T0237946

Communications
in Computer and Information Science   444

Anne Laurent   Olivier Strauss
Bernadette Bouchon-Meunier
Ronald R. Yager (Eds.)

Information Processing and Management of Uncertainty in Knowledge-Based Systems

15th International Conference, IPMU 2014
Montpellier, France, July 15-19, 2014
Proceedings, Part III

 Springer

Volume Editors

Anne Laurent
Université Montpellier 2, LIRMM, France
E-mail: laurent@lirmm.fr

Olivier Strauss
Université Montpellier 2, LIRMM, France
E-mail: olivier.strauss@lirmm.fr

Bernadette Bouchon-Meunier
Sorbonne Universités, UPMC Paris 6, France
E-mail: bernadette.bouchon-meunier@lip6.fr

Ronald R. Yager
Iona College, New Rochelle, NY, USA
E-mail: ryager@iona.edu

ISSN 1865-0929   e-ISSN 1865-0937
ISBN 978-3-319-08851-8   e-ISBN 978-3-319-08852-5
DOI 10.1007/978-3-319-08852-5
Springer Cham Heidelberg New York Dordrecht London

Library of Congress Control Number: 2014942459

Typesetting: Camera-ready by author, data conversion by Scientific Publishing Services, Chennai, India

Printed on acid-free paper

Springer is part of Springer Science+Business Media (www.springer.com)

Preface

Here we provide the proceedings of the 15th International Conference on Information Processing and Management of Uncertainty in Knowledge-based Systems, IPMU 2014, held in Montpellier, France, during July 15–19, 2014. The IPMU conference is organized every two years with the focus of bringing together scientists working on methods for the management of uncertainty and aggregation of information in intelligent systems.

This conference provides a medium for the exchange of ideas between theoreticians and practitioners working on the latest developments in these and other related areas. This was the 15th edition of the IPMU conference, which started in 1986 and has been held every two years in the following locations in Europe: Paris (1986), Urbino (1988), Paris (1990), Palma de Mallorca (1992), Paris (1994), Granada (1996), Paris (1998), Madrid (2000), Annecy (2002), Perugia (2004), Malaga (2008), Dortmund (2010) and Catania (2012).

Among the plenary speakers at past IPMU conferences, there have been three Nobel Prize winners: Kenneth Arrow, Daniel Kahneman, and Ilya Prigogine. An important feature of the IPMU Conference is the presentation of the Kampé de Fériet Award for outstanding contributions to the field of uncertainty. This year, the recipient was Vladimir N. Vapnik. Past winners of this prestigious award were Lotfi A. Zadeh (1992), Ilya Prigogine (1994), Toshiro Terano (1996), Kenneth Arrow (1998), Richard Jeffrey (2000), Arthur Dempster (2002), Janos Aczel (2004), Daniel Kahneman (2006), Enric Trillas (2008), James Bezdek (2010), Michio Sugeno (2012).

The program of the IPMU 2014 conference consisted of 5 invited academic talks together with 180 contributed papers, authored by researchers from 46 countries, including the regular track and 19 special sessions. The invited academic talks were given by the following distinguished researchers: Vladimir N. Vapnik (NEC Laboratories, USA), Stuart Russell (University of California, Berkeley, USA and University Pierre et Marie Curie, Paris, France), Inés Couso (University of Oviedo, Spain), Nadia Berthouze (University College London, United Kingdom) and Marcin Detyniecki (University Pierre and Marie Curie, Paris, France).

Industrial talks were given in complement of academic talks and highlighted the necessary collaboration we all have to foster in order to deal with current challenges from the real world such as Big Data for dealing with massive and complex data.

The success of IPMU 2014 was due to the hard work and dedication of a large number of people, and the collaboration of several institutions. We want to acknowledge the industrial sponsors, the help of the members of the International Program Committee, the reviewers of papers, the organizers of special sessions, the Local Organizing Committee, and the volunteer students. Most of all, we appreciate the work and effort of those who contributed papers to the conference. All of them deserve many thanks for having helped to attain the goal of providing a high quality conference in a pleasant environment.

May 2014

Bernadette Bouchon-Meunier
Anne Laurent
Olivier Strauss
Ronald R. Yager

Conference Committee

General Chairs

Anne Laurent   Université Montpellier 2, France
Olivier Strauss   Université Montpellier 2, France

Executive Directors

Bernadette Bouchon-Meunier   CNRS-UPMC, France
Ronald R. Yager   Iona College, USA

Web Chair

Yuan Lin   INRA, SupAgro, France

Proceedings Chair

Jérôme Fortin   Université Montpellier 2, France

International Advisory Board

Giulianella Coletti (Italy)
Miguel Delgado (Spain)
Mario Fedrizzi (Italy)
Laurent Foulloy (France)
Salvatore Greco (Italy)
Julio Gutierrez-Rios (Spain)
Eyke Hüllermeier(Germany)
Luis Magdalena (Spain)
Christophe Marsala (France)

Benedetto Matarazzo (Italy)
Manuel Ojeda-Aciego (Spain)
Maria Rifqi (France)
Lorenza Saitta (Italy)
Enric Trillas (Spain)
Llorenç Valverde (Spain)
José Luis Verdegay (Spain)
Maria-Amparo Vila (Spain)
Lotfi A. Zadeh (USA)

Special Session Organizers

Jose M. Alonso   European Centre for Soft Computing, Spain
Michal Baczynski   University of Silesia, Poland
Edurne Barrenechea   Universidad Pública de Navarra, Spain
Zohra Bellahsene   University Montpellier 2, France
Patrice Buche   INRA, France
Thomas Burger   CEA, France
Tomasa Calvo   Universidad de Alcalá, Spain

Brigitte Charnomordic	INRA, France
Didier Coquin	University of Savoie, France
Cécile Coulon-Leroy	École Supérieure d'Agriculture d'Angers, France
Sébastien Destercke	CNRS, Heudiasyc Lab., France
Susana Irene Diaz Rodriguez	University of Oviedo, Spain
Luka Eciolaza	European Centre for Soft Computing, Spain
Francesc Esteva	IIIA - CSIC, Spain
Tommaso Flaminio	University of Insubria, Italy
Brunella Gerla	University of Insubria, Italy
Manuel González Hidalgo	University of the Balearic Islands, Spain
Michel Grabisch	University of Paris Sorbonne, France
Serge Guillaume	Irstea, France
Anne Laurent	University Montpellier 2, France
Christophe Labreuche	Thales Research and Technology, France
Kevin Loquin	University Montpellier 2, France
Nicolás Marín	University of Granada, Spain
Trevor Martin	University of Bristol, UK
Sebastia Massanet	University of the Balearic Islands, Spain
Juan Miguel Medina	University of Granada, Spain
Enrique Miranda	University of Oviedo, Spain
Javier Montero	Complutense University of Madrid, Spain
Jesús Medina Moreno	University of Cádiz, Spain
Manuel Ojeda Aciego	University of Málaga, Spain
Martin Pereira Farina	Universidad de Santiago de Compostela, Spain
Olga Pons	University of Granada, Spain
Dragan Radojevic	Serbia
Anca L. Ralescu	University of Cincinnati, USA
François Scharffe	University Montpellier 2, France
Rudolf Seising	European Centre for Soft Computing, Spain
Marco Elio Tabacchi	Università degli Studi di Palermo, Italy
Tadanari Taniguchi	Tokai University, Japan
Charles Tijus	LUTIN-CHArt, Université Paris 8, France
Konstantin Todorov	University Montpellier 2, France
Lionel Valet	University of Savoie, France

Program Committee

Michal Baczynski	Ulrich Bodenhofer	Rita Casadio
Gleb Beliakov	P. Bonissone	Yurilev Chalco-Cano
Radim Belohlavek	Bernadette Bouchon-Meunier	Brigitte Charnomordic
Salem Benferhat		Guoqing Chen
H. Berenji	Patrice Buche	Carlos A. Coello Coello
Isabelle Bloch	Humberto Bustince	Giulianella Coletti

Oscar Cordón
Inés Couso
Keeley Crockett
Bernard De Baets
Sébastien Destercke
Marcin Detyniecki
Antonio Di Nola
Gerard Dray
Dimiter Driankov
Didier Dubois
Francesc Esteva
Mario Fedrizzi
Janos Fodor
David Fogel
Jérôme Fortin
Laurent Foulloy
Sylvie Galichet
Patrick Gallinari
Maria Angeles Gil
Siegfried Gottwald
Salvatore Greco
Steve Grossberg
Serge Guillaume
Lawrence Hall
Francisco Herrera
Enrique Herrera-Viedma
Kaoru Hirota
Janusz Kacprzyk
A. Kandel
James M. Keller
E.P. Klement
Laszlo T. Koczy
Vladik Kreinovich
Christophe Labreuche

Jérôme Lang
Henrik L. Larsen
Anne Laurent
Marie-Jeanne Lesot
Churn-Jung Liau
Yuan Lin
Honghai Liu
Weldon Lodwick
Kevin Loquin
Thomas Lukasiewicz
Luis Magdalena
Jean-Luc Marichal
Nicolás Marín
Christophe Marsala
Trevor Martin
Mylène Masson
Silvia Massruha
Gilles Mauris
Gaspar Mayor
Juan Miguel Medina
Jerry Mendel
Enrique Miranda
Pedro Miranda
Paul-André Monney
Franco Montagna
Javier Montero
Jacky Montmain
Serafín Moral
Yusuke Nojima
Vilem Novak
Hannu Nurmi
S. Ovchinnikov
Nikhil R. Pal
Endre Pap

Simon Parsons
Gabriella Pasi
W. Pedrycz
Fred Petry
Vincenzo Piuri
Olivier Pivert
Henri Prade
Anca Ralescu
Dan Ralescu
Mohammed Ramdani
Marek Reformat
Agnes Rico
Maria Rifqi
Enrique H. Ruspini
Glen Shafer
J. Shawe-Taylor
P. Shenoy
P. Sobrevilla
Umberto Straccia
Olivier Strauss
M. Sugeno
Kay-Chen Tan
S. Termini
Konstantin Todorov
Vicenc Torra
I. Burhan Turksen
Barbara Vantaggi
José Luis Verdegay
Thomas Vetterlein
Marco Zaffalon
Hans-Jürgen
  Zimmermann
Jacek Zurada

Additional Members of the Reviewing Committee

Iris Ada
Jesús Alcalá-Fernandez
Cristina Alcalde
José María Alonso
Alberto Alvarez-Alvarez
Leila Amgoud
Massih-Reza Amini

Derek Anderson
Plamen Angelov
Violaine Antoine
Alessandro Antonucci
Alain Appriou
Arnaud Martin
Jamal Atif

Rosangela Ballini
Carlos Barranco
Edurne Barrenechea
Gleb Beliakov
Zohra Bellahsene
Alessio Benavoli
Eric Benoit

Table of Contents – Part III

Intelligent Databases and Information Systems

Theory of Evidence

Aggregation Functions

Big Data - The Role of Fuzzy Methods

Imprecise Probabilities: From Foundations to Applications

Intelligent Measurement and Control for Nonlinear Systems

Aggregation Functions

Robust Selection of Domain-Specific Semantic Similarity Measures from Uncertain Expertise

Stefan Janaqi[*], Sébastien Harispe, Sylvie Ranwez, and Jacky Montmain

LGI2P/ENSMA Research Centre, Parc Scientifique G. Besse,
30035 Nîmes Cedex 1, France
firstname.name@mines-ales.fr

Abstract. Knowledge-based semantic measures are cornerstone to exploit ontologies not only for exact inferences or retrieval processes, but also for data analyses and inexact searches. Abstract theoretical frameworks have recently been proposed in order to study the large diversity of measures available; they demonstrate that groups of measures are particular instantiations of general parameterized functions. In this paper, we study how such frameworks can be used to support the selection/design of measures. Based on (i) a theoretical framework unifying the measures, (ii) a software solution implementing this framework and (iii) a domain-specific benchmark, we define a semi-supervised learning technique to distinguish best measures for a concrete application. Next, considering uncertainty in both experts' judgments and measures' selection process, we extend this proposal for robust selection of semantic measures that best resists to these uncertainties. We illustrate our approach through a real use case in the biomedical domain.

Keywords: semantic similarity measures, ontologies, unifying semantic similarity measures framework, measure robustness, uncertain expertise.

1   Introduction

Formal knowledge representations (KRs) can be used to express domain-specific knowledge in a computer-readable and understandable form. These KRs, generally called ontologies, are commonly defined as formal, explicit and shared conceptualizations [3]. They bridge the gap between domain-specific expertise and computer resources by enabling a partial transfer of expertise to computer systems. Intelligence can further be simulated by developing reasoners which will process KRs w.r.t. the semantics of the KR language used for their definition (e.g., OWL, RDFS).

From gene analysis to recommendation systems, knowledge-based systems are today the backbones of numerous business and research projects. They are extensively used for the task of classification or more generally to answer exact queries w.r.t. domain-specific knowledge. They also play an important role for knowledge discovery, which, contrary to knowledge inferences, relies on inexact search techniques. Cornerstones of such algorithms are *semantic measures*, functions used to estimate the degree of likeness of concepts defined in a KR [12, 15]. They are for instance

A. Laurent et al. (Eds.): IPMU 2014, Part III, CCIS 444, pp. 1–10, 2014.

used to estimate the semantic proximity of resources (e.g., diseases) indexed by concepts (e.g., syndromes) [9]. Among the large diversity of semantic measures (refer to [5] for a survey), knowledge-based semantic similarity measures (SSM) can be used to compare how similar the meanings of concepts are according to taxonomical evidences formalized in a KR. Among their broad range of practical applications, SSMs are used to disambiguate texts, to design information retrieval algorithms, to suggest drug repositioning, or to analyse genes products [5]. Nevertheless, most SSMs were designed in an *ad hoc* manner and only few domain-specific comparisons involving restricted subsets of measures have been made. It is therefore difficult to select a SSM for a specific usage. Moreover, in a larger extent, it is the study of SSM and all knowledge-based systems relying on SSMs which are hampered by this diversity of proposals since it is difficult today to distinguish the benefits of using a particular measure.

In continuation of several contributions which studied similitudes between measures [2, 13, 15], a theoretical unifying framework of SSMs has recently been proposed [4]. It enables the decomposition of SSMs through a small set of intuitive core elements and parameters. This highlights the fact that most SSMs can be instantiated from general similarity formula. This result is not only relevant for theoretical studies; it opens interesting perspectives for practical applications of measures, such as the definition of new measures or the characterization of measures best performing in specific application contexts. To this end, the proposal of the theoretical framework has been completed by the development of the *Semantic Measures Library* (SML) [6][1], an open source generic software solution dedicated to semantic measures.

Both the theoretical framework of SSMs and the SML can be used to tackle fundamental questions regarding SSMs: which measure should be considered for a concrete application? Are there strong implications associated to the selection of a specific SSM (e.g., in term of accuracy)? Do some of the measures have similar behaviours? Is the accuracy of a measure one of its intrinsic properties, i.e., are there measures which always better perform? Is a given measure sensitive w.r.t. uncertainty in data or selection process? Can the impact of these uncertainties be estimated?

In this paper, we focus on the uncertainty relative to the selection of a SSM in a particular context of use. Considering that a representative set of pairs of concepts (x, y) and expected similarities $sim(x, y)$ have been furnished by domain experts, we propose to use the aforementioned framework and its implementation, to study SSMs and to support context-specific selection of measures. Applying semi-supervised learning techniques, we propose to '*learn*' the core elements and the parameters of general measures in order to fit domain-expert's knowledge. Moreover, we focus on the impact of experts' uncertainty on the selection of measures that is: to which extent can a given SSM be reliably used knowing that similarities provided by experts are inherently approximate assessments? We define the robustness of a SSM as its ability to remain a reliable model in presence of uncertainties in the learning dataset.

---

[1] http://www.semantic-measures-library.org

The rest of the paper is organized as follows. Section 2 presents the unifying framework of SSMs. Section 3 describes a learning procedure for choosing the core elements and parameters on the basis of expert's knowledge. A simple, yet realistic, model of uncertainty for experts' assessments and indicators of robustness are introduced. Section 4 illustrates the practical application of our proposal in the biomedical domain. Finally, section 5 provides conclusions as well as some lines of future work.

2   Semantic Similarity Measures and Their Unification

This section briefly introduces SSMs for the comparison of two concepts defined in a KR, more details are provided in [5]. We present how the unifying framework recently introduced in [4] can be used to define parametric measures from which existing and new SSMs can be expressed. We focus on semantic similarity which is estimated based on the taxonomical relationships between concepts; we therefore focus on the taxonomy of concepts of a KR. This taxonomy is a rooted direct acyclic graph denoted G; it defines a partial order $\preccurlyeq$ among the set of concepts C. We use these notations:

- $A(x)$: the ancestors of the concept x, i.e., $A(x) = \{c \mid x \preccurlyeq c\}$.
- $depth(x)$: the length of the shortest-path linking x to the root of the taxonomy.
- $LCA(x,y)$: the Least Common Ancestor of concepts x and y, i.e. their deeper common ancestor.
- $IC(x)$: the information content of x, e.g., originally defined as $-\log p(x)$, with $p(x)$ the probability of usage of concept x, e.g., in a text corpus. Topological alternatives have also been proposed (see [5, 15]).
- $MICA(x,y)$ the Most Informative Common Ancestor of concepts x and y.

In [4], it is shown that a large diversity of SSMs can easily be expressed from a small set of primitive components. The framework is composed of two main elements: a set of primitive elements used to compose measures, and general measure expressions which define how these primitives can be associated to build a concrete measure. With $\mathbb{K}$ a domain containing any subset of the taxonomy (e.g., $A(x)$, G_x^+ the subgraph of G induced by $A(x)$), the primitives used by the framework are:

- **Semantic representation** (ρ): Canonical form adopted to represent a concept and derive evidences of semantic similarity, with $\rho: C \to \mathbb{K}$. We note $\rho(x)$ or $\tilde{x}$, the semantic representation of the concept x. As an example, a concept can be seen as the set of senses it encompasses, i.e., $A(x)$ its subsumers in G.
- **Specificity of a concept** (θ): The specificity of a concept u is estimated by a function $\theta: C \to \mathbb{R}^+$, with $x \preccurlyeq y \Rightarrow \theta(x) \geq \theta(y)$, e.g., IC, depth of a concept.
- **Specificity of a concept representation** (Θ): The specificity of a concept representation $\tilde{x}$, $\Theta(\tilde{x})$, is estimated by a function: $\Theta: \mathbb{K} \to \mathbb{R}^+$. Θ decreases from the leaves to the root of G, i.e., $x \preccurlyeq y \Rightarrow \Theta(\tilde{x}) \geq \Theta(\tilde{y})$.

- **Commonality (Ψ):** The commonality of two concepts (x, y) according to their semantic representations $(\tilde{x}, \tilde{y})$ is given by a function $\Psi(\tilde{x}, \tilde{y})$, $\Psi: \mathbb{K} \times \mathbb{K} \to \mathbb{R}^+$.
- **Difference (Φ):** The amount of knowledge represented in $\tilde{x}$ not found in $\tilde{y}$ is estimated using a function $\Phi(\tilde{x}, \tilde{y})$: $\Phi: \mathbb{K} \times \mathbb{K} \to \mathbb{R}^+$.

The abstract functions ρ, Ψ, Φ are the core elements of most SSMs, numerous general measures can be expressed from them; we here present sim_{RM}, an abstract formulation of the *ratio model* introduced by Tversky [16]:

$$sim_{RM}(x, y) = \frac{\Psi(\tilde{x}, \tilde{y})}{\alpha\,\Phi(\tilde{x}, \tilde{y}) + \beta\,\Phi(\tilde{y}, \tilde{x}) + \Psi(\tilde{x}, \tilde{y})}$$

Note that α, β are positive, bounded, and that important α, β values have no appeal in our case. sim_{RM} can be used to express numerous SSMs [4]; using the core elements of Table 1, some well known SSMs instantiations of sim_{RM} are provided in Table 2 (in all cases, $\Phi(\tilde{x}, \tilde{y}) = \Theta(\tilde{x}) - \Psi(\tilde{x}, \tilde{y})$). Please refer to [4, 5] for citations and for a more extensive list of both abstract measures and instantiations.

Table 1. Examples of instantiations of the primitive components defined by the framework

Case	1	2	3	4
$\rho(u) = \tilde{u}$	G_u^+	$A(u)$	$A(u)$	$A(u)$
$\Theta(\tilde{u})$	$depth(u)$	$IC(u)$	$\sum_{c \in A(u)} IC(c)$	$|A(u)|$
$\Psi(\tilde{u}, \tilde{v})$	$depth(LCA(u,v))$	$IC(MICA(u,v))$	$\sum_{c \in A(u) \cap A(v)} IC(c)$	$|A(u) \cap A(v)|$

Table 2. Instantiations of SSMs which can be used to assess the similarity of concepts

Measures	Case	Parameters
$sim_{Wu\,\&\,Palmer}(u, v) = \dfrac{2\,depth(LCA(u,v))}{depth(u)+depth(v)}$	1	$\alpha = 0.5,\ \beta = 0.5$
$sim_{Lin}(u, v) = \dfrac{2\,IC(MICA(u,v))}{IC(u)+IC(v)}$	2	$\alpha = 0.5,\ \beta = 0.5$
$sim_{Faith}(u, v) = \dfrac{IC(MICA(u,v))}{IC(u)+IC(v)-IC(MICA(u,v))}$	2	$\alpha = 1,\quad \beta = 1$
$sim_{Mazandu}(u, v) = \dfrac{2\,\sum_{c \in A(u) \cap A(v)} IC(c)}{\sum_{c \in A(u)} IC(c)+\sum_{c \in A(v)} IC(c)}$	3	$\alpha = 0.5,\quad \beta = 0.5$
$sim_{CMatch}(u, v) = \dfrac{|A(u) \cap A(v)|}{|A(u) \cup A(v)|}$	4	$\alpha = 1,\quad \beta = 1$

3   Learning Measures Tuning from Expert Knowledge

Considering a particular abstract expression of a measure, here sim_{RM}, the objective is to define the "right" combination of parameters $(\rho, \Theta, \Psi, \Phi, \alpha, \beta)$. Following the framework presented in section 2, this choice proceeds through two steps:

- *Step 1*: Define a finite list $\Pi = \{\pi_l \mid \pi_l = (\rho_l, \Theta_l, \Psi_l, \Phi_l), l = 1, ..., L\}$ of possible instantiations of the core elements $(\rho, \Theta, \Psi, \Phi)$, see Table 1. This choice can be guided by semantic concerns and application constraints, e.g., based on: (i) the analysis of the assumptions on which rely specific instantiations of measures [5], (ii) on the will to respect particular mathematical properties such as the identity of the indiscernibles or (iii) the computational complexity of measures.

- *Step 2*: Choose the couples of parameters $(\alpha_l, \beta_l), l = 1, ..., L$ to be associated to π_l in sim_{RM}. A couple (α_l, β_l) may be selected in an *ad hoc* manner from a finite list of well known instantiations (see Table 2), e.g., based on the *heavy* assumption that measures performing correctly in other benchmarks are suited for our specific use case. Alternatively, knowing the expected similarities $sim(x, y)$ furnished by domain experts on a learning dataset, (α_l, β_l) can also be obtained from a continuous optimization process over this dataset. This latter issue is developed hereafter.

For any instantiation $(\pi_l, \alpha_l, \beta_l)$ of the abstract measure sim_{RM}, let us denote, $s_l(x, y) \equiv sim_{RM(\pi_l, \alpha_l, \beta_l)}(x, y)$ for any couple of concepts (x, y). Suppose now that experts have given the expected similarities $s_k = s(x_k, y_k), k = 1, ..., N$ for a subset of N couples of concepts (x_k, y_k). Let $\mathbf{s} = [s_1, ..., s_N]^T$ be the vector of these expected similarity values. It is then possible to estimate the quality of a particular SSM tuning s_l through the value of a fitting function. We denote $\mathbf{s}_l$ the vector which contains the similarities obtained by s_l for each pair of concepts evaluated to build $\mathbf{s}$, with: $\mathbf{s}_l = [s_l(x_k, y_k)]_{k=1,...,N}$. Given π_l, the similarities $s_l(x_k, y_k)$ only depend on (α_l, β_l); it is thus possible to find the optimal (α_l^0, β_l^0) values that optimize a fitting function, e.g. the correlation between $\mathbf{s}$ and $\mathbf{s}_l$:

$$\begin{cases} \max_{\alpha, \beta} \; corr(\mathbf{s}, \mathbf{s}_l) \\ \quad 0 \le \alpha, \beta \le M \end{cases} \tag{1}$$

The bound constraint of this optimization problem (1) is reasonable since the case $\alpha \to \infty$ or $\beta \to \infty$ should imply $sim_{RM}(x, y) = 0$ which have no appeal for us.

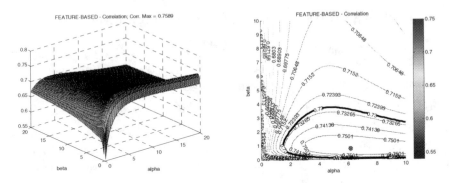

Fig. 1. The fitting function $corr(\mathbf{s}, \mathbf{s}_l(\alpha, \beta))$ and its level lines for choice 4 (see Table 1)

Fig. 1 presents an experiment which has been made to distinguish the optimal (α_l^0, β_l^0) parameters for instantiation of sim_{RM} using case 4, and considering a bio-medical benchmark (discussed in section 4). The maximal correlation value is 0.759, for $\alpha_l^0 = 6.17$ and $\beta_l^0 = 0.77$ (the dot in the right image). The strong asymmetry in the contour lines is a consequence of $\Phi(\tilde{u}, \tilde{v}) \neq \Phi(\tilde{v}, \tilde{u})$. This approach is efficient to derive the optimal configuration considering a given vector of similarities (**s**). Nevertheless it does not take into account the fact that expert assessments of similarity are inherently marred by uncertainty.

In our application domain, expected similarities are often provided in a finite ordinal linear scale of type $v_i = v_0 + i\Delta, i = 1 \ldots V$ (e.g., $v_i \in \{1, 2, 3, 4\}$ in the next section). If Δ denotes the difference between two contiguous levels of the scale, then we assume in this case that $\varepsilon_k \in \{-\Delta, 0, \Delta\}$ with probability $p(\varepsilon_k = 0) = q, p(\varepsilon_k = -\Delta) = p(\varepsilon_k = \Delta) = \frac{1-q}{2}$. This model of uncertainty merely means that expert assessment errors cannot exceed $\pm\Delta$. In addition, it allows computing the probability distributions of the optimal couples $(\alpha_l(\varepsilon), \beta_l(\varepsilon)) \sim D_{\alpha_l, \beta_l}^u$ with $(\alpha_l(\varepsilon), \beta_l(\varepsilon))$ being the solution of problem (1) with $\mathbf{t} = \mathbf{s} + \varepsilon$ instead of $\mathbf{s}$ as inputs, and u the uncertainty parameter ($u = q$) (note that $(\alpha_l^0, \beta_l^0) = (\alpha(0), \beta(0))$).

The aim is to quantify the impact of uncertainty in expert assessments on the selection of a measure instantiation, i.e. selection of $(\pi_l, \alpha_l, \beta_l)$. We are interested in the evaluation of SSM robustness w.r.t. expert uncertainty, and we more particularly focus on the relevance to consider (α_l^0, β_l^0) in case of uncertain expert assessments. Finding a robust solution to an optimization problem without knowing the probability density of data is well known [1, 7]. In our case, we do not use any hypothesis on the distribution of $\mathbf{s} - \mathbf{s}_l$. We define a set of *near optimal* (α_l, β_l) using a threshold value r (domain-specific). The near optimal solutions are those in the level set: $L_r = \{(\alpha_l, \beta_l) \mid corr(\mathbf{s}, \mathbf{s}_l) \geq r\}$, i.e. the set of pairs of parameters α, β giving good correlation values w.r.t the threshold r. The robustness is therefore given by: $(u) = \iint_{L_r} D_{\alpha_l, \beta_l}^u d\alpha_l d\beta_l$; the bigger R, the more robust the model (α_l^0, β_l^0) is. Nevertheless, given that analytical form for the distribution $D_{\alpha, \beta}^u$ cannot be established, even in the normal case $\varepsilon \sim N(0, \Sigma)$, estimation techniques are used, e.g., Monte Carlo method.

The computation of $D^u_{\alpha_l, \beta_l}$ allows identifying a robust couple (α_l, β_l) for a given uncertainty level u. An approximation of this point, named (α^*_l, β^*_l), is given by the median of points $(\alpha_l(\varepsilon), \beta_l(\varepsilon))$ generated by Monte Carlo method. Note that (α^*_l, β^*_l) coincides with (α^0_l, β^0_l) for $u = 0$ or little values of u. Therefore (α^*_l, β^*_l) remains inside L_r for most u and is significantly different from (α^0_l, β^0_l) when u increases.

4 Practical Choice of a Robust Semantic Similarity Measure

Most algorithms and treatments based on SSMs require measures to be highly corre-lated with human judgement of similarity [10–12]. SSMs are thus commonly evalu-ated regarding their ability to mimic human appreciation of similarity between domain-specific concepts. In our experiment, we considered a benchmark commonly used in biomedicine to evaluate SSMs according to similarities provided by medical experts [12]. The benchmark contains 29 pairs of medical terms which have been associated to pairs of concepts defined in the Medical Subject Headings (MeSH) the-saurus [14]. Despite its reduced size – due to the fact that benchmarks are hard to produce – we used this benchmark since it is commonly used in the biomedical do-main to evaluate SSM accuracy. The average of expert similarities is given for each pair of concepts; initial ratings are of the form $s_k \in \{1, 2, 3, 4\}$. As a consequence, the uncertainty is best modelled defining $\varepsilon_k \in \{-1, 0, 1\}$ with probability distribution: $p(\varepsilon_k = 0) = q, \; p(\varepsilon_k = -1) = p(\varepsilon_k = 1) = \frac{1-q}{2}$.

The approach described in Section 3 is applied with instantiations π_l presented in Table 2. Optimal (α, β) were found by resolving Problem (1); SSMs computations were performed by the Semantic Measures Library [6]. Table 3 shows that one of the optimal configuration is obtained using Case 2 expressions (with $z = MICA(u, v)$):

$$sim_{RM-C2}(u, v) = \frac{IC(z)}{18.62\,(IC(u) - IC(z)) + 4.23(IC(v) - IC(z)) + IC(z)}$$

Table 3. Best correlations of parametric SSMs, refer to Table 2 for case details

	Case 1 (C1)	Case 2 (C2)	Case 3 (C3)	Case 4 (C4)
Max. Correlation	0.719	**0.768**	0.736	0.759
Optimal (α^0_l, β^0_l)	(9.89,1.36)	(18.62,4.23)	(7.26,0.40)	(6.17,0.77)

It can be observed from the analysis of the results that, naturally, *the choice of core elements affects the maximal correlation*; instantiations C2 and C4 always resulted in the highest correlations (results are considered to be significant since the accuracies of C2 and C4 are around 5% better than C1 and C3). Another interesting aspect of the results is that *asymmetrical measures provide best results:* all experiments provided the best correlations by tuning the measures with asymmetric contributions of α and

β parameters – which can be linked to the results obtained in [4, 8]. Note that best tunings and (α, β) ratio vary depending on the core elements considered.

Setting a threshold of correlation at 0.75, we now focus on the instantiations C2 and C4; they have comparable results (respectively 0.768/0.759, $\Delta = 0.009$). The aim is to evaluate their robustness according to the framework introduced in section 3. Considering inter-agreements between pools of experts reported in [12] (0.68/0.78), the level of near optimality (L_r) is fixed to $r = 0.73$. We also choose uncertainty values $q \in \{0.9, 0.8, 0.7, 0.6, 0.5\}$. The probability for the expert(s) to give erroneous values, i.e. their uncertainty, is $1 - q \in \{0.1, 0.2, 0.3, 0.4, 0.5\}$. For each q-value, a large number of ε-vectors are generated to derive (α_l^*, β_l^*). Estimated values of the robustness $R(u)$ and (α_l^*, β_l^*) for measures C2 and C4 are given in Table 4 and are illustrated in Fig. 2.

Table 4. Robustness of parametric SSMs for case 2 (C2) and case 4 (C4) measures

	$1 - q = 0.1$	$1 - q = 0.2$	$1 - q = 0.3$	$1 - q = 0.4$	$1 - q = 0.5$
$R_{C2}(u)$	0.83	0.70	0.56	0.49	0.39
$(\alpha_{C2}^*, \beta_{C2}^*)$	(18.62, 4.23)	(18.62, 4.23)	(15.31, 4.23)	(16.70, 4.07)	(13.71, 4.02)
$R_{C4}(u)$	0.76	0.54	0.46	0.39	0.35
$(\alpha_{C4}^*, \beta_{C4}^*)$	(6.17, 0.77)	(6.17, 0.76)	(5.52, 0.71)	(5.12, 0.64)	(4.06, 0.70)

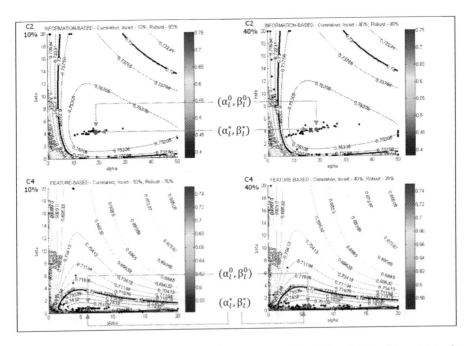

Fig. 2. Robustness of case 2 (C2) and case 4 (C4) measures for 10% and 40% of uncertainty. In each figure the solutions (α_l^0, β_l^0), (α_l^*, β_l^*) and L_r are plotted.

Fig. 2 shows the spread of the couples $(\alpha(\varepsilon), \beta(\varepsilon))$ for measures C2 and C4 considering the levels of uncertainty set to 10% and 40%. An interesting aspect of the results is that robustness is significantly different depending on the case considered, 83% for C2 and 76% for C4. Therefore, despite their correlations were comparable ($\Delta = 0.009$), C2 is less sensible to uncertainty w.r.t. to the learning dataset used to distinguish best-suited parameters Indeed, only based on correlation analysis, SSM users will most of the time prefer measures which have been derived from C2 since their computational complexity is lower than those derived from C4 (computations of the IC and the MICA are more complex). Nevertheless, C4 appears to be a more risky choice considering the robustness of the measures and the uncertainty inherently associated to expert evaluations. In this case, one can reasonably conclude that (α_l^0, β_l^0) of C2 is robust for uncertainty lower than 10% ($1 - q = 0.1$; $R(u) = 0.83$). The size of the level set L_r is also a relevant feature for the selection of SSMs; it indicates the size of the set of parameters (α_l, β_l) that give high correlations considering imprecise human expectations. Therefore, an analytical (and its graphical counterpart) estimator of robustness is introduced. Another interesting finding of this study is that, even if human observations are marred by uncertainty and the semantic choice of measures parameters, $\pi_l = (\rho_l, \Theta_l, \Psi_l, \Phi_l)$, is not a precise process, the resulting SSM *is not so sensitive* to all these uncertainty factors.

5   Conclusions and Future Work

Considering the large diversity of measures available, an important contribution for end-users of SSMs would be to provide tools to select best-suited measures for domain-specific usages. Our approach paves the way to the development of such tool and can more generally be used to perform detailed evaluations of SSMs in other contexts and applications (i.e., other knowledge domains, ontologies and training data). In this paper, we used the SSM unifying framework established in [4] and a well-established benchmark in order to design a SSM that fits the objectives of a practitioner/designer in a given application context.

We particularly focused on the fact that the selection of the best-suited SSM is affected by uncertainties, in particular due to the uncertainty associated to the ratings of human experts used to evaluate the measures, etc. To our knowledge, we are the first to propose an approach that finds/creates a best-suited SM which is robust to these uncertainties. Indeed, contrary to most of existing studies which only compare measures based on their correlation with expected scores of similarity (e.g., human appreciation of similarity), our study highlights the fact that robustness of measures is an essential criteria to better understand measures' behaviour and therefore drive their comparison and selection. We therefore propose an analytical estimator of robustness, and its graphical counterpart, which can be used to characterize this important property of SSM. Therefore, by putting into light the limit of existing estimator of measures' accuracy, especially when uncertainty is regularly impacting measures (evaluation and definition), we are convinced that our proposals open interesting perspectives for measure characterization and will therefore ease their accurate selection for domain specific studies.

In addition, results of the real-world example used to illustrate our approach give us the opportunity to capture new insights about specific types of measures (i.e. particular instantiation of an abstract measure, sim_{RM}). Nevertheless, the observations made in this experiment have been made based on the analysis of specific configurations of measures, using a single ontology and a unique (unfortunately reduced) benchmark. More benchmarks have thus to be studied to derive more general conclusions and the sensitivity of our approach w.r.t the benchmark properties (e.g. size) have to be discussed. This will help to better understanding SSMs and more particularly to better analyse the role and connexions between abstract measures expressions, core elements instantiations and extra parameters (e.g. α, β) regarding the precision and the robustness of SSMs. Finally, note that a similar approach can be used to study SSMs expressed from abstract formulae other than sim_{RM}, and therefore study the robustness of a large diversity of measures not presented in this study.

References

1. Ben-Tal, A., et al.: Robust optimization. Princeton series in applied mathematics. Priceton University Press (2009)
2. Blanchard, E., et al.: A generic framework for comparing semantic similarities on a subsumption hierarchy. In: 18th Eur. Conf. Artif. Intell., pp. 20–24 (2008)
3. Gruber, T.: A translation approach to portable ontology specifications. Knowl. Acquis. 5(2), 199–220 (1993)
4. Harispe, S., et al.: A Framework for Unifying Ontology-based Semantic Similarity Measures: a Study in the Biomedical Domain. J. Biomed. Inform. (in press, 2013)
5. Harispe, S., et al.: Semantic Measures for the Comparison of Units of Language, Concepts or Entities from Text and Knowledge Base Analysis. ArXiv.1310.1285 (2013)
6. Harispe, S., et al.: The Semantic Measures Library and Toolkit: fast computation of semantic similarity and relatedness using biomedical ontologies. Bioinformatics 30(5), 740–742 (2013)
7. Janaqi, S., et al.: Robust real-time optimization for the linear oil blending. RAIRO - Oper. Res. 47, 465–479 (2013)
8. Lesot, M.-J., Rifqi, M.: Order-based equivalence degrees for similarity and distance measures. In: Hüllermeier, E., Kruse, R., Hoffmann, F. (eds.) IPMU 2010. LNCS (LNAI), vol. 6178, pp. 19–28. Springer, Heidelberg (2010)
9. Mathur, S., Dinakarpandian, D.: Finding disease similarity based on implicit semantic similarity. J. Biomed. Inform. 45(2), 363–371 (2012)
10. Pakhomov, S., et al.: Semantic Similarity and Relatedness between Clinical Terms: An Experimental Study. In: AMIA Annu. Symp. Proc. 2010, pp. 572–576 (2010)
11. Pakhomov, S.V.S., et al.: Towards a framework for developing semantic relatedness reference standards. J. Biomed. Inform. 44(2), 251–265 (2011)
12. Pedersen, T., et al.: Measures of semantic similarity and relatedness in the biomedical domain. J. Biomed. Inform. 40(3), 288–299 (2007)
13. Pirró, G., Euzenat, J.: A Feature and Information Theoretic Framework for Semantic Similarity and Relatedness. In: Patel-Schneider, P.F., Pan, Y., Hitzler, P., Mika, P., Zhang, L., Pan, J.Z., Horrocks, I., Glimm, B. (eds.) ISWC 2010, Part I. LNCS, vol. 6496, pp. 615–630. Springer, Heidelberg (2010)
14. Rogers, F.B.: Medical subject headings. Bull. Med. Libr. Assoc. 51, 114–116 (1963)
15. Sánchez, D., Batet, M.: Semantic similarity estimation in the biomedical domain: An ontology-based information-theoretic perspective. J. Biomed. Inform. 44(5), 749–759 (2011)
16. Tversky, A.: Features of similarity. Psychol. Rev. 84(4), 327–352 (1977)

Coping with Imprecision During
a Semi-automatic Conceptual Indexing Process

Nicolas Fiorini[1], Sylvie Ranwez[1], Jacky Montmain[1], and Vincent Ranwez[2]

[1] Centre de recherche LGI2P de l'école des mines d'Alès,
Parc Scientifique Georges Besse, F-30 035 Nîmes cedex 1, France
{nicolas.fiorini,sylvie.ranwez,jacky.montmain}@mines-ales.fr
[2] Montpellier SupAgro, UMR AGAP, F-34 060 Montpellier, France
vincent.ranwez@supagro.inra.fr

Abstract. Concept-based information retrieval is known to be a powerful and reliable process. It relies on a semantically annotated corpus, *i.e.* resources indexed by concepts organized within a domain ontology. The conception and enlargement of such index is a tedious task, which is often a bottleneck due to the lack of (semi-)automated solutions. In this paper, we first introduce a solution to assist experts during the indexing process thanks to semantic annotation propagation. The idea is to let them position the new resource on a semantic map, containing already indexed resources and to propose an indexation of this new resource based on those of its neighbors. To further help users, we then introduce indicators to estimate the robustness of the indexation with respect to the indicated position and to the annotation homogeneity of nearby resources. By computing these values before any interaction, it is possible to visually inform users on their margins of error, therefore reducing the risk of having a non-optimal, thus unsatisfying, annotation.

Keywords: conceptual indexing, semantic annotation propagation, imprecision management, semantic similarity, man-machine interaction.

1 Introduction

Nowadays, information technologies allow to handle huge quantity of resources of various types (images, videos, text documents, etc.). While they provide technical solutions to store and transfer almost infinite quantity of data, they also induce new challenges related to collecting, organizing, curating or analyzing information. Information retrieval (IR) is one of them and is a key process for scientific research or technology watch. Most information retrieval systems (IRS) rely on NLP (natural language processing) algorithms and on indexes — of both resources and queries to catch them — based on potentially weighted keywords. Nevertheless, those approaches are hampered by ambiguity of keywords in the indexes. There is also no structure acknowledged among the terms, *e.g.*, "car" is not considered to have anything in common with "vehicle". Concept-based information retrieval overcomes these issues when a semantically annotated corpus is

A. Laurent et al. (Eds.): IPMU 2014, Part III, CCIS 444, pp. 11–20, 2014.

available [1]. The vocabulary used to describe the resources is therefore controlled and structured as it is limited to the concepts available in the retained domain ontology. The latter also provides a framework to calculate semantic similarities to assess the relatedness between two concepts or between two groups of concepts. For instance, the relevance score between a query and a document can be computed using such measures, thus taking into account specialization and generalization relationships defined in the ontology. However, the requirement of an annotated corpus is a limit of such semantic based approach because augmenting a concept-based index is a tedious and time-consuming task which needs a high level of expertise. Here we propose an assisted semantic indexing process build upon a solution that we initially developed for resources indexed by a series of numerical values. Indeed, few years ago we proposed a semi-automatic indexing method to ease and support the indexing process of music pieces [2]. The goal of this approach was to infer a new annotation thanks to existing ones: what we called annotation propagation. A representative sample of the corpus is displayed on a map to let the expert point out (e.g. by clicking) the expected location of a new item in a neighborhood of similar resources. An indexation is then proposed to characterize the new item by summarizing those of nearby resources. In practice, this is done by taking the mean of the annotations of the closest neighbors, weighted by their distances to the click location. According to experts (DJs), the generated annotations were satisfying and our method greatly sped up their annotation process (the vector contained 23 independent dimensions and more than 20 songs were indexed in less than five minutes by an expert). We also successfully applied it in another context, to annotate photographs [3].

Here we propose a solution to extend this approach to a concept-based indexing process. Indeed, using indexation propagation is not straightforward in the semantic framework since concepts indexing a resource are not necessarily independent and the *average* of a set of concepts is not as obvious as the average of a set of numerical values. Summarizing the annotations of the closest neighbors cannot be done using a mere barycenter anymore. In our approach, we rely on semantic similarity measures in order to define the new item's annotation, thanks to its neighbors' ones. The exact location of the user click may have more or less impact on the proposed indexation depending on the resources' indexation and density in the selected area. To further support the user during the indexation process, we thus propose to analyze the semantic map to give him a visual indication of this impact. The user can then pay more attention to his selection when needed and go faster when possible.

After having described the annotation propagation related works in section 2, section 3 introduces our framework. We explain the objective function leading to the annotation and the impact of expert's imprecision withing the annotation propagation process. Section 4 focuses on the management of this imprecision and more specifically on solutions to help the user to take it into account within the indexing process. Results are discussed in section 5 with a special care of visual rendering proposed to assist the end-user.

2   Related Work

The need of associating metadata to increasingly numerous documents led the community to conceive approaches easing and supporting this task. Some of them infer new annotations thanks to existing ones [4]. The propagation of annotations then relies on existing links among entities to determine the propagation rules. For instance, [5] proposes to propagate metadata on the Internet by using the webpages hyperlinks, while [6] uses document citations. However, such propagation strategies only rely on application specific links and not on the underlying semantics of entities. A document cited as a counter example will thus inherit from metadata which are possibly irrelevant.

Some propagation methods have been specifically designed to handle media-specific resources not easily applicable in our context, *e.g.* picture annotation. They generally rely on image analysis of other pictures to define, thanks to probabilistic approaches, which terms to pick in order to annotate the new one. Some of them are based on maximal entropy [7], n-grams, bayesian inference [8] or SVM — Support Vector Machine. [9] proposed to propagate WordNet[1] based indexations rather than mere terms. [4] proposed a different approach in which resources are not necessarily pictures and their associated annotations are based on concepts. During the manual indexing process, they propose to assist the end user by propagating annotation from a newly annotated entity to the rest of the corpus. This led to a very fast solution, able to deal with millions of images, that assign the exact same annotation to several resources without any expert validation of the propagated information. We propose to use the opposite strategy and to assist the user by providing him with an initial annotation of the new resource based on existing ones. Indeed we believe this strategy is better adapted for the numerous contexts in which a smaller corpus is used but a high quality annotation is crucial. Each annotation should thus be specific and validated by a human expert. Maintaining and augmenting an index is always mentioned as a tedious task for the experts in the literature [9]. This issue is yet more problematic when the annotation is made of concepts because it requires an expertise of the whole knowledge model (*e.g.* a domain ontology). This means that, besides high-level knowledge of a specific domain, the user is also supposed to perfectly control the concepts and their structure, *i.e.* the model that analysts impose on him. While it is difficult to imagine a fully automatic annotation process (probably including NLP, clustering, etc.), we propose a semi-automatic approach supporting the expert's task.

3   A New Semantic Annotation Propagation Framework

Since it is difficult to visualize, analyze and understand the whole corpus at once (even with zoom and pan tools), we suggest presenting a semantic map displaying an excerpt of entities from the indexed corpus to the user (see Fig. 1). When having a new item to index, the expert is asked to identify, by clicking

---

[1] http://wordnet.princeton.edu

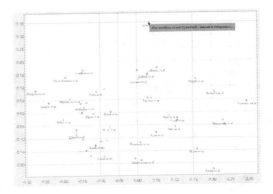

Fig. 1. The interactive map presented to the user

on this map, the location where this item should be placed. This clicked place is inherently a source of information: when the expert indicates a location on the map, he tries to assess which entities are related to the new one and which ones differ from it. By knowing the description of the closest items, the system may infer a new annotation and make use of the data structure they rely on to refine the proposal. The construction of the semantic map presented to the user requires i) to identify a subset of relevant resources and ii) to organize them in a visual and meaningful way. The first step is obviously crucial and can be tackled using NLP preprocessing, cross-citations or co-authorship. As this is an open question in itself and not the core of this contribution, we do not consider this point hereafter and assume that the excerpt is an input. Given an input set of relevant resources, we chose to use the MDS (Multi Dimensional Scaling) to display them on a semantic map so that resource closeness on the map reflects as much as possible their semantic relatedness.

3.1 Propagation of Semantic Annotations

The annotation propagation starts with the selection of the set N of the k closest neighbors of the click. We make a first raw annotation A_0, which is the union of all annotations of N, so $A_0 = \bigcup_{n_i \in N} Annotation(n_i)$. A_0 may be a specific and extensive annotation for the new entity, however, it is rarely a concise one. Indeed, during our tests (where $k = 10$) it leads to an annotation of 70 concepts on average. As we want to propose this to the user, the suggestion must contain fewer concepts that properly summarize the information in the neighborhood. To do so, we defined an objective function which, when maximized, gives an annotation $A^* \subset A_0$ that is the median of those of the elements of N, *i.e.*:

$$A^* = \arg\max_{A \subseteq A_0}\{score(A)\}, \quad score(A) = \sum_{n_i \in N} sim(A, Annotation(n_i)) \quad (1)$$

Where $sim(A, Annotation(n_i))$ denotes the similarity between two groups of concepts, respectively A and $Annotation(n_i)$. To assess it, we used the Lin pairwise semantic similarity measure [10] agregated by BMA [11]. This subset can not be found using a brute force approach as there are $2^{|A_0|}$ solutions. Therefore, the computation relies on a greedy heuristic starting from A_0 and deleting concepts one by one. The concept to be deleted at each step is the one leading to the greatest improvement of the objective function. When there is no possible improvement, the algorithm stops and returns a locally optimal A^* (see Algorithm 1).

Algorithm 1. Annotate$(M, (x, y))$

Input: M (an excerpt of) a semantic map; (x, y) a localization on M
Output: a semantic annotation for position (x, y) of M

```
N ← the 10 nearest neighbors of (x, y) on M
```
$A^* \leftarrow \bigcup_{n_i \in N} Annotation(n_i)$
```
repeat
    best_A ← A*
    local_optimum ← true
    for all c ∈ A* do
```
      $A \leftarrow A^* \setminus c$
```
        if score(A) > score(best_A) then
            best_A ← A
            local_optimum ← false
        end if
    end for
```
    $A^* \leftarrow best\_A$
```
until local_optimum is true
return  A*
```

3.2   Towards Assessment of Imprecision Impact

A misplacement of a new item on a map may have more or less impact on its proposed annotation depending on the local map content. To illustrate and better understand this point we studied the impact of misplacing a scientific article on a map containing papers annotated by MeSH terms. The PubMed annotation of those articles is considered to be the true one, and we examined two extreme cases. In the first one we focused on a central item on a homogeneous map (*i.e.* containing papers with highly similar annotations). Whereas in the second case we focused on an item surrounded by numerous resources displayed on a heterogeneous map (*i.e.* where entity annotations vary much more). We re-annotated the selected paper and computed the *annotation score*, which is the similarity between the computed annotation and the one proposed by PubMed. We did this for the correct location (*i.e.* the one corresponding to the projection of the PubMed annotation) and for 400 random locations.

We plot for each map the variation of the score with respect to the distance between the random location and the correct place (Fig. 2). As expected in both cases the annotation accuracy tends to decrease when the distance to the

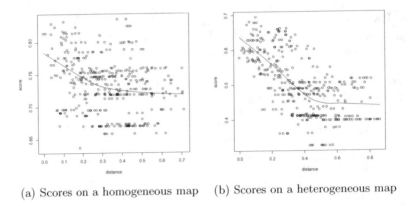

(a) Scores on a homogeneous map   (b) Scores on a heterogeneous map

Fig. 2. Impact of a misplacement on the annotation of a resource embedded in a map containing rather homogeneous (a) or rather heterogeneous (b) resources. The x-axis represents the distance from the correct location and the y-axis shows the similarity score between the correct annotation (from PubMed) and the propagated one.

correct position increases, and this decrease is much faster in the heterogeneous map than in the homogeneous one. Note also that in both cases a plateau is reached, which reflects the fact that both maps are meaningful excerpt that contain documents related to the one that is re-annotated.

4   Coping with Imprecision During Semantic Annotation Propagation

The definition of the neighborhood is a key step of the annotation propagation. Indeed, including too few neighbors will lead to an incomplete annotation whereas including too many or too semantically distant neighbors will lead to a less accurate annotation. Two main causes may affect the neighborhood definition. The first one is related to the creation of the map, where only relevant items should be displayed to the user. As this relevance problem is very context-specific, we do not tackle this potential source of error and we consider that the map is correct. The second one is directly related to one of the advantages of the method: the user interaction. While it is crucial to rely on the human expertise in such a cognitive process, imprecision inherent to any expert assessment cannot be neglected in the control of the annotation's accuracy. As illustrated in the previous section, the impact of such imprecision depends on the local map content. We hence propose to estimate the robustness of the proposed annotation with respect to the click position to help the user focusing on difficult cases while going faster on easier ones. Therefore, the user needs to approximately know the impact of a misplacement of an item on its suggested annotation. We compute an annotation stability indicator prior to display the map and visually help users by letting them know their margin of error when clicking. On a zone where this indicator is high the annotation is robust to a misplacement of a new resource because all elements' annotations are rather similar. Whereas if this indicator is low, the annotation variability associated to a misplaced click is high. Computing a single indicator for the whole map would not be very handy because the annotation variability may vary depending on the position on the map.

To efficiently compute those annotation stability indicators, we first split the map in n^2 smaller elementary pieces and generate the annotation of their center thanks to Algorithm 1. Using those pre-computed indexations we then assess the robustness of each submaps of M by identifying the number of connected elementary submaps sharing a similar annotation. By default, we split the map in 400 pieces ($n = 20$) and consider the set of elementary submaps with annotation having at least 95% of similarity with the considered one (*e.g.*, a divergence threshold of 5%).

5   Results, Tests and Discussions

5.1   Application Protocol

Our application protocol is based on scientific paper annotation. The corpus essentially deals with tumor-related papers and has been conceived by ITMO[2] Cancer to gather people working on tumors in France. As the documents are annotated via the MeSH ontology, we rely on this structure in our application. We need to set a groupwise semantic similarity measure to assess the similarity between pairs of documents (represented as groups of concepts). As explained in [12], there are plenty of measures available but many are context-specific. [13] propose a framework generalizing all these measures and implement most of them in a Java library, SML[3]. We decide to use the generic measure introduced by Lin [10], which is available in this library. It relies on an information content (IC) measure we want generic too so we pick the IC function using the ontology structure proposed in [14]. Finally, Lin is a pairwise semantic measure and we need to compare groups of concepts. Therefore, we use the Best Match Average (BMA) to aggregate such pairwise measures into a groupwise measure [11].

5.2   Objective Function and Its Heuristic Optimization

We rely on the protocol described hereafter to evaluate the heuristic and objective function we introduced in this paper to propagate semantic annotations. Since we have a benchmark corpus with 38,000 manually indexed documents — from PubMed —, we re-annotate some of these documents using the Algorithm 1 and estimate the quality of the resulting annotations with respect to several criteria. Firstly, its accuracy, that can be easily assessed by comparing it to the original PubMed annotation. Secondly, its concision, which is crucial to favor the expert validation of this suggested annotation and can be easily measured

---

[2] French multi-organism thematic institute initiated by the AvieSan program.
[3] Semantic Measures Library, http://www.semantic-measures-library.org/

as its number of concepts. Finally, the execution time which is crucial to have a reactive system favoring human-machine interaction. We compare our proposed solution with two simpler ones which are (i) the union of annotations of the 10 closest neighbors — denoted by A_0 — and (ii) the annotation of the closest neighbor.

Table 1. Comparison of 3 strategies to propagate semantic annotations. The first one relies on Algorithm 1, the second one considers the union of closest neighbors' annotation and the third one simply replicates the closest neighbor's annotation. Comparison is made through 3 criteria: the similarity with the reference annotation of PubMed, the annotation size and the execution time.

	Average similarity with PubMed	Average annotation size	Average execution time (ms)
A^* (Algorithm 1)	0.843	6.71	791
A_0 (union)	0.787	61.72	10
Closest neighbor	0.792	17.90	4

The performances of the 3 strategies are compared on the re-annotation of the same excerpt of 1,500 scientific papers randomly chosen from our above described corpus. Therefore, strategy performances were compared using a paired Student's test. Detailed results are provided in Table 1. The average similarity with PubMed is significantly better when the annotation is computed thanks to our Algorithm 1 than when computed with either the union of neighbors' annotations (p-value $= 8.93 * 10^{-8}$) or the annotation of the closest neighbor (p-value $= 8.443 * 10^{-6}$). The size of annotations is concise enough when using Algorithm 1 or the annotation obtained by duplicating that of the closest point; whereas it contains far too much concepts when considering the union of neighbors' annotation (A_0). Finally, even though the execution time of our algorithm is significantly higher than with other protocols, the average duration remains below 1 second ensuring a good system responsiveness. We observed the same trends for different k values (k being the number of considered neighbors) and except for very small values of k (< 4), there is no major variation in the method performances.

Note that the map construction time, which requires around 10 seconds is not included in computing times provided in Table 1. We excluded them since they are not impacted by the chosen annotation method and are done only once for several annotation. Indeed, we do not recompute the map each time users annotate a new resource, otherwise the map would constantly change impeding the construction of a mental representation of the data.

5.3   Visually Assist the User

In order to enhance the human-machine interaction and to improve the efficiency of the indexation process, we propose to give visual hints to the end-user about the impact of a misplacement. To that aim we color the area of the map surrounding the current mouse position that will led to similar annotations. More precisely, the colored area is such that positioning the item anywhere in this area will lead to an indexation similar to the one obtained by positioning the item at the current mouse position. This approach we think meaningful reminds the brush selection in graphics editing program: with an oversized brush, it is difficult to precisely color an area even if the graphic artist is very talented. Here, a big grey zone shows that the click position does not really matter on the final annotation, whereas a small grey area means that precision is expected. Figure 3 shows a representation of such zones on different parts of the same map.

By default, the accepted imprecision is 5% so that clicking in the colored area or at the current position will provide annotation with at least 95% of similarity. This default threshold can easily be changed through the graphic user interface. The definition of the zone to color for a pointed location can be done in real-time thanks to the pre-process (see Section 4).

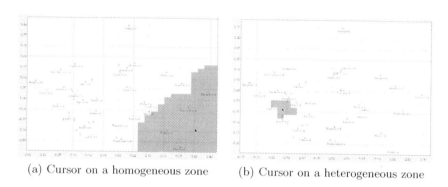

(a) Cursor on a homogeneous zone   (b) Cursor on a heterogeneous zone

Fig. 3. Visual hints of position deviation impact. The cursor is surrounded by a grey area indicating positions that would lead to similar annotation. A large area means that the user can be confident when clicking whereas a small one means that precision is important.

6   Conclusion

In this paper, we present a way to propagate semantic annotation by positioning new resources on a semantic map and to provide visual hints of the impact of a misplacement of the resource on this map. Our method i) greatly simplifies the otherwise tedious annotation process, ii) is significantly more accurate than 2 naive propagation strategies and iii) conserves fast running time below the second. We considered the identification of relevant documents to be displayed on the map as an independent task out of the scope of this paper. This is nonetheless a key step when the corpus is too big to be fully displayed. In the case of scientific papers, several strategies can be adopted to identify those relevant documents based on NLP techniques, cross-citations or co-authoring. In order to compare our method with more classic classifiers, we are planning to evaluate this approach with precision/recall, even though those values are not fully satisfactory in a semantic context.

An advantage of having a semi-automatic annotation process is that the underlying system handle the ontology-based knowledge representation and its impact on semantic similarities. Experts doing the annotation can thus focus on resources similarities without the need of having an extra knowledge of the ontology structure. This does not only simplify their task but also limit the subjectivity of their annotations. This method paves the way for deeply needed assisted semantic annotation tools.

References

1. Haav, H.M., Lubi, T.L.: A survey of concept-based information retrieval tools on the web. In: Proceedings of the 5th East-European Conference ADBIS, vol. 2, pp. 29–41 (2001)
2. Crampes, M., Ranwez, S., Velickovski, F., Mooney, C., Mille, N.: An integrated visual approach for music indexing and dynamic playlist composition. In: Electronic Imaging 2006, p. 607103. International Society for Optics and Photonics (2006)
3. Crampes, M., de Oliveira-Kumar, J., Ranwez, S., Villerd, J.: Visualizing social photos on a hasse diagram for eliciting relations and indexing new photos. IEEE Transactions on Visualization and Computer Graphics 15(6), 985–992 (2009)
4. Lazaridis, M., Axenopoulos, A., Rafailidis, D., Daras, P.: Multimedia search and retrieval using multimodal annotation propagation and indexing techniques. Signal Processing: Image Communication 28(4), 351–367 (2013)
5. Marchiori, M.: The limits of web metadata, and beyond. Computer Networks and ISDN Systems 30(1), 1–9 (1998)
6. Abrouk, L., Gouaïch, A.: Automatic annotation using citation links and co-citation measure: Application to the water information system. In: Mizoguchi, R., Shi, Z.-Z., Giunchiglia, F. (eds.) ASWC 2006. LNCS, vol. 4185, pp. 44–57. Springer, Heidelberg (2006)
7. Jeon, J., Manmatha, R.: Using maximum entropy for automatic image annotation. In: Enser, P.G.B., Kompatsiaris, Y., O'Connor, N.E., Smeaton, A.F., Smeulders, A.W.M. (eds.) CIVR 2004. LNCS, vol. 3115, pp. 24–32. Springer, Heidelberg (2004)
8. Zhang, H., Su, Z.: Improving cbir by semantic propagation and cross modality query expansion. In: Proc. of the Int. Workshop on Multimedia Content-based Indexing and Retrieval, Brescia, pp. 19–21 (2001)
9. Shevade, B., Sundaram, H.: Vidya: an experiential annotation system. In: Proceedings of the 2003 ACM SIGMM Workshop on Experiential Telepresence, pp. 91–98. ACM (2003)
10. Lin, D.: An information-theoretic definition of similarity. In: ICML, vol. 98, pp. 296–304 (1998)
11. Schlicker, A., Domingues, F.S., Rahnenführer, J., Lengauer, T.: A new measure for functional similarity of gene products based on gene ontology. BMC Bioinformatics 7(1), 302 (2006)
12. Harispe, S., Janaqi, S., Ranwez, S., Montmain, J.: From theoretical framework to generic semantic measures library. In: Demey, Y.T., Panetto, H. (eds.) OTM 2013 Workshops. LNCS, vol. 8186, pp. 739–742. Springer, Heidelberg (2013)
13. Harispe, S., Ranwez, S., Janaqi, S., Montmain, J.: The Semantic Measures Library and Toolkit: fast computation of semantic similarity and relatedness using biomedical ontologies (2013), doi:10.1093/bioinformatics
14. Seco, N., Veale, T., Hayes, J.: An intrinsic information content metric for semantic similarity in wordnet. In: ECAI, vol. 16, p. 1089. Citeseer (2004)

A Fuzzy R Code Similarity Detection Algorithm

Maciej Bartoszuk[1] and Marek Gagolewski[2,3]

[1] Interdisciplinary PhD Studies Program,
Systems Research Institute, Polish Academy of Sciences, Poland
m.bartoszuk@phd.ipipan.waw.pl
[2] Systems Research Institute, Polish Academy of Sciences,
ul. Newelska 6, 01-447 Warsaw, Poland
gagolews@ibspan.waw.pl
[3] Faculty of Mathematics and Information Science, Warsaw University of Technology,
ul. Koszykowa 75, 00-662 Warsaw, Poland

Abstract. R is a programming language and software environment for performing statistical computations and applying data analysis that increasingly gains popularity among practitioners and scientists. In this paper we present a preliminary version of a system to detect pairs of similar R code blocks among a given set of routines, which bases on a proper aggregation of the output of three different $[0, 1]$-valued (fuzzy) proximity degree estimation algorithms. Its analysis on empirical data indicates that the system may in future be successfully applied in practice in order e.g. to detect plagiarism among students' homework submissions or to perform an analysis of code recycling or code cloning in R's open source packages repositories.

Keywords: R, antiplagiarism detection, code cloning, fuzzy proximity relations, aggregation.

1 Introduction

The R [16] programming language and environment is used by many practitioners and scientists in the fields of statistical computing, data analysis, data mining, machine learning and bioinformatics. One of the notable features of R is the availability of a centralized archive of software packages called CRAN – the Comprehensive R Archive Network. Although it is not the only source of extensions, it is currently the largest one, featuring 5505 packages as of May 3, 2014. This repository is a stock of very interesting data, which may provide a good test bed for modern soft computing, data mining, and aggregation methods. For example, some of its aspects can be examined by using the impact functions aiming to measure the performance of packages' authors, see [6], not only by means of the software quality (indicated e.g. by the number of dependencies between packages or their downloads) but also their creators' productivity.

To perform sensible analyses, we need to cleanse the data set. For example, some experts hypothesize that a considerable number of contributors treat open

A. Laurent et al. (Eds.): IPMU 2014, Part III, CCIS 444, pp. 21–30, 2014.

source-ness too liberally and do not "cite" the packages providing required facilities, i.e. while developing a package, they sometimes do not state that its code formally depends on some facilities provided by a third-party library. Instead, they just copy-paste the code needed, especially when its size is small. In order to detect such situations, reliable code similarity detection algorithms are needed.

Moreover, it may be observed that R is being more and more eagerly taught at universities. In order to guarantee the high quality of the education processes, automated methods for plagiarism detection, e.g. in students' homework submissions, are of high importance.

The very nature of the R language is quite different from the other ones. Although R's syntax resembles that of C/C++ to some degree, it is a functional language with its own unique features. It may be observed (see Sec. 4) that existing plagiarism detection software, like MOSS [1] or JPlag [13], fails to identify similarities between R functions' source codes correctly. Thus, the aim of this paper is to present a preliminary version of a tool of interest. It is widely known from machine learning that no single method has perfect performance in every possible case: when dealing with individual heuristic methods one investigates only selected aspects of what he/she thinks plagiarism is in its nature, and does not obtain a "global view" on the subject. Thus, the proposed algorithm bases on a proper aggregation of (currently) three different fuzzy proximity degree estimation procedures (two based on the literature and one is our own proposal) in order to obtain a wider perspective of the data set. Such a synthesis is quite challenging, as different methods may give incomparable estimates that should be calibrated prior to their aggregation. An empirical analysis performed on an exemplary benchmark set indicates that our approach is highly promising and definitely worth further research.

The paper is structured as follows. Sec. 2 describes 3 fuzzy proximity measures which may be used to compare two functions' source codes. Sec. 3 discusses the choice of an aggregation method that shall be used to combine the output of the aforementioned measures. In Sec. 4 we present an empirical study of the algorithm's discrimination performance. Finally, Sec. 5 concludes the paper.

2   Three Code Similarity Measures

Assume we are given a set of n functions' source codes $\mathcal{F} = \{f_1, \ldots, f_n\}$, where f_i is a character string, i.e. $f_i \in \bigcup_{k=1}^{\infty} \Sigma^k$, where Σ is a set of e.g. ASCII-encoded characters. Each f_i should be properly normalized by i.a. removing unnecessary comments and redundant white spaces, as well as by applying the same indentation style. In R, if f represents source code (character vector), this may be easily done by calling f <- deparse(parse(text=f)).

We are interested in creating a $[0, 1]$-valued (fuzzy) proximity relation, cf. e.g. [5], defined by a membership function $\mu : \mathcal{F}^2 \to [0, 1]$ such that it is at least:

– reflexive ($\mu(f_i, f_i) = 1$),
– symmetric ($\mu(f_i, f_j) = \mu(f_j, f_i)$).

Intuitively, we have $\mu(f_i, f_j) = 0$ iff f_i, f_j are entirely distinct (according to some criterion), $\mu(f_i, f_j) = 1$ iff they are identical, and immediate values of $\mu(f_i, f_j)$ describe the degree of their "partial proximity". As this is a preliminary study, we omit the discussion on the transitivity (formally, T-transitivity for some t-norm T) of μ and leave it for further research.

Measuring code similarity is always a task that bases on some heuristics. Below we present three different methods μ_1, μ_2, μ_3 to compare two R functions. μ_1 computes the Levenshtein distance [10] between source codes. This is a perfect method to detect verbatim copies, but small modifications (like changing names of variables) should also be easily revealed. Further on, μ_2 bases on our own proposal that takes into account the counts of the number of calls to other functions as well as the names of these functions. This approach is potentially fruitful, as each R statement corresponds to a call to some function. For example:

`x <- y * z`	is equivalent to	`'<-'(x,`
		`    '*'(y, z)`
		`)`

and

`for (i in 1:10)`
`{ x <- x+i }`
	is in fact	`'for'(i,`
		`    ':'(1, 10),`
		`    '{'(`
		`       '<-'(x,`
		`           '+'(x, i)`
		`) ) )`

For instance, a function which calls `sqrt()` twice and `pnorm()` five times is certainly different from the one that calls `readLines()` three times. Last method, μ_3, is based on tokens' analysis, i.e. utilizes the code syntax tree. Here two token sequences are compared by the well-known "greedy string tiling" algorithm, cf. [18].

2.1   Levenshtein Distance

The first proximity relation is based on the Levenshtein distance [10] between two character strings. Intuitively, the Levenshtein distance between two sequences is the minimum number of single-character edits (insertions, deletions, substitutions) required to obtain one string from the other. More formally, the Levenshtein distance between two strings a, b is given by $\mathrm{lev}_{a,b}(|a|, |b|)$ where:

$$\mathrm{lev}_{a,b}(i, j) = \begin{cases} \max(i, j) & \text{if } \min(i, j) = 0, \\ \min \begin{cases} \mathrm{lev}_{a,b}(i - 1, j) + 1 \\ \mathrm{lev}_{a,b}(i, j - 1) + 1 \\ \mathrm{lev}_{a,b}(i - 1, j - 1) + \mathbf{I}_{(a_i \neq b_j)} \end{cases} & \text{otherwise,} \end{cases}$$

where $\mathbf{I}_{(a_i \neq b_j)} = 1$ iff $a_i \neq b_j$ and 0 otherwise, and $|a|$ is the length of a.

A fast implementation of the Levenshtein distance uses a dynamic programming technique. Here we rely on the `adist()` function in the *utils* package for R.

Of course, we get $\text{lev}_{a,b}(i, j) = 0$ if the strings are equal. As we require the proximity degree to be in the interval $[0, 1]$, we ought to normalize the result somehow. In our case, we assume:

$$\mu_1(f_i, f_j) = 1 - \frac{\text{lev}_{f_i, f_j}(|f_i|, |f_j|)}{\max(|f_i|, |f_j|)}.$$

It is easily seen that for a pair of identical strings we obtain value of 1. On the other hand, for "abc" and "defghi" we get 0, as $\text{lev}_{\text{"abc", "defghi"}}(3, 6) = 6$.

2.2 Function Calls Counts

If a function calls some mathematical routines many times and another one rather more often utilizes some data processing methods, we may presume that these two functions are rather not similar. On the other hand, if two functions call `sort()` twice each, then we cannot be certain that there are the same. We see that this heuristic may provide us with a "necessary", but not a "sufficient" proximity estimate. Note that – as it has been stated earlier on – R is a programming language in which function calls play a central role.

Let $\mathcal{R}$ denote the set of names of all possible R functions and $c_i(g)$ be equal to the number of calls of $g \in \mathcal{R}$ within f_i. Our second proximity estimation method is defined by:

$$\mu_2(f_i, f_j) = \frac{\sum_{g \in \mathcal{R}} 2\left(c_i(g) \wedge c_j(g)\right)}{\sum_{g \in \mathcal{R}} \left(c_i(g) + c_j(g)\right)}.$$

2.3 Longest Common Token Substrings

The idea of the third algorithm has been inspired by the method presented in [14], cf. also [2]. It is quite invulnerable – at least in theory – to such methods used by plagiarists as changing the variable names, swapping two large fragments of code or modifications of numeric constants. It operates in two phases. First, we parse all the functions that are to be compared and convert them into token strings. Transformation $T : \mathcal{F} \to \mathcal{F}'$ gives a token string f_i' calculated from f_i. Then we compare the token strings in pairs to estimate the similarity of each pair. This task is based on finding the longest common token substrings.

Conversion of Functions' Source Codes to Token Strings. As a rule, a tokenization should be performed in such a way that it depicts the highly abstract "essence" of a function (which is not easy to alter by a plagiarist). In the R language, creation of token strings is very easy: the `getParseData()` function from the *utils* package takes some parsed code as an argument and returns i.a. the information we are looking for. Interestingly, each token is represented internally by an integer number, and a token string is in fact an integer vector.

Example 1. Consider the following source code.

```
1      f <- function(x)
2      {
3          stopifnot(is.numeric(x))
4          y <- sum(x)
5          y
6      }
```

The tokenization procedure results in:

```
1   expr, SYMBOL, expr, LEFT_ASSIGN, expr, FUNCTION, '(',
        SYMBOL_FORMALS, ')', expr,
2   '{',
3   expr, SYMBOL_FUNCTION_CALL, expr, '(', expr,
        SYMBOL_FUNCTION_CALL, expr, '(', SYMBOL, expr, ')', ')'
4   expr, SYMBOL, expr, LEFT_ASSIGN, expr, SYMBOL_FUNCTION_CALL
        expr, '(', SYMBOL, expr, ')',
5   SYMBOL, expr,
6   '}'
```

Comparison of Two Token Strings. The "greedy string tiling" algorithm, as described in [18], is a method used to compare the similarity of two token strings. Interestingly, such a method is also used in JPlag [13].

The pseudo-code of the algorithm is presented below. The $\oplus$ operator in line 14 means that we add a match to a matches set if and only if it does not overlap with one of the matches already in this set. The triple $match(a, b, l)$ denotes an association between corresponding substrings of A and B, starting at positions A_a and B_b respectively, of length l.

```
1   Greedy_String_Tiling(String A, String B) {
2       tiles = {};
3       do {
4           maxmatch = MinimumMatchLength;
5           matches = {};
6           Forall unmarked tokens A_a in A {
7               Forall unmarked tokens B_b in B {
8                   j=0;
9                   while (A_{a+j}==B_{b+j} &&
10                          unmarked(A_{a+j}) &&
11                          unmarked(B_{b+j}))
12                       j++;
13                   if (j==maxmatch)
14                       matches = matches ⊕ match(a,b,j);
15                   else if (j > maxmatch){
16                       matches = {match(a,b,j)};
17                       maxmatch = j;
18                   }
```

```
19                        }
20                    }
21                Forall match(a,b,maxmatch) ∈ matches {
22                    for j=0...(maxmatch−1){
23                        mark(A_{a+j});
24                        mark(B_{b+j});
25                    }
26                    tiles = tiles ∪ match(a,b,maxmatch);
27                }
28            } while(maxmatch > MinimumMatchLength);
29            return tiles;
30  }
```

The last task consists of computing a proximity degree of two token strings f'_i and f'_j. In [14], the following formula was proposed:

$$\mu_3(f'_i, f'_j) = \frac{2\text{coverage}(\textit{tiles})}{|f'_i| + |f'_j|},$$

where

$$\text{coverage}(\textit{tiles}) = \sum_{\text{match}(a,b,length)\in\textit{tiles}} length,$$

and $\textit{tiles} = \text{Greedy\_String\_Tiling}(f'_i, f'_j)$.

3 Aggregation and Defuzzification of Proximity Measures

Of course, each of the three proximity relations described above may be represented as a square matrix $M^{(k)}$, where $m_{ij}^{(k)}$ equals to $\mu_k(f_i, f_j)$, $k = 1, 2, 3$. When they are obtained, special attention should be paid to their proper aggregation, and then defuzzification, i.e. projection to $\{0, 1\}$.

The most straightforward approach could consist of determining

$$\mu(f_i, f_j) = A(\mu_1(f_i, f_j), \mu_2(f_i, f_j), \mu_3(f_i, f_j))$$

using an aggregation function $A : [0, 1]^3 \to [0, 1]$, i.e. a function at least fulfilling the following conditions: (a) $A(0, 0, 0) = 0$, (b) $A(1, 1, 1) = 1$, (c) A is a non-decreasing in each variable, cf. [7]. For example, one could study the family of weighted arithmetic means, given by 3 free parameters a_1, a_2, a_3:

$$\mu(f_i, f_j) = a_1\mu_1(f_i, f_j) + a_2\mu_2(f_i, f_j) + a_3\mu_3(f_i, f_j),$$

where $a_1 + a_2 + a_3 = 1$ and $a_1, a_2, a_3 \geq 0$. After computing μ, it should be defuzzified in order to obtain a crisp proximity relation P. This may be done by setting $P(f_i, f_j) = \mathbf{I}_{\mu(f_i, f_j) \geq \alpha}$ for some $\alpha \in [0, 1]$.

However, we may note that the values of μ_1, μ_2, and μ_3 are not necessarily comparable – they are not of the same order of magnitude. In other words,

$\mu_1(f_i, f_j) = 0.6$ may not describe the same "objective" proximity degree as $\mu_2(f_i, f_j) = 0.6$. This is illustrated in Fig. 1a: in our exemplary empirical analysis μ_2 seems to give much larger values than the other relations. Thus, their values should be normalized somehow before aggregation. Nevertheless, in our case we will apply a different approach based on logistic regression, which is described in the next section.

4 Experimental Results

In order to verify the proposed method, we have created a benchmark set of R functions. The data set consists of students' homework submissions (8 different tasks). A few of them were evidently classified by the tutors as cases of plagiarisms (which was then confirmed by the very students). Moreover, we have added some straightforward modifications of the analyzed routines, e.g. ones with just the variable names changed or those with a more "expanded" forms, like:

```
1   sort_list <- function (x, f) {
2       o <- order (unlist (lapply (x, f)))
3       x[o]
4   }
```

being decomposed to:

```
1   sort_list <- function (x, f) {
2       v1 <- lapply (x, f)
3       v2 <- unlist (v1)
4       o  <- order (v2)
5       x[o]
6   }
```

Table 1. Performance of the systems considered

	FP	FN	t [s]
Proposed approach	3	7	3
JPlag - > 0.1 - plain text - original code	39	55	≈ 10
JPlag - > 0.1 - plain text - normalized code	63	57	≈ 10
JPlag - > 0.4 - plain text - original code	0	107	≈ 10
JPlag - > 0.4 - plain text - normalized code	0	110	≈ 10
MOSS - plain text - normalized code	31	89	≈ 3
MOSS - plain text - original code	29	84	≈ 3
MOSS - C - original code	1	94	≈ 3
MOSS - C - normalized code	1	94	≈ 3
MOSS - Haskell - normalized code	22	66	≈ 3
MOSS - Haskell - original code	14	72	≈ 3

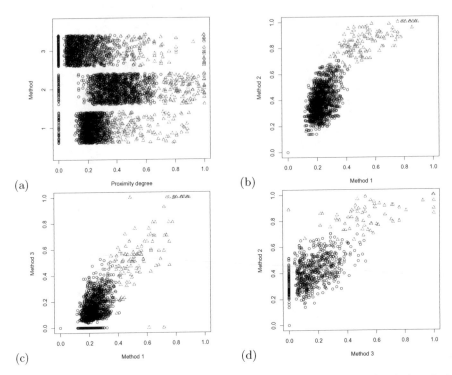

Fig. 1. Values of proximity measures μ_1, μ_2, μ_3 for non-similar pairs (circles) and plagiarisms (triangles)

In result, we obtained a set of 58 functions. 138 pairs, i.e. ca. 4%, are classified a priori as cases of plagiarisms. The whole procedure was computed within just a few seconds on a PC with Intel Core i5-2500K CPU 3.30GHz and 8 GB of RAM (the greedy string tiling was implemented in C++). Notably, similar timings have been obtained by analyzing the same data set with MOSS [1] and JPlag [13].

Figure 1bcd depicts the relationships between the three proximity measures. The measures are strongly positively correlated. The Spearman ρ is equal to about 0.8 for (μ_1, μ_2) and (μ_1, μ_3) and 0.75 for (μ_2, μ_3). Moreover from Fig. 1a we see that the plagiarism and non-plagiarism case cannot be discriminated so easily with just one measure. The best solution was obtained for α-cut of the form $\mathbf{I}_{\mu_1(f_i,f_j)\geq 0.38}$ (5 false positives (**FP**), i.e. cases in which we classify as plagiarisms non-similar pairs of functions, and 16 false negatives (**FN**), i.e. cases in which we fail to detect plagiarisms), $\mathbf{I}_{\mu_2(f_i,f_j)\geq 0.65}$ (6 and 8, respectively), and $\mathbf{I}_{\mu_3(f_i,f_j)\geq 0.41}$ (15 and 33, respectively). Thus, the intuition behind combining multiple proximity relations seems to be valid.

In order to discriminate between the observations from two classes we may e.g. apply discrimination analysis. A fitted logistic regression model which gives 3 false positives and 7 false negatives is of the form:

$$P(f_i, f_j) = \mathbf{I}\left(\frac{\exp\left(-46.9 + 26\mu_1(f_i, f_j) + 59.1\mu_2(f_i, f_j) - 5.5\mu_3(f_i, f_j)\right)}{1 + \exp\left(-46.9 + 26\mu_1(f_i, f_j) + 59.1\mu_2(f_i, f_j) - 5.5\mu_3(f_i, f_j)\right)} \geq 0.5 \right).$$

The performance of our method was compared with JPlag and MOSS, see Table 1 for a summary. As none of them has built-in support for R syntax, we analyzed two versions of our benchmark set: one "as is" and the other one normalized with deparse() and parse(). Moreover, we tried to determine which language (-l) flag leads to best result in case of MOSS. Quite interestingly, the functional language Haskell gave the smallest number of incorrect hits. On the other hand, even though JPlag supports Java, C/C++, Scheme, and C#, it treated R sources as plain text (note that μ_3, however, uses a JPlag-like approach). In this case, the user may decide which similarity level pair of function should be displayed. Here we have chosen a very liberal 10% level and a more conservative 40% level.

5   Conclusions

The preliminary version of our plagiarism detection system seems to be quite fast and accurate. As far as the analyzed data set is concerned, it correctly classified most of the suspicious similarities. Moreover, the system is – at least theoretically – not vulnerable to typical attacks, like changing names of variables, swapping code places and function composition/decomposition.

Of course, there is still much to be done to make our algorithm more reliable. First of all, we should investigate some kind of transitivity notion in our proximity relation. This is needed to detect fuzzy cliques of functions (e.g. in order to detect groups of plagiarists). Moreover, we could think of a more valid normalization of the relations' values or applying different modern data mining techniques for data classification.

Even more, there are still many other well-known groups of methods which can be aggregated. For example, a plagiarism-detection method based on the Program Dependency Graph (PDG) [4,11,15] represents the so-called control and data dependencies between the statements in a program. This approach aims to be immune to changing variables' names, swapping small fragments of code, inserting or deleting single lines of code and substituting while loops for their equivalents.

There is still an open question if the Levenshtein distance is the most proper plain text metric. There are many other measures, such as the Damerau-Levenshtein distance [3], Hamming distance [8], Jaro-Winkler distance [17], Lee distance [9] which should be examined, see also [12] for a survey on the topic.

Lastly, we should collect and analyze a more extensive data set of R functions, so that our method could be calibrated more properly, perhaps even with a form of human feedback.

Acknowledgments. M. Bartoszuk acknowledges the support by the European Union from resources of the European Social Fund, Project "Information technologies: Research and their interdisciplinary applications", agreement UDA-POKL.04.01.01-00-051/10-00 via the Interdisciplinary PhD Studies Program.

References

1. Aiken, A.: MOSS (Measure of software similarity) plagiarism detection system, `http://theory.stanford.edu/~aiken/moss/`
2. Chilowicz, M., Duris, E., Roussel, G.: Viewing functions as token sequences to highlight similarities in source code. Science of Computer Programming 78, 1871–1891 (2013)
3. Damerau, F.J.: A technique for computer detection and correction of spelling errors. Communications of the ACM 7(3), 171–176 (1964)
4. Ferrante, J., Ottenstein, K.J., Warren, J.D.: The program dependence graph and its use in optimization. ACM Trans. Program Lang. Syst. 9(3), 319–349 (1987)
5. Fodor, J., Roubens, M.: Fuzzy Preference Modelling and Multicriteria Decision Support. Springer (1994)
6. Gagolewski, M., Grzegorzewski, P.: Possibilistic analysis of arity-monotonic aggregation operators and its relation to bibliometric impact assessment of individuals. International Journal of Approximate Reasoning 52(9), 1312–1324 (2011)
7. Grabisch, M., Marichal, J.L., Mesiar, R., Pap, E.: Aggregation functions. Cambridge University Press (2009)
8. Hamming, R.W.: Error detecting and error correcting codes. Bell System Technical Journal 29(2), 147–160 (1950)
9. Lee, C.Y.: Some properties of nonbinary error-correcting codes. IRE Transactions on Information Theory 4(2), 77–82 (1958)
10. Levenshtein, I.: Binary codes capable of correcting deletions, insertions, and reversals. Soviet Physics Doklady 10(8), 707–710 (1966)
11. Liu, C., Chen, C., Han, J., Yu, P.S.: GPLAG: Detection of Software Plagiarism by Program Dependence Graph Analysis. In: Proc. 12th ACM SIGKDD Intl. Conf. Knowledge Discovery and Data Mining (KDD 2006), pp. 872–881 (2006)
12. Navarro, G.: A guided tour to approximate string matching. ACM Computing Surveys 33(1), 31–88 (2001)
13. Prechelt, L., Malpohl, G., Philippsen, M.: Finding plagiarisms among a set of programs with JPlag. Journal of Universal Computer Science 8(11), 1016–1038 (2002)
14. Prechelt, L., Malpohl, G., Phlippsen, M.: JPlag: Finding plagiarisms among a set of programs. Tech. rep. (2000)
15. Qu, W., Jia, Y., Jiang, M.: Pattern mining of cloned codes in software systems. Information Sciences 259, 544–554 (2014)
16. R Core Team: R: A Language and Environment for Statistical Computing. R Foundation for Statistical Computing, Vienna, Austria (2014), `http://www.R-project.org/`
17. Winkler, W.E.: String Comparator Metrics and Enhanced Decision Rules in the Fellegi-Sunter Model of Record Linkage. In: Proc. Section on Survey Research Methods (ASA), pp. 354–359 (1990)
18. Wise, M.J.: String similarity via greedy string tiling and running Karp-Rabin matching. Tech. rep., Dept. of Computer Science, University of Sydney (1993)

L-Fuzzy Context Sequences
on Complete Lattices

Cristina Alcalde[1] and Ana Burusco[2]

[1] Dept. Matemática Aplicada, Escuela Universitaria Politécnica,
UPV/EHU Plaza de Europa 1, 20018 San Sebastián, Spain
c.alcalde@ehu.es
[2] Departamento de Automática y Computación, Universidad Pública de Navarra,
Campus de Arrosadía, 31006 Pamplona, Spain
burusco@unavarra.es

Abstract. This work studies the L-fuzzy context sequences when L is a complete lattice extending the results obtained in previous works with $L = [0, 1]$. To do this, we will use n-ary OWA operators on complete lattices. With the aid of these operators, we will study the different contexts values of the sequence using some new relations. As a particular case, we have the study when $L = \mathcal{J}([0, 1])$. Finally, we illustrate all the results by means of an example.

Keywords: L-fuzzy context, L-fuzzy concept, L-fuzzy context sequences, n-ary OWA operators.

1 Introduction

The *L-Fuzzy Concept Analysis* studies the information from an *L*-fuzzy context by means of the *L*-fuzzy concepts. These *L*-fuzzy contexts are tuples (L, X, Y, R), with L a complete lattice, X and Y sets of objects and attributes, and $R \in L^{X \times Y}$ an *L*-fuzzy relation between the objects and the attributes.

In some situations, we have a sequence formed by the *L*-fuzzy contexts (L, X, Y, R_i), $i = \{1, \ldots, n\}$, $n \in \mathbb{N}$, where R_i is the ith relation between the objects of X and the attributes of Y. The study of these *L*-fuzzy context sequences will be the main target of this work.

A particular case of this sequence is when it represents an evolution in time of an *L*-fuzzy context.

To start, we will see some important results about the *L*-Fuzzy Concept Analysis.

2 *L*-Fuzzy Contexts

The Formal Concept Analysis of R. Wille [23] extracts information from a binary table that represents a Formal context (X, Y, R) with X and Y finite sets of objects and attributes respectively and $R \subseteq X \times Y$. The hidden information

A. Laurent et al. (Eds.): IPMU 2014, Part III, CCIS 444, pp. 31–40, 2014.

consists of pairs (A, B) with $A \subseteq X$ and $B \subseteq Y$, called Formal concepts, verifying $A^* = B$ and $B^* = A$, where $(\cdot)^*$ is a derivation operator that associates the attributes related to the elements of A with every object set A, and the objects related to the attributes of B with every attribute set B. These Formal Concepts can be interpreted as a group of objects A that shares the attributes of B.

In previous works [8,9] we have defined the L-fuzzy contexts (L, X, Y, R), with L a complete lattice, X and Y sets of objects and attributes respectively and $R \in L^{X \times Y}$ a fuzzy relation between the objects and the attributes.

In our case, to work with these L-fuzzy contexts, we have defined the derivation operators $(\cdot)_1$ and $(\cdot)_2$ given by means of these expressions:

$$\forall A \in L^X, \forall B \in L^Y, \ A_1(y) = \inf_{x \in X}\{\mathcal{I}(A(x), R(x, y))\}$$
$$B_2(x) = \inf_{y \in Y}\{\mathcal{I}(B(y), R(x, y))\}$$

with $\mathcal{I}$ a fuzzy implication operator defined in the lattice $(L, \leq)$.

Some authors use a residuated implication operator in their definitions of derivation operators [7,19,20].

The information stored in the context is visualized by means of the L-fuzzy concepts that are pairs $(A, B) \in (L^X \times L^Y)$ fulfilling $A_1 = B$ and $B_2 = A$. These pairs, whose first and second components are said to be the fuzzy extension and intension respectively, represent a group of objects that share a group of attributes in a fuzzy way.

On the other hand, given $A \in L^X$, (or $B \in L^Y$) we can obtain the associated L-fuzzy concept applying twice the derivation operators. In the case of using a residuated implication, as we do in this work, the associated L-fuzzy concept is (A_{12}, A_1) (or (B_2, B_{21})).

Other important results about this theory are in [6,10,7,2,19,15,16,20].

3   L-Fuzzy Context Sequences

In this section, we are interested in the study of the L-fuzzy context sequences where L is a complete lattice. We have analyzed these sequences when $L = [0, 1]$ in [4,5]. To do this, we have given the formal definition:

Definition 1. *An L-fuzzy context sequence is a tuple $(L, X, Y, R_i), i = \{1, \ldots, n\}$, $n \in \mathbb{N}$, with L a complete lattice, X and Y sets of objects and attributes respectively and $R_i \in L^{X \times Y}, \forall i = \{1, \ldots, n\}$, a family of L-fuzzy relations between X and Y.*

In previous works [12,11], we have done some studies in order to aggregate the information of different contexts with the same set of objects and attributes. The use of weighted averages [13,14] (with $L=[0,1]$) in order to summarize the information stored in the different relations allows us to associate different weights to the L-fuzzy contexts highlighting some of them. However, it is possible that some observations of an L-fuzzy context of the sequence are interesting whereas others not so much. For instance, in [3] we study that the used methods to obtain the L-fuzzy concepts do not give good results when we have very low values in some relations. Moreover, we want now to do different studies based on different exigency levels. This is one of the new contributions of this work.

In order to introduce this subject, let us see the following example.

Example 1. Let $(L, X, Y, R_i), i = \{1, \ldots, n\}$, be an L-fuzzy context sequence that represents the sales of sports articles (X) in some establishments (Y) throughout a period of time (I), and we want to study the places where the main sales hold taking into account that there are seasonal sporting goods (for instance skies, bathing suits) and of a certain zone (it is more possible to sale skies in Colorado than in Florida).

In this case, the weighted average model is not valid since it is very difficult to associate a weight to an L-fuzzy context (in some months more bath suits are sold whereas, in others, skies are). To analyze this situation, it could be interesting the use of the OWA operators [21,17] with the most of the weights near the largest values. In this way, we give more relevance to the largest observations, independently of the moment when they have taken place and, on the other hand, we would avoid some small values in the resulting relations (that can give problems in the calculation of the L-fuzzy concepts as was studied in [3]).

The next section summarizes the main results about these operators.

4 $n-$ary OWA Operators

This is the definition of these operators given by Yager [21]:

Definition 2. *A mapping F from $L^n \longrightarrow L$, where $L = [0,1]$ is called an OWA operator of dimension n if associated with F is a weighting $n-$tuple $W = (w_1, w_2, \ldots, w_n)$ such that $w_i \in [0,1]$ and $\sum_{1 \leq i \leq n} w_i = 1$, where $F(a_1, a_2, \ldots, a_n) = w_1.b_1 + w_2.b_2 + \cdots + w_n.b_n$, with b_i the ith largest element in the collection $a_1, a_2, \ldots, a_n$.*

To study the fuzzy context sequence, we are interested in the use of operators close to *or*. To measure this proximity we can use the orness degree [21].

However, Yager's OWA operators are not easy to be extended to any complete lattice L. The main difficult is that Yager's construction is based on a previous arrangement of the real values which have to be aggregated, which is not always possible in a partially ordered set. In order to overcome this problem Lizasoain and Moreno [18] have built an ordered vector for each given vector in the lattice. This construction allows to define the n-ary OWA operator on any complete lattice which has Yager's OWA operator as a particular case.

Their contribution involves the construction, for each vector $(a_1, \ldots, a_n) \in L^n$ of a totally ordered vector $(b_1, \ldots, b_n)$ as shown in the following proposition:

Proposition 1. *Let $(L, \leq_L)$ be a complete lattice. For any $(a_1, a_2, \ldots, a_n) \in L^n$, consider the values*

- $b_1 = a_1 \vee \cdots \vee a_n \in L$
- $b_2 = [(a_1 \wedge a_2) \vee \cdots \vee (a_1 \wedge a_n)] \vee [(a_2 \wedge a_3) \vee \cdots \vee (a_2 \wedge a_n)] \vee \cdots \vee [a_{n-1} \wedge a_n] \in L$

$\vdots$

- $b_k = \bigvee \{a_{j_1} \wedge \cdots \wedge a_{j_k} \mid \{j_1, \ldots, j_k\} \subseteq \{1, \ldots, n\}\} \in L$

$\vdots$

- $b_n = a_1 \wedge \cdots \wedge a_n \in L$

Then $a_1 \wedge \cdots \wedge a_n = b_n \leq_L b_{n-1} \leq \cdots \leq_L b_1 = a_1 \vee \cdots \vee a_n$.

Moreover, if the set $\{a_1, \ldots, a_n\}$ *is totally ordered, then the vector* $(b_1, \ldots, b_n)$ *agrees with* $(a_{\sigma(1)}, \ldots, a_{\sigma(n)})$ *for some permutation* σ *of* $\{1, \ldots, n\}$.

On the other hand, it is very easy to see that if $\{a_1, \ldots, a_n\}$ is a chain, b_k is the k-th order statistic.

This proposition allows us to generalize Yager's n-ary OWA operators from $[0, 1]$ to any complete lattice. To do this, Lizasoain and Moreno give the following:

Definition 3. *Let* $(L, \leq_L, T, S)$ *be a complete lattice endowed with a t-norm* T *and a t-conorm* S. *We will say that* $(\alpha_1, \alpha_2, \ldots, \alpha_n) \in L^n$ *is a*

(i) weighting vector in $(L, \leq_L, T, S)$ *if* $S(\alpha_1, \ldots, \alpha_n) = 1_L$ *and*
(ii) distributive weighting vector in $(L, \leq_L, T, S)$ *if it also satisfies that* $a = T(a, S(\alpha_1, \ldots, \alpha_n)) = S(T(a, \alpha_1), \ldots T(a, \alpha_n))$ *for any* $a \in L$.

Definition 4. *Let* $(\alpha_1, \ldots, \alpha_n) \in L^n$ *be a distributive weighting vector in* $(L, \leq_L, T, S)$. *For each* $(a_1, \ldots, a_n) \in L^n$, *call* $(b_1, \ldots, b_n)$ *the totally ordered vector constructed in Proposition 1. The function* $F_\alpha : L^n \longrightarrow L$ *given by*

$$F_\alpha(a_1, \ldots, a_n) = S(T(\alpha_1, b_1), \ldots, T(\alpha_n, b_n)),$$

$(a_1, \ldots, a_n) \in L^n$, *is called* n-ary OWA operator.

We will use these n-ary OWA operators in the following sections.

5   *L*-Fuzzy Context Sequences General Study

Returning to the initial situation, we can give a definition that summarizes the information stored in the L-fuzzy context sequence:

Definition 5. *Let* $(L, \leq_L, T, S)$ *be a complete lattice endowed with a t-norm* T *and a t-conorm* S. *Let* $(L, X, Y, R_i), i = \{1, \ldots, n\}$, *be the L-fuzzy context sequence,* $\alpha = (\alpha_1, \alpha_2, \ldots, \alpha_n)$ *a distributive weighting vector and* F_α *the* n-ary OWA *operator associated with* α. *We can define an L-fuzzy relation* R_{F_α} *that aggregates the information of the different L-fuzzy contexts by means of this expression:*

$$R_{F_\alpha}(x, y) = F_\alpha(R_1(x, y), R_2(x, y), \ldots, R_n(x, y)) =$$
$$= S(T(\alpha_1, b_1(x, y)), T(\alpha_2, b_2(x, y)), \ldots, T(\alpha_n, b_n(x, y)),$$
$$\forall x \in X, y \in Y$$

with $(b_1(x, y), b_2(x, y), \ldots, b_n(x, y))$ *the totally ordered vector constructed in Proposition 1 for* $(R_1(x, y), R_2(x, y), \ldots, R_n(x, y))$.

In general, the election of the distributive weighting vector will be very important in order to obtain different results.

On the other hand, we want also to establish different demand levels for a more exhaustive study of the L-fuzzy context sequence. To do this, we are going to define n relations using $n-$ary OWA operators where the distributive weighting vector α has just one non-null value $\alpha_k = 1$, for a certain $k \leq n$.

Relevant Case 1. Let $(L, X, Y, R_i), i = \{1, \ldots, n\}$, be an L-fuzzy context sequence with $(L, \leq_L, T, S)$ a complete lattice, X and Y sets of objects and attributes respectively and $R_i \in L^{X \times Y}, \forall i = \{1, \ldots, n\}$, and consider $k \in \mathbb{N}, k \leq n$. We define the relation $R_{F_{\alpha^k}}$ using the $n-$ary OWA operator F_{α^k} with the distributive weighting vector $\alpha^k = (\alpha_1, \alpha_2, \ldots, \alpha_n)$ such that $\alpha_k^k = 1_L$ and $\alpha_i^k = 0_L, \forall i \neq k$:

$$\forall x \in X, y \in Y, \quad R_{F_{\alpha^k}}(x, y) = F_{\alpha^k}(R_1(x, y), R_2(x, y), \ldots, R_n(x, y)) =$$
$$= S(T(0_L, b_1(x, y)), T(0_L, b_2(x, y)), \ldots, T(1_L, b_k(x, y)), \ldots, T(0_L, b_n(x, y))$$

The value $R_{F_{\alpha^k}}(x, y)$ is the minimum of the k largest values associated with the pair (x, y) in the relations R_i. So, this new relation measures the degree in which the object x is at least k times related with the attribute y.

Observe that the defined α^k is a distributive weighting vector. Moreover, notice that if $L = [0, 1]$, then we are using step-OWA operators [22].

Proposition 2. For any t-norm T and a t-conorm S, it is verified that

$$R_{F_{\alpha^k}}(x, y) = b_k(x, y), \forall(x, y) \in X \times Y.$$

Proof. Taking into account the basic properties of a t-norm T and a t-conorm S, and the definition of the distributive weighting vector.   □

Another interesting case is obtained when we want to analyze the average of the k largest values associated with the pair (x, y) in the relations R_i. In order to do it, we can consider the following relation:

Relevant Case 2. If $L = [0, 1]$, $T(a, b) = ab$ and $S(a, b) = \min\{a+b, 1\}, \forall a, b \in [0, 1]$, using a distributive weighting vector $\widehat{\alpha}^k$ such that $\widehat{\alpha}_i^k = 1/k, i \leq k$ and $\widehat{\alpha}_i^k = 0, i > k$, the obtained relation $R_{F_{\widehat{\alpha}^k}}$ is given by:

$$R_{F_{\widehat{\alpha}^k}}(x, y) = \sum_{i=1}^{k} \frac{b_i(x, y)}{k}, \quad \forall(x, y) \in X \times Y \tag{1}$$

Remark 1. It is immediate to prove that if $h \leq k$, then $R_{F_{\alpha^h}} \geq R_{F_{\alpha^k}}$ and $R_{F_{\widehat{\alpha}^h}} \geq R_{F_{\widehat{\alpha}^k}}$.

The observation of the L-fuzzy contexts defined from these new relations gives the idea for the following propositions:

Proposition 3. *Consider $k \in \mathbb{N}$, such that $k \leq n$. If (A, B) is an L-fuzzy concept of the L-fuzzy context $(L, X, Y, R_{F_{\alpha k}})$, then $\forall h \in \mathbb{N}, h \leq k$, there exists an L-fuzzy concept (C, D) of the L-fuzzy context $(L, X, Y, R_{F_{\alpha h}})$ such that $A \leq C$ and $B \leq D$.*

Proof. If $h \leq k$, then $R_{F_{\alpha k}}(x, y) \leq R_{F_{\alpha h}}(x, y) \quad \forall (x, y) \in X \times Y$.

Let (A, B) be an L-fuzzy concept of the L-fuzzy context $(L, X, Y, R_{F_{\alpha k}})$. Then, as any implication operator is increasing on its second argument,

$$\forall y \in Y, \;\; B(y) = \inf_{x \in X}\{\mathcal{I}(A(x), R_{F_{\alpha k}}(x, y))\} \leq \inf_{x \in X}\{\mathcal{I}(A(x), R_{F_{\alpha h}}(x, y))\} = D(y)$$

Thus, the L-fuzzy set B derived from A in $(L, X, Y, R_{F_{\alpha k}})$ is a subset of the L-fuzzy set D derived from A in $(L, X, Y, R_{F_{\alpha h}})$. Therefore, $B \leq D$.

As the used implication operator $\mathcal{I}$ is residuated, if we derive the set D in $(L, X, Y, R_{F_{\alpha h}})$, we obtain the set $C = D_2$ and the pair (C, D) is an L-fuzzy concept of the L-fuzzy context $(L, X, Y, R_{F_{\alpha h}})$. Now, applying the properties of this closure operator formed by the composition of the derivation operators in $(L, X, Y, R_{F_{\alpha h}})$ [6]: $A \leq A_{12} = D_2 = C$. Therefore, the other inequality also holds.

The proposition is analogously proved using the relations $R_{F_{\hat{\alpha} k}}$ and $R_{F_{\hat{\alpha} h}}$. $\quad\square$

The following result sets up relations between the L-fuzzy concepts associated with the same starting set (see section 2) in the different L-fuzzy contexts.

Proposition 4. *Consider $h, k \in \mathbb{N}$, such that $h \leq k \leq n$, and $A \in L^X$. If the L-fuzzy concepts associated with the set A in the contexts $(L, X, Y, R_{F_{\alpha k}})$ and $(L, X, Y, R_{F_{\alpha h}})$ are denoted by (A^k, B^k) and (A^h, B^h), then $B^k \leq B^h$.*

The same result is obtained if we consider the L-fuzzy contexts associated with the relations $R_{F_{\hat{\alpha} k}}$ and $R_{F_{\hat{\alpha} h}}$.

Proof. Consider $A \in L^X$ and the L-fuzzy contexts associated with the relations $R_{F_{\alpha k}}$ and $R_{F_{\alpha h}}$. Unfolding the fuzzy extensions of both L-fuzzy concepts, and taking into account that a fuzzy implication operator is increasing on its second argument, $\forall y \in Y$:

$$B^k(y) = \inf_{x \in X}\{\mathcal{I}(A(x), R_{F_{\alpha k}}(x, y))\} \leq \inf_{x \in X}\{\mathcal{I}(A(x), R_{F_{\alpha h}}(x, y))\} = B^h(y)$$

This inequality holds for every $A \in L^X$ and for every implication $\mathcal{I}$.

The result can be similarly proved considering the L-fuzzy contexts associated with the relations $R_{F_{\hat{\alpha} k}}$ and $R_{F_{\hat{\alpha} h}}$. $\quad\square$

6   L-Fuzzy Context Sequences on $\mathcal{J}([0, 1])$

One of the most interesting situations is when we use interval-valued L-Fuzzy contexts. We have previously published some works [11,2] in which the chosen lattice is $L = \mathcal{J}([0, 1])$.

In this case, notice that $(\mathcal{J}([0, 1]), \leq)$ with the usual order ($[a_1, c_1] \leq [a_2, c_2] \iff a_1 \leq a_2$ and $c_1 \leq c_2$) is a complete but not totally ordered lattice.

Then, we can give the following definition:

Definition 6. *Let* $(\mathcal{J}([0,1]), \leq, \boldsymbol{T}, \boldsymbol{S})$ *be the complete lattice of the closed intervals in* $[0,1]$ *endowed with the t-norm* $\boldsymbol{T}$ *and the t-conorm* $\boldsymbol{S}$ *and consider the sequence of interval-valued L-fuzzy contexts* $(\mathcal{J}([0,1]), X, Y, R_i), i = \{1, \ldots, n\}$. *If* $[\alpha, \beta] = ([\alpha_1, \beta_1], [\alpha_2, \beta_2], \ldots, [\alpha_n, \beta_n])$ *is a distributive weighting vector of intervals and* $F_{[\alpha, \beta]}$ *the n-ary OWA operator associated with* $[\alpha, \beta]$, *then the interval-valued L-fuzzy relation* $R_{F_{[\alpha, \beta]}}$ *that aggregates the information of the different L-fuzzy contexts can be defined* $\forall(x, y) \in X \times Y$ *as:*

$$R_{F_{[\alpha, \beta]}}(x, y) = F_{[\alpha, \beta]}(R_1(x, y), R_2(x, y), \ldots, R_n(x, y)) =$$
$$= \boldsymbol{S}(\boldsymbol{T}([\alpha_1, \beta_1], [b_1(x, y), d_1(x, y)]), \ldots, \boldsymbol{T}([\alpha_n, \beta_n], [b_n(x, y), d_n(x, y)])$$

where $([b_1(x, y), d_1(x, y)], [b_2(x, y), d_2(x, y)], \ldots, [b_n(x, y), d_n(x, y)])$ *is the totally ordered vector constructed from* $(R_1(x, y), R_2(x, y), \ldots, R_n(x, y))$.

Also in this case two relevant situations can be highlighted. In the first one we will establish an exigence level k and in order to measure the degree in which an object is at least k times related to an attribute we will use the following relation:

Relevant Case 3. *Consider* $k \in \mathbb{N}$ *such that* $k \leq n$. *If we represent by* $[\![\alpha]\!]^k$ *the distributive weighting vector* $[\![\alpha]\!]^k = ([\alpha_1, \beta_1], [\alpha_2, \beta_2], \ldots, [\alpha_n, \beta_n])$ *such that* $[\alpha_k, \beta_k] = [1, 1]$, *and* $[\alpha_i, \beta_i] = [0, 0], \forall i \neq k$, *we can define the relation* $R_{F_{[\![\alpha]\!]^k}}$ *as:*

$$\forall(x, y) \in X \times Y,$$
$$R_{F_{[\![\alpha]\!]^k}}(x, y) = \boldsymbol{S}(\boldsymbol{T}([0, 0], [b_1(x, y), d_1(x, y)]), \ldots,$$
$$\boldsymbol{T}([1, 1], [b_k(x, y), d_k(x, y)]), \ldots, \boldsymbol{T}([0, 0], [b_n(x, y), d_n(x, y)])$$

It is immediate to prove that, also in this case, for any t-norm $\boldsymbol{T}$ and t-conorm $\boldsymbol{S}$, the following proposition holds:

Proposition 5. $R_{F_{[\![\alpha]\!]^k}}(x, y) = [b_k(x, y), d_k(x, y)], \ \forall(x, y) \in X \times Y.$

The second interesting family of relations is associated with the average of the observations:

Relevant Case 4. *In the complete lattice* $(\mathcal{J}([0,1]), \leq)$, *consider the t-norm* $\boldsymbol{T}$ *and the t-conorm* $\boldsymbol{S}$ *given for any* $[a_1, c_1], [a_2, c_2] \in \mathcal{J}([0,1])$ *by*

$$\boldsymbol{T}([a_1, c_1], [a_2, c_2]) = [a_1 a_2, c_1 c_2]$$
$$\boldsymbol{S}([a_1, c_1], [a_2, c_2]) = [min\{a_1 + a_2, 1\}, min\{c_1 + c_2, 1\}]$$

If $k \leq n$ *and we use the weighting vector* $[\![\hat{\alpha}]\!]^k = ([\alpha_1, \beta_1], \ldots, [\alpha_n, \beta_n]) \in \mathcal{J}([0,1])^n$ *verifying* $[\alpha_i, \beta_i] = [\frac{1}{k}, \frac{1}{k}]$ *for every* $i \leq k$, *and* $[\alpha_i, \beta_i] = [0, 0]$ *for every* $i > k$, *we can define the relation* $R_{F_{[\![\hat{\alpha}]\!]^k}}$ *as follows:*

$$R_{F_{[\![\hat{\alpha}]\!]^k}}(x, y) = \left[\sum_{i=1}^{k} \frac{b_i(x, y)}{k}, \sum_{i=1}^{k} \frac{d_i(x, y)}{k}\right], \quad \forall(x, y) \in X \times Y$$

Let see an example to better understand the difference of using both ways of aggregated information in $L = \mathcal{J}([0,1])$:

Example 2. We come back to the L-fuzzy context sequence $(L, X, Y, R_i), i = \{1, \ldots, 5\}$, of Example 1 that represents the sports articles $X = \{x_1, x_2, x_3\}$ sales in some establishments $Y = \{y_1, y_2, y_3\}$ during 5 months. Every interval-valued observation of the relations $R_i \in \mathcal{J}([0,1])^{X \times Y}$, represents the variation of the percentage of the daily product sales in each establishment along a month.

$$R_1 = \begin{pmatrix} [0.7, 0.8] & [1, 1] & [0.8, 1] \\ [0, 0] & [0.1, 0.4] & [0.1, 0.3] \\ [0, 0.2] & [0.1, 0.3] & [0, 0.6] \end{pmatrix} R_2 = \begin{pmatrix} [1, 1] & [0.8, 1] & [1, 1] \\ [0.8, 0.9] & [0.4, 0.5] & [0.1, 0.3] \\ [0, 0] & [0, 0.2] & [0.2, 0.4] \end{pmatrix}$$

$$R_3 = \begin{pmatrix} [1, 1] & [1, 1] & [1, 1] \\ [0.6, 0.8] & [0.5, 0.5] & [0.7, 0.8] \\ [0, 0] & [0.1, 0.2] & [0.2, 0.4] \end{pmatrix} R_4 = \begin{pmatrix} [0.5, 0.5] & [0.4, 0.6] & [0.6, 0.8] \\ [0.1, 0.3] & [0.5, 0.6] & [0.3, 0.5] \\ [0.6, 0.6] & [0.8, 0.9] & [0.8, 1] \end{pmatrix}$$

$$R_5 = \begin{pmatrix} [0.1, 0.4] & [0, 0.2] & [0, 0.2] \\ [0, 0] & [0.1, 0.3] & [0, 0.2] \\ [0.8, 1] & [1, 1] & [0.9, 0.9] \end{pmatrix}$$

We want to study in what establishments are the highest sales for each product, no matter when the sale has been carried out, taking into account that there are seasonal sporting goods that are sold in certain periods of time and not in others (skies, bathing suits ...).

If we fix the demand level, for instance to $k = 3$, and we want to analyze if the products have been sold at least during three months, then associated with the distributive weighting vector $[\![\alpha]\!]^3$, we have the relation:

$$R_{F_{[\![\alpha]\!]^3}} = \begin{pmatrix} [0.7, 0.8] & [0.8, 1] & [0.8, 1] \\ [0.1, 0.3] & [0.4, 0.5] & [0.1, 0.3] \\ [0, 0.2] & [0.1, 0.3] & [0.2, 0.6] \end{pmatrix}$$

Now, we take the L-fuzzy context $(L, X, Y, R_{F_{[\![\alpha]\!]^3}})$ and obtain the interval-valued L-fuzzy concept derived from the crisp singleton $\{x_2\}$ using the interval-valued implication operator defined from the Brouwer-Gödel implication ($\mathcal{I}(a, b) = 1, a \leq b$ and $\mathcal{I}(a, b) = b$ in other case) [1]:

$(\{x_1/[1, 1], x_2/[1, 1], x_3/[0, 0.2]\}, \{y_1/[0.1, 0.3], y_2/[0.4, 0.5], y_3/[0.1, 0.3]\})$

In this case, we can say that x_1 and x_2 have been important sales mainly in establishment y_2 at least during three months.

On the other hand, we can analyze the average sale of each article in the three months with highest sales. To do this, we will use the weighting vector $[\![\hat{\alpha}]\!]^3$ and the obtained relation is:

$$R_{F_{[\![\hat{\alpha}]\!]^3}} = \begin{pmatrix} [0.9, 0.93] & [0.93, 1] & [0.93, 1] \\ [0.5, 0.67] & [0.46, 0.53] & [0.36, 0.53] \\ [0.46, 0.6] & [0.63, 0.73] & [0.63, 0.83] \end{pmatrix}$$

In this case, we will take the interval-valued implication operator obtained from the fuzzy implication $\mathcal{I}(a,b) = \min\{1, b/a\}$ associated with the t-norm $T(a,b) = ab$ [1]. Then, taking as a starting point x_2, we obtain the interval-valued L-fuzzy concept:

$$(\{x_1/[1,1], x_2/[1,1], x_3/[0.89, 0.89]\}, \{y_1/[0.5, 0.67], y_2/[0.46, 0.53], y_3/[0.36, 0.53]\})$$

We can say that, taking the average of the sales in the three months with highest sales, all the articles have been acceptable sales in all the establishments (the sales in y_1 and y_3 that were not important with the previous definition, now are because they are compensated using the values of the three months).

The use of the $n-$ary OWA operators allow us to ignore the small values of the relations (the sales of a non-seasonal sporting goods are close to 0) since, in this case, if we take the average of all the relations, the results will be biased.

7   Conclusions and Future Work

In this work, we have used OWA operators to study the L-fuzzy context sequence and the derived information by means of the L-fuzzy contexts.

A more complete study can be done when we work with L-fuzzy context sequences that represent the evolution in time of an L-fuzzy context.

On the other hand, these L-fuzzy contexts that evolve in time can be generalize if we study L-fuzzy contexts where the observations are other L-fuzzy contexts. This is the task that we will study in the future.

Acknowledgments. This paper is supported by the Research Group "Intelligent Systems and Energy (SI+E)" of the Basque Government, under Grant IT677-13, and by the Research Group "Artificial Intelligence and Approximate Reasoning" of the Public University of Navarra.

References

1. Alcalde, C., Burusco, A., Fuentes-González, R.: A constructive method for the definition of interval-valued fuzzy implication operators. Fuzzy Sets and Systems 153, 211–227 (2005)
2. Alcalde, C., Burusco, A., Fuentes-González, R., Zubia, I.: Treatment of L-fuzzy contexts with absent values. Information Sciences 179(1-2), 1–15 (2009)
3. Alcalde, C., Burusco, A., Fuentes-González, R., Zubia, I.: The use of linguistic variables and fuzzy propositions in the L-Fuzzy Concept Theory. Computers and Mathematics with Applications 62, 3111–3122 (2011)
4. Alcalde, C., Burusco, A., Fuentes-González, R.: The study of fuzzy context sequences. International Journal of Computational Intelligence Systems 6(3), 518–529 (2013)
5. Alcalde, C., Burusco, A., Fuentes-González, R.: Application of OWA Operators in the L-Fuzzy Concept Analysis. In: Bustince, H., Fernandez, J., Mesiar, R., Calvo, T. (eds.) Aggregation Functions in Theory and in Practise. AISC, vol. 228, pp. 133–146. Springer, Heidelberg (2013)

6. Bělohlávek, R.: Fuzzy Galois Connections. Math. Logic Quarterly 45(4), 497–504 (1999)
7. Belohlavek, R.: What is a fuzzy concept lattice? II. In: Kuznetsov, S.O., Ślęzak, D., Hepting, D.H., Mirkin, B.G. (eds.) RSFDGrC 2011. LNCS, vol. 6743, pp. 19–26. Springer, Heidelberg (2011)
8. Burusco, A., Fuentes-González, R.: The Study of the L-fuzzy Concept Lattice. Mathware and Soft Computing 1(3), 209–218 (1994)
9. Burusco, A., Fuentes-González, R.: Construction of the L-fuzzy Concept Lattice. Fuzzy Sets and Systems 97(1), 109–114 (1998)
10. Burusco, A., Fuentes-González, R.: Concept lattices defined from implication operators. Fuzzy Sets and Systems 114(1), 431–436 (2000)
11. Burusco, A., Fuentes-González, R.: The study of the interval-valued contexts. Fuzzy Sets and Systems 121, 69–82 (2001)
12. Burusco, A., Fuentes-González, R.: Contexts with multiple weighted values. The International Journal of Uncertainty, Fuzziness and Knowledge-based Systems 9(3), 355–368 (2001)
13. Calvo, T., Mesiar, R.: Weighted triangular norms-based aggregation operators. Fuzzy Sets and Systems 137, 3–10 (2003)
14. Calvo, T., Mesiar, R.: Aggregation operators: ordering and bounds. Fuzzy Sets and Systems 139, 685–697 (2003)
15. Djouadi, Y., Prade, H.: Interval-Valued Fuzzy Galois Connections: Algebraic Requirements and Concept Lattice Construction. Fundamenta Informaticae 99(2), 169–186 (2010)
16. Djouadi, Y., Prade, H.: Possibility- theoretic extension of derivation operators in formal concept analysis over fuzzy lattices. FODM 10(4), 287–309 (2011)
17. Fodor, J., Marichal, J.L., Roubens, M.: Characterization of the Ordered Weighted Averaging Operators. IEEE Transactions on Fuzzy Systems 3(2), 236–240 (1995)
18. Lizasoain, I., Moreno, C.: OWA operators defined on complete lattices. Fuzzy Sets and Systems 224, 36–52 (2013)
19. Medina, J., Ojeda-Aciego, M.: Multi-adjoint t-concept lattices. Information Sciences 180(5), 712–725 (2010)
20. Medina, J.: Multi-adjoint property-oriented concept lattices. Information Sciences 190, 95–106 (2012)
21. Yager, R.R.: On ordered weighted averaging aggregation operators in multi-criteria decision making. IEEE Transactions on Systems, Man and Cibernetics 18, 183–190 (1988)
22. Yager, R.R.: Families of OWA operators. Fuzzy Sets and Systems 59, 125–148 (1993)
23. Wille, R.: Restructuring lattice theory: an approach based on hierarchies of concepts. In: Rival, I. (ed.) Ordered Sets, pp. 445–470. Reidel, Dordrecht-Boston (1982)

Variable-Range Approximate Systems Induced by Many-Valued *L*-Relations

Aleksandrs Eļkins[1], Sang-Eon Han[3], and Alexander Šostak[1,2,⋆]

[1] Department of Mathematics, University of Latvia, LV-1002, Riga, Latvia
[2] Institute of Mathematics and CS, University of Latia, LV-1459, Riga, Latvia
[3] Faculty of Liberal Education, Institute of Pure and Applied Mathematics
Chonbuk National University, Jeonju-City, Jeonbuk 561-756, Republic of Korea

Abstract. The concept of a many-valued *L*-relation is introduced and studied. Many-valued *L*-relations are used to induce variable-range quasi-approximate systems defined on the lines of the paper (A. Šostak, Towards the theory of approximate systems: variable-range categories. Proceedings of ICTA 2011, Cambridge Univ. Publ. (2012) 265–284.) Such variable-range (quasi-)approximate systems can be realized as special families of *L*-fuzzy rough sets indexed by elements of a complete lattice.

Keywords: Variable-range (quasi-)approximate system, many-valued *L*-relation, *L*-fuzzy rough set.

1   Introduction

The concept of an approximate system was first introduced in [19] and further thoroughly studied in [20]. Among the most important examples of approximate systems of essentially different nature are approximate systems generated by *L*-(fuzzy) topologies [6], [9], or, more generally, by *L*-(fuzzy) ditopologies [4], [5] and approximate systems induced by reflexive symmetric transitive *L*-(fuzzy) relations. The last ones are approximate systems which are closely related to *L*-fuzzy rough sets, see e.g. [7], [23], [24], see also [10] concerning relations between approximate systems and *L*-fuzzy rough sets. Since the properties of reflexivity, symmetry and transitivity in many considerations are too restrictive, and on the other hand, in some cases they are not very crucial, in our paper [11] we introduced concepts more general, than the one of an approximate system - namely quasi-approximate and pseudo-approximate systems. Among other advantages of such systems is that they suit also for operating with *L*-relations missing one or several of the properties of reflexivity, transitivity and symmetry.

Aiming to develop an approach which would allow us to consider approximate systems with different ranges at the same time, we introduced the category of variable-range approximate systems in [21]. Basic properties of this category and some of its subcategories were studied there, too.

⋆ The support of the ESF project 2013/0024/1DP/1.1.1.2.0/13/APIA/VIAA/045 is kindly announced.

A. Laurent et al. (Eds.): IPMU 2014, Part III, CCIS 444, pp. 41–50, 2014.
© Springer International Publishing Switzerland 2014

It was the primary aim of this work to study variable-range approximate systems induced by many-valued L-relations and to apply them for the study of what could be called variable-range L-fuzzy rough sets interpreted as special approximate systems. However, how it was mentioned above, if an L-relation is missing one of the properties of reflexivity, symmetry or transitivity, then the resulting structure fails to be an approximate system as it is defined in [20] but is only a quasi-approximate system. Therefore, in order to present our research in a sufficiently broad context, first we introduce the concepts related to quasi-approximate systems (Section 2), and in the next, Section 3, develop the basics of the theory of *variable-range quasi-approximate systems*. Here to a certain extent we follow [21]. Further, we proceed with considering the category of many-valued L-relations (Section 4) and then study variable-range approximate systems induced by many-valued L-relations (Section 5). In Section 6 we construct a functor embedding the category of many-valued L-relations into the category of variable-range quasi-approximate systems and study some properties of this embedding. Finally, in the last Section 7, Conclusions, we briefly review the results of this work, in particular, interpreting them from the viewpoint of L-fuzzy rough set theory, and indicate some directions for future work.

2   Quasi-Approximate, Pseudo-Approximate and Approximate Systems

Let $\mathbb{L}$ be a complete infinitely distributive lattice whose bottom and top elements are $0_{\mathbb{L}}$ and $1_{\mathbb{L}}$ respectively where $0_{\mathbb{L}} \neq 1_{\mathbb{L}}$, and let $\mathbb{M}$ be any complete lattice.

Definition 1. [11] An *upper quasi $\mathbb{M}$-approximation operator* on a lattice $\mathbb{L}$ is a mapping $u : \mathbb{L} \times \mathbb{M} \to \mathbb{L}$ such that

(1u) $u(0_{\mathbb{L}}, \alpha) = 0_{\mathbb{L}} \ \forall \alpha \in \mathbb{M}$,
(2u) $u(a \vee b, \alpha) = u(a, \alpha) \vee u(b, \alpha) \ \forall a, b \in \mathbb{L}, \ \forall \alpha \in \mathbb{M}$, and
(mu) $\alpha \leq \beta \implies u(a, \alpha) \leq u(a, \beta) \ \forall a \in \mathbb{L}$ and $\forall \alpha, \beta \in \mathbb{M}$.

An upper quasi approximation operator $u : \mathbb{L} \times \mathbb{M} \to \mathbb{L}$ is called *an upper pseudo $\mathbb{M}$-approximation operator* if

(3'u) $u(u(a, \alpha), \alpha) \leq u(a, \alpha) \ \forall a \in \mathbb{L}, \forall \alpha \in \mathbb{M}$.

An upper pseudo $\mathbb{M}$-approximation operator $u : \mathbb{L} \times \mathbb{M} \to \mathbb{L}$ is called *an upper $\mathbb{M}$-approximation operator* (cf. [20, Definition 3.1]) if

(4u) $a \leq u(a, \alpha) \ \forall a \in \mathbb{L}, \forall \alpha \in \mathbb{M}$

Definition 2. [11] *A lower quasi $\mathbb{M}$-approximation operator on $\mathbb{L}$ is a mapping* $l : \mathbb{L} \times \mathbb{M} \to \mathbb{L}$ *such that*

(1l) $l(1_{\mathbb{L}}, \alpha) = 1_{\mathbb{L}} \ \forall \alpha \in \mathbb{M}$,
(2l) $l(a \wedge b, \alpha) = l(a, \alpha) \wedge l(b, \alpha) \ \forall a, b \in \mathbb{L}, \forall \alpha \in \mathbb{M}$, *and*
(ml) $\alpha \leq \beta \implies l(a, \alpha) \geq l(a, \beta) \ \forall a \in \mathbb{L}, \forall \alpha, \beta \in \mathbb{M}$.

A *lower quasi* $\mathbb{M}$-*approximation operator* $l : \mathbb{L} \times \mathbb{M} \to \mathbb{L}$ *is called a lower pseudo* $\mathbb{M}$-*approximation operator if*

(3'l) $l(l(a,\alpha),\alpha) \geq l(a,\alpha) \ \forall a \in \mathbb{L}, \forall \alpha \in \mathbb{M}.$

A *lower pseudo* $\mathbb{M}$-*approximation operator* $l : \mathbb{L} \times \mathbb{M} \to \mathbb{L}$ *is called a lower* $\mathbb{M}$-*approximation operator* (cf. [20, Definition 3.1]) *if*

(4l) $a \geq l(a,\alpha) \ \forall a \in \mathbb{L}, \forall \alpha \in \mathbb{M}.$

Definition 3. [11] A quadruple $(\mathbb{L}, \mathbb{M}, u, l)$ is called:

- a *quasi* $\mathbb{M}$-*approximate system on* $\mathbb{L}$ if u and l are quasi $\mathbb{M}$-approximation operators on $\mathbb{L}$;
- *strongly quasi* $\mathbb{M}$-*approximate* if it is a quasi $\mathbb{M}$-approximate and $l \leq u$;
- *pseudo* $\mathbb{M}$-*approximate* if u and l are pseudo $\mathbb{M}$-approximation operators;
- *an* $\mathbb{M}$-*approximate system* if u and l are $\mathbb{M}$-approximation operators.

3 Category QAS of Variable-Range Quasi-Approximate Systems and Its Subcategories

Let **QAS** (**PAS**, **AS** respectively) be the family of quasi-approximate (pseudo-approximate, approximate, resp.) systems $(\mathbb{L}, \mathbb{M}, u, l)$. To consider **QAS** as a category whose class of objects are quasi-approximate systems $(\mathbb{L}, \mathbb{M}, u, l)$ (and respectively **PAS**, **AS** as its full subcategories) we have to specify its morphisms. Given two quasi-approximate systems $(\mathbb{L}_1, \mathbb{M}_1, u_1, l_1), (\mathbb{L}_2, \mathbb{M}_2, u_2, l_2)$ by a morphism $F : (\mathbb{L}_1, \mathbb{M}_1, u_1, l_1) \to (\mathbb{L}_2, \mathbb{M}_2, u_2, l_2)$ we call a pair $(f, \varphi) = F$ such that

(1mor) $f : \mathbb{L}_2 \to \mathbb{L}_1$ is a morphism in the category **IDCLAT** of infinitely distributive complete lattices;
(2mor) $\varphi : \mathbb{M}_2 \to \mathbb{M}_1$ is a morphism in the category **CLAT** of complete lattices;
(3mor) $u_1(f(b), \varphi(\beta)) \leq f(u_2(b, \beta)) \ \forall b \in \mathbb{L}_2, \ \forall \beta \in \mathbb{M}_2$;
(4mor) $f(l_2(b, \beta)) \leq l_1(f(b), \varphi(\beta)) \ \forall b \in \mathbb{L}_2, \forall \beta \in \mathbb{M}_2$

Theorem 1. **QAS** *thus obtained is indeed a category.*

Proof Let $F = (f, \varphi) : (\mathbb{M}_1, \mathbb{L}_1, u_1, l_1) \to (\mathbb{M}_2, \mathbb{L}_2, u_2, l_2)$ and $G = (g, \psi) : (\mathbb{M}_2, \mathbb{L}_2, u_2, l_2) \to (\mathbb{M}_3, \mathbb{L}_3, u_3, l_3)$ be defined as above and let $G \circ F = (f \circ g, \varphi \circ \psi) : (\mathbb{M}_1, \mathbb{L}_1, u_1, l) \to (\mathbb{M}_3, \mathbb{L}_3, u_3, l_3)$. We have to verify that $f \circ g$ and $\psi \circ \varphi$ satisfy conditions (3mor) and (4mor) above. Let $c \in \mathbb{L}_3$ and $\gamma \in \mathbb{M}_3$. Then $(u_1(g \circ f)(c), (\psi \circ \varphi)(\gamma)) = u_1(f(g(c)), \varphi(\psi(\gamma))) \leq f(u_2(g(c), \psi(\gamma))) \leq f(g(u_3(c), \gamma)) = (g \circ f)(u_3(c, \gamma))$ and hence condition (3mor) holds. In a similar way we verify condition (4mor). We conclude the proof by noticing that the pair $F = (id_\mathbb{L}, id_\mathbb{M}) : (\mathbb{L}, \mathbb{M}, u, l) \to (\mathbb{L}, \mathbb{M}, u, l)$ is obviously a morphism.

Remark 1. Let $\mathbb{M}$ be fixed. Then the subcategory $\textbf{QAS}^{\mathbb{M}}$ of the category **QAS**, whose morphisms are of the form $F = (f, id_\mathbb{M})$ and its full subcategories $\textbf{PAS}^{\mathbb{M}}$ and $\textbf{AS}^{\mathbb{M}}$ can be identified with categories considered in [11], the last one was thoroughly studied also in [20].

Theorem 2. *Every source $F : (\mathbb{L}_1, \mathbb{M}_1) \to (\mathbb{L}_2, \mathbb{M}_2, u_2, l_2)$ has a unique initial lift $F : (\mathbb{L}_1, \mathbb{M}_1, u_1, l_1) \to (\mathbb{L}_2, \mathbb{M}_2, u_2, l_2)$ in category* **QAS** *and in its subcategories* **PAS** *and* **AS**.

Proof   Let $F = (f, \varphi) : (\mathbb{L}_1, \mathbb{M}_1) \to (\mathbb{L}_2, \mathbb{M}_2, u_2, l_2)$. We define an upper approximation operator $u_1 : \mathbb{L}_1 \times \mathbb{M}_1 \to \mathbb{L}_1$ by

$$u_1(a, \alpha) = \bigwedge \{ f(u_2(b, \beta)) \mid f(b) \geq a, \varphi(\beta) \geq \alpha, b \in \mathbb{L}_2, \beta \in \mathbb{M}_2 \}.$$

We verify that $u_1 : \mathbb{L}_1 \times \mathbb{M}_1 \to \mathbb{L}_1$ is indeed an upper $\mathbb{M}_1$-approximate operator. The first property is obvious from the definition of u_1. To verify property (2u) let $a_1, a_2 \in \mathbb{L}_1, \alpha \in \mathbb{M}_1$, then
$u_1(a_1 \vee a_2, \alpha) = \bigwedge \{ f(u_2(b, \beta)) \mid f(b) \geq a_1 \vee a_2, \varphi(\beta) \geq \alpha, b \in \mathbb{L}_2, \beta \in \mathbb{M}_2 \}$
$\leq \bigwedge \{ f(u_2(b_1 \vee b_2), \beta) \mid f(b_1) \geq a_1, f(b_2) \geq a_2, \varphi(\beta) \geq \alpha, b_1, b_2 \in \mathbb{L}_2, \beta \in \mathbb{M}_2 \}$
$= \bigwedge \{ f(u_2(b_1, \beta)) \vee f(u_2(b_2, \beta)) \mid f(b_1) \geq a_1, f(b_2) \geq a_2, \varphi(\beta) \geq \alpha, b_1, b_2 \in \mathbb{L}_2, \beta \in \mathbb{M}_2 \} = (\bigwedge \{ f(u_2(b_1, \beta)) \mid f(b_1) \geq a_1, \varphi(\beta) \geq \alpha, b_1 \in \mathbb{L}_2, \beta \in \mathbb{M}_2 \}) \vee (\bigwedge \{ f(u_2(b_2, \beta)) \mid f(b_2) \geq a_2, \varphi(\beta) \geq \alpha, b_2 \in \mathbb{L}_2, \beta \in \mathbb{M}_2 \}) = u_1(a_1, \alpha) \vee u_1(a_2, \alpha)$.
The converse inequality is obvious.
To verify property (mu) let $\alpha_1 \leq \alpha_2 \in \mathbb{M}_1$ Then
$u_1(a, \alpha_1) = \bigwedge \{ f(u_2(b, \beta) \mid f(b) \geq a, \varphi(\beta) \geq \alpha_1, b \in \mathbb{L}_2, \beta \in \mathbb{M}_2 \} \leq \bigwedge \{ f(u_2(b, \beta) \mid f(b) \geq a, \varphi(\beta) \geq \alpha_2, b \in \mathbb{L}_2, \beta \in \mathbb{M}_2 \} = u_2(a, \alpha_2)$.
To verify condition (3'u) in case u_2 satisfies it, notice that $u_1(u_1(a)), \alpha) \leq$
$\bigwedge_{f(b) \geq a, \varphi(\beta) \geq \alpha} u_1 (f(u_2(b, \beta)), \alpha) \leq \bigwedge_{f(b) \geq a, \varphi(\beta) \geq \alpha} u_1 (f(u_2(b, \beta)), \varphi(\beta))$
$\leq \bigwedge_{f(b) \geq a, \varphi(\beta \geq \alpha} f(u_2(u_2(b, \beta), \beta))) = \bigwedge_{f(b) \geq a, \varphi(\beta) \geq \alpha} f(u_2(b, \beta)) = u_1(a, \alpha)$.
The validity of property (4u) for u_2 is obvious from its definition whenever u_1 satisfies this condition.
    We define the lower quasi M-approximation operator $l_1 : \mathbb{L}_1 \times \mathbb{M}_1 \to \mathbb{L}_1$ by

$$l_1(a, \alpha) = \bigvee \{ f(l_2(b, \beta)) \mid f(b) \leq a, \varphi(\beta) \geq \alpha, b \in \mathbb{L}_2, \beta \in \mathbb{M}_2 \}.$$

The validity of the first condition for l_1 follows from the corresponding property of l_2. To verify the second condition let $a_1, a_2 \in \mathbb{L}_1$, and $\alpha \in \mathbb{M}_1$. Then
$l_1(a_1, \alpha) \wedge l_1(a_2, \alpha) = (\bigvee \{ f(l_2(b_1, \beta)) \mid f(b_1)) \leq a_1, \varphi(\beta) \geq \alpha, b_1 \in \mathbb{L}_2, \beta \in \mathbb{M}_2 \}) \wedge (\bigvee \{ f(l_2(b_2, \beta)) \mid f((b_2)) \leq a_2, \varphi(\beta) \geq \alpha, b_2 \in \mathbb{L}_2, \beta \in \mathbb{M}_2 \})$
$= \bigvee \{ f(l_2(b_1, \beta)) \wedge f(l_2(b_2, \beta)) \mid f(b_1) \leq a_1, f(b_2) \leq a_2, \varphi(\beta) \geq \alpha, b_1, b_2 \in \mathbb{L}_2, \} \leq \bigvee \{ f(l_2(b_1 \wedge b_2, \beta) \mid f(b_1) \wedge f(b_2) \leq a_1 \wedge a_2, \varphi(\beta) \geq \alpha, b_1, b_2 \in \mathbb{L}_2, \beta \in \mathbb{M}_2 \}$
$= \bigvee \{ f(l_2(b, \beta)) \mid f(b) \leq a_1 \wedge a_2, \varphi(\beta) \geq \alpha, b_1, b_2 \in \mathbb{L}_2, \beta \in \mathbb{M}_2 \} = l_1(a_1 \wedge a_2, \alpha)$.
The converse inequality is obvious.
    To verify property (ml) for l_1 let $\alpha_1 \leq \alpha_2, \in \mathbb{M}_1$. Then
$l_1(a, \alpha_1) = \bigvee \{ f(l_2(b, \beta) \mid f(b) \geq a, \varphi(\beta) \geq \alpha_1, b \in \mathbb{L}_2, \beta \in \mathbb{M}_2 \} \geq \bigvee \{ f(l_2(b, \beta) \mid f(b) \leq a, \varphi(\beta) \geq \alpha_2, b \in \mathbb{L}_2, \beta \in \mathbb{M}_2 \} = l_2(a, \alpha_2)$.
    Property (3'u) of the operator $l_1 : \mathbb{L}_1 \times \mathbb{M}_1 \to \mathbb{L}_1$ (in case l_2 has this property) is established as follows. Given $a \in \mathbb{L}_1$ we have
$l_1(l_1(a, \alpha), \alpha) = l_1 \left(\bigvee_{f(b) \leq a, \varphi(\beta) \geq \alpha} f(l_2(b, \beta), \alpha) \right) \geq \bigvee_{f(b) \leq a, \varphi(\beta) \geq \alpha} l_1(f(l_2, \beta), \alpha)$
$\geq \bigvee_{f(b) \leq a, \varphi(\beta) \geq \alpha} l_1(f(l_2, \beta), \varphi(\beta)) = \bigvee_{f(b) \leq a, \varphi(\beta) \geq \alpha} f(l_2(b, \beta)) = l_1(a, \alpha)$.
The validity of property (4l) for l_2 is obvious from its definition whenever l_1 satisfies this condition.

To complete the proof, let $G = (g, \psi) : (\mathbb{L}_3, \mathbb{M}_3, u_3, l_3) \to (\mathbb{L}_2, \mathbb{M}_2, u_2, l_2)$ be a morphism in **QAS** and let $H = (h, \chi)$ where $h : \mathbb{L}_1 \to \mathbb{L}_3$, $\chi : \mathbb{M}_1 \to \mathbb{M}_3$ are morphisms in categories **IDCLAT** and **CLAT** respectively, such that $f \circ h = g$ and $\varphi \circ \chi = \psi$, see the diagram below:

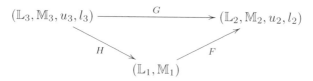

Then from the construction it is clear that $u_3(h(a), \chi(\alpha)) \leq h(u_1(a, \alpha))$ and $h(l_1(a, \alpha)) \leq l_3(h(a), \chi(\alpha))$ for every $a \in \mathbb{L}_1$ and every $\alpha \in \mathbb{M}$. Hence $H : (\mathbb{L}_3, \mathbb{M}_3, u_3, l_3) \to (L_1, \mathbb{M}_1, u_1, l_1)$ is a morphism in **QAS**. Thus $F : (\mathbb{L}_1, \mathbb{M}_1, u_1, l_1) \to (\mathbb{L}_2, \mathbb{M}_2, u_2, l_2)$ is indeed the initial lift of the source $F : (\mathbb{L}_1, \mathbb{M}_1) \to (\mathbb{L}_2, \mathbb{M}_2, u_2, l_2)$. The uniqueness of the lift is obvious.

Remark 2. As shown in [21] in case of the category **AS** of approximate systems the statement of the previous theorem holds also for a source consisting of arbitrary many morphisms. Unfortunately, we do not know whether such statement is valid also for quasi-approximate, or, at least for pseudo-approximate systems.

From Theorem 2 by duality principle we have

Theorem 3. *Every sink containing a single morphism* $F : (\mathbb{L}_1, \mathbb{M}_1, u_1, l_1) \to (\mathbb{L}_2, \mathbb{M}_2)$ *has a unique final lift* $F : (\mathbb{L}_1, \mathbb{M}_1, u_1, l_1) \to (\mathbb{L}_2, \mathbb{M}_2, u_2, l_2)$ *in category* **QAS** *and in its subcategories* **PAS** *and* **AS**.

The upper and lower quasi-approximate operators on $(\mathbb{L}_2, \mathbb{M}_2)$ efficiently can be defined as follows:

$$u_2(b, \beta) = \bigwedge \{c \in \mathbb{L}_2 \mid c \geq b, f(c) \geq u_1(f(b), \varphi(\beta))\}.$$

$$l_2(b, \beta) = \bigvee \{c \in \mathbb{L}_2 \mid c \leq b, f(c) \leq l_1(f(b), \varphi(\beta))\}.$$

4   Many-Valued L-Relations

The concept of an L-relation was first introduced in Zadeh's paper [25] and then redefined, under various assumptions on the range lattice L, by different authors, see, e.g. [22] [2], [3], [7], et al. In this section we extend the concept of an L-relation, in order to get a concept well coordinated with variable range quasi-approximate systems considered above. In the result we come to what we call a many-valued L-relation. In order to define it we have to enrich lattice L with a further binary operation $* : L \times L \to L$. Namely, in the sequel we assume that $(L, \leq, \wedge, \vee, *)$ is a cl-monoid.

Definition 4. [1] *A cl-monoid is a complete lattice* $(L, \leq, \wedge, \vee)$ *enriched with a binary commutative associative monotone operation* $* : L \times L \to L$ *such that* $a * 1_L = a$, $a * 0_L = 0_L$ *for each* $a \in L$ *and* $*$ *distributes over arbitrary joins, that is* $a * (\bigvee_i b_i) = \bigvee_i (a * b_i)$ $\forall a \in L$, $\forall \{b_i \mid i \in I\} \subseteq \mathbb{L}$.

Note that a cl-monoid can be defined also as an integral commutative quantale in the sense of K.I. Rosenthal [15].

Extending known properties of L-relations (see e.g.[23], [24], [14]) to the case of many-valued L-relations we come to the following

Definition 5. *A many-valued L-relation on a set X, or an $\mathbb{M}$-valued L-relation for short, is a non-decreasing on the last coordinate mapping $R : X \times X \times \mathbb{M} \to L$.*

(c) R is right-connected, (resp. left-connected) if for each $x \in X$ there exists $x' \in X$ (resp. $x'' \in X$) such that $R(x, x', \alpha) = 1_L$ (resp. $R(x'', x, \alpha) = 1_L$) $\forall \alpha \in \mathbb{M}$; an L-relation is connected if it is both left- and right-connected;

(r) R is reflexive, if $R(x, x, \alpha) = 1_L$ for each $x \in X$ and each $\alpha \in \mathbb{M}$.

(s) R is symmetric, if $R(x, y, \alpha) = R(y, x, \alpha)$ for all $x, y \in X$ and each $\alpha \in \mathbb{M}$;

(t) R is transitive, if $R(x, y, \alpha) * R(y, z, \beta) \leq R(x, z, \alpha \wedge \beta)$ for all $x, y, z \in X$ and all $\alpha, \beta \in \mathbb{M}$.

Let a cl-monoid L be fixed and let $\mathbf{REL}(L)$ denote the set of all triples $(X, \mathbb{M}, R)$ where $R : X \times X \times \mathbb{M} \to L$ is an $\mathbb{M}$-valued L-relation on it. To realize $\mathbf{REL}(L)$ as a category we specify its morphisms as pairs $(f, \varphi) : (X_1, \mathbb{M}_1, R_1) \to (X_2, \mathbb{M}_2, R_2)$ such that $R_1(x, y, \varphi(\beta)) \leq R_2(f(x), f(y), \beta)$ for all $x, y \in X$ and all $\beta \in \mathbb{M}_2$.

Let $\mathbf{REL}_c(L)$, $\mathbf{REL}_r(L)$, $\mathbf{REL}_s(L)$, $\mathbf{REL}_t(L)$ denote the full subcategories of $\mathbf{REL}(L)$, the L-relations in which are right-connected, reflexive, symmetric and transitive respectively. The notations like $\mathbf{REL}_{st}(L)$, $\mathbf{REL}_{rt}(L)$, etc., should be clear. In the next section we shall discuss connections between these categories and the corresponding categories of variable-range approximate-type systems.

5   Quasi-Approximate, Pseudo-Approximative, and Approximate Systems Generated by Many-Valued L-Relations

Let a set X and a cl-monoid $L = (L, \leq, \wedge, \vee, *)$ be fixed. Given an $\mathbb{M}$-valued L-relation $R : X \times X \times \mathbb{M} \to L$, we define an upper and a lower $\mathbb{M}$-quasi-approximation operators $u_R : L^X \times \mathbb{M} \to L^X$ and $l_R : L^X \times \mathbb{M} \to L^X$ by

$$u_R(A)(x, \alpha) = \sup_{x'} (R(x, x', \alpha) * A(x')) \ \forall A \in L^X, \ \forall x \in X, \forall \alpha \in \mathbb{M};$$

$$l_R(A)(x, \alpha) = \inf_{x'} (R(x, x', \alpha) \mapsto A(x')) \ \forall A \in L^X \ \forall x \in X, \forall \alpha \in \mathbb{M}$$

where $\mapsto: L \times L \to L$ is the residuation induced by operation $* : L \times L \to L$, see e.g. [18], [12].

Let $\mathfrak{S}(L) = \{(L^X, u_R, l_R, \mathbb{M}) \mid X \in \mathbf{SET}, \mathbb{M} \in \mathbf{CLAT}\}$ that is $\mathfrak{S}(L)$ is the family of all quasi-approximate systems induced by many-valued L-relations. (Note that cl-monoid L in these considerations is fixed and the L-power-set lattice L^X corresponds to the lattice $\mathbb{L}$ in sections 2 and 3!)

Theorem 4. *Every $\Sigma_{(X,\mathbb{M},R)} = (L^X, \mathbb{M}, u_R, l_R) \in \mathfrak{S}(X, L)$ is an $\mathbb{M}$-quasi-approximate system. If R is right-connected, then $u_R \geq l_R$, and hence $\Sigma_{(X,\mathbb{M},R)} = (L^X, \mathbb{M}, u_R, l_R)$ is strongly quasi-approximate. If R is transitive, then $\Sigma_{(X,\mathbb{M},R)} = (L^X, \mathbb{M}, u_R, l_R)$ is pseudo-approximate. Finally, if R is transitive and reflexive, then $\Sigma_{(X,\mathbb{M},R)} = (L^X, \mathbb{M}, u_R, l_R)$ is an $\mathbb{M}$-approximate system.*

Proof The validity of conditions (1u) and (1l) from definitions 1 and 2 for u_R and l_R follows from the properties of the lattices L^X and $\mathbb{M}$. What concerns properties (2u) and (2l), we shall prove stronger, namely infinite versions of these conditions:

(2*u) $u_R((\bigvee_i A_i), \alpha) = \bigvee_i u_R(A_i, \alpha) \quad \forall \{A_i \mid i \in I\} \subseteq L^X, \forall \alpha \in \mathbb{M};$
(2*l) $l_R((\bigwedge_i A_i), \alpha) = \bigwedge_i l_R(A_i, \alpha) \quad \forall \{A_i \mid i \in I\} \subseteq L^X, \forall \alpha \in \mathbb{M}.$

We get these properties as follows:
$u_R(\bigvee_i A_i, \alpha)(x) = \sup_{x'}(R(x, x', \alpha)(\bigvee_i A_i(x'), \alpha)) = \sup_{x'}(\bigvee_i R(x, x', \alpha) * A_i(x'))$
$= \bigvee_i(\sup_{x'}(R(x, x', \alpha) * A_i(x'))) = \bigvee_i(u_R(A_i)(x), \alpha) = (\bigvee_i(u_R(A_i, \alpha))(x);$
$l_R(\bigwedge_i A_i, \alpha)(x) = \inf_{x'}(R(x, x', \alpha) \mapsto (\bigwedge_i A_i, \alpha)(x'))$
$= \inf_{x'}(\bigwedge_i(R(x, x', \alpha) \mapsto A_i(x'))) = \bigwedge_i(\inf_{x'}(R(x, x', \alpha) \mapsto A_i(x'))$
$= (\bigwedge_i l_R(A_i, \alpha))(x).$
Assume now that the L-relation R is right-connected and for a given $x \in X$ choose $y \in X$ such that $R(x, y, \alpha) = 1$ for all $\alpha \in \mathbb{M}$. Then
$u_R(A, \alpha)(x) = \sup_{x'}(R(x, x', \alpha) * A(x')) \geq R(x, y, \alpha) * A(y) = A(y);$
$l_R(A, \alpha)(x) = \inf_{x'}(R(x, x', \alpha) \mapsto A(x')) \leq R(x, y, \alpha) \mapsto A(y) = A(y),$
that is $l_R(A, \alpha) \leq u_R(A, \alpha)$ and hence the system $\Sigma_{(X,\mathbb{M},R)}$ is strongly quasi-approximate. In case R is reflexive, in the above inequalities we may take $y = x$ and obtain $l_R(A, \alpha)(x) \leq A(x) \leq u_R(A, \alpha)(x) \ \forall A \in L^X, \ \forall x \in X$, that is $l_R(A) \leq A \leq u_R(A)$.
To prove that $\Sigma_{(X,\mathbb{M},R)}$ is a pseudo-approximate system in case when R is a transitive L-relation, we show that $u_R(u_R(A)) \leq u_R(A)$ and $l_R(l_R(A)) \geq l_R(A)$ for each $A \in L^X$. We get the first one of these inequalities referring to the transitivity of the relation R, as follows: $u(u(A, \alpha), \alpha)(x) = \sup_y\big(\sup_z(A(z) * R(x, z, \alpha)) * R(y, z, \alpha))\big) = \sup_z\big(\sup_y(A(z) * (R(x, z, \alpha) * R(y, z, \alpha)))\big)$
$\leq \sup_z\big(A(z) * R(x, z, \alpha)\big) = u(A, \alpha)(x).$
    The next series of inequalities, which is also justified by the transitivity of the relation R, establishes the second inequality:
$l_R(l_R(A, \alpha))(x) = \inf_y(R(x, y, \alpha) \mapsto (\inf_z(R(y, z, \alpha) \mapsto A(z))))$
$= \inf_y \inf_z\big((R(x, y, \alpha) * R(y, z, \alpha)) \mapsto A(z)\big)$
$= \inf_z \inf_y\big((R(x, y, \alpha) * R(y, z, \alpha)) \mapsto A(z)\big)$
$\geq \inf_z\big(R(x, z, \alpha) \mapsto A(z)\big) = l_R(A, \alpha)(x).$
To complete the proof, it is sufficient to notice that in case when R is both transitive and reflexive, all conditions (1u) - (4u) and (1l) - (4l) are satisfied and hence $\Sigma_{(X,\mathbb{M},R)}$ is an approximate system.

Remark 3. By the definition of the quasi approximation operators u_R and l_R, it is clear that $R \leq R'$ implies that $u_R \leq u_{R'}$ and $l_R \geq l_{R'}$. Moreover, note that if $R_1, R_2 : X \times X \to L$ are $\mathbb{M}$-valued L-relations on the same set X and $R_1 \neq R_2$,

then $u_{R_1} \neq u_{R_2}$ and hence $\Sigma_{R_1} \neq \Sigma_{R_2}$. Indeed, let $x_1, x_2 \in X$ and $\alpha \in \mathbb{M}$ be such that $R_1(x_1, x_2, \alpha) \neq R_2(x_1, x_2, \alpha)$ and let $A = \{x_2\}$ (as usual we identify a crisp set with its characteristic function). Then
$u_{R_1}(A)(x_1, \alpha) = \sup_{x' \in X}\big(R_1(x, x', \alpha) * A(x')\big) = R_1(x_1, x_2, \alpha); \; u_{R_2}(A)(x_1, \alpha) = \sup_{x' \in X}\big(R_2(x, x', \alpha) * A(x')\big) = R_2(x_1, x_2, \alpha),$
and hence $u_{R_1} \neq u_{R_2}$ and $\Sigma_{(X,\mathbb{M},R_1)} \neq \Sigma_{(X,\mathbb{M},R_2)}$. Note that in this case also $l_{R_1}(A)(x_1, \alpha) = R_1(x_1, x_2, \alpha) \neq R_2(x_1, x_2, \alpha) = l_{R_2}(A)(x_1, \alpha)$.

6 Category of Quasi-Approximate Systems Induced by Many-Valued L-Relations and Its Subcategories

Let a cl-monoid L be fixed and let $\mathbf{QASMR}(L)$ be the category of variable-range quasi approximate systems induced by many-valued L-relations on non-empty sets. Further, let $\mathbf{QASMRt}(L), \mathbf{QASMRc}(L), \mathbf{QASMRtr}(L)$ be its subcategories induced respectively by transitive, right-connected and transitive reflexive many-valued L-relations.

To construct a functor Φ from the category $\mathbf{REL}(L)$ where L is fixed into the category $\mathbf{QAS}$ we first define this functor on objects. Let $(X, \mathbb{M}, R) \in Ob(\mathbf{REL}(L))$, then we set

$$\Phi(X, \mathbb{M}, R) = \Sigma_{(X,\mathbb{M},R)}(= (L^X, \mathbb{M}, u_R, l_R))$$

where u_R and l_R are respectively the upper and lower approximation operators induced by the many-valued relation $R : X \times X \times \mathbb{M} \to L$. Further, given a morphism $(f, \varphi) : (X_1, \mathbb{M}_1, R_1) \to (X_2, \mathbb{M}_2, R_2)$ in $\mathbf{REL}(L)$ let

$$\Phi(f, \varphi) = (f_L^{\leftarrow}, \varphi) : \Sigma_{(X_1,\mathbb{M}_1,R_1)} \to \Sigma_{(X_2,\mathbb{M}_2,R_2)}$$

where $f_L^{\leftarrow} : L^{X_2} \to L^{X_1}$ is the backward operator induced by f, see [17], that is $f^{\leftarrow}(B) = B \circ f \in L^{X_1}$ for each $B \in L^{X_2}$. Let $F = \Phi(f, \varphi)$. We show that $F : \Sigma_{(X_1,R_1,\mathbb{M}_1)} \to \Sigma_{(X_2,R_2,\mathbb{M}_2)}$ thus defined is a morphism in the category $\mathbf{QAS}$.

Let $\mathbb{L}_1 = L^{X_1}$, $\mathbb{L}_2 = L^{X_2}$. Then it is well-known (and can be easily seen) that $f_L^{\leftarrow} : \mathbb{L}_2 \to \mathbb{L}_1$ is a morphism in the category of $\mathbf{IDCLAT}$ of infinitely distributive complete lattices, and hence condition (1mor) for F holds. The validity of condition (2mor) is obvious. To verify condition (3mor) for F let $B \in \mathbb{L}_2$ and $x \in X_1$. Then
$u_1(f_L^{\leftarrow}(B), \varphi(\beta))(x) = \sup_{x'}\big(R_1(x, x', \beta) * f_L^{\leftarrow}(B)(x)\big)$
$\leq \sup_{x'}\big(R_2(f(x), f(x'), \beta) * B(f(x'))\big) \leq \sup_{y'}\big(R_2(f(x), y', \beta) * B(f(x))\big).$
On the other hand,
$f_L^{\leftarrow}(u_2(B), \beta)(x) = u_2(B, \beta)(f(x)) = \big(R_2(f(x), y', \beta) * B(f(x)))\big),$
and hence $u_1(f_L^{\leftarrow}(B), \varphi(\beta)) \leq f_L^{\leftarrow}(u_2(B, \beta))$, so that the condition (3mor) holds.

To verify condition (4mor), let again $B \in \mathbb{L}_2$ and $x \in X_1$. Then
$f_L^{\leftarrow}(l_2(B, \beta))(x) = l_2(B, \beta)(f(x)) = \inf_{y'}\big(R_2(f(x), y', \beta) \mapsto B(f(x)).$
On the other hand, $l_1\big(f_L^{\leftarrow}(B), \varphi(\beta)\big)(x) = \inf_{x'}\big(R_1(x, x'), \varphi(\beta) \mapsto f_L^{\leftarrow}(B)(x)\big)$
$\geq \inf_{x'}\big(R_2(f(x), f(x'), \beta) \mapsto B(f(x)))\big) \geq \inf_{y'}\big(R_2(y', f(x), \beta) \mapsto B(f(x)).$
Hence $f_L^{\leftarrow}(l_2(B, \beta)) \leq l_1(f_L^{\leftarrow}(B), \beta)$ and so the condition (4mor) holds.

Combining the obtained information and taking into account Theorem 4 and Remark 3 we come to the following

Theorem 5. *By assigning to $(X, R, \mathbb{M}) \in Ob(\mathbf{REL}(L))$ a quasi-approximate system $\Sigma_R = (L^X, \mathbb{M}, u_R, l_R)$ and assigning to a morphism $f : (X_1, R_1, \mathbb{M}_1) \to (X_2, R_2, \mathbb{M}_2)$ of $\mathbf{REL}(L)$ the morphism $F = (f_L^\leftarrow, \varphi) : \Sigma_{R_1} \to \Sigma_{R_2}$ of $\mathbf{QAS}$ we obtain an isomorphism $\Phi : \mathbf{REL}(L) \to \mathbf{QASMR}(L)$ from $\mathbf{REL}(L)$ onto the sub-category $\mathbf{QASMR}(L)$ of $\mathbf{QAS}$. The restrictions of this functor to $\mathbf{REL}_t(L)$ and $\mathbf{REL}_{tr}(L)$ are isomorphisms of these categories and the categories $\mathbf{QASMR}_t(L)$ and $\mathbf{QASMR}_{rt}(L)$ respectively.*

7 Conclusion

We introduced the concept of a many-valued L-relation which in essence is a well-coordinated family of L-relations indexed by elements of a complete lattice. To develop the theory of many-valued L-relations we extended the category of variable-range approximate systems and defined less restricted categories of variable-range quasi-approximate and pseudo-approximate systems These categories include also systems induced by L-relations which either miss properties of reflexivity, symmetry and transitivity or satisfy weaken versions of these axioms.

As a by-product of the theory worked out here we obtain a many-valued viewpoint on the subject of the theory of L-fuzzy rough sets. Specifically, in the framework of the approach developed in this work, one can study lattices of well-coordinated families of rough approximations of an L-fuzzy set $A \in L^X$, with the triple $(A, 1_{L^X}, 0_{L^X})$ as (possibly) the coarsest L-rough approximation and the triple (A, A, A) as (possibly) the finest L-rough approximation.

Turning to the prospectives of the theory initiated here, we see both purely theoretical issues as well as its possible applications. Concerning theoretical prospectives as the first challenge we see a further study of the lattice characteristics and the categorical properties of variable-range (quasi-)approximate systems generated by many-valued L-relations (see e.g. the unsolved problem in Remark 2; its clarification is essential, in particular, for the operation of product for variable-range quasi-approximate systems). Considering possible applications, we expect that the concepts introduced here and the obtained results, in particular, realized in the context of many-valued L-fuzzy rough sets can be helpful, specifically, in the study of classification and clusterization problems in case when different unrelated sets of parameters should be taken into account.

Acknowledgment. The authors are grateful to the anonymous referees for reading the paper carefully and making comments which allowed to improve the exposition.

References

1. Birkhoff, G.: Lattice Theory. AMS, Providence (1995)
2. Bodenhofer, U.: Ordering of fuzzy sets based on fuzzy orderings. I: The basic approach. Mathware Soft Comput. 15, 201–218 (2008)

3. Bodenhofer, U.: Ordering of fuzzy sets based on fuzzy orderings. II: Generalizations. Mathware Soft Comput. 15, 219–249 (2008)

4. Brown, L.M., Ertürk, R., Dost, Ş.: Ditopological texture spaces and fuzzy topology, I. Basic concepts. Fuzzy Sets and Syst. 110, 227–236 (2000)

5. Brown, L.M., Ertürk, R., Dost, Ş.: Ditopological texture spaces and fuzzy topology, II. Topological considerations. Fuzzy Sets and Syst. 147, 171–199 (2004)

6. Chang, C.L.: Fuzzy topological spaces. J. Math. Anal. Appl. 24, 182–190 (1968)

7. Dubois, D., Prade, H.: Rough fuzzy sets and fuzzy rough sets. Internat. J. General Syst. 17, 191–209 (1990)

8. Gierz, G., Hoffman, K.H., Keimel, K., Lawson, J.D., Mislove, M.W., Scott, D.S.: Continuous Lattices and Domains. Cambridge University Press, Cambridge (2003)

9. Goguen, J.A.: The fuzzy Tychonoff theorem. J. Math. Anal. Appl. 43, 734–742 (1973)

10. Elkins, A., Šostak, A.: On some categories of approximate systems generated by L-relations. In: 3rd Rough Sets Theory Workshop, Milan, Italy, pp. 14–19 (2011)

11. Sang-Eon, H., Soo, K.I., Šostak, A.: On approximate-type systems generated by L-relations. Inf. Sci. (to appear)

12. Klement, E.P., Mesiar, R., Pap, E.: Triangular Norms. Kluwer Acad. Publ. (2000)

13. Pawlak, Z.: Rough sets. Intern. J. of Computer and Inform. Sci. 11, 341–356 (1982)

14. Radzikowska, A.M., Kerre, E.E.: A comparative study of fuzzy rough sets. Fuzzy Sets and Syst. 126, 137–155 (2002)

15. Rosenthal, K.I.: Quantales and Their Applications. Pirman Research Notes in Mathematics 234. Longman Scientific & Technical (1990)

16. Rodabaugh, S.E.: Powers-set operator based foundations for point-set lattice-theoretic (poslat) fuzzy set theories and topologies. Quaest. Math. 20, 463–530 (1997)

17. Rodabaugh, S.E.: Power-set operator foundations for poslat fuzzy set theories and topologies. In: Höhle, U., Rodabaugh, S.E. (eds.) Mathematics of Fuzzy Sets: Logic, Topology and Measure Theory, pp. 91–116. Kluwer Acad. Publ. (1999)

18. Schweitzer, B., Sklar, A.: Probabilistic Metric Spaces. North Holland, New York (1983)

19. Šostak, A.: On approximative operators and approximative systems. In: Proceedings of Congress of the International Fuzzy System Association (IFSA 2009), Lisbon, Portugal, July 20-24, pp. 1061–1066 (2009)

20. Šostak, A.: Towards the theory of M-approximate systems: Fundamentals and examples. Fuzzy Sets and Syst. 161, 2440–2461 (2010)

21. Šostak, A.: Towards the theory of approximate systems: variable range categories. In: Proceedings of ICTA 2011, Islamabad, Pakistan, pp. 265–284. Cambridge University Publ. (2012)

22. Valverde, L.: On the structure of F-indistinguishability operators. Fuzzy Sets and Syst. 17, 313–328 (1985)

23. Yao, Y.Y.: A comparative study of fuzzy sets and rough sets. Inf. Sci. 109, 227–242 (1998)

24. Yao, Y.Y.: On generalizing Pawlak approximation operators. In: Polkowski, L., Skowron, A. (eds.) RSCTC 1998. LNCS (LNAI), vol. 1424, pp. 298–307. Springer, Heidelberg (1998)

25. Zadeh, L.: Similarity relations and fuzzy orderings. Inf. Sci. 3, 177–200 (1971)

Introducing Similarity Relations in a Framework for Modelling Real-World Fuzzy Knowledge

Víctor Pablos-Ceruelo and Susana Muñoz-Hernández[*]

The Babel Research Group
Facultad de Informática
Universidad Politécnica de Madrid, Spain
{vpablos,susana}@babel.ls.fi.upm.es
http://babel.ls.fi.upm.es

Abstract. There is no need for justifying the use of fuzzy logic (FL) to model the real-world knowledge. Bi-valued logic cannot conclude if a real-world sentence like "the restaurant is close to the city center" is true or false because it is neither true nor false. Letting apart paradoxes' sentences[1], there are sentences (as the previous one) that are not true nor false but true up to some degree of truth or true at least to some degree of truth. In order to represent the truth or falsity of such sentences we need FL.

Similarity is a relation between real-world concepts. As in the representation of the truth of the first sentence, the representation of the similarity between two (fuzzy or not) concepts can be true, false or true up to (or at least to) some degree. We present syntactic constructions (and their semantics) for modelling such relation between concepts. The interest is in, for example, obtaining "spanish food restaurants" when asking for "mediterranean food restaurants" (only if the similarity between spanish and mediterranean food is explicitly stated in the program file). We hope this allows to represent in a better way the real-world knowledge, specially the concepts that are defined just by their similarity relations to some other concepts.

Keywords: fuzzy logic, framework, similarity relations.

1 Introduction

From the beginning the human being has tried to create machines with the capability to understand the real world as he does and help him to carry out tasks that he does not like to do.

[*] This work is partially supported by research projects DESAFIOS10 (TIN2009-14599-C03-00) funded by Ministerio Ciencia e Innovación of Spain, PROMETIDOS (P2009/TIC-1465) funded by Comunidad Autónoma de Madrid and Research Staff Training Program (BES-2008-008320) funded by the Spanish Ministry of Science and Innovation. It is partially supported too by the Universidad Politécnica de Madrid entities Departamento de Lenguajes, Sistemas Informáticos e Ingeniería de Software and Facultad de Informática.

[1] Paradoxes' sentences are sentences like "the only barber in the town shaves anyone that does not shave himself" (attributed to Bertrand Russell), which have self-references that do not let us to assign them any truth value because using logic inference we always conclude the opposite truth value.

A. Laurent et al. (Eds.): IPMU 2014, Part III, CCIS 444, pp. 51–60, 2014.
© Springer International Publishing Switzerland 2014

When representing the world as the human being understands it and how he takes decisions and interacts with the first one we encounter the problem of representing fuzzy characteristics (it is hot), fuzzy rules (if it is hot, turn on the fan) and fuzzy actions (since it is not too hot, turn on the fan at medium speed). So, a machine needs all this information (or knowledge) if we want it to understand the world as the human being does and take decisions as the human being does.

One of the most successful programming languages for representing knowledge in computer science is Prolog, whose main advantage with respect to the other ones is being a more declarative programming language[2]. Prolog is based on logic. It is usual to identify logic with bi-valued logic and assume that the only available values are "yes" and "no" (or "true" and "false"), but logic is much more than bi-valued logic. In fact we use fuzzy logic (FL), a subset of logic that allow us to represent not only if an individual belongs or not to a set, but the grade in which it belongs. Supposing a database with the contents shown in Fig. 1, the definition for the function "close" in Fig. 1 and the question "Is restaurant X close to the center?" with FL we can deduce that Il tempietto is "definitely" close to the center, Tapasbar is "almost" close, Ni Hao is "hardly" close and Kenzo is "not" close to the center. We highlight the words "definitely", "almost", "hardly" and "not" because the usual answers for the query are "1", "0.9", "0.1" and "0" for the individuals Il tempietto, Tapasbar, Ni Hao and Kenzo and the humanization of the crisp values is done in a subsequent step by defuzzification.

name	distance	price avg.	food type
Il_tempietto	100	30	italian
Tapasbar	300	20	spanish
Ni Hao	900	10	chinese
Kenzo	1200	40	japanese

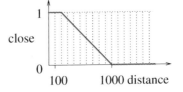

Fig. 1. Restaurants database and close fuzzification function

Modelling the real-world knowledge as the human being does is not an easy task. The human mind tends to determine if two concepts have something in common. If they have it he usually memorizes the most general one or a new one that comprises the common parts and the differences between this one and the other one(s). By learning by heart only this the human mind can store many concepts with a very low storage cost and have a very fast process time (the less concepts we have the faster we retrieve them). Besides, it works in the same way for rules that explain the world behaviour or how to act in each situation: he learns the general rule and/or the exceptions to the general rule and what they depend on.

This way of storing information (or knowledge) and reasoning with it is what we try to allow the programmer to represent in programs. Due to its complexity we are here

---

[2] We say that it is a more declarative programming language because it removes the necessity to specify the flow control in most cases, but the programmer still needs to know if the interpreter or compiler implements depth or breadth-first search strategy and left-to-right or any other literal selection rule.

concerned only with allowing him/her to represent similarity between concepts. This allows to save space and time when modelling the real-world knowledge, at the same time that we get a little bit closer to the human being way of understanding the world.

To introduce similarity in fuzzy logic we could use any of the existing frameworks for representing fuzzy knowledge. Leaving apart the theoretical frameworks, as [20], we know about the Prolog-Elf system [7], the FRIL Prolog system [1], the F-Prolog language [8], the FuzzyDL reasoner [2], the Fuzzy Logic Programming Environment for Research (FLOPER) [15], the Fuzzy Prolog system [19,6], or Rfuzzy [16]. All of them implement in some way the fuzzy set theory introduced by Lotfi Zadeh in 1965 ([22]), and all of them allow you to extend the base language with your own modifications. We choose Rfuzzy with priorities [17,18] because we need the capability to define that the results provided by some rule are preferred to the ones provided by some other rule, no matter if the last one provides a higher truth value.

To our knowledge, the works similar to ours are [21,5,3,4]. The main differences between our work and this ones are (1) that we do not force the similarity relation to be reflexive, symmetric and transitive, i.e., an equivalence relation. As some of they mention, this is too restrictive for real-world applications. And (2) that we do not try to measure the closeness (or similarity) between two fuzzy propositions. Our work goes in the other direction: we take the similarity value computed and return the elements considered to be similar to the one we are looking for.

The paper is structured as follows: an introduction to the syntax we use goes first (sec. 2). A little bit of background on FL with priorities goes after (sec. 3) and the syntax and semantics we propose for representing similarity just after it (sec. 4). Conclusions and current work go in last place (sec. 5), as usual.

2   Syntax

We will use a signature Σ of function symbols and a set of variables V to "build" the *term universe* $TU_{\Sigma,V}$ (whose elements are the *terms*). It is the minimal set such that each variable is a term and terms are closed under Σ-operations. In particular, constant symbols are terms. Similarly, we use a signature Π of predicate symbols to define the *term base* $TB_{\Pi,\Sigma,V}$ (whose elements are called *atoms*). Atoms are predicates whose arguments are elements of $TU_{\Sigma,V}$. Atoms and terms are called *ground* if they do not contain variables. As usual, the *Herbrand universe* **HU** is the set of all ground terms, and the *Herbrand base* **HB** is the set of all atoms with arguments from the Herbrand universe. A substitution σ or ξ is (as usual) a mapping from variables from V to terms from $TU_{\Sigma,V}$ and can be represented in suffix ($(Term)\sigma$) or in prefix notation ($\sigma(Term)$).

To capture different interdependencies between predicates, we will make use of a signature Ω of *many-valued connectives* formed by *conjunctions* $\&_1, \&_2, ..., \&_k$, *disjunctions* $\vee_1, \vee_2, ..., \vee_l$, *implications* $\leftarrow_1, \leftarrow_2, ..., \leftarrow_m$, *aggregations* $@_1, @_2, ..., @_n$ and tuples of real numbers in the interval $[0, 1]$ represented by (p, v).

While Ω denotes the set of connective symbols, $\hat{\Omega}$ denotes the set of their respective associated truth functions. Instances of connective symbols and truth functions are denoted by $\&_i$ and $\hat{\&}_i$ for conjunctors, $\vee_i$ and $\hat{\vee}_i$ for disjunctors, $\leftarrow_i$ and $\hat{\leftarrow}_i$ for implicators, $@_i$ and $\hat{@}_i$ for aggregators and (p, v) and $(\hat{p}, v)$ for the tuples.

Truth functions for the connectives are then defined as $\&: [0,1]^2 \to [0,1]$ monotone[3] and non-decreasing in both coordinates, $\hat{\vee}: [0,1]^2 \to [0,1]$ monotone in both coordinates, $\hat{\leftarrow}: [0,1]^2 \to [0,1]$ non-increasing in the first and non-decreasing in the second coordinate, $\hat{@}: [0,1]^n \to [0,1]$ as a function that verifies $\hat{@}(0, \ldots, 0) = 0$ and $\hat{@}(1, \ldots, 1) = 1$ and $(\mathsf{p}, \mathsf{v}) \in \Omega^{(0)}$ are functions of arity 0 (constants) that coincide with the connectives.

Immediate examples for connectives that come to mind for conjunctors are: in Łukasiewicz logic ($\hat{F}(x,y) = max(0, x + y - 1)$), in Gödel logic ($\hat{F}(x,y) = min(x,y)$), in product logic ($\hat{F}(x,y) = x \cdot y$), for disjunctors: in Łukasiewicz logic ($\hat{F}(x,y) = min(1, x + y)$), in Gödel logic ($\hat{F}(x,y) = max(x,y)$), in product logic ($\hat{F}(x,y) = x \cdot y$), for implicators: in Łukasiewicz logic ($\hat{F}(x,y) = min(1, 1 - x + y)$), in Gödel logic ($\hat{F}(x,y) = y$ if $x > y$ else 1), in product logic ($\hat{F}(x,y) = x \cdot y$) and for aggregation operators[4]: arithmetic mean, weighted sum or a monotone function learned from data.

3   A Small Revision on FL with Priorities

FL with priorities has basically the same properties that FL, but instead of having a truth value $\mathsf{v} \in [0,1]$ has a tuple of real numbers between 0 and 1, $(\mathsf{p}, \mathsf{v}) \in \Omega^{(0)}$, where $\mathsf{p} \in [0,1]$ denotes the (accumulated) priority. This simple change implies two more changes, one in the classical ordering definition of the truth values and the other one in the connectives meaning. We include first a brief introduction to the structure that gives meaning to our FL programs, a particular case of the multi-adjoint algebra (more info can be found in [12,9,10,11,13,14]), and to the modifications needed to manage priorities (more info can be found in [17,18]). The strong point of using this structure is that we can obtain the credibility for the rules that we write from real-world data (in an automatic way), although this time we do not focus in that advantage.

The multi-adjoint semantics of fuzzy logic programs is based on a maximum operator. Since we have now a tuple $(\mathsf{p}, \mathsf{v}) \in \Omega^{(0)}$ instead of the truth value $\mathsf{v} \in [0,1]$, we need to define the ordering between two or more tuples. The usual representation (p, v) is sometimes changed into (pv) to highlight that the variable is only one and it can take the value $\perp$ (no answer), and the set of all possible values is symbolized by **KT**.

Definition 1 ($\preccurlyeq$ **KT**).

$$\perp \preccurlyeq_{\mathbf{KT}} \perp \preccurlyeq_{\mathbf{KT}} (\mathsf{p}, \mathsf{v})$$

$$(\mathsf{p}_1, \mathsf{v}_1) \preccurlyeq_{\mathbf{KT}} (\mathsf{p}_2, \mathsf{v}_2) \quad \leftrightarrow \quad (\mathsf{p}_1 < \mathsf{p}_2) \; or \; (\mathsf{p}_1 = \mathsf{p}_2 \; and \; \mathsf{v}_1 \leq \mathsf{v}_2) \quad (1)$$

where $<$ is defined as usually (v_i and p_j are just real numbers between 0 and 1).

---

[3] As usually, a n-ary function $\hat{F}$ is called *monotonic in the* i-th argument ($i \leq n$), if $x \leq x'$ implies $\hat{F}(x_1, \ldots, x_{i-1}, x, x_{i+1}, \ldots, x_n) \leq \hat{F}(x_1, \ldots, x_{i-1}, x', x_{i+1}, \ldots, x_n)$ and a function is called *monotonic* if it is monotonic in all arguments.

[4] Note that the above definition of aggregation operators subsumes all kinds of minimum, maximum or mean operators.

We define now the syntax of the multi-adjoint logic programs, their valuations and interpretations, the operator needed by the connectives to manage the value of p_j in the tuples and the satisfaction and model. All this conforms the syntax and semantics of our programs.

Definition 2 (Multi-Adjoint Logic Program). *A multi-adjoint logic program is a set of clauses of the form*

$$A \xleftarrow{(\mathsf{p},\,\mathsf{v}).\,\&_i} @_j (B_1, \ldots, B_k, \ldots, B_n) \quad \textit{if COND} \tag{2}$$

where $(\mathsf{p}, \mathsf{v}) \in \mathbf{KT}$, $\&_i$ *is a conjunctor,* $@_j$ *an aggregator[5], A and* B_k, $k \in [1..n]$, *are atoms and COND is a first-order formula (basically a bi-valued condition) formed by the predicates in* $\mathrm{TB}_{\Pi,\Sigma,V}$, *the predicates* $=, \neq, \geq, \leq, >$ *and* $<$ *restricted to terms from* $\mathrm{TU}_{\Sigma,V}$, *the symbol true and the conjunction* $\wedge$ *and disjunction* $\vee$ *in their usual meaning.*

Definition 3 (Valuation, Interpretation). *A* valuation *or* instantiation $\sigma : V \to \mathbf{HU}$ *is an assignment of ground terms to variables and uniquely constitutes a mapping* $\hat{\sigma} : \mathrm{TB}_{\Pi,\Sigma,V} \to \mathbf{HB}$ *that is defined in the obvious way.*

A fuzzy Herbrand interpretation (or short, interpretation) of a fuzzy logic program is a mapping $I : \mathbf{HB} \to \mathbf{KT}$ *that assigns an element in our lattice to ground atoms[6].*

It is possible to extend uniquely the mapping I defined on $\mathbf{HB}$ *to the set of all ground formulas of the language by using the unique homomorphic extension. This extension is denoted* $\hat{I}$ *and the set of all interpretations of the formulas in a program* $\mathbf{P}$ *is denoted* $I_{\mathbf{P}}$.

Definition 4 (The operator $\circ$). *The application of some conjunctor* $\bar{\&}$ *(resp. implicator* $\bar{\leftarrow}$, *aggregator* $\bar{@}$ *) to elements* $(\mathsf{p}, \mathsf{v}) \in \mathbf{KT} \setminus \{\bot\}$ *refers to the application of the truth function* $\hat{\&}$ *(resp.* $\hat{\leftarrow}$, $\hat{@}$ *) to the second elements of the tuples while* $\circ_{\&}$ *(resp.* $\circ_{\leftarrow}$, $\circ_{\&}$ *) is the one applied to the first ones. The operator* $\circ$ *is defined by*

$$x \circ_{\&} y = \frac{x + y}{2} \quad \textit{and} \quad z \circ_{\leftarrow} y = 2 * z - y.$$

Definition 5 (Satisfaction, Model). *Let* $\mathbf{P}$ *be a multi-adjoint logic program,* $I \in I_{\mathbf{P}}$ *an interpretation and* $A \in \mathbf{HB}$ *a ground atom. We say that a clause* $Cl_i \in \mathbf{P}$ *of the form shown in eq. 2 is satisfied by I or I is a model of the clause* Cl_i *(* $I \Vdash Cl_i$ *) if and only if (iff) for all ground atoms* $A \in \mathbf{HB}$ *and for all instantiations* σ *for which* $B\sigma \in \mathbf{HB}$ *(note that* σ *can be the empty substitution) it is true that*

$$\hat{I}(A) \succcurlyeq_{\mathbf{KT}} (\mathsf{p}, \mathsf{v}) \ \bar{\&}_i \ \bar{@}_i(\hat{I}(B_1\sigma), \ldots, \hat{I}(B_n\sigma)) \tag{3}$$

whenever COND is satisfied (true). Finally, we say that I is a model of the program $\mathbf{P}$ *and write* $I \Vdash \mathbf{P}$ *iff* $I \Vdash Cl_i$ *for all clauses in our multi-adjoint logic program* $\mathbf{P}$.

---

[5] Unnecessary if $k \in [1..1]$ or $n = 1$

[6] The *domain* of an interpretation is the set of all atoms in the Herbrand Base (interpretations are total functions), although for readability reasons we present interpretations as sets of pairs $(A, (\mathsf{p}, \mathsf{v}))$ where $A \in \mathbf{HB}$ and $(\mathsf{p}, \mathsf{v}) \in \mathbf{KT} \setminus \{\bot\}$ (we omit those atoms whose interpretation is the truth value $\bot$).

4   Syntax and Semantics for the New Similarity Constructions

Now that we have introduced the basics of our formal semantics we introduce the syntax and semantics of the similarity constructions that we propose. Since this constructions must live with some others that were previously defined in the framework, we first include a brief revision of them and refer to the contributions [17,18] for more details.

The syntactical constructions defined in [18] are basically eight. One of them serves to map the contents of a database into concepts that we can use in our programs, three of them act as tail of the remaining four (and modify slightly the meaning of these four when they are used) and the last four are for defining fuzzy characteristics from the non-fuzzy data stored in the database. Due to lack of space we only include the syntax and semantics of the first one, the common part of the semantics of the last four and how the three tails affect the meaning (or semantics) of those four.

The construction used to map the contents of a database into concepts that we can use in our programs is shown in eq. 4. We provide an example in eq. 5 to clarify, in which the restaurant vdbt has four columns: the first for the unique identifier given to each restaurant (its name), the second for the distance to the city center from that restaurant, the third for the restaurant's price average and the last one for the food type served there.

$$define\_database(pT/pA, [(pN, pT')]) \tag{4}$$

$$define\_database(\ restaurant/5, \quad (id,\ string\_type),$$
$$(distance\_to\_the\_city\_center,\ integer\_type),$$
$$(price\_average,\ integer\_type),\ (food\_type,\ enum\_type)]). \tag{5}$$

The three tails' constructions that serve to slightly modify the meaning of the remaining four are shown in eqs. 6, 7, 8. If they appear the programmer wants, respectively, (eq. 6) to limit the set of elements in our database for which he wants to use the fuzzy clause or rule or (eq. 7) to define a personalized rule, one that only applies when the name of the user logged in and the user name in the rule are the same one or (eq. 8) to (re)define the credibility of the construction in which it appears as tail, and the operator used to compute the resultant truth value. The changes produced in the semantics of any clause with the form of one of the four constructions when any of this three constructions (or a combination of them) appears as tail of the first ones is summarized in the table in Fig. 3.

The four remaining constructions are used to define fuzzy characteristics of the elements in our database from the data stored in the database. All of them get their semantics by translating the syntax proposed into the syntax of eq. 2. The only difference between them is the values given to the variables appearing in that construction, which are $fPredName(Individual)$[7] for A and the values shown in the table in Fig. 2 for the variables p, v, $\&_i$, $@_j$ $(B_1, \ldots, B_n)$ and $COND$.

---

[7] $fPredName$ is the name of the fuzzy predicate we are defining and individual is a variable for introducing the element of the database to which we want to obtain the fuzzy value or a variable in case we want to obtain the results for all the individuals in our database (by using Prolog's backtracking).

$$if(pN(pT) \ comp \ value). \tag{6}$$

$$only\_for\_user \ 'UserName' \tag{7}$$

$$with\_credibility(credOp, \ credVal) \tag{8}$$

construction	p	v	$\&_i$	$@_j (B_1,\ldots,B_n)$	COND
fuzzy value	0.8	1	product	TV	true
fuzzification function	0.6	1	product	$pN(Individual) * \frac{(valOut\_2 - valOut\_1)}{(valIn\_2 - valIn\_1)}$	$(valIn\_1 < pN(Individual) \le valIn\_2)$
fuzzy rule	0.4	1	product	$@_j (B_1,\ldots,B_n)$	true
default fuzzy value	0	1	product	TV	true

Fig. 2. Summary of the values given to the variables p, v, $\&_i$, $@_j (B_1,\ldots,B_n)$ and COND

tail construction	p	v	$\&_i$	COND
eq. 6	p + 0.05	v	$\&_i$	$COND \wedge (pN(Individual) \ comp \ value)$
eq. 7	p + 0.1	v	$\&_i$	$COND \wedge currentUser(Me) \wedge Me = 'UserName'$
eq. 8	p	credVal	credOp	COND

Fig. 3. Changes in the values given to the variables p, v, $\&_i$ and COND when the tails' constructions in eqs. 6, 7, 8 are used

The syntactical constructions we propose for modelling similarity are shown in eqs. 9 and 13. The necessity for two constructions is justified by the existence of two kinds of similarity: between attributes (eq. 9) and between fuzzy predicates (eq. 13). Illustrative examples are "the food type mediterranean is 0.7 similar to the spanish food"[8] (eq. 10) or "unexpensive is similar to (or a synonym of) cheap" (eq. 14). In the syntax shown pT is the name of the virtual database table[9] (vdbt), pN is the name assigned to a column of the vdbt named pT, $V1$ and $V2$ are possible values for the column pN of the vdbt named pT (column that must be of type $enum\_type$) and TV is a truth value (a float number between 0 and 1).

The semantics of the constructions are presented in eqs. 11 and 15 and the ones for the examples in eqs. 12 and 16. In eqs. 11 and 15 the values for the variables p, v, $\&_i$, $@_j (B_1,\ldots,B_n)$ and COND are the ones in the table in Fig. 4. The syntactical structures for similarity can be followed by the tails' constructions in eqs. 6, 7, 8 and, when this occurs, their semantics change as the semantics of the other syntactical constructions (see the table in Fig. 3).

---

[8] Be careful, we are not saying that the spanish food is 0.7 similar to the mediterranean one. You need to add another clause with that information if you wanna say that too.

[9] In [18] the authors use the "virtual database table" concept to highlight that the structure of the database used in programs can be different to the real structure of the database.

construction	p	v	$\&_i$	$@_j (B_1,\ldots,B_n)$	$COND$
similarity between attributes	0.8	1	product	TV	true
similarity between fuzzy predicates	0	1	product	$fPredName2(Individual)$	true

Fig. 4. Summary of the values given to the variables p, v, $\&_i$ and $COND$ for similarity

$$similarity\_between(pT,\ pN(V1),\ pN(V2),\ TV) \tag{9}$$

$$similarity\_between(restaurant,$$
$$food\_type(mediterranean),\ food\_type(spanish),\ 0.7) \tag{10}$$

$$similarity(pT(pN(V1,V2))) \xleftarrow{(p,\ v),\ \&_i} TV\ \text{if}\ COND \tag{11}$$

$$similarity(restaurant(food\_type(mediterranean, spanish))) \xleftarrow{(0.8,\ 1),\ prod} 0.7\ \text{if}\ true \tag{12}$$

$$fPredName(pT) :\sim synonym\_of(fPredName2(pT),\ credOp,\ credVal) \tag{13}$$

$$unexpensive(restaurant) : synonym\_of(cheap(restaurant),\ prod,\ 1). \tag{14}$$

$$fPredName(Individual) \xleftarrow{(p,\ v),\ \&_i} fPredName2(Individual)\ \text{if}\ COND \tag{15}$$

$$unexpensive(Individual) \xleftarrow{(0,\ 1),\ prod} cheap(Individual)\ \text{if}\ true \tag{16}$$

The introduction of this new syntactical constructions needs a little bit of explanation. The first construction, the one in eq. 11, serves to define the similarity between two attributes. Allowing to define this similarity is not the goal, but only the means to get in our framework the information needed for answering questions of the form "give me all the elements in the database whose attribute X is similar to Y". This allows us, for example, to define the pink color similar to the red one and get all the pink cars when asking for cars with a color similar to red. The second construction, the one in eq. 15, serves to define fuzzy predicates from other fuzzy predicates, when the programmer considers that the first one is a synonym of the last one. The interest in allowing to do this is, for example, in having a richer vocabulary without the cost of defining and storing the definitions of all of the new words. We can, for example, define "unexpensive" from "cheap", "wet" from "damp" or "gorgeous" and "handsome" from "beautiful". In this way we can answer the query "unexpensive cars" from the definition of "cheap cars" and the similarity link between "unexpensive" and "cheap" cars.

The inclusion of the last construction forced us to slightly modify the semantics of the "default fuzzy values" construction. The reason for this is that we want the predicate defined as similar to other one to be modifiable in case of necessity. So, we could for example define "gorgeous" person and "handsome" person from "beautiful" person, but redefine "gorgeous" to have a value 0 when the person is a man. It is, since some men (male humans) feel uncomfortable if we say that they are "gorgeous" we avoid using this qualification for them, but for women it keeps the original meaning given by "beautiful". To get this behaviour, we needed the results provided by the similarity rule to have the smallest priority value. Due to the fact that the "default fuzzy values" construction had the lowest priority value (0), we have modified it to 0.2, so the construction with the lowest priority is now the similarity construction.

5 Conclusions

We have presented the syntax and semantics of two constructions for representing similarity relations. Our goal with this constructions is allowing the representation of real-world similarity relations while modelling the real world using fuzzy logic. The advantages of doing it are mainly two: (1) reuse the definitions of concepts that are similar to the ones we are defining and (2) get the programming language we use a little bit more close to the human way of thinking. When we say that we get it a little bit closer to the human way of thinking we mean that the human being tends to group the concepts by using similarity relations, saving just the differences between concepts, and the similarity constructions allows him/her to code just as his/her mind is "programmed". By copying this behaviour we facilitate him/her the process of representing what he has in his/her mind. It is obvious that we still have a lot of work to do in this line, but this is one step more for achieving the final goal of transferring the human knowledge to a machine. Links to a beta version of our implementation and to a web application that we have developed for testing existing example programs (with the possibility to upload new ones) is available at our web page.

Our current research focus on deriving similarity relations from the modelization of a problem in the framework's language. In this way we could, for example, derive from the RGB composition of two colors their similarity relation.

References

1. Baldwin, J.F., Martin, T.P., Pilsworth, B.W.: Fril-Fuzzy and Evidential Reasoning in Artificial Intelligence. John Wiley & Sons, Inc., New York (1995)
2. Bobillo, F., Straccia, U.: fuzzydl: An expressive fuzzy description logic reasoner. In: 2008 International Conference on Fuzzy Systems (FUZZ 2008), pp. 923–930. IEEE Computer Society (2008)
3. Dubois, D., Prade, H.: Comparison of two fuzzy set-based logics: similarity logic and possibilistic logic. In: Proceedings of 1995 IEEE Int. Fuzzy Systems, International Joint Conference of the Fourth IEEE International Conference on Fuzzy Systems and The Second International Fuzzy Engineering Symposium, vol. 3, pp. 1219–1226 (1995)
4. Esteva, F., Garcia, P., Godo, L., Ruspini, E., Valverde, L.: On similarity logic and the generalized modus ponens. In: Proceedings of the Third IEEE Conference on Computational Intelligence, Fuzzy Systems, IEEE World Congress on Computational Intelligence, vol. 2, pp. 1423–1427 (1994)
5. Godo, L., Rodriguez, R.O.: A fuzzy modal logic for similarity reasoning. In: Cai, K.-Y., Chen, G., Ying, M. (eds.) Fuzzy Logic And Soft Computing. Kluwer Academic (1999)
6. Guadarrama, S., Muñoz-Hernández, S., Vaucheret, C.: Fuzzy prolog: a new approach using soft constraints propagation. Fuzzy Sets and Systems 144(1), 127–150 (2004)
7. Ishizuka, M., Kanai, N.: Prolog-elf incorporating fuzzy logic. In: IJCAI 1985: Proceedings of the 9th International Joint Conference on Artificial Intelligence, pp. 701–703. Morgan Kaufmann Publishers Inc., San Francisco (1985)

8. Li, D., Liu, D.: A fuzzy Prolog database system. John Wiley & Sons, Inc., New York (1990)
9. Medina, J., Ojeda-Aciego, M., Vojtáš, P.: A completeness theorem for multi-adjoint logic programming. In: FUZZ-IEEE, pp. 1031–1034 (2001)
10. Medina, J., Ojeda-Aciego, M., Vojtáš, P.: Multi-adjoint logic programming with continuous semantics. In: Eiter, T., Faber, W., Truszczyński, M. (eds.) LPNMR 2001. LNCS (LNAI), vol. 2173, pp. 351–364. Springer, Heidelberg (2001)
11. Medina, J., Ojeda-Aciego, M., Vojtáš, P.: A procedural semantics for multi-adjoint logic programming. In: Brazdil, P., Jorge, A. (eds.) EPIA 2001. LNCS (LNAI), vol. 2258, pp. 290–297. Springer, Heidelberg (2001)
12. Medina, J., Ojeda-Aciego, M., Vojtáš, P.: A multi-adjoint approach to similarity-based unification. Electronic Notes in Theoretical Computer Science 66(5), 70–85 (2002), UNCL'2002, Unification in Non-Classical Logics (ICALP 2002 Satellite Workshop)
13. Medina, J., Ojeda-Aciego, M., Vojtáš, P.: Similarity-based unification: a multi-adjoint approach. Fuzzy Sets and Systems 146(1), 43–62 (2004)
14. Moreno, J.M., Ojeda-Aciego, M.: On first-order multi-adjoint logic programming. In: 11th Spanish Congress on Fuzzy Logic and Technology (2002)
15. Morcillo, P.J., Moreno, G.: Floper, a fuzzy logic programming environment for research. In: Fundación Universidad de Oviedo (ed.) Proceedings of VIII Jornadas sobre Programación y Lenguajes (PROLE 2008), Gijón, Spain, pp. 259–263 (october 2008)
16. Muñoz-Hernández, S., Pablos-Ceruelo, V., Strass, H.: Rfuzzy: Syntax, semantics and implementation details of a simple and expressive fuzzy tool over prolog. Information Sciences 181(10), 1951–1970 (2011), Special Issue on Information Engineering Applications Based on Lattices
17. Pablos-Ceruelo, V., Muñoz-Hernández, S.: Introducing priorities in rfuzzy: Syntax and semantics. In: CMMSE 2011: Proceedings of the 11th International Conference on Mathematical Methods in Science and Engineering, Benidorm, Alicante, Spain, vol. 3, pp. 918–929 (June 2011)
18. Pablos-Ceruelo, V., Muñoz-Hernández, S.: Getting answers to fuzzy and flexible searches by easy modelling of real-world knowledge. In: FCTA 2013: Proceedings of the 5th International Conference on Fuzzy Computation Theory and Applications (2013)
19. Vaucheret, C., Guadarrama, S., Muñoz-Hernández, S.: Fuzzy prolog: A simple general implementation using CLP(R). In: Baaz, M., Voronkov, A. (eds.) LPAR 2002. LNCS (LNAI), vol. 2514, pp. 450–464. Springer, Heidelberg (2002)
20. Vojtáš, P.: Fuzzy logic programming. Fuzzy Sets and Systems 124(3), 361–370 (2001)
21. Wang, J.-B., Xu, Z.-Q., Wang, N.-C.: A fuzzy logic with similarity. In: Proceedings of the 2002 International Conference on Machine Learning and Cybernetics, vol. 3, pp. 1178–1183 (2002)
22. Zadeh, L.A.: Fuzzy sets. Information and Control 8(3), 338–353 (1965)

An Approach Based on Rough Sets to Possibilistic Information

Michinori Nakata[1] and Hiroshi Sakai[2]

[1] Faculty of Management and Information Science,
Josai International University
1 Gumyo, Togane, Chiba, 283-8555, Japan
nakatam@ieee.org

[2] Department of Mathematics and Computer Aided Sciences,
Faculty of Engineering, Kyushu Institute of Technology,
Tobata, Kitakyushu, 804-8550, Japan
sakai@mns.kyutech.ac.jp

Abstract. Rough approximations, which consist of lower and upper approximations, are described under objects characterized by possibilistic information that is expressed by a normal possibility distribution. Concepts of not only possibility but also certainty are used to construct an indiscernibility relation. First, rough approximations are shown for a set of discernible objects by using the indiscernibility relation. Next, a set of objects characterized by possibilistic information is approximated. Consequently, rough approximations consist of objects with a degree expressed by an interval value where lower and upper degrees mean the lower and the upper bounds of the actual degree. This leads to the complementarity property linked with lower and upper approximations in the case of a set of discernible objects, as is valid under complete information. Furthermore, a criterion is introduced to judge whether or not an object is regarded as supporting rules. By using the criterion, we can select only objects that are regarded as inducing rules.

Keywords: Rough sets, Incomplete Information, Possibilistic information, Indiscernibility relation, Lower and upper approximations.

1 Introduction

Possibilistic information systems consist of objects whose attribute values are described by normal possibility distributions. Possibility distributions can be used to express fuzzy terms [12]. For example, "about 50" is expressed by the possibility distribution $\{(47, 0.3), (48, 0.7), (49, 1), (50, 1), (51, 1), (52, 0.7), (53, 0.3)\}_p$ in the sentence "his age is about 50." Such a fuzzy term is ubiquitous in natural languages that we use in daily life. We live in a flood of fuzzy terms. Therefore, possibilistic information systems are suitable for dealing with information obtained from our daily life.

The framework of rough sets, proposed by Pawlak [9], is used as an effective tool for data analysis in various fields such as pattern recognition, machine

A. Laurent et al. (Eds.): IPMU 2014, Part III, CCIS 444, pp. 61–70, 2014.
© Springer International Publishing Switzerland 2014

learning, data mining, and so on. The rough sets are based on indiscernibility of objects whose characteristic values are indistinguishable. The fundamental framework is specified by rough approximations, which consist of lower and upper approximations, under indiscernibility relations obtained from information tables containing only complete information.

The framework requires some extensions to deal with possibilistic information. Słowiński and Stefanowski introduce the concept of possible indiscernibility between objects [11]. Nakata and Sakai express rough approximations by using possible equivalence classes [6]. The rough approximations coincide with those obtained from the approach based on possible worlds. Couso and Dubois express rough approximations by using the degree of possibility that objects belong to the same equivalence class under indiscernibility relations [1]. These approaches consider only the possibility that objects are indistinguishable. Therefore, the rough approximations obtained from the approaches are possible ones. As a result, the complementarity property linked with lower and upper approximations does not hold, although it is valid under complete information.

In the field of databases dealing with information that is not complete, it is well-known that the actual answer to a query cannot be obtained in query processing. Two types of sets, which mean certain and possible answers, are obtained [3,4]. The certain answer is included in the actual answer while the possible answer contains the actual answer. Similarly, results of query processing in possibilistic databases show that an object has two degrees to which it certainly and possibly satisfies given conditions [10]. Recently Nakata and Sakai have examined rough approximations by using possible equivalence classes in the case of information tables containing missing values [8]. Their work shows that rough approximations are not unique, but consist of lower and upper bounds, called certain and possible rough approximations. Therefore, from the viewpoint of not only possibility but also certainty rough sets should be examined for information tables containing possibilistic information.

In this paper, we formulate rough approximations from the viewpoint of certainty and possibility to deal with possibilistic information that includes partly-known values and missing values as special cases. We extend rough approximations by directly using an indiscernibility relation, as is shown in fuzzy rough sets [2], although our previous work is based on possible equivalence classes obtained from the indiscernibility relation [6,7]. This is because the number of possible equivalence classes exponentially increases as the number of values that are not complete increases.

The paper is organized as follows. In section 2, an approach based on rough sets is briefly addressed under complete information. In section 3, we develop an approach based on indiscernibility relations under possibilistic information from the viewpoint of certainty and possibility. In section 4, conclusions are addressed.

2   Rough Sets in Complete Information Systems

A data set is represented as a table, called an information table, where each row and each column represent an object and an attribute, respectively. A mathematical model of an information table with complete information is called a complete information system. The complete information system is a triplet expressed by $(U, AT, \{D(a_i) \mid a_i \in AT\})$. U is a non-empty finite set of objects called the universe, AT is a non-empty finite set of attributes such that $a_i : U \rightarrow D(a_i)$ for every $a_i \in AT$ where $D(a_i)$ is the domain of attribute a_i. Binary relation R_{a_i} for indiscernibility of objects on attribute $a_i \in AT$, which called the indiscernibility relation for a_i, is:

$$R_{a_i} = \{(o, o') \in U \times U \mid a_i(o) = a_i(o')\}, \tag{1}$$

where $a_i(o)$ is the value for attribute a_i of object o. From the indiscernibility relation, equivalence class $[o]_{a_i}$ for object o is obtained:

$$[o]_{a_i} = \{o' \mid (o, o') \in R_{a_i}\}. \tag{2}$$

Finally, family $\mathcal{E}_{a_i}{}^1$ of equivalence classes on a_i is:

$$\mathcal{E}_{a_i} = \{[o]_{a_i} \mid o \in U\}. \tag{3}$$

Using the family of equivalence classes on a_i, lower approximation $\underline{apr}_{a_i}(\mathcal{O})$ and upper approximation $\overline{apr}_{a_i}(\mathcal{O})$ of set $\mathcal{O}$ of indiscernible objects are:

$$\underline{apr}_{a_i}(\mathcal{O}) = \{o \mid [o]_{a_i} \in \mathcal{E}_{a_i} \wedge \forall_{o' \in [o]_{a_i}} o' \in \mathcal{O}\}, \tag{4}$$

$$\overline{apr}_{a_i}(\mathcal{O}) = \{o \mid [o]_{a_i} \in \mathcal{E}_{a_i} \wedge \exists_{o' \in [o]_{a_i}} o' \in \mathcal{O}\}. \tag{5}$$

Lower and upper approximations are not independent, but are linked with each other. The relationship between lower and upper approximations, called complementarity property, is:

$$\underline{apr}_{a_i}(\mathcal{O}) = U - \overline{apr}_{a_i}(U - \mathcal{O}). \tag{6}$$

When objects are characterized by values for a set of attributes, a set of objects being approximated is partitioned by equivalence classes on the set of attributes. Thus, to approximate a set of objects is to approximate the family of equivalence classes that is derived from the set. Let $\mathcal{E}_{a_j}(\mathcal{O})$ be the family of equivalence classes derived from $\mathcal{O}$ on attribute a_j. Lower approximation $\underline{apr}_{a_i}(\mathcal{O}/a_j)$ and upper approximation $\overline{apr}_{a_i}(\mathcal{O}/a_j)$ of set $\mathcal{O}$ of objects that are characterized by values for a_j are obtained on a_i:

$$\underline{apr}_{a_i}(\mathcal{O}/a_j) = \{o \mid o \in \underline{apr}_{a_i}(\mathcal{O}') \wedge \mathcal{O}' \in \mathcal{E}_{a_j}(\mathcal{O})\}, \tag{7}$$

$$\overline{apr}_{a_i}(\mathcal{O}/a_j) = \{o \mid o \in \overline{apr}_{a_i}(\mathcal{O}') \wedge \mathcal{O}' \in \mathcal{E}_{a_j}(\mathcal{O})\}. \tag{8}$$

For formulae on sets A and B of attributes whose individual attributes are denoted by a_i and a_j,

$$R_A = \cap_{a_i \in A} R_{a_i}, \tag{9}$$

---

[1] $\mathcal{E}_{a_i}$ is formally $\mathcal{E}_{a_i}(U)$. (U) is usually omitted.

$$[o]_A = \{o' \mid (o, o') \in R_A\} = \cap_{a_i \in A}[o]_{a_i}, \tag{10}$$

$$\mathcal{E}_A = \{[o]_A \mid o \in U\} = \{E \mid o \in U \wedge E = \cap_{a_i \in A}[o]_{a_i}\}, \tag{11}$$

$$\underline{apr}_A(\mathcal{O}) = \{o \mid [o]_A \in \mathcal{E}_A \wedge \forall_{o' \in [o]_A} o' \in \mathcal{O}\}, \tag{12}$$

$$\overline{apr}_A(\mathcal{O}) = \{o \mid [o]_A \in \mathcal{E}_A \wedge \exists_{o' \in [o]_A} o' \in \mathcal{O}\}, \tag{13}$$

$$\underline{apr}_A(\mathcal{O}/a_j) = \{o \mid o \in \underline{apr}_A(\mathcal{O}') \wedge \mathcal{O}' \in \mathcal{E}_{a_j}(\mathcal{O})\}, \tag{14}$$

$$\overline{apr}_A(\mathcal{O}/a_j) = \{o \mid o \in \overline{apr}_A(\mathcal{O}') \wedge \mathcal{O}' \in \mathcal{E}_{a_j}(\mathcal{O})\}, \tag{15}$$

$$\underline{apr}_A(\mathcal{O}/B) = \cap_{a_j \in B}\{o \mid o \in \underline{apr}_A(\mathcal{O}/a_j)\}, \tag{16}$$

$$\overline{apr}_A(\mathcal{O}/B) = \cap_{a_j \in B}\{o \mid o \in \overline{Apr}_A(\mathcal{O}/a_j)\}. \tag{17}$$

3   Rough Sets in Possibilistic Information Systems

In possibilistic information systems, $a_i : U \to \pi_{a_i}$ for every $a_i \in AT$ where π_{a_i} is the set of all normal possibility distributions over domain $D(a_i)$ of attribute a_i. When value $a_i(o)$ for attribute a_i of object o is expressed by a normal possibility distribution $\{(v, \pi_{a_i(o)}(v)) \mid v \in D(a_i) \wedge \pi_{a_i(o)}(v) > 0\}_p$, $\pi_{a_i(o)}(v)$ denotes the possibilistic degree that $a_i(o)$ has value $v \in D(a_i)$ for attribute a_i.

Indiscernibility relations in a possibilistic information system are expressed by using indiscernibility degrees. An indiscernibility degree of two objects is expressed by not a single value, but two values that means degrees for certainty and possibility. This point is different from fuzzy rough sets [2]. Indiscernibility degree $\mu_{R_{a_i}}(o_k, o_l)$ of two objects o_k and o_l for attribute a_i is expressed by a pair of degrees $C\mu_{R_{a_i}}(o_k, o_l)$ and $P\mu_{R_{a_i}}(o_k, o_l)$ that mean degrees for certainty and for possibility. They are calculated by:

$$P\mu_{R_{a_i}}(o_k, o_l) = \begin{cases} 1 & \text{if } k = l, \\ \max_u \min(\pi_{a_i(o_k)}(u), \pi_{a_i(o_l)}(u)) & \text{otherwise,} \end{cases} \tag{18}$$

$$C\mu_{R_{a_i}}(o_k, o_l) = \begin{cases} 1 & \text{if } k = l, \\ 1 - \max_{u \neq v} \min(\pi_{a_i(o_k)}(u), \pi_{a_i(o_l)}(v)) & \text{otherwise.} \end{cases} \tag{19}$$

The two degrees are reflexive and symmetric, but not max-min transitive.

Example 3.1
Let information table T be obtained as follows:

T

U	a_1	a_2
1	$\{(x, 1)\}_p$	$\{(c, 1), (d, 0.2)\}_p$
2	$\{(x, 1), (y, 0.2)\}_p$	$\{(a, 0.9), (b, 1)\}_p$
3	$\{(y, 1)\}_p$	$\{(b, 1)\}_p$
4	$\{(y, 1), (z, 1)\}_p$	$\{(b, 1)\}_p$
5	$\{(x, 0.4), (w, 1)\}_p$	$\{(c, 1), (d, 0.7)\}_p$

In information table T, $U = \{o_1, o_2, o_3, o_4, o_5\}$, where domains $D(a_1)$ and $D(a_2)$ of attributes a_1 and a_2 are $\{w, x, y, z\}$ and $\{a, b, c, d\}$, respectively. Using formula

(18) and (19), indiscernibility degrees in the indiscernibility relation for a_1 in T are:

$$\mu_{R_{a_1}}(o_k, o_l) = [C\mu_{R_{a_1}}(o_k, o_l), P\mu_{R_{a_1}}(o_k, o_l)] =$$

$$\begin{pmatrix}
[1,1] & [0.8,1] & [0,0] & [0,0] & [0,0.4] \\
[0.8,1] & [1,1] & [0,0.2] & [0,0.2] & [0,0.4] \\
[0,0] & [0,0.2] & [1,1] & [0,1] & [0,0] \\
[0,0] & [0,0.2] & [0,1] & [1,1] & [0,0] \\
[0,0.4] & [0,0.4] & [0,0] & [0,0] & [1,1]
\end{pmatrix}$$

We cannot obtain the actual membership degree to which an object belongs to rough approximations, because the information that characterizes objects is not complete and is expressed by possibility distributions. This is different from fuzzy rough sets [2]. We obtain lower and upper bounds of the actual membership degree, called certain and possible membership degrees.

Let $\mathcal{O}$ be a set of discernible objects. Possible membership degree $P\mu_{\underline{apr}_{a_i}(\mathcal{O})}(o)$ to which object o possibly belongs to lower approximation $\underline{apr}_{a_i}(\mathcal{O})$ is:

$$P\mu_{\underline{apr}_{a_i}(\mathcal{O})}(o) = \min_{o' \in U} \max(1 - C\mu_{R_{a_i}}(o, o'), \mu_{\mathcal{O}}(o')), \qquad (20)$$

where $\mu_{\mathcal{O}}(o') = 1$ if $o' \in \mathcal{O}$, 0 otherwise.

Certain membership degree $C\mu_{\underline{apr}_{a_i}(\mathcal{O})}(o)$ to which object o certainly belongs to lower approximation $\underline{apr}_{a_i}(\mathcal{O})$ is:

$$C\mu_{\underline{apr}_{a_i}(\mathcal{O})}(o) = \min_{o' \in U} \max(1 - P\mu_{R_{a_i}}(o, o'), \mu_{\mathcal{O}}(o')). \qquad (21)$$

Proposition 3.1
$\forall o \in U \; C\mu_{\underline{apr}_{a_i}(\mathcal{O})}(o) \leq P\mu_{\underline{apr}_{a_i}(\mathcal{O})}(o).$

Similarly, possible membership degree $P\mu_{\overline{apr}_{a_i}(\mathcal{O})}(o)$ to which object o possibly belongs to upper approximation $\overline{apr}_{a_i}(\mathcal{O})$ is:

$$P\mu_{\overline{apr}_{a_i}(\mathcal{O})}(o) = \max_{o' \in U} \min(P\mu_{R_{a_i}}(o, o'), \mu_{\mathcal{O}}(o')). \qquad (22)$$

Certain membership degree $C\mu_{\overline{apr}_{a_i}(\mathcal{O})}(o)$ to which object o certainly belongs to upper approximation $\overline{apr}_{a_i}(\mathcal{O})$ is:

$$C\mu_{\overline{apr}_{a_i}(\mathcal{O})}(o) = \max_{o' \in U} \min(C\mu_{R_{a_i}}(o, o'), \mu_{\mathcal{O}}(o')). \qquad (23)$$

Proposition 3.2
$\forall o \in U \; C\mu_{\overline{apr}_{a_i}(\mathcal{O})}(o) \leq P\mu_{\overline{apr}_{a_i}(\mathcal{O})}(o).$

Proposition 3.3

$\forall o \in U \; C\mu_{\underline{apr}_{a_i}}(\mathcal{O})(o) \leq C\mu_{\overline{apr}_{a_i}}(\mathcal{O})(o)$ and $\forall o \in U \; P\mu_{\underline{apr}_{a_i}}(\mathcal{O})(o) \leq P\mu_{\overline{apr}_{a_i}}(\mathcal{O})(o)$.

Proposition 3.4

$\forall o \in U \; C\mu_{\underline{apr}_{a_i}}(\mathcal{O})(o) \leq P\mu_{\underline{apr}_{a_i}}(\mathcal{O})(o) \leq C\mu_{\overline{apr}_{a_i}}(\mathcal{O})(o) \leq P\mu_{\overline{apr}_{a_i}}(\mathcal{O})(o)$.

Four membership degrees are linked with each other.

Proposition 3.5

$P\mu_{\underline{apr}_{a_i}}(\mathcal{O})(o) = 1 - C\mu_{\overline{apr}_{a_i}}(U-\mathcal{O})(o)$ and $C\mu_{\underline{apr}_{a_i}}(\mathcal{O})(o) = 1 - P\mu_{\overline{apr}_{a_i}}(U-\mathcal{O})(o)$.

Each object has degrees of membership for these four approximations denoted by formulae (20) - (23). Using these degrees, rough approximations are expressed as follows:

$$\underline{apr}_{a_i}(\mathcal{O}) = \{(o, [C\mu_{\underline{apr}_{a_i}}(\mathcal{O})(o), P\mu_{\underline{apr}_{a_i}}(\mathcal{O})(o)]) \mid P\mu_{\underline{apr}_{a_i}}(\mathcal{O})(o) > 0\}, \quad (24)$$

$$\overline{apr}_{a_i}(\mathcal{O}) = \{(o, [C\mu_{\overline{apr}_{a_i}}(\mathcal{O})(o), P\mu_{\overline{apr}_{a_i}}(\mathcal{O})(o)]) \mid P\mu_{\overline{apr}_{a_i}}(\mathcal{O})(o) > 0\}. \quad (25)$$

These formulae show that each object has membership degrees expressed by not a single, but an interval value for lower and upper approximations, which is essential in possibilistic information systems. Degrees of imprecision for the membership degrees of o are evaluated by $P\mu_{\underline{apr}_{a_i}}(\mathcal{O})(o) - C\mu_{\underline{apr}_{a_i}}(\mathcal{O})(o)$ and $P\mu_{\overline{apr}_{a_i}}(\mathcal{O})(o) - C\mu_{\overline{apr}_{a}}(\mathcal{O})(o)$ in $\underline{apr}_{a_i}(\mathcal{O})$ and $\overline{apr}_{a_i}(\mathcal{O})$, respectively. The two approximations depend on each other; namely, the comprementarity property linked with lower and upper approximations holds, as is so in complete information systems.

Proposition 3.6

$$\underline{apr}_{a_i}(\mathcal{O}) = U - \overline{apr}_{a_i}(U - \mathcal{O}),$$

where

$$1 - [C\mu_{\overline{apr}_{a_i}}(U-\mathcal{O})(o), P\mu_{\overline{apr}_{a_i}}(U-\mathcal{O})(o)] = [1 - P\mu_{\overline{apr}_{a_i}}(U-\mathcal{O})(o), 1 - C\mu_{\overline{apr}_{a_i}}(U-\mathcal{O})(o)].$$

Example 3.2

Let us go back to Example 3.1. Let a set $\mathcal{O}$ of discernible objects be $\{o_2, o_3, o_4\}$. Using formulae (20)-(23), for object o_1,

$$C\mu_{\underline{apr}_{a_1}}(\mathcal{O})(o_1) = 0, P\mu_{\underline{apr}_{a_1}}(\mathcal{O})(o_1) = 0, C\mu_{\overline{apr}_{a_1}}(\mathcal{O})(o_1) = 0.8, P\mu_{\overline{apr}_{a_1}}(\mathcal{O})(o_1) = 1.$$

Similarly, calculating membership degrees for the other objects,

$$\underline{apr}_{a_1}(\mathcal{O}) = \{(o_2, [0, 0.2]), (o_3, [1, 1]), (o_4, [1, 1])\},$$

$$\overline{apr}_{a_1}(\mathcal{O}) = \{(o_1, [0.8, 1]), (o_2, [1, 1]), (o_3, [1, 1]), (o_4, [1, 1]), (o_5, [0, 0.4])\}.$$

Subsequently, we describe the case where a set of objects characterized by possibilistic information is approximated by objects with complete information.

Let objects in U have complete information for a_i and $\mathcal{O}$ be characterized by a_j with possibilistic information. Membership degrees for four approximations are:

$$P\mu_{\underline{apr}_{a_i}}(\mathcal{O}/a_j)(o) = \max_{o'' \in \mathcal{O}} \min_{o' \in [o]_{a_i}} \min(P\mu_{R_{a_j}}(o', o''), \mu_{\mathcal{O}}(o')), \qquad (26)$$

$$C\mu_{\underline{apr}_{a_i}}(\mathcal{O}/a_j)(o) = \max_{o'' \in \mathcal{O}} \min_{o' \in [o]_{a_i}} \min(C\mu_{R_{a_j}}(o', o''), \mu_{\mathcal{O}}(o')), \qquad (27)$$

$$P\mu_{\overline{apr}_{a_i}}(\mathcal{O}/a_j)(o) = \max_{o'' \in \mathcal{O}} \max_{o' \in [o]_{a_i}} \min(P\mu_{R_{a_j}}(o', o''), \mu_{\mathcal{O}}(o')), \qquad (28)$$

$$C\mu_{\overline{apr}_{a_i}}(\mathcal{O}/a_j)(o) = \max_{o'' \in \mathcal{O}} \max_{o' \in [o]_{a_i}} \min(C\mu_{R_{a_j}}(o', o''), \mu_{\mathcal{O}}(o')). \qquad (29)$$

Combining the above two cases, we can obtain membership degrees in the case where both objects used to approximate and objects approximated are characterized by attributes with possibilistic information. Possible membership degree $P\mu_{\underline{apr}_{a_i}}(\mathcal{O}/a_j)(o)$ to which object o possibly belongs to lower approximation $\underline{apr}_{a_i}(\mathcal{O}/a_j)$ is:

$$P\mu_{\underline{apr}_{a_i}}(\mathcal{O}/a_j)(o) = \max_{o'' \in \mathcal{O}} \min_{o' \in U} \max(1 - C\mu_{R_{a_i}}(o, o'),$$

$$\min(P\mu_{R_{a_j}}(o', o''), \mu_{\mathcal{O}}(o'))). \quad (30)$$

Certain membership degree $C\mu_{\underline{apr}_{a_i}}(\mathcal{O}/a_j)(o)$ to which object o certainly belongs to lower approximation $\underline{apr}_{a_i}(\mathcal{O}/a_j)$ is:

$$C\mu_{\underline{apr}_{a_i}}(\mathcal{O}/a_j)(o) = \max_{o'' \in \mathcal{O}} \min_{o' \in U} \max(1 - P\mu_{R_{a_i}}(o, o')$$

$$\min(C\mu_{R_{a_j}}(o', o''), \mu_{\mathcal{O}}(o'))). \quad (31)$$

Proposition 3.7
$\forall o \in U \; C\mu_{\underline{apr}_{a_i}}(\mathcal{O}/a_j)(o) \leq P\mu_{\underline{apr}_{a_i}}(\mathcal{O}/a_j)(o).$

Similarly, possible membership degree $P\mu_{\overline{apr}_{a_i}}(\mathcal{O}/a_j)(o)$ to which object o possibly belongs to upper approximation $\overline{apr}_{a_i}(\mathcal{O}/a_j)$ is:

$$P\mu_{\overline{apr}_{a_i}}(\mathcal{O}/a_j)(o) = \max_{o'' \in \mathcal{O}} \max_{o' \in U} \min(P\mu_{R_{a_i}}(o, o'), P\mu_{a_j}(o', o''), \mu_{\mathcal{O}}(o')). \quad (32)$$

Certain membership degree $C\mu_{\overline{apr}_{a_i}}(\mathcal{O}/a_j)(o)$ to which object o certainly belongs to upper approximation $\overline{apr}_{a_i}(\mathcal{O}/a_j)$ is:

$$C\mu_{\overline{apr}_{a_i}}(\mathcal{O}/a_j)(o) = \max_{o'' \in \mathcal{O}} \max_{o' \in U} \min(C\mu_{R_{a_i}}(o, o'), C\mu_{a_j}(o', o''), \mu_{\mathcal{O}}(o')). \quad (33)$$

Proposition 3.8
$\forall o \in U \; C\mu_{\overline{apr}_{a_i}}(\mathcal{O}/a_j)(o) \leq P\mu_{\overline{apr}_{a_i}}(\mathcal{O}/a_j)(o).$

Proposition 3.9

$\forall o \in U \; C\mu_{\underline{apr}_{a_i}}(\mathcal{O}/a_j)(o) \leq C\mu_{\overline{apr}_{a_i}}(\mathcal{O}/a_j)(o)$ and $\forall o \in U \; P\mu_{\underline{apr}_{a_i}}(\mathcal{O}/a_j)(o) \leq P\mu_{\overline{apr}_{a_i}}(\mathcal{O}/a_j)(o)$.

Proposition 3.10

$\forall o \in U \; C\mu_{\underline{apr}_{a_i}}(\mathcal{O}/a_j)(o) \leq P\mu_{\underline{apr}_{a_i}}(\mathcal{O}/a_j)(o) \leq C\mu_{\overline{apr}_{a_i}}(\mathcal{O}/a_j)(o) \leq P\mu_{\overline{apr}_{a_i}}(\mathcal{O}/a_j)(o)$.

Example 3.3

Let us go back to information table T in Example 3.1. Let $\mathcal{O}$ be $\{o_2, o_3, o_4\}$ that is characterized by values of attribute a_2. Using formulae (30)-(33),

$$\underline{apr}_{a_1}(\mathcal{O}/a_2) = \{(o_2, [0, 0.2]), (o_3, [0.8, 1]), (o_4, [0.8, 1])\},$$

$$\overline{apr}_{a_1}(\mathcal{O}/a_2) = \{(o_1, [0.8, 1]), (o_2, [1, 1]), (o_3, [1, 1]), (o_4, [1, 1]), (o_5, [0, 0.4])\}.$$

It is significant to focus on the membership degree that an object has in lower and upper approximations. From the viewpoint of rule induction, objects in lower and upper approximations consistently and inconsistently support rules with degrees, respectively. The degrees are closely related with the membership degrees that the objects have in the lower and upper approximations.

Object o_2 has a low degree denoted by $[0, 0.2]$ for $\underline{apr}_{a_1}(\mathcal{O}/a_2)$ in Example 3.3. Can an object with such a low degree be regarded as actually supporting rules? To solve this problem, we introduce a criterion for an object regarded as supporting rules, as is used in possibilistic databases [5]. Now, we obtain certain and possible membership degrees of an object belonging to lower and upper approximations. By using them, we can derive membership degrees $[\neg C\mu_{\underline{apr}_{a_i}}(\mathcal{O}/a_j)(o),$ $\neg P\mu_{\underline{apr}_{a_i}}(\mathcal{O}/a_j)(o)]$ and $[\neg C\mu_{\overline{apr}_{a_i}}(\mathcal{O}/a_j)(o), \neg P\mu_{\overline{apr}_{a_i}}(\mathcal{O}/a_j)(o)]$ with which the object does not belong to the lower and upper approximations, where

$$\neg C\mu_{\underline{apr}_{a_i}}(\mathcal{O}/a_j)(o) = 1 - P\mu_{\underline{apr}_{a_i}}(\mathcal{O}/a_j)(o),$$

$$\neg P\mu_{\underline{apr}_{a_i}}(\mathcal{O}/a_j)(o) = 1 - C\mu_{\underline{apr}_{a_i}}(\mathcal{O}/a_j)(o),$$

$$\neg C\mu_{\overline{apr}_{a_i}}(\mathcal{O}/a_j)(o) = 1 - P\mu_{\overline{apr}_{a_i}}(\mathcal{O}/a_j)(o),$$

$$\neg P\mu_{\overline{apr}_{a_i}}(\mathcal{O}/a_j)(o) = 1 - C\mu_{\overline{apr}_{a_i}}(\mathcal{O}/a_j)(o).$$

We introduce the following criterion for an object regarded as supporting rules:

The membership degree with which an object belongs to rough approximations is larger than or equal to that with which the object does not so.

This is expressed for lower and upper approximations as follows:

$$[C\mu_{\underline{apr}_{a_i}}(\mathcal{O}/a_j)(o), P\mu_{\underline{apr}_{a_i}}(\mathcal{O}/a_j)(o)] \geq [\neg C\mu_{\underline{apr}_{a_i}}(\mathcal{O}/a_j)(o), \neg P\mu_{\underline{apr}_{a_i}}(\mathcal{O}/a_j)(o)],$$

$$[C\mu_{\overline{apr}_{a_i}}(\mathcal{O}/a_j)(o), P\mu_{\overline{apr}_{a_i}}(\mathcal{O}/a_j)(o)] \geq [\neg C\mu_{\overline{apr}_{a_i}}(\mathcal{O}/a_j)(o), \neg P\mu_{\overline{apr}_{a_i}}(\mathcal{O}/a_j)(o)].$$

From these formulae,

$$C\mu_{\underline{apr}_{a_i}}(O/a_j)(o) \geq \neg C\mu_{\underline{apr}_{a_i}}(O/a_j)(o),$$

$$P\mu_{\underline{apr}_{a_i}}(O/a_j)(o) \geq \neg P\mu_{\underline{apr}_{a_i}}(O/a_j)(o),$$

$$C\mu_{\overline{apr}_{a_i}}(O/a_j)(o) \geq \neg C\mu_{\overline{apr}_{a_i}}(O/a_j)(o),$$

$$P\mu_{\overline{apr}_{a_i}}(O/a_j)(o) \geq \neg P\mu_{\overline{apr}_{a_i}}(O/a_j)(o).$$

Therefore, the criteria for lower and upper approximations are equivalent to:

$$C\mu_{\underline{apr}_{a_i}}(O/a_j)(o) + P\mu_{\underline{apr}_{a_i}}(O/a_j)(o) \geq 1, \tag{34}$$

$$C\mu_{\overline{apr}_{a_i}}(O/a_j)(o) + P\mu_{\overline{apr}_{a_i}}(O/a_j)(o) \geq 1. \tag{35}$$

By using these criteria, we can select only objects that is regarded as supporting rules.

Example 3.4

Let us go back to lower and upper approximations in Example 3.3. By using the criterion (34), objects that are regarded as supporting rules consistently are o_3 with the degree $[0.8, 1]$ and o_4 with the degree $[0.8, 1]$. By using the criterion (35), objects that are regarded as suproting rules inconsistently are o_1 with $[0.8, 1]$, o_2 with $[1, 1]$, o_3 with $[1, 1]$, and o_4 with $[1, 1]$.

4   Conclusions

We have examined rough approximations in a possibilistic information system. An attribute value is expressed by a normal possibility distribution in the possibilistic information system. We deal with possibilistic information under indiscernibility relations from the viewpoint of certainty and possibility. First, We have shown rough approximations for the case where only objects used to approximate are characterized by attributes with possibilistic information. Second, we have shown rough approximations in the case where only objects in a set approximated have possibilistic information. Finally, rough approximations have been shown the case where both objects used to approximate and objects approximated are characterized by attributes with possibilistic information.

An object has certain and possible membership degrees to which the object certainly and possibly belongs to rough approximations. The certain and possible membership degrees are lower and upper bounds of the actual membership degrees of the object for rough approximations, respectively. Therefore, with what degree an object belongs to rough approximations is expressed by not a single, but an interval value. This is essential in possibilistic information systems. As a result, the complementarity property linked with lower and upper approximations holds for a set of discernible objects, as is valid under complete information.

Objects whose membership degrees are low for rough approximations appear in possibilistic information systems. It does not seem that the objects support rules. We can naturally introduce a criterion for an object regarded as supporting rules, because we use certain and possible membership degrees of the object for the rough approximations. By using the criterion, we can select only objects that are regarded as supporting rules.

Acknowledgment. This work has been partially supported by the Grant-in-Aid for Scientific Research (C), Japan Society for the Promotion of Science, No. 26330277.

References

1. Couso, I., Dubois, D.: Rough Sets, Coverings and Incomplete Information. Fundamenta Informaticae 108(3-4), 223–247 (2011)
2. Dubois, D., Prade, H.: Rough Fuzzy Sets and Fuzzy Rough Sets. International Journal of General Systems 17, 191–209 (1990)
3. Lipski, W.: On Semantics Issues Connected with Incomplete Information Databases. ACM Transactions on Database Systems 4, 262–296 (1979)
4. Lipski, W.: On Databases with Incomplete Information. Journal of the ACM 28, 41–70 (1981)
5. Nakata, M.: Unacceptable components in fuzzy relational databases. International Journal of Intelligent Systems 11(9), 633–647 (1996)
6. Nakata, M., Sakai, H.: Lower and Upper Approximations in Data Tables Containing Possibilistic Information. In: Peters, J.F., Skowron, A., Marek, V.W., Orłowska, E., Słowiński, R., Ziarko, W.P. (eds.) Transactions on Rough Sets VII. LNCS, vol. 4400, pp. 170–189. Springer, Heidelberg (2007)
7. Nakata, M., Sakai, H.: Rule Induction Based on Rough Sets from Information Tables Containing Possibilistic Information. In: Proceedings of the 2013 Joint IFSA World Congress and NAFIPS Annual Meeting (IFSA/NAFIPS), pp. 91–96. IEEE Press (2013)
8. Nakata, M., Sakai, H.: Twofold rough approximations under incomplete information. International Journal of General Systems 42, 546–571 (2013)
9. Pawlak, Z.: Rough Sets: Theoretical Aspects of Reasoning about Data. Kluwer Academic Publishers (1991)
10. Prade, H., Testemale, C.: Generalizing Database Relational Algebra for the Treatment of Incomplete or Uncertain Information and Vague Queries. Information Sciences 34, 115–143 (1984)
11. Słowiński, R., Stefanowski, J.: Rough Classification in Incomplete Information Systems. Mathematical and Computer Modelling 12, 1347–1357 (1989)
12. Zadeh, L.A.: Fuzzy Sets as a Basis for a Theory of Possibility. Fuzzy Sets and Systems 1, 3–28 (1978)

Antitone L-bonds[*]

Jan Konecny

Data Analysis and Modeling Lab, Department of Computer Science
Faculty of Science, Palacky University, Olomouc, Czech Republic
jan.konecny@upol.cz

Abstract. L-bonds represent relationships between fuzzy formal contexts. We study these intercontextual structures w.r.t. antitone Galois connections in fuzzy setting. Furthermore, we define direct ◁-product and ▷-product of two formal fuzzy contexts and show conditions under which a fuzzy bond can be obtained as an intent of the product. This extents our previous work on isotone fuzzy bonds.

1 Introduction

Formal Concept Analysis (FCA) [10] is an exploratory method of analysis of relational data. The method identifies some interesting clusters (formal concepts) in a collection of objects and their attributes (formal context) and organizes them into a structure called concept lattice. Formal Concept Analysis in fuzzy setting [3] allows us to work with graded data.

In the present paper, we deal with intercontextual relationships in FCA in fuzzy setting. Particularly, our approach originated in relation to [16] on the notion of Chu correspondences between formal contexts, which led to obtaining information about the structure of **L**-bonds. In [15] we studied properties of **L**-bonds w.r.t. isotone concept-forming operators.

The present paper concerns with **L**-bonds with antitone character; We describe their properties and explain how these **L**-bonds relate to the structures studied in [16]. In addition, we also focus on the direct products of two formal fuzzy contexts and show conditions under which a bond can be obtained as an intent of the product.

The paper is structured as follows: in Section 2 we recollect some notions used in this paper; in Section 3 we define the **L**-bonds and direct products, and describe their properties. Our conclusions and related further research are summarized in Section 4.

2 Preliminaries

In this section, we recall some basic notions used in the paper.

---

[*] Supported by the ESF project No. CZ.1.07/2.3.00/20.0059, the project is cofinanced by the European Social Fund and the state budget of the Czech Republic.

A. Laurent et al. (Eds.): IPMU 2014, Part III, CCIS 444, pp. 71–80, 2014.

2.1   Residuated Lattices and Fuzzy Sets

We use complete residuated lattices as basic structures of truth-degrees. A complete residuated lattice [3,12,21] is a structure $\mathbf{L} = \langle L, \wedge, \vee, \otimes, \rightarrow, 0, 1 \rangle$ such that

(i) $\langle L, \wedge, \vee, 0, 1 \rangle$ is a complete lattice, i.e. a partially ordered set in which arbitrary infima and suprema exist;

(ii) $\langle L, \otimes, 1 \rangle$ is a commutative monoid, i.e. $\otimes$ is a binary operation which is commutative, associative, and $a \otimes 1 = a$ for each $a \in L$;

(iii) $\otimes$ and $\rightarrow$ satisfy adjointness, i.e. $a \otimes b \leqslant c$ iff $a \leqslant b \rightarrow c$.

0 and 1 denote the least and greatest elements. The partial order of $\mathbf{L}$ is denoted by $\leqslant$. Throughout this paper, $\mathbf{L}$ denotes an arbitrary complete residuated lattice.

Elements a of L are called truth degrees. Operations $\otimes$ (multiplication) and $\rightarrow$ (residuum) play the role of a (truth functions of) "fuzzy conjunction" and "fuzzy implication". Furthermore, we define the complement of $a \in L$ as

$$\neg a = a \rightarrow 0. \tag{1}$$

L-*sets and* **L**-*relations* An **L**-set (or fuzzy set) A in a universe set X is a mapping assigning to each $x \in X$ some truth degree $A(x) \in L$. The set of all **L**-sets in a universe X is denoted L^X, or $\mathbf{L}^X$ if the structure of $\mathbf{L}$ is to be emphasized.

The operations with **L**-sets are defined componentwise. For instance, the intersection of **L**-sets $A, B \in L^X$ is an **L**-set $A \cap B$ in X such that $(A \cap B)(x) = A(x) \wedge B(x)$ for each $x \in X$, etc. An **L**-set $A \in L^X$ is also denoted $\{^{A(x)}/x \mid x \in X\}$. If for all $y \in X$ distinct from $x_1, x_2, \dots, x_n$ we have $A(y) = 0$, we also write

$$\{^{A(x_1)}/x_1, {}^{A(x_2)}/x_1, \dots, {}^{A(x_n)}/x_n\}.$$

An **L**-set $A \in L^X$ is called crisp if $A(x) \in \{0, 1\}$ for each $x \in X$. Crisp **L**-sets can be identified with ordinary sets. For a crisp A, we also write $x \in A$ for $A(x) = 1$ and $x \notin A$ for $A(x) = 0$. An **L**-set $A \in L^X$ is called empty (denoted by $\varnothing$) if $A(x) = 0$ for each $x \in X$. For $a \in L$ and $A \in L^X$, the **L**-sets $a \otimes A, a \rightarrow A, A \rightarrow a$, and $\neg A$ in X are defined by

$$(a \otimes A)(x) = a \otimes A(x), \tag{2}$$
$$(a \rightarrow A)(x) = a \rightarrow A(x), \tag{3}$$
$$(A \rightarrow a)(x) = A(x) \rightarrow a, \tag{4}$$
$$\neg A(x) = A(x) \rightarrow 0. \tag{5}$$

An a-complement is an **L**-set A which satisfies $(A \rightarrow a) \rightarrow a = A$.

Binary **L**-relations (binary fuzzy relations) between X and Y can be thought of as **L**-sets in the universe $X \times Y$. That is, a binary **L**-relation $I \in L^{X \times Y}$ between a set X and a set Y is a mapping assigning to each $x \in X$ and each $y \in Y$ a truth degree $I(x, y) \in L$ (a degree to which x and y are related by I).

$V \subseteq L^X$ is called an **L**-*closure system* if

– V is closed under left $\rightarrow$-multiplication (or $\rightarrow$-shift), i.e. for every $a \in L$ and $C \in V$ we have $a \rightarrow C \in V$,
– V is closed under intersection, i.e. for $C_j \in V$ $(j \in J)$ we have $\bigcap_{j \in J} C_j \in V$.

$V \subseteq L^X$ is called an **L**-*interior system* if

– V is closed under left $\otimes$-multiplication, i.e. for every $a \in L$ and $C \in V$ we have $a \otimes C \in V$,
– V is closed under union, i.e. for $C_j \in V$ $(j \in J)$ we have $\bigcup_{j \in J} C_j \in V$.

Relational products We use three relational product operators, $\circ$, $\lhd$, and $\rhd$, and consider the corresponding products $R = S \circ T$, $R = S \lhd T$, and $R = S \rhd T$ (for $R \in L^{X \times Z}, S \in L^{X \times Y}, T \in L^{Y \times Z}$). In the compositions, $R(x, z)$ is interpreted as the degree to which the object x has the attribute z; $S(x, y)$ as the degree to which the factor y applies to the object x; $T(y, z)$ as the degree to which the attribute z is a manifestation (one of possibly several manifestations) of the factor y. The composition operators are defined by

$$(S \circ T)(x, z) = \bigvee_{y \in Y} S(x, y) \otimes T(y, z), \tag{6}$$

$$(S \lhd T)(x, z) = \bigwedge_{y \in Y} S(x, y) \rightarrow T(y, z), \tag{7}$$

$$(S \rhd T)(x, z) = \bigwedge_{y \in Y} T(y, z) \rightarrow S(x, y). \tag{8}$$

Note that these operators were extensively studied by Bandler and Kohout, see e.g. [13]. They have natural verbal descriptions. For instance, $(S \circ T)(x, z)$ is the truth degree of the proposition "there is factor y such that y applies to object x and attribute z is a manifestation of y"; $(S \lhd T)(x, z)$ is the truth degree of "for every factor y, if y applies to object x then attribute z is a manifestation of y". Note also that for $L = \{0, 1\}$, $S \circ T$ coincides with the well-known composition of binary relations.

We will need following lemma.

Lemma 1 ([3]). *For $R \in L^{W \times X}, S \in L^{X \times Y}, T \in L^{Y \times Z}$ we have*

$$R \lhd (S \lhd T) = (R \circ S) \lhd T \quad and \quad R \rhd (S \circ T) = (R \rhd S) \rhd T.$$

2.2   Formal Concept Analysis in the Fuzzy Setting

An **L**-context is a triplet $\langle X, Y, I \rangle$ where X and Y are (ordinary) sets and $I \in L^{X \times Y}$ is an **L**-relation between X and Y. Elements of X are called objects, elements of Y are called attributes, I is called an incidence relation. $I(x, y) = a$ is read: "The object x has the attribute y to degree a." An **L**-context is usually depicted as a table whose rows correspond to objects and whose columns correspond to attributes; entries of the table contain the degrees $I(x, y)$.

Concept-forming operators induced by an **L**-context $\langle X, Y, I \rangle$ are the following operators: First, the pair $\langle \uparrow, \downarrow \rangle$ of operators $\uparrow : L^X \to L^Y$ and $\downarrow : L^Y \to L^X$ is defined by

$$A^\uparrow(y) = \bigwedge_{x \in X} A(x) \to I(x, y), \quad B^\downarrow(x) = \bigwedge_{y \in Y} B(y) \to I(x, y). \tag{9}$$

Second, the pair $\langle \cap, \cup \rangle$ of operators $\cap : L^X \to L^Y$ and $\cup : L^Y \to L^X$ is defined by

$$A^\cap(y) = \bigvee_{x \in X} A(x) \otimes I(x, y), \quad B^\cup(x) = \bigwedge_{y \in Y} I(x, y) \to B(y), \tag{10}$$

Third, the pair $\langle \wedge, \vee \rangle$ of operators $\wedge : L^X \to L^Y$ and $\vee : L^Y \to L^X$ is defined by

$$A^\wedge(y) = \bigwedge_{x \in X} I(x, y) \to A(x), \quad B^\vee(x) = \bigvee_{y \in Y} B(y) \otimes I(x, y), \tag{11}$$

for $A \in L^X$, $B \in L^Y$. When we need to emphasize that a pair of concept-forming operators is induced by a particular **L**-relation we write it as a subscript, for instance we write $\uparrow_I$ instead of just $\uparrow$.

Furthermore, denote the corresponding sets of fixed points by $\mathcal{B}^{\uparrow\downarrow}(X, Y, I)$, $\mathcal{B}^{\cap\cup}(X, Y, I)$, and $\mathcal{B}^{\wedge\vee}(X, Y, I)$, i.e.

$$\mathcal{B}^{\uparrow\downarrow}(X, Y, I) = \{\langle A, B \rangle \in L^X \times L^Y \mid A^\uparrow = B, B^\downarrow = A\},$$
$$\mathcal{B}^{\cap\cup}(X, Y, I) = \{\langle A, B \rangle \in L^X \times L^Y \mid A^\cap = B, B^\cup = A\},$$
$$\mathcal{B}^{\wedge\vee}(X, Y, I) = \{\langle A, B \rangle \in L^X \times L^Y \mid A^\wedge = B, B^\vee = A\}.$$

The sets of fixpoints are complete lattices [1,11,20], called **L**-concept lattices associated to I, and their elements are called formal concepts.

For a concept lattice $\mathcal{B}^{\Delta\nabla}(X, Y, I)$, where $\mathcal{B}^{\Delta\nabla}$ is either of $\mathcal{B}^{\uparrow\downarrow}$, $\mathcal{B}^{\cap\cup}$, or $\mathcal{B}^{\wedge\vee}$, denote the corresponding sets of extents and intents by $\mathrm{Ext}^{\Delta\nabla}(X, Y, I)$ and $\mathrm{Int}^{\Delta\nabla}(X, Y, I)$. That is,

$$\mathrm{Ext}^{\Delta\nabla}(X, Y, I) = \{A \in L^X \mid \langle A, B \rangle \in \mathcal{B}^{\Delta\nabla}(X, Y, I) \text{ for some } B\},$$
$$\mathrm{Int}^{\Delta\nabla}(X, Y, I) = \{B \in L^Y \mid \langle A, B \rangle \in \mathcal{B}^{\Delta\nabla}(X, Y, I) \text{ for some } A\}.$$

The operators induced by an **L**-context and their sets of fixpoints have been extensively studied, see e.g. [1,2,4,11,20].

3   L-bonds

This section introduces antitone **L**-bonds, namely a-bonds and c-bonds, and describes their properties.

Definition 1. *(a)* An a-bond *from* **L**-context $\mathbb{K}_1 = \langle X_1, Y_1, I_1 \rangle$ *to* **L**-context $\mathbb{K}_2 = \langle X_2, Y_2, I_2 \rangle$ *is an* **L**-relation $\beta \in L^{X_1 \times Y_2}$ *s.t.*

$$\mathrm{Ext}^{\uparrow\downarrow}(X_1, Y_2, \beta) \subseteq \mathrm{Ext}^{\cap\cup}(X_1, Y_1, I_1) \text{ and } \mathrm{Int}^{\uparrow\downarrow}(X_1, Y_2, \beta) \subseteq \mathrm{Int}^{\uparrow\downarrow}(X_2, Y_2, I_2).$$

(b) A c-bond *from* **L**-context $\mathbb{K}_1 = \langle X_1, Y_1, I_1 \rangle$ *to* **L**-context $\mathbb{K}_2 = \langle X_2, Y_2, I_2 \rangle$ *is an* **L**-relation $\beta \in L^{X_1 \times Y_2}$ *s.t.*

$$\mathrm{Ext}^{\uparrow\downarrow}(X_1, Y_2, \beta) \subseteq \mathrm{Ext}^{\uparrow\downarrow}(X_1, Y_1, I_1) \ and \ \mathrm{Int}^{\uparrow\downarrow}(X_1, Y_2, \beta) \subseteq \mathrm{Int}^{\wedge\vee}(X_2, Y_2, I_2).$$

Remark 1. 1) The terms—a-bond and c-bond—were chosen to match with notions of a-morphism and c-morphism [7,14,9]. We show in Theorem 2 that the a-bonds and c-bonds are in one-to-one correspondence of a-morphisms and c-morphisms, respectively, on sets of intents of associated concept lattices.

2) Note that all considered sets of extents and intents in Definition 1 are **L**-closure systems. From this point of view, the condition of subsethood is natural.

Theorem 1. *(a)* $\beta \in L^{X_1 \times Y_2}$ *is an a-bond between* $\mathbb{K}_1 = \langle X_1, Y_1, I_1 \rangle$ *and* $\mathbb{K}_2 = \langle X_2, Y_2, I_2 \rangle$ *iff there exist* **L**-*relations* $S_i \in L^{Y_1 \times Y_2}$ *and* $S_e \in L^{X_1 \times X_2}$, *such that*

$$\beta = I_1 \lhd S_i = S_e \lhd I_2. \tag{12}$$

(b) $\beta \in L^{X_1 \times Y_2}$ *is a c-bond between* $\mathbb{K}_1 = \langle X_1, Y_1, I_1 \rangle$ *and* $\mathbb{K}_2 = \langle X_2, Y_2, I_2 \rangle$ *iff there exist* **L**-*relations* $S_i \in L^{Y_1 \times Y_2}$ *and* $S_e \in L^{X_1 \times X_2}$, *such that*

$$\beta = I_1 \rhd S_i = S_e \rhd I_2. \tag{13}$$

Proof. Follows from results in [9].   □

3.1   Morphisms

This section explains correspondence of **L**-bonds with morphisms of **L**-interior/**L**-closure spaces. First, we recall notions of c-morphisms and a-morphisms. These morphisms were previously studied in [7,9,14].

Definition 2. *(a) A mapping* $h : V \to W$ *from an* **L**-*interior system* $V \subseteq L^X$ *into an* **L**-*closure system* $W \subseteq L^Y$ *is called an* a-morphism *if*

- $h(a \otimes C) = a \to h(C)$ *for each* $a \in L$ *and* $C \in V$;
- $h(\bigvee_{k \in K} C_k) = \bigwedge_{k \in K} h(C_k)$ *for every collection of* $C_k \in V$.

An a-morphism $h : V \to W$ *is called an* extendable a-morphism *if* h *can be extended to an a-morphism of* L^X *to* L^Y, *i.e. if there exists an a-morphism* $h' : L^X \to L^Y$ *such that for every* $C \in V$ *we have* $h'(C) = h(C)$.

(b) A mapping $h : V \to W$ *from an* **L**-*closure system* $V \subseteq L^X$ *into an* **L**-*closure system* $W \subseteq L^Y$ *is called a* c-morphism *if it is a* $\to$- *and* $\bigwedge$-*morphism and it preserves a-complements, i.e. if*

- $h(a \to C) = a \to h(C)$ *for each* $a \in L$ *and* $C \in V$;
- $h(\bigwedge_{k \in K} C_k) = \bigwedge_{k \in K} h(C_k)$ *for every collection of* $C_k \in V$ $(k \in K)$;
- *if* C *is an a-complement then* $h(C)$ *is an a-complement.*

A c-morphism $h : V \to W$ is called an extendable c-morphism *if h can be extended to a c-morphism of L^X to L^Y, i.e. if there exists a c-morphism $h' : L^X \to L^Y$ such that for every $C \in V$ we have $h'(C) = h(C)$.*

In this paper we consider only extendable {a,c}-morphims.

Theorem 2. *(a) The a-bonds between $\mathbb{K}_1 = \langle X_1, Y_1, I_1 \rangle$ and $\mathbb{K}_2 = \langle X_2, Y_2, I_2 \rangle$ are in one-to-one correspondence with*

- *a-morphisms from $\mathrm{Int}^{\cap\cup}(X_1, Y_1, I_1)$ to $\mathrm{Int}^{\uparrow\downarrow}(X_2, Y_2, I_2)$;*
- *c-morphisms from $\mathrm{Ext}^{\uparrow\downarrow}(X_2, Y_2, I_2)$ to $\mathrm{Ext}^{\cap\cup}(X_1, Y_1, I_1)$.*

(b) The c-bonds between $\mathbb{K}_1 = \langle X_1, Y_1, I_1 \rangle$ and $\mathbb{K}_2 = \langle X_2, Y_2, I_2 \rangle$ are in one-to-one correspondence with

- *c-morphisms from $\mathrm{Int}^{\uparrow\downarrow}(X_1, Y_1, I_1)$ to $\mathrm{Int}^{\wedge\vee}(X_2, Y_2, I_2)$;*
- *a-morphisms from $\mathrm{Ext}^{\wedge\vee}(X_2, Y_2, I_2)$ to $\mathrm{Ext}^{\uparrow\downarrow}(X_1, Y_1, I_1)$.*

Proof. Follows from Theorem 1 and results in [9,14].

Theorem 3. *(a) The system of all a-bonds is an $\mathbf{L}$-closure system.*
(b) The system of all c-bonds is an $\mathbf{L}$-closure system.

Proof. (a) Consider a collection of a-bonds β_i. By Theorem 1 the β_is are in the form $\beta_i = I_1 \vartriangleleft S_i = S_e \vartriangleleft I_2$. We have

$$\bigcap_{j \in J} \beta_j = \bigcap_{j \in J}(I_1 \vartriangleleft S_{ij}) = I_1 \vartriangleleft \left(\bigcap_{j \in J} S_{ij} \right)$$
$$= \bigcap_{j \in J}(S_{ej} \vartriangleleft I_2) = \left(\bigcup_{j \in J} S_{ej} \right) \vartriangleleft I_2;$$
$$a \to \beta = a \to (I_1 \vartriangleleft S_i) = I_1 \vartriangleleft (a \to S_i)$$
$$= a \to (S_e \vartriangleleft I_2) = (a \otimes S_e) \vartriangleleft I_2.$$

Thus, $\bigcap_{j \in J} \beta_j$ and $a \to \beta$ are a-bonds. Proof of (b) is similar. $\square$

3.2   Direct Products

In this part, we focus on direct products of $\mathbf{L}$-contexts related to a-bonds and c-bonds.

Definition 3. *Let $\mathbb{K}_1 = \langle X_1, Y_1, I_1 \rangle, \mathbb{K}_2 = \langle X_2, Y_2, I_2 \rangle$ be $\mathbf{L}$-contexts.*

(a) A direct $\vartriangleleft$-product of $\mathbb{K}_1$ and $\mathbb{K}_2$ is defined as the $\mathbf{L}$-context $\mathbb{K}_1 \boxminus \mathbb{K}_2 = \langle X_2 \times Y_1, X_1 \times Y_2, \Delta \rangle$ with $\Delta(\langle x_2, y_1 \rangle, \langle x_1, y_2 \rangle) = I_1(x_1, y_1) \to I_2(x_2, y_2)$ for all $x_1 \in X_1, x_2 \in X_2, y_1 \in Y_1, y_2 \in Y_2$.

(b) A direct $\vartriangleright$-product of $\mathbb{K}_1$ and $\mathbb{K}_2$ is defined as the $\mathbf{L}$-context $\mathbb{K}_1 \boxminus \mathbb{K}_2 = \langle X_2 \times Y_1, X_1 \times Y_2, \Delta \rangle$ with $\Delta(\langle x_2, y_1 \rangle, \langle x_1, y_2 \rangle) = I_2(x_2, y_2) \to I_1(x_1, y_1)$ for all $x_1 \in X_1, x_2 \in X_2, y_1 \in Y_1, y_2 \in Y_2$.

The following theorem shows that $\mathbb{K}_1 \boxminus \mathbb{K}_2$ (resp. $\mathbb{K}_1 \boxminus \mathbb{K}_2$) induces a-bonds (resp. c-bonds) as its intents.

Theorem 4. *(a) The intents of $\mathbb{K}_1 \boxminus \mathbb{K}_2$ w.r.t $\langle \uparrow, \downarrow \rangle$ are a-bonds from $\mathbb{K}_1$ to $\mathbb{K}_2$, i.e for each $\phi \in L^{X_2 \times Y_1}$, $\phi^{\uparrow}$ is an a-bond from $\mathbb{K}_1$ to $\mathbb{K}_2$.*
(b) The intents of $\mathbb{K}_1 \boxminus \mathbb{K}_2$ w.r.t $\langle \uparrow, \downarrow \rangle$ are c-bonds from $\mathbb{K}_1$ to $\mathbb{K}_2$, i.e for each $\phi \in L^{X_2 \times Y_1}$, $\phi^{\uparrow}$ is a c-bond from $\mathbb{K}_1$ to $\mathbb{K}_2$.

Proof. (a) For $\phi \in L^{X_2 \times Y_1}$ we have

$$\phi^{\uparrow}(x_1, y_2) = \bigwedge_{\langle x_2, y_1 \rangle \in X_2 \times Y_1} \phi(x_2, y_1) \to \Delta(\langle x_2, y_1 \rangle, \langle x_1, y_2 \rangle)$$

$$= \bigwedge_{x_2 \in X_2} \bigwedge_{y_1 \in Y_1} \phi(x_2, y_1) \to (I_1(x_1, y_1) \to I_2(x_2, y_2))$$

$$= \bigwedge_{x_2 \in X_2} \bigwedge_{y_1 \in Y_1} (I_1(x_1, y_1) \to (\phi(x_2, y_1) \to I_2(x_2, y_2)))$$

$$= \bigwedge_{y_1 \in Y_1} (I_1(x_1, y_1) \to \bigwedge_{x_2 \in X_2} (\phi(x_2, y_1) \to I_2(x_2, y_2)))$$

$$= \bigwedge_{y_1 \in Y_1} (I_1(x_1, y_1) \to \bigwedge_{x_2 \in X_2} (\phi^{\mathrm{T}}(y_1, x_2) \to I_2(x_2, y_2)))$$

$$= \bigwedge_{y_1 \in Y_1} I_1(x_1, y_1) \to (\phi^{\mathrm{T}} \triangleleft I_2)(y_1, y_2)$$

$$= (I_1 \triangleleft (\phi^{\mathrm{T}} \triangleleft I_2))(x_1, y_2)$$

$$= ((I_1 \circ \phi^{\mathrm{T}}) \triangleleft I_2)(x_1, y_2).$$

Thus $\phi^{\uparrow}$ is an a-bond by Theorem 1. Proof of (b) is similar.   □

Not all a-bonds are intents of the direct product as the following examples shows.

Example 1. Consider **L**-context $\mathbb{K} = \langle \{x\}, \{y\}, \{^{0.5}/\langle x, y \rangle\} \rangle$ with **L** being the three-element Lukasiewicz chain. Obviously, $\{^{0.5}/\langle x, y \rangle\}$ is an a-bond from $\mathbb{K}$ to $\mathbb{K}$. We have $\mathbb{K} \boxminus \mathbb{K} = \langle \{\langle x, y \rangle\}, \{\langle x, y \rangle\}, \{\langle x, y \rangle, \langle x, y \rangle\} \rangle$. The only intent of $\mathbb{K} \boxminus \mathbb{K}$ is $\{\langle x, y \rangle\}$; thus the a-bond $\{^{0.5}/\langle x, y \rangle\}$ is not among its intents.

Example 2. Consider following **L**-context with **L** being three-element Lukasiewicz chain.

$$\mathbb{K}_1 = \begin{vmatrix} 0 & 0 & 0 & \frac{1}{2} \\ 1 & 0 & \frac{1}{2} & \frac{1}{2} \end{vmatrix} \quad \mathbb{K}_2 = \begin{vmatrix} 0 & 1 & 1 \\ 1 & 1 & 1 \\ \frac{1}{2} & \frac{1}{2} & 1 \end{vmatrix}.$$

There are 11 a-bonds from $\mathbb{K}_1$ to $\mathbb{K}_2$, but $\mathbb{K}_1 \boxminus \mathbb{K}_2$ has only 9 concepts; see Figure 1.

Since the definition of direct $\triangleleft$-product and direct $\triangleright$-product differ only in the direction of residuum, we can make the following corollary.

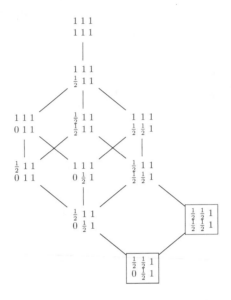

Fig. 1. System of a-bonds between $\mathbb{K}_1$ and $\mathbb{K}_2$ from Example 2. Boxed a-bonds are those which are not intents of $\mathbb{K}_1 \boxminus \mathbb{K}_2$.

Corollary 1. *(a) The extents of the direct $\lhd$-product of $\langle X_1, Y_1, I_1 \rangle$ and $\langle X_2, Y_2, I_2 \rangle$ are a-bonds from $\langle X_2, Y_2, I_2 \rangle$ to $\langle X_1, Y_1, I_1 \rangle$.*
(b) The extents of the direct $\rhd$-product of $\langle X_1, Y_1, I_1 \rangle$ and $\langle X_2, Y_2, I_2 \rangle$ are c-bonds from $\langle X_2, Y_2, I_2 \rangle$ to $\langle X_1, Y_1, I_1 \rangle$.

3.3   Strong Antitone L-bonds

As classical bonds connect contexts with antitone Galois connections, we also consider **L**-bonds from $\langle X_1, Y_1, I_1 \rangle$ to $\langle X_2, Y_2, I_2 \rangle$ defined as **L**-relations $J \in L^{X_1 \times Y_2}$ such that

$$\mathrm{Ext}^{\uparrow\downarrow}(X_1, Y_2, J) \subseteq \mathrm{Ext}^{\uparrow\downarrow}(X_1, Y_1, I_1) \quad \text{and} \quad \mathrm{Int}^{\uparrow\downarrow}(X_1, Y_2, J) \subseteq \mathrm{Int}^{\uparrow\downarrow}(X_2, Y_2, I_2). \tag{14}$$

In what follows, we call the **L**-relations defined by (14) *strong antitone **L**-bonds*.

Using the double negation law. If the double negation law holds true in L, each pair of concept-forming operators (9)–(11) is definable by any of other two. As a consequence, we have

$$\mathcal{B}^{\uparrow\downarrow}(X, Y, I) \text{ and } \mathcal{B}^{\cap\cup}(X, Y, \neg I) \text{ are isomorphic as lattices} \tag{15}$$

with $\langle A, B \rangle \mapsto \langle A, \neg B \rangle$ being an isomorphism. In addition, we have

$$\mathrm{Ext}^{\uparrow\downarrow}(X, Y, \neg I) = \mathrm{Ext}^{\cap\cup}(X, Y, I) \text{ and, dually, } \mathrm{Int}^{\uparrow\downarrow}(X, Y, \neg I) = \mathrm{Int}^{\wedge\vee}(X, Y, I).$$

Theorem 5. *Let the double negation law hold true in* **L**. *the strong antitone* **L**-*bonds from* $\langle X_1, Y_1, I_1 \rangle$ *to* $\langle X_2, Y_2, I_2 \rangle$ *are exactly a-bonds from* $\langle X_1, Y_1, \neg I_1 \rangle$ *to* $\langle X_2, Y_2, I_2 \rangle$; *and c-bonds from* $\langle X_1, Y_1, I_1 \rangle$ *to* $\langle X_2, Y_2, \neg I_2 \rangle$.

Note that the incidence relation Δ in direct product $\mathbb{K}_1 \boxminus \mathbb{K}_2$ then becomes

$$\Delta(\langle x_2, y_1 \rangle, \langle x_1, y_2 \rangle) = \neg I_1(x_1, y_1) \to I_2(x_2, y_2);$$

that is in agreement with results in [16]. Similarly, the incidence relation Δ in direct product $\mathbb{K}_1 \boxminus \mathbb{K}_2$ becomes

$$\Delta(\langle x_2, y_1 \rangle, \langle x_1, y_2 \rangle) = I_1(x_1, y_1) \to \neg I_2(x_2, y_2).$$

Using an alternative notion of complement. The mutual reducibility of concept-forming operators (9)–(11) does not hold generally. In [8], we proposed a new notion of complement of **L**-relation to overcome that. Using this notion we showed that each for each $I \in L^{X \times Y}$, one can define $\ulcorner I \in L^{X \times (Y \times L)}$ as

$$\ulcorner I(x, \langle y, a \rangle) = I(x, y) \to a,$$

and obtain

$$\mathrm{Ext}^{\uparrow\downarrow}(X, Y \times L, \ulcorner I) = \mathrm{Ext}^{\cap\cup}(X, Y, I)$$

and, similarly,

$$\mathrm{Int}^{\uparrow\downarrow}(X, Y \times L, \ulcorner I) = \mathrm{Int}^{\wedge\vee}(X \times L, Y, (\ulcorner I^{\mathrm{T}})^{\mathrm{T}})$$

Unfortunately, the opposite direction holds true only for those **L**-contexts $\langle X, Y, I \rangle$ whose set $\mathrm{Ext}^{\uparrow\downarrow}(X, Y, I)$ (resp. $\mathrm{Int}^{\uparrow\downarrow}(X, Y, I)$) is a c-closure system [7]; i.e. an **L**-closure system generated by a system of all a-complements of some $\mathcal{T} \subseteq L^X$.

Theorem 6. *If* $\mathrm{Ext}^{\uparrow\downarrow}(X_1, Y_1, I_1)$ *is a c-closure system, the strong antitone* **L**-*bonds from* $\langle X_1, Y_1, I_1 \rangle$ *to* $\langle X_2, Y_2, I_2 \rangle$ *are exactly a-bonds from* $\langle X_1, Y_1 \times L, \ulcorner I_1 \rangle$ *to* $\langle X_2, Y_2, I_2 \rangle$. *If* $\mathrm{Int}^{\uparrow\downarrow}(X_2, Y_2, I_2)$ *is a c-closure system, the antitone* **L**-*bonds from* $\langle X_1, Y_1, I_1 \rangle$ *to* $\langle X_2, Y_2, I_2 \rangle$ *are exactly c-bonds from* $\langle X_1, Y_1, I_1 \rangle$ *to* $\langle X_2 \times L, Y_2, (\ulcorner I_2^{\mathrm{T}})^{\mathrm{T}} \rangle$.

We omit further details due to the lack of space.

4 Conclusions and Further Research

We studied bonds between fuzzy contexts related to mutually different types of concept-forming operators and their relationship to antitone fuzzy bonds.

Our future research includes:

- Covering the **L**-bonds described above and isotone **L**-bonds in [15] by a general framework. The isotone and antitone concept-forming operators are one type of operators in [6,5]; also in [18].
- Generalizing the described theory to bond **L**-contexts which each use different residuated lattice as the structure of truth-degrees. Results described in [17] seem to be promising for this goal.

References

1. Belohlavek, R.: Fuzzy Galois connections. Math. Logic Quarterly 45(6), 497–504 (1999)
2. Belohlavek, R.: Fuzzy closure operators. Journal of Mathematical Analysis and Applications 262(2), 473–489 (2001)
3. Belohlavek, R.: Fuzzy Relational Systems: Foundations and Principles. Kluwer Academic Publishers, Norwell (2002)
4. Belohlavek, R.: Concept lattices and order in fuzzy logic. Ann. Pure Appl. Log. 128(1-3), 277–298 (2004)
5. Belohlavek, R.: Optimal decompositions of matrices with entries from residuated lattices. Submitted to J. Logic and Computation (2009)
6. Belohlavek, R.: Sup-t-norm and inf-residuum are one type of relational product: Unifying framework and consequences. Fuzzy Sets Syst. 197, 45–58 (2012)
7. Belohlavek, R., Konecny, J.: Closure spaces of isotone Galois connections and their morphisms. In: Wang, D., Reynolds, M. (eds.) AI 2011. LNCS, vol. 7106, pp. 182–191. Springer, Heidelberg (2011)
8. Belohlavek, R., Konecny, J.: Concept lattices of isotone vs. antitone Galois connections in graded setting: Mutual reducibility revisited. Information Sciences 199, 133–137 (2012)
9. Belohlavek, R., Konecny, J.: Row and column spaces of matrices over residuated lattices. Fundam. Inform. 115(4), 279–295 (2012)
10. Ganter, B., Wille, R.: Formal Concept Analysis – Mathematical Foundations. Springer (1999)
11. Georgescu, G., Popescu, A.: Non-dual fuzzy connections. Arch. Math. Log. 43(8), 1009–1039 (2004)
12. Hájek, P.: Metamathematics of Fuzzy Logic (Trends in Logic). Springer (November 2001)
13. Kohout, L.J., Bandler, W.: Relational-product architectures for information processing. Information Sciences 37(1-3), 25–37 (1985)
14. Konecny, J.: Closure and Interior Structures in Relational Data Analysis and Their Morphisms. PhD thesis, Palacky University (2012)
15. Konecny, J., Ojeda-Aciego, M.: Isotone L-bonds. In: Ojeda-Aciego, Outrata (eds.) [19], pp. 153–162
16. Krídlo, O., Krajči, S., Ojeda-Aciego, M.: The category of l-chu correspondences and the structure of L-bonds. Fundam. Inf. 115(4), 297–325 (2012)
17. Kridlo, O., Ojeda-Aciego, M.: CRL-Chu correspondences. In: Ojeda-Aciego, Outrata (eds.) [19], pp. 105–116
18. Medina, J., Ojeda-Aciego, M., Ruiz-Calviño, J.: Formal concept analysis via multi-adjoint concept lattices. Fuzzy Sets Syst. 160(2), 130–144 (2009)
19. Ojeda-Aciego, M., Outrata, J. (eds.): Proceedings of the Tenth International Conference on Concept Lattices and Their Applications, La Rochelle, France, October 15-18. CEUR Workshop Proceedings, vol. 1062. CEUR-WS.org (2013)
20. Pollandt, S.: Fuzzy Begriffe: Formale Begriffsanalyse von unscharfen Daten. Springer, Heidelberg (1997)
21. Ward, M., Dilworth, R.P.: Residuated lattices. Transactions of the American Mathematical Society 45, 335–354 (1939)

Minimal Solutions of Fuzzy Relation Equations with General Operators on the Unit Interval

Jesús Medina[1], Esko Turunen[2], Eduard Bartl[3],
and Juan Carlos Díaz-Moreno[1]

[1] Department of Mathematics, University of Cádiz, Spain
{jesus.medina,juancarlos.diaz}@uca.es
[2] Research Unit Computational Logic,
Vienna University of Technology, Wien, Austria
esko.turunen@tut.fi
[3] Department of Computer Science, Palacky University, Olomouc, Czech Republic
eduard.bartl@upol.cz

Abstract. Fuzzy relation equations arise as a mechanism to solve problems in several frameworks, such as in fuzzy logic. Moreover, the solvability of these equations has been related to fuzzy property-oriented concept lattices.

This paper studies a procedure to obtain the minimal solutions of fuzzy relation equations $R \circ X = T$, with an isotone binary operation associated with a (left or right) residuated implication on the unit interval. From this study several results, based on the covering problem, are introduced generalizing other ones given in the literature.

Keywords: Fuzzy relation equations, minimal solutions, residual structures.

1 Introduction

E. Sanchez [18] introduced fuzzy relation equations in the seventies, in order to investigate theoretical and applicational aspects of fuzzy set theory [7]. Several generalizations of the original equations have been introduced, such as [2,4,10]. These and other many papers study the existence of solutions of these equations [1,3,6,17,20], and, in the affirmative case, they prove that the greatest solution is $R \Rightarrow T$, where $\Rightarrow$ is the residuated implication of $\circ$.

The solvability of these equations has recently been related to a particular fuzzy property-oriented concept lattice [10,11] and they have been used in [8,9] to solve several problems arisen in fuzzy logic programming [15].

Find out minimal solutions is the next step, since the existence of these solutions offer important properties of the considered fuzzy relation equation, such as, the complete set of solutions of the fuzzy relation equation can be built. Different papers study how these solutions can be obtained, such as in [5,13,19,21], but the considered frameworks are very restrictive. This paper is a further study of the previous ones which consider a more general setting, in which neither

A. Laurent et al. (Eds.): IPMU 2014, Part III, CCIS 444, pp. 81–90, 2014.

commutative nor associative operators can be considered, satisfying that they are inf-preserving mappings on the right argument. This assumption is not very restrictive as we will explain later. The only restriction needed is the linearity of the carrier in which the values are considered.

The plan of this paper is the following: a preliminary notions are introduced in Section 2. Section 3 introduces a mechanism to obtain the minimal solutions of the general fuzzy relation equations introduced. A detailed example is given in Section 4 and the paper ends with several conclusions and prospects for future work.

2 Preliminaries

The operator used to define the fuzzy relation equation is $\odot\colon [0,1] \times [0,1] \to [0,1]$, such that it is order preserving and there exists $\to\colon [0,1] \times [0,1] \to [0,1]$, satisfying one of the following two adjoint properties with $\odot$:

1. $x \odot y \leq z$ if and only if $y \leq x \to z$
2. $x \odot y \leq z$ if and only if $x \leq y \to z$

for each $x, y, z \in [0,1]$. From now on, we will assume a pair $(\odot, \to)$ satisfying Condition 1. A similar theory can be developed with Condition 2.

One important notion, also needed along this paper, is the definition of *covering*.

Definition 1. *Given an ordered set $(A, \leq)$ and subsets $S_1, \ldots, S_n \in \mathcal{P}(A)$, where $\mathcal{P}(A)$ is the powerset of A, the element $a \in A$ covers $\{S_1, \ldots, S_n\}$, if for each $i \in \{1, \ldots, n\}$, then there exists $s_i \in S_i$ such that $s_i \leq a$.*

A cover $b \in A$ is called minimal *if for any $d \in A$ satisfying $d < b$, then d is not a cover of $\{S_1, \ldots, S_n\}$.*

Note that when $(A, \leq)$ is a complete lattice, minimal covers always exist in A.

Given the pair $(\odot, \to)$, a fuzzy relation equation in the environment of this paper is the equation

$$R \circ X = T \tag{1}$$

in which R, X, T are *finite* fuzzy relations, i.e. the $[0,1]$-valued relations $R\colon U \times V \to [0,1]$, $X\colon V \times W \to [0,1]$, $T\colon U \times W \to [0,1]$ can be expressed by $n \times k$, $k \times m$ and $n \times m$ matrices, respectively, with $n, m, k \in \mathbb{N}$; and $R \circ X$ is defined, for each $u \in U$, $w \in W$, as

$$(R \circ X)\langle u, w \rangle = \bigvee \{ R\langle u, v \rangle \odot X\langle v, w \rangle \mid v \in U \} \tag{2}$$

Therefore, this is a $[0,1]$-fuzzy relation which can also be written as a matrix.

As in the usual case, this equation may not have a solution. The following theorem shows a result similar to the one given to the max-min relation equation [20].

Theorem 1. *Given the pair* $(\odot, \rightarrow)$, *a fuzzy relation equation defined as in Expresion 1 has a solution if and only if*

$$(R \Rightarrow T)\langle v, w \rangle = \bigwedge \{R\langle u, v \rangle \rightarrow T\langle u, w \rangle \mid u \in U\}$$

is a solution and, in that case, it is the greatest solution.

The following section presents a mechanism to obtain the minimal solutions of the introduced fuzzy relation equations.

3   Minimal Solutions Generated by a Given Solution

The algebraic procedure to obtain minimal solutions of a fuzzy relation equation will be based on the following observations.

In order to verify the equality

$$\bigvee \{R\langle u_i, v \rangle \odot (R \Rightarrow T)\langle v, w_j \rangle \mid v \in V\} = T\langle u_i, w_j \rangle, \tag{3}$$

we have that several elements in V are redundant and so, they are not needed to obtain the value $T\langle u_i, w_j \rangle$ and, therefore, they can be omitted. Hence the guiding principle to find out minimal solutions is to replace all the unnecessary values in the matrix representation of $R \Rightarrow T$ by 0 and possibly reduce some other values $(R \Rightarrow T)\langle v, w_j \rangle$; this is done column by column. Specifically, by the finiteness and linearity assumptions, for each fixed i, j there is at least one $v_0 \in V$ such that

$$R\langle u_i, v_0 \rangle \odot (R \Rightarrow T)\langle v_0, w_j \rangle = T\langle u_i, w_j \rangle \tag{4}$$

and the other $v \in V \setminus \{v_0\}$ are irrelevant and so, $(R \Rightarrow T)\langle v, w_j \rangle$ can be substituted by 0. Moreover, in some cases the value $(R \Rightarrow T)\langle v_0, w_j \rangle$ can even be reduced to a smaller value. Therefore, for any v_0 satisfying Equation (4), we can consider the set $V' = \{v_0\}$, which defines a $k \times 1$ column that can be used in the matrix representation of minimal solutions.

There exist a variety of composition operators of fuzzy relations [12,14,16,21], several of them are very restrictive since, for example, the commutativity and associativity properties of the considered conjunctors is needed. Moreover, in order to obtain the minimal solutions extra properties, as the *zero-or-greatest property* [13,19], are assumed. However, based on the above examples, we will show that the composition 'o' does not need to satisfy strong properties.

This section is focused on finding out the minimal solutions of a general fuzzy relation equation, in which we only need to assume a linear complete lattice $[0, 1]$ endowed with an increasing operation $\odot$ that has a residuum $\rightarrow$ and satisfies the following equality:

$$a \odot \bigwedge \{b_i \mid i \in \Gamma\} = \bigwedge \{a \odot b_i \mid i \in \Gamma\} \tag{5}$$

for all elements $a \in [0, 1]$ and all non-empty subsets $\{b_i \mid i \in \Gamma\} \subseteq [0, 1]$. This last condition will be called *IPNE-Condition*, making reference to that $\odot$ is Infimum

Preserving of arbitrary Non-Empty sets. Several properties and observations of this condition will be given in the next section. Note that this property give us the possibility of consider infinite lattices.

Therefore, let us consider a general solvable fuzzy relation equation (1), where R, X, T are finite, $U = \{u_1, \ldots, u_n\}$, $W = \{w_1, \ldots, w_m\}$, and $\odot$ satisfies the IPNE-Condition.

The first result characterizes the solutions of a solvable fuzzy relation equation $R \circ X = T$ by the covering elements of a family of subsets S_{ij}. Next, these sets are defined. First of all, the auxiliary sets V_{ij} need to be introduced, which are associated with the elements u_i, w_j and the greatest solution $R \Rightarrow T$. Since for each $j = 1, \ldots, m$, $i = 1, \ldots, n$

$$\bigvee\{R\langle u_i, v\rangle \odot (R \Rightarrow T)\langle v, w_j\rangle \mid v \in V\} = T\langle u_i, w_j\rangle, \tag{6}$$

$[0, 1]$ is linear and V is finite, then there exists, at least one, v_s validating the equation

$$R\langle u_i, v_s\rangle \odot (R \Rightarrow T)\langle v_s, w_j\rangle = T\langle u_i, w_j\rangle. \tag{7}$$

Therefore, the set $V_{ij} = \{v_s \in V \mid R\langle u_i, v_s\rangle \odot (R \Rightarrow T)\langle v_s, w_j\rangle = T\langle u_i, w_j\rangle\}$ is not empty and, for all $v \notin V_{ij}$, the strict inequality $R\langle u_i, v\rangle \odot (R \Rightarrow T)\langle v_s, w_j\rangle < T\langle u_i, w_j\rangle$ holds.

Each v_s in V_{ij} will provide a fuzzy subset S_{ijs} as follows: Given $v_s \in V_{ij}$, we have that

$$\{d \in [0, 1] \mid R\langle u_i, v_s\rangle \odot d = T\langle u_i, w_j\rangle\} \neq \varnothing$$

and the infimum $\bigwedge\{d \in [0, 1] \mid R\langle u_i, v_s\rangle \odot d = T\langle u_i, w_j\rangle\} = e_s$ also satisfies the equality

$$R\langle u_i, v_s\rangle \odot e_s = T\langle u_i, w_j\rangle$$

by the IPNE-Condition. These elements are used to define the characteristic fuzzy subsets $S_{ijs} \colon V \to [0, 1]$ of V, defined by

$$S_{ijs}(v) = \begin{cases} e_s & \text{if } v = v_s \\ 0 & \text{otherwise} \end{cases}$$

which form the set S_{ij}, that is $S_{ij} = \{S_{ijs} \mid v_s \in V_{ij}\}$, for each $i = 1, \ldots, n$, $j = 1, \ldots, m$. These sets will be used to characterize the set of solutions of Equation (1) by the notion of *covering*[1]. The proof of this result is not include here due to the lack of space and, mainly, it follows the idea behind the procedure considered in the example detailed in Section 4.

Theorem 2. *The $[0, 1]$-fuzzy relation $X \colon V \times W \to [0, 1]$ is a solution of a solvable Equation (1) if and only if $X \leq (R \Rightarrow T)$ and, for each $j = 1, \ldots, m$, the fuzzy subset $C_j \colon V \to [0, 1]$, defined by $C_j(v) = X\langle v, w_j\rangle$, is a cover of $\{S_{1j}, \ldots, S_{nj}\}$.*

---

[1] The definition of cover (Definition 1) is inspired by [14], see also [12] and [13], however our approach is more general and it does not depend on the special type of the composition 'o' of fuzzy relations.

As a consequence, the minimal solutions are characterized by the minimal covers.

Corollary 1. *A fuzzy binary relation* $X \colon V \times W \to [0,1]$ *is a minimal solution of Equation (1) if and only if, for each* $j = 1, \ldots, m$, $C_j \colon V \to [0,1]$, *defined by* $C_j(v) = X\langle v, w_j \rangle$, *is a minimal cover of* $\{S_{1j}, \ldots, S_{nj}\}$.

Hence, from the corollary above, minimal solutions of the fuzzy relation equation (1) are obtained from $R \Rightarrow T$ as follows:

Procedure to Obtain Minimal Solutions of Equation (1)

For each $j \in \{1, \ldots, m\}$ (jth element $w_j \in W$) the following steps are applied:

1. For each $i \in \{1, \ldots, n\}$ (ith element $u_i \in U$) compute the set $V_{ij} \subseteq V$.
2. Construct the set S_{ij} of corresponding characteristic mappings S_{ijs}.
3. Compute the minimal cover(s) C_j of the set $\{S_{1j}, \ldots, S_{nj}\}$.
4. Define the $[0,1]$-fuzzy matrix $X \colon V \times W \to [0,1]$ as $X\langle v, w_j \rangle = C_j(v)$.

4 Worked Out Example

Let us assume the standard MV–algebra, that is, the unit interval, the Łukasiewicz operator and its residuated implication.

Given $U = \{u_1, u_2, u_3\}$, $V = \{v_1, v_2, v_3\}$ $W = \{w_1, w_2, w_3\}$ and the fuzzy binary relation, defined from the following tables

R	v_1	v_2	v_3
u_1	0.9	0.5	0.9
u_2	0.2	0.9	0.7
u_3	0.8	0.6	0.9

and

T	w_1	w_2	w_3
u_1	0.8	0.4	0.7
u_2	0.6	0.7	0.3
u_3	0.8	0.4	0.6

direct computation shows that the relation $R \Rightarrow T$, defined from the table

$R \Rightarrow T$	w_1	w_2	w_3
v_1	0.9	0.5	0.8
v_2	0.7	0.8	0.4
v_3	0.9	0.5	0.6

is the greatest solution of Equation (1). During the verification we go through the following calculations:
When computing $R \circ (R \Rightarrow T)\langle u_1, w_1 \rangle = 0.8$, we consider the maximum of

$$R\langle u_1, v_1 \rangle \odot (R \Rightarrow T)\langle v_1, w_1 \rangle = 0.9 + 0.9 - 1 = 0.8$$
$$R\langle u_1, v_2 \rangle \odot (R \Rightarrow T)\langle v_2, w_1 \rangle = 0.5 + 0.7 - 1 = 0.2$$
$$R\langle u_1, v_3 \rangle \odot (R \Rightarrow T)\langle v_3, w_1 \rangle = 0.9 + 0.9 - 1 = 0.8$$

Notice that from v_1 and v_3 we get the maximum. Hence, in order to obtain this maximum, we only need to consider $\{v_1\}$ or $\{v_3\}$, hence $V_{11} = \{v_1, v_3\}$.

Moreover, the values 0.9 associated with v_1 and 0.9 associated with v_3 cannot be decreased since, with this assumption, a value less than 0.8 will be obtained in the computation and we do not get a solution. Therefore, the characteristic fuzzy subsets considered are:

$$S_{111}(v) = \begin{cases} 0.9 & \text{if } v = v_1 \\ 0 & \text{otherwise} \end{cases} \qquad S_{113}(v) = \begin{cases} 0.9 & \text{if } v = v_3 \\ 0 & \text{otherwise} \end{cases}$$

and so, $S_{11} = \{S_{111}, S_{113}\}$. Now, we need to compute S_{21} and S_{31} associated with the first column of T.

The equality $R \circ (R \Rightarrow T)\langle u_2, w_1 \rangle = 0.6$ is similarly studied. The value $R \circ (R \Rightarrow T)\langle u_2, w_1 \rangle$ is the maximum of the values

$$R\langle u_2, v_1 \rangle \odot (R \Rightarrow T)\langle v_1, w_1 \rangle = 0.2 + 0.9 - 1 = 0.1$$
$$R\langle u_2, v_2 \rangle \odot (R \Rightarrow T)\langle v_2, w_1 \rangle = 0.9 + 0.7 - 1 = 0.6$$
$$R\langle u_2, v_3 \rangle \odot (R \Rightarrow T)\langle v_3, w_1 \rangle = 0.7 + 0.9 - 1 = 0.6$$

for which $\{v_2\}$ or $\{v_3\}$ is only necessary and so, $V_{21} = \{v_2, v_3\}$ and the fuzzy subsets are:

$$S_{212}(v) = \begin{cases} 0.7 & \text{if } v = v_2 \\ 0 & \text{otherwise} \end{cases} \qquad S_{213}(v) = \begin{cases} 0.9 & \text{if } v = v_3 \\ 0 & \text{otherwise} \end{cases}$$

Therefore, $S_{21} = \{S_{212}, S_{213}\}$. Finally, when computing $R \circ (R \Rightarrow T)\langle u_3, w_1 \rangle = 0.8$ we pass by

$$R\langle u_3, v_1 \rangle \odot (R \Rightarrow T)\langle v_1, w_1 \rangle = 0.8 + 0.9 - 1 = 0.7$$
$$R\langle u_3, v_2 \rangle \odot (R \Rightarrow T)\langle v_2, w_1 \rangle = 0.6 + 0.7 - 1 = 0.3$$
$$R\langle u_3, v_3 \rangle \odot (R \Rightarrow T)\langle v_3, w_1 \rangle = 0.9 + 0.9 - 1 = 0.8$$

In this case, only v_3 is necessary and one fuzzy subset is only considered $S_{31} = \{S_{313}\}$, where

$$S_{313}(v) = \begin{cases} 0.9 & \text{if } v = v_3 \\ 0 & \text{otherwise} \end{cases}$$

We observe that

$$C_1(v) = \begin{cases} 0.9 & \text{if } v = v_3 \\ 0 & \text{otherwise} \end{cases}$$

is in $S_{11} \cap S_{21} \cap S_{31}$ and, therefore, X_1 is (the only) minimal fuzzy subset that covers the set $\{S_{11}, S_{21}, S_{31}\}$. Moreover, we conclude that a fuzzy relation X_1, defined as

X_1	w_1	w_2	w_3
v_1	0	0.5	0.8
v_2	0	0.8	0.4
v_3	0.9	0.5	0.6

solves the fuzzy relation equation (1). This is a first approximation in order to obtain the minimal solutions, for this goal, the rest of columns need to be considered.

Next we consider the second column of $R \Rightarrow T$, which provides a different case. For $R \circ (R \Rightarrow T)\langle u_1, w_2 \rangle = 0.4$ we have

$$R\langle u_1, v_1 \rangle \odot (R \Rightarrow T)\langle v_1, w_2 \rangle = 0.9 + 0.5 - 1 = 0.4$$
$$R\langle u_1, v_2 \rangle \odot (R \Rightarrow T)\langle v_2, w_2 \rangle = 0.5 + 0.8 - 1 = 0.3$$
$$R\langle u_1, v_3 \rangle \odot (R \Rightarrow T)\langle v_3, w_2 \rangle = 0.9 + 0.5 - 1 = 0.4$$

Hence, the maximum is obtained from v_1 or v_3 and, therefore, $V_{12} = \{v_1, v_3\}$ and $S_{12} = \{S_{121}, S_{123}\}$, where

$$S_{121}(v) = \begin{cases} 0.5 & \text{if } v = v_1 \\ 0 & \text{otherwise} \end{cases} \qquad S_{123}(v) = \begin{cases} 0.5 & \text{if } v = v_3 \\ 0 & \text{otherwise} \end{cases}$$

For $R \circ (R \Rightarrow T)\langle u_2, w_2 \rangle = 0.7$ we have

$$R\langle u_2, v_1 \rangle \odot (R \Rightarrow T)\langle v_1, w_2 \rangle = 0$$
$$R\langle u_2, v_2 \rangle \odot (R \Rightarrow T)\langle v_2, w_2 \rangle = 0.9 + 0.8 - 1 = 0.7$$
$$R\langle u_2, v_3 \rangle \odot (R \Rightarrow T)\langle v_3, w_2 \rangle = 0.7 + 0.5 - 1 = 0.2$$

Consequently, the set obtained are $V_{22} = \{v_2\}$ and $S_{22} = \{S_{222}\}$, where

$$S_{222}(v) = \begin{cases} 0.8 & \text{if } v = v_2 \\ 0 & \text{otherwise} \end{cases}$$

For $R \circ (R \Rightarrow T)\langle u_3, w_2 \rangle = 0.4$ we have

$$R\langle u_3, v_1 \rangle \odot (R \Rightarrow T)\langle v_1, w_2 \rangle = 0.8 + 0.5 - 1 = 0.3$$
$$R\langle u_3, v_2 \rangle \odot (R \Rightarrow T)\langle v_2, w_2 \rangle = 0.6 + 0.8 - 1 = 0.4$$
$$R\langle u_3, v_3 \rangle \odot (R \Rightarrow T)\langle v_3, w_2 \rangle = 0.9 + 0.5 - 1 = 0.4$$

Hence, in v_2 and v_3 the maximum is obtained and so, $V_{32} = \{v_2, v_3\}$ and the fuzzy subsets

$$S_{322}(v) = \begin{cases} 0.8 & \text{if } v = v_2 \\ 0 & \text{otherwise} \end{cases} \qquad S_{323}(v) = \begin{cases} 0.5 & \text{if } v = v_3 \\ 0 & \text{otherwise} \end{cases}$$

form $S_{32} = \{S_{322}, S_{323}\}$. In this case, since $S_{12} \cap S_{22} \cap S_{32} = \varnothing$, a minimal cover of $\{S_{12}, S_{22}, S_{32}\}$ cannot be obtained from the intersection. As $S_{22} = \{S_{222}\}$ and $S_{222} = S_{322} \in S_{32}$, then a minimal cover only need to be greater than S_{222} and S_{121}, or S_{222} and S_{123}. Therefore,

$$C_2(v) = \begin{cases} 0.8 & \text{if } v = v_2 \\ 0.5 & \text{if } v = v_3 \\ 0 & \text{otherwise} \end{cases} \qquad C_2'(v) = \begin{cases} 0.5 & \text{if } v = v_1 \\ 0.8 & \text{if } v = v_2 \\ 0 & \text{otherwise} \end{cases}$$

are (the only) minimal fuzzy subsets that cover the set $\{S_{12}, S_{22}, S_{32}\}$. Finally, the values in the third column of $R \Rightarrow T$ are reduced.

For $R \circ (R \Rightarrow T)\langle u_1, w_3 \rangle = 0.7$, we compute

$$R\langle u_1, v_1 \rangle \odot (R \Rightarrow T)\langle v_1, w_3 \rangle = 0.9 + 0.8 - 1 = 0.7$$
$$R\langle u_1, v_2 \rangle \odot (R \Rightarrow T)\langle v_2, w_3 \rangle = 0$$
$$R\langle u_1, v_3 \rangle \odot (R \Rightarrow T)\langle v_3, w_3 \rangle = 0.9 + 0.6 - 1 = 0.5$$

Hence, $V_{13} = \{v_1\}$ and $S_{13} = \{S_{131}\}$, where

$$S_{131}(v) = \begin{cases} 0.8 & \text{if } v = v_1 \\ 0 & \text{otherwise} \end{cases}$$

For $R \circ (R \Rightarrow T)\langle u_2, w_3 \rangle = 0.3$, we have

$$R\langle u_2, v_1 \rangle \odot (R \Rightarrow T)\langle v_1, w_3 \rangle = 0.2 + 0.8 - 1 = 0$$
$$R\langle u_2, v_2 \rangle \odot (R \Rightarrow T)\langle v_2, w_3 \rangle = 0.9 + 0.4 - 1 = 0.3$$
$$R\langle u_2, v_3 \rangle \odot (R \Rightarrow T)\langle v_3, w_3 \rangle = 0.7 + 0.6 - 1 = 0.3$$

two possibilities providing two fuzzy subsets: $S_{23} = \{S_{232}, S_{233}\}$,

$$S_{231}(v) = \begin{cases} 0.4 & \text{if } v = v_2 \\ 0 & \text{otherwise} \end{cases} \quad S_{233}(v) = \begin{cases} 0.6 & \text{if } v = v_3 \\ 0 & \text{otherwise} \end{cases}$$

For $R \circ (R \Rightarrow T)\langle u_3, w_3 \rangle = 0.6$ we have

$$R\langle u_3, v_1 \rangle \odot (R \Rightarrow T)\langle v_1, w_3 \rangle = 0.8 + 0.8 - 1 = 0.6$$
$$R\langle u_3, v_2 \rangle \odot (R \Rightarrow T)\langle v_2, w_3 \rangle = 0.6 + 0.4 - 1 = 0$$
$$R\langle u_3, v_3 \rangle \odot (R \Rightarrow T)\langle v_3, w_3 \rangle = 0.9 + 0.6 - 1 = 0.5$$

Therefore, only v_1 is needed to obtain the maximum and so, $S_{33} = \{S_{231}\}$, with

$$S_{231}(v) = \begin{cases} 0.8 & \text{if } v = v_1 \\ 0 & \text{otherwise} \end{cases}$$

In this case, $S_{13} = \{S_{131}\} = S_{33}$ and, therefore, the minimal covers must be greater than S_{131} and S_{231}, or S_{131} and S_{233}. Therefore, there also are two minimal fuzzy subsets that cover the set $\{S_{13}, S_{23}, S_{33}\}$:

$$C_3(v) = \begin{cases} 0.8 & \text{if } v = v_1 \\ 0.4 & \text{if } v = v_2 \\ 0 & \text{otherwise} \end{cases} \quad C_3'(v) = \begin{cases} 0.8 & \text{if } v = v_1 \\ 0.6 & \text{if } v = v_3 \\ 0 & \text{otherwise} \end{cases}$$

Thus, by Theorem 2, the minimal solutions are given by the $[0, 1]$-fuzzy relations $X \colon V \times W \to [0, 1]$, defined by $X\langle v, w_j \rangle = C_j(v)$, where C_j are the previous covers, with $j \in \{1, 2, 3\}$. This yields to four fuzzy relations, defined as following:

X_1	w_1	w_2	w_3
v_1	0	0	0.8
v_2	0	0.8	0.4
v_3	0.9	0.5	0

X_2	w_1	w_2	w_3
v_1	0	0	0.8
v_2	0	0.8	0
v_3	0.9	0.5	0.6

X_3	w_1	w_2	w_3
v_1	0	0.5	0.8
v_2	0	0.8	0.4
v_3	0.9	0	0

X_4	w_1	w_2	w_3
v_1	0	0.5	0.8
v_2	0	0.8	0
v_3	0.9	0	0.6

that solve the fuzzy relation equation $R \circ X = T$.

5   Conclusions and Future Work

A deterministic procedure has been introduced to obtain the minimal solutions of a solvable general fuzzy relation equation. This procedure has been based on a relation between the solutions of the equation and the covering elements of a set of characteristic fuzzy subsets, which generalizes other ones given in the literature.

In the future, an extension of the results with conjunctors in a general carrier set will be studied. The main goal will be to consider a general lattice as values set.

Acknowledgements. Jesús Medina and Juan Carlos Díaz were partially supported by the Spanish Science Ministry projects TIN2009-14562-C05-03 and TIN2012-39353-C04-04, and by Junta de Andalucía project P09-FQM-5233. Esko Turunen was supported by the Czech Technical University in Prague under project SGS12/187/OHK3/3T/13. The authors are grateful for M. Navara and R. Horčík for their comments to improve the final version of this research.

References

1. Bandler, W., Kohout, L.: Semantics of implication operators and fuzzy relational products. Int. J. Man-Machine Studies 12, 89–116 (1980)
2. Bartl, E.: Fuzzy Relational Equation. Phd Dissertation. PhD thesis, Faculty of Science, Palacky University Olomouc (2013)
3. Bělohlávek, R.: Fuzzy Relational Systems: Foundations and Principles. Kluwer Academic Publishers (2002)
4. Bělohlávek, R.: Sup-t-norm and inf-residuum are one type of relational product: Unifying framework and consequences. Fuzzy Sets and Systems 197, 45–58 (2012)
5. Chen, L., Wang, P.: Fuzzy relation equations (ii): The branch-point-solutions and the categorized minimal solutions. Soft Computing - A Fusion of Foundations, Methodologies and Applications 11, 33–40 (2007)
6. De Baets, B.: Analytical solution methods for fuzzy relation equations. In: Dubois, D., Prade, H. (eds.) The Handbooks of Fuzzy Sets Series, vol. 1, pp. 291–340. Kluwer, Dordrecht (1999)
7. Di Nola, A., Sanchez, E., Pedrycz, W., Sessa, S.: Fuzzy Relation Equations and Their Applications to Knowledge Engineering. Kluwer Academic Publishers, Norwell (1989)
8. Díaz, J., Medina, J.: Applying multi-adjoint relation equations to fuzzy logic programming. In: XV Congreso Español sobre Tecnologías y Lógica Fuzzy, ESTYLF (2014)
9. Díaz, J.C., Medina, J.: Concept lattices in fuzzy relation equations. In: The 8th International Conference on Concept Lattices and Their Applications, pp. 75–86 (2011)
10. Díaz, J.C., Medina, J.: Multi-adjoint relation equations: Definition, properties and solutions using concept lattices. Information Sciences 253, 100–109 (2013)
11. Díaz, J.C., Medina, J.: Solving systems of fuzzy relation equations by fuzzy property-oriented concepts. Information Sciences 222, 405–412 (2013)

12. Lin, J.-L.: On the relation between fuzzy max-archimedean t-norm relational equations and the covering problem. Fuzzy Sets and Systems 160(16), 2328–2344 (2009)
13. Lin, J.-L., Wu, Y.-K., Guu, S.-M.: On fuzzy relational equations and the covering problem. Information Sciences 181(14), 2951–2963 (2011)
14. Markovskii, A.: On the relation between equations with max-product composition and the covering problem. Fuzzy Sets and Systems 153(2), 261–273 (2005)
15. Medina, J., Ojeda-Aciego, M., Vojtáš, P.: Multi-adjoint logic programming with continuous semantics. In: Eiter, T., Faber, W., Truszczyński, M. (eds.) LPNMR 2001. LNCS (LNAI), vol. 2173, pp. 351–364. Springer, Heidelberg (2001)
16. Peeva, K.: Resolution of fuzzy relational equations: Method, algorithm and software with applications. Information Sciences 234, 44–63 (2013)
17. Perfilieva, I., Nosková, L.: System of fuzzy relation equations with inf-→ composition: Complete set of solutions. Fuzzy Sets and Systems 159(17), 2256–2271 (2008)
18. Sanchez, E.: Resolution of composite fuzzy relation equations. Information and Control 30(1), 38–48 (1976)
19. Shieh, B.-S.: Solution to the covering problem. Information Sciences 222, 626–633 (2013)
20. Turunen, E.: On generalized fuzzy relation equations: necessary and sufficient conditions for the existence of solutions. Acta Universitatis Carolinae. Mathematica et Physica 028(1), 33–37 (1987)
21. Yeh, C.-T.: On the minimal solutions of max-min fuzzy relational equations. Fuzzy Sets and Systems 159(1), 23–39 (2008)

Generating Isotone Galois Connections on an Unstructured Codomain

Francisca García-Pardo, Inma P. Cabrera, Pablo Cordero,
Manuel Ojeda-Aciego, and Francisco J. Rodríguez

Universidad de Málaga, Spain*

Abstract. Given a mapping $f\colon A \to B$ from a partially ordered set A into an unstructured set B, we study the problem of defining a suitable partial ordering relation on B such that there exists a mapping $g\colon B \to A$ such that the pair of mappings (f, g) forms an isotone Galois connection between partially ordered sets.

1 Introduction

Galois connections were introduced by Ore [25] as a pair of antitone mappings satisfying certain conditions which generalize Birkhoff's theory of polarities to apply to complete lattices. Later, Kan [19] introduced the notion of *adjunction* in a categorical context which, after instantiating to partially ordered sets turned out to be the isotone version of the notion of Galois connection.

In the recent years there has been a notable increase in the number of publications concerning Galois connections, both isotone and antitone. On the one hand, one can find lots of papers on theoretical developments or theoretical applications [7, 9, 20]; on the other hand, of course, there exist as well a lot of applications to computer science, see [23] for a first survey on applications, although more specific references on certain topics can be found, for instance, to programming [24], data analysis [23], logic [12, 18], etc.

Two research topics that have benefitted recently from the use of the theory of Galois connections is that of approximate reasoning using rough sets [13, 17, 26], and Formal Concept Analysis (FCA), either theoretically [1, 3, 6, 22] or applicatively [10, 11]. It is not surprising to see so many works dealing with both Galois connections and FCA, since the derivation operators used to define the concepts form a (antitone) Galois connection.

A number of results can be found in the literature concerning sufficient or necessary conditions for a Galois connection between ordered structures to exist.

---

* This work is partially supported by the Spanish research projects TIN2009-14562-C05-01, TIN2011-28084 and TIN2012-39353-C04-01, and Junta de Andalucía project P09-FQM-5233.

A. Laurent et al. (Eds.): IPMU 2014, Part III, CCIS 444, pp. 91–99, 2014.

The main result of this paper is related to the existence and construction of the adjoint pair to a given mapping f, but *in a more general framework*.

Our initial setting is to consider a mapping $f \colon A \to B$ from a partially ordered set A into an unstructured set B, and then characterize those situations in which the set B can be partially ordered and an isotone mapping $g \colon B \to A$ can be built such that the pair (f, g) is an isotone Galois connection.

The structure of the paper is as follows: in Section 2 we introduce the preliminary definitions and results; then, in Section 3, given $f \colon A \to B$ we focus on the case in which the domain A has a poset structure, the necessary and sufficient conditions for the existence of a unique ordering on B and a mapping g such that (f, g) is an adjunction are given; Finally, in Section 4, we draw some conclusions and discuss future work.

2   Preliminaries

We assume basic knowledge of the properties and constructions related to a partially ordered set. For the sake of self-completion, we include below the formal definitions of the main concepts to be used in this section.

Definition 1. *Given a partially ordered set $\mathbb{A} = (A, \leq_A)$, $X \subseteq A$, and $a \in A$.*

- *Element a is said to be the* maximum *of X, denoted $\max X$, if $a \in X$ and $x \leq a$ for all $x \in X$.*
- *The* downset *$a^{\downarrow}$ of a is defined as $a^{\downarrow} = \{x \in A \mid x \leq_A a\}$.*
- *The* upset *$a^{\uparrow}$ of a is defined as $a^{\uparrow} = \{x \in A \mid x \geq_A a\}$.*

A mapping $f \colon (A, \leq_A) \to (B, \leq_B)$ between partially ordered sets is said to be

- *isotone if $a_1 \leq_A a_2$ implies $f(a_1) \leq_B f(a_2)$, for all $a_1, a_2 \in A$.*
- *antitone if $a_1 \leq_A a_2$ implies $f(a_2) \leq_B f(a_1)$, for all $a_1, a_2 \in A$.*

As usual, f^{-1} is the inverse image of f, that is, $f^{-1}(b) = \{a \in A \mid f(a) = b\}$. In the particular case in which $A = B$,

- *f is inflationary (also called extensive) if $a \leq_A f(a)$ for all $a \in A$.*
- *f is deflationary if $f(a) \leq_A a$ for all $a \in A$.*

As we are including the necessary definitions for the development of the construction of isotone Galois connections (hereafter, for brevity, termed *adjunctions*) between posets, we state below the definition of adjunction we will be working with.

Definition 2. *Let $\mathbb{A} = (A, \leq_A)$ and $\mathbb{B} = (B, \leq_B)$ be posets, $f \colon A \to B$ and $g \colon B \to A$ be two mappings. The pair (f, g) is said to be an **adjunction between $\mathbb{A}$ and $\mathbb{B}$**, denoted by $(f, g) \colon \mathbb{A} \leftrightharpoons \mathbb{B}$, whenever for all $a \in A$ and $b \in B$ we have that*

$$f(a) \leq_B b \qquad \text{if and only if} \qquad a \leq_A g(b)$$

The mapping f is called left adjoint *and g is called* right adjoint.

The following theorem states equivalent definitions of adjunction between posets that can be found in the literature, see for instance [5, 16].

Theorem 1. *Let* $\mathbb{A} = (A, \leq_A), \mathbb{B} = (B, \leq_B)$ *be two posets,* $f: \mathbb{A} \to \mathbb{B}$ *and* $g: \mathbb{B} \to \mathbb{A}$ *be two mappings. The following statements are equivalent:*

1. $(f, g): \mathbb{A} \leftrightarrows \mathbb{B}$.
2. f *and* g *are isotone,* $g \circ f$ *is inflationary, and* $f \circ g$ *is deflationary.*
3. $f(a)^\uparrow = g^{-1}(a^\uparrow)$ *for all* $a \in A$.
4. $g(b)^\downarrow = f^{-1}(b^\downarrow)$ *for all* $b \in B$.
5. f *is isotone and* $g(b) = \max f^{-1}(b^\downarrow)$ *for all* $b \in B$.
6. g *is isotone and* $f(a) = \min g^{-1}(a^\uparrow)$ *for each* $a \in A$.

We introduce the technical lemma below which shows that, in some case, it is possible to get rid of the downsets (as used in item 5 of the previous theorem).

Lemma 1. *Let* $(A, \leq_A)$ *and* $(B, \leq_B)$ *be posets and* $f: A \to B$ *an isotone mapping. If* $\max f^{-1}(b^\downarrow)$ *exists for some* $b \in f(A)$, *then* $\max f^{-1}(b)$ *exists and* $\max f^{-1}(b^\downarrow) = \max f^{-1}(b)$.

Proof. Let us denote $m = \max f^{-1}(b^\downarrow)$ and we will prove that $a \leq_A m$, for all $a \in f^{-1}(b)$, and $m \in f^{-1}(b)$, in order to have $m = \max f^{-1}(b)$.

Consider $a \in f^{-1}(b)$, then $f(a) = b \in b^\downarrow$ and $a \in f^{-1}(b^\downarrow)$, hence $a \leq_A m$.

Now, isotonicity of f shows that $f(a) = b \leq_B f(m)$. For the other inequality, simply consider that $m = \max f^{-1}(b^\downarrow)$ implies $m \in f^{-1}(b^\downarrow)$, which means $f(m) \leq_B b$. Therefore, $f(m) = b$ because of antisymmetry of $\leq_B$. □

3   Building Adjunctions between Partially Ordered Sets

With the general aim of finding conditions for a mapping from a poset $(A, \leq_A)$ to an unstructured set B, in order to construct an adjunction we will naturally consider the canonical decomposition of $f: A \to B$ through A_f, the quotient set of A wrt the **kernel relation** $\equiv_f$, defined as $a \equiv_f b$ if and only if $f(a) = f(b)$:

$$
\begin{array}{ccc}
A & \xrightarrow{\ f\ } & B \\
\pi \big\downarrow & & \big\uparrow i \\
A_f & \dashrightarrow_{\varphi} & f(A)
\end{array}
$$

In general, given a poset $(A \leq_A)$ together with an equivalence relation $\sim$ on A, it is customary to consider the set $A_\sim = A/\sim$, the quotient set of A wrt $\sim$, and the natural projection $\pi: A \to A_\sim$. As usual, the equivalence class of an element $a \in A$ is denoted $[a]$ and, then, $\pi(a) = [a]$.

The following lemma provides sufficient conditions for π being the left component of an adjunction.

Lemma 2. *Let $(A, \leq_A)$ be a poset and $\sim$ an equivalence relation on A. Suppose that the following conditions hold*

1. *There exists $\max([a])$, for all $a \in A$.*
2. *If $a_1 \leq_A a_2$ then $\max([a_1]) \leq_A \max([a_2])$, for all $a_1, a_2 \in A$.*

Then, the relation $\leq_{A_\sim}$ defined by $[a_1] \leq_{A_\sim} [a_2]$ if only if $a_1 \leq_A \max([a_2])$ is an ordering in $A_\sim$ and, moreover, the pair $(\pi, \max)$ is an adjunction.

Proof. To begin with, the relation $\leq_{A_\sim}$ is well defined since, by the first hypothesis, $\max([a])$ exists for all $a \in A$.

Reflexivity. Obvious, since $[a] \leq_{A_\sim} [a]$ if and only if $a \leq_A \max([a])$, and the latter holds for all $a \in A$.

Transitivity. Assume $[a_1] \leq_{A_\sim} [a_2]$ and $[a_2] \leq_{A_\sim} [a_3]$.
From $[a_1] \leq_{A_\sim} [a_2]$, by definition, we have $a_1 \leq_A \max([a_2])$. Now, from $[a_2] \leq_{A_\sim} [a_3]$ we obtain, by definition of the ordering and the second hypothesis that $\max([a_2]) \leq_A \max([a_3])$. As a result, we obtain $[a_1] \leq_{A_\sim} \max([a_3])$, that is, $[a_1] \leq_{A_\sim} [a_3]$.

Antisymmetry. Assume $a_1, a_2 \in A$ such that $[a_1] \leq_{A_\sim} [a_2]$ and $[a_2] \leq_{A_\sim} [a_1]$. By hypothesis, we have that $a_1 \leq_A \max([a_2])$ then $\max([a_1]) \leq_A \max([a_2])$, and $a_2 \leq_A \max([a_1])$ then $\max([a_2]) \leq_A \max([a_1])$. Since $\leq_A$ is antisymmetric, then $\max([a_1]) = \max([a_2])$; now, we have that the intersection of the two classes $[a_1]$ and $[a_2]$ is non-empty, therefore $[a_1] = [a_2]$.

Once again by the first hypothesis, max can be seen as a mapping $A_\sim \to A$. Now, the adjunction follows by the definition of π and the ordering:

$$\pi(a_1) \leq_{A_\sim} [a_2] \text{ if and only if } [a_1] \leq_{A_\sim} [a_2]$$
$$\text{if and only if } a_1 \leq_A \max([a_2])$$

$\square$

The previous lemma gave sufficient conditions for π being a left adjoint; the following result states that the conditions are also necessary, and that the ordering relation and the right adjoint are uniquely defined.

Lemma 3. *Let $(A \leq_A)$ be a poset and $\sim$ an equivalence relation on A. Let $A_\sim = A/\sim$ be the quotient set of A wrt $\sim$, and $\pi \colon A \to A_\sim$ the natural projection. If there exists an ordering relation $\leq_{A_\sim}$ in $A_\sim$ and $g \colon A_\sim \to A$ such that $(\pi, g) \colon A \leftrightharpoons A_\sim$ then,*

1. *$g([a]) = \max([a])$ for all $a \in A$.*
2. *$[a_1] \leq_{A_\sim} [a_2]$ if and only if $a_1 \leq_A \max([a_2])$ for all $a_1, a_2 \in A$.*
3. *If $a_1 \leq_A a_2$ then $\max([a_1]) \leq_A \max([a_2])$ for all $a_1, a_2 \in A$.*

Proof.

1. By Theorem 1, we have $g([a]) = \max \pi^{-1}([a]^{\downarrow})$. Now, Lemma 1 leads to $\max \pi^{-1}([a]^{\downarrow}) = \max \pi^{-1}([a]) = \max([a])$.

   There is a slight abuse of notation in that $[a]$ is sometimes considered as a single element, i.e. one equivalence class of the quotient set, and sometimes as the set of elements of the equivalence class. The context helps to clarify which meaning is intended in each case.

2. By the adjointness of (π, g), definition of π, and the previous item we have the following chain of equivalences

$$[a_1] \leq_{A_\sim} [a_2] \text{ if and only if } \pi(a_1) \leq_{A_f} [a_2]$$
$$\text{if and only if } a_1 \leq_A g([a_2])$$
$$\text{if and only if } a_1 \leq_A \max([a_2])$$

3. Finally, since π and g are isotone maps, $a_1 \leq_A a_2$ implies $[a_1] \leq_{A_f} [a_2]$, and $g([a_1]) \leq_A g([a_2])$, therefore $\max([a_1]) \leq_A \max([a_2])$ by item 1 above. □

Continuing with the analysis of the decomposition, we naturally arrive to the following result.

Lemma 4. *Consider a poset $(A, \leq_A)$ and a bijective mapping $\varphi \colon A \to B$, then there exists a unique ordering relation in B, which is defined as $b \leq_B b'$ if and only if $\varphi^{-1}(b) \leq_A \varphi^{-1}(b')$, such that $(\varphi, \varphi^{-1}) \colon A \leftrightharpoons B$.*

Proof. Straightforward.   □

As a consequence of the previous results, we have established necessary and sufficient conditions ensuring the existence and uniqueness of right adjoint for any surjective mapping f from a poset A to an unstructured set B.

The third part of this section is devoted to considering the case in which f is not surjective. In this case, in general, there are several possible orderings on B which allow to define the right adjoint. The crux of the construction is related to the definition of an order-embedding of the image into the codomain set.

More generally, the idea is to extend an ordering defined just on a subset of a set to the whole set.

Definition 3. *Given a subset $X \subseteq B$, and a fixed element $m \in X$, any preordering $\leq_X$ in X can be extended to a preordering $\leq_m$ on B, defined as the reflexive and transitive closure of the relation $\leq_X \cup \{(m, y) \mid y \notin X\}$.*

Note that the relation above can be described as, for all $x, y \in B$, $x \leq_m y$ if and only if some of the following holds:

(a) $x, y \in X$ and $x \leq_X y$

(b) $x \in X, y \notin X$ and $x \leq_X m$

(c) $x, y \notin X$ and $x = y$

It is not difficult to check that if the initial relation $\leq_X$ is an ordering relation, then $\leq_m$ is an ordering as well. Formally, we have

Lemma 5. *Given a subset $X \subseteq B$, and a fixed element $m \in X$, then $\leq_X$ is an ordering in X if and only if $\leq_m$ is an ordering on B.*

Proof. Just some routine computations are needed to check that $\leq_m$ is antisymmetric using the properties of $\leq_X$.

Conversely, if $\leq_m$ is an ordering, then $\leq_X$ is an ordering as well, since it is a restriction of $\leq_m$. □

Lemma 6. *Let X be a subset of B, consider a fixed element $m \in X$, and an ordering $\leq_X$ in X. Define the mapping $j_m \colon (B, \leq_m) \to (X, \leq_X)$ as*

$$j_m(x) = \begin{cases} x & \text{if } x \in X \\ m & \text{if } x \notin X \end{cases}$$

Then, $(i, j_m) \colon (X, \leq_X) \leftrightharpoons (B, \leq_m)$ where i denotes the inclusion $X \hookrightarrow B$.

Proof. It follows easily by routine computation. □

Theorem 2. *Given a poset $(A, \leq_A)$ and a map $f \colon A \to B$, let $\equiv_f$ be the kernel relation. Then, there exists an ordering $\leq_B$ in B and a map $g \colon B \to A$ such that $(f, g) \colon A \leftrightharpoons B$ if and only if*

1. *There exists $\max([a])$ for all $a \in A$.*
2. *For all $a_1, a_2 \in A$, $a_1 \leq_A a_2$ implies $\max([a_1]) \leq_A \max([a_2])$.*

Proof. Assume that there exists an adjunction $(f, g) \colon A \leftrightharpoons B$ and let us prove items 1 and 2.

Given $a \in A$, item 1 holds because of the following chain of equalities, where the first equality follows from Theorem 1, the second one follows from Lemma 1, and the third because of the definition of $[a]$:

$$g(f(a)) = \max f^{-1}(f(a)^\downarrow) = \max f^{-1}(f(a)) = \max([a]) \tag{1}$$

Now, item 2 is straightforward, because if $a_1 \leq_A a_2$ then, by isotonicity, $f(a_1) \leq_B f(a_2)$ and $g(f(a_1)) \leq_A g(f(a_2))$. Therefore, by Equation (1) above, $\max([a_1]) \leq_A \max([a_2])$.

Conversely, given $(A, \leq_A)$ and $f \colon A \to B$ and items 1 and 2, let us prove that f is the left adjoint of a mapping $g \colon B \to A$. To begin with, consider the canonical decomposition of f through the quotient set A_f of A wrt $\equiv_f$, see below, where $\pi\colon A \to A_f$ is the natural projection, $\pi(a) = [a]$, $\varphi([a]) = f(a)$, and $i(b) = b$ is the inclusion mapping.

$$g = \max \circ \varphi^{-1} \circ j_m$$

$$
\begin{array}{ccc}
A & \xrightarrow{\ f\ } & B \\
\max \left\uparrow\downarrow\right\pi & & i \left\uparrow\downarrow\right j_m \\
A_f & \xleftarrow[\varphi^{-1}]{\varphi} & f(A)
\end{array}
$$

Firstly, by Lemma 2, using conditions 1 and 2, and the fact that $[a] = \pi(a)$, we obtain that $(\pi, \max)\colon A \leftrightharpoons A_f$.

Moreover, since the mapping $\varphi\colon A_f \to f(A)$ is bijective, we can apply Lemma 4 in order to induce an ordering $\leq_{f(A)}$ on $f(A)$ such that we have another adjunction, the pair $(\varphi, \varphi^{-1})\colon A_f \leftrightharpoons f(A)$.

Then, considering an arbitrary element $m \in f(A)$, the ordering $\leq_{f(A)}$ also induces an ordering $\leq_m$ on B, as stated in Lemma 5, and a map $j_m\colon B \to f(A)$ such that $(i, j_m)\colon f(A) \leftrightharpoons B$.

Finally, the composition $g = \max \circ \varphi^{-1} \circ j_m\colon B \to A$ is such that (f, g) is an adjunction.   □

We end this section with two counterexamples showing that the conditions in the theorem cannot be removed.

Let $A = \{a, b, c\}$ and $B = \{d, e\}$ be two sets and $f\colon A \to B$ defined as $f(a) = d$ and $f(b) = f(c) = e$.

Condition 1 cannot be removed: Consider $(A, \leq)$ where $a \leq b, a \leq c$ and b, c not related. Then $[b] = \{b, c\}$ and there does not exist $\max([b])$.

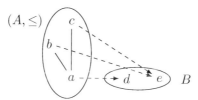

The right adjoint does not exist because $\max f^{-1}(e^{\downarrow})$ would not be defined for any ordering in B.

Condition 2 cannot be removed: Consider $(A, \leq)$, where $b \leq a \leq c$.

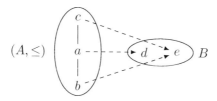

In this case, Condition 1 holds, since there exist both max[a] = a and max[b] = c, but Condition 2 clearly does not. Again, the right adjoint does not exist because f will never be isotone in any possible ordering defined in B.

4 Conclusions

Given a mapping $f\colon A \to B$ from a partially ordered set A into an unstructured set B, we have obtained necessary and sufficient conditions which allow us for defining a suitable partial ordering relation on B such that there exists a mapping $g\colon B \to A$ such that the pair of mappings (f, g) forms an adjunction between partially ordered sets. The results obtained in Theorem 2, regardless of the fact that the proof is not exactly straightforward, are in consonance with the intuition and the well-known facts about Galois connections.

A first source of future work is to consider A to be a preordered set, and try to find an isotone Galois connection between preorders. In this context, there are no clear candidate conditions for the existence of the preorder relation in B, since the notion of maximum is not unique in a preordered setting due to the absence of antisymmetry.

Another topic for future work is related to obtaining a fuzzy version of the obtained result, in the sense of considering either fuzzy Galois connections [2,4, 15,21] or considering the framework of fuzzy posets and fuzzy preorders.

References

1. Antoni, L., Krajči, S., Krídlo, O., Macek, B., Pisková, L.: On heterogeneous formal contexts. Fuzzy Sets and Systems 234, 22–33 (2014)
2. Bělohlávek, R.: Fuzzy Galois connections. Math. Logic Q. 45(4), 497–504 (1999)
3. Bělohlávek, R., Konečný, J.: Concept lattices of isotone vs. antitone Galois connections in graded setting: Mutual reducibility revisited. Information Sciences 199, 133–137 (2012)
4. Bělohlávek, R., Osička, P.: Triadic fuzzy Galois connections as ordinary connections. In: IEEE Intl Conf. on Fuzzy Systems (2012)
5. Blyth, T.S.: Lattices and Ordered Algebraic Structures. Springer (2005)
6. Butka, P., Pócs, J., Pócsová, J.: On equivalence of conceptual scaling and generalized one-sided concept lattices. Information Sciences 259, 57–70 (2014)
7. Castellini, G., Koslowski, J., Strecker, G.: Closure operators and polarities. Annals of the New York Academy of Sciences 704, 38–52 (1993)
8. Cohen, D.A., Creed, P., Jeavons, P.G., Živný, S.: An Algebraic theory of complexity for valued constraints: Establishing a Galois Connection. In: Murlak, F., Sankowski, P. (eds.) MFCS 2011. LNCS, vol. 6907, pp. 231–242. Springer, Heidelberg (2011)
9. Denecke, K., Erné, M., Wismath, S.L.: Galois connections and applications, vol. 565. Springer (2004)

10. Díaz, J.C., Medina, J.: Multi-adjoint relation equations: Definition, properties and solutions using concept lattices. Information Sciences 253, 100–109 (2013)
11. Dubois, D., Prade, H.: Possibility theory and formal concept analysis: Characterizing independent sub-contexts. Fuzzy Sets and Systems 196, 4–16 (2012)
12. Dzik, W., Järvinen, J., Kondo, M.: Intuitionistic propositional logic with Galois connections. Logic Journal of the IGPL 18(6), 837–858 (2010)
13. Dzik, W., Järvinen, J., Kondo, M.: Representing expansions of bounded distributive lattices with Galois connections in terms of rough sets. International Journal of Approximate Reasoning 55(1), 427–435 (2014)
14. Erné, M., Koslowski, J., Melton, A., Strecker, G.E.: A primer on Galois connections. Annals of the New York Academy of Sciences 704, 103–125 (1993)
15. Frascella, A.: Fuzzy Galois connections under weak conditions. Fuzzy Sets and Systems 172(1), 33–50 (2011)
16. García-Pardo, F., Cabrera, I.P., Cordero, P., Ojeda-Aciego, M.: On Galois Connections and Soft Computing. In: Rojas, I., Joya, G., Cabestany, J. (eds.) IWANN 2013, Part II. LNCS, vol. 7903, pp. 224–235. Springer, Heidelberg (2013)
17. Järvinen, J.: Pawlak's information systems in terms of Galois connections and functional dependencies. Fundamenta Informaticae 75, 315–330 (2007)
18. Järvinen, J., Kondo, M., Kortelainen, J.: Logics from Galois connections. Int. J. Approx. Reasoning 49(3), 595–606 (2008)
19. Kan, D.M.: Adjoint functors. Transactions of the American Mathematical Society 87(2), 294–329 (1958)
20. Kerkhoff, S.: A general Galois theory for operations and relations in arbitrary categories. Algebra Universalis 68(3), 325–352 (2012)
21. Konecny, J.: Isotone fuzzy Galois connections with hedges. Information Sciences 181(10), 1804–1817 (2011)
22. Medina, J.: Multi-adjoint property-oriented and object-oriented concept lattices. Information Sciences 190, 95–106 (2012)
23. Melton, A., Schmidt, D.A., Strecker, G.E.: Galois connections and computer science applications. In: Poigné, A., Pitt, D., Rydeheard, D., Abramsky, S. (eds.) Category Theory and Computer Programming. LNCS, vol. 240, pp. 299–312. Springer, Heidelberg (1986)
24. Mu, S.-C., Oliveira, J.N.: Programming from Galois connections. The Journal of Logic and Algebraic Programming 81(6), 680–704 (2012)
25. Ore, Ø.: Galois connections. Trans. Amer. Math. Soc. 55, 493–513 (1944)
26. Restrepo, M., Cornelis, C., Gómez, J.: Duality, conjugacy and adjointness of approximation operators in covering-based rough sets. International Journal of Approximate Reasoning 55(1), 469–485 (2014)
27. Wolski, M.: Galois connections and data analysis. Fundamenta Informaticae 60, 401–415 (2004)

Fuzzy Core DBScan Clustering Algorithm

Gloria Bordogna[1] and Dino Ienco[2]

[1] CNR IdA detached at IREA, Via Bassini 15, Milano, Italy
bordogna.g@irea.cnr.it
[2] Irstea, UMR TETIS, Montpellier, France
LIRMM, Montpellier, France
dino.ienco@irstea.fr

Abstract. In this work we propose an extension of the *DBSCAN* algorithm to generate clusters with fuzzy density characteristics. The original version of *DBSCAN* requires two parameters (*minPts* and ϵ) to determine if a point lies in a dense area or not. Merging different dense areas results into clusters that fit the underlined dataset densities. In this approach, a single density threshold is employed for all the datasets of points while the distinct or the same set of points can exhibit different densities. In order to deal with this issue, we propose *Approx Fuzzy Core DBSCAN* that applies a soft constraint to model different densities, thus relaxing the rigid assumption used in the original algorithm. The proposal is compared with the classic *DBSCAN*. Some results are discussed on synthetic data.

1 Introduction

Density based clustering algorithms have a wide applicability in data mining. They apply a local criterion to group objects: clusters are regarded as regions in the data space where the objects are dense, and which are separated by regions of low object density (noise). Among the density based clustering algorithms *DBSCAN* is very popular due both to its low complexity and its ability to detect clusters of any shape, which is a desired characteristics when one does not have any knowledge of the possible clusters' shapes, or when the objects are distributed heterogenously such as along paths of a graph or a road network. Nevertheless, to drive the process, this algorithm needs two numeric input parameters, *minPts* and ϵ which together define the desired density characteristics of the generated clusters. Specifically, *minPts* is a positive integer specifying the minimum number of objects that must exist within a maximum distance ϵ of the data space in order for an object to belong to a cluster.

Since *DBSCAN* is very sensible to the setting of these input parameters they must be chosen with great accuracy [2] by considering both the scale of the dataset and the closeness of the objects in order not to affect too much both the speed of the algorithm and the effectiveness of the results. To fix the right values of these parameters one generally engages an exploration phase of trials and errors in which the clustering is run several times with distinct values of the parameters.

A. Laurent et al. (Eds.): IPMU 2014, Part III, CCIS 444, pp. 100–109, 2014.

In fact, a common drawback of all crisp flat clustering algorithms used to group objects whose distribution has a faint and smooth density profile is that they draw crisp boundaries to separate clusters, which are often somewhat arbitrary.

There are applications in which the positions of the objects is ill-known, such as in the case of databases of moving objects, whose locations are recorded at fixed timestamps with uncertainty on their intermediate positions, or in the case of objects appearing in remote sensing images having a coarse spatial resolution so that a pixel is much greater than the object dimension, and thus uncertainty is implied when one has to detect the exact position of the object within an area. Last but not least, when one has to detect communities of users in a social network, while one can specify easily that the users must have at most a given number of degrees of separation on the network, it my be questionable to define the precise minimum number of elements defining a social community.

In this contribution we investigate an extension of the *DBSCAN* algorithm defined within the framework of fuzzy set theory whose aim is to detect fuzzy clusters with approximate density characteristics. In the literature a fuzzy extension of *DBSCAN* has been proposed, named FN-SBSCAN, with the objective of allowing the specification of an approximate distance between objects instead of a precise ϵ value [6]. Our proposal is dual since we want to leverage the setting of the precise value $minPts$ by allowing the specification of an approximate minimum number of objects for defining a cluster. In our proposal a user drives the clustering algorithm by specifying a soft condition that can be expressed as follow: *group at least approximately $minPts_{min} - minPts_{max}$ objects which are within a maximum distance ϵ "*.

The two values $minPts_{min}, minPts_{max}$ indicate the approximate number of objects defining the density of the clusters and define a soft constraint with a not decreasing membership function on the basic domain of positive integers. The algorithm uses this approximate input to generate clusters with a fuzzy core, i.e., clusters whose elements are associated with a numeric membership degree in [0,1]. Having fuzzy clusters allows several advantages: with a single run of the clustering it is possible to perform a sensitivity analysis by generating several distinct crisp partions obtained by specifying distinct thresholds on the membership degrees of the objects to the clusters. This allows an easy exploration of the spatial distribution of the objects without the need of several runs of the clustering as it occurres when one uses the classic *DBSCAN*. The contribution first recalls the classic *DBSCAN* algorithm; then, the *Approx Fuzzy Core DBSCAN* is defined. Section 4. discusses the results obtained by the extended algorithm; section 5 compared the proposal with related literature, and the conclusions summarize the main achievements.

2   Classic DBScan Algorithm

For sake of clarity in the following we will consider a set of objects represented by distinct points defined in a multidimensional domain. These objects can be either actual entities located on the bidimensional spatial domain such as cars, taxi cabs, airplanes, or virtual entities, such as web pages and tweets represented in the virtual n-dimensional space of the terms they contain. DBSCAN can be applied to group these objects based on their local densities in the space. This makes it possible to identify traffic jams of cars on the roads, or to identify groups of web pages and tweets that deal with same topics.

DBSCAN assigns points of a spatial domain defined on RxR to particular clusters or designates them as statistical noise if they are not sufficiently close to other points. DBSCAN determines cluster assignments by assessing the local density at each point using two parameters: distance (ϵ) and minimum number of points ($minPts$). A single point which meets the minimum density criterion, namely that there are $minPts$ located within distance ϵ, is designated a core point. Formally, Given a set P of N points $p_i = (x_{i_1}, x_{i_2}, ..., x_{i_n})$ with x_{i_j} defined on the n-dimensional domain R^n. $p \in P$ is a core point if at least a minimum number $minPts$ of points $p_1, , p_{minPts} \in P \exists s.t ||p_j - p|| < \epsilon$, Two core points p_i and p_j with $i, j s.t ||p_i - p_j|| < \epsilon$ define a cluster c, $p_i, p_j \in c$ and are core points of c, i.e., $p_j, p_j \in core(c)$ All not core points within the maximum distance ϵ from a core point are considered non-core members of a cluster, and are boundary or border points: $p \notin core(c)$ is a boundary point of c if $\exists p_i \in core(c)$ with $||p - p_i|| < \epsilon$. Finally, points that are not part of a cluster are considered noise: $p \notin core(c)$ are noise if $\forall c, \nexists p_i \in core(c)$ with $||p - p_i|| < \epsilon$. In the following the classic DBSCAN algorithm is described:

Algorithm 1. $DBSCAN(D, \epsilon, MinPts)$

Require: P: dataset of points
Require: ϵ: the maximum distance around a point defining the point neighbourhood
Require: $MinPts$: density, in points, around a point to be considered a core point
  1. $C = 0$
  2. $Clusters = \emptyset$
  3. **for all** $p \in P$ s.t. p is unvisited **do**
  4.   mark p as visited
  5.   neighborsPts = regionQuery(p, ϵ)
  6.   **if** $(sizeof(neighborsPts) <= MinPts)$ **then**
  7.   mark p as NOISE
  8.   **else**
  9. C = next cluster
 10. $Clusters = Clusters \cup expandCluster(p, neighborsPts, C, \epsilon, MinPts)$
 11.   **end if**
 12. **end for**
 13. **return** $Clusters$

---

3   Generating Clusters with Fuzzy Cores

The extension of the classic $DBSCAN$ algorithm we propose, named fuzzy core DBSCAN, is obtained by considering crisp the distance, as in the classic approach, and by introducing an approximate value of the desired cardinality of the neighborhood of a point $minPts$. This can be done by substituting the numeric value $minPts$ with a soft constraint defined by a non decreasing membership function on the domain of the positive integers. This soft constraint specifies

---

Algorithm 2. $expandCluster(p, neighborsPts, C, \epsilon, MinPts)$

Require: p: the point just marked as visited
Require: $neighborsPts$: the neighborhood of p
Require: C: the actual cluster
Require: ϵ the distance around a point to compute its density
Require: $MinPts$: density, in points, defining the minimum cardinality of the neighborhood of a
    point to be considered a core point
  1. add p to cluster C
  2. **for all** $p' \in neighborsPts$ **do**
  3.   **if** p' is not visited **then**
  4.   mark p' as visited
  5. $neighborsPts' = $ regionQuery(p', ϵ)
  6.   **if** $sizeof(neighborsPts') > MinPts$ **then**
  7. $neighborsPts = neighborsPts \cup neighborsPts'$
  8.   **end if**
  9.   **end if**
10.   **if** p' is not yet member of any cluster **then**
11.   add p' to cluster C
12.   **end if**
13. **end for**
14. **return** C

---

the approximate number of points that are required in the neighborhood of a point for generating a fuzzy core of a cluster. Let us define the piecewise linear membership function as follows:

$$\mu_{minP}(x) \begin{cases} 1, & \text{if } x \geq Mpts_{Max} \\ \frac{x - Mpts_{Min}}{Mpts_{Max} - Mpts_{Min}}, & \text{if } Mpts_{Min} < x < Mpts_{Max} \\ 0, & \text{if } x \leq Mpts_{Min} \end{cases} \quad (1)$$

This membership function gives the value 1 when the number x of elements in the neighbourhood of a point is greater than $Mpts_{Max}$, a value 0 when x is below $Mpts_{Min}$ and intermediate values when x is in between $Mpts_{Min}$ and pts_{Max}.

Since users may find it difficult to specify the two values $Mpts_{Min}$ and $Mpts_{Max}$ when they are not aware of the total number of objects involved in the process, they can specify two percentage values, $\%Mpts_{Min}$ and $\%Mpts_{Max}$ which are then converted into $Mpts_{Min}$ and $Mpts_{Max}$ as follows:

$Mpts_{Min} = $ round$(\%Mpts_{Min} * N$ and $pts_{Max} = $round$(\%Mpts_{Min} * N$, in which N is the total number of objects and $round(m)$ returns the closest integer to m.

Let us redefine the fuzzy core . Given a set P of N objects represented by N points in the n-dimensional domain R^n $p_1, p_2, ...p_N$, where p_i has the coordinates $x_{i_1}, x_{i_2}, ..., x_{i_n}$.

Given a point $p \in P$, if x points $p_i \ \exists$ in the neighbourhood of point p , i.e., with $\|p_i - p\| < \epsilon$, s.t. $\mu_{minP}(x) > 0$ the p is a fuzzy core point with membership degree to the fuzzy core given by $Fuzzycore(p) = \mu_{MinP}(x)$ If two fuzzy core points p_i, p_j ($Fuzzycore(p_i) > 0$ and $Fuzzycore(p_j) > 0$) $\exists$ with $i \neq j$ s.t. $\|p_i - p\| < \epsilon$ then they define a cluster c, $p_i, p_j \in c$, and are fuzzy core points of c, i.e., $p_i, p_j \in fuzzycore(c)$ with membership degrees $Fuzzycore_c(p_i)$ and $Fuzzycore_c(p_j)$.

A point p of a cluster that is not a fuzzy core point is a boundary or border point if it satifies the following: Given $p \notin fuzzycore(c)$ if $\exists p_i \in fuzzycore(c)$, i.e., with membership degree $fuzzycore_c(p_i) > 0$, s.t. $\|p_i - p\| < \epsilon$ then p gets a membership degree to c defined as: $\mu_c(p) = max_{p_i \in fuzzycore(c)} fuzzycore_c(p_i)$

Finally, points p that are not part of a cluster are considered noise: $\forall c$ if $\nexists p_i \in fuzzycore(c)$ s.t. $\|p_i - p\| < \epsilon$, then p is noise.

Notice that the points belonging to a cluster c gets distinct membership values to the cluster reflecting the number of their neighbours within a maximum distance ϵ. This definition allows generating fuzzy clusters with a fuzzy core, where the membership degrees represent the variable cluster density.

Moreover, a boundary point p can partially belong to a single cluster c since its membership degree is upperbounded by the maximum membership degree of its neighbouring fuzzy core points. Notice, that this algorithm does not generate overlapping fuzzy clusters, but the support of the fuzzy clusters is still a crisp partition as in the classic DBSCAN:

$c_i \cap c_j = \emptyset$

Further property, the fuzzy core DBSCAN reduces to the classic DBSCAN when the input values $MinPts_{Min} = MinPts_{Max}$: in this case the fuzzy core DBSCAN produces the same results of the classic DBSCAN with $minPts = MinPts_{Min} = MinPts_{Max}$ and same distance ϵ. In fact, the level based soft condition imposed by μ_{minP} is indeed a crisp condition $\mu_{MinP}(x) \in 0, 1$ on the minimum number of points defining the local density of the neighbourhood: $\mu_{minP} = 0$ when the number of points within a maximum distance ϵ of any point p is less than $minPts = MinPts_{Min} = MinPts_{Max}$, on the contrary $\mu_{minP} = 1$. In this case, the membership degrees of all fuzzy core points is 1, and thus the fuzzy core reduces to a crisp core as in the classic DBSCAN.

The border points are thus defined as in the classic approach too, since their membership degree is the maximum of their closest core points, i.e., it is always 1.

The Fuzzy procedure is sketched in Algorithms 3 and 4. Considering the outer loop of the process (Algorithm 3), the difference with the original version (Algorithm 1) lies at line 6.

In the fuzzy version, a point is marked as *NOISE* if its neighborhood size is less than or equal to $MinPts_{Min}$ otherwise it will be a fuzzy core point with a given membership value. Once the point is recognized as fuzzy core point the procedure $expandClusterFuzzyCore$ is called (Algorithm 4).

As in the classical *DBSCAN*, this procedure is devoted to find all the reachable points from p and to mark them as core or border points. In the original version the assignment of the point p is crisp while we introduce a fuzzy assignment (line 1) modelled by the fuzzy function $\mu_{MinP}()$. The same function is employed when a new fuzzy core point is detected (line 8). Also in this case, firstly we verify the density around a given point p' w.r.t. $MinPts_{Min}$ and then, if the point verifies the soft constraint, we add the point to the fuzzy core of cluster C with its associated membership value. Differently from the original version, line 10 is only devoted to detect border points not yet assigned to any cluster.

Algorithm 3. *Approx Fuzzy Core DBSCAN*($D,\epsilon,MinPts_{M}in,MinPts_{Max}$)

Require: P: dataset of points
Require: ϵ: the maximum distance around a point defining the point neighbourhood
Require: $MinPts_{Min}, MinPts_{Max}$: soft constraint interval for the density around a point to be considered a core point to a degree
1. $C = 0$
2. $Clusters = \emptyset$
3. **for all** $p \in P$ s.t. p is unvisited **do**
4.   mark p as visited
5.   neighborsPts = regionQuery(p,ϵ)
6.   **if** $\big(sizeof(neighborsPts) \leq MinPts_{Min}\big)$ **then**
7.   mark p as NOISE
8.   **else**
9. C = next cluster
10. $Clusters = Clusters \cup expandClusterFuzzyCore(p, neighborsPts, C, \epsilon, MinPts_{Min}, MinPts_{Max})$
11.   **end if**
12. **end for**
13. **return** $Clusters$

Algorithm 4. $expandClusterFuzzyCore(p, neighborsPts, C, \epsilon, MinPts_{Min}, MinPts_{Max})$

Require: p: the point just marked as visited
Require: $neighborsPts$: the points in the neighbourhood of p
Require: C: the actual cluster
Require: ϵ the distance around a point to compute its density
Require: $MinPts_{Min}, MinPts_{Max}$: soft constraint interval for the density around a point to be considered a core point
1. add p to C with membership $Fuzzycore(p) = \mu_{MinP}(|neighborsPts|)$
2. **for all** $p' \in neighborsPts$ **do**
3.   **if** p' is not visited **then**
4.   mark p' as visited
5. $neighborsPts' = $ regionQuery(p',ϵ)
6.   **if** $sizeof(neighborsPts') > MinPts_{Min}$ **then**
7. $neighborsPts = neighborsPts \cup neighborsPts'$
8.   add p' to C with membership $Fuzzycore(p') = \mu_{MinP}(|neighborsPts'|)$
9.   **end if**
10.   **if** p' is not yet member of any cluster **then**
11.   add p' to C (as border point)
12.   **end if**
13.   **end if**
14. **end for**
15. **return** C

---

4   Experiments

In this section we discuss the properties of our proposed algorithm by showing the results we can obtain when applying the fuzzy core DBSCAN to a synthetic data set in a bidimensional domain. Figure 1 (a) and 1(b) depict with distinct colors the clusters identified by the classic DBSCAN and the fuzzy core DBSCAN algorithms respectively. It can be noticed that the classic approach provided in input with the parameters $minPts = 9$ and $\epsilon = 12$ divides into two distinct clusters, cluster 1 and cluster 6, and also cluster 5 and cluster 3, while the fuzzy core clustering identifies as two clusters, cluster 2 and cluster 3 respectively. It can also be noticed that with the fuzzy DBSCAN the obtained conflated clusters have a variable density profile.

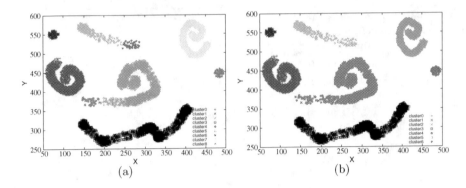

(a)   (b)

Fig. 1. Results of a) *DBSCAN* b) *Approx Fuzzy Core DBSCAN*. We set Mpts = 9 and $\epsilon = 12$ while for *Approx Fuzzy Core DBSCAN* the soft constraint over the minimum number of points ranges from 7 to 12 and ϵ is always equal to 12.

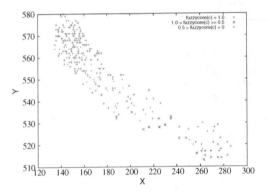

Fig. 2. Inspection of a cluster generated with the *Approx Fuzzy Core DBSCAN* approach (*Mpts*=(9,12), ϵ=12]). For the light blue cluster (Cluster2) shown in Figure 1b we visualize the fuzzy core points grouped in three category: fuzzy cores with membership equal to 1 (red cross), fuzzy cores with membership lesser than 1 and greater or equal to 0.5 (blue X) and fuzzy cores with membership lesser than 0.5 and bigger than 0 (green star).

This happens in the classic approach since the input parameters were not chosen accurately. To obtain a correct partition we would have to choose a smaller value *minPts* < 9. Nevertheless, to find the correct parameters we might have to re-run the classic DBSCAN several times before obtaining a satisfactory solution. Conversely, just with a single run of *Approx Fuzzy Core DBSCAN* we can identify the appropriate clusters, also the ones, cluster 2 and cluster 3, that are characterised by a variable density of their core.

Moreover, having the membership degrees associated with each pair *object − cluster* we can inspect the partitions that we can obtain by associating with each range of membership degrees a distinct color. An example of this analysis is depicted in Figure 2. We plot the fuzzy core points of cluster 2 of Figure 1(b).

We avoid to plot border point as they are not influenced by the fuzzification process. Cluster 2 has the profile of a comet with a dense nucleous on the upper left side and a faint tail on the lower right side. The membership degrees to the fuzzy core are discretised into three bins: fuzzy core points with a membership value equals to 1.0, fuzzy core points with a membership in the range (1.0,0.5] and fuzzy core points with a membership in the interval (0.5,0). This way we quantify how the density distribution varies, having all full core points in the nucleus, and partial fuzzy core points on the tail.

If we apply a threshold so as to detect only the points with full membership 1 to the fuzzy core, we can observe that the tail of the comet splits into two parts as with the classic DBSCAN result in Figure 1(a).

5   Related Work

Clustering algorithms can be grouped into five main categories such as hierarchical, partition-based, grid-based, density based , and model-based and, furthermore, one can distinguish among crisp and soft (fuzzy or probabilistic) clustering according to the fact that elements belong to clusters with a full or a partial membership degree, in this latter case with the possibility for and element to simultaneously belong to several clusters. Among the partitional density based clustering algorithms, $DBSCAN$ is the most popular one due to its ability to detect irregularly shaped clusters by copying with noise data sets. Nevertheless it performances are strongly dependent of the parameters setting and this is the main reason that lead to its soft extensions. In the literature there have been a few extensions of the $DBSCAN$ algorithm in order to detect fuzzy clusters, as also discussed in the recent paper [3] in which the authors report a survey of the main density based clustering algorithms. The most cited paper [6] proposes a fuzzy extension of the $DBSCAN$, named $FN - DBSCAN$ (fuzzy neighborhood $DBSCAN$), whose main characteristic is to use a fuzzy neighborhood relation. In this approach the authors address the difficulty of the user in setting the values of the input parameters when the distances of the points are in distinct scales. Thus, they first normalize the distances between all points in [0,1], and then they allow specifying distinct membership functions on the distance to delimit the neighborhood of points, i.e., the decaying of the membership degree as a function of the distance. Then, they select as belonging to the fuzzy neighbourhood of a point only those points having a minimum membership degree greater than zero. This extension of $DBSCAN$ uses a level-based neighborhood set instead of a distance-based neighborhood set, and it uses the concept of fuzzy cardinality instead of classical cardinality for identifying core points. This last choice causes the creation (with the same run of the algorithm) of both fuzzy clusters with cores having many sparse points and fuzzy clusters with cores having only a few close points. Thus the density characteristic of the generated clusters is very heterogeneous. Furthermore in this extensions points can belong to several clusters with distinct membership degree. This extension of the $DBSCAN$ can be considered dual to our proposal since we fuzzify the minimum number of points in the neighbourhood of a point to be considered as part of the fuzzy core, while the maximum distance ϵ is still crisp, thus generating fuzzy non overlapping clusters. This kind of fuzziness of clusters reflects the variable densities of the clusters' cores: as a consequence the membership degree of a point to the fuzzy core depends on the number of points in its crisp neighborhood, and thus a point is assigned to only one cluster, the one with the greatest number of core points in its neighbourhood. With this semantics of fuzziness we want to favor the growing of denser clusters with respect to fainter ones. The utility of the fuzzy $DBSCAN$ is pointed out in the paper [4] where the authors use FN-DBSCAN in conjunction with the computation of the convex hull of the generated fuzzy clusters to derive connected footprints of entities with arbitrary shape. Having fuzzy clusters allows generating isolines footprints. An efficient implementation is proposed in [2]. It tackles the problem of clustering a huge number of objects strongly affected by noise when the scale distributions of objects are heterogeneous. To remove noise they first map the distance of any point from its k-neighbours and rank the distance values in decreasing order; then they determine the threshold θ on the distance which corresponds to the first minimum on the ordered values. All points in the first ranked positions having a distance above the thresholds θ are noise points and are removed, while the remaining will belong to a cluster. These latter points are clustered with the classic $DBSCAN$ by providing as input parameters $minPts = K$ and $\epsilon = \theta$. Another motivation of defining fuzzy $DBSCAN$ is to cluster objects whose position is ill-known, as in the paper [5] where the authors propose the FDBSCAN algorithm in which a fuzzy distance measure is defined as the probability that an object is directly density-reachable from another objects. This problem could be modeled in our approach by allowing the neighbouhood of any object as consisting of an approximate number of other objects, thus capturing the uncertainty on the positions of the moving objects, which could be inside or outside the radius ϵ. This way, the grouping of moving objects could be modeled with a simpler approach with respect to the proposal [5] that use fuzzy distance and probability distributions. Finally, the most recent soft extension of $DBSCAN$ has been proposed in [1] where the authors combine the classic $DBSCAN$ with the fuzzy C-means algorithm. They detect seeds points by the classic DBSCAN and in a second phase they compute the degrees of membership to the clusters around the seeds by relying on the fuzzy C-means clustering algorithm. Nevertheless, this extension has the objective of determining seeds to feed the Fuzzy C-Means, like the approximate clustering algorithm based on the mountain method [7], which is different from our proposal since we do not grow the boundary by applying the fuzzy C-means, but rely on the DBSCAN. The result is not a fuzzy partition but is still a crip partition of the elements into distinct clusters, even if each element can belong to a single cluster with a distinct degree. In fact our aim is tofold: firstly of all we want to subsume several runs of the classic DBSCAN with different paramanater settings with a single run of our algorithm, and secondly we want to give a preference to the grouth of clusters with denser core with respect to clusters with fainter core.

6   Conclusion

In this work we present a new fuzzy clustering algorithm *Approx Fuzzy Core DBSCAN* that extends the original *DBSCAN* method. The main characteristics of this algorithm is to introduce a soft constraint to specify the approximate number of points that must exists in the neighbourhood of a point for generating a cluster. Specifically, *Approx Fuzzy Core DBSCAN* allows assigning a core point to a cluster with a membership value so clusters can contain core points with different membership values representing this way the distinct local densities. Beside leveraging the specification of the precise input, the proposal supplies with a single run of the clustering algorithm a solution that summarises multiple runs of the original classic DBSCAN algorithm: specifically, all the runs with a value of *MinPoints* belonging to the support of the soft constraint. The visual quality of the results yielded by *Approx Fuzzy Core DBSCAN* is tested in the experimental section. A synthetic dataset is employed to highlight the benefit of representing the fuzziness of the local density around points in order to obtain meaningful results. As a future direction we would employ soft constraints also over the maximum distance ϵ and, then, over both parameters of the *DBSCAN* simultaneously in order to be more flexible over data exhibit heterogenous densities.

References

1. Smiti, A., Eloudi, Z.: Soft dbscan: Improving dbscan clustering method using fuzzy set theory. Human System Interaction 1, 380–385 (2013)
2. Ester, M., Kriegel, H.P., Sander, J., Xu, X.: A density-based algorithm for discovering clusters in large spatial databases with noise. In: KDD, vol. 160, pp. 226–231 (1996)
3. Ulutagay, G., Nasibov, E.N.: Fuzzy and crisp clustering methods based on the neighborhood concept: A comprehensive review. Journal of Intelligent and Fuzzy Systems 23, 1–11 (2012)
4. Parker, J.K., Downs, J.A.: Footprint generation using fuzzy-neighborhood clustering. Geoinformatica 17, 283–299 (2013)
5. Kriegel, H.P., Pfeifle, M.: Density-based clustering of uncertain data. In: KDD 2005, vol. 17, pp. 672–677 (2005)
6. Nasibov, E.N., Ulutagay, G.: Robustness of density-based clustering methods with various neighborhood relations. Fuzzy Sets and Systems 160(24), 3601–3615 (2009)
7. Yager, R.R., Filev, D.P.: Approximate clustering via the mountain method. IEEE Transactions on Systems, Man and Cybernetics 24(8), 1279–1284 (1994)

Estimating Null Values in Relational Databases Using Analogical Proportions

William Correa Beltran, Hélène Jaudoin, and Olivier Pivert

University of Rennes 1, Irisa
Technopole Anticipa 22305 Lannion Cedex France
william.correa_beltran@irisa.fr, {jaudoin,pivert}@enssat.fr

Abstract. This paper presents a novel approach to the prediction of null values in relational databases, based on the notion of analogical proportion. We show in particular how an algorithm initially proposed in a classification context can be adapted to this purpose. This work focuses on the case of a transactional database, where attributes are Boolean. The experimental results reported here, even though preliminary, are encouraging since the approach yields a better precision, on average, than the classical nearest neighbors technique.

1 Introduction

In this paper, we propose a novel solution to a classical database problem that consists in estimating null (i.e., unknown) values in incomplete relational databases. Many approaches have been proposed to tackle this issue, both in the database community and in the machine learning community (based on functional dependencies [2], association rules [13,14], classification rules [5], clustering techniques [4], etc). See also [8,9].

Here, we investigate a new idea, that comes from artificial intelligence and consists in exploiting analogical proportions [11]. An analogical proportion is a statement of the form "A is to B as C is to D". As emphasized in [12], analogy is not a mere question of similarity between two objects (or situations) but rather a matter of proportion or relation between objects. An analogical proportion equates a relation between two objects with the relation between two other objects. These relations can be considered as a symbolic counterpart to the case where the ratio or the difference between two similar things is a matter of degree or number. As such, an analogical proportion of the form "A is to B as C is to D" poses an analogy of proportionality by (implicitly) stating that the way two objects A and B, otherwise similar, differ is the same way as the two objects C and D, which are similar in some respects, differ. In other words, transformations applied to A (resp. B) in order to obtain B (resp. A) are the same as that applied to C (resp. D) in order to obtain D (resp. C).

Up to now, the notion of analogical proportion has been studied mainly in artificial intelligence, notably for classification purposes (see, e.g., [1]). Our objective is to exploit it in a database context in order to predict the null values

A. Laurent et al. (Eds.): IPMU 2014, Part III, CCIS 444, pp. 110–119, 2014.

in a tuple t by finding quadruples of items (including t) that are linked by an analogical proportion.

The remainder of the paper is organized as follows. In Section 2, we provide a refresher on the notion of analogical proportion. Section 3 presents the general principle of the approach we propose for estimating null values, inspired by the classification method proposed in [1,6]. Section 4 reports on an experimentation aimed at assessing the performances of the approach and at comparing its results with those obtained using the classical k nearest neighbors technique. Finally, Section 5 recalls the main contributions and outlines perspectives for future work.

2   Refresher on Analogical Proportions

The following presentation is mainly drawn from [7]. An analogical proportion is a statement of the form "A is to B as C is to D". This will be denoted by $(A : B :: C : D)$. In this particular form of analogy, the objects A, B, C, and D usually correspond to descriptions of items under the form of objects such as sets, multisets, vectors, strings or trees. In the following, if the objects A, B, C, and D are tuples having n attributes, i.e., $A = \langle a_1, \ldots, a_n \rangle, \ldots, D = \langle d_1, \ldots, d_n \rangle$, we shall say that A, B, C, and D are in analogical proportion if and only if for each component i an analogical proportion "a_i is to b_i as c_i is to d_i" holds.

We now have to specify what kind of relation an analogical proportion may mean. Intuitively speaking, we have to understand how to interpret "is to" and "as" in "A is to B as C is to D". A may be similar (or identical) to B in some respects, and differ in other respects. The way C differs from D should be the same as A differs from B, while C and D may be similar in some other respects, if we want the analogical proportion to hold. This view is enough for justifying three postulates that date back to Aristotle's time:

- (ID) $(A : B :: A : B)$
- (S) $(A : B :: C : D) \Leftrightarrow (C : D :: A : B)$
- (CP) $(A : B :: C : D) \Leftrightarrow (A : C :: B : D)$.

(ID) and (S) express reflexivity and symmetry for the comparison "as", while (CP) allows for a central permutation.

A *logical proportion* [10] is a particular type of Boolean expression $T(a, b, c, d)$ involving four variables a, b, c, d, whose truth values belong to $\mathbb{B} = \{0, 1\}$. It is made of the conjunction of two distinct equivalences, involving a conjunction of variables a, b on one side, and a conjunction of variables c, d on the other side of $\equiv$, where each variable may be negated. Analogical proportion is a special case of a logical proportion, and it expression is [7]: $(a\bar{b} \equiv c\bar{d}) \wedge (\bar{a}b \equiv \bar{c}d)$. The six valuations leading to truth value 1 are thus $(0, 0, 0, 0)$, $(1, 1, 1, 1)$, $(0, 0, 1, 1)$, $(1, 1, 0, 0)$, $(0, 1, 0, 1)$ and $(1, 0, 1, 0)$.

As noted in [12], the idea of proportion is closely related to the idea of extrapolation, i.e., to guess/compute a new value on the ground of existing values, which is precisely what we intend to do. In other words, if for whatever reason, it is assumed or known that a logical proportion holds between four binary elements, three being known, then one may try to infer the value of the fourth one.

3   Principle of the Approach

3.1   General Idea

The approach we propose is inspired by a method of "classification by analogy" introduced in [1] where the authors describe an algorithm named FADANA (FAst search of the least Dissimilar ANAlogy). This algorithm uses a measure of *analogical dissimilarity* between four objects, which estimates how far these objects are from being in analogical proportion. Roughly speaking, the analogical dissimilarity ad between four Boolean values is the minimum number of bits that have to be switched to get a proper analogy. For instance $ad(1,0,1,0) = 0$, $ad(1,0,1,1) = 1$ and $ad(1,0,0,1) = 2$. Thus, denoting by $\mathcal{A}$ the relation of analogical proportion, we have $\mathcal{A}(a,\ b,\ c,\ d) \Leftrightarrow ad(a,b,c,d) = 0$. One has for instance $ad(0,\ 0,\ 0,\ 1) = 1$ and $ad(0,\ 1,\ 1,\ 0) = 2$.

When, instead of having four Boolean values, we deal with four Boolean vectors in $\mathbb{B}^n$, we add the ad evaluations componentwise to get the analogical dissimilarity, which leads to an integer in the interval $[0,\ 2n]$. This principle has been used in [1] to implement a classification algorithm that takes as an input a training set S of classified items, a new item d to be classified, and an integer k. The algorithm proceeds as follows:

1. For every triple $(a,\ b,\ c)$ of S^3, compute $ad(a,\ b,\ c,\ d)$.
2. Sort these n triples by increasing value of their ad when associated with d.
3. If the k-th triple has the integer value p for ad, then let k' be the greatest integer such that the k'-th triple has the value p.
4. Solve the k' analogical equations on the label of the class. take the winner of the k' votes and allocate this winner as the class of d.

Example 1. Let S be a training set composed of four labelled objects. The set of objects in S are showed in Table 1, where the first column indicates their number or *id*, the columns A_1, A_2, and A_3 their attribute values, and the column cl gives the class they belong to.

Table 1. Training set

id	A_1	A_2	A_3	cl
1	0	0	0	0
2	0	1	0	1
3	0	1	1	1
4	1	1	1	1

Table 2. Computation of **ad**

id	A_1	A_2	A_3	
1	0	0	0	
2	0	1	0	
3	0	1	1	
x	1	0	0	
ad	1	2	1	= 4

Now, let $x \notin S$ be an object to be classified, defined as $A_1 = 1$, $A_2 = 0$, $A_3 = 0$. One first has to compute the **ad** value between x and every possible triple of objects from S. Table 2 shows the **ad** value obtained with the triple (1, 2, 3). Table 3 shows the list of the first 10 triples (ranked according to **ad**).

Table 3. Triples ranked according to **ad**

Combination	a	b	c	d	ad
1)	3	1	4	x	0
2)	3	4	1	x	0
3)	2	3	4	x	1
4)	3	4	2	x	1
5)	2	4	1	x	1
6)	2	1	4	x	1
7)	3	1	2	x	2
8)	3	2	1	x	2
9)	2	1	3	x	2
10)	4	1	3	x	2

Let $k = 5$; all the triples such that their associated **ad** value equals at most that of the 5th tuple (here, 1), are chosen. The triples 1 to 6 are then used to find the class of x. The six corresponding analogical equations are then solved. For instance, combination 2) yields the equation $1 : 1 :: 0 : cl$, leading to $cl=0$. Finally, the class that gets the most votes is retained for d. ◇

3.2   Application to the Prediction of Missing Values

This method may be adapted to the case of null value prediction in a transactional database as follows. Let r be a relation of schema $(A_1, \ldots, A_m)$ and t a tuple of r involving a missing value for attribute A_i: $t[A_i] = NULL$. In order to estimate the value of $t[A_i]$ — that is 0 or 1 in the case of a transactional database — , one applies the previous algorithm considering that A_i corresponds to the class cl to be determined. The training set S corresponds to a sample (of a predefined size) of tuples from r (minus attribute A_i that does not intervene in the calculus of **ad** but represents the "class") involving no missing values. Besides, the attributes A_h, $h \neq i$ such that $t[A_h] = NULL$ are ignored during the computation aimed at predicting $t[A_i]$.

4 Preliminary Experimentation

The main objective of the experimentation was to compare the results obtained using this technique with those produced by other approaches (in particular the "nearest neighbors" technique), thus to extimate its relative effectiveness in terms of *precision* (i.e., of percentage of values correctly predicted).

Let us first emphasize that the performance aspect (in terms of execution time) is not so crucial here, provided of course that the class of complexity remains reasonable. Indeed, the prediction of missing values is to be performed offline. However, this aspect will be tackled in the conclusion of this section, and we will see that different optimization techniques make it possible to significantly improve the efficiency of the algorithm.

4.1 Experimental Results

In order to test the approach, six datasets from the Frequent Itemset Mining Implementations Repository[1], namely *accidents* (30 attributes), *chess* (20 attributes), *mushroom* (32 attributes), *pumsb* (268 attributes), and *connect* (26 attributes)) and one from the UCI machine learning repository[2], *solarFlare* (16 attributes) have been used. These datasets contain categorical attributes that have been binarized. For each dataset, a sample E has been extracted by randomly choosing 1000 tuples (300 in the case of *solarFlare*) from the relation. A subset M of E has been modified, i.e., a certain percentage of values of each of its tuples has been replaced by *NULL*. Then, the FADANA algorithm has been run so as to predict the missing values: for each tuple d involving at least a missing value, a random sample D of $E \backslash M$ (thus made of complete tuples) has been chosen. This sample D is used for running the algorithm inspired from FADANA, detailed in Section 3. Each time, the k nearest neighbors (kNN) technique has also been taken as a reference, and has been run with the same number of tuples and the same value of k. Let us recall that the kNN approach is based on a distance computation between the tuple to be completed and the tuples from the training set (one keeps only the k closest tuples, and a voting procedure similar to that involved in FADANA, is used to predict the missing values).

In a first step, we also tested a simple probabilistic approach that consists, for a given attribute, in computing the proportion of missing values equal to 1 (resp., to 0), and to use it as a probability when predicting a missing value. For instance, if there exist, in the dataset, 60 % of values equal to 1 (hence 40 % of values equal to 0) for attribute A_i, then the probability that the missing value be 1 equals 0.6, and 0.4 for 0. As the performances of this method appeared to be rather poor, it has not been retained further as a reference.

We also evaluated the proportion of values correctly predicted by FADANA and not by kNN, and reciprocally. In the following tables, No-FADANA (resp. No-kNN) means "incorrectly predicted by FADANA (resp. by kNN)", whereas

---

[1] http://fimi.ua.ac.be/data/
[2] http://http://archive.ics.uci.edu/ml/datasets.html

"Fadana & kNN" means "correctly predicted both by FADANA and by kNN". The results corresponding to the dataset *connect* and the average values computed over the six datasets are both presented.

Table 4. Precision wrt the number of modified values (set to NULL) per tuple (proportion of modified tuples: 70%, $k = 40$, size of the training set: 40)

%	5	10	20	40	60	80
FADANA	93.38	93.84	93.56	91.95	90.18	86.68
kNN	83.12	83.78	83.49	84.94	83.9	82.36
FADANA & kNN	77.83	77.35	77.54	78.45	75.45	74.70
FADANA & No-kNN	14.91	15.22	14.41	11.41	12.00	9.30
No-FADANA & kNN	2.33	2.03	2.04	2,95	4,01	4,14
No-FADANA & No-kNN	3.39	3.67	4.27	5.12	6.53	9.86

Table 5. Precision wrt the number of modified values (set to NULL) per tuple (proportion of modified tuples: 70%, $k = 40$, size of the training set: 40) for the dataset *connect*

%	5	10	20	40	60	80
FADANA	94.83	94.5	94.33	92.5	90.5	89.17
kNN	86.83	86.5	86.83	86.33	86.6	86.5
FADANA & kNN	83.66	83.66	83.66	82.83	82.5	83
FADANA & No-kNN	10	9.5	9.33	8.33	6.33	4.33
No-FADANA & kNN	1.83	1.5	1.66	1.83	2.66	1.83
No-FADANA & No-kNN	3.16	3.5	3.66	5	6.5	8.5

Tables 4 and 5 show how precision evolves with the number of modified values per tuple. One can see that both FADANA and kNN are rather robust in this respect, since even with 80% of values modified, precision remains over 80% (and even close to 90% most of the time for FADANA).

Tables 6 and 7 show how precision evolves with the value of k used in the algorithm. Again, one notices a remarkable stability, both for FADANA and for kNN, and one can see that even with a relatively small value of k, the vote leads to correct results in most of the cases.

Tables 8 and 9 describe the impact of the size of the training set on the precision of the algorithm. Unsurprisingly, one can see that a too small set negatively affects the effectiveness of the approach, but as soon as the set is large enough (30 or 40 tuples), FADANA reaches a precision level that is almost optimal (around 92%). Let us emphasize that it would be illusory to expect a 100% precision since there exist, in general, tuples that do not participate in any (sufficiently valid) analogical relation.

Table 6. Precision wrt the value of k (proportion of modified tuples: 70%, proportion of modified values per tuple: 40%, size of the training set: 40)

k	1	5	10	15	30	50	100
FADANA	92.54	92.35	92.61	92.61	92.31	91.88	92.32
kNN	88.73	87.98	86.72	85.53	84.23	84.14	83.85
FADANA & kNN	82.91	80.33	79.08	77.37	75.72	75.97	75.91
FADANA & No-kNN	7.42	9.40	11.19	12.59	13.25	13.01	13.62
No-FADANA & kNN	3.58	3.36	3.30	3.05	3.36	3.30	3.04
No-FADANA & No-kNN	4.04	4.87	4.37	4.74	4.50	5.92	5.50

Table 7. Precision wrt the value of k (proportion of modified tuples: 70%, proportion of modified values per tuple: 40%, size of the training set: 40) for the dataset connect

k	1	5	10	15	30	50	100
FADANA	93.83	93	92.83	93	92.5	92.66	93.33
kNN	91.33	88.83	88	87.5	86.5	86.5	86.5
FADANA & kNN	87.33	84.66	83.83	84	82.5	83	83.33
FADANA & No-kNN	4.833	6.66	7.66	7.66	8.66	8.166	8.33
No-FADANA & kNN	2.5	2.5	2.5	2.16	2.33	2.16	1.66
No-FADANA & No-kNN	2.833	4	3.83	4	4.66	5	4.83

Globally, a remarkable result is that the precision of FADANA is better than that of kNN, on average. In order to prove that the differences observed are statistically significant, we followed the recommendations of [3], which suggests a set of non-parametric tests for statistical comparisons of classifiers. When comparing only two classifiers, the use of the Wilcoxon signed ranks test is proposed. This method works as follows: assuming that two classifiers c_1 and c_2 are being compared, it starts by computing their difference in precision for each data set, then ranking these differences in increasing order. Then, let R^+ be the sum of ranks associated with the cases where c_1 outperformed c_2, and R^- the sum of ranks for the opposite situation, the minimum of those two values i.e., $\min(R^+, R^-)$ is used to determine if the difference of performance of the two classifiers is statistically significant (along with the number of datasets evaluated and some confidence level p, the most frequently used being 0.05).

We only used this method when evaluating the precision with respect to the value of k, specifically when $k \in (1, 5, 10, 15)$, since these are the only cases for which there is no unanimity with respect to the winner over the six datasets. For the other evaluations, FADANA always outperform kNN, except when the size of the training set is equal to 5, case in which it is kNN that outperforms FADANA. In all of the cases that we evaluated, i.e, $k \in (1, 5, 10, 15)$, according to the Wilcoxon signed-rank test, the results appear significant with a confidence value $p \leq 0.05$.

Table 8. Precision wrt the size of the training set (proportion of modified tuples: 70%, $k = 40$, proportion of modified values per tuple: 40%)

$|TS|$	5	10	20	30	40
FADANA	72.58	86.5	90.35	92.12	92.19
kNN	81.77	82.96	83.47	84.22	84.09
FADANA & kNN	56.89	72.33	74.82	76.48	76.20
FADANA & No-kNN	10.09	10.85	12.77	13.16	13.75
No-FADANA & kNN	19.35	5.83	4.00	3.19	3.41
No-FADANA & No-kNN	11.61	8.91	6.49	4.91	4.79

Table 9. Precision wrt the size of the training set (proportion of modified tuples: 70%, $k = 40$, proportion of maodified values per tuple: 40%) for the dataset *connect*

$|TS|$	5	10	20	30	40
FADANA	78.33	87.5	90.66	92.66	92.5
kNN	86.66	86.33	86.66	86.66	86.16
FADANA & kNN	72.16	80	82.33	83.16	82.33
FADANA & No-kNN	4.83	5.83	7	8.16	8.33
No-FADANA & kNN	13	4.66	2.66	2	2.66
No-FADANA & No-kNN	8	7.37	5.83	4.83	4.66

4.2   Optimization Aspects

As mentioned in the preamble, temporal performances of the approach are not so crucial since the prediction process is to be executed offline. However, it is interesting to study the extent to which the calculus could be optimized. With the base algorithm presented in Section 3, complexity is in $\theta(N^3)$ for the prediction of a missing value, where N denotes the size of the training set TS (indeed, an incomplete tuple has to be associated with every triple that can be built from this set, the analogical relation being quaternary). An interesting idea consists in detecting *a priori* the triples from TS that are the most "useful" for the considered task, i.e., those the most likely to take part in a sufficiently valid analogical relation. For doing so, one just has to run the algorithm on a small subset of the database containing artificially introduced missing values, and to count, for each triple of the training set, the number of k-lists in which it appears as well as the average number of cases in which the prediction is correct. One can then keep the sole N' triples that appear the most frequently with a good rate of success, and use them to predict the missing values in the entire database. Complexity is then in $\theta(N')$ for estimating a given missing value.

We ran this optimized algorithm several times, with k varying between 20 and 40, the size of the training set between 20 and 40, and N' between 100 and 1000. For a total of 3000 incomplete tuples, the basic algorithm was run over the first 500, treating the others with the optimized method.

While the precision of the regular FADANA algorithm is 91% on average, that of the optimized method is about 84%, i.e., there is a difference of about 7 percents (whereas the precision of the kNN method over the same dataset is about 85%). On the other hand, the optimized method is much more efficient: it is 1300 times faster than the regular FADANA algorithm when the size of the training set equals 40, and 25 times faster when it equals 20.

These results show that this method does not imply a huge loss of precision, but leads to a very significant reduction of the overall processing time. Further experiments and analyses are needed, though, in order to determine which properties make a triple more "effective" than others.

Let us mention that another optimization axis would consist in parallelizing the calculus on the basis of a vertical partitioning of the relation involved, which would make it possible to assign a subset of attributes to each processor, the intermediate results being summed in order to obtain the final value of the analogical dissimilarity ad.

5   Conclusion

In this paper, we have presented a novel approach to the estimation of missing values in relational databases, that exploits the notion of analogical proportion. We have shown how an algorithm proposed in the context of classification could be adapted to this end. The results obtained, even though preliminary, appear very encouraging since the approach yields a significantly better precision than the classical nearest neighbors technique.

Among the many perspectives opened by this work, let us mention the following four ones. Future work should i) compare the analogy-based prediction approach with a larger sample of approaches — beyond kNN — but one can be reasonably optimistic since, on the one hand, many approaches from the literature are, as kNN, based on the notion of distance and, on the other hand, since no other approach exploits the type of "regularity" that we do, namely analogical proportion, one may reasonably hope that a combination of our algorithm with another approach would improve the effectiveness of both; ii) deal in a more sophisticated way with categorical attributes by taking into account notions such as synonymy, hyponymy/hypernymy, etc. iii) study the possibility of building an *optimal* training set, on the basis of the remark made in Subsection 4.2; iv) study the way predicted values must be handled, in particular during the database querying process. This will imply using an uncertain database model (for instance of a probabilistic nature) inasmuch as an estimated value remains tainted with uncertainty, even if the prediction process has a good level of reliability.

References

1. Bayoudh, S., Miclet, L., Delhay, A.: Learning by analogy: A classification rule for binary and nominal data. In: Veloso, M.M. (ed.) IJCAI, pp. 678–683 (2007)

2. Chen, S.M., Chang, S.T.: Estimating null values in relational database systems having negative dependency relationships between attributes. Cybernetics and Systems 40(2), 146–159 (2009)

3. Demšar, J.: Statistical comparisons of classifiers over multiple data sets. J. Mach. Learn. Res. 7, 1–30 (2006)

4. Fujikawa, Y., Ho, T.-B.: Cluster-based algorithms for dealing with missing values. In: Chen, M.-S., Yu, P.S., Liu, B. (eds.) PAKDD 2002. LNCS (LNAI), vol. 2336, pp. 549–554. Springer, Heidelberg (2002)

5. Liu, W.Z., White, A.P., Thompson, S.G., Bramer, M.A.: Techniques for dealing with missing values in classification. In: Liu, X., Cohen, P., Berthold, M. (eds.) IDA 1997. LNCS, vol. 1280, pp. 527–536. Springer, Heidelberg (1997)

6. Miclet, L., Bayoudh, S., Delhay, A.: Analogical dissimilarity: Definition, algorithms and two experiments in machine learning. J. Artif. Intell. Res. (JAIR) 32, 793–824 (2008)

7. Miclet, L., Prade, H.: Handling analogical proportions in classical logic and fuzzy logics settings. In: Sossai, C., Chemello, G. (eds.) ECSQARU 2009. LNCS, vol. 5590, pp. 638–650. Springer, Heidelberg (2009)

8. Myrtveit, I., Stensrud, E., Olsson, U.H.: Analyzing data sets with missing data: An empirical evaluation of imputation methods and likelihood-based methods. IEEE Trans. Software Eng. 27(11), 999–1013 (2001)

9. Nogueira, B.M., Santos, T.R.A., Zárate, L.E.: Comparison of classifiers efficiency on missing values recovering: Application in a marketing database with massive missing data. In: CIDM, pp. 66–72. IEEE (2007)

10. Prade, H., Richard, G.: Reasoning with logical proportions. In: Lin, F., Sattler, U., Truszczynski, M. (eds.) KR. AAAI Press (2010)

11. Prade, H., Richard, G.: Homogeneous logical proportions: Their uniqueness and their role in similarity-based prediction. In: Brewka, G., Eiter, T., McIlraith, S.A. (eds.) KR. AAAI Press (2012)

12. Prade, H., Richard, G.: Analogical proportions and multiple-valued logics. In: van der Gaag, L.C. (ed.) ECSQARU 2013. LNCS, vol. 7958, pp. 497–509. Springer, Heidelberg (2013)

13. Ragel, A.: Preprocessing of missing values using robust association rules. In: Żytkow, J.M., Quafafou, M. (eds.) PKDD 1998. LNCS, vol. 1510, pp. 414–422. Springer, Heidelberg (1998)

14. Shen, J.J., Chen, M.T.: A recycle technique of association rule for missing value completion. In: AINA, pp. 526–529. IEEE Computer Society (2003)

Exception-Tolerant Skyline Queries

Hélène Jaudoin, Olivier Pivert, and Daniel Rocacher

Université de Rennes 1, Irisa
Technopole Anticipa 22305 Lannion Cedex, France
{jaudoin,pivert,rocacher}@enssat.fr

Abstract. This paper presents an approach aimed at reducing the impact of exceptional points/outliers when computing skyline queries. The phenomenon that one wants to avoid is that noisy or suspect elements "hide" some more interesting answers just because they dominate them in the sense of Pareto. The approach we propose is based on the fuzzy notion of typicality and makes it possible to distinguish between genuinely interesting points and potential anomalies in the skyline obtained.

Keywords: skyline query, exception, gradual approach.

1 Introduction

In this paper, a qualitative view of preference queries is chosen, namely the *Skyline* approach introduced by Börzsönyi *et al.* [2]. Given a set of points in a space, a skyline query retrieves those points that are not dominated by any other in the sense of Pareto order. When the number of dimensions on which preferences are expressed gets high, many tuples may become incomparable. Several approaches have been proposed to define an order for two incomparable tuples, based on the number of other tuples that each of the two tuples dominates (notion of k-representative dominance proposed by Lin *et al.* [12]), on a preference order of the attributes (see for instance the notions of k-dominance and k-frequency introduced by Chan *et al.* [3,4]), or on a notion of representativity ([14] redefines the approach proposed by [12] and proposes to return only the more representative points of the skyline, i.e., a point among those present in each cluster of the skyline points). Other approaches fuzzify the concept of skyline in different ways, see e.g. [9]. Here, we are concerned with a different problem, namely that of the possible presence of *exceptional* points, aka outliers, in the dataset over which the skyline is computed. Such exceptions may correspond to noise or to the presence of *nontypical* points in the collection considered. The impact of such points on the skyline may obviously be important if they dominate some other, more representative ones.

Two strategies can be considered to handle exceptions. The former consists in removing anomalies by adopting cleaning procedures or defining data entry constraints. However, the task of automatically distinguishing between odd points and simply exceptional points is not always easy. Another solution is to define an approach that is *tolerant to exceptions*, that highlights representative points of the database and that points out the possible outliers. In this paper, we present

A. Laurent et al. (Eds.): IPMU 2014, Part III, CCIS 444, pp. 120–129, 2014.

such an approach based on the fuzzy notion of typicality [15]. We revisit the definition of a skyline and show that it i) makes it possible to retrieve the dominant points without discarding other potentially interesting ones, and ii) constitutes a flexible tool for visualizing the answers.

The remainder of the paper is structured as follows. Section 2 provides a refresher about skyline queries and motivates the approach. Section 3 presents the principle of exception-tolerant skyline queries, based on the fuzzy concept of typicality. Section 4 gives the main implementation elements of the approach whereas Section 5 presents preliminary experimental results obtained on a real-world dataset. Finally, Section 6 recalls the main contributions and outlines perspectives for future work.

2 Refresher about Skyline Queries and Motivations

Let $\mathcal{D} = \{D_1, \ldots, D_d\}$ be a set of d dimensions. Let us denote by $dom(D_i)$ the domain associated with dimension D_i. Let $\mathcal{S} \subseteq dom(D_1) \times \ldots \times dom(D_d)$, p and q two points of $\mathcal{S}$, and $\succ_i$ an order on D_i. One says that p *dominates* q on $\mathcal{D}$ (p is better than q according to Pareto order), denoted by $p \succ_{\mathcal{D}} q$, iff

$$\forall i \in [1, d] : p_i \succeq_i q_i \text{ and } \exists j \in [1, d] : p_i \succ_i q_i.$$

A skyline query on $\mathcal{D}$ applied to a set of points $\mathcal{S}$, whose result is denoted by $\text{SKY}_{\mathcal{D}}(\mathcal{S})$, according to order relations $\succ_i$, produces the set of points that are not dominated by any other point of $\mathcal{S}$:

$$\text{SKY}_{\mathcal{D}}(\mathcal{S}) = \{p \in \mathcal{S} \mid \nexists q \in \mathcal{S} : q \succ_{\mathcal{D}} p\}$$

Depending on the context, one may try, for instance, to maximize or minimize the values of $dom(D_i)$, assuming that $dom(D_i)$ is a numerical domain.

In order to illustrate the principle of the approach we propose, let us consider the dataset *Iris* [8], graphically represented in Figure 1.

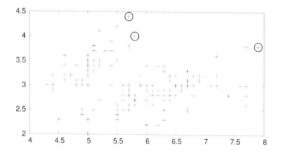

Fig. 1. The dataset *Iris*

The vertical axis corresponds to the attribute *sepal width* whereas the horizontal axis is associated with *sepal length*. The skyline query:

```
select * from iris
skyline of sepallength max, sepalwidth max
```

looks for those points that maximize the dimensions *length* and *width* of the sepals (the circled points in Figure 1).

In this dataset, the points form two groups that respectively correspond to the intervals [4, 5.5] and [5.5, 7] on attribute *length*. By definition, the skyline points are on the border of the region that includes the points of the dataset. However, these points are very distant from the areas corresponding to the two groups and are thus not very representative of the dataset. It could then be interesting for a user to be able to visualize the points that are "almost dominant", closer to the clusters, then more representative of the dataset. The notion of typicality discussed in the next section makes it possible to reach that goal.

2.1   Computing a Fuzzy Set of Typical Values

The *typicality* of an element in a set indicates the extent to which this element is similar to many other points from the set. The notion of fuzzy typicality has been much studied in the contexts of data summaries and approximate reasoning. Zadeh [15] states that x is a typical element of a fuzzy set A iff i) x has a high membership degree to A and ii) *most of the elements of A are similar to x*. In the case where A is a crisp set — as it will be the case in the following —, the definition becomes: x is in A and most of the elements of A are similar to x.

In [7], the authors define a typicality index based on frequency and similarity. We adapt their definition as follows. Let us consider a set $\mathcal{E}$ of points. We say that a point is all the more typical as it is close to many other points. The proximity relation considered is based on Euclidean distance. We consider that two points p_1 and p_2 are close to each other if $d(p_1, p_2) \leq \tau$ where τ is a predefined threshold. In the experiment performed on the dataset *Iris*, we used $\tau = 0.5$. The frequency of a point is defined as:

$$F(p) = \frac{|\{p_i \in \mathcal{E}, d(p, p_i) \leq \tau\}| - 1}{|\mathcal{E}|}. \tag{1}$$

This degree is then normalized into a typicality degree in [0, 1]:

$$typ(p) = \frac{F(p)}{\max_{p_i \in \mathcal{E}}\{F(p_i)\}}.$$

We will also use the following notations:

$$\text{TYP}(\mathcal{E}) = \{typ(p)/p \mid p \in \mathcal{E}\}$$

$$\text{TYP}_\gamma(\mathcal{E}) = \{p \mid p \in \mathcal{E} \text{ and } typ(p) \geq \gamma\}.$$

$\text{TYP}(\mathcal{E})$ represents the fuzzy set of points that are somewhat typical of the set $\mathcal{E}$ while $\text{TYP}_\gamma(\mathcal{E})$ gathers the points of $\mathcal{E}$ whose typicality is over the threshold γ. An excerpt of the typicality degrees computed over the *Iris* dataset is presented in Table 1.

Table 1. Excerpt of the *Iris* dataset with associated typicality degrees

length	width	frequency	typicality
7.4	2.8	0.0600	0.187
7.9	3.8	0.0133	0.0417
6.4	2.8	0.253	0.792
6.3	2.8	0.287	0.896
6.1	2.6	0.253	0.792
7.7	3.0	0.0467	0.146
6.3	3.4	0.153	0.479
6.4	3.1	0.293	0.917
6.0	3.0	0.320	1.000

3   Principle of the Exception-Tolerant Skyline

As explained in the introduction, our goal is to revisit the definition of the skyline so as to take into account the typicality of the points in the database, in order to control the impact of exceptions or anomalies.

3.1   Boolean View

A first idea is to restrict the computation of the skyline to a subset of $\mathcal{E}$ that corresponds to sufficiently typical points. The corresponding definition is:

$$\text{SKY}_{\mathcal{D}}(\text{TYP}_\gamma(\mathcal{S})) = \{p \in \text{TYP}_\gamma(\mathcal{S}) \mid \not\exists q \in \text{TYP}_\gamma(\mathcal{S}) \text{ such that } q \succ_{\mathcal{D}} p\} \quad (2)$$

Such an approach obviously reduces the cost of the processing since only the points that are typical at least to the degree γ are considered in the calculus. However, this definition does not make it possible to discriminate the points of the result according to their degree of typicality since the skyline obtained is a crisp set. Figure 2 illustrates this behavior and shows the maxima (circled points) obtained when considering the points that are typical to a degree ≥ 0.7 (represented by crosses).

Another drawback of this definition is to exclude the nontypical points altogether, even though some of them could be interesting answers. A more cautious definition consists in keeping the nontypical points while computing the skyline and transform Equation (2) into:

$$\text{SKY}_{\mathcal{D}}(\text{TYP}_\gamma(\mathcal{S})) = \{p \in \mathcal{S} \mid \not\exists q \in \text{TYP}_\gamma(\mathcal{S}) \text{ such that } q \succ_{\mathcal{D}} p\} \quad (3)$$

Figure 3 illustrates this alternative solution. It represents (circled points) the objets from the *Iris* dataset that are not dominated by any item typical to the degree $\gamma = 0.7$ at least (represented by crosses).

With Equation (2), the nontypical points are discarded, whereas with Equation (3), the skyline is larger and includes nontypical extrema. This approaches relaxes skyline queries in such a way that the result obtained is not a line anymore but a stripe composed of the regular skyline elements completed with

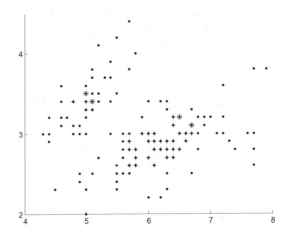

Fig. 2. Skyline of the Iris points whose typicality degree is ≥ 0.7

possible "substitutes". However, the main drawbacks of this definition are: i) the potentially large number of points returned, ii) the impossibility to distinguish, among the skyline points, those that are not at all dominated from those that *are* dominated (by weakly typical points).

3.2   Gradual View

A third version makes it possible to compute a *graded* skyline, seen as a fuzzy set, that preserves the gradual nature of the concept of typicality. By doing so, no threshold (γ) is applied to typicality degrees. The definition is as follows:

$$\text{SKY}_{\mathcal{D}}(\text{TYP}(\mathcal{S})) =$$
$$= \{\mu/p \mid p \in \mathcal{S} \wedge \mu = \min_{q \in \mathcal{S}} \left(\max(1 - \mu_{\text{TYP}}(q), \ deg(\neg(q \succ_{\mathcal{D}} p))) \right)\} \quad (4)$$

where $deg(\neg(q \succ_{\mathcal{D}} p)) = 1$ if q does not dominate p (i.e., $(q \succ_{\mathcal{D}} p)$ is false), 0 otherwise. A point totally belongs to the skyline (membership degree equal to 1) if it is dominated by no other point. A point does not belong at all to the skyline (membership degree equal to 0) if it is dominated by at least one totally typical point. In the case where p is dominated by somewhat (but not totally) typical points, its degree of membership to the skyline depends on the typicality of these points. Equation (4) may be rewritten as follows:

$$\text{SKY}_{\mathcal{D}}(\text{TYP}(\mathcal{S})) = \{\mu/p \mid p \in \mathcal{S} \wedge \mu = 1 - \max_{q \in \mathcal{S} \mid q \succ_{\mathcal{D}} p} (\mu_{\text{TYP}}(q))\}. \quad (5)$$

With the *Iris* dataset, one gets the result presented in Figure 4, where the degree of membership to the skyline corresponds to the z axis. As expected, the points of the classical skyline totally belong to the graded skyline, along with some additional answers that more or less belong to it. This approach appears

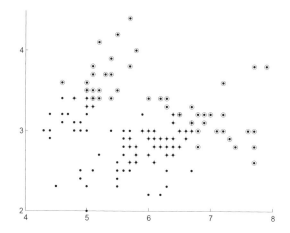

Fig. 3. Points that are not dominated by any other whose typicality degree is ≥ 0.7

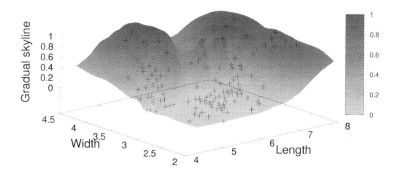

Fig. 4. Graded skyline obtained with the *Iris* dataset

interesting in terms of visualization. Indeed, the score associated with each point makes it possible to focus on different α-cuts of the skyline. In Figure 4, one may notice a slope from the optimal points towards the less typical or completely dominated ones. The user may select points that are not necessarily optimal but that represent good alternatives to the regular skyline answers (in the case, for instance, where the latter look "too good to be true"). Finally, an element of the graded skyline is associated with two scores: a degree of membership to the skyline, and a typicality degree (that expresses the extent to which it is *not* exceptional). One may imagine different ways of navigating inside areas in order to explore the set of answers: a simple scan for displaying the characteristics of the points, the use of different filters aimed, for instance, at optimizing diversity on certain attributes, etc.

4   Implementation Aspects

Two steps are necessary for obtaining the graded result: i) computation of the typicality degrees, and ii) computation of the skyline. Many algorithms have been proposed for processing skyline queries: Block-Nested-Loops(BNL) [2]; Divide and Conquer [2]; a technique exploiting a B-tree or an R-tree [2]; an algorithm based on bitmap indices [13]; an improvement of the BNL method, named Sort-Filter-Skyline [5,6], and a strategy proposed in [1] that relies on a preordering of the tuples aimed to limit the number of elements to be accessed and compared. We have based our implementation on the approach proposed in [13] with which Formula (5) appears the easiest to evaluate.

The data structure underlying the algorithm described in [13] is an array of Boolean values or *bitmap*. A bitmap index is defined for each dimension of the skyline: every column corresponds to a possible value in the dimension considered, and every row references a tuple from the database. Value 1 at the intersection of row l and column c means that the tuple referenced in row l has the value corresponding to column c. Then, every point p of the database S is tested in order to determine if it belongs to the skyline or not. For doing so, two other data structures are created. The first one, denoted by A, gathers *the tuples that are as good as p* on every dimension, the second one, denoted by B, contains *the tuples that are better than p* on at least one dimension. A and B are defined as two-dimensional tables of Booleans whose columns are associated with the tuples of S. They are initialized using the bitmap indices.

Algorithm 1 constitutes the heart of our prototype and follows the principle described above. We have also used three tables ($\mathcal{T}$, Sky_{grad}, A') where each column corresponds to a tuple of S, that contain real numbers in [0, 1]. In $\mathcal{T}$, these numbers correspond to typicality degrees whereas in Sky_{grad} they represent degrees of membership to the graded skyline. AND corresponds to the logical conjunction between the pairs of values ($A[i]$, $B[i]$) so as to check whether the point i is both as good as p on all dimensions and better than p on one dimension. If it is the case, then this point dominates p. $MULT$ is used to compute the product of the values $A[i]$ and $\mathcal{T}[i]$, which makes it possible to associate a point i with its typicality degree if it dominates the point p considered. Finally, MAX returns the maximal value of the array A. The membership degree of p to the graded skyline is then obtained by means of Formula (5).

The computation of the typicality degrees uses a threshold on the distance that depends on the attributes involved in the skyline query. The complexity of the computation is obviously in $\theta(n^2)$ since one has to compute the Euclidean distance between each pair of points of the dataset.

One may think of two ways for computing typicality: either on demand, on the attributes specified in the *skyline* clause of the query, or beforehand, on different relevant sets of attributes. The first method implies an additional cost to the evaluation of skyline queries, whereas the second one makes it necessary to update the typicality degrees when the dataset evolves. In any case, indices such as those used for retrieving k nearest neighbors (for instance kdtrees) may be exploited. Furthermore, since in general the computation of the graded

Algorithm 1. Main algorithm for computing the graded skyline

Require: d distance, n cardinality of the dataset $\mathcal{S}$, the points of the dataset $p \in \mathcal{S}$, the set of dimensions $\{d_i\}$

Ensure: graded skyline: $\forall p \in \mathcal{S},\ Sky_{grad}(p)$

  Preprocessing: creation of the bitmap indices on the d_i's

  Preprocessing: computation of the typicality of the points $\mathcal{T}$: $\forall p \in \mathcal{S},\ Typ(p)$

  for all $p \in \mathcal{S}$ **do**

    // Search for those points that dominate p

    Creation of A

    Creation of B

    $A := A$ AND B

    $A' := A$ MULT $\mathcal{T}$

    $Sky_{grad}(p) := 1 - Max(A')$

  end for

---

skyline concerns only a small fragment of the database, the extra cost related to typicality should not be too high.

It is worth emphasizing that the algorithm could be parallelized by partitioning the arrays A, B, A' and $\mathcal{T}$. Similarly, the creation of the structures A and B may be parallelized, provided that the bitmap indices and the typicality degrees are distributed or shared.

Table 2. Excerpt of the database and associated degrees (skyline and typicality)

id	price	km	skyline	typicality
1156771	6000	700	1	0.247
1596085	5800	162643	1	0.005
1211574	7000	500	1	0.352
1054357	1800	118000	1	0
1333992	500	220000	1	0
1380340	800	190000	1	0
891125	1000	170000	1	0
1229833	5990	10000	1	0.126
1276388	1300	135000	1	0
916264	5990	2514000	0.874	0
1674045	6000	3500	0.753	0.315

5 Experimental Results

The approach has been tested using a subset of the database of 845810 ads about second hand cars from the website *Le bon coin* [1] from 2012. The skyline query used hereafter as an example aims at minimizing both the price and the mileage. In the query considered, we focus on small urban cars with a regular (non-diesel) engine, which corresponds to 441 ads. Figure 5 shows the result obtained. In dark

---

[1] www.leboncoin.fr

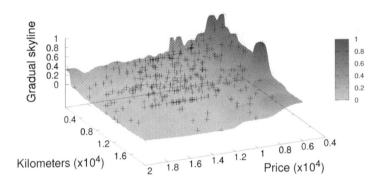

Fig. 5. 3D representation of the graded skyline

Table 3. Excerpt of the area [0.6, 0.8]

id	price	km	skyline	typicality
870279	6900	1000	0.716	0.358
981939	6500	4000	0.637	0.363
1022586	6500	7200	0.637	0.258
1166077	7750	2214	0.642	0.532
1208620	6500	3300	0.716	0.363
1267726	6500	100000	0.637	0
1334605	10500	500	0.647	0.642
1366336	7490	4250	0.637	0.516
1529678	7980	650	0.647	0.458
1635437	9900	590	0.647	0.621
1685854	7890	1000	0.642	0.458

grey are the points that belong the most to the skyline (membership degree between 0.8 and 1). These points are detailed in Table 2. According to the definition used, points dominated by others that are not totally typical belong to the result. It is the case for instance of ad number 916264 that is dominated by ads numbered 1054357 and 1229833. The identifiers in bold correspond to the points that belong to the regular skyline. One may observe that the points from Table 2 (area [0.8, 1]) are not very (or even not at all) typical. Moreover, certain features may not satisfy the user (the mileage can be very high, the price can be very low) and may look suspicious. On the other hand, Table 3, which shows an excerpt of the 0.6-cut of the graded skyline, contains more typical – thus more credible – points whose overall satisfaction remains high. Let us also mention that the time taken by the precomputation of the typicality degrees associated with the selected elements is twice as large (around 0.54 second) as the time devoted to the computation of the graded skyline (about 0.22 second). However, this result must taken carefully as the computation of typicality has not been optimized in the prototype yet.

6   Conclusion

In this paper, we have proposed a graded version of skyline queries aimed at controlling the impact of exceptions on the result (so as to prevent interesting points to be hidden because they are dominated by an exceptional one). An improvement could consist in using more sophisticated techniques for characterizing the points according to their level of representativity as a typicality-based clustering approach [11] or statistical methods for detecting outliers [10]. As a short-term perspective, we intend to carry out a parallel implementation of the algorithm and to use indexes for reducing the processing time devoted to the computation of typicality degrees.

References

1. Bartolini, I., Ciaccia, P., Patella, M.: Efficient sort-based skyline evaluation. ACM Trans. Database Syst. 33(4), 1–49 (2008)
2. Börzsönyi, S., Kossmann, D., Stocker, K.: The skyline operator. In: Proc. of ICDE 2001, pp. 421–430 (2001)
3. Chan, C., Jagadish, H., Tan, K., Tung, A., Zhang, Z.: Finding k-dominant skylines in high dimensional space. In: Proc. of SIGMOD 2006, pp. 503–514 (2006)
4. Chan, C.-Y., Jagadish, H.V., Tan, K.-L., Tung, A.K.H., Zhang, Z.: On high dimensional skylines. In: Ioannidis, Y., Scholl, M.H., Schmidt, J.W., Matthes, F., Hatzopoulos, M., Böhm, K., Kemper, A., Grust, T., Böhm, C. (eds.) EDBT 2006. LNCS, vol. 3896, pp. 478–495. Springer, Heidelberg (2006)
5. Chomicki, J., Godfrey, P., Gryz, J., Liang, D.: Skyline with presorting. In: Proc. of ICDE 2003, pp. 717–719 (2003)
6. Chomicki, J., Godfrey, P., Gryz, J., Liang, D.: Skyline with presorting: Theory and optimizations. In: Proc. of IIS 2005, pp. 595–604 (2005)
7. Dubois, D., Prade, H.: On data summarization with fuzzy sets. In: Proc. of IFSA 1993, pp. 465–468 (1984)
8. Fisher, R.A.: The use of multiple measurements in taxonomic problems. Annals of Eugenics 7(2), 179–188 (1936)
9. Hadjali, A., Pivert, O., Prade, H.: On different types of fuzzy skylines. In: Kryszkiewicz, M., Rybinski, H., Skowron, A., Raś, Z.W. (eds.) ISMIS 2011. LNCS (LNAI), vol. 6804, pp. 581–591. Springer, Heidelberg (2011)
10. Hodge, V., Austin, J.: A survey of outlier detection methodologies. Artificial Intelligence Review 22(2), 85–126 (2004)
11. Lesot, M.: Typicality-based clustering. Int. J. of Information Technology and Intelligent Computing 12, 279–292 (2006)
12. Lin, X., Yuan, Y., Zhang, Q., Zhang, Y.: Selecting stars: the k most representative skyline operator. In: Proc. of the ICDE 2007, pp. 86–95 (2007)
13. Tan, K.L., Eng, P.K., Ooi, B.C.: Efficient progressive skyline computation. In: Proc. of VLDB 2001, pp. 301–310 (2001)
14. Tao, Y., Ding, L., Lin, X., Pei, J.: Distance-based representative skyline. In: Ioannidis, Y.E., Lee, D.L., Ng, R.T. (eds.) ICDE, pp. 892–903. IEEE (2009)
15. Zadeh, L.A.: A computational theory of dispositions. In: Wilks, Y. (ed.) COLING, pp. 312–318. ACL (1984)

Towards a Gradual QCL Model
for Database Querying

Ludovic Liétard[1], Allel Hadjali[2], and Daniel Rocacher[1]

[1] IRISA/ENSSAT, Rue de Kérampont BP 80518 Lannion, France
{ludovic.lietard,daniel.rocacher}@univ-rennes1.fr
[2] LIAS/ENSMA, Téléport 2 - 1 Avenue Clément Ader - BP 40109, 86961
FUTUROSCOPE CHASSENEUIL Cedex, France
allel.hadjali@ensma.fr

Abstract. The Qualitative Choice Logic (QCL) is devoted to a logic expressing preferences for Boolean alternatives. This paper puts the first foundations to extend QCL to fuzzy alternatives. In particular, some relationships between QCL and the bipolar expression of preferences queries are emphasized. A new type of bipolar conditions is defined in the Boolean context to express QCL statements. This new type of bipolar conditions is extended to graduality and it is shown that this extension can be the basis to define a gradual QCL model.

Keywords: database, preference query, qualitative choice logic, bipolarity.

1 Introduction

This paper is devoted to the integration of user's preferences inside queries addressed to relational databases (see as examples, [1,2,3]). More precisely, we consider fuzzy bipolar conditions to model sophisticated user preferences [4,5,6,7,8,9]. In this context, a fuzzy bipolar condition is made of two fuzzy predicates (two poles), the first one expresses a constraint to define the elements to be retrieved (its negation is the set of values to be rejected), the other one expresses a more restrictive attitude to define the best elements (among the ones satisfying the constraint). The advantage of this type of condition can be illustrated by the querying of a database of a travel agency. A fuzzy bipolar condition can be useful to take into consideration two aspects: the mandatory requirement of the client (a trip in Italy with a cheap price) and the less important requirements (a trip in august including Roma and Pisa). Here, it is not only a matter of importance but also a matter of obligation. A trip which does not satisfy the mandatory requirements (not in Italy or without a cheap price) is rejected.

Independently, a Qualitative Choice Logic (QCL) has been proposed where a new connective in propositional logic has been defined ($A \overrightarrow{\times} B$) to express ordered disjunctions (if A, but if A is impossible then at least B). This new connective is closely related to the "or else" type of fuzzy bipolar conditions.

A. Laurent et al. (Eds.): IPMU 2014, Part III, CCIS 444, pp. 130–139, 2014.

This paper shows some relationships between fuzzy bipolar conditions and QCL statements. We define the first basis of a gradual QCL and the ultimate aim is to enrich bipolar queries by the use of this logic.

Section 2 provides a very brief summary of the relational algebra defined in [6] for fuzzy bipolar conditions. A presentation of QCL is introduced in Section 3. Section 4 discusses a comparison of both theories. Finally, Section 5 shows the basis for the definition of a gradual QCL.

2   A Relational Algebra for Fuzzy Bipolar Condition

A bipolar condition is an association of a negative condition (negative pole) and positive condition (positive pole). In this paper, a bipolar condition is made of two conditions defined on the same universe: i) a constraint c, which describes the set of acceptable elements, ii) a wish w which defines the set of desired or wished elements. The negation of c is the set of rejected elements since it describes non-acceptable elements. Since it is not coherent to wish a rejected element, the following property of coherence holds: $w \subseteq c$.

In addition, condition c is mandatory because an element which does not satisfy c is rejected; $\neg c$ is then considered as the negative pole of the bipolar condition. Condition w is optional because its non-satisfaction does not automatically mean the rejection. But condition w describes the best elements and w is then considered as the positive pole of the bipolar condition.

If c and w are boolean conditions, the satisfaction with respect to (c, w) is an ordered pair from $\{0, 1\}^2$. When querying a database with such a condition, tuples satisfying the constraint and the wish are returned in priority to the user. They are the top answers. Tuples satisfying only the constraint are delivered but are ranked after the top answers. If there is no top answers, they are considered as the best answers it is possible to provide to the user.

If c and w are fuzzy conditions (defined on the universe U), the property of coherence becomes: $\forall u \in U, \mu_w(u) \leq \mu_c(u)$. The satisfaction is a pair of degrees where $\mu_w(u)$ expresses the optimality while $\mu_c(u)$ expresses the non rejection. When dealing with such conditions, two different attitudes can be considered. The first one is to assess that the non rejection is the most important pole and the fuzzy bipolar condition is an and-if-possible condition (to satisfy c and if possible to satisfy w). The idea is then "to be not rejected and if possible to be optimal". The satisfaction with respect to such a fuzzy bipolar condition is denoted $(\mu_c(u), \mu_w(u))$. The second attitude is the opposite one, it considers that the optimality is the most important pole and the fuzzy bipolar condition is an or-else condition (to satisfy w or else to satisfy c). The idea is then "to be optimal or else to be non rejected". The satisfaction with respect to such a fuzzy bipolar condition is denoted $[\mu_w(u), \mu_c(u)]$. Both types of conditions are different because it is possible to show [6] that the same values of satisfaction for w and c does not lead to a same ordering whether it is an and-if-possible condition or an or-else condition. The satisfactions with respect of both types of conditions can be sorted using the lexicographical order and the minimum and the maximum on the lexicographical order can respectively be used to define the conjunction and disjunction of fuzzy bipolar conditions [6]. The negation of an and-if-possible condition is an or-else condition (and vice-versa) and an algebraic framework can be defined [6] to handle fuzzy bipolar conditions. Another definition for the negation can be found in [10] but this negation cannot be interpreted as a fuzzy bipolar condition.

3   A Qualitative Choice Logic

The Qualitative Choice Logic [11] (QCL) defines a new connective in propositional logic $(A \vec{\times} B)$ to express ordered disjunctions (if A, but if A is impossible then at least B).

In QCL each logical expression is associated to an integer stating its level of satisfaction (it is called degree by the authors). The level 1 is the full satisfaction and the higher this level, the less satisfied the expression. The rules defining the Qualitative Choice Logic are the following ones (where I is an interpretation, A, A_i are propositional rules and P and Q ordered disjunctions):

(1) $I \models_k (A_1 \vec{\times} \ldots \vec{\times} A_n)$ iff $I \models (A_1 \vee \ldots \vee A_n)$ and $k = min\{j \mid I \models A_j\}$;

(2) $I \models_k A$ iff $k = 1$ and $A \in I$ (for propositional atom A);

(3) $I \models_k (P \wedge Q)$ iff $I \models_m P$ and $I \models_n Q$ $k = max(m, n)$;

(4) $I \models_k (P \vee Q)$ iff $I \models_m P$ or $I \models_n Q$ and $k = min\{r \mid I \models_r P$ or $I \models_r Q\}$;

(5) $I \models_k \neg A$ iff $k = 1$ and $A \notin I$ (for propositional atom A);

(6) $I \models_k \neg(A_1 \vec{\times} \ldots \vec{\times} A_n)$ iff $I \models_k (\neg A_1 \vec{\times} \ldots \vec{\times} \neg A_n)$.

Rules (1)-(4) are obvious and comes from the definition of QCL [11]. Rule (5) is nothing but the negation used for propositional atoms. Rule (6) is a little bit more complex since it defines the negation of an ordered disjunction. It is not the original definition proposed in [11] but a more recent one introduced in [12] in the context of *Prioritized* Qualitative Choice Logic (PQCL). The idea of PQCL is to overcome the drawbacks of the original proposition for the negation and to introduce a prioritized disjunction and conjunction (which are not commutative, see [12] for more details).

Finally, a last rule (from the original QCL [11]) allows a construct of ordered disjunctions from ordered disjunctions:

(7) $I \models_k (P \vec{\times} Q)$ iff $I \models_k P$ or $(I \models_1 \neg P$ and $I \models_m Q$ and $k = m + opt(P))$

where $opt(P)$ is the number of satisfaction levels for P.

This rule (7) states that if I satisfies P, at a certain level, then it satisfies $P \vec{\times} Q$ at the same level. When it does not satisfy P, its satisfaction for $P \vec{\times} Q$ is given by the one from Q but with a penalization (of $opt(P)$) due to its non satisfaction to P.

However, it is possible to show that the rule (6) defining the negation does not fully satisfy the requirement for a negation (see Example 1). As a consequence, we ignore this rule in this paper and we suggest in section 4 (definition (6')) an alternative definition for the negation of QCL statements.

Example 1. A user is looking for a trip and his preference is "if possible *Air France* (AF) and if *Air France* is impossible then *KLM* and if *KLM* is impossible then *British Airways* (BA)". This preference is expressed by the ordered disjunction:

$$AF \vec{\times} KLM \vec{\times} BA.$$

An interpretation $I_1 = \{AF\}$ receives the level 1 of satisfaction, an interpretation $I_2 = \{KLM\}$ receives the level 2 of satisfaction, and an interpretation $I_3 = \{BA\}$ receives the level 3 of satisfaction. As a consequence, I_1 is preferred to I_2 which is preferred to I_3. According to (6), the negation is:

$$\neg AF \vec{\times} \neg KLM \vec{\times} \neg BA,$$

and then both interpretations I_2 and I_3 receive level 1 while interpretation I_1 receives level 2. This behaviour looks difficult to be justified for database querying because we expect I_3 to be preferred to I_2 to be preferred to I_1 for the negation (we expect a negation to reverse the order).

4   Fuzzy Bipolar Conditions and QCL

This section compares fuzzy bipolar conditions and QCL. Since QCL is defined in a Boolean context, we should consider the restriction of fuzzy bipolar conditions to Boolean predicates (bipolar conditions). Furthermore, such (boolean) bipolar conditions should be extended to several components.

The first subsection (4.1) introduces this new type of bipolar conditions (Boolean multipolar conditions or BMC for short). The next section (4.2) shows that QCL is in adequation with the semantics conveyed by this new type of bipolar conditions.

4.1   Boolean Multipolar Conditions

We consider the extension of the fuzzy bipolar condition to several arguments to define BMC's made of n arguments (an example of a multipolar or-else condition is "C_1 or else C_2 or else ... or else C_n"). We consider the case of Boolean arguments and we propose an interpretation for these conditions, a disjunction, a conjunction and a negation. The original bipolar conditions [6] restricted to boolean conditions is a particular case of the results introduced here (it is the case where $n=2$).

In the following C_1, C_2, ..., C_n are n boolean conditions such that $\forall i, C_i \Rightarrow C_{i+1}$. It means that condition C_{i+1} is a relaxation of condition C_i (C_i is included in C_{i+1}). This property is called normalization.

Definition. A (multipolar) "or else" condition is noted $[C_1, C_2, \ldots, C_n]$ and expresses "to satisfy C_1 or else to satisfy C_2 or else ... or else to satisfy C_n", while a (multipolar) "and if possible" condition is noted $(C_n, C_{(n-1)}, \ldots, C_1)$ and expresses "to satisfy C_n and if possible to satisfy $C_{(n-1)}$ and if possible ... and if possible to satisfy C_1".

It is very important to keep the property : $\forall i, C_i \Rightarrow C_{i+1}$. Condition C_{i+1} is a relaxation of condition C_i and the "or else" has a precise meaning: *to satisfy a condition* or else *a relaxation of this condition*. Similarly, the "and if possible" condition has a precise meaning: *to satisfy a condition* and if possible *a more restrictive variante of this condition.*

The satisfaction of different elements with respect of both types of conditions can be represented by a vector of n values from $\{0, 1\}$. The different vectors can be ranked using the lexicographical order. The highest the ranking, the more preferred the element. In case of an "or else" condition, the first positions of this vector are the 0 values while the last values are 1 values (if such values exists). One can observe that the vector is sorted in increasing order. In case of an "and if possible" condition, the first positions of this vector are the 1 values (if such values exists) while the last values are 0 values. One can observe that the vector is sorted in decreasing order. We can now define the following properties (the proof is obvious).

Property 1. An "or else" condition is entirely false when the associated vector is made of 0 (it has no rank). Otherwise, the ranking is the position of the first value 1 in the vector. In other words, the ranking is (k'+1) where k' is the number of 0 values in the vector (due to the normalization).

Property 2. An "and if possible" condition is entirely false when the associated vector is made of 0 (it has no rank). Otherwise, the ranking is (k'+1) where k' is the number of 0 values in the vector.

It means that, since we deal with Boolean arguments, the ranking is the same whether we deal with "C_1 or else C_2 or else ... or else C_n" or "C_n and if possible $C_{(n-1)}$ and if possible ... and if possible C_1" conditions. As a consequence, when dealing with Boolean predicates, multipolar "or else" conditions and multipolar "and if possible" conditions lead to same results (a similar conclusion is made [6] when dealing with only two conditions). Mutipolar Boolean conditions of type "or else" and "if possible" are then equivalent (due to the Boolean context).

As a consequence, in the following we only deal with multipolar "or else" conditions but the obtained results are also valid for multipolar "and if possible" conditions. Furthermore, we consider the level of ranking in the lexicographical order as the level of satisfaction of a bipolar condition (the smaller the level, the more preferred it is).

We now define the negation, the conjunction and the disjunction of bipolar conditions.

Negation. The negation of $[C_1, C_2, \ldots, C_n]$ is $[\neg C_n, \neg C_{(n-1)}, \ldots, \neg C_1]$.

Proof. It is obvioulsy idempotent, and it reverses the order (since the ranking of a condition is the number of 0). The vector made of 0's is turned into the one made of 1's and vice-versa.

One may remark that the normalization is kept by the negation. Let A and B be two "or else" conditions:

Conjunction. $A \wedge B$ has no level of satisfaction when A *or* B is entirely false. Otherwise, the level of satisfaction for $A \wedge B$ is $max(k, k')$ where k is the level of satisfaction of A and k' the one for B.

Disjunction. $A \vee B$ has no level of satisfaction when A *and* B are entirely false. When only A (resp. B) is entirely false, the level of satisfaction for $A \vee B$ is that of B (resp. A). Otherwise, the level of satisfaction for $A \vee B$ is $min(k, k')$ where k is the level of satisfaction for A and k' the one for B.

It is obvious to show that these two definitions satisfy the properties of a conjunction (extended t-norm) and disjunction (extended t-conorm).

MBC's can then be handled using extended algebraic operators (conjunction, disjunction and negation). On may remark that a Boolean condition C is rewritten $[C, C, \ldots, C]$ or equivalently $(C, C, \ldots, C)$ and its vector representing its satisfaction is either the vector made only with values 1 or only with values 0.

4.2   QCL and Boolean Multipolar Conditions

First of all, in order to compare QCL and BMC's it is necessary to limit the application of QCL to a particular form of formulas. This form is that of normalized formulas defined by:

$$A_1 \overrightarrow{\times} \ldots \overrightarrow{\times} A_n \text{ with } \forall i, A_i \Rightarrow A_{i+1}.$$

This is not a real limitation of QCL because any QCL expression can be normalized[1] and it does not affect the meaning of QCL and its rules.

Result. The negation excepted, QCL statements and rules can be expressed by MBC's.

Proof. The obtained ranking for $(A_1 \overrightarrow{\times} \ldots \overrightarrow{\times} A_n)$ in QCL and $[A_1, \ldots, A_n]$ (even if it is not normalized) are the same (it is the position of the first value 1). The disjunction and the conjunction are similar for QCL and MBC's. Rules (1),(3)

---

[1] Each QCL expression $C_1 \overrightarrow{\times} \ldots \overrightarrow{\times} C_n$ is turned into $A_1 \overrightarrow{\times} \ldots \overrightarrow{\times} A_n$ with $A_i = C_1 \vee C_2 \vee \ldots C_i$.

and (4) can then be defined by MBC's. Rules (2) and (5) (dealing with propositional atoms) can also be expressed by MBC's because a Boolean condition C is defined as "C or else C or else C ..." and its level is 1 when it is satisfied. However, it can be observed that the negation (6) is not the one proposed for MBC's. **EndProof.**

In order to have a complete matching between QCL and MBC's, it is necessary to replace negation (6) of QCL by a new definition of the negation. This new definition is given by (6'):

(6') $I \models_k \neg(A_1 \vec{\times} \ldots \vec{\times} A_n)$ iff $I \models_k (\neg A_n \vec{\times} \ldots \vec{\times} \neg A_1)$.

It is possible to show that the negation defined by (6') satisfies the order-reversing property:

Properties

- $I \models_k (A_1 \vec{\times} \ldots \vec{\times} A_n)$ (with $k \neq 1$) $\Leftrightarrow I \models_{k'} \neg(A_1 \vec{\times} \ldots \vec{\times} A_n)$ with $k' = n + 2 - k$;
- $I \models_1 (A_1 \vec{\times} \ldots \vec{\times} A_n) \Leftrightarrow I$ is not considered for $\neg(A_1 \vec{\times} \ldots \vec{\times} A_n)$ (i.e. $\neg \exists k$ such that $I \models_k \neg(A_1 \vec{\times} \ldots \vec{\times} A_n)$);
- $I \models_1 \neg(A_1 \vec{\times} \ldots \vec{\times} A_n) \Leftrightarrow I$ is not considered for $(A_1 \vec{\times} \ldots \vec{\times} A_n)$ (i.e. $\neg \exists k$ such that $I \models_k (A_1 \vec{\times} \ldots \vec{\times} A_n)$).

Proof

- We consider $I \models_k A_1 \vec{\times} \ldots \vec{\times} A_n$ (with $k \neq 1$). We get $I \models (A_1 \vee \ldots \vee A_n)$ and $k = min\{j \mid I \models A_j\}$. Since k is a minimum, then:
  - $\forall 1 \leq j \leq (k - 1)$, $I \models \neg A_j$ (if it not holds, as example for k = 5 and for j=2, we have $I \models A_2$ and then the minimum is smaller or equal to 2 and it cannot be k),
  - $I \models A_k$ and, due to the normalization, $\forall j, k \leq j \leq n, I \models A_j$.
  As a consequence we obtain $I \models_{k'} (\neg A_n \vec{\times} \ldots \vec{\times} \neg A_1)$ with $k' = n + 2 - k$. From (6') we get $I \models_{k'} \neg(A_1 \vec{\times} \ldots \vec{\times} A_n)$ with $k' = n + 2 - k$.
- When $I \models_1 A_1 \vec{\times} \ldots \vec{\times} A_n \Leftrightarrow I \models A_1 \Leftrightarrow$ due to the normalization, $\forall j, 1 \leq j \leq n, I \models A_j \Leftrightarrow \neg \exists k$ such that $I \models_k \neg(A_1 \vec{\times} \ldots \vec{\times} A_n)$.
- When $I \models_1 \neg(A_1 \vec{\times} \ldots \vec{\times} A_n) \Leftrightarrow I \models_1 (\neg A_n \vec{\times} \ldots \vec{\times} \neg A_1) \Leftrightarrow I \models \neg A_n \Leftrightarrow$ due to the normalization, $\forall j, 1 \leq j \leq n, I \models \neg A_j \Leftrightarrow \neg \exists k$ such that $I \models_k (A_1 \vec{\times} \ldots \vec{\times} A_n)$.

Example 2. We reconsider Example 1 already given for the negation. A user is looking for a trip and his preference is "if possible *Air France* and if *Air France* is impossible then *KLM* and if *KLM* is impossible then *British Airways*". This preference is expressed by the (normalized) ordered disjunction:

$$AF \vec{\times} (AF \vee KLM) \vec{\times} (AF \vee KLM \vee BA).$$

An interpretation $I_1 = \{AF\}$ receives the level 1 of satisfaction, an interpretation $I_2 = \{KLM\}$ receives the level 2 of satisfaction, and an interpretation $I_3 = \{BA\}$ receives the level 3 of satisfaction. As a consequence, I_1 is preferred to I_2 which is preferred to I_3. The negation is:

$$\neg(AF \vee KLM \vee BA)\overset{\rightarrow}{\times}\neg(AF \vee KLM)\overset{\rightarrow}{\times}\neg AF.$$

Then, interpretation I_1 is not considered ($\neg\exists k$ such that $I_1 \models_k \neg(AF \vee KLM \vee BA)\overset{\rightarrow}{\times}\neg(AF \vee KLM)\overset{\rightarrow}{\times}\neg AF$), interpretation I_2 receives level 3 while interpretation I_3 receives level 2. The order is reversed and interpretation I_1 is not considered by the negation since it is the best one for the (not negated) ordered disjunction. Furthermore, if we consider interpretation $I_4 = \{AL\}$ which sells tickets only with *Alitalia*, this interpretation is not considered by $AF\overset{\rightarrow}{\times}(AF \vee KLM)\overset{\rightarrow}{\times}(AF \vee KLM \vee BA)$ ($\neg\exists k$ such that $I_1 \models_k AF\overset{\rightarrow}{\times}(AF \vee KLM)\overset{\rightarrow}{\times}(AF \vee KLM \vee BA)$) but it has level 1 for its negation.

5   Towards a Gradual Qualitative Choice Logic

In this section, we consider fuzzy conditions defined on a symbolic scale of m values $\{w_1 = \text{best}, w_2 = \text{very-good}, w_3 = \text{good}, \ldots, w_m = \text{rejected}\}$ instead of the unit interval $[0, 1]$. This symbolic scale is such that $(w_i > w_{i+1})$ and $(\forall i,\ \bar{w}_i = 1 - w_i = w_{m-i+1})$. When comparing with the unit interval $[0, 1]$, such a scale is easier to be understood by an and user and it does not change the obtained results. Furthermore, in the following $C_1, C_2, \ldots, C_n$ are n fuzzy conditions such that $\forall i, \forall u, \mu_{C_i}(u) \leq \mu_{C_{i+1}}(u)$. Condition C_{i+1} is a relaxation of condition C_i.

Definitions. A fuzzy (multipolar) "or else" condition is noted $[C_1, C_2, \ldots, C_n]$ and expresses "to satisfy C_1 or else to satisfy C_2 or else ... or else to satisfy C_n" while a fuzzy (multipolar) "and if possible" condition is noted $(C_n, C_{(n-1)}, \ldots, C_1)$ and expresses "to satisfy C_n and if possible to satisfy $C_{(n-1)}$ and if possible ...and if possible to satisfy C_1".

Here again, it is very important to keep the property such that condition C_{i+1} is a relaxation of condition C_i and an "or else" condition has a precise meaning: *to satisfy a condition* or else *a relaxation of this condition*. A fuzzy multipolar "or else" condition can be illustrated by the following example: "find the trips having a *cheap* price and a *short* flight time *or else* a *cheap* price". Similarly, an "and if possible" condition has a precise meaning: *to satisfy a condition* and if possible *a more restrictive version of this condition*.

The satisfaction of different elements with respect of both types of conditions can be represented by a vector of n values from $\{w_1, w_2, w_3, \ldots, w_m\}$. The different vectors can be ranked using the lexicographical order. The higher the ranking, the more preferred the element. We observe that, in case of an "or else" condition, the vector is sorted in increasing order. We observe that, in case of an "and if possible" condition, the vector is sorted in decreasing order. Furthermore, we consider the level of ranking in the lexicographical order as the level of satisfaction of a bipolar condition (the smaller the level, the more preferred it is). The rank of one vector can be computed using the indexes of its components (the proof of the following property is omitted due to a lack of space). We recall that a vector made of 0 values has no rank because it corresponds to an entirely false bipolar condition.

Property 3. We consider a fuzzy multipolar condition of type $[C_1, C_2, \ldots, C_n]$ or $(C_n, C_{(n-1)}, \ldots, C_1)$. The ranking position k in the lexicographical order is:

$$k = \sum_{j=1}^{(n-1)} m^{n-j}(\lambda_j - 1) + \lambda_n,$$

where m is the number of symbols in the symbolic scale, and the vector expressing the satisfaction of the bipolar condition is noted $(\omega_{\lambda_1}, \omega_{\lambda_2}, \ldots, \omega_{\lambda_n})$.

The conjunction and the disjunction of bipolar conditions are the same as the ones shown in 4.1. It is obvious to show that these two definitions satisfy the properties of a conjunction (extended t-norm) and disjunction (extended t-conorm). We now define the negation:

Negation. The negation of $[C_1, C_2, \ldots, C_n]$ is $(\neg C_1, \neg C_2, \ldots, \neg C_n)$ and the negation of $(C_n, C_{(n-1)}, \ldots, C_1)$ is $[\neg C_n, \neg C_{(n-1)}, \ldots, \neg C_1]$.

Proof. The proof is similar to the one introduced in [6] where bipolar conditions made of two components are considered.

One may remark that the normalization is kept by the negation. In addition, this definition is in accordance with the one introduced in section 4.1 in the case of Boolean condition C_i's. The negation of $[C_1, C_2, \ldots, C_n]$ is $(\neg C_1, \neg C_2, \ldots, \neg C_n)$ which is equivalent to $[\neg C_n, \neg C_{(n-1)}, \ldots, \neg C_1]$ in the Boolean case.

Fuzzy multipolar conditions can then be handled using extended algebraic operators (conjunction, disjunction and negation). One may remark that a fuzzy condition C is rewritten $[C, C, \ldots, C]$ or equivalently $(C, C, \ldots, C)$ and the vector representing its satisfaction is the vector made only with its satisfaction value.

The extension of QCL rules to fuzzy alternatives can be based on these definitions. For this purpose, it becomes necessary to consider the *ordered conjunction* in addition to the *ordered disjunction*. More precisely, fuzzy multipolar conditions of type "and if possible" are defining *ordered conjunctions* while fuzzy multipolar conditions of type "or else" are defining *ordered disjunctions*.

6   Conclusion

This paper has considered the Qualitative Choice Logic (QCL) and a new type of bipolar expression for preference queries: Boolean multipolar conditions to express conditions like "C_1 or else ... or else C_n" or like "C_n and if possible ... and if possible C_1" where each Boolean condition C_{i+1} is a relaxation of the Boolean condition C_i. The conjunction, disjunction and negation have been defined for such conditions and it has been shown how QCL can be expressed using these new types of bipolar conditions. The Boolean multipolar conditions have been extended to fuzzy predicates (each C_i is then a fuzzy predicate) and definitions for the conjunction, the disjunction and the negation of such conditions have been proposed. This extension provides the basis for a gradual QCL model in order to consider fuzzy alternatives in QCL. Future works would aim at defining the complete specification of a gradual QCL and its application to database querying.

References

1. Chomicki, J.: Preference formulas in relational queries. ACM Transactions on Database Systems 28(4), 427–466 (2003)
2. Kießling, W., Köstler, G.: Preference SQL - design, implementation, experiences. In: Proceedings of the 28th VLDB Conference, Hong Kong, China (2002)
3. Bosc, P., Pivert, O.: SQLf: A relational database language for fuzzy querying. IEEE Transactions on Fuzzy Systems 3(1), 1–17 (1995)
4. Dubois, D., Prade, H.: An introduction to bipolar representations of information and preference. International Journal of Intelligent Systems 23, 174–182 (2008)
5. Dubois, D., Prade, H.: Handling bipolar queries in fuzzy information processing. In: Galindo, J. (ed.) Handbook of Research on Fuzzy Information Processing in Databases, pp. 97–114. Information Science Reference, Hershey (2008)
6. Liétard, L., Rocacher, D., Tamani, N.: A relational algebra for generalized fuzzy bipolar conditions. In: Pivert, O., Zadrożny, S. (eds.) Flexible Approaches in Data, Information and Knowledge Management. SCI, vol. 497, pp. 45–69. Springer, Heidelberg (2013)
7. Dubois, D., Prade, H.: Modeling and if possible" and or at least": different forms of bipolarity in flexible querying. In: Pivert, O., Zadrożny, S. (eds.) Flexible Approaches in Data, Information and Knowledge Management. SCI, vol. 497, pp. 3–44. Springer, Heidelberg (2013)
8. De Tré, G., Zadrożny, S., Matthé, T., Kacprzyk, J., Bronselaer, A.: Dealing with positive and negative query criteria in fuzzy database querying bipolar satisfaction degrees. In: Andreasen, T., Yager, R.R., Bulskov, H., Christiansen, H., Larsen, H.L. (eds.) FQAS 2009. LNCS, vol. 5822, pp. 593–604. Springer, Heidelberg (2009)
9. Zadrożny, S., Kacprzyk, J.: Bipolar queries: A way to enhance the flexibility of database queries. In: Ras, Z.W., Dardzinska, A. (eds.) Advances in Data Management. SCI, vol. 223, pp. 49–66. Springer, Heidelberg (2009)
10. Bosc, P., Pivert, O.: On a fuzzy bipolar relational algebra. Information Sciences 219, 1–16 (2013)
11. Brewka, G., Benferhat, S., Berre, D.L.: Qualitative choice logic. Artificial Intelligence 157, 203–237 (2004)
12. Benferhat, S., Sedki, K.: A revised qualitative choice logic for handling prioritized preferences. In: Mellouli, K. (ed.) ECSQARU 2007. LNCS (LNAI), vol. 4724, pp. 635–647. Springer, Heidelberg (2007)

A Vocabulary Revision Method
Based on Modality Splitting

Grégory Smits[1], Olivier Pivert[1], and Marie-Jeanne Lesot[2,3]

[1] IRISA - University of Rennes 1, UMR 6074, Lannion, France
{gregory.smits,olivier.pivert}@irisa.fr
[2] Sorbonne Universités, UPMC Univ Paris 06, UMR 7606,
LIP6, F-75005, Paris, France
[3] CNRS, UMR 7606, LIP6, F-75005, Paris, France
marie-jeanne.lesot@lip6.fr

Abstract. Using linguistic fuzzy variables to describe data improves the interpretability of data querying systems and thus their quality, under the condition that the considered modalities induce an indistinguishability relation in adequacy with the underlying data structure. This paper proposes a method to identify and split too general modalities so as to finally obtain a more appropriate vocabulary wrt. the data structure.

Keywords: interpretability, indistinguishability, linguistic variables, adequacy.

1 Introduction

The use of linguistic variables leads to interpretable descriptions of data that can be easily understood by human beings, making it possible to improve the quality of data querying processes. Furthermore, the choice of the considered linguistic terms can be left to the user, e.g. an applicative context expert, offering the possibility for personalization, improving further the system quality.

Formally, fuzzy linguistic variables can be used to faithfully model the imprecise and gradual nature of the terms. They induce indistinguishability relations insofar as objects with different numerical values cannot be distinguished if they are described with the same linguistic labels. This indistinguishability is legitimate as it corresponds to objects that are equally preferred by the expert.

However, the vocabulary is defined by an expert for a given applicative context but not a particular data set, and it is necessary to check whether the indistinguishability relation does not hide the data specificity and the existence of subgroups that should be differentiated. This issue can be illustrated by the fictitious example of a tourism office expert who possesses a vocabulary to describe and query hotels in Paris, France. If he is asked to deal with hotels in Sophia, Bulgaria, where the accommodation market is globally cheaper, the modality used to characterize *cheap* hotels may not be appropriate anymore, as illustrated in Fig. 1: most of the different groups of Bulgarian hotels, $\{C_1, C_2, C_3, C_4\}$, fully satisfy the modality *cheap* and are thus indistinguishable on their price.

A. Laurent et al. (Eds.): IPMU 2014, Part III, CCIS 444, pp. 140–149, 2014.

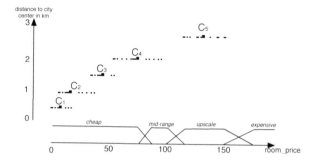

Fig. 1. Inappropriate vocabulary to describe Bulgarian hotels (fictitious values)

This paper proposes to address this issue through the detection of inadequate modalities and the suggestion of local modifications through modality splitting, leading to a specific case of vocabulary revision to better adapt to a given data set. For the illustrative example, the proposed methodology for instance decomposes the modality *cheap* to finally have a more refined linguistic description of the hotels in the Bulgarian data set.

It must be underlined that inappropriate modalities are interpreted here as too general terms that inadequately cover several data subgroups. Too detailed terms that may introduce distinctions within natural data subgroups, are not considered as problematic: they can be exploited using term disjunctions and do not lead to misleading data interpretations as too general terms do.

The paper is organized as follows: Section 2 describes the considered task more formally and discusses related works. Sections 3 and 4 present the two main steps of the proposed method, namely the identification of problematic modalities and the revision process based on splitting operations. Section 5 describes experimental results obtained on artificial data and Section 6 concludes and draws perspectives for future works.

2   Formalization

This section presents the two spaces that can be used to describe the data, respectively based on numerical attributes and on linguistic variables. It then describes the proposed interpretation of adequacy in terms of indistinguishability matching and discusses some related works.

2.1   Numerical Description

The considered data set, denoted by $\mathcal{D} = \{x_1, x_2, ..., x_n\}$, is described by m numerical attributes $A_1, A_2, ..., A_m$, respectively defined on domain $\mathcal{D}_j, j = 1..m$.

The data underlying structure, defined by the subgroups of similar data, can be extracted automatically, applying a clustering algorithm [1]. In the considered task, three requirements must be taken into account when selecting an appropriate algorithm: it must be scalable to process large data sets, as may arise in the data base querying context. It must also be able to automatically determine the appropriate number of clusters: the aim is to identify the underlying data structure, it is not justified to assume that the data expert knows how many clusters should be identified. Third, each cluster must be associated to a representative, called its center: the proposed methodology relies on the distinguishability of these cluster summaries when they are described using the considered vocabulary.

It must be underlined that the proposed methodology aims at adjusting the vocabulary with respect to the structure identified through the clustering step, trusting the clustering algorithm and not questioning the results it yields.

2.2   Linguistic Description

The vocabulary whose appropriateness must be measured is defined by linguistic variables, associating each attribute with a set of linguistic labels and a fuzzy Ruspini partition [2]: formally, for attribute A_j, $j = 1..m$, a_j denotes the number of associated modalities and $V_j = \{v_{j1}, \ldots v_{ja_j}\}$ their associated fuzzy sets. The Ruspini property imposes that $\forall j = 1..m$, $\forall x \in \mathcal{D}_j$, $\sum_{k=1}^{a_j} \mu_{v_{jk}}(x) = 1$.

This paper considers the case of trapezoidal modalities, represented by quadruplets (a, b, c, d) where $]a, d[$ denotes the fuzzy set support and $]b, c[$ its core.

An object $x \in \mathcal{D}$ can then be rewritten as a vector of $\sum_{k=1}^{m} a_k$ membership degrees $\langle \mu_{v_{11}}(x.A_1), \ldots, \mu_{v_{1a_1}}(x.A_1), \ldots, \mu_{v_{m1}}(x.A_m), \ldots, \mu_{v_{ma_m}}(x.A_m) \rangle$ where $x.A$ denotes the value taken by attribute A for data point x. Due to the Ruspini partition property, each object can partially satisfy up to two modalities for a given attribute: the above vector has at most $2m$ non zero components.

As discussed in the introduction, a data representation based on fuzzy linguistic variables induces an indistinguishability relation insofar as it does not allow to differentiate between objects having the same membership degrees, in particular the objects within the core of each modality.

2.3   Confronting Separability and Indistinguishability

The adequacy of the two description spaces can be measured by confronting the separability of the clusters and the indistinguishability of the linguistic descriptions. It thus corresponds to the accuracy issue of a fuzzy partition system, related to its capability of faithfully and precisely describing the clusters using the words defined in the expert vocabulary: clusters, corresponding to separated groups of similar objects, must be linguistically described differently. It has been shown in various applicative contexts, from fuzzy rule-based systems inference [3] or fuzzy partitions generation [4,5] to fuzzy decision tree revision [6], that these two notions of accuracy and interpretability are generally contradictory properties: a trade-off has to be found so as to avoid accurate but useless systems or symmetrically interpretable systems returning irrelevant results.

2.4   Related Works and Characteristics of the Proposed Method

There exist many criteria to measure the quality of a cluster decomposition [7,8] or the global adequacy between two partitions [9,10]. However, these methods usually compare only numerical descriptions of the data and do not aim at matching different types of data representations.

In [11], the adequacy between a numerical and a linguistic representation is quantified, in a global approach that evaluates the whole set of linguistic variables altogether. In this paper, we propose a local approach, considering each modality in turn: it makes it possible to identify inappropriate modalities and to suggest modifications for the latter.

There exist methods to elicit fuzzy sets from numerical data descriptions [5,3], e.g. for inducing fuzzy decision trees [4,6]. In this paper we propose to start from the user defined vocabulary and to perform local revision, so as to increase interpretability and to preserve the personalization capability of the system.

In these approaches that automatically generate fuzzy partitions or fuzzy rules, the interpretability of the system is e.g. quantified by the standard deviation of the number of partition elements [3] or of their density [5]. In [3], it is also pointed out that the use of normalized partitions that cover the whole domain of each attribute without huge overlapping and with distinguishable prototype elements for each partition element improves the overall interpretability: the latter mainly depends on the distinguishability of the most representative elements [3], when using the vocabulary terms.

This justifies the use of Ruspini partitions to represent the expert vocabulary and the focus on cluster centers distinguishability during the vocabulary revision process. Moreover, considering centers only, as a summary of the cluster instead of all its members, offers a scalability property. Finally, as a subjective notion, interpretability is also related to the fact that the proposed method performs local revision and reduces the modifications of the expert vocabulary.

3   Identification of Problematic Modalities

3.1   Principle

As described and justified in Section 2.1, given a data set and an expert vocabulary, the preliminary step of the proposed methodology consists in identifying the cluster decomposition of the data.

The first step then consists in identifying inappropriate fuzzy modalities, that induce an indistinguishability relation that does not match structural distinguishability: the adequacy of a given modality is derived from its ability to distinguish between well separated cluster centers.

Indeed, its adequacy cannot be defined solely from its ability to distinguish all cluster centers, as illustrated on Figure 2: both in the left and right cases, C_i and C_j are described with the modality v and thus indistinguishable. However, in the right case, they are not distinguished in the numerical description space either: the considered attribute is not relevant to characterize these clusters. Thus they should not imply that v is problematic, contrary to the left case.

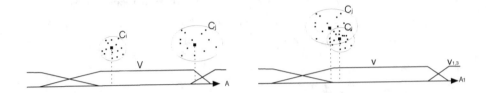

Fig. 2. (Left) Problematic indistinguishability, (right) justified indistinguishability

3.2   Proposed Method

The adequacy of a modality $v = (a, b, c, d)$ associated to an attribute A is assessed using the following 4-step procedure:

1. Identify the relevant clusters C, defined as clusters mainly represented by v on attribute A, i.e. such that $\mu_v(c.A) \geq 0.5$ where c is C center.
   Indeed it can be considered that clusters mainly represented by other modalities are not significant to assess v adequacy.
2. Sort the relevant clusters according to the value of their center on A.
3. Measure the relative numerical distance for all pairs of adjacent relevant centers (c_i, c_j), as their distance according to attribute A compared to v support: $dis_v(c_i, c_j) = (c_j.A - c_i.A)/(d - a)$. The support of v thus defines a distance scale, making the definition of adequacy given below relative to each modality. The lower dis_v, the more indistinguishable the two adjacent centers.
4. Measure the linguistic indistinguishability for all pairs of successive relevant centers $ind_v(c_i, c_j) = 1 - 2 \times |\mu_v(c_i.A) - \mu_v(c_j.A)|$. The factor 2 is used to obtain a value in $[0, 1]$ as $\mu_v(c_i.A) \geq 0.5$ and $\mu_v(c_j.A) \geq 0.5$.

The adequacy issue can then be visualized in the two-dimensional space illustrated by Fig 3. A point in this space represents a triplet composed of a modality and two adjacent centers that are mainly rewritten by the considered modality. Four regions can be considered: the upper left part corresponds to centers that are relatively far apart and have a low linguistic indistinguishability, i.e. that have satisfyingly distinct rewritten forms. They thus indicate an adequate modality. Symmetrically, the lower right part corresponds to numerically close centers with high linguistic indistinguishability. On the contrary, the upper right part corresponds to centers that are relatively far apart but have the same rewritten form for the considered modality. It thus indicates too large modalities that are inadequate and may be solved by a splitting operation, as presented in the next section. The lower left region corresponds to centers that are close in the numerical space, mainly rewritten by the same modality but with different degrees. They may be made more indistinguishable by adjusting the slopes of the fuzzy modalities. However this less problematic kind of local revision of the vocabulary is left for future work.

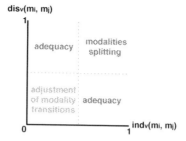

Fig. 3. Confronting numerical center distance and linguistic indistinguishability

4   Suggestion of Local Revisions

4.1   Priority Order for Revision

In order to process first the most serious cases of indistinguishability, i.e. the ones from the upper right part of Fig. 3, the two previous criteria are aggregated into a degree of seriousness defined as

$$ds(v, c_i, c_j) = \min(dis_v(c_i, c_j), ind_v(c_i, c_j)).$$

where c_i and c_j are two relevant adjacent centers, i.e. adjacent centers mainly rewritten by v. These values are then aggregated at the modality level as

$$DS(v) = \frac{1}{M} \sum ds(v, c_i, c_j),$$

where the sum applies to all pairs of relevant adjacent centers and M is the number of such pairs.

The modalities can then be processed in decreasing order of DS and for a given modality, the problematic pairs of centers in decreasing order of ds. Revisions may be suggested only for the modalities whose degree is higher than a predefined threshold, or iteratively for all problematic modalities as long as the expert validates them.

4.2   Modality Split

The revision of a modality defined on attribute A and represented by $v = (a, b, c, d)$ due to the insufficient indistinguishability of two relevant adjacent centers c_1 and c_2 is performed through a splitting operation that leads to 2 new trapezoidal modalities, respectively defined as $v' = (a, b', \gamma, \delta)$ and $v'' = (\gamma, \delta, c', d)$. Thus the modality support outer bounds a and d are kept unchanged.

The core outer bounds of the 2 new modalities v' and v'' are possibly expanded into $b' = \min(b, c_1.A)$ and $c' = \max(c, c_2.A)$: these definitions guarantee that each cluster center totally belongs to one modality. This may lead to modifying the modalities adjacent to v so as to preserve the Ruspini partition property.

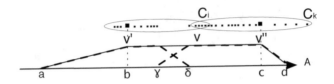

Fig. 4. Modality splitting to improve the cluster interpretability

The inner bounds γ and δ then determine the indistinguishability regions induced by v' and v''. It is not justified to define them so as to try to optimize the separation between the cluster members as the latter can be justified by dimensions others than A. This is illustrated in Fig. 4: when restricted to the sole attribute A, the two clusters overlap, making it impossible to justify a splitting position. Therefore we propose to define γ and δ in adequacy with the distribution of the whole data set in the interval $[b, c]$, so that their difference corresponds to the maximal gap between successive values observed on A: indexing the data $x \in \mathcal{D}$ so that they are sorted according to their A values, we set

$$l^* = \arg\max_l (x_{l+1}.A - x_l.A) \qquad \gamma = x_{l^*} \quad \delta = x_{l^*+1}$$

This favours splitting in low density regions, which leads to the desired matching between the revised partition and the data distribution. It can be observed that, on the other hand, it may lead to modalities with low degree of fuzziness if the local data density is high. Finally, the system asks the expert to linguistically qualify the two modalities resulting from the split operation.

5   Experimental Results

5.1   Considered Data Set

Experiments to assess the relevance of the proposed methodology are carried out on a small representative 2D artificial data set, generated as a mixture of 3 well-separated Gaussian distributions. It is illustrated on the left part of Fig. 5 that also shows the considered initial vocabularies. For the x attribute, 6 partitions are considered, ordered by increasing number of modalities. $P1$ contains a single, too generic, modality that makes all data indistinguishable. $P2$ is a correct partition with the two expected modalities. It must be underlined that it actually leads to the same rewriting form for the centers of the two left subgroups. $P2a$ also contains 2 modalities, but the position of the transition between them is inappropriate wrt. the data distribution. $P3$ and $P3l$ contain 3 modalities making the central part of the space indistinguishable and bridging the gap between the two lower subgroups. They differ by the size of the overlapping region. Partition $P4$ represents a too detailed partition with too many modalities. For the y-attribute, a single linguistic variable is considered, defined by a partition that matches the data distribution.

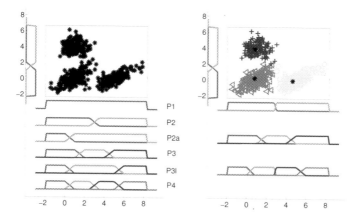

Fig. 5. (Left) Considered data and vocabularies, (right) clustering results (* denotes the identified cluster centers) and revised vocabularies

The underlying structure of the data set is automatically extracted applying the *l-fcmed-select* algorithm [12], that satisfies the requirements described in Section 2.1. As can be seen on the upper right part of Fig. 5, it succeeds in identifying 3 clusters, only a few incorrect assignments occur.

5.2   Obtained Results

Revised Modalities. The proposed methodology is applied on this data set with the threshold for the degree of seriousness DS being set to 0.2. The obtained revised modalities are shown on the right part of Fig. 5: the partitions $P2$, $P3$ and $P4$ are left unchanged, as well as the partition defined on the y-attribute, whereas $P1$, $P2a$ and $P3l$ are modified.

Indeed, for the partition defined on the y-attribute, the first modality, denoted by v_{y1} leads to the same rewritten form of the lower cluster centers denoted by c_2 and c_3: $ind_{v_{y1}}(c_2, c_3) = 1$. Yet these centers are very close in the numerical space and $dis_{v_{y1}}(c_2, c_3) = 0.07$. Thus $ds(v_{y1}, c_2, c_3) < 0.2$. For the second modality, there is no pair of centers mainly represented by v_{y2}, thus it does not need rewriting either. The same principle applies to partition $P2$ on the x-attribute: for the first modality v_{x1}, $ind_{v_{x1}}(c_1, c_2) = 1$ but $dis_{v_{x1}}(c_1, c_2) = 0.006$, thus its degree of seriousness is below the revision threshold: the identical rewriting form for the cluster centers is justified by their low difference in the numerical domain, not needing revision.

For partition $P4$, all centers have distinct rewriting forms, so no pair of centers is mainly represented by a common modality. This is compatible with the principle stated in the introduction: $P4$ e.g. allows to describe the rightmost cluster as the disjunction "v_{x3} or v_{x4}". Therefore the too detailed partition does

Table 1. Average and standard deviation of the vocabulary quality measured by the q_{XB} criterion [11] over 1000 initializations of the clustering algorithm

Partition	Before revision	After revision
P1	Inf	0.04 ± 0.04
P2	0.04 ± 0.03	0.04 ± 0.04
P2a	Inf	0.24 ± 0.03
P3	0.20 ± 0.03	0.21 ± 0.05
P3l	Inf	0.36 ± 0.07
P4	0.33 ± 0.05	0.34 ± 0.06

not misleadingly ignore the underlying data distribution and does not need revision. Similarly for partition $P3$, all centers have distinct rewriting forms and no revision is necessary: although the central modality bridges the gap between the two lower clusters, it is not considered as too general with respect to their centers. This result is compatible with the aim of minimizing the amount of modification to the expert vocabulary, the scale for determining whether a split is necessary depending on the data summary through the cluster centers.

On the other hand, for partition $P3l$, the central modality is so large that it reaches the cluster centers, making them indistinguishable and leading to revision. Indeed, $dis_{v_{x2}}(c_2, c_3) = 0.64$ and $ind_{v_{x2}}(c_2, c_3) = 0.63$, leading to $DS(v_{x2}) = 0.63$. Splitting this modality leads to a partition similar to $P4$.

Partition $P1$ which is indeed much too general is split as expected. The maximal gap between successive data on the x-attribute corresponds to the low density region that defines the limit between the two clusters; as a result, the obtained revised $P1$ is very similar to the correct partition $P2$. Likewise, the revision of $P2a$ leads to split the too large modality, leading to a right modality that indeed characterizes the rightmost cluster and a global result similar to $P3$.

Numerical Evaluation. In order to numerically assess the quality of the revised modalities, we apply the Xie-Beni vocabulary adequacy criterion q_{XB} [11] that must be minimized: it is defined as the quotient between cluster compactness and separability, where clusters are identified from the numerical data representation and compactness and separability are measured from the rewritten representation, based on a distance between vectors of membership degrees.

Table 1 shows the values obtained for each vocabulary before and after revision. The low standard deviations over 1000 initializations of the clustering algorithm show that on this data set, the *l-fcmed-select* algorithm is stable.

The revised partitions initially have infinite q_{XB}: when cluster centers are indistinguishable, their separability is zero. It can be noted that it may be the case that partitions with initial finite q_{XB} undergo revision: q_{XB} is infinite only if two centers exactly have the same rewritten form, for all attributes and modalities, whereas revision takes place independently for each modality. As discussed when looking at the obtained revised partitions, q_{XB} value for the the revised $P1$ is comparable that of $P2$, $P2a$ becomes comparable to $P3$ and $P3l$ to $P4$.

6 Conclusion and Perspectives

In the perspective of developing natural and possibly personalized data processing and querying systems, this paper addressed the crucial issue of adapting a user-defined vocabulary to the specificity of the data set it must describe. It proposes a method to balance two contradictory aims: on one hand the preservation of the user subjectivity, expressed by the linguistic terms he defines and, on the other hand, its adequacy to the underlying data distribution. It focused on the revision of too general terms that do not allow to make the difference between distinct subgroups of the data and lead to misleading representation of the data content, hiding the data specificity. The proposed method first identifies problematic modalities and then splits them in adequacy to the data distribution.

Ongoing works aim at performing a more thorough analysis of the proposed method behaviour, in particular to study its scalability property and to process more complex data, real data offering the possibility of a user-based subjective evaluation. Other perspectives include the study of other vocabulary revision principles, in particular to locally adjust the slope of the modalities.

References

1. Jain, A., Murty, M., Flynn, P.: Data clustering: a review. ACM Computing Survey 31(3), 264–323 (1999)
2. Ruspini, E.H.: A new approach to clustering. Information and Control 15(1), 22–32 (1969)
3. Gacto, M.J., Alcalá, R., Herrera, F.: Interpretability of linguistic fuzzy rule-based systems: An overview of interpretability measures. Information Sciences 181(20), 4340–4360 (2011)
4. Marsala, C.: Fuzzy partition inference over a set of numerical values. In: Proc. of the IEEE Int. Conf. on Fuzzy Systems, pp. 1512–1517 (1995)
5. Guillaume, S., Charnomordic, B.: Generating an interpretable family of fuzzy partitions from data. IEEE Transactions on Fuzzy Systems 12(3), 324–335 (2004)
6. Marsala, C.: Incremental tuning of fuzzy decision trees. In: Proc. of the 6th IEEE Int. Conf. on Soft Computing and Intelligent Systems, SCIS, pp. 2061–2064 (2012)
7. Meila, M.: Comparing clustering, an axiomatic view. In: Proc. of the 22nd Int. Conf. on Machine Learning (2005)
8. Le Capitaine, H., Frélicot, C.: A cluster-validity index combining an overlap measure and a separation measure based on fuzzy-aggregation operators. IEEE Trans. on Fuzzy Systems 19(3), 580–588 (2011)
9. Valet, L., Mauris, G., Bolon, P., Keskes, N.: A fuzzy linguistic-based software tool for seismic image interpretation. IEEE Trans. on Instrumentation and Measurement 52(3), 675–680 (2003)
10. Raschia, G., Mouaddib, N.: Évaluation de la qualité des partitions de concepts dans un processus de résumés de bases de données. In: Proc. of LFA (2000)
11. Lesot, M.J., Smits, G., Pivert, O.: Adequacy of a user-defined vocabulary to the data structure. In: Proc. of the IEEE Int. Conf. on Fuzzy Systems (2013)
12. Lesot, M.-J., Revault d'Allonnes, A.: Credit-card fraud profiling using a hybrid incremental clustering methodology. In: Hüllermeier, E., Link, S., Fober, T., Seeger, B. (eds.) SUM 2012. LNCS, vol. 7520, pp. 325–336. Springer, Heidelberg (2012)

Dealing with Aggregate Queries in an Uncertain Database Model Based on Possibilistic Certainty

Olivier Pivert[1] and Henri Prade[2]

[1] University of Rennes 1, Irisa, Lannion, France
[2] CNRS/IRIT, University of Toulouse, Toulouse, France
pivert@enssat.fr, prade@irit.fr

Abstract. This paper deals with the evaluation of aggregate queries in the framework of an uncertain database model where the notion of necessity is used to qualify the certainty that an ill-known piece of data takes a given value or belongs to a given subset. Two facets of the problem are considered, that correspond to: i) the nature of the data (certain or uncertain), and ii) the nature of the query (crisp or fuzzy) that specifies the relation over which the aggregate has to be computed.

1 Introduction

Uncertain information may appear in various contexts, such as data warehouses that collect information coming from different sources, automated recognition of objects, sensor networks, forecasts or archives where only partial information is known for sure. In the database community, the last ten years have witnessed a growing interest in uncertain databases, see e.g. [2, 15, 1, 5]. Let us note, however, that the early works on the topic are much older and date back to the late 70s and early 80s [17, 21, 16]). Most authors promote a probabilistic modeling of uncertainty (see [20] for a detailed overview), but a few alternative works [4] rather favor a qualitative modeling of uncertainty through possibility theory, which is our case here. In contrast with probability theory, one expects the following advantages when using possibility theory:

- the qualitative nature of the model simplifies the elicitation of the degrees attached to candidate values (a symbolic scale may be used, in particular);
- in probability theory, the fact that the sum of the degrees from a distribution must equal 1 makes it difficult to deal with incompletely known distributions.

In this latter category of approaches, a pioneering work is that by Prade and Testemale [19] who introduced in the early 80s what may be called a "full-possibilistic" database model (an ill-known attribute value is represented by a possibility distribution of candidate values, and the result of a query is a relation where each tuple is associated with a possibility and a necessity degree). More recently, Bosc and Pivert [3] refined this model so as to make it a *strong*

A. Laurent et al. (Eds.): IPMU 2014, Part III, CCIS 444, pp. 150–159, 2014.

representation system [16] for a significant subset of relational algebra, i.e., a model such that the uncertain relation obtained by a direct evaluation of a query (involving authorized operators only) is equivalent to the set of classical relations obtained when evaluating the query on each possible instance of the uncertain database.

More recently, Bosc *et al.* [6, 7] introduced a new model based on *possibilistic certainty*. The idea is to use the notion of necessity from possibility theory to qualify the certainty that an ill-known piece of data takes a given value (or belongs to a given subset). In contrast with both probabilistic databases *and* possibilistic ones in the sense of [19, 3], the main advantage of the certainty-based model lies in the fact that operations from relational algebra can be extended in a simple way and with a data complexity that is the same as in a classical database context (i.e., where all data are certain).

Here, our objective is to enrich the query language by introducing aggregation operators (*avg, sum, min,* and *max* — the case of *count*, which is more complex, will be only briefly discussed here). Let us recall that in SQL, aggregation operators can be used either in the *select* clause (for instance to express a query such as "find the average salary of the employees for each department": select avg(*salary*) from *Emp* group by #*dep*) or in a *having* clause (for instance in a query as "find the departments where the average salary is higher than \$3000": select #dep from Emp group by #dep having avg(*salary*) > 3000). Two facets of the problem will be considered, that correspond to: i) the nature of the data (certain or uncertain), and ii) the nature of the query (crisp or fuzzy) that produces the relation over which the aggregate has to be computed.

The remainder of the paper is structured as follows. Section 2 presents the main features of the certainty-based model. Section 3 investigates the simple case where the aggregate is precise but the data are uncertain. Section 4 deals with the situation where the aggregate is imprecise (because a fuzzy condition has first been applied) and the data are certain. Section 5 deals with the most general case, i.e., that where the aggregate is imprecise and the data are uncertain. Some difficulties raised by the computation of the aggregation operator *count* in this framework are briefly discussed in Section 6. Finally, Section 7 recalls the main contributions and outlines perspectives for future work.

2   A Short Overview of the Certainty-Based Model

As that described in [3], the model introduced in [6] is based on possibility theory [22, 8], but it represents the values that are more or less certain instead of those which are more or less possible. This corresponds to the most important part of information (in this approach, a possibility distribution is approximated by keeping its most plausible elements, associated with a certainty level). The idea is to attach a certainty level to each piece of data (by default, a piece of data has certainty 1). Certainty is modeled as a lower bound of a necessity measure. For instance, $\langle 037, John, (40, \alpha) \rangle$ denotes the existence of a person named John, whose age is 40 with certainty α. Then the possibility that his age differs from 40 is upper bounded by $1 - \alpha$ without further information on the respective possibility degrees of other possible values. More generally, the underlying possibility distribution associated with an uncertain attribute value (a, α) is $\{1/a, (1 - \alpha)/\omega\}$ where $1/a$ means that a is a completely possible $(\pi = 1)$ candidate value and ω denotes $domain(A) - \{a\}$, A being the attribute considered. This is due to the duality necessity (certainty) / possibility: $N(a) \geq \alpha \Leftrightarrow \Pi(a) \leq 1 - \alpha$ [11]. For instance, let us assume that the domain of attribute *City* is $\{Boston, Newton, Quincy\}$. The uncertain attribute value $(Boston, \alpha)$ is assumed to correspond to the possibility distribution $\{1/Boston, (1-\alpha)/Newton, (1-\alpha)/Quincy\}$. The model can also deal with disjunctive uncertain values, and the underlying possibility distributions π are of the form $\pi(u) = \max(S(u), 1-\alpha)$ where S is an α-certain subset of the attribute domain and $S(u)$ equals 1 if $u \in S$, 0 otherwise.

   Moreover, since some operations (e.g., the selection) may create "maybe tuples", each tuple t from an uncertain relation r has to be associated with a degree N expressing the certainty that t exists in r. It will be denoted by N/t.

Example 1. Let us consider the relation r of schema $(\#id, Name, City)$ containing tuple $t_1 = \langle 1, John, (Quincy, 0.8) \rangle$, and the query "find the persons who live in Quincy". Let the domain of attribute *City* be $\{Boston, Newton, Quincy\}$. The answer contains $0.8/t_1$ since it is 0.8 certain that t_1 satisfies the requirement, while the result of the query "find the persons who live in Boston, Newton or Quincy" contains $1/t_1$ since it is totally certain that t_1 satisfies the condition (only cities in the attribute domain are somewhat possible).◇

To sum up, a tuple $\beta/\langle 037, John, (Quincy, \alpha) \rangle$ from relation r means that it is β certain that person 037 exists in the relation, and that it is α certain that 037 lives in Quincy (independently from the fact that it is or not in relation r). In the following, for the sake of readability of the formulas, we will assume that N equals 1 for every tuple of the relation concerned. However, the case where the computation of the aggregate involves maybe tuples (i.e., tuples such that $N < 1$) does not raise any particular difficulty: one just has to replace an uncertain value (a, α) coming from a tuple whose degree N equals β by $(a, \min(\alpha, \beta))$ when computing the aggregate. Moreover, beyond its value, the existence of the computed aggregate may be uncertain. As soon as at least one of the aggregated values has certainty 1, then the computed aggregate exists with certainty 1. If there is no fully certain tuple to be aggregated, then the certainty of the existence of the result should be taken as equal to the maximum of the certainty degrees of the involved tuples. However, in the case of the aggregation function *count*, this latter situation cannot take place since the cardinality always exists for sure (being possibly equal to 0).

   Given a query, we only look for answers that are somewhat certain. Consider the relations r and s given in Table 1 and a query asking for the persons who live in a city where there is a flea market), then *John* will be retrieved with a certainty level equal to $min(\alpha, \beta)$ (in agreement with the calculus of necessity measures [11]).

Table 1. Relations r (left) and s (right)

#id	Name	City	N		City	Flea Market	N
1	John	(Boston, α)	1		Boston	(yes, β)	1
2	Mary	(Newton, δ)	1		Newton	(no, γ)	1

As mentioned above, it is also possible to accommodate cases of disjunctive information in this setting. For instance, the tuple $\langle 3,\ Peter,\ (Newton \vee Quincy,\ 0.8)\rangle$ represents the fact that it is 0.8-certain that the person number 3 named Peter lives in Newton or in Quincy.

3   Precise Aggregates of Uncertain Data

The most simple case is when each uncertain value that has to be aggregated is represented by a singleton associated with a certainty degree, i.e., is of the form (a_i, α_i). This means that it is α_i-certain that the value is a_i, i.e., $N(a_i) \geq \alpha_i$, where N is a necessity measure, as in possibilistic logic [11–13].

Then, the formula for computing the aggregation function f (where the a_i's are numerical attribute values, and f stands for avg, sum, min, or max) is:

$$f((a_1,\alpha_1),\ \ldots,\ (a_n,\ \alpha_n)) = (f(a_1,\ \ldots,\ a_n),\ \min_i \alpha_i). \tag{1}$$

This expression is a particular case of the extension principle applied for computing $f(A_1\ \ldots,\ A_n)$ where A_i is a fuzzy quantity (here $A_i(u_i) = \max(\mu_{a_i}(u),\ 1 - \alpha_i)$ where $\mu_{a_i}(u) = 1$ if $u = a_i$ and $\mu_{a_i}(u) = 0$ otherwise), in agreement with possibility theory [8]:

$$f(A_1\ \ldots,\ A_n)(v) = \sup_{u_1,\ \ldots,\ u_n : f(u_1,\ \ldots,\ u_n) = v} \min(A_1(u_1),\ldots,A_n(u_n)).$$

Indeed, it is clear that $f(A_1\ \ldots,\ A_n)(v) = 1$ only if $v = f(a_1,\ \ldots,\ a_n)$, and that for any other value of v, $f(A_1\ \ldots,\ A_n)(v)$ is *upper bounded* by $\max_i 1 - \alpha_i = 1 - \min_i \alpha_i$. Hence, $N(f(a_1,\ \ldots,\ a_n)) \geq \min_i \alpha_i$. Note that when the domain of the considered attribute is the whole real line, $f(A_1\ \ldots,\ A_n)(v)$ reaches the upper bound [8], but this is not always the case when we deal with strict subparts of the real line.

In the case where uncertain values are a disjunction of possible values (let us denote by c_i the disjunctive subset representing the set of possible values for the considered attribute in the considered tuple t_i), the formula is (in agreement with the extension principle, and the lower bound interpretation of the certainty pairs):

$$f((c_1,\alpha_1),\ \ldots,\ (c_n,\ \alpha_n)) = (\bigvee_{a_{i_1} \in c_1,\ \ldots,\ a_{i_n} \in c_n} f(a_{i_1},\ \ldots,\ a_{i_n}),\ \min_i \alpha_i). \tag{2}$$

This case is illustrated in Example 2 below. Obviously, when the relation contains many different disjunctions for the attribute considered, the aggregate obtained can be a quite large disjunction. Assume now that an ill-known value is represented by an interval I_i with certainty α_i. The formula is:

$$f((I_1, \alpha_1), \ldots, (I_i, \alpha_i), \ldots, (I_n, \alpha_n)) = (f(I_1, \ldots, I_i, \ldots, I_n), \min_i \alpha_i). \quad (3)$$

An interesting point is that since f is a monotonic function in the computation of aggregates, only the bounds of the intervals have to be considered in the calculus [8].

Example 2. Let us consider the data from Table 2 and the query searching for the average age of the employees. The result is:

Table 2. Relation *Employee*

#id	Name	Age	N
1	John	(35, 0.8)	1
2	Mary	(22 $\vee$ 32, 0.7)	1
3	Paul	(45 $\vee$ 54, 0.4)	1

$(\frac{35+22+45}{3} \vee \frac{35+32+45}{3} \vee \frac{35+22+54}{3} \vee \frac{35+32+54}{3}, \min(0.8, 0.7, 0.4))$
$= (34 \vee 37.3 \vee 37 \vee 40.3, 0.4)$.

Now if we had $([22, 32], 0.7)$ for Mary's age and $([45, 54], 0.4)$ for Paul's, the result would simply be $([34, 40.3], 0.4)$.◇

Remark 1. When some ill-known values are represented by disjunctions of intervals, the formula to be used is a straightforward mix of (2) and (3) — we do not give it here as it is rather cumbersome — and one gets a disjunction of intervals as a result.

Example 3. Let us now assume that the aggregation operation is the minimum. Let us consider two values $v_1 = (12 \vee 14, 1)$ and $v_2 = (8 \vee 13, 0.8)$. Then,

$$\min(v_1, v_2) = (8 \vee 12 \vee 13, \min(1, 0.8))$$

as a result of the application of Equation 2:

$$\min(v_1, v_2) = \min(12, 8) \vee \min(12, 13) \vee \min(14, 8) \vee \min(14, 13)$$

which corresponds to the union of the partial results considering each possible value in the disjunction separately.

Now, assuming $v_1 = ([12, 14], 1)$ and $v_2 = ([8, 13], 0.8)$, one gets by applying Equation 3:

$$\min(v_1, v_2) = ([\min(12, 8), \min(14, 13)], \min(1, 0.8)) = ([8, 13], 0.8).$$

Similarly, with $v_2' = ([8, 15], 0.8)$, one would get $\min(v_1, v_2') = ([8, 14], 0.8)$. Note that the minimum of two interval values is not always one of the two interval values, as shown in this example. ◇

Remark 2. It should be noticed that the kind of aggregates described above remains cautious, and in some sense incomplete as illustrated by the following example. Let us suppose that we are asking about the minimum salary among employees in some department, and that we have several agents with salary 1500 at certainty 1, and one agent with salary 1300 at certainty $\alpha < 1$, all other agents having a salary larger than 1500 for sure. Then the result of the minimum is $(1300, \alpha)$. This result may be considered as providing an incomplete view of the situation since it comes from the only uncertain piece of data, while *all the other* pieces of data (thus forgetting the piece of data responsible for the result) would yield together $(1500, 1)$. In such a case, the imprecise aggregate $([1300, 1500], 1)$ may be seen as a useful complement, or as a valuable substitute, which may be even completed by the fact that there is a possibility $1 - \alpha$ that the result is 1500 (instead of 1300). More generally, there is an issue here of balancing an imprecise but certain aggregate result against a precise but uncertain aggregate result: think of computing the average salary in a 1000 employees department where for 999 ones we have $(1500, 1)$, and for one the information is $(1300, \alpha)$ (with α small), against making the average of 999 times 1500 and one time the *whole interval* of the salary domain (which is a fully certain piece of information for the last employee). In the former case, we get a precise result with low certainty, while in the latter the result is fully certain but still rather precise (i.e., it is a narrow interval). However, we leave such potential refinements aside in the rest of the paper.

4   Imprecise Aggregates of Certain Data

When we ask "what is the average age of well-paid people" (in face of a database with fully certain values), the implicit question is to know to what extent the answer *varies* with our understanding of *well-paid*; see [14] for a similar view in the case of fuzzy association rules for which the confidence in the rule should not vary too much when the understanding of the fuzzy sets appearing in the rule varies. An alternative view leading to a scalar evaluation would be to compute some expected value as in [9]. In the following, we give preference to an evaluation mode keeping track of the variations if any. The following formula gives the basis of the proposed evaluation as *a fuzzy set of values* (where the membership degree is before the /):

$$(\bigcup_\beta \beta / f([a_1, \ldots, a_i, \ldots, a_n]_\beta), 1) \tag{4}$$

where $[a_1, \ldots, a_i, \ldots, a_n]_\beta$ denotes the β-cut of the fuzzy constraint (*well-paid* in the example), i.e., the set of (uncertain) age values a_i attached to the tuples whose satisfaction degree with respect to *well-paid* is $\geq \beta$. The computation of $f([a_1, \ldots, a_i, \ldots, a_n]_\beta)$ follows the same principle as in Section 3. Here, the aggregate value obtained is a (completely certain) fuzzy set.

Example 4. Let us consider Table 3 and assume that the satisfaction degrees related to the fuzzy predicate *well-paid* are 0.8 for John, 0.3 for Mary and 1 for

Table 3. Relation *Employee*

#id	Name	Age	Salary
1	John	35	3000
2	Mary	22	2000
3	Paul	45	4000

Paul. The result of the query "what is the average age of well-paid people" is:
$(\{1/45, 0.8/(\frac{45+35}{2}), 0.3/(\frac{45+35+22}{3})\}, 1) = (\{1/45, 0.8/40, 0.3/34\}, 1).\diamond$

5   General Case

When the data are uncertain and the aggregation operator applies to a fuzzy set of objects, the formula that serves as a basis for the evaluation, and combines the ideas presented in the two previous sections, is as follows:

$$(\{(f(I_1, \ldots, I_i, \ldots, I_n)_\beta, \min_i \alpha_i)\}_{\beta \in [0,1]}, 1). \tag{5}$$

In other words, the result is a (completely certain) fuzzy set (where the degrees correspond to the level cuts of the fuzzy constraint) of more or less certain evaluations. Each more or less certain evaluation $(f(I_1, \ldots, I_i, \ldots, I_n)_{\beta_j}, \min_i \alpha_i)$ may be viewed itself as a fuzzy set F_j. One can then apply the canonic reduction of a fuzzy set of fuzzy sets to a fuzzy set, according to the following transformation:

$$\{\beta_j/F_j \mid \beta_j \in [0, 1]\} \rightarrow \max_j \min(\beta_j, F_j). \tag{6}$$

Note that the fuzziness of the result is due to the following facts:

- the data are imprecise and / or uncertain as reflected in the F_j's (this corresponds to the necessity-based modeling of the uncertainty);
- the constraint is fuzzy, which leads to consider that different answers may be more or less guaranteed as being possible (according to the β_j's; this would correspond to a guaranteed possibility in possibility theory terms [10]).

Table 4. Relation *Employee*

#id	Name	Age	Salary	N
1	John	(35, 0.4)	3000	1
2	Mary	(22, 0.6)	2000	1
3	Paul	(45, 0.7)	4000	1
4	James	(58, 1)	1500	1

Example 5. Let us consider Table 4 and assume that the satisfaction degrees related to the fuzzy predicate *well-paid* are 0.8 for John, 0.3 for Mary, 1 for Paul, and 0.2 for James. The result of the query "what is the average age of well-paid people" is:

$$(\{1/(45, 0.7), 0.8/(40, \min(0.7, 0.4)),$$
$$0.3/(34, \min(0.7, 0.4, 0.6)), 0.2/(40, \min(0.7, 0.4, 0.6, 1))\}, 1)$$
$$= (\{1/(45, 0.7), 0.8/(40, 0.4), 0.3/(34, 0.4), 0.2/(40, 0.4)\}, 1)$$

which can be transformed into:

$$(\{0.7/45, 0.4/40, 0.3/34\}, 1).$$

Notice that the degree 0.4 associated with value 40 is the result of

$$\max(\min(0, 8, 0.4), \min(0.2, 0.4))$$

according to Formula (6).◇

Remark 3. Extending the model to accommodate such entities does not seem to raise any difficulty since the only difference between such a fuzzy set and a regular disjunction stands in the degrees associated with the candidate values. The good properties of the model (in particular the fact that we do not need any lineage mechanism as in, e.g., [2]) resides in the fact that one is only interested in those answers that are somewhat *certain* (and not only possible), and nothing changes in this respect here.

6   Difficulties Raised by the Aggregate *Count*

Let us now make some brief comments concerning the aggregate function *count*, whose purpose is to compute the cardinality of a relation. Let us first consider, as a simple example, the query "how many employees have a salary at least equal to \$2000" addressed to a database where salary values are precise (i.e., represented by a singleton) but not necessarily certain. One may get results such as: "(at least 5, certainty $= 1$), (at least 7, certainty ≥ 0.8), ..., (at least 12, certainty ≥ 0.2)". Note that such a result has a format that is more complex than those of the aggregation operators considered previously since it involves different more or less certain values (whereas in Example 2, for instance, one only had to handle a more or less certain disjunction of values). Using the data from Table 4, the result of the previous query would be:

$$\langle(\textit{at least } 1, 0.7), (\textit{at least } 2, 0.6), (\textit{at least } 3, 0.4)\rangle.$$

The format of such a result appears too complex to be easily usable. Again as in Remark 2, this may be a matter of choosing to deliver a count that is not too imprecise, but still sufficiently certain.

The situation gets even worse for a query such as "how many employees at most have a salary at least equal to \$2000" when the database includes imprecise values (i.e. values represented by a disjunction or an interval). Then, some values (for instance [1800, 2100]) may overlap with the condition $salary \geq 2000$, which means that some possible values may not be certain. And of course, it does not get any simpler when the aggregate function applies to a referential of fuzzy objects, as in "how many young employees have a salary over \$2000"...

7   Conclusion

In this paper, we have shown how classical aggregation operators (min, max, avg, sum) could be interpreted in the framework of an uncertain database model based on the concept of possibilistic certainty. Three cases have been considered: i) precise aggregate over uncertain data, ii) imprecise aggregate over data that are certain (which corresponds to the case where a fuzzy condition has been applied first), iii) imprecise aggregate over uncertain data. Depending on the situation, the aggregate may take two forms: either an uncertain value (or an uncertain disjunction of values or intervals) or a fuzzy set. It has been pointed out that the aggregation operator $count$ raises specific issues, and its thorough study is left for future work.

It is worth emphasizing that the data complexity (in the database sense) of aggregate query processing is as in the classical database case, i.e., linear. The only extra cost is related to the possible presence of disjunctions or intervals but this does not impact the number of accesses to the database (which is the crucial factor in terms of performances). On the other hand, the problem of evaluating aggregate queries is much more complex in a probabilistic database context, see e.g. [18], where approximations are necessary if one wants to avoid exponentially sized results.

Among perspectives for future work, let us mention i) an in-depth study of the aggregation operator $count$, ii) the computation of proper summaries with balanced imprecision and uncertainty in a meaningful way as discussed in Remark 2, iii) an extension of the model so as to deal with uncertain fuzzy values, in order to have a fully compositional framework (but, as explained in Remark 3, this should not be too problematic), iv) the implementation of a DBMS prototype based on the model and query language described in [7] augmented with the aggregate queries investigated here.

References

1. Aggarwal, C.C., Yu, P.S.: A survey of uncertain data algorithms and applications. IEEE Trans. Knowl. Data Eng. 21(5), 609–623 (2009)
2. Benjelloun, O., Das Sarma, A., Halevy, A., Theobald, M., Widom, J.: Databases with uncertainty and lineage. VLDB Journal 17(2), 243–264 (2008)
3. Bosc, P., Pivert, O.: About projection-selection-join queries addressed to possibilistic relational databases. IEEE Trans. on Fuzzy Systems 13(1), 124–139 (2005)

4. Bosc, P., Prade, H.: An introduction to the fuzzy set and possibility theory-based treatment of soft queries and uncertain of imprecise databases. In: Smets, P., Motro, A. (eds.) Uncertainty Management in Information Systems: From Needs to Solutions, pp. 285–324. Kluwer, Dordrecht (1997)
5. Bosc, P., Pivert, O.: Modeling and querying uncertain relational databases: a survey of approaches based on the possible worlds semantics. International Journal of Uncertainty, Fuzziness and Knowledge-Based Systems 18(5), 565–603 (2010)
6. Bosc, P., Pivert, O., Prade, H.: A model based on possibilistic certainty levels for incomplete databases. In: Godo, L., Pugliese, A. (eds.) SUM 2009. LNCS, vol. 5785, pp. 80–94. Springer, Heidelberg (2009)
7. Bosc, P., Pivert, O., Prade, H.: An uncertain database model and a query algebra based on possibilistic certainty. In: Martin, T.P., Muda, A.K., Abraham, A., Prade, H., Laurent, A., Laurent, D., Sans, V. (eds.) SoCPaR, pp. 63–68. IEEE (2010)
8. Dubois, D., Prade, H.: Possibility Theory: An Approach to Computerized Processing of Uncertainty. Plenum Press, New York (1988), with the collaboration of Farreny, H., Martin-Clouaire, R., Testemale, C.
9. Dubois, D., Prade, H.: Measuring properties of fuzzy sets: A general technique and its use in fuzzy query evaluation. Fuzzy Sets Syst. 38(2), 137–152 (1990)
10. Dubois, D., Prade, H.: Possibility theory: qualitative and quantitative aspects. In: Gabbay, D.M., Smets, P. (eds.) Quantified Representation of Uncertainty and Imprecision, Handbook of Defeasible Reasoning and Uncertainty Management Systems, vol. 1, pp. 169–226. Kluwer Acad. Publ. (1998)
11. Dubois, D., Prade, H.: Necessity measures and the resolution principle. IEEE Trans. on Systems, Man and Cybernetics 17(3), 474–478 (1987)
12. Dubois, D., Prade, H.: Processing fuzzy temporal knowledge. IEEE Trans. on Systems, Man and Cybernetics 19(4), 729–744 (1989)
13. Dubois, D., Prade, H.: On the ranking of ill-known values in possibility theory. Fuzzy Sets and Systems, Aggregation and Best Choices of Imprecise Opinions 43, 311–317 (1991)
14. Dubois, D., Prade, H., Sudkamp, T.: On the representation, measurement, and discovery of fuzzy associations. IEEE T. Fuzzy Systems 13(2), 250–262 (2005)
15. Haas, P.J., Suciu, D.: Special issue on uncertain and probabilistic databases. VLDB J. 18(5), 987–988 (2009)
16. Imielinski, T., Lipski, W.: Incomplete information in relational databases. J. of the ACM 31(4), 761–791 (1984)
17. Lipski, W.: Semantic issues connected with incomplete information databases. ACM Trans. on Database Syst. 4(3), 262–296 (1979)
18. Murthy, R., Ikeda, R., Widom, J.: Making aggregation work in uncertain and probabilistic databases. IEEE Trans. Knowl. Data Eng. 23(8), 1261–1273 (2011)
19. Prade, H., Testemale, C.: Generalizing database relational algebra for the treatment of incomplete/uncertain information and vague queries. Information Sciences 34(2), 115–143 (1984)
20. Suciu, D., Olteanu, D., Ré, C., Koch, C.: Probabilistic Databases. Synthesis Lectures on Data Management. Morgan & Claypool Publishers (2011)
21. Wong, E.: A statistical approach to incomplete information in database systems. ACM Trans. Database Syst. 7(3), 470–488 (1982)
22. Zadeh, L.: Fuzzy sets as a basis for a theory of possibility. Fuzzy Sets and Systems 1(1), 3–28 (1978)

Bipolar Comparison of 3D Ear Models

Guy De Tré[1], Dirk Vandermeulen[2], Jeroen Hermans[2], Peter Claeys[2],
Joachim Nielandt[1], and Antoon Bronselaer[1]

[1] Department of Telecommunications and Information Processing,
Ghent University, Sint-Pietersnieuwstraat 41, B-9000 Ghent, Belgium
{Guy.DeTre,Joachim.Nielandt,Antoon.Bronselaer}@UGent.be
[2] Department of Electrical Engineering (ESAT),
KU Leuven, Kasteelpark Arenberg 10, box 2440, B-3001 Leuven, Belgium
{dirk.vandermeulen,peter.claes}@esat.kuleuven.be,
jeroen.hermans@uzleuven.be

Abstract. Comparing ear photographs is considered to be an important
aspect of victim identification. In this paper we study how automated
ear comparison can be improved with soft computing techniques. More
specifically we describe and illustrate how bipolar data modelling techniques can be used for handling data imperfections more adequately. In
order to minimise rescaling and reorientation problems, we start with
3D ear models that are obtained from 2D ear photographs. To compare two 3D models, we compute and aggregate the similarities between
corresponding points. Hereby, a novel bipolar similarity measure is proposed. This measure is based on Euclidian distance, but explicitly deals
with hesitation caused by bad data quality. Comparison results are expressed using bipolar satisfaction degrees which, compared to traditional
approaches, provide a semantically richer description of the extent to
which two ear photographs match.

Keywords: Ear comparison, data quality, bipolarity, similarity.

1 Introduction

Ear biometrics are considered to be a reliable source for disaster victim identification. Indeed, ears are relatively immune to variation due to ageing [9] and the
external ear anatomy constitutes unique characteristic features [13]. Moreover,
ears are often among the intact parts of found bodies, automated comparison of
photographs is in general faster and cheaper than DNA analysis and collecting
ante mortem photographs is considered to be a humane process for relatives.

Although there is currently no hard evidence that ears are unique, there is
neither evidence that they are not. Experiments comparing over ten thousand
ears revealed that no two ears were indistinguishable [13,5] and another study
revealed that fraternal and identical twins have a similar but still clearly distinguishable ear structure. More research is needed to examine the validity of
uniqueness but, despite of that a match or mismatch of ear biometrics can provide forensic experts with useful information in identification tasks. This makes
research on the comparison of ear photographs relevant and interesting.

A. Laurent et al. (Eds.): IPMU 2014, Part III, CCIS 444, pp. 160–169, 2014.

When considering a missing person and the found body of a victim, ear identification practically boils down to a comparison of a set of ear photographs of
the missing person with a set of ear photographs of the victim. Ear pictures of a
victim are taken in post mortem conditions and hence referred to as post mortem
(PM) pictures. Pictures of a missing person are always taken ante mortem and
therefore called ante mortem (AM) pictures. PM pictures are assumed to be of
good quality because they are usually taken by forensic experts under controlled
conditions: high resolution, correct angle, uniform lighting, with the ear completely exposed. AM photos are often of lower, unprofessional quality. They are
not taken with the purpose of ear identification and in most cases are provided
by relatives or social media. Because we have no control over the conditions in
which these pictures were taken, we can only hope to retrieve the best we can.
Moreover, parts of the ear might be obscured by hair, headgear or other objects.
The ear can also be deformed by glasses, earrings or piercings. Efficiently coping
with all these aspects that have a negative impact on the data quality and hence
also on the comparison is a research challenge and the subject of this work.

A considerable part of related work focusses on comparisons where an ear
photo from a given set of photos is compared to all photos in this set (e.g,
[25,12,21]). This is a simplified case because matches between identical photos
are searched for. The work in this paper is more general because it involves
the matching of identical ears on different photos. An important step of each
automated ear comparison process is the ear recognition step during which corresponding extracted features from two ears are compared in order to decide
whether the ears match or not. Related work on ear recognition can be cate-
gorised based on the feature extraction scheme used. Intensity based methods
use techniques like principal component analysis, independent component analysis and linear discriminant analysis for the comparison (e.g., [26,22]). Other
categories of methods are based on force field transformations (e.g., [3]), 2D ear
curves geometry (e.g., [8]), Fourier descriptors [1], wavelet transformation (e.g.,
[11]), Gabor filters (e.g., [18]) or scale-invariant feature transformation (e.g.,
[15]). A last category of comparison techniques are based on 3D shape features.
Most approaches use an iterative closest point algorithm for ear recognition (e.g.,
[7,23,14,6]). In [24] both point-to-point and point-to-surface matching schemes
are used, whereas the method in [20] is based on the extraction and comparison
of a compact biometric signature. An elaborate survey on ear recognition is [2].

Current approaches for ear recognition cannot adequately handle, measure
and reflect data quality issues. Nevertheless, efficiently coping with aspects that
have a negative impact on correct ear detection and ear recognition is recognised
to be an important research challenge. Ear identification methods should not only
support the annotation of areas of bad data quality in an ear photo, but also be
able to quantify these and reflect their impact on the results of ear comparisons.
Indeed, forensic experts would benefit from extra information expressing the
quality of data on which comparisons are based. For example, the case where
AM photo A and PM photo P only partially match, but both having sufficient
quality, clearly differs from the case where A and P partially match but A is of

low quality. In this work we investigate if and how soft computing techniques can be used to explicitly cope with AM data of bad quality in ear recognition. We use a 3D ear model on which we apply a point-to-point comparison method. Bipolar data modelling techniques are applied to denote and quantify areas of bad data quality. A novel bipolar similarity measure is proposed and used for the comparison. Ear comparison results are expressed using bipolar satisfaction degrees [16] which quantify hesitation about each result caused by bad data quality of the AM photos.

The remainder of the paper is structured as follows. In Section 2 some preliminaries are given. Some general issues on bipolarity in ear comparison are explained. Next, some basic concepts and definitions of bipolar satisfaction degrees are described. In Section 3, the 3D ear model is described. Section 4 deals with ear recognition and comprises the main contribution of the paper. It consecutively describes how corresponding points in two ear models can be compared, proposes a novel bipolar similarity measure, describes how this bipolar similarity measure can be used for the comparison of two 3D ear models and discusses the interpretation of comparison results in a bipolar setting. Some conclusions and plans for related research are reported in Section 5.

2   Preliminaries

2.1   Bipolarity Issues in Ear Comparison

In the context of information handling the term *bipolarity* is, among others, used to denote that information can be of a positive or negative nature [16]. Positive information describes what is true, correct, preferred. Oppositely, negative information describes what is false, incorrect, not preferred. In most situations, especially in a scientific context, positive and negative information complement each other. This is called *symmetric bipolarity* [10]. Boolean logic and probability theory are examples of mathematical frameworks where symmetric bipolarity is assumed. So-called *dual bipolarity* is assumed in possibility theory where positive and negative information are dually related to each other and measured on different scales based on the same source of knowledge. The most general form of bipolarity is *heterogeneous bipolarity*. Two separate knowledge sources provide positive and negative information which are independent and hence do not have to complement each other. In the remainder of the paper, heterogeneous bipolarity is assumed.

Victim identification by ear biometrics can be seen as a pattern recognition process where PM ear photos of a victim are reduced to a set of features that is subsequently compared with the feature sets that are obtained from the AM photos of missing persons in order to help determine the identity of the victim on the basis of the best match. The following steps are hereby distinguished:

1. *Ear detection.* Hereby, ears are positioned and extracted from the photos.
2. *Ear normalisation and enhancement.* Detected ears are transformed to a consistent ear model using, e.g., geometrical and photometric corrections.

3. *Feature extraction.* Representative features are extracted from the ear model.
4. *Ear recognition.* Feature sets of AM and PM ears are compared. A matching score indicating the similarity between the ears is computed.
5. *Decision.* The matching scores are ranked and used to render an answer that supports forensic experts in their decision making.

Errors in the first three steps can undermine the utility of the process. So, features that are obtained from bad quality data should be handled with care. For that reason, we consider that a feature set provides us with heterogeneous bipolar information: some features are obtained from reliable data and positively contribute in the identification process, other features might turn out to be unreliable and might have a negative impact which should be avoided, while for still other features there can be hesitation about whether they are useful or not.

2.2   Bipolar Satisfaction Degrees

To efficiently handle heterogeneous bipolarity in the comparison process, bipolar satisfaction degrees are used [16]. A bipolar satisfaction degree (BSD) is a couple

$$(s, d) \in [0, 1]^2 \tag{1}$$

where s is the *satisfaction degree* and d is the *dissatisfaction degree*. Both s and d take their values in the unit interval $[0, 1]$ reflecting to what extent the BSD represents satisfied, resp. dissatisfied. The extreme values are 0 ('not at all'), and 1 ('fully'). The values s and d are independent of each other. A BSD can be used to express the result of a comparison in which case s (resp. d) denotes to which extent the comparison condition is accomplished (resp. not accomplished).

Three cases are distinguished:

1. If $s + d = 1$, then the BSD is *fully specified*. This situation corresponds to traditional involutive reasoning.
2. If $s + d < 1$, then the BSD is *underspecified*. In this case, the difference $h = 1 - s - d$ reflects the *hesitation* about the accomplishment of the comparison. This situation corresponds to membership and non-membership degrees in intuitionistic fuzzy sets [4].
3. If $s + d > 1$, then the BSD is *overspecified*. In this case, the difference $c = s + d - 1$ reflects the *conflict* in the comparison results.

With the understanding that i denotes a t-norm (e.g., min) and u denotes its associated t-conorm (e.g., max), the basic operations for BSDs (s_1, d_1) and (s_2, d_2) are [16]:

– *Conjunction.* $(s_1, d_1) \wedge (s_2, d_2) = (i(s_1, s_2), u(d_1, d_2))$
– *Disjunction.* $(s_1, d_1) \vee (s_2, d_2) = (u(s_1, s_2), i(d_1, d_2))$
– *Negation.* $\neg(s_1, d_1) = (d_1, s_1)$.

3   3D Ear Model

In our previous work, we used 2D ear images for accomplishing ear recognition [19]. Imperfect geometrical and photometric transformations of 2D AM photos put a limit on the quality of the results. To improve this approach we now use a 3D ear model. This 3D ear model is obtained by estimating the parameters of a mathematical shape function such that the resulting shape optimally fits the images of the ear. For a PM ear, a 3D camera image can be used, whereas for an AM ear usually a set of 2D photos is used. The description of this fitting process is outside the scope of this paper. At this point it is sufficient to assume that for each ear we obtained a 3D model that captures the three dimensional details of the ear surface as shown in Fig. 1 (left and middle).

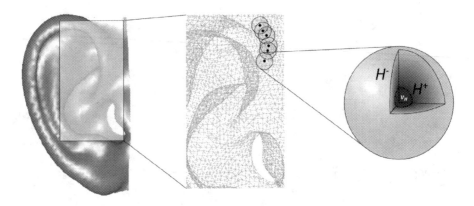

Fig. 1. 3D ear model (with hesitation spheres)

The 3D ear model is normalised for all ears, so all ear models have the same resolution and scale. However, unclear parts of 2D AM ear photos might decrease the quality of (parts of) a 3D AM ear model. Indeed, if parts of the 2D ears are inadequately visible or unreliable then their corresponding parts in the 3D model will as well be unreliable. To cope with this, unreliable parts of 3D ear models are indicated by so-called *hesitation spheres*. As illustrated in Fig. 1 (right), a hesitation sphere H is defined by two concentric spheres H^+ and H^-. All points p inside or on the surface of the inner sphere H^+ have a fixed associated hesitation value $h_H(p) = v_H \in]0,1]$. For points on the surface or outside the outer sphere H^- the hesitation is 0, whereas for points between both spheres the hesitation is gradually decreasing from v_H to 0, depending on their distance from H^+, i.e.,

$$h_H(p) = v_H \cdot \left(1 - \frac{d(H^+, p)}{d(H^+, H^-)}\right) \qquad (2)$$

where d denotes the Euclidean distance. In general, forensic experts can manually assign as many hesitation spheres as required to indicate unreliable parts in the

3D model. This assignment process is subject for (semi-)automation in future work. In the presence of $k > 1$ hesitation spheres H_k, the overall hesitation about the quality of a point p is computed by

$$h(p) = \max_k h_{H_k}(p). \tag{3}$$

Thus, the maximal hesitation assigned to the point is taken.

Feature extraction boils down to selecting n representative points of the 3D ear model. The more points that are considered, the better the matching results, but also the longer the computation time. For normalisation purposes, a fixed list $L_S = [p_1^S, \ldots, p_n^S]$ of n points is selected on a standard, reference ear model S. Ear fitting, i.e., determining the optimal parameters for the shape function, will transform L_S into a list $L_E = [p_1^E, \ldots, p_n^E]$ of n points of the best fitting 3D ear model E. Hereby, each point p_i^S corresponds to the point p_i^E ($i = 1, \ldots, n$). Moreover, using the same ear model S and the same list L_S for fitting two different ear models A and P guarantees that each point p_i^A of L_A corresponds to the point p_i^P of L_P ($i = 1, \ldots, n$).

4   Ear Recognition

A basic step in ear recognition is the comparison of two left (or two right) ears. As such, in victim identification a set of AM photos of one ear have to be compared with a set of PM photos of the other ear. Using the 3D ear modelling technique explained in the previous section, the feature list L_A of the ear model A of the AM photos has to be compared with the feature list L_P of the ear model P of the PM photos. To reflect data quality issues, the hesitation $h(p)$ of each point in the lists L_A and L_P has to be taken into account.

4.1   Similarity of Corresponding Features

A commonly used comparison technique for corresponding points of two feature lists is to use the Euclidean distance. In the 3D space defined by the three orthogonal X, Y and Z-axes, the Euclidean distance between a point p^A of L_A and its corresponding point p^P in L_P is given by:

$$d(p^A, p^P) = \sqrt{((p^A)_x - (p^P)_x)^2 + ((p^A)_y - (p^P)_y)^2 + ((p^A)_z - (p^P)_z)^2} \tag{4}$$

where $(.)_x$, $(.)_y$ and $(.)_z$ denote the x, y and z coordinates of the point.

The similarity between the points is then obtained by applying a similarity function to their distance. This similarity function μ_{Sim} can generally be defined by a fuzzy set Sim over the domain of distances, e.g.,

$$\mu_{Sim} : [0, +\infty[\rightarrow [0, 1] \tag{5}$$
$$d \mapsto 1, \text{ iff } d \leq \epsilon_1$$
$$d \mapsto 0, \text{ iff } d \geq \epsilon_0$$
$$d \mapsto 1 - \frac{d - \epsilon_1}{\epsilon_0 - \epsilon_1}, \text{ iff } \epsilon_1 < d < \epsilon_0$$

where $0 \le \epsilon_1 \le \epsilon_0$. Hence, if the distance $d < \epsilon_1$ then the similarity between the points is considered to be 1, if $d > \epsilon_0$, the similarity is 0, and for distances d between ϵ_1 and ϵ_0 the similarity is gradually decreasing from 1 to 0.

Hence the similarity between two points p^A and p^P yields

$$\mu_{Sim}(d(p^A, p^P)) \in [0, 1]. \tag{6}$$

4.2 Bipolar Similarity

The similarity function μ_{Sim} is not taking into account any hesitation that might exist about the points p^A and p^P. For that reason, the following novel similarity measure, based on both μ_{Sim} and the overall hesitation h (cf. Eq. (3)), is proposed.

$$f_{Bsim} : \mathbb{P} \times \mathbb{P} \to [0, 1]^2 \tag{7}$$
$$(p^A, p^P) \mapsto (s, d)$$

where $\mathbb{P} \subseteq \mathbb{R}^3$ denotes the 3D space in which the ear model is defined and (s, d) is the BSD expressing the result of comparing p^A and p^P as described in the preliminaries. The BSD (s, d) is defined by

$$s = (1 - \max(h(p^A), h(p^P))) \cdot \mu_{Sim}(d(p^A, p^P)) \tag{8}$$

and

$$d = (1 - \max(h(p^A), h(p^P))) \cdot (1 - \mu_{Sim}(d(p^A, p^P))). \tag{9}$$

Herewith it is reflected that we consider a consistent situation where $h = 1 - s - d$ and we consider s (resp. d) to be the proportion of $1 - h$ that corresponds with the similarity (resp. dissimilarity) between p^A and p^B.

Remark that, with the former equations, the hesitation h represented by the BSD (s, d) becomes $h = 1 - s - d = \max(h(p^A), h(p^P)) \in [0, 1]$. Also $s + d = 1 - h \le 1$, such that the resulting BSD can either be fully specified, or be underspecified. So, Eq. 7 is consistent with the semantics of BSDs.

4.3 Comparing 3D Ear Models

The comparison of an AM ear model A and a PM ear model P is based on the comparison of all features p_i^A and p_i^P in their respective feature lists L_A and L_P ($i = 1, \ldots, n$). More specifically, the comparison results of all n corresponding AM and PM points should be aggregated to an overall similarity and hesitation indication. Because these overall similarity and hesitation have to reflect the global similarity and hesitation of all points under consideration, the arithmetic mean can be used as an aggregator for the n similarities of the corresponding points in the feature lists. Therefore we propose the following similarity measure for feature lists of 3D ear models.

$$f_{bsim}^* : \mathbb{P}^n \times \mathbb{P}^n \to [0, 1]^2 \tag{10}$$
$$([p_1^A, \ldots, p_n^A], [p_1^P, \ldots, p_n^P]) \mapsto (s, d)$$

where $\mathbb{P}^n$ denotes the set of all feature lists consisting of n points of $\mathbb{P}$ and (s, d) expresses the result of comparing the feature lists $L_A = [p_1^A, \ldots, p_n^A]$ and $L_P = [p_1^P, \ldots, p_n^P]$. The BSD (s, d) is defined by

$$s = \left(1 - \frac{\sum_{i=1}^{n} \max(h(p_i^A), h(p_i^P))}{n}\right) \cdot \frac{\sum_{i=1}^{n} \mu_{Sim}(d(p_i^A, p_i^P))}{n} \tag{11}$$

and

$$d = \left(1 - \frac{\sum_{i=1}^{n} \max(h(p_i^A), h(p_i^P))}{n}\right) \cdot \left(1 - \frac{\sum_{i=1}^{n} \mu_{Sim}(d(p_i^A, p_i^P))}{n}\right). \tag{12}$$

The hesitation h in the BSD (s, d) is $h = 1 - s - d = \frac{\sum_{i=1}^{n} \max(h(p_i^A), h(p_i^P))}{n} \in [0, 1]$ and again $s + d = 1 - h \leq 1$.

4.4 Interpreting the Results

In a typical victim identification search, a PM 3D ear model is compared with a set of m AM 3D ear models taken from a database with missing persons. Each of these comparisons results in a BSD (s_i, d_i), $i = 1, \ldots, m$, from which the hesitation $h_i = 1 - s_i - d_i$ about the result can be derived. Hence, the information provided from the comparison is the following:

1. s_i ($\in [0, 1]$): denotes how satisfied/convinced the method is about the matching of both ear models.
2. d_i ($\in [0, 1]$): denotes how dissatisfied/unconvinced the method is.
3. h_i ($\in [0, 1]$): expresses the overall hesitation about the comparison results (due to inadequate data quality).

In practice, forensic experts will be interested in the top-k matches for a given PM 3D ear model. For that purpose, the resulting BSDs (s_i, d_i), $i = 1, \ldots, m$, have to be ranked. In the given context, the best ear matches are those where s_i is as high as possible and h_i is as low as possible. Therefore, considering that $h_i = 1 - s_i - d_i$, the ranking function

$$r : [0, 1]^2 \to [0, 1] \tag{13}$$

$$(s, d) \mapsto \frac{s + (1 - d)}{2}$$

can be used. This function computes a single ranking value $r((s_i, d_i))$ for each BSD (s_i, d_i), which can then be used to rank order the comparison results and select the top-k among them. Other ranking functions are possible and discussed in [17].

Another option is to work with two threshold values $\delta_s, \delta_h \in [0, 1]$. In such a case, only ear models for which the resulting BSD (s, d) satisfies $s \geq \delta_s$ and $h \leq \delta_h$ (or $1 - s - d \leq \delta_h$) are kept in the comparison result.

5 Conclusions and Future Work

In this paper, we described some theoretical aspects of a novel, bipolar approach for comparing 3D ear models. Soft computing techniques based on heterogeneous bipolar satisfaction degrees (BSDs) support explicitly coping with the hesitation that occurs when low(er) quality ear photos have to be compared with other ear photos (taken in different position, on a different time, ...). The use of BSDs allows to provide user with extra quantitative information about the overall hesitation on the comparison results (due to inadequate data quality).

The focus in the paper is on the ear recognition and decision processes of an ear identification approach. The presented technique departs from a 3D ear model that is obtained from ear detection, normalisation and enhancement processes. On this model, parts of low quality are annotated using a set of so-called hesitation spheres. From each 3D ear model a feature list is extracted. Feature lists are compared with each other by using a novel bipolar similarity measure, which provides quantitative information on the similarity of two ears and on the overall hesitation about (the quality of) the data involved in the comparison.

Up to now the method has only been tested on synthetically modified ear models. Experiments with models of real ears are necessary for parameter fine-tuning and validation purposes and are planned in the near future.

Acknowledgements. This work is supported by the Flemish Fund for Scientific Research.

References

1. Abate, A., Nappi, N., Riccio, D., Ricciardi, R.: Ear recognition by means of a rotation invariant descriptor. In: 18th IEEE International Conference on Pattern Recognition (ICPR), pp. 437–440 (2006)
2. Abaza, A., Ross, A., Hebert, C., Harrison, M.A., Nixon, M.S.: A Survey on Ear Biometrics. ACM Computing Surveys 45(2), article 22 (2013)
3. Abdel-Mottaleb, M., Zhou, J.: Human ear recognition from face profile images. In: Zhang, D., Jain, A.K. (eds.) ICB 2006. LNCS, vol. 3832, pp. 786–792. Springer, Heidelberg (2006)
4. Atanassov, K.: Intuitionistic fuzzy sets. Fuzzy Sets and Systems 20, 87–96 (1986)
5. Burge, M., Burger, W.: Ear biometrics. In: Jain, A., et al. (eds.) Biometrics: Personal Identification in Networked Society. Kluwer Academic Publishers (1998)
6. Cadavid, S., Abdel-Mottaleb, M.: 3D ear modeling and recognition from video sequences using shape from shading. In: 19th IEEE International Conference on Pattern Recognition (ICPR), pp. 1–4 (2008)
7. Chen, H., Bhanu, B.: Contour matching for 3D ear recognition. In: IEEE Workshops on Application of Computer Vision (WACV), pp. 123–128 (2005)
8. Choras, M., Choras, R.: Geometrical algorithms of ear contour shape representation and feature extraction. In: 6th IEEE International Conference on Intelligent Systems Design and Applications (ISDA), pp. 123–128 (2006)
9. Cummings, A.H., Nixon, M.S., Carter, J.N.: A Novel Ray Analogy for Enrolment of Ear Biometrics. In: 4th IEEE Conference on Biometrics: Theory, Applications and Systems, pp. 302–308 (2010)

10. Dubois, D., Prade, H.: An introduction to bipolar representations of information and preference. International Journal of Intelligent Systems 23, 866–877 (2008)

11. Feng, J., Mu, Z.: Texture analysis for ear recognition using local feature descriptor and transform filter. In: SPIE (MIPPR 2009: Pattern Recognition and Computer Vision, vol. 7496, Session 1 (2009)

12. Gutierrez, L., Melin, P., Lopez, M.: Modular neural network integrator for human recognition from ear images. In: International Joint Conference on Neural Networks, pp. 1–5 (2010)

13. Iannarelli, A.: Ear Identification. Forensic Identification Series. Paramount Publishing Company, California (1989)

14. Islam, S., Bennamoun, M., Davies, R.: Fast and fully automatic ear detection using cascaded adaboost. In: IEEE Workshop on Applications of Computer Vision, pp. 1–6 (2008)

15. Kisku, D.R., Mehrotra, H., Gupta, P., Sing, J.K.: SIFT-Based ear recognition by fusion of detected key-points from color similarity slice regions. In: IEEE International Conference on Advances in Computational Tools for Engineering Applications (ACTEA), pp. 380–385 (2009)

16. Matthé, T., De Tré, G., Zadrozny, S., Kacprzyk, J., Bronselaer, A.: Bipolar Database Querying Using Bipolar Satisfaction Degrees. International Journal of Intelligent Systems 26(10), 890–910 (2011)

17. Matthé, T., De Tré, G.: Ranking of bipolar satisfaction degrees. In: Greco, S., Bouchon-Meunier, B., Coletti, G., Fedrizzi, M., Matarazzo, B., Yager, R.R. (eds.) IPMU 2012, Part II. CCIS, vol. 298, pp. 461–470. Springer, Heidelberg (2012)

18. Nanni, L., Lumini, A.: A supervised method to discriminate between impostors and genuine in biometry. Expert Systems with Applications 36(7), 10401–10407 (2009)

19. Nielandt, J., Bronselaer, A., Matthé, T., De Tré, G.: Bipolarity in ear biometrics. In: European Society for Fuzzy Logic and Technology Conference and les rencontres francophones sur la Logique Floue et ses Applications (EUSFLAT/LFA), pp. 409–415 (2011)

20. Passalis, G., Kakadiaris, I., Theoharis, T., Toderici, G., Papaioannou, T.: Towards fast 3D ear recognition for real-life biometric applications. In: IEEE Conference on Advanced Video and Signal Based Surveillance, pp. 39–44 (2007)

21. Sánchez, D., Melin, P.: Optimization of modular granular neural networks using hierarchical genetic algorithms for human recognition using the ear biometric measure. Engineering Applications of Artificial Intelligence 27, 41–56 (2014)

22. Xie, Z., Mu, Z.: Ear recognition using lle and idlle algorithm. In: 19th IEEE International Conference on Pattern Recognition (ICPR), pp. 1–4 (2008)

23. Yan, P., Bowyer, K., Chang, K.: ICP-Based approaches for 3D ear recognition. In: SPIE (Biometric Technology for Human Identification II), pp. 282–291 (2005)

24. Yan, P., Bowyer, K.: Biometric recognition using 3D ear shape. IEEE Transactions on Pattern Analysis and Machine Intelligence 29(8), 1297–1308 (2007)

25. Yuizono, T., Wang, Y., Satoh, K., Nakayama, S.: Study on individual recognition for ear images by using genetic local search. In: IEEE Congress on Evolutionary Computation (CEC), pp. 237–242 (2002)

26. Zhang, H., Mu, Z., Qu, W., Liu, L., Zhang, C.: A novel approach for ear recognition based on ICA and RBF network. In: 4th IEEE International Conference on Machine Learning and Cybernetics, pp. 4511–4515 (2005)

A Logical Version of the Belief Function Theory

Laurence Cholvy

ONERA
2 Avenue Edouard Belin
31055 France
Laurence.Cholvy@onera.fr

Abstract. This paper presents a version of the belief function theory in which masses are assigned to propositional formulas. It also provides two combination rules which apply within this framework. Finally, it proves that the belief function theory and its extension to non-exclusive hypotheses are two particular cases of this logical framework.

Keywords: Belief function theory, logic.

1 Introduction

This paper focuses on a particular framework used to model uncertainty, the belief function theory.

This theory, as initially defined in [2] and [9], starts from a frame of discernment which is a finite set of elements called hypotheses. That comes to consider alternatives (hypotheses) which are exhaustive and exclusive. In this setting, generally called Dempster-Shafer's theory, masses i.e., real numbers belonging to [0, 1] are assigned to sets of hypotheses so that 0 is assigned to $\emptyset$ and the sum of these masses is equal to 1. More recently, [3] extends this theory by relaxing the assumption of hypotheses exclusivity. There, a frame of discernment is a finite set of hypotheses which may intersect. Masses are assigned to sets of hypotheses or to intersections of sets of hypotheses (notice that here, intersections of sets are not sets).

[1] showed that (i) in both previous cases, the frame of discernment can be seen as a propositional language; (ii) under the assumption of hypotheses exclusivity, expressions on which masses are assigned are equivalent to propositional positive clauses; (iii) when relaxing this assumption, expressions on which masses are assigned are equivalent to some particular kind of conjunctions of positive clauses.

The question we ask in this present paper is the following: is it possible to be more general and assign masses to any kind of propositional formulas ? This question is motivated by the following example.

Example 1. We consider the example given by Shafer in [10]: my friend Betty reports her observation about something which fell on my car.

A. Laurent et al. (Eds.): IPMU 2014, Part III, CCIS 444, pp. 170–179, 2014.

- (a) Betty tells me that a limb fell on my car. Suppose that from this testimony alone, I can justify a 0.9 degree of belief that a limb fell on my car[1]. Modelling this leads to consider a frame of discernment with two exclusive hypotheses $\{limb, nolimb\}$ and to consider the mass function: $m(\{limb\}) = 0.9,\ m(\{limb, nolimb\}) = 0.1$.
- (b) Assume now that Betty can distinguish between oaks, ashes and birches and that she tells me that an oak limb fell on my car. Modelling this within Dempster-Shafer theory can be done by considering 4 exclusive hypotheses: H_1 which represents the fact that an oak limb fell, H_2 which represents the fact that a ash limb fell, H_3 which represents the fact that a birch limb fell, H_4 which represents the fact that no limb fell. Then, my beliefs are modelled by the mass function: $m(\{H_1\}) = 0.9,\ m(\{H_1, H_2, H_4, H_4\}) = 0.1$. But we can also use the model of [3] and consider the frame of discernment $\{limb, nolimb, oak, ash, birch\}$ in which the pairs of exclusive hypotheses are: $(limb, nolimb)$, (oak, ash), $(oak, birch)$, $(ash, birch)$. Then, my beliefs are modelled by: $m(\{limb\} \cap \{oak\}) = 0.9,\ m(\{limb, nolimb, oak, ash, birch\}) = 0.1$.
- (c) Consider finally that Betty tells me that a limb fell on my car and it was not an oak limb. Modelling this within Dempster-Shafer theory leads to the following mass function: $m(\{H_2, H_3\}) = 0.9,\ m(\{H_1, H_2, H_3, H_4\}) = 0.1$. Modelling this within the model of [3] leads to consider the mass function: $m(\{limb\} \cap \{ash, birch\}) = 0.9,\ m(\{limb, nolimb, oak, ash, birch\}) = 0.1$.

Let us come back to case (c). We can notice that in both models, the modeler has to reformulate the information to be modelled. Indeed, the information Betty tells me is that *a non-oak limb fell*. Within the first model, this information is reformulated as: *an ash limb fell or a birch limb fell* (i.e., $\{H_2, H_3\}$). Within the second model, this information is reformulated as: *a limb of a tree, which is an ash or a birch, fell* (i.e., $\{limb\} \cap \{ash, birch\}$).

Our suggestion is to offer the modeller a language which allows it to express this information without any reformulation. For doing so, we will abandon the set theory for expressing information and offer propositional logic instead. More precisely, in this paper we will allow the modeler to express information by means of a propositional language. This will lead him/her to express his/her beliefs by *any kind of propositional formulas*. In the previous example, this will lead to consider the propositional language whose letters are $limb, oak, ash, birch$. The first focal element will then be modelled by $limb \wedge \neg oak$ which is the very information reported by Betty. Notice that this propositional formula is a conjunction of two atomic clauses, the second one being negative.

This paper is organized as follows. Section 2 quickly presents the propositional logic, the belief functions theory and its extension to non-exclusive hypotheses. Section 3 describes our proposal, i.e., a version of the belief functions theory in

---

[1] Let us mention that, according to Shafer [10], this degree is a consequence of the fact that my subjective probability that Betty is reliable is 0.9. However, in this paper, we do not discuss the meaning of such values nor do we discuss the intuition behind the combination rules.

which masses are assigned to propositional formulas. It also presents two combination rules. Section 4 shows that the belief function theory and its extension to non-exhaustive hypotheses are two particular cases of this framework. Finally, section 5 concludes the paper.

2 Propositional Logic and Belief Function Theory

2.1 Propositional Logic

Let us first recall some definitions and results that will be useful in the rest of the paper.

A *propositional language* Θ is defined by a set of propositional letters, connectives $\neg, \wedge, \vee, \rightarrow, \leftrightarrow$ and parentheses. In what follows, it is sufficient to consider a *finite language* i.e a language composed of a finite set of letters.

The set of *formulas*, denoted $FORM$, is the smallest set of words built on this alphabet such that: if a is a letter, then a is a formula; $\neg A$ is a formula if A is a formula; $A \wedge B$ is a formula if A and B are formulas. Other formulas are defined by abbreviation. More precisely, $A \vee B$ denotes $\neg(\neg A \wedge \neg B)$; $A \rightarrow B$ denotes $\neg A \vee B$; $A \leftrightarrow B$ denotes $(A \rightarrow B) \wedge (B \rightarrow A)$.

A *literal* is a letter or the negation of a letter. In the first case it is called a *positive litteral*, in the second it is called a *negative litteral*.

A *clause* is a disjunction of literals.

A *positive clause* is a clause whose literals are positive.

A clause C_1 *subsumes* a clause C_2 iff the literals of C_1 are literals of C_2.

A formula is $+mcnf$ if it is a conjunction of positive clauses such that no clause subsumes another one.

An *interpretation* i is a mapping from the set of letters to the set of truth values $\{0, 1\}$. An interpretation i can be extended to the set of formulas by: $i(\neg A) = 1$ iff $i(A) = 0$; $i(A \wedge B) = 1$ iff $i(A) = 1$ and $i(B) = 1$. Consequently, $i(A \vee B) = 1$ iff $i(A) = 1$ or $i(B) = 1$, and $i(A \rightarrow B) = 1$ iff $i(A) = 0$ or $i(B) = 1$; $i(A \leftrightarrow B) = 1$ iff $i(A) = i(B)$.

The *set of interpretations* of the language Θ will be denoted I_Θ.

The interpretation i is *a model* of formula A iff $i(A) = 1$. We say that i *satisfies* A.

The set of models of formula A is denoted $Mod(A)$.

A is *satisfiable* iff $Mod(A) \neq \emptyset$.

A is *a tautology* iff $Mod(A) = I_\Theta$. Tautologies are denoted by *true*.

A is *a contradiction* iff $Mod(A) = \emptyset$. Contradictions are denoted by *false*.

Formula B is a *logical consequence* of formula A iff $Mod(A) \subseteq Mod(B)$. It is denoted $A \models B$.

Formula A and formula B are *logically equivalent* (or *equivalent*) iff $Mod(A) = Mod(B)$. It is denoted $\models A \leftrightarrow B$.

Any propositional formula is equivalent to a conjunction of clauses in which no clause is subsumed.

Let σ be a satisfiable formula. An equivalence relation denoted $\overset{\sigma}{\leftrightarrow}$ is defined on $FORM$ by: $A \overset{\sigma}{\leftrightarrow} B$ iff $\sigma \models A \leftrightarrow B$. Thus, two formulas are related by relation $\overset{\sigma}{\leftrightarrow}$ iff they are equivalent when σ is true.

The following convention is made from now: We will consider that all the formulas of a given equivalence class of $\overset{\sigma}{\leftrightarrow}$ are *identical*. For instance, if $\sigma = \neg(a \wedge b)$ then $c \vee (a \wedge b)$ and c are identical; $a \wedge b$ and $b \wedge a$ are identical to *false*.

With the previous convention, if Θ is finite and if σ is a satisfiable formula, then we can consider that the set of formulas is finite. It is denoted $FORM^\sigma$.

CL^σ is the finite set of clauses that can be built if one considers σ. I.e., $CL^\sigma = \{A : A$ is a clause and there is a formula ϕ in $FORM^\sigma$ such that A is identical to $\phi\}$.

$+MCNF^\sigma$ is the finite set of formulas which are $+mcnf$. I.e., $+MCNF^\sigma = \{A : A \in FORM^\sigma$ and A is $+mcnf\}$.

2.2 Belief Function Theory

Belief function theory considers *a finite frame of discernment* $\Theta = \{\theta_1, ... \theta_n\}$ whose elements, called hypotheses, are *exhaustive and exclusive*. A *basic belief assignment (or mass function)* is a function $m : 2^\Theta \to [0,1]$ such that: $m(\emptyset) = 0$ and $\sum_{A \subseteq \Theta} m(A) = 1$. Given a mass function m, one can define *a belief function* $Bel : 2^\Theta \to [0,1]$ by: $Bel(A) = \sum_{B \subseteq A} m(B)$. One can also define *a plausibility function* $Pl : 2^\Theta \to [0,1]$ such that $Pl(A) = 1 - Bel(\overline{A})$.

Let m_1 and m_2 be two mass functions on the frame Θ. *Dempster's combination rule* defines a mass function denoted $m_1 \oplus m_2$, by: $m_1 \oplus m_2(\emptyset) = 0$ and for any $C \neq \emptyset$

$$m_1 \oplus m_2(C) = \frac{\sum_{A \cap B = C} m_1(A).m_2(B)}{\sum_{A \cap B \neq \emptyset} m_1(A).m_2(B)} \quad \text{for any } C \neq \emptyset$$

if $\sum_{A \cap B \neq \emptyset} m_1(A).m_2(B) \neq 0$

2.3 Extension of Belief Function Theory to Non-exclusive Hypotheses

The formalism described by [3] assumes that frames of discernment are finite sets of hypotheses which are exhaustive but *not necessarily exclusive*. It has then been extended in [11] by considering *integrity constraints*. Here, we summarize this extended version.

Let $\Theta = \{\theta_1, ... \theta_n\}$ be a frame of discernment. We say that an expression is under *reduced conjunctive normal form*, iff it is an intersection of unions of hypotheses of Θ such that no union contains another one. The *hyper-power-set*, D^Θ, is the set of all the expressions under reduced conjunctive normal form.

Integrity constraints are represented by IC, an expression of the form $E = \emptyset$, where $E \in D^\Theta$.

Taking integrity constraints into account comes down to considering a restriction of the hyper power set D^Θ. This restricted set contains fewer elements than in the general case and we will denote it D^Θ_{IC} to indicate the fact that this new set depends on IC. For instance, if $\Theta = \{\theta_1, \theta_2\}$ then $D^\Theta = \{\emptyset, \theta_1, \theta_2, \theta_1 \cap \theta_2, \theta_1 \cup \theta_2\}$. Consider now the constraint $IC = (\theta_1 \cap \theta_2 = \emptyset)$. Then we get $D^\Theta_{IC} = 2^\Theta = \{\emptyset, \theta_1, \theta_2, \theta_1 \cup \theta_2\}$.

The mass functions are then defined on elements of the hyper-power-set D^Θ_{IC}. Definitions of belief functions and plausibility functions are unchanged.

Finally, several combination rules have been defined in this theory in particular, the so called Proportional Conflict redistribution Rules (PCR) which redistribute the mass related to conflicting beliefs proportionally to expressions responsible of this conflict. They are several ways for redistributing. In this paper, we do not discuss the intuition behind the combination rules and we focus on the fifth rule called PCR5. This rule takes two mass functions m_1 and m_2, and builds a third mass function: $m_1 \oplus_5 m_2 : D^\Theta_{IC} \to [0, 1]$ defined by:

$$m_1 \oplus_5 m_2(\emptyset) = 0$$

And for any $X \neq \emptyset$, $m_1 \oplus_5 m_2(X) = m_{12}(X) + m'_{12}(X)$
with

$$m_{12}(X) = \sum_{\substack{X_1, X_2 \in D^\Theta_{IC} \\ X_1 \cap X_2 = X}} m_1(X_1).m_2(X_2)$$

and

$$m'_{12}(X) = \sum_{\substack{Y \in D^\Theta_{IC} \\ X \cap Y = \emptyset}} \left(\frac{m_1(X)^2 m_2(Y)}{m_1(X) + m_2(Y)} + \frac{m_2(X)^2 m_1(Y)}{m_2(X) + m_1(Y)} \right)$$

(where any fraction whose denominator is 0 is discarded)

3   A Logical Version of Belief Function Theory

In this section, we present a formalism that allows a modeler to represent beliefs by assigning masses to propositional formulas.

3.1   Logical Mass Functions, Logical Belief Functions, Logical Plausibility Functions

We introduce new definitions for mass functions, belief functions and plausibility functions in order to apply them to propositional formulas.

Let Θ be a finite propositional language, σ be a satisfiable formula.

Definition 1. *A logical mass function is a function* $m : FORM^\sigma \to [0, 1]$ *such that:* $m(false) = 0$ *and* $\sum_{A \in FORM^\sigma} m(A) = 1$.

Definition 2. *Given a logical mass function, the logical belief function Bel is defined by:* $Bel : FORM^\sigma \to [0, 1]$ *by:*

$$Bel(A) = \sum_{\substack{B \in FORM^\sigma \\ \sigma \models B \to A}} m(B)$$

Definition 3. *Given a logical mass function, the logical plausibility function Pl is defined by:* $Pl : FORM^\sigma \to [0, 1]$ *such that* $Pl(A) = 1 - Bel(\neg A)$.

Theorem 1.

$$Pl(A) = \sum_{\substack{B \in FORM^\sigma \\ (\sigma \wedge B \wedge A) \ is \ satisfiable}} m(B)$$

Example 2. Let Θ be the propositional language whose letters are: $limb, oak, ash,$ $birch$ respectively meaning "what fell is a limb", "what fell comes from an oak", "what fell comes from an ash", "what fell comes from a birch". Consider $\sigma = limb \to (oak \vee ash \vee birch) \wedge \neg(oak \wedge ash) \wedge \neg(oak \wedge birch) \wedge \neg(ash \wedge birch)$ expressing that if a limb fell then it is comes from an oak, an ash or a birch, and these three types of trees are different.

Consider the following logical mass function: $m(limb \wedge \neg oak) = 0.8$, $m(ash) = 0.2$. This expresses that my degree of belief in the piece of information " a non-oak limb fell" is 0.8 and my degree of belief in the piece of information "what fell comes from an ash" is 0.2.

The logical belief function is then defined by: $Bel(limb \wedge \neg oak) = 0.8$, $Bel(ash) = 0.2$, $Bel(ash \vee birch) = 1$, $Bel(\neg oak) = 1$, $Bel(\neg birch) = 0.2$, $Bel(birch) = 0$, $Bel(limb) = 0.8$, $Bel(\neg limb) = 0$, etc....

The logical plausibility function is then defined by: $Pl(limb \wedge \neg oak) = 1$, $Pl(ash) = 1$, $Pl(ash \vee birch) = 1$, $Pl(\neg oak) = 1$, $Pl(\neg birch) = 1$, $Pl(birch) = 0.8$, $Pl(limb) = 1$, $Pl(\neg limb) = 0.2$, etc....

3.2   Combination Rules

Let us now address the question of combining two logical mass functions. As it is done the belief function theory community, we could define plenty of rules according to the meanings of the masses or according to the properties that the combination rule must fulfill. But the purpose of this paper is not to define a new combination rule nor to comment the drawbacks or advantages of such or such existing rule. This is why we arbitrarily propose two rules for combining two logical mass functions. The first one is called "logical DS rule", denoted $\oplus_L$; the second one is called "logical PCR5 rule" or "logical Proportional Conflict Redistribution Rule 5" and denoted $\oplus_{5L}$. They are defined by the two following definitions and illustrated on example 3.

Definition 4. *Let m_1 and m_2 be two logical mass functions. The logical DS rule, denoted $\oplus_L$, defines the logical mass function $m_1 \oplus_L m_2 : FORM^\sigma \to [0,1]$ by: $m_1 \oplus_L m_2(false) = 0$ and for any $C \not\overset{\sigma}{\to} false$*

$$m_1 \oplus_L m_2(C) = \frac{\sum_{C \overset{\sigma}{\leftrightarrow}(A \wedge B)} m_1(A).m_2(B)}{\sum_{\sigma \wedge (A \wedge B) \ satisfiable} m_1(A).m_2(B)}$$

$$\text{if} \quad \sum_{\sigma \wedge (A \wedge B) \ satisfiable} m_1(A).m_2(B) \quad \neq 0$$

Thus, according to this definition, the logical DS rule defines a logical mass function $m_1 \oplus_L m_2$ such that the mass assigned to any formula C which is not a contradiction, is the normalized sum of the product $m_1(A).m_2(B)$ where $A \wedge B$ and C are identical (i.e., equivalent when σ is true).

Definition 5. *Let m_1 and m_2 be two logical mass functions. The logical PCR5 rule defines the logical mass function $m_1 \oplus_{5L} m_2 : FORM^\sigma \to [0,1]$ by:*

$$m_{5L}(false) = 0$$

$$m_{5L}(X) = m_{12}(X) + m'_{12}(X) \quad \text{if } X \not\overset{\sigma}{\to} false$$

with

$$m_{12}(X) = \sum_{\substack{X_1.X_2 \in FORM^\sigma \\ X_1 \wedge X_2 \overset{\sigma}{\leftrightarrow} X}} m_1(X_1).m_2(X_2)$$

and

$$m'_{12}(X) = \sum_{\substack{Y \in FORM^\sigma \\ (\sigma \wedge X \wedge Y) \ unsatisfiable}} \left(\frac{m_1(X)^2 m_2(Y)}{m_1(X) + m_2(Y)} + \frac{m_2(X)^2 m_1(Y)}{m_2(X) + m_1(Y)} \right)$$

(where any fraction whose denominator is 0 is discarded)

Thus, according to this definition, the logical PCR5 rule defines a logical mass function $m_1 \oplus_{5L} m_2$ such that the mass assigned to any formula X which is not a contradiction, is the sum of two numbers: the first number is the sum of the products $m_1(X_1).m_2(X_2)$ where $X_1 \wedge X_2$ and X are identical (i.e., equivalent when σ is true); the second number depends on the masses of the propositions which (under σ) contradict X.

Example 3. Let us consider Θ and σ as in example 2. We consider here two witnesses: Betty and Sally. Suppose that Betty tells that a limb fell on my car but it was not an oak limb and that this testimony alone implies that I have the following degrees of belief: $m_1(limb \wedge \neg oak) = 2/3$, $m_1(true) = 1/3$. Suppose that Sally tells that what fell on my car was not a limb but it comes down from an ash and that this testimony alone implies that I have the following degrees of belief: $m_2(\neg limb \wedge ash) = 2/3$, $m_2(true) = 1/3$

1. Let us see what the logical DS rule provides. First, notice that $\sigma \wedge (limb \wedge \neg oak) \wedge (\neg limb \wedge ash)$ is not satisfiable. Thus $N = 5/9$. Consequently, the logical DS rule provides the following logical mass function: $m_1 \oplus_L m_2(limb \wedge \neg oak) = 2/5$, $m_1 \oplus_L m_2(\neg limb \wedge ash) = 2/5$, $m_1 \oplus_L m_2(true) = 1/5$.
   Thus, the logical belief function Bel, associated with $m_1 \oplus_L m_2$ is so that: $Bel(limb) = 2/5$, $Bel(\neg limb) = 2/5$, $Bel(oak) = 0$, $Bel(\neg oak) = 4/5$, $Bel(ash) = 2/5$, $Bel(\neg ash) = 0$ $Bel(birch) = 0$, $Bel(\neg birch) = 2/5$ etc.
   And the logical plausibility function Pl associated with $m_1 \oplus_L m_2$ is so that: $Pl(limb) = 3/5$, $Pl(\neg limb) = 3/5$, $Pl(oak) = 1/5$, $Pl(\neg oak) = 1$, $Pl(ash) = 1$, $Pl(\neg ash) = 3/5$ $Pl(birch) = 3/5$, $Pl(\neg birch) = 1$ etc.
2. Let us consider now the logical PCR5 rule. This rule provides the following logical mass function: $m_1 \oplus_{5L} m_2(limb \wedge \neg oak) = 4/9$, $m_1 \oplus_{5L} m_2(\neg limb \wedge ash) = 4/9$, $m_1 \oplus_{5L} m_2(true) = 1/9$.
   Thus the logical belief function which is associated with $m_1 \oplus_{5L} m_2$ is so that: $Bel(limb) = 4/9$, $Bel(\neg limb) = 4/9$, $Bel(ash) = 4/9$, $Bel(\neg ash) = 0$, $Bel(oak) = 0$, $Bel(\neg oak) = 8/9$, $Bel(birch) = 0$, $Bel(\neg birch) = 4/9$, $Bel(limb \wedge \neg oak) = 4/9$, $Bel(ash \vee birch) = 8/9$, etc....
   And the logical plausibility function which is associated with $m_1 \oplus_{5L} m_2$ is so that: $Pl(limb) = 5/9$, $Pl(\neg limb) = 5/9$, $Pl(oak) = 5/9$, $Pl(\neg oak) = 1$, $Pl(ash) = 1$, $Pl(\neg ash) = 5/9$ $Pl(birch) = 5/9$, $Pl(\neg birch) = 1$ etc.

4   Particular Cases

In this section, we consider two particular cases of the logical framework previously introduced. We first show that if we consider a particular σ which expresses that hypotheses are exclusive and exhaustive and restrict formulas to clauses, then the logical framework reduces to the belief function theory. Then we show that if we consider another particular σ and restrict formulas to those of $+MCNF^\sigma$, then the logical framework reduces to the extension of the belief function theory to non-exclusive hypotheses. *In both cases, we consider a frame of discernment which is isomorphic to the propositional language.*

4.1   First Particular Case

We consider a first particular case of our logical proposal in which $\sigma = (H_1 \vee ... \vee H_n) \wedge \bigwedge_{i \neq j} \neg(H_i \wedge H_j)$, where $H_1, ... H_n$ are the letters of the propositional language Θ. Thus, considering σ as true leads to consider that one and only one H_i is true. Furthermore, in this section, logical mass functions are restricted to those whose focal elements are clauses only. I.e., here, *a logical mass function is a function* $m : CL^\sigma \to [0, 1]$ *such that:* $m(false) = 0$ *and* $\sum_{A \in CL^\sigma} m(A) = 1$.

Definition 6. *With Θ, we can associate a frame of discernment with n exhaustive and exclusive hypotheses. These hypotheses are denoted $f(H_1), ... f(H_n)$ and the frame is denoted $f(\Theta)$.*

Theorem 2. *Under the previous assumptions:*

1. *Any clause C of CL^σ can be associated with an unique expression of $2^{f(\Theta)}$.*
2. *Any logical mass function m whose focal elements are $C_1, ...C_k$ can be associated with an unique mass function denoted $f(m)$ whose focal elements are $f(C_1), ...f(C_k)$ and so that each $f(C_i)$ is given the mass $m(C_i)$.*
3. *Given two logical mass functions m_1 and m_2, then for any clause C in CL^σ, we have $m_1 \oplus_L m_2(C) = f(m_1) \oplus f(m_2)(f(C))$*

4.2 Second Particular Case

We consider a second particular case of our logical proposal in which $\sigma = (H_1 \vee ... \vee H_n) \wedge IC$ where $H_1, ...H_n$ are the letters of the propositional language Θ and IC a propositional formula different from $\neg(H_1 \vee ... \vee H_n)$. Moreover, logical mass functions are restricted to those whose focal elements are $+$mcnf formulas only. I.e., here, *a logical mass function* is a function $m : +MCNF^\sigma \to [0,1]$ so that: $m(false) = 0$ and $\sum_{A \in +MCNF^\sigma} m(A) = 1$.

Definition 7. *With Θ, we can associate a frame of discernment with n exhaustive hypotheses. These hypotheses are denoted $g(H_1), ..., g(H_n)$ and the frame is denoted $g(\Theta)$.*

Theorem 3. *Under the previous assumptions:*

1. *Any formula F of $+MCNF^\sigma$ can be associated with an unique expression of $D_g(\Sigma)^{g(\Theta)}$.*
2. *Any logical mass function m whose focal elements are $F_1, ...F_k$ can be associated with an unique mass function denoted $g(m)$ whose focal elements are $g(F_1), ...g(F_k)$ and so that each $g(F_i)$ is given the mass $m(F_i)$.*
3. *Given two logical mass functions m_1 and m_2, then for any $+$mcnf formula F in $+MCNF^\sigma$, we have $m_1 \oplus_{5L} m_2(F) = g(m_1) \oplus_5 g(m_2)(g(F))$*

5 Conclusion

This paper presented a version of the belief function theory in which masses are assigned to propositional formulas. In this logical framework, we can assume that some formula (denoted σ before) is true and two combination rules have been defined. Finally, its has been shown that the belief function theory and its extension to non-exclusive hypotheses are two particular cases of this framework.

We think that this work is original and we have not found in the literature any work proposing this logical extension or something equivalent. However, the relation between focal elements of the belief function theory and propositional clauses is known for a long time. This link was early mentioned in Shafer's book (p. 37) but this remark did not lead to any improvement. In [7], Provan proposed a logical view of Demspter-Shafer theory in which focal elements are modeled by propositional clauses. This corresponds to the result given in section

4.1. In [5], Kwisthout and Dastani wondered if the belief function theory can be applied to the beliefs of an agent when they are represented by logical formulas of an agent programming language. So their motivation was very close to ours. However, instead of defining a logical version of belief function theory as we do here, they translated the problem in the "classical" belief function theory. In [6], Lehmann was interested in reducing the cost of computations in the belief function theory. He proposed a compact representation of focal elements based on bitstrings. For doing so, he considers hypotheses which are interpretations of propositional formulas. And he obviously shows that, instead of assigning masses to unions of interpretations one can assign masses to formulas. However, even if Lehmann's motivation for finding a more compact representation meets ours (see example in 1), he did not address the question of combination rules in this compact representation. Finally, notice that the logical framework we introduced is *not* a logic and cannot be compared with the belief-function logic introduced in [8] nor with the fuzzy modal logic for belief functions defined by [4].

As a future work, we aim to extend this logical framework to first order logic beliefs. This would allow one to assign masses to first order formulas. We could then have logical mass functions like: $m(\forall x \; bird(x) \rightarrow fly(x)) = 0.1$ and $m(true) = 0.9$ which would express my high uncertainty about the fact that all birds fly.

Acknowledgments. I thank the anonymous reviewers whose questions and comments helped me to improve this paper.

References

1. Cholvy, L.: Non-exclusive hypotheses in Dempster-Shafer Theory. Journal of Approximate Reasoning 53, 493–501 (2012)
2. Dempster, A.P.: A generalization of bayesian inference. Journal of the Royal Statistical Society, Series B 30, 205–247 (1968)
3. Dezert, J.: Foundations for a new theory of plausible and paradoxical reasoning. Information and Security. An International Journal 9, 90–95 (2002)
4. Godo, L., Hájek, P., Esteva, F.: A fuzzy modal logic for belief-functions. Fundamenta Informaticae XXI, 1001–1020 (2001)
5. Kwisthout, J., Dastani, M.: Modelling uncertainty in agent programming. In: Baldoni, M., Endriss, U., Omicini, A., Torroni, P. (eds.) DALT 2005. LNCS (LNAI), vol. 3904, pp. 17–32. Springer, Heidelberg (2006)
6. Lehmann, N.: Shared ordered binary decision diagrams for Dempster-Shafer Theory. In: Mellouli, K. (ed.) ECSQARU 2007. LNCS (LNAI), vol. 4724, pp. 320–331. Springer, Heidelberg (2007)
7. Provan, G.: A logic-based analysis of Dempster-Shafer Theory. International Journal of Approximate Reasoning 4, 451–495 (1990)
8. Saffiotti, A.: A belief-function logic. In: Proceedings of the 10th AAAI Conference, San Jose, CA, pp. 642–647 (1992)
9. Shafer, G.: A mathematical theory of evidence. Princeton University Press (1976)
10. Shafer, G.: Perspectives on the theory and practice of belief functions. Int. Journal of Approximate Reasoning 4, 323–362 (1990)
11. Smarandache, F., Dezert, J.: An introduction to the DSm Theory for the combination of paradoxical, uncertain and imprecise sources of information (2006), http://www.gallup.unm.edu/~smarandache/DSmT-basic-short.pdf

Evidential-EM Algorithm Applied to Progressively Censored Observations

Kuang Zhou[1,2], Arnaud Martin[2], and Quan Pan[1]

[1] School of Automation, Northwestern Polytechnical University,
Xi'an, Shaanxi 710072, P.R. China
[2] IRISA, University of Rennes 1, Rue E. Branly, 22300 Lannion, France
kzhoumath@163.com, Arnaud.Martin@univ-rennes1.fr, quanpan@nwpu.edu.cn

Abstract. Evidential-EM (E2M) algorithm is an effective approach for computing maximum likelihood estimations under finite mixture models, especially when there is uncertain information about data. In this paper we present an extension of the E2M method in a particular case of incomplete data, where the loss of information is due to both mixture models and censored observations. The prior uncertain information is expressed by belief functions, while the pseudo-likelihood function is derived based on imprecise observations and prior knowledge. Then E2M method is evoked to maximize the generalized likelihood function to obtain the optimal estimation of parameters. Numerical examples show that the proposed method could effectively integrate the uncertain prior information with the current imprecise knowledge conveyed by the observed data.

Keywords: Belief function theory, Evidential-EM, Mixed-distribution, Uncertainty, Reliability analysis.

1 Introduction

In life-testing experiments, the data are often censored. A datum T_i is said to be right-censored if the event occurs at a time after a right bound, but we do not exactly know when. The only information we have is this right bound. Two most common right censoring schemes are termed as Type-I and Type-II censoring. The experiments using these test schemes have the drawback that they do not allow removal of samples at time points other than the terminal of the experiment. The progressively censoring scheme, which possesses this advantage, has become very popular in the life tests in the last few years [1]. The censored data provide some kind of imprecise information for reliability analysis.

It is interesting to evaluate the reliability performance for items with mixture distributions. When the population is composed of several subpopulations, an instance in the data set is expected to have a label which represents the origin, that is, the subpopulation from which the data is observed. In real-world data, observed labels may carry only partial information about the origins of samples. Thus there are concurrent imprecision and uncertainty for the censored data from mixture distributions. The Evidential-EM (E2M) method, proposed

A. Laurent et al. (Eds.): IPMU 2014, Part III, CCIS 444, pp. 180–189, 2014.

by Denœux [4,3], is an effective approach for computing maximum likelihood estimates for the mixture problem, especially when there is both imprecise and uncertain knowledge about the data. However, it has not been used for reliability analysis and the censored life tests.

This paper considers a special kind of incomplete data in life tests, where the loss of information is due simultaneously to the mixture problem and to censored observations. The data set analysed in this paper is merged by samples from different classes. Some uncertain information about class values of these unlabeled data is expressed by belief functions. The pseudo-likelihood function is obtained based on the imprecise observations and uncertain prior information, and then E2M method is invoked to maximize the generalized likelihood function. The simulation studies show that the proposed method could take advantages of using the partial labels, and thus incorporates more information than traditional EM algorithms.

2 Theoretical Analysis

Progressively censoring scheme has attracted considerable attention in recent years, since it has the flexibility of allowing removal of units at points other than the terminal point of the experiment [1]. The theory of belief functions is first described by Dempster [2] with the study of upper and lower probabilities and extended by Shafer later [6]. This section will give a brief description of these two concepts.

2.1 The Type-II Progressively Censoring Scheme

The model of Type-II progressively censoring scheme (PCS) is described as follows [1]. Suppose n independent identical items are placed on a life-test with the corresponding lifetimes $X_1, X_2, \cdots, X_n$ being identically distributed. We assume that X_i ($i = 1, 2, \cdots, n$) are i.i.d. with probability density function (pdf) $f(x; \theta)$ and cumulative distribution function (cdf) $F(x; \theta)$. The integer $J < n$ is fixed at the beginning of the experiment. The values $R_1, R_2, \cdots, R_J$ are J pre-fixed satisfying $R_1 + R_2 + \cdots + R_J + J = n$. During the experiment, the jth failure is observed and immediately after the failure, R_j functioning items are randomly removed from the test. We denote the time of the jth failure by $X_{j:J:n}$, where J and n describe the censored scheme used in the experiment, that is, there are n test units and the experiment stops after J failures are observed. Therefore, in the presence of Type-II progressively censoring schemes, we have the observations $\{X_{1:J:n}, \cdots, X_{J:J:n}\}$. The likelihood function can be given by

$$L(\theta; x_{1:J:n}, \cdots, x_{J:J:n}) = C \prod_{i=1}^{J} f(x_{i:J:n}; \theta)[1 - F(x_{i:J:n}; \theta)]^{R_i}, \qquad (1)$$

where $C = n(n-1-R_1)(n-2-R_1-R_2) \cdots (n-J+1-R_1-R_2-\cdots-R_{J-1})$.

2.2 Theory of Belief Functions

Let $\Theta = \{\theta_1, \theta_2, \ldots, \theta_N\}$ be the finite domain of X, called the discernment frame. The mass function is defined on the power set $2^\Theta = \{A : A \subseteq \Theta\}$. The function $m : 2^\Theta \rightarrow [0, 1]$ is said to be the basic belief assignment (bba) on 2^Θ, if it satisfies:

$$\sum_{A \subseteq \Theta} m(A) = 1. \tag{2}$$

Every $A \in 2^{\Theta}$ such that $m(A) > 0$ is called a focal element. The credibility and plausibility functions are defined in Eq. (3) and Eq. (4).

$$Bel(A) = \sum_{\emptyset \neq B \subseteq A} m(B), \forall A \subseteq \Theta, \tag{3}$$

$$Pl(A) = \sum_{B \cap A \neq \emptyset} m(B), \forall A \subseteq \Theta. \tag{4}$$

Each quantity $Bel(A)$ denotes the degree to which the evidence supports A, while $Pl(A)$ can be interpreted as an upper bound on the degree of support that could be assigned to A if more specific information became available [7]. The function $pl : \Theta \to [0, 1]$ such that $pl(\theta) = Pl(\{\theta\})$ is called the contour function associated to m.

If m has a single focal element A, it is said to be categorical and denoted as m_A. If all focal elements of m are singletons, then m is said to be Bayesian. Bayesian mass functions are equivalent to probability distributions.

If there are two distinct pieces of evidences (bba) on the same frame, they can be combined using Dempster's rule [6] to form a new bba:

$$m_{1 \oplus 2}(C) = \frac{\sum_{A_i \cap B_j = C} m_1(A_i) m_2(B_j)}{1 - k} \qquad \forall C \subseteq \Theta, C \neq \emptyset \tag{5}$$

If m_1 is Bayesian mass function, and its corresponding contour function is p_1. Let m_2 be an arbitrary mass function with contour function pl_2. The combination of m_1 and m_2 yields a Bayesian mass function $m_1 \oplus m_2$ with contour function $p_1 \oplus pl_2$ defined by

$$p_1 \oplus pl_2 = \frac{p_1(\omega) pl_2(\omega)}{\sum_{\omega' \in \Omega} p_1(\omega') pl_2(\omega')}. \tag{6}$$

The conflict between p_1 and pl_2 is $k = 1 - \sum_{\omega' \in \Omega} p_1(\omega') pl_2(\omega')$. It equals one minus the expectation of pl_2 with respect to p_1.

3   The E2M Algorithm for Type-II PCS

3.1   The Generalized Likelihood Function and E2M Algorithm

E2M algorithm, similar to the EM method, is an iterative optimization tactics to obtain the maximum of the observed likelihood function [4,3]. However, the data applied to E2M model can be imprecise and uncertain. The imprecision may be brought by missing information or hidden variables, and this problem can be solved by the EM approach. The uncertainty may be due to the unreliable sensors, the errors caused by the measuring or estimation methods and so on. In the E2M model, the uncertainty is represented by belief functions.

Let X be a discrete variable defined on Ω_X and the probability density function is $p_X(\cdot; \theta)$. If x is an observation sample of X, the likelihood function can be expressed as:

$$L(\theta; x) = p_X(x; \theta). \tag{7}$$

If x is not completely observed, and what we only know is that $x \in A, A \subseteq \Omega_X$, then the likelihood function becomes:

$$L(\theta; A) = \sum_{x \in A} p_X(x; \theta). \tag{8}$$

If there is some uncertain information about x, for example, the experts may give their belief about x in the form of mass functions: $m(A_i), i = 1, 2, \cdots, r$, $A_i \subseteq \Omega_X$, then the likelihood becomes:

$$L(\theta; m) = \sum_{i=1}^{r} m(A_i) L(\theta; A_i) = \sum_{x \in \Omega_x} p_X(x; \theta) pl(x). \tag{9}$$

It can be seen from Eq. (9) that the likelihood $L(\theta; m)$ only depends on m through its associated contour function pl. Thus we could write indifferently $L(\theta; m)$ or $L(\theta; pl)$.

Let $W = (X, Z)$ be the complete variable set. Set X is the observable data while Z is unobservable but with some uncertain knowledge in the form of pl_Z. The log-likelihood based on the complete sample is $\log L(\theta; W)$. In E2M, the observe-data log likelihood is $\log L(\theta; X, pl_Z)$.

In the E-step of the E2M algorithm, the pseudo-likelihood function should be calculated as:

$$Q(\theta, \theta^k) = E_{\theta^k}[\log L(\theta; W)|X, pl_Z; \theta^k], \tag{10}$$

where pl_Z is the contour function describing our uncertainty on Z, and θ^k is the parameter vector obtained at the kth step. E_{θ^k} represents the expectation with respect to the following density:

$$\gamma'(Z = j|X, pl_Z; \theta^k) \triangleq \gamma(Z = j|X; \theta^k) \oplus pl_Z. \tag{11}$$

Function γ' could be regarded as a combination of conditional probability density $\gamma(Z = j|X; \theta^k) = p_Z(Z = j|X; \theta^k)$ and the contour function pl_Z. It depicts the current information based on the observation X and the prior uncertain information on Z, thus this combination is similar to the Bayes rule.

According to the Dempster combination rule and Eq. (9), we can get:

$$\gamma'(Z = j|X, pl_Z; \theta^k) = \frac{r(Z = j|X; \theta^k)pl_Z(Z = j)}{\sum_j r(Z = j|X; \theta^k)pl_Z(Z = j)}. \tag{12}$$

Therefore, the pseudo-likelihood is:

$$Q(\theta, \theta^k) = \frac{\sum_j r(Z = j|X; \theta^k)pl(Z = j)\log L(\theta; W)}{L(\theta^k; X, pl_Z)}. \tag{13}$$

The M-step is the same as EM and requires the maximization of $Q(\theta, \theta^k)$ with respect to θ. The E2M algorithm alternately repeats the E- and M-steps above until the increase of general observed-data likelihood becomes smaller than a given threshold.

3.2   Mixed-Distributed Progressively Censored Data

Here, we present a special type of incomplete data, where the imperfection of information is due both to the mixed-distribution and to some censored observations. Let Y denote the lifetime of test samples. The n test samples can de divided into two parts, *i.e.* Y_1, Y_2, where Y_1 is the set of observed data, while Y_2 is the censored data set. Let Z be the class labels and $W = (Y, Z)$ represent the complete data.

Assume that Y is from mixed-distribution with p.d.f.

$$f_Y(y; \theta) = \sum_{z=1}^{p} \lambda_z f(y; \xi_z), \tag{14}$$

where $\theta = (\lambda_1, \cdots, \lambda_p, \xi_1, \cdots, \xi_p)$. The complete data distribution of W is given by $P(Z = z) = \lambda_z$ and $P(Y|Z = z) = f(y; \xi_z)$. Variable Z is hidden but we can have a prior knowledge about it. This kind of prior uncertain information of Z can be described in the form of belief functions:

$$pl_Z(Z = j) = pl_j, j = 1, 2, \cdots, p. \tag{15}$$

The likelihood of the complete data is:

$$L^c(\theta; Y, Z) = \prod_{j=1}^{n} f(y_j, z_j; \theta), \tag{16}$$

and the pseudo-likelihood function is:

$$Q(\theta, \theta^k) = \mathrm{E}_{\theta^k}[\log L^c(\theta; Y, Z)|Y^*, pl_Z; \theta^k], \tag{17}$$

where $\mathrm{E}_{\theta^k}[\cdot|Y^*, pl_Z; \theta^k]$ denotes expectation with respect to the conditional distribution of W given the observation Y^* and the uncertain information pl_Z.

Theorem 1. *For (y_j, z_j) are complete and censored, $f_{YZ}(y_j, z_j|y_j^*; \theta^k)$ can be calculated according to Eq. (18) and Eq. (19) respectively. Let y_j^* be the jth observation. If the jth sample is completely observed, $y_j = y_j^*$; Otherwise $y_j \geq y_j^*$.*

$$f_{YZ}^1(y_j, z_j|y_j^*; \theta^k) = \mathrm{I}_{\{y_j=y_j^*\}} P_{1jz}^k, \tag{18}$$

$$f_{YZ}^2(y_j, z_j|y_j^*; \theta^k) = \mathrm{I}_{\{y_j>y_j^*\}} P_{2jz}^k \frac{f(y_j; \xi_z^k)}{\overline{F}(y_j^*; \xi_z^k)}. \tag{19}$$

where P^k_{1jz} and P^k_{2jz} are shown in Eq. (20).

$$P^k_{jz}(z_j = z|Y^*; \theta) = \begin{cases} P^k_{1jz}(z_j = z|y^*_j; \theta^k) & \text{for the completely observed data} \\ P^k_{2jz}(z_j = z|y^*_j; \theta^k) & \text{for the censored data} \end{cases}$$

(20)

where

$$P^k_{1jz}(z_j = z|y^*_j; \theta^k) = \frac{f(y^*_j; \xi^k_z)\lambda^k_z}{\sum_z f(y^*_j; \xi^k_z)\lambda^k_z},$$

(21)

$$P^k_{2jz}(z_j = z|y^*_j; \theta^k) = \frac{\overline{F}(y^*_j; \xi^k_z)\lambda^k_z}{\sum_z \overline{F}(y^*_j; \xi^k_z)\lambda^k_z}.$$

(22)

Proof. If (y_j, z_j) are completely observed,

$$f^1_{yz}(y_j, z_j|y^*_j; \theta^k) = P^k_{1jz} f(y_j|y^*_j = y_j, Z_j = z; \theta^k),$$

we obtain Eq. (18).

If (y_j, z_j) are censored,

$$f^2_{yz}(y_j, z_j|y^*_j; \theta^k) = P^k_{2jz} f(y_j|y^*_j < y_j, Z_j = z; \theta^k),$$

From the theorem in [5],

$$f(y_j|y^*_j < y_j, Z_j = z; \theta^k) = \frac{f(y_j; \xi^k_z)}{\overline{F}(y^*_j; \xi^k_z)} I_{\{y_j > y^*_j\}},$$

we can get Eq. (19).

This completes this proof.

From the above theorem, the pseudo-likelihood function can be written as:

$$\begin{aligned} Q(\theta, \theta^k) &= E_{\theta^k}[\log f^c(Y, Z)|Y^*, pl_Z; \theta^k] \\ &= \sum_{j=1}^{n} E_{\theta^k}[\log \lambda_z + \log f(y_j|\xi_z)|Y^*, pl_Z; \theta^k] \\ &= \sum_{y_j \in Y_1} \sum_z P'^k_{1jz} \log \lambda_z + \sum_{y_j \in Y_2} \sum_z P'^k_{2jz} \log \lambda_z \\ &+ \sum_{y_j \in Y_1} \sum_z P'^k_{1jz} \log f(y^*_j|\xi_z) \\ &+ \sum_{y_j \in Y_2} \sum_z P'^k_{2jz} \int_{y^*_j}^{+\infty} \log f(x|\xi_z) \frac{f(x|\xi^k_z)}{\overline{F}(y^*_j; \xi^k_z)} dx, \end{aligned}$$

(23)

where

$$P'^k_{ijz}(z_j = z|y^*_j, pl_{Z_j}; \theta^k) = P^k_{ijz}(z_j = z|y^*_j; \theta^k) \oplus pl_{Z_j}, i = 1, 2.$$

It can be seen that $P'^k_{ijz}(z_j = z|y^*_j, pl_{Z_j}; \theta^k)$ is a Dempster combination of the prior and the observed information.

Assume that the data is from the mixed-Rayleigh distribution without loss of generality, the p.d.f. is shown in Eq. (24):

$$f_X(x; \lambda, \xi) = \sum_{j=1}^{p} \lambda_j g_X(x; \xi_j) = \sum_{j=1}^{p} \lambda_j \xi_j^2 \exp\{-\frac{1}{2}\xi_j^2 x^2\}, \tag{24}$$

After the kth iteration and $\theta^k = \lambda^k$ is got, the $(k+1)^{th}$ step of E2M algorithm is shown as follows:

1. **E-step:** For $j = 1, 2, \cdots, n$, $z = 1, 2 \cdots, p$, use Eq. (23) to obtain the conditional p.d.f. of $\log L^c(\theta; W)$ based on the observed data, the prior uncertain information and the current parameters.
2. **M-step:** Maximize $Q(\theta|\theta^k)$ and update the parameters:

$$\lambda_z^{k+1} = \frac{1}{n} \left(\sum_{y_j \in Y_1} P_{1jz}^{\prime k} + \sum_{y_j \in Y_2} P_{2jz}^{\prime k} \right), \tag{25}$$

$$(\xi_z^{k+1})^2 = \frac{2 \left(\sum_{y_j \in Y_1} P_{1jz}^{\prime k} + \sum_{y_j \in Y_2} P_{2jz}^{\prime k} \right)}{\sum_{y_j \in Y_1} P_{1jz}^{\prime k} y_j^{*2} + \sum_{y_j \in Y_2} P_{2jz}^{\prime k} (y_j^{*2} + 2/(\xi_z^k)^2)}. \tag{26}$$

It should be pointed out that the maximize of $Q(\theta, \theta^k)$ is conditioned on $\sum_{i=1}^{p} \lambda_i = 1$. By Lagrange multipliers method we have the new objective function:

$$Q(\theta, \theta^k) - \alpha(\sum_{i=1}^{p} \lambda_i - 1).$$

4   Numerical Results

In this section, we will use Monte-Carlo method to test the proposed method. The simulated data set in this section is drawn from mixed Rayleigh distribution as shown in Eq. (24) with $p = 3$, $\lambda = (1/3, 1/3, 1/3)$ and $\xi = (4, 0.5, 0.8)$. The test scheme is $n = 500$, $m = n * 0.6$, $R = (0, 0, \cdots, n - m)_{1 \times m}$. Let the initial values be $\lambda^0 = (1/3, 1/3, 1/3)$ and $\xi^0 = (4, 0.5, 0.8) - 0.01$. As mentioned before, usually there is no information about the subclass labels of the data, which is the case of unsupervised learning. But in real life, we may get some prior uncertain knowledge from the experts or experience. These partial information is assumed to be in the form of belief functions here.

To simulate the uncertainty on the labels of the data, the original generated datasets are corrupted as follows. For each data j, an error probability q_j is drawn randomly from a beta distribution with mean ρ and standard deviation 0.2. The value q_j expresses the doubt by experts on the class of sample j. With probability q_j, the label of sample j is changed to any (three) class (denoted by z_j^*) with equal probabilities. The plausibilities are then determined as

$$pl_{Z_j}(z_j) = \begin{cases} \frac{q_j}{3} & if \ z_j \neq z_j^*, \\ \frac{q_j}{3} + 1 - q_j & if \ z_j = z_j^* \end{cases}. \tag{27}$$

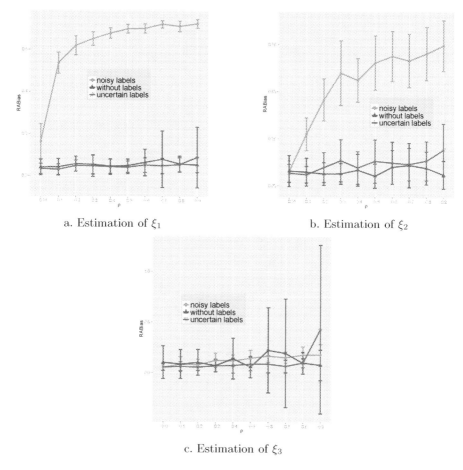

a. Estimation of ξ_1 b. Estimation of ξ_2

c. Estimation of ξ_3

Fig. 1. Average RABias values (plus and minus one standard deviation) for 20 repeated experiments, as a function of the error probability ρ for the simulated labels

The results of our approach with uncertain labels are compared with the cases of noisy labels and no information on labels. The former case with noisy labels is like supervised learning, while the latter is the traditional EM algorithm applied to progressively censored data. In each case, the E2M (or EM) algorithm is run 20 times. The estimations of parameters are compared to their real value using absolute relative bias (RABias). We recall that this commonly used measure equals 0 for the absolutely exact estimation $\hat{\theta} = \theta$.

The results are shown graphically in Figure 1. As expected, a degradation of the estimation performance is observed when the error probability ρ increases using noisy and uncertain labels. But our solution based on soft labels does not suffer as much that using noisy labels, and it clearly outperforms the supervised learning with noisy labels. The estimations for ξ_1 and ξ_3 by our approach (uncertain labels) are better than the unsupervised learning with unknown labels.

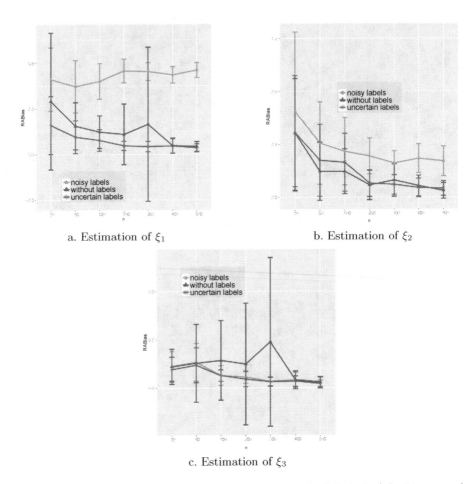

a. Estimation of ξ_1 b. Estimation of ξ_2

c. Estimation of ξ_3

Fig. 2. Average RABias values (plus and minus one standard deviation) for 20 repeated experiments, as a function of the sample numbers n

Although the estimation result for ξ_2 using uncertain labels seems not better than that by traditional EM algorithm when ρ is large, it still indicates that our approach is able to exploit additional information on data uncertainty when such information is available as the case when ρ is small.

In the following experiment, we will test the algorithm with different sample numbers n. In order to illustrate the different behavior of the approach with respect to n, we consider a fixed censored scheme with $(m =)$ 60% of samples are censored. With a given n, the test scheme is as follows: $m = n * 0.6$, $R = (0, 0, \cdots, n - m)_{1 \times m}$. Let the error probability be $\rho = 0.1$. Also we will compare our method using uncertain labels with those by noisy labels and without using any information of labels. The RABias for the results with different methods is shown in Figure 2. We can get similar conclusions as before that uncertainty on class labels appears to be successfully exploited by the proposed approach. Moreover, as n increases, the RABias decreases, which indicates the large sample properties of the maximum-likelihood estimation.

5   Conclusion

In this paper, we investigate how to apply E2M algorithm to progressively censored data analysis. From the numerical results we can see that the proposed method based on E2M algorithm has a better behavior in terms of the RABias of the parameter estimations as it could take advantage of the available data uncertainty. Thus the belief function theory is an effective tool to represent and deal with the uncertain information in reliability evaluation. The Monte-Carlo simulations show that the RABiases decreases with the increase of n for all cases. The method does improve for large sample size.

The mixture distribution is widely used in reliability project. Engineers find that there are often failures of tubes or other devices at the early stage, but the failure rate will remain stable or continue to raise with the increase of time. From the view of statistics, these products should be regarded to come from mixed distributions. Besides, when the reliability evaluation of these complex products is performed, there is often not enough priori information. Therefore, the application of the proposed method is of practical meaning in this case.

References

1. Balakrishnan, N.: Progressive censoring methodology: an appraisal. TEST 16, 211–259 (2007)
2. Dempster, A.P.: Upper and lower probabilities induced by a multivalued mapping. Annals of Mathematical Statistics 38, 325–328 (1967)
3. Denœux, T.: Maximum likelihood estimation from uncertain data in the belief function framework. IEEE Transactions on Knowledge and Data Engineering 25(1), 119–130 (2013)
4. Denœux, T.: Maximum likelihood from evidential data: An extension of the EM algorithm. In: Borgelt, C., González-Rodríguez, G., Trutschnig, W., Lubiano, M.A., Gil, M.Á., Grzegorzewski, P., Hryniewicz, O. (eds.) Combining Soft Computing and Statistical Methods in Data Analysis. AISC, vol. 77, pp. 181–188. Springer, Heidelberg (2010)
5. Ng, H., Chan, P., Balakrishnan, N.: Estimation of parameters from progressively censored data using EM algorithm. Computational Statistics and Data Analysis 39(4), 371–386 (2002)
6. Shafer, G.: A mathematical theory of evidence. Princeton University Press (1976)
7. Smets, P., Kennes, R.: The transferable belief model. Artificial Intelligence 66(2), 191–234 (1994)

A First Inquiry into Simpson's Paradox with Belief Functions

François Delmotte, David Mercier, and Frédéric Pichon

Univ. Lille Nord de France, F-59000 Lille, France
UArtois, LGI2A, F-62400, Béthune, France
firstname.lastname@univ-artois.fr

Abstract. Simpson's paradox, also known as the Yule-Simpson effect, is a statistical paradox which plays a major role in causality modelling and decision making. It may appear when marginalizing data: an effect can be positive for all subgroups of a population, but it can be negative when aggregating all the subgroups. This paper explores what happens if data are considered in the framework of belief functions instead of classical probability theory. In particular, the co-occurrence of the paradox with both the probabilistic approach and our belief function approach is studied.

Keywords: Simpson's paradox, Belief functions, Marginalization, Decision Making.

1   Introduction

First relations about Simpson's paradox, also known as the Yule-Simpson effect or the reversal paradox, were discovered one century ago [12], and more studies were published from the fifties [11]. It concerns statistical results that are very strange to the common sense: the influence of one variable can be positive for every subgroups of a population, but it may be negative for the whole population. For instance a medical treatment can be effective when considering the gender of patients, being effective for both males and females separately. But when studying the population as a whole, when the gender is ignored by marginalizing this variable, the treatment becomes inefficient. Real life examples are numerous. In [12] several examples are given concerning batting averages in baseball, kidney stone treatment, or the Berkeley sex bias case when studying the admittance at this university depending on the gender of the applicants.

Simpson's paradox plays a major role in causality modelling, when it is ignored if a variable has an influence on the others, or not. Since it has been shown that given a set of data involving several variables, any relation may be reversed when marginalizing (see [7] for a graphical proof), Simpson's paradox has a direct role in the adjustment problem, when somebody tries to know what variables have to be taken into account in order to obtain a model of the process. A proposed solution to this paradox does not lie in the original data, but involves extraneous information, provided by experts [9], called causal relations.

A. Laurent et al. (Eds.): IPMU 2014, Part III, CCIS 444, pp. 190–199, 2014.

So far Simpson's paradox phenomenon has been described with probability measures. However, in the last 40 years, other models of uncertainty have been defined, and in particular belief functions. Introduced by Shafer [13], they have wider and wider domains of application (see in particular [10, Appendix A]).

It would be interesting to know if such reversal of the decisions also occurs with belief functions when marginalizations are involved. Through simulations, a first inquiry into that question is presented in this paper.

This paper is organized as follows. A few details about the paradox are given in Section 2 and basic concepts on belief functions are exposed in Section 3. Then, in Section 4 is presented a belief function approach to handle data at the origin of the paradox with an academic example. Then a Monte Carlo experiment is provided in Section 5 to provide the (co) occurences of the paradox. Section 6 concludes this paper.

2   Simpson's Paradox

Introducing Simpson's paradox may be done purely arithmetically with 8 numbers such that:

$$a/b < A/B \tag{1}$$

$$c/d < C/D \tag{2}$$

$$(a+c)/(b+d) <> (A+C)/(B+D) \tag{3}$$

In the last equation (Eq. 3), it is not known what terms is the biggest. In other words, (Eq. 1) and (Eq. 2) are insufficient to deduce any order in (Eq. 3). For example, the next relations [12] hold:

$$\frac{1}{5} < \frac{2}{8} \text{ (Eq. 1)}, \ \frac{6}{8} < \frac{4}{5} \text{ (Eq. 2), and } \frac{7}{13} > \frac{6}{13} \text{ (Eq. 3)}. \tag{4}$$

When considering probabilities, these sets of inequalities can be seen as:

$$P(A \mid B, C) < P(A \mid \overline{B}, C) \tag{5}$$

$$P(A \mid B, \overline{C}) < P(A \mid \overline{B}, \overline{C}) \tag{6}$$

$$P(A \mid B) <> P(A \mid \overline{B}) \tag{7}$$

This is where Simpson's paradox appears [9]: a property may be true in every subgroups of a population while being false for the whole population (In the previous example two subgroups corresponding to C and $\overline{C}$ are considered).

Real life examples of such discrepancies are numerous. They are even encountered in evolutionary games involving populations of rats and lemmings [2,7].

Such an example [9,12] about a medical treatment is presented in Table 1. It can be observed that the treatment is effective when the variable Gender is taken into account. Indeed:

$$P(S \mid A, m) \simeq .93 > P(S \mid B, m) \simeq .87 , \qquad (8)$$
$$P(S \mid A, f) \simeq .73 > P(S \mid B, f) \simeq .69 . \qquad (9)$$

Table 1. Probabilities of success (S) and failure (F) of a treatment knowing it has been given (A) or not (B) when a male m is encountered versus a female f, a number of 700 cases being considered

	\multicolumn{2}{c}{m}		\multicolumn{2}{c}{f}	
	A	B	A	B
S	$81/87 \simeq 0.93$	$234/270 \simeq 0.87$	$192/263 \simeq 0.73$	$55/80 \simeq 0.69$
F	$6/87 \simeq 0.07$	$36/270 \simeq 0.13$	$71/263 \simeq 0.27$	$25/80 \simeq 0.31$

However, when the gender is marginalized, for the whole population, no treatment becomes a better option. Indeed, as it can be seen in Table 2, inequalities are reversed:

$$P(S \mid A) \simeq .78 < P(S \mid B) \simeq .83 . \qquad (10)$$

Table 2. Probability of success and failure obtained from Table 2 when the gender is marginalized

	A (with treatment)	B (no treatment)
S	$P_{11} = 273/350 \simeq 0.78$	$P_{12} = 289/350 \simeq 0.83$
F	$P_{21} = 77/350 \simeq 0.22$	$P_{22} = 61/350 \simeq 0.17$

Let us also note that changing numbers while keeping ratios constant, with $a/b = (\alpha a)/(\alpha b)$ and α free, may alter the paradox. Indeed ratios in (Eq. 1) and (Eq. 2) are constant, as is the right term of (Eq. 3), but the left term of the latter now depends on α. For example consider $\alpha = 10$, then equation 4 becomes $10/50 < 2/8$, $6/8 < 4/5$ and $16/58 < 6/13$, *i.e,* the opposite conclusion is reached.

A solution proposed by Pearl [9] consists in using extraneous data, called causal information, that are different from the raw data. Causal information is the knowledge of the influences of some variables on some other ones. They enable one to know if one must reason with the full contingency table, or the reduced one. So the conclusions drawn from each table cannot be compared, and so the paradox becomes impossible. However, causal information must be provided by experts and therefore may be unavailable or difficult to obtain.

3   Belief Functions

This section recalls some basic concepts on belief functions which are used in this paper.

3.1   Main Functions

Let $\Omega = \{\omega_1, \ldots, \omega_n\}$ be a finite set called the *frame of discernment*. The quantity $m(A) \in [0,1]$ with $A \subseteq \Omega$ is the part of belief supporting A that, due to a lack of information, cannot be given to any strict subset of A [14]. Mass function m (or basic belief assignment) has to satisfy:

$$\sum_{A \subseteq \Omega} m^{\Omega}(A) = 1 . \tag{11}$$

Throughout this article, 2^{Ω} represents all the subsets of Ω. Plausibility $Pl^{\Omega}(A)$ represents the total amount of belief that may be given to A with further pieces of evidence:

$$Pl^{\Omega}(A) = \sum_{A \cap B \neq \emptyset} m^{\Omega}(B) . \tag{12}$$

These functions are in one-to-one correspondence [13], so they are used indifferently with the same term **belief function**. A set A such that $m(A) > 0$ is called a *focal element* of m. The vacuous belief function is defined by $m(\Omega) = 1$. It represents the total lack of information.

3.2   Refinement, Vacuous Extension and Marginalization

Let R be a mapping from 2^{Θ} to 2^{Ω} such that every singleton $\{\theta\}$, with $\theta \in \Theta$ is mapped into one or several elements of Ω, and such that all images $R(\{\theta\}) \subseteq \Omega$ form a partition of Ω. Such a mapping R is called a *refining*, Ω is called a *refinement* of Θ and Θ is called a *coarsening* of Ω [13].

Any belief function m^{Θ} defined on Θ can be extended to Ω. This operation is called the vacuous extension of m^{Θ} on Ω, it is denoted by $m^{\Theta \uparrow \Omega}$ and is defined by:

$$m^{\Theta \uparrow \Omega}(A) = \begin{cases} m^{\Theta}(B) & \text{if } A = R(B) \\ 0 & \text{otherwise} \end{cases} \tag{13}$$

The coarsening operation is the opposite step. Starting from a belief function defined on Ω, a belief function on Θ is defined trough a mapping R defined as previously. The problem is that R is generally not an onto mapping: usually there will be some focal elements of m^{Ω} that are not the images of sets of Θ by R. In this case Shafer [13, chapter 6, page 117] has introduced two envelopes, called *inner* and *outer reductions*.

The inner reduction (or lower envelop) $\underline{\Theta}$ and the outer reduction (or upper envelop) $\overline{\Theta}$ are mappings respectively defined, from 2^{Ω} to 2^{Θ}, for all $A \subseteq \Omega$ by:

$$\underline{\Theta}(A) = \{\theta \in \Theta, \ R(\{\theta\}) \subseteq A\} \ , \tag{14}$$

and

$$\overline{\Theta}(A) = \{\theta \in \Theta, \ R(\{\theta\}) \cap A \neq \emptyset\} \ . \tag{15}$$

The *inner reduction* and *outer reduction* on Θ of a belief function m^{Ω} defined on Ω are then respectively given for all $B \subseteq \Theta$ by:

$$\underline{m}^{\Theta}(B) = \sum_{A \subseteq \Omega, \ \underline{\Theta}(A) = B} m^{\Omega}(A) \ , \tag{16}$$

and

$$\overline{m}^{\Theta}(B) = \sum_{A \subseteq \Omega, \ \overline{\Theta}(A) = B} m^{\Omega}(A) \ . \tag{17}$$

Conceptually, *marginalization* is a special case of coarsening where $\Omega = X \times Y$ and either $\Theta = X$ or $\Theta = Y$ [13].

3.3   Pignistic Transformation

Decisions from belief functions can be made using the pignistic transformation $BetP$ justified in [14,15] based on rationality requirements and axioms. It is defined, for all ω of Ω, by:

$$BetP^{\Omega}(\{\omega\}) = \sum_{\omega \in A} \frac{m^{\Omega}(A)}{|A|\,(1 - m(\emptyset))} \ . \tag{18}$$

The converse operation of the pignistic rule has also been defined. It actually yields the least specific mass function whose pignistic transformation is the probability measure P^{Ω} [6]. Let $m^{\Omega}_{LSP(P^{\Omega})}$ denote this mass function. It has n nested focal elements :

$$m^{\Omega}_{LSP(P^{\Omega})}(\{\omega_i, \ldots, \omega_n\}) = (n - i + 1)(p_i - p_{i-1}), \quad \forall i \in \{1, \ldots, n\}, \tag{19}$$

with $0 < P^{\Omega}(\{\omega_1\}) = p_1 < \ldots < p_n$ and $p_0 = 0$. If some p_i are equal, the result is the same whatever the order.

4   Inferring Decisions from Contingency Tables with Belief Functions

4.1   Proposed Approach Outline

The whole approach used in this article is summarized in Figure 1.

From initial data on a 3 binary variables contingency table $\Omega = X \times Y \times Z$, a probability measure $\hat{P}^{\Omega}$ is estimated. From the same data a belief function

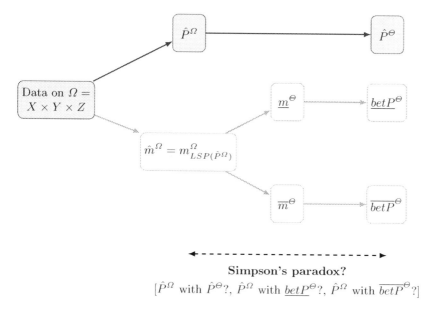

Fig. 1. Comparison of the Simpson's paradox when using probabilities and the proposed approach based on belief functions

$\hat{m}^{\Omega} = m^{\Omega}_{LSP(\hat{P}^{\Omega})}$ is estimated as the least specific belief function whose pignistic transformation gives $\hat{P}^{\Omega}$. It is obtained using Equation 19.

Then the probability measure $\hat{P}^{\Omega}$ and the belief function $\hat{m}^{\Omega}$ are marginalized on $\Theta = X \times Y$.

With belief functions, two reductions for this marginalization are considered: the inner reduction $\underline{m}^{\Theta}$ of $\hat{m}^{\Omega}$ (Eq. 16) and the outer reduction $\overline{m}^{\Theta}$ of $\hat{m}^{\Omega}$ (Eq. 17).

Finally, decisions based on the full contingency table are compared to decisions taken on the reduced space Θ to detect Simpson's paradoxes.

4.2 An Academic Example

In this section, an example is given to illustrate the approach with the inner reduction. Values in Tables 1 and 2 are used again.

Let us consider the spaces $\Omega = \{S, F\} \times \{A, B\} \times \{m, f\}$ and $\Theta = \{S, F\} \times \{A, B\}$ when the variable gender is marginalized.

In Table 3 are sorted out the numbers of cases of Table 1 to compute the least specific belief isopignistic to $\hat{P}^{\Omega}$.

Table 3. Computing $\hat{P}^{\Omega}$ from Table 1

Singleton	$\hat{P}^{\Omega}$
$FAm = (F, A, m)$	$\frac{6}{700} \simeq .01 = p_1$
FBf	$\frac{25}{700} \simeq .04 = p_2$
FBm	$\frac{36}{700} \simeq .05 = p_3$
SBf	$\frac{55}{700} \simeq .08 = p_4$
FAf	$\frac{71}{700} \simeq .10 = p_5$
SAm	$\frac{81}{700} \simeq .12 = p_6$
SAf	$\frac{192}{700} \simeq .27 = p_7$
SBm	$\frac{234}{700} \simeq .33 = p_8$

Using Equation 19, the least specific belief $\hat{m}^{\Omega}$ obtained from Table 3 can be computed:

$$
\begin{aligned}
m^{\Omega}_{LSP(\hat{P})}(\Omega) &= 8p_1 &= x_1 \\
m^{\Omega}_{LSP(\hat{P})}(\Omega \setminus \{FAm\}) &= 7(p_2 - p_1) &= x_2 \\
m^{\Omega}_{LSP(\hat{P})}(\Omega \setminus \{FBf, FAm\}) &= 6(p_3 - p_2) &= x_3 \\
m^{\Omega}_{LSP(\hat{P})}(\Omega \setminus \{FBm, FBf, FAm\}) &= 5(p_4 - p_3) &= x_4 \\
m^{\Omega}_{LSP(\hat{P})}(\{SBm, SAf, SAm, FAf\}) &= 4(p_5 - p_4) &= x_5 \\
m^{\Omega}_{LSP(\hat{P})}(\{SBm, SAf, SAm\}) &= 3(p_6 - p_5) &= x_6 \\
m^{\Omega}_{LSP(\hat{P})}(\{SBm, SAf\}) &= 2(p_7 - p_6) &= x_7 \\
m^{\Omega}_{LSP(\hat{P})}(\{SBm\}) &= p_8 - p_7 &= x_8
\end{aligned}
\tag{20}
$$

Mapping $\underline{\Theta}$ (Eq. 14) of $\Theta = \{S, F\} \times \{A, B\}$ is defined by:

$$
\begin{aligned}
\underline{\Theta}(\Omega) &= \{SA, SB, FA, FB\} \\
\underline{\Theta}(\Omega \setminus FAm) &= \{SA, SB, FB\} \\
\underline{\Theta}(\Omega \setminus \{FBf, FAm\}) &= \{SA, SB\} \\
\underline{\Theta}(\Omega \setminus \{FBm, FBf, FAm\}) &= \{SA, SB\} \\
\underline{\Theta}(\{SBm, SAf, SAm, FAf\}) &= \{SA\} \\
\underline{\Theta}(\{SBm, SAf, SAm\}) &= \{SA\} \\
\underline{\Theta}(\{SBm, SAf\}) &= \emptyset \\
\underline{\Theta}(\{SBm\}) &= \emptyset
\end{aligned}
\tag{21}
$$

So the inner reduction $\underline{m}^\Theta$ of $\hat{m}^\Omega$ on the space Θ is given by (Eq. 16):

$$\underline{m}^\Theta : \begin{cases} \{SA, SB, FA, FB\} & \mapsto x_1 \\ \{SA, SB, FB\} & \mapsto x_2 \\ \{SA, SB\} & \mapsto x_3 + x_4 \\ \{SA\} & \mapsto x_5 + x_6 \\ \emptyset & \mapsto x_7 + x_8 \end{cases} \tag{22}$$

and pignistic values are the followings (Eq. 18):

$$\begin{array}{ll} \underline{BetP}^\Theta(\{SA\}) = k(\frac{x_1}{4} + \frac{x_2}{3} + \frac{x_3+x_4}{2} + x_5 + x_6) \simeq 0.51 \\ \underline{BetP}^\Theta(\{SB\}) = k(\frac{x_1}{4} + \frac{x_2}{3} + \frac{x_3+x_4}{2}) \simeq 0.31 \\ \underline{BetP}^\Theta(\{FA\}) = k(\frac{x_1}{4}) \simeq 0.03 \\ \underline{BetP}^\Theta(\{FB\}) = k(\frac{x_1}{4} + \frac{x_2}{3}) \simeq 0.15 , \end{array} \tag{23}$$

with $k = \frac{1}{1 - x_7 - x_8}$.

So $\underline{BetP}(S \mid A) \simeq \frac{.51}{.51 + .03} \simeq 0.95$ and $\underline{BetP}(S \mid B) \simeq 0.68$, and unlike the probability case (Eq. 10) $\underline{BetP}(S \mid A) > \underline{BetP}(S \mid B)$, which leads to the same decision as the one obtained on the whole space $\{S, F\} \times \{A, B\} \times \{m, f\}$ (Eq. 8 and 9).

In this example, there is no longer a Simpson's paradox when considering belief functions and the inner reduction. However, as shown in next section, the paradox may also occur with belief functions, depending on the reduction, and without being triggered necessarily in the same time as with a Bayesian approach.

5   Numerical Simulations

In this section, the results of the following experiment are given:

- 10^8 contingency tables of 3 binary variables composed of numbers defined randomly in the interval $[1, 10^3]$ are built. Frequencies about the paradox are then observed for the three approaches: probability case, belief functions with inner and outer reduction cases.
- the preceding point is repeated 10 times in order to obtain mean values and variances of the obtained frequencies.

Table 4 provides the mean and standard deviation values obtained for the eight frequencies of the paradox triggering in the experiment. Each frequency corresponds to a case of appearance considering the probability approach ($Proba$) and the approaches with belief functions marginalized with the inner and outer reductions ($Inner$, $Outer$). A "1" means a paradox.

A first remark when looking at these results is that Simpson's paradox can be triggered for the three approaches, and all combinations of occurrences are possible.

Table 4. Appearance frequencies of Simpson's paradox over 10^8 contingency tables composed of numbers randomly chosen between 1 and 1000. "1" means a paradox. The first number is the mean value, the second one the standard deviation.

	$Proba = 0$		$Proba = 1$	
	$Outer = 0$	$Outer = 1$	$Outer = 0$	$Outer = 1$
$Inner = 0$	94%	3.77%	0.04%	0.9%
	0.003%	0.0012%	$1.5.10^{-4}\%$	$7.10^{-4}\%$
$Inner = 1$	1.2%	0.004%	0.013%	0.0046%
	0.0014%	$9.10^{-5}\%$	$1.4.10^{-4}\%$	$9.10^{-5}\%$

In most cases (94%), the paradox is absent for all the approaches.

Simpson's paradox occurs in this experiment for the probability approach with a 0.9576% frequency[1] $(0.04 + 0.9 + 0.013 + 0.0046)$ which is lower than for the other cases (inner and outer reductions).

At last it can be observed that the chance (0.9%) of observing a paradox for both the probabilistic and outer reduction approaches, is much more important than that of observing a paradox for the inner reduction approach with any or both of the other approaches. Taking into account the marginal chances of observing a paradox for each approach, we may further remark that the probabilistic approach is actually somewhat included in the outer reduction approach (when a paradox is observed for the probabilistic approach, a paradox will in general be observed for the outer reduction), whereas the inner reduction approach is almost disjoint from the other ones.

6   Conclusion and Discussion

In this article, similarly to the probability case, frequencies of appearance of Simpson's paradox have been studied with belief functions. The marginalization step is shown to impact it, since the inner and outer reductions have different behaviours. It has been shown than a paradox can occur in each approach and each combination of approaches.

To complete this study, instead of the least specific isopignistic belief function, it might be interesting to investigate on other estimators since they may yield different results than those obtained with consonant beliefs. Indeed, for instance, the Bayesian belief function identified to the probability measure, leads to different conclusions than those based on consonant beliefs.

Another parameter that influences the paradox is the decision rule. In this paper the pignistic rule is used. But other decision rules exist, and among them

---

[1] The same frequency was valued at 1.67% in another experiment [8]. In this article, the experiment is based on 7 independent proportions, the eight one summing up to one. This discrepancy between the two results may be due to the fact that proportions may shift the triggering of the paradox, as recalled in the penultimate paragraph of Section 2.

the maximum of plausibility emerges. However, the analysis of the full decision chain based on this rule is more complex (for instance, differences between decisions obtained under this rule and under the pignistic rule already appear on the set Ω) and is left for further work.

Lastly, it would be interesting to investigate whether causality as addressed in belief function theory [1], could bring a solution to the paradox, as is in the probabilistic case.

References

1. Boukhris, I., Elouedi, Z., Benferhat, S.: Dealing with external actions in belief causal networks. International Journal of Approximate Reasoning 54(8), 978–999 (2013)
2. Chater, N., Vlaev, I., Grinberg, M.: A new consequence of Simpson's Paradox: Stable co-operation in one-shot Prisoner's Dilemma from populations of individualistic learning agents, University College London/New Bulgarian University, http://else.econ.ucl.ac.uk/papers/uploaded/210.pdf
3. Cobb, B.R., Shenoy, P.P.: On the plausibility transformation method for translating belief function models to probability models. International Journal of Approximate Reasoning 41(3), 314–330 (2006)
4. Dubois, D., Prade, H.: A set-theoretic view of belief functions: logical operations and approximations by fuzzy sets. International Journal of General Systems 12(3), 193–226 (1986)
5. Dubois, D., Moral, S., Prade, H.: Belief Change Rules in Ordinal and Numerical Uncertainty Theories. In: Handbook of Defeasible Reasoning and Uncertainty Management Systems, vol. 3, pp. 311–392 (1998)
6. Dubois, D., Prade, H., Smets, P.: A Definition of Subjective Possibility. International Journal of Approximate Reasoning 48, 352–364 (2008)
7. Malinas, G.: Simpson's Paradox, Stanford encyclopedia of philosophy, http://plato.stanford.edu/entries/paradox-simpson/ (first published February 2, 2004)
8. Pavlides, M.G., Perlman, M.D.: How likely is Simpson's paradox? The American Statistician 63(3), 226–233 (2009)
9. Pearl, J.: Causality: Models, Reasoning, and Inference. Cambridge University, New York (2000)
10. Sentz, K., Ferson, S.: Combination of evidence in Dempster–Shafer theory, Technical report, SANDIA National Laboratories, 96 pages (2002)
11. Simpson, E.H.: The interpretation of interaction in contingency tables. Journal of the Royal Statistical Society, Ser. B 13, 238–241 (1951)
12. Simpson's paradox (2008), http://en.wikipedia.org/wiki/Simpson's_paradox
13. Shafer, G.: A mathematical theory of evidence. Princeton University Press, Princeton (1976)
14. Smets, P., Kennes, R.: The transferable belief model. Artificial Intelligence 66, 191–234 (1994)
15. Smets, P.: Decision Making in the TBM: the Necessity of the Pignistic Transformation. International Journal of Approximate Reasoning 38(2), 133–147 (2005)
16. Approaches to Causal Structure Learning, Workshop, UAI 2012-2013, http://www.stat.washington.edu/tsr/uai-causal-structure-learning-workshop/, and all papers available online

Comparison of Credal Assignment Algorithms in Kinematic Data Tracking Context

Samir Hachour, François Delmotte, and David Mercier

Univ. Lille Nord de France, UArtois, EA 3926 LGI2A, Béthune, France
samir_hachour@ens.univ-artois.fr,
{francois.delmotte,david.mercier}@univ-artois.fr

Abstract. This paper compares several assignment algorithms in a multi-target tracking context, namely: the optimal Global Nearest Neighbor algorithm (GNN) and a few based on belief functions. The robustness of the algorithms are tested in different situations, such as: nearby targets tracking, targets appearances management. It is shown that the algorithms performances are sensitive to some design parameters. It is shown that, for kinematic data based assignment problem, the credal assignment algorithms do not outperform the standard GNN algorithm.

Keywords: multi-target tracking, optimal assignment, credal assignment, appearance management.

1 Introduction

Multiple target tracking task consists of the estimation of some random targets state vectors, which are generally composed of kinematic data (e.g. position, velocity). Based on some measured data (e.g. position in x and y directions), the state estimation can be ensured by: Kalman filters, particles filters, Interacting Multiple Model algorithm which are used in this article and so on. Targets state estimation quality depends on how accurately the measured data are assigned to the tracked targets. In fact the assignment task is quite hard as far as the measured data are imperfect.

This paper focuses on distance optimization based assignment, where, the well known optimal Global Nearest Neighbor algorithm (GNN) [1] is compared with some, recently developed, equivalent credal solutions, namely: the works of Denoeux et al. [2], Mercier et al. [3], Fayad and Hamadeh [4] and Lauffenberger et al. [5]. Discussions with some other approaches are included [6].

This paper highlights drawbacks of turning distances into mass functions, in the credal algorithms. Simulation examples show the difficulties to correctly define the parameters of all the methods, including the appearances and disappearances management. In particular, it is shown that the performance criteria is linked to two distinct errors, namely: miss-assignment of two nearby targets and false decision about new targets. A method to define the most accurate mass function, allowing the credal algorithms to get the same performance as the GNN, is presented. It appears that in the described mono-sensor experiments

A. Laurent et al. (Eds.): IPMU 2014, Part III, CCIS 444, pp. 200–211, 2014.

based on kinematic data, credal assignment algorithms do not outperform the standard GNN algorithm.

In this paper, the problem of conflicting assignment situation and the proposed solutions are described in Section 2. A relation between the algorithms parameters is presented in Section 3. Some tests and results in tracking assignment context are depicted in Section 4. Section 5 concludes the paper.

2 Assignment Problem and Related Solutions

In multi target tracking contexts, updating the state estimations is much more complex than in a mono target framework. Indeed the first task is to assign the observations to the known objets.

Let us illustrate the problem in Fig. 1, where at a given time k, three targets $T = \{T_1, T_2, T_3\}$ are known and four observations $O = \{O_1, O_2, O_3, O_4\}$ are received. The question is: how to assign the observations to the known targets and taking into account the appearances and disappearances?

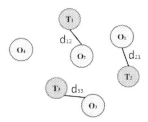

Fig. 1. Distance based assignment

Global Nearest Neighbor (GNN) solution: GNN algorithm is one of the firstly proposed solutions to the assignment problem in multi-target tracking context. It provides an optimal solution, in the sense where global distance between known targets and observations is minimized. Let $r_{i,j} \in \{0, 1\}$ be the relation that object T_i is associated or not associated with observation O_j ($r_{i,j} = 1$ means that observation O_j is assigned to object T_i, $r_{i,j} = 0$ otherwise). The objective function of such problem is formalized as follows:

$$min \sum_{i=1}^{n} \sum_{j=1}^{m} d_{i,j} r_{i,j} , \tag{1}$$

where,

$$\sum_{i=1}^{n} r_{i,j} = 1 , \tag{2}$$

$$\sum_{j=1}^{m} r_{i,j} \leq 1 , \tag{3}$$

where $d_{i,j}$ represents the Mahalanobis distance between the known target T_i and the observation O_j.

The generalized distances matrix $[d_{i,j}]$ for the example given in Fig. 1, can have the following form:

	O_1	O_2	O_3	O_4
T_1	$d_{1,1}$	$d_{1,2}$	$d_{1,3}$	$d_{1,4}$
T_2	$d_{2,1}$	$d_{2,2}$	$d_{2,3}$	$d_{2,4}$
T_3	$d_{3,1}$	$d_{3,2}$	$d_{3,3}$	$d_{3,4}$
NT_1	λ	∞	∞	∞
NT_2	∞	λ	∞	∞
NT_3	∞	∞	λ	∞
NT_4	∞	∞	∞	λ

When they are not assigned to existing targets, observations initiate new targets noted NT. If the probability p that an observation is generated by an existing target is known, the threshold λ can be derived from the χ^2 table as far as the Mahalanobis[1] distance follows a χ^2 distribution [7]:

$$P(d_{i,j} < \lambda) = p, \tag{4}$$

otherwise, it must be trained to lower the false decisions rate.

2.1   Denoeux et al.'s Solution [8]

In Denœux et al.'s approach as in most credal approaches, available evidence on the relation between objects T_i and O_j is assumed to be given for each couple (T_i, O_j) by a mass function $m_{i,j}$ expressed on the frame $\{0,1\}$ and calculated in the following manner:

$$\begin{cases} m_{i,j}(\{1\}) = & \alpha_{i,j}, & \text{supporting } r_{i,j} = 1. \\ m_{i,j}(\{0\}) = & \beta_{i,j}, & \text{supporting } r_{i,j} = 0. \\ m_{i,j}(\{0,1\}) = 1 - \alpha_{i,j} - \beta_{i,j}, & \text{ignorance on the assignment of } O_j \text{ to } T_i. \end{cases} \tag{5}$$

With R the set of all possible relations between objects T_i and O_j, $R_{i,j}$ denotes the set of relations matching object T_i with observation O_j:

$$R_{i,j} = \{r \in R | r_{i,j} = 1\}. \tag{6}$$

Each mass function $m_{i,j}$ is then extended to R by transferring masses $m_{i,j}(\{1\})$ to $R_{i,j}$, $m_{i,j}(\{0\})$ to $\overline{R_{i,j}}$ and $m_{i,j}(\{0,1\})$ to R. For all $r \in R$, associated plausibility function $Pl_{i,j}$ verifies:

$$Pl_{i,j}(\{r\}) = (1 - \beta_{i,j})^{r_{i,j}} (1 - \alpha_{i,j})^{1-r_{i,j}}. \tag{7}$$

---

[1] For fair tests, all the algorithms are based on Mahalanobis distances.

After combining all the $m_{i,j}$ by Dempster's rule, the obtained global plausibility function Pl is shown to be proportional to the $Pl_{i,j}$ and given for all $r \in R$ by:

$$Pl(\{r\}) \propto \prod_{i,j} (1 - \beta_{i,j})^{r_{i,j}} (1 - \alpha_{i,j})^{1 - r_{i,j}}. \tag{8}$$

Finally, the calculation of the logarithm function of (8), $\beta_{i,j}$ and $\alpha_{i,j}$ being all considered strictly lower than 1, allows the authors to express the search of the most plausible relation as a linear programming problem defined as follows with n objects T_i, m observations O_j and $w_{i,j} = \ln(1 - \beta_{i,j}) - \ln(1 - \alpha_{i,j})$:

$$max \sum_{i,j} w_{i,j} r_{i,j}, \quad i = \{1, ..., n\}, j = \{1, ..., m\}, \tag{9}$$

with

$$\sum_{i}^{n} r_{i,j} \leq 1 , \tag{10}$$

$$\sum_{j}^{m} r_{i,j} \leq 1 , \tag{11}$$

$$r_{i,j} \in \{0, 1\}, \quad \forall i \in \{1, ..., n\}, \forall j \in \{1, ..., m\} . \tag{12}$$

This problem can be solved using Hungarian or Munkres algorithms [9]. More specifically, the authors also propose to solve an equivalent algorithm by considering, instead of (9), the following objective function:

$$max \sum_{i,j} w'_{i,j} r'_{i,j}, \quad i = \{1, ..., n\}, j = \{1, ..., m\}, \tag{13}$$

with $w'_{i,j} = max(0, w'_{i,j})$.

To experiment this algorithm with kinematic data based assignments, weights $\alpha_{i,j}$ and $\beta_{i,j}$ are computed in [8] as follows:

$$\begin{cases} m_{i,j}(\{1\}) = & \alpha_{i,j} & = & \sigma \exp(-\gamma d_{i,j}) \\ m_{i,j}(\{0\}) = & \beta_{i,j} & = \sigma(1 - \exp(-\gamma d_{i,j})) \\ m_{i,j}(\{0, 1\}) = 1 - \alpha_{i,j} - \beta_{i,j} = & & 1 - \sigma \end{cases} \tag{14}$$

where $d_{i,j}$ is the distance between object T_i and observation O_j and γ is a weighting parameter. Parameter σ is used to discount the information according to the sensor reliability [10].

In this article, all sensors have an equal perfect reliability, and so $\sigma = 0.9$. Moreover this parameter could be used in the same manner for all credal algorithms. On the contrary, parameter γ will be optimized. Although appealing, set of equations 14 remains empirical.

Mercier et al.'s solution [3]: mass functions in the works of Mercier et al. are calculated as in Equation (14), they are extended (vacuous extension [10]) to the frame of discernment $T^* = \{T_1, T_2, T_3, *\}$ or $O^* = \{O_1, O_2, O_3, *\}$ depending on if we want to associate the observations to the targets or the targets to the observations. Element $(*)$ models the non-detection or new target appearance. Once the mass functions are all expressed on a common frame of discernment, they are conjunctively combined [10] and then a mass function is obtained for each element O_j or T_i, according to the assignment point of view. Finally, The mass functions are transformed to pignistic probabilities [10]. The assignment decision is made by taking the maximum pignistic probabilities among the possible relations.

It is shown that this method is asymmetric when it comes to manage targets appearances and disappearances: assigning observations to targets is different than targets to observations. In this paper, only observations points of view are considered.

Lauffenberger et al.'s solution [5]: at the credal level, this method is almost similar to Mercier et al's method. To avoid the asymmetry problem, the authors, propose a different decision making strategy. For a given realization of distances, they perform the previous algorithm in both sides and obtain two pignistic probability matrices, which are not normalized since the weight on the empty set resulting from the conjunctive combination is isolated and used for a decision making purpose. A dual pignistic matrix is calculated by performing an element-wise product of calculated two pignistic probabilities matrices. The maximum pignistic probability is retained for each target (each row of the dual matrix), if this pignistic probability is greater than a given threshold, the target is associated with the column corresponding element, else it is considered as non-detected. The same procedure is performed for the observations (column elements of the dual matrix). The probabilities are also compared to the conflict weight generated by the conjunctive combination. If the conflict weight is greater that the dual matrix (rows/columns) probabilities, no assignment decision is made.

Fayad and Hamadeh's solution [4]: mass functions calculations in this method are different of the one adopted by the previous methods. For each observation O_j, in Fig. 1, for example, a mass function over the set of known targets $T^* = \{T_1, T_2, T_3, *\}$ is calculated. The element $(*)$ is a virtual target for which assigned observations are considered as new targets. Distances between known targets and each observation are sorted (minimum distance to maximum distance) and the mass function weights are calculated by inverting the distances and then normalizing the weighs in order to get a representative mass functions. Once mass functions of all observations are calculated, they are combined and expressed on the set of all possible hypotheses: $[\{(O_1, *), (O_2, *), (O_3, *), (O_4, *)\}, \{(O_1, T_1), (O_2, *), (O_3, *), (O_4, *)\}, ...]$. The combination is done by means of a cautious rule based on the "*min*" operator. This method becomes quickly intractable when the number of observations and/or targets gets over 3.

3   Relation between Parameters γ and λ

These two parameters can be optimized through a training. Moreover, if one is known the other can be deduced from it. To illustrate the necessity of choosing adequate parameters, let us consider that $d_{3,3}$, in Fig. 1, is equal to 1. The weights concerning the assignment of O_3 to T_3 for two different values of $\gamma = \{0.6, 0.8\}$ are: $\alpha_{i,j} = \exp(-0.6) = 0.55$ and $\beta_{i,j} = 1 - \exp(-0.6) = 0.45$. The parameter $\alpha_{i,j}$, in this case, is greater that $\beta_{i,j}$ so O_3 is associated to T_3. In another hand, if $\gamma = 0.8$, $\alpha_{i,j} = \exp(-0.8) = 0.45$ and $\beta_{i,j} = 1 - \exp(-0.8) = 0.55$, this means that T_3 is non-detected and O_3 is considered as a new target.

Fig. 2 represents the evolution of functions $\alpha_{i,j}$ and $\beta_{i,j}$.

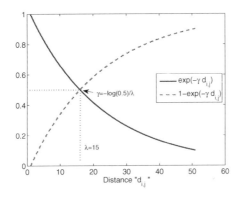

Fig. 2. Parameter "γ" determination

The fail-over distance (confirming/refuting the assignment) depends on the value given to γ. It can be chosen in such a way to get exactly the same fail-over distance as for the algorithm GNN, namely the parameter λ in Equation (4): it can be seen in Fig. 2 that the fail-over weight is given by the junction of the two curves ($\alpha_{i,j} = \beta_{i,j}$), if we want to impose λ as a fail-over distance, we just have to put $\exp(-\gamma\lambda) = 1 - \exp(-\gamma\lambda)$ and then deduce the value of γ using the following relation:

$$\gamma = \frac{-\log(0.5)}{\lambda}. \tag{15}$$

4   Assignment Tests in Tracking Context

Targets are evolving according to linear constant velocity models originally defined for aircrafts [1]:

$$x_i(k) = Ax_i(k) + Bu(k) + w(k) , \tag{16}$$

where:

$$A = \begin{bmatrix} 1 & \Delta T & 0 & 0 \\ 0 & 1 & 0 & 0 \\ 0 & 0 & 1 & \Delta T \\ 0 & 0 & 0 & 1 \end{bmatrix}, \quad B = \begin{bmatrix} (\Delta T)^2/2 & 0 \\ \Delta T & 0 \\ 0 & (\Delta T)^2/2 \\ 0 & \Delta T \end{bmatrix}, \quad (17)$$

where ΔT represents the sampling time and w represents a Gaussian state noise. Input matrix is represented by B, where B' is matrix B transpose. Vector $u = [a_x \;\; a_y]^T$ in Equation (16) represents a given acceleration mode in x, y or both x and y directions.

Sensor measurements are modeled by:

$$O_i(k) = Hx_i(k) + v(k), \quad (18)$$

where,

$$H = \begin{bmatrix} 1 & 0 & 0 & 0 \\ 0 & 0 & 1 & 0 \end{bmatrix}, \quad (19)$$

and v a Gaussian measurement noise.

Let us start by giving a numerical example showing the effect of an arbitrary choice of the parameter γ, for example, in the concerned credal assignment methods.

Two time steps illustrating example: let $D(k) = [d_{i,j}]$ be a distances matrix at time step k, it is calculated based on 3 known targets $T_1(k)$, $T_2(k)$, $T_3(k)$ and 3 observations $O_1(k)$, $O_2(k)$, $O_3(k)$:

$$D(k) = \begin{bmatrix} 6.9 & 8.1 & 7.1 \\ 9.9 & 6.9 & 9.1 \\ 10.3 & 11.2 & 6.4 \end{bmatrix}. \quad (20)$$

The resolution of this matrix using GNN and Fayad's algorithms gives the following solution:

$$S_{G,F}(k) = \begin{bmatrix} 1 & 0 & 0 \\ 0 & 1 & 0 \\ 0 & 0 & 1 \end{bmatrix}. \quad (21)$$

The matrices $\alpha = [\alpha_{i,j}]$ and $\beta = [\beta_{i,j}]$, corresponding to the transformation of the distances into mass functions are given by:

$$\alpha(k) = \begin{bmatrix} 0.4514 & 0.4004 & 0.4425 \\ 0.3344 & 0.4514 & 0.3623 \\ 0.3213 & 0.2937 & 0.4746 \end{bmatrix}, \beta(k) = \begin{bmatrix} 0.4486 & 0.4996 & 0.4575 \\ 0.5656 & 0.4486 & 0.5377 \\ 0.5787 & 0.6063 & 0.4254 \end{bmatrix}, \quad (22)$$

For these matrices, the credal algorithms (except Lauffenberger et al.'s algorithm) gives the same solution as the GNN (see Equation (21)). Lauffenberger et al.'s algorithm gives the following solution:

$$S_L(k) = \begin{bmatrix} 0 & 0 & 0 \\ 0 & 0 & 0 \\ 0 & 0 & 0 \end{bmatrix}. \quad (23)$$

This means that all the know targets are disappeared and all the observations are considered as new targets. This illustrates the limits of the algorithm in conflicting situation (nearby targets, given that the cross-distances are almost equal). This is due to the fact that the assignment decision is made based on the conflict generated by the mass functions combination, and when targets are close to each other, the conflict is high and then considerably influence the assignment decision.

At time step $k + 1$, the measurements $O_1(k + 1)$, $O_2(k + 1)$, $O_3(k + 1)$ of the same known targets are affected by the sensor noise, which leads to a different distance matrix $D(k + 1)$:

$$D(k + 1) = D(k) + noise = \begin{bmatrix} 7.8 & 9.4 & 8.5 \\ 10 & 7.8 & 11 \\ 10.2 & 12 & 7.9 \end{bmatrix}, \qquad (24)$$

The obtained solution in the GNN and Fayad's algorithms is the same as in Equation (21). In order to get the other credal algorithms solutions, these distances are transformed into mass functions in the following matrices:

$$\alpha(k+1) = \begin{bmatrix} 0.4126 & 0.3516 & 0.3847 \\ 0.3311 & 0.4126 & 0.2996 \\ 0.3245 & 0.2711 & 0.4085 \end{bmatrix}, \beta(k+1) = \begin{bmatrix} 0.4874 & 0.5484 & 0.5153 \\ 0.5689 & 0.4874 & 0.6004 \\ 0.5755 & 0.6289 & 0.4915 \end{bmatrix}, \quad (25)$$

The algorithms depending on γ give the following solution:

$$S_{D,M,L}(k + 1) = \begin{bmatrix} 0 & 0 & 0 \\ 0 & 0 & 0 \\ 0 & 0 & 0 \end{bmatrix}. \qquad (26)$$

This solution means that all the known targets are not detected and all the acquired measurements are considered as new targets which is a false decision.

Let us now, consider the conflicting scenario of two nearby target, in (a) of Fig. 3 and compare the performances of the assignment algorithms which are given in (b) of the same figure.

The results in (b) Fig. 3 confirms the dependency of the algorithms on their respective parameters. In this simulation the parameters λ and γ are linked by the relation in Equation (15), therefore, λ depending algorithms and γ depending ones have almost the same performances. For a given value of their parameters, they supply the same performances, with the minimum error rate (10%) depending only on the scenario (noises and so on).

The results in Fig. 4 represent a robustness test of a λ depending algorithm, namely, GNN algorithm and a γ depending algorithm, namely, Denoeux's algorithm. The results shows that the algorithms are almost equivalent and similarly dependent on their respective parameters and sensor noise. Another robustness test is added in Fig. 5.

It can be seen in Fig. 5 that algorithms performances are sensitive and proportional to the modeling error which is simulated by state noise variation. In this simulation algorithms use the optimal values of their respective parameters.

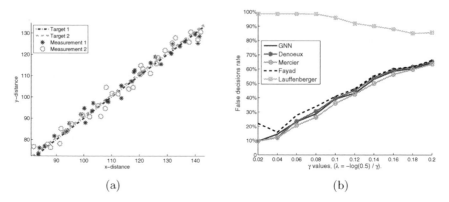

(a)   (b)

Fig. 3. (a): Conflicting scenario, (b): False assignments rates with the variation of the algorithms parameters

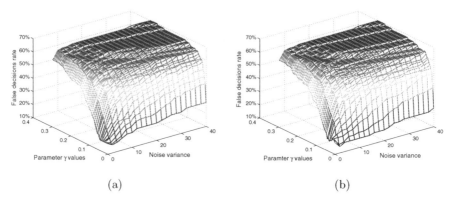

(a)   (b)

Fig. 4. (a): GNN algorithm robustness test, (b): Denoeux's algorithm robustness test

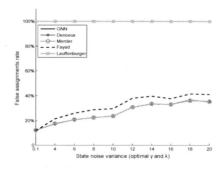

Fig. 5. Robustness against modeling error

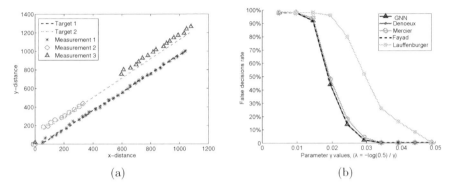

Fig. 6. (a): Target appearance scenario, (b): False decisions rate depending on the parameter λ for GNN and Fayad's algorithms and γ for Denoeux, Mercier and Lauffenberger's algorithms.

The following simulation (Fig. 6) confirms the almost equal performances on the second kind of errors about new targets appearances. This second simulation aims to calculate the false decisions rates, which means how often the newly detected target "Target 3", in (a) Fig. 6, is erroneously assigned to a previously non-detected one "Target 2".

It can be seen in (b) Fig. 6, that the false decisions rate depends on the parameter λ for the GNN and Fayad's algorithms, and depends on parameter γ for the other algorithms. The result shows that the algorithms reach equal performances for given values of λ and γ. When the probability p is known, λ can be determined according to Equation (2) and γ can be deduce using Equation (15).

A last simulation including the two precedent tests is added in the following. It tries to train the optimal parameters λ and γ on the scenario of Fig. 7, without any a priori knowledge.

Results of this simulation are obtained separately for λ depending algorithms (GNN and Fayad's algorithms) and γ depending algorithms (Denœux, Mercier and Lauffenberger's algorithms). They are depicted in Fig. 8.

As expected this last results show the necessity to adequately choose the algorithms parameters for a tracking assignment purpose. They also confirms that the trained optimal parameters $\lambda = 46$ and $\gamma = 0.015$ are linked by the relation of Equation (15) presented in Section 3, since $0.015 \simeq -log(0.5)/46$.

A final test is added to give an idea on the computational complexity of the studied algorithms. Computational times for an increasing number of observations are depicted in Fig. 9.

Test in Fig. 9 shows that Fayad's algorithm is the most computationally demanding. It is followed by Mercier's algorithm. GNN and Deneoux's algorithms seem to be the less complex algorithms.

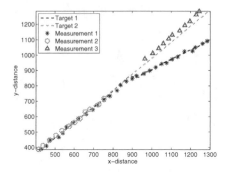

Fig. 7. Test on both false assignment and targets appearances management

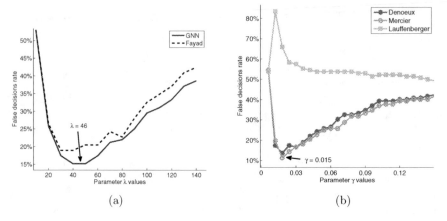

(a)   (b)

Fig. 8. (a): Performances of the algorithms depending on λ, (b): Performances of the algorithms depending on γ

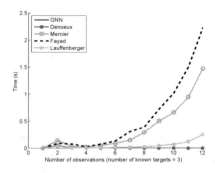

Fig. 9. Calculation time for an increasing number of data

5   Conclusion

This paper proposes a comparison study of various assignment algorithms in a context of multi-target tracking based on kinematic data. These algorithms depends on parameters that must be trained, otherwise, the performances are decreased. Contrarily to previous articles, it is shown here that the standard GNN algorithm with optimized parameters provides the same best performances than other algorithms. It is also less time-consuming. It is shown that there exists a relation between the optimized design parameters λ and γ.

It can be noticed that Lauffenberger's algorithm makes wrong decisions in conflicting scenarios. This is a priori due to the use of a decision making process based on conflict, where generated conflict in such situation is high and then refutes all assignments.

In the future, we will tackle the possible benefits of using belief functions in multi-sensors cases.

Acknowledgments. The authors are very grateful to Prof. T. Denœux for having shared the MatlabTM code of his assignment algorithm.

References

1. Blackman, S.S., Popoli, R.: Design and analysis of modern tracking systems. Artech House, Norwood (1999)
2. El Zoghby, N., Cherfaoui, V., Denoeux, T.: Optimal object association from pairwise evidential mass functions. In: Proceedings of the 16th International Conference on Information Fusion (2013)
3. Mercier, D., Lefèvre, É., Jolly, D.: Object association with belief functions, an application with vehicles. Information Sciences 181(24), 5485–5500 (2011)
4. Fayad, F., Hamadeh, K.: Object-to-track association in a multisensor fusion system under the tbm framework. In: In 11th International Conference on Information Sciences, Signal Processing and their Applications (ISSPA 2012), Montreal, Quebec, Canada, pp. 1001–1006 (2012)
5. Lauffenberger, J.-P., Daniel, J., Saif, O.: Object-to-track association in a multisensor fusion system under the tbm framework. In: In IFAC Workshop on Advances in Control and Automation Theory for Transportation Applications (ACATTA 2013), Istanbul, Turkey (2013)
6. Dallil, A., Oussalah, M., Ouldali, A.: Evidential data association filter. In: Hüllermeier, E., Kruse, R., Hoffmann, F. (eds.) IPMU 2010. CCIS, vol. 80, pp. 209–217. Springer, Heidelberg (2010)
7. McLachlan, G.J.: Mahalanobis distance. Resonance 4(6), 20–26 (1999)
8. Denœux, T., El Zoghby, N., Cherfaoui, V., Jouglet, A.: Optimal object association in the dempster-shafer framework. IEEE Transactions on Cybernetics
9. Bourgeois, F., Lassalle, J.-C.: An extension of the Munkres algorithm for the assignment problem to rectangular matrices. Communications of the ACM 14(12), 802–804 (1971)
10. Smets, P., Kennes, R.: The Transferable Belief Model. Artificial Intelligence 66(2), 191–234 (1994)

Towards a Conflicting Part of a Belief Function

Milan Daniel⋆

Institute of Computer Science, Academy of Sciences of the Czech Republic
Pod vodárenskou věží 2, CZ - 187 02, Prague 8, Czech Republic
milan.daniel@cs.cas.cz

Abstract. Belief functions usually contain some internal conflict. Based on Hájek-Valdés algebraic analysis of belief functions, a unique decomposition of a belief function into its conflicting and non-conflicting part was introduced at ISIPTA'11 symposium for belief functions defined on a two-element frame of discernment.

This contribution studies the conditions under which such a decomposition exists for belief functions defined on a three-element frame. A generalisation of important Hájek-Valdés homomorphism f of semigroup of belief functions onto its subsemigroup of indecisive belief functions is found and presented. A class of quasi-Bayesian belief functions, for which the decomposition into conflicting and non-conflicting parts exists is specified. A series of other steps towards a conflicting part of a belief function are presented. Several open problems from algebra of belief functions which are related to the investigated topic and are necessary for general solution of the issue of decomposition are formulated.

Keywords: Belief function, Dempster-Shafer theory, Dempster's semigroup, conflict between belief functions, uncertainty, non-conflicting part of belief function, conflicting part of belief function.

1 Introduction

Belief functions represent one of the widely used formalisms for uncertainty representation and processing; they enable representation of incomplete and uncertain knowledge, belief updating, and combination of evidence [20].

When combining belief functions (BFs) by the conjunctive rules of combination, conflicts often appear which are assigned to $\emptyset$ by the non-normalised conjunctive rule $\bigodot$ or normalised by Dempster's rule of combination $\oplus$. Combination of conflicting BFs and interpretation of conflicts is often questionable in real applications; hence a series of alternative combination rules were suggested and a series of papers on conflicting BFs were published, e.g., [13, 16–18, 22].

In [5, 10, 11], new ideas concerning interpretation, definition, and measurement of conflicts of BFs were introduced. We presented three new approaches to interpretation and computation of conflicts: combinational conflict, plausibility conflict, and comparative conflict. Later, pignistic conflict was introduced [11].

⋆ This research is supported by the grant P202/10/1826 of the Czech Science Foundation (GA ČR). and partially by the institutional support of RVO: 67985807.

A. Laurent et al. (Eds.): IPMU 2014, Part III, CCIS 444, pp. 212–222, 2014.

When analyzing mathematical properties of the three approaches to conflicts of BFs, there appears a possibility of expression of a BF Bel as Dempster's sum of a non-conflicting BF Bel_0 with the same plausibility decisional support as the original BF Bel has and of an indecisive BF Bel_S which does not prefer any of the elements of frame of discernment. A new measure of conflict of BFs based on conflicting and non-conflicting parts of BFs is recently under development.

A unique decomposition to such BFs Bel_0 and Bel_S was demonstrated for BFs on 2-element frame of discernment in [6]. The present study analyses possibility of its generalisation and presents three classes of BFs on a 3-element frame for which such decomposition exists; it remains an open problem for other BFs.

2   Preliminaries

2.1   General Primer on Belief Functions

We assume classic definitions of basic notions from theory of *belief functions* (BFs) [20] on finite frames of discernment $\Omega_n = \{\omega_1, \omega_2, ..., \omega_n\}$, see also [4–9]. A *basic belief assignment (bba)* is a mapping $m : \mathcal{P}(\Omega) \longrightarrow [0,1]$ such that $\sum_{A \subseteq \Omega} m(A) = 1$; its values are called *basic belief masses (bbm)*. $m(\emptyset) = 0$ is usually assumed, if it holds, we speak about *normalised bba*. A *belief function (BF)* is a mapping $Bel : \mathcal{P}(\Omega) \longrightarrow [0,1]$, $Bel(A) = \sum_{\emptyset \neq X \subseteq A} m(X)$. A *plausibility function* $Pl(A) = \sum_{\emptyset \neq A \cap X} m(X)$. There is a unique correspondence between m and the corresponding Bel and Pl; thus we often speak about m as a BF.

A *focal element* is $X \subseteq \Omega$, such that $m(X) > 0$. If all of the focal elements are *singletons* (i.e. one-element subsets of Ω), this is what we call a *Bayesian belief function* (BBF). If all of the focal elements are either singletons or whole Ω (i.e. $|X| = 1$ or $|X| = |\Omega|$), this is what we call a *quasi-Bayesian belief function* (qBBF). If all focal elements have non-empty intersections (or all are nested), we call this a *consistent BF* (or a *consonant BF*, also a *possibility measure*).

Dempster's (conjunctive) rule of combination $\oplus$ is given as $(m_1 \oplus m_2)(A) = \sum_{X \cap Y = A} K m_1(X) m_2(Y)$ for $A \neq \emptyset$, where $K = \frac{1}{1-\kappa}$, $\kappa = \sum_{X \cap Y = \emptyset} m_1(X) m_2(Y)$, and $(m_1 \oplus m_2)(\emptyset) = 0$ [20]; putting $K = 1$ and $(m_1 \oplus m_2)(\emptyset) = \kappa$ we obtain the *non-normalised conjunctive rule of combination* $\odot$, see, e. g., [21].

We say that BF Bel is *non-conflicting* (or conflict free, i.e., it has no internal conflict), when it is consistent; i.e., whenever $Pl(\omega_i) = 1$ for some $\omega_i \in \Omega_n$. Otherwise, BF is *conflicting*, i.e., it contains an internal conflict [5]. We can observe that Bel is non-conflicting if and only if the conjunctive combination of Bel with itself does not produce any conflicting belief masses[1] (when $(Bel \odot Bel)(\emptyset) = 0$).

U_n is the *uniform Bayesian belief function*[2] [5], i.e., the uniform probability distribution on Ω_n. The *normalised plausibility of singletons*[3] of Bel is the BBF (prob. distrib.) $Pl\_P(Bel)$ such, that $(Pl\_P(Bel))(\omega_i) = \frac{Pl(\{\omega_i\})}{\sum_{\omega \in \Omega} Pl(\{\omega\})}$ [1, 4].

---

[1] Martin calls $(m \odot m)(\emptyset)$ autoconflict of the BF [18].

[2] U_n which is idempotent w.r.t. Dempster's rule $\oplus$, and moreover neutral on the set of all BBFs, is denoted as $_nD0'$ in [4], $0'$ comes from studies by Hájek & Valdés.

[3] Plausibility of singletons is called the *contour function* by Shafer [20], thus $Pl\_P(Bel)$ is in fact a normalisation of the contour function.

Let us define an *indecisive (indifferent) BF* as a BF which does not prefer any $\omega_i \in \Omega_n$, i.e., a BF which gives no decisional support for any ω_i, i.e., a BF such that $h(Bel) = Bel \oplus U_n = U_n$, i.e., $Pl(\{\omega_i\}) = const.$, that is, $(Pl\_P(Bel))(\{\omega_i\}) = \frac{1}{n}$. Let us further define an *exclusive BF* as a BF Bel such[4] that $Pl(X) = 0$ for a certain $\emptyset \neq X \subset \Omega$; the BF is otherwise *non-exclusive*.

2.2   Belief Functions on a Two-Element Frame of Discernment; Dempster's Semigroup

Let us suppose that the reader is slightly familiar with basic algebraic notions like *a group, semigroup, homomorphism*, etc. (Otherwise, see e.g., [3, 14, 15].)

We assume $\Omega_2 = \{\omega_1, \omega_2\}$, in this subsection. We can represent any BF on Ω_2 by a couple (a, b), i.e., by enumeration of its m-values $a = m(\{\omega_1\}), b = m(\{\omega_2\})$, where $m(\{\omega_1, \omega_2\}) = 1 - a - b$. This is called *Dempster's pair* or simply *d-pair* [14, 15] (it is a pair of reals such that $0 \leq a, b \leq 1, a + b \leq 1$).

The set of all non-extremal d-pairs (i.e., d-pairs different from $(1, 0)$ and $(0, 1)$) is denoted by D_0; the set of all non-extremal *Bayesian d-pairs* (where $a + b = 1$) is denoted by G; the set of d-pairs such that $a = b$ is denoted by S (set of indecisive d-pairs); the set where $b = 0$ by S_1 ($a = 0$ by S_2), simple support BFs. Vacuous BF is denoted as $0 = (0, 0)$ and $0' = U_2 = (\frac{1}{2}, \frac{1}{2})$, see Figure 1.

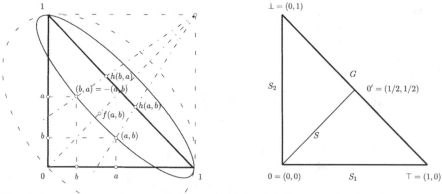

Fig. 1. Dempster's semigroup D_0. Homomorphism h is here a projection of the triangle D_0 to G along the straight lines running through $(1, 1)$. All of the d-pairs lying on the same ellipse are mapped by homomorphism f to the same d-pair in semigroup S.

The *(conjunctive) Dempster's semigroup* $\mathbf{D_0} = (D_0, \oplus, 0, 0')$ is the set D_0 endowed with the binary operation $\oplus$ (i.e., with Dempster's rule) and two distinguished elements 0 and $0'$. Dempster's rule can be expressed by the formula $(a, b) \oplus (c, d) = (1 - \frac{(1-a)(1-c)}{1-(ad+bc)}, 1 - \frac{(1-b)(1-d)}{1-(ad+bc)})$ for d-pairs [14]. In D_0 it is defined further: $-(a, b) = (b, a)$, $h(a, b) = (a, b) \oplus 0' = (\frac{1-b}{2-a-b}, \frac{1-a}{2-a-b})$, $h_1(a, b) = \frac{1-b}{2-a-b}$, $f(a, b) = (a, b) \oplus (b, a) = (\frac{a+b-a^2-b^2-ab}{1-a^2-b^2}, \frac{a+b-a^2-b^2-ab}{1-a^2-b^2})$; $(a, b) \leq (c, d)$ iff $[h_1(a, b) < h_1(c, d)$ or $h_1(a, b) = h_1(c, d)$ and $a \leq c]$ [5].

---

[4] BF Bel excludes all ω_i such, that $Pl(\{\omega_i\}) = 0$.

[5] Note that $h(a, b)$ is an abbreviation for $h((a, b))$, similarly for $h_1(a, b)$ and $f(a, b)$.

Theorem 1. *(i)* $\mathbf{G} = (G, \oplus, -, 0', \leq)$ *is an ordered Abelian group, isomorphic to the additive group of reals with the usual ordering.* $G^{\leq 0'}$, $G^{\geq 0'}$ *are its cones.*
(ii) *The sets* S, S_1, S_2 *with operation* $\oplus$ *and the ordering* $\leq$ *form ordered commutative semigroups with neutral element* 0; *all isomorphic to* $(Re, +, -, 0, \leq)^{\geq 0}$.
(iii) h *is ordered homomorphism:* $\mathbf{D}_0 \longrightarrow \mathbf{G}$; $h(Bel) = Bel \oplus 0' = Pl_-P(Bel)$.
(iv) f *is homomorphism:* $(D_0, \oplus, -, 0, 0') \longrightarrow (S, \oplus, -, 0)$; *(but, not ordered).*
(v) *Mapping* $- : \mathbf{D}_0 \longrightarrow \mathbf{D}_0$, $-(a, b) = (b, a)$ *is automorphism of* $\mathbf{D}_0$.

2.3   Dempster's Semigroup on a 3-Element Frame of Discernment

Analogously to d-pairs we can represent BFs by six-tuples $(d_1, d_2, d_3, d_{12}, d_{13}, d_{23}) = (m(\{\omega_1\}), m(\{\omega_2\}), m(\{\omega_3\}), m(\{\omega_1, \omega_2\}), m(\{\omega_1, \omega_3\}), m(\{\omega_2, \omega_3\}))$, i.e. by enumeration of its $2^3 - 2$ values, where the $(2^3 - 1)$-th value $m(\Omega_3) = 1 - \sum_i d_i$. Thus there is a significant increase of complexity considering 3-element frame of discernment Ω_3. While we can represent BFs on Ω_2 by a 2-dimensional triangle, we need a 6-dimensional simplex in the case of Ω_3. Further, all the dimensions are not equal: there are 3 independent dimensions corresponding to singletons from Ω_3, but there are other 3 dimensions corresponding to 2-element subsets of Ω_3, each of them somehow related to 2 dimensions corresponding to singletons.

Dempster's semigroup $\mathbf{D}_3$ on Ω_3 is defined analogously to $\mathbf{D}_0$. First algebraic results on $\mathbf{D}_3$ were presented at IPMU'12 [8] (a quasi-Bayesian case $\mathbf{D}_{3-0}$, the dimensions related to singletons only, see Figure 2) and a general case in [9].

Let us briefly recall the following results on $\mathbf{D}_3$ which are related to our topic.

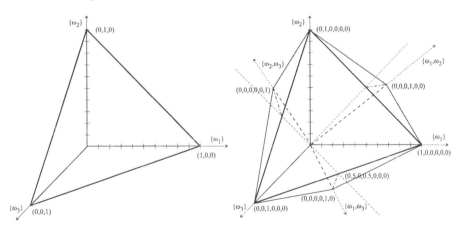

Fig. 2. Quasi-Bayesian BFs on Ω_3 **Fig. 3.** General BFs on 3-elem. frame Ω_3

Theorem 2. *(i)* $\mathbf{D}_{3-0} = (D_{3-0}, \oplus, 0, U_3)$ *is subalgebra of* $\mathbf{D}_3 = (D_3, \oplus, 0, U_3)$, *where* D_{3-0} *is set of non-exclusive qBBFs* $\mathbf{D}_{3-0} = \{(a, b, c, 0, 0, 0)\}$, D_3 *is set of all non-exclusive BFs on* Ω_3, $0 = (0, 0, 0, 0, 0, 0)$, *and* $U_3 = (\frac{1}{3}, \frac{1}{3}, \frac{1}{3}, 0, 0, 0)$.
(ii) $G_3 = (\{(a, b, c, 0, 0, 0) \mid a+b+c = 1; 0 < a, b, c\}, \oplus, "-", U_3)$ *is a subgroup of* $\mathbf{D}_3$, *where* "$-$" *is given by* $-(a, b, c, 0, 0, 0) = (\frac{bc}{ab+ac+bc}, \frac{ac}{ab+ac+bc}, \frac{ab}{ab+ac+bc}, 0, 0, 0)$.

(iii a) $S_0 = (\{(a,a,a,\ 0,0,0)\,|\,0 \le a \le \frac{1}{3}\}, \oplus, 0)$, $S_1 = (\{(a,0,0,0,0,0)\,|\,0 \le a <$ $1\}, \oplus, 0)$, S_2, S_3, *are monoids with neutral element* 0, *all are isomorphic to the positive cone of the additive group of reals* $\mathbf{Re}^{\ge 0}$ *(S_0 to $\mathbf{Re}^{\ge 0+}$ with* ∞*)*.

(iii b) *Monoids* $S = (\{(a,a,a,b,b,b) \in D_3\}, \oplus, 0)$ *and* $S_{Pl} = (\{(d_1,d_2,...,d_{23}) \in D_3\,|$ $Pl(d_1,d_2,...,d_{23}) = U_3\}, \oplus, 0)$ *are alternative generalisations of Hájek-Valdés S, both with neutral idempotent* 0 *and absorbing one* U_3. *(note that set of BFs* $\{(a,a,a,a,a,a) \in D_3\}$ *is not closed under* $\oplus$, *thus it does not form a semigroup).*

(iv) *Mapping* h *is homomorphism:* $(D_3, \oplus, 0, U_3) \longrightarrow (G_3, \oplus, " - ", U_3)$; $h(Bel) =$ $Bel \oplus U_3 = Pl\_P(Bel)$ *i.e., the normalised plausibility of singletons.*

See Theorems 2 and 3 in [8] and [9], assertion (iv) already as Theorem 3 in [6]. Unfortunately, a full generalisation either of $-$ or of f was not yet found [8, 9].

3   State of the Art

3.1   Non-conflicting and Conflicting Parts of Belief Functions on Ω_2

Using algebraic properties of group G, of semigroup S (including 'Dempster's subtraction' $(s, s) \oplus (x, x) = (s', s')$, and 'Dempster's half' $(x, x) \oplus (x, x) = (s, s)$, see [6]) and homomorphisms f and h we obtain the following theorem for BFs on Ω_2 (for detail and proofs see [6]):

Theorem 3. *Any BF* (a, b) *on a 2-element frame of discernment* Ω_2 *is Dempster's sum of its unique* non-conflicting *part* $(a_0, b_0) \in S_1 \cup S_2$ *and of its unique* conflicting *part* $(s, s) \in S$, *which does not prefer any element of* Ω_2, *that is,* $(a, b) = (a_0, b_0) \oplus (s, s)$. *It holds true that* $s = \frac{b(1-a)}{1-2a+b-ab+a^2} = \frac{b(1-b)}{1-a+ab-b^2}$ *and* $(a_0, b_0) = (\frac{a-b}{1-b}, 0) \oplus (s, s)$ *for* $a \ge b$; *and similarly that* $s = \frac{a(1-b)}{1+a-2b-ab+b^2} =$ $\frac{a(1-a)}{1-b+ab-a^2}$ *and* $(a_0, b_0) = (0, \frac{b-a}{1-a}) \oplus (s, s)$ *for* $a \le b$.   (See Theorem 2 in [6].)

3.2   Non-Conflicting Part of BFs on General Finite Frames

We would like to verify that Theorem 3 holds true also for general finite frames:

Hypothesis 1 *We can represent any BF Bel on an* n*-element frame of discernment* $\Omega_n = \{\omega_1, ..., \omega_n\}$ *as Dempster's sum* $Bel = Bel_0 \oplus Bel_S$ *of non-conflicting BF* Bel_0 *and of indecisive conflicting BF* Bel_S *which has no decisional support, i.e. which does not prefer any element of* Ω_n *to the others, see Figure 4.*

Using algebraic properties of Bayesian BFs and homomorphic properties of h we have a partial generalisation of mapping "$-$" to sets of Bayesian and consonant BFs, thus we have $-h(Bel)$ and $-Bel_0$.

Theorem 4. *(i) For any BF Bel on* Ω_n *there exists unique consonant BF* Bel_0 *such that,* $h(Bel_0 \oplus Bel_S) = h(Bel)$ *for any* Bel_S *such that* $Bel_S \oplus U_n = U_n$.

(ii) If for $h(Bel) = (h_1, h_2, ..., h_n, 0, 0, ..., 0)$ *holds true that,* $0 < h_i < 1$, *then unique BF* $-Bel_0$ *and* $-h(Bel)$ *exist, such that,* $(h(-Bel_0 \oplus Bel_s) = -h(Bel)$ *and* $h(Bel_0) \oplus$ $h(-Bel_0) = U_n$ (See Theorem 4 in [6].)

Corollary 1. *(i) For any consonant BF Bel such that $Pl(\{\omega_i\}) > 0$ there exists a unique BF $-Bel$; $-Bel$ is consonant in this case.*

Let us notice the importance of the consonance here: a stronger statement for general non-conflicting BFs does not hold true on Ω_3, for detail see [6].

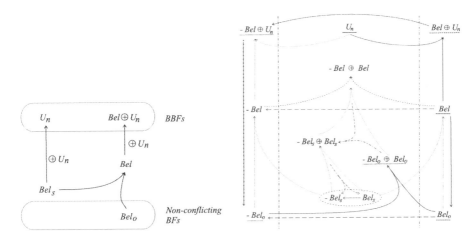

Fig. 4. Schema of Hypothesis 1

Fig. 5. Detailed schema of a decomposition of BF Bel

Including Theorem 4 into the schema of decomposition we obtain Figure 5. We have still partial results, as we have only underlined BFs; to complete the diagram, we need a definition of $-Bel$ for general BFs on Ω_3 to compute $Bel \oplus -Bel$; we further need an analysis of indecisive BFs to compute $Bel_S \oplus -Bel_S$ and resulting Bel_S and to specify conditions under which a unique Bel_S exists.

4   Towards Conflicting Parts of BFs on Ω_3

4.1   A General Idea

An introduction to the algebra of BFs on a 3-element frame was performed, but a full generalisation of basic homomorphisms of Dempster's semigroup – and f is still missing [6–9]. We need $f(Bel) = -Bel \oplus Bel$ to complete the decomposition diagram (Figure 5) according to the original idea from [6].

Let us forget for a moment a meaning of $'-'$ and its relation to group 'minus' in subgroups G and G_3; and look at its construction $-(a, b) = (b, a)$. It is a simple transposition of m-values of ω_1 and ω_2 in fact. Generally on Ω_3 we have:

Lemma 1. *Any transposition τ of a 3-element frame of discernment Ω_3 is an automorphism of D_3. $\tau_{12}(\omega_1, \omega_2, \omega_3) = (\omega_2, \omega_1, \omega_3)$, $\tau_{23}(\omega_1, \omega_2, \omega_3) = (\omega_1, \omega_3, \omega_2)$, $\tau_{13}(\omega_1, \omega_2, \omega_3) = (\omega_3, \omega_2, \omega_1)$.*

Theorem 5. *Any permutation π of a 3-element frame of discernment Ω_3 is an automorphism of D_3.*

For proofs of statements in this section see [12] (Lems 2–5 and Thms 6–9).

Considering function '$-$' as transposition (permutation), we have $f(a,b) = (a,b) \oplus (b,a)$ a Dempster's sum of all permutations of Bel given by (a,b) on Ω_2. Analogously we can define

$$f(Bel) = \bigoplus_{\pi \in \Pi_3} \pi(Bel),$$

where $\Pi_3 = \{\pi_{123}, \pi_{213}, \pi_{231}, \pi_{132}, \pi_{312}, \pi_{321}\}$, i.e., $f(a,b,c,d,e,f;g) = \bigoplus_{\pi \in \Pi_3} \pi(a,b,c,d,e,f;g) = (a,b,c,d,e,f;g) \oplus (b,a,c,d,f,e;g) \oplus (b,c,a,f,d,e;g) \oplus (a,c,b,e,d,f;g) \oplus (c,a,b,e,f,d;g) \oplus (c,b,a,f,e,d;g)$.

Theorem 6. *Function $f : D_3 \longrightarrow S$, $f(Bel) = \bigoplus_{\pi \in \Pi_3} \pi(Bel)$ is homomorphism of Dempster's semigroup $\mathbf{D}_3$ to its subsemigroup $S = (\{(a,a,a,b,b,b;1-3a-3b)\}, \oplus)$.*

Having homomorphism f, we can leave a question of existence $-Bel$ such that $h(-Bel) = -h(Bel)$, where '$-$' from group of BBFs G_3 is used on the right hand side. Unfortunately, we have not an isomorphism of S subsemigroup of $\mathbf{D}_3$ to the additive group of reals as in the case of semigroup S of $\mathbf{D}_0$, thus we still have an open question of subtraction there. Let us focus, at first, on the subsemigroup of quasi-Bayesian BFs for simplification.

4.2   Towards Conflicting Parts of Quasi-Bayesian BFs on Ω_3

Let us consider qBBFs $(a,b,c,0,0,0;1-a-b-c) \in \mathbf{D}_{3-0}$ in this section. Following Theorem 6 we obtain the following formulation for qBBFs:

Theorem 7. *Function $f : D_{3-0} \longrightarrow S$, $f(Bel) = \bigoplus_{\pi \in \Pi_3} \pi(Bel)$ is homomorphism of Dempster's semigroup $\mathbf{D}_{3-0}$ to its subsemigroup $S_0 = (\{(a,a,a,0,0,0;1-3a)\}, \oplus)$.*

S_0 is isomorphic to the positive cone of the additive group of reals, see Theorem 2, thus there is subtraction which is necessary for completion of diagram from Figure 5. Utilising isomorphism with reals, we have also existence of 'Dempster's sixth'[6] which is needed to obtain preimage of $f(Bel)$ in S_0. Supposing Hypothesis 1, $Bel = Bel_0 \oplus Bel_S$, thus $f(Bel) = \bigoplus_{\pi \in \Pi_3} \pi(Bel_0) \oplus \bigoplus_{\pi \in \Pi_3} \pi(Bel_S)$; all 6 permutations are equal to identity for any qBBF $Bel_S \in S_0$), thus we have $f(Bel) = \bigoplus_{\pi \in \Pi_3} \pi(Bel_0) \oplus \bigoplus_{(6\text{-}times)} Bel_S$ in our case):

Lemma 2. *'Dempster's sixth'.*
Having $f(Bel_S)$ in S_0, there is unique $f^{-1}(f(Bel_S)) \in S_0$, such that $\bigoplus_{(6\text{-}times)} f^{-1}(f(Bel_S)) = f(Bel_S)$. If $Bel_S \in S_0$ then $f^{-1}(f(Bel_S)) = Bel_S$.

On the other hand there is a complication considering qBBFs on Ω_3 that their non-conflicting part is a consonant BF frequently out of $\mathbf{D}_{3-0}$. Hence we can simply use the advantage of properties of S_0 only for qBBFs with singleton simple support non-conflicting parts.

---

[6] Analogously we can show existence of general 'Dempster's k-th' for any natural k and any BF Bel from S_0, but we are interested in 'Dempster's sixth' in our case.

Lemma 3. *Quasi-Bayesian belief functions which have quasi-Bayesian non-conflicting part are just BFs from the following sets* $Q_1 = \{(a, b, b, 0, 0, 0) \mid a \geq b\}$, $Q_2 = \{(b, a, b, 0, 0, 0) \mid a \geq b\}$, $Q_3 = \{(b, b, a, 0, 0, 0) \mid a \geq b\}$. Q_1, Q_2, Q_3 *with oplus are subsemigroups of* $\mathbf{D}_{3-0}$; *their union* $Q = Q_1 \cup Q_2 \cup Q_3$ *is not closed w.r.t.* $\oplus$. $(Q_1, \oplus)$ *is further subsemigroup of* $D_{1-2-3} = (\{(d_1, d_2, d_2, 0, 0, 0)\}, \oplus, 0, U_3)$, *for detail on* D_{1-2-3} *see [8, 9], following this, we can denote* $(Q_i, \oplus)$ *as* $D_{i-j=k}^{i \geq j = k}$.

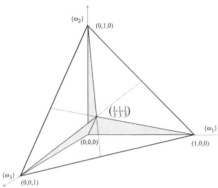

For quasi-Bayesian BFs out of Q (i.e. BFs from $\mathbf{D}_{3-0} \setminus Q$) we have not yet decomposition into conflicting and non-conflicting part according to Hypothesis 1, as we have not $f(Bel_0) \in S_0$ and have not subtraction in S in general. BFs from $\mathbf{D}_{3-0} \setminus Q$ either have their conflicting part in $S_{Pl} \setminus S_0$ or in $S_{Pl} \setminus S$ or have not conflicting part according to Hypothesis 1 (i.e. their conflicting part is a pseudo belief function out of D_3). Solution of the problem is related to a question of subtraction in subsemigroups S and S_{Pl}, as $f(Bel_0)$ is not in S_0 but in $S \setminus S_0$ for qBBFs out of Q. Thus we have to study these qBBFs together with general BFs from the point of view of their conflicting parts.

Fig. 6. Quasi-Bayesian BFs with unique decomposition into $Bel_0 \oplus Bel_S$

Theorem 8. *Belief functions* Bel *from* $Q = D_{1-2=3}^{1 \geq 2=3} \cup D_{2-1=3}^{2 \geq 1=3} \cup D_{3-1=2}^{3 \geq 1=2}$ *have unique decomposition into their conflicting part* $Bel_S \in S_0$ *and non-conflicting part in* S_1 *(S_2 or S_3 respectively).*

4.3   Towards Conflicting Parts of General Belief Functions on Ω_3

There is a special class of general BFs with singleton simple support non-conflicting part, i.e. BFs with $f(Bel_0) \in S_0$. Nevertheless due to the generality of Bel, we have $f(Bel) \in S$ in general, thus there is a different special type of belief 'subtraction' ($(a, a, a, b, b, b) \ominus (c, c, c, 0, 0, 0)$).

Following the idea from Figure 5, what do we already have?
We have the entire right part: given Bel, $Bel \oplus U_3$, and non-conflicting part Bel_0 (Theorem 4 (i)); in the left part we have $-Bel \oplus U_3 = -(Bel \oplus U_3)$ using G_3 group '$-$' (Theorem 2 (ii)) and $-Bel_0 = (-Bel \oplus U_3)_0$ (a non-conflicting part of $-Bel \oplus U_3$). In the central part of the figure, we only have U_3 and $-Bel_0 \oplus Bel_0$ in fact. As we have not $-Bel$ we have not $-Bel \oplus Bel$, we use $f(Bel) = \bigoplus_{\pi \in \Pi_3} \pi(Bel)$ instead of it; $f(Bel) \in S$ in general.

We can compute $-Bel_0 \oplus Bel_0$; is it equal to $f(Bel_0)$? We can quite easily find a counter-example, see [12]. Thus neither $-Bel \oplus Bel$ is equal to $f(Bel) = \bigoplus_{\pi \in \Pi_3} \pi(Bel)$ in general. What is a relation of these two approaches? What is their relation to the decomposition of Bel?

Lemma 4. $-Bel_0 \oplus Bel_0$ *is not equal to* $\bigoplus_{\pi \in \Pi_3} \pi(Bel)$ *in general. Thus there are two different generalisations of homomorphism* f *to* $\mathbf{D}_3$.

Learning this, we can update the diagram of decomposition of a BF *Bel* into its conflicting and non-conflicting part as it is in Figure 8.

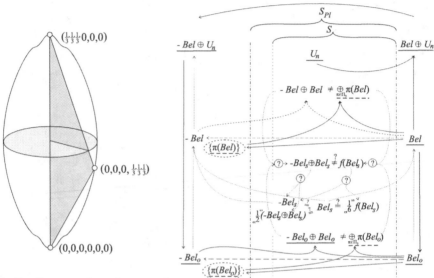

Fig. 7. S_{Pl} — subsemigroup of general indecisive belief functions   **Fig. 8.** Updated schema of decomposition of *Bel*

5   Open Problems for a Future Research

There are three main general open problems from the previous section:
- Elaboration of algebraic analysis, especially of sugbroup S_{Pl} (indecisive BFs).
- What are the properties of two different generalisations of homomorphism f; what is their relationship?
- Principal question of the study: verification of Hypothesis 1; otherwise a specification of sets of BFs which are or are not decomposable ($Bel_0 \oplus Bel_S$).

Open is also a question of relationship to and of possible utilisation of Cuzzolin's geometry [2] and Quaeghebeur-deCooman extreme lower probabilities [19].

6   Summary and Conclusions

New approach to understanding operation '−' and homomorphism f from $\mathbf{D}_0$ (a transposition of elements instead of some operation related to group 'minus' of G, G_3) is introduced in this study.

The first complete generalisation of Hájek-Valdés important homomorphism f is presented. Specification of several classes of BFs (on Ω_3) which are decomposable into $Bel_0 \oplus Bel_S$, and several other partial results were obtained.

The presented results improve general understanding of conflicts of BFs and of the entire nature of BFs. These results can be also used as one one of the mile-stones to further study of conflicts between BFs. Correct understanding of conflicts may consequently improve a combination of conflicting belief functions.

References

1. Cobb, B.R., Shenoy, P.P.: A Comparison of Methods for Transforming Belief Function Models to Probability Models. In: Nielsen, T.D., Zhang, N.L. (eds.) ECSQARU 2003. LNCS (LNAI), vol. 2711, pp. 255–266. Springer, Heidelberg (2003)
2. Cuzzolin, F.: The geometry of consonant belief functions: Simplicial complexes of necessity measures. Fuzzy Sets and Systems 161(10), 1459–1479 (2010)
3. Daniel, M.: Algebraic structures related to Dempster-Shafer theory. In: Bouchon-Meunier, B., Yager, R.R., Zadeh, L.A. (eds.) IPMU 1994. LNCS, vol. 945, pp. 51–61. Springer, Heidelberg (1995)
4. Daniel, M.: Probabilistic Transformations of Belief Functions. In: Godo, L. (ed.) ECSQARU 2005. LNCS (LNAI), vol. 3571, pp. 539–551. Springer, Heidelberg (2005)
5. Daniel, M.: Conflicts within and between Belief Functions. In: Hüllermeier, E., Kruse, R., Hoffmann, F. (eds.) IPMU 2010. LNCS (LNAI), vol. 6178, pp. 696–705. Springer, Heidelberg (2010)
6. Daniel, M.: Non-conflicting and Conflicting Parts of Belief Functions. In: Coolen, F., de Cooman, G., Fetz, T., Oberguggenberger, M. (eds.) ISIPTA 2011, pp. 149–158, Studia Universitätsverlag, Innsbruck (2011)
7. Daniel, M.: Morphisms of Dempster's Semigroup: A Rev. and Interpret. In: Barták, R. (ed.) Proceedings of 14th Czech-Japan Seminar on Data Analysis and Decision Making Under Uncertainty, CJS 2011, pp. 26–34. Matfyzpress, Prague (2011)
8. Daniel, M.: Introduction to an Algebra of Belief Functions on Three-Element Frame of Discernment — A Quasi Bayesian Case. In: Greco, S., Bouchon-Meunier, B., Coletti, G., Fedrizzi, M., Matarazzo, B., Yager, R.R. (eds.) IPMU 2012, Part III. CCIS, vol. 299, pp. 532–542. Springer, Heidelberg (2012)
9. Daniel, M.: Introduction to Algebra of Belief Functions on Three-element Frame of Discernment - A General Case. In: Kroupa, T., Vejnarová, J. (eds.) WUPES 2012. Proceedings of the 9th Workshop on Uncertainty Processing, pp. 46–57. University of Economics, Prague (2012)
10. Daniel, M.: Properties of Plausibility Conflict of Belief Functions. In: Rutkowski, L., Korytkowski, M., Scherer, R., Tadeusiewicz, R., Zadeh, L.A., Zurada, J.M. (eds.) ICAISC 2013, Part I. LNCS (LNAI), vol. 7894, pp. 235–246. Springer, Heidelberg (2013)
11. Daniel, M.: Belief Functions: A Revision of Plausibility Conflict and Pignistic Conflict. In: Liu, W., Subrahmanian, V.S., Wijsen, J. (eds.) SUM 2013. LNCS (LNAI), vol. 8078, pp. 190–203. Springer, Heidelberg (2013)
12. Daniel, M.: Steps Towards a Conflicting Part of a Belief Function. Technical report V-1179, ICS AS CR, Prague (2013)
13. Destercke, S., Burger, T.: Revisiting the Notion of Conflicting Belief Functions. In: Denœux, T., Masson, M.-H. (eds.) Belief Functions: Theory & Appl. AISC, vol. 164, pp. 153–160. Springer, Heidelberg (2012)
14. Hájek, P., Havránek, T., Jiroušek, R.: Uncertain Information Processing in Expert Systems. CRC Press, Boca Raton (1992)
15. Hájek, P., Valdés, J.J.: Generalized algebraic foundations of uncertainty processing in rule-based expert systems (dempsteroids). Computers and Artificial Intelligence 10(1), 29–42 (1991)
16. Lefèvre, É., Elouedi, Z., Mercier, D.: Towards an Alarm for Opposition Conflict in a Conjunctive Combination of Belief Functions. In: Liu, W. (ed.) ECSQARU 2011. LNCS (LNAI), vol. 6717, pp. 314–325. Springer, Heidelberg (2011)

17. Liu, W.: Analysing the degree of conflict among belief functions. Artificial Intelligence 170, 909–924 (2006)
18. Martin, A.: About Conflict in the Theory of Belief Functions. In: Denœux, T., Masson, M.-H. (eds.) Belief Functions: Theory & Appl. AISC, vol. 164, pp. 161–168. Springer, Heidelberg (2012)
19. Quaeghebeur, E., de Cooman, G.: Extreme lower probabilities. Fuzzy Sets and Systems 159(16), 2163–2175 (1990)
20. Shafer, G.: A Mathematical Theory of Evidence. Princeton University Press, Princeton (1976)
21. Smets, P.: The combination of evidence in the transferable belief model. IEEE-Pattern Analysis and Machine Intelligence 12, 447–458 (1990)
22. Smets, P.: Analyzing the combination of conflicting belief functions. Information Fusion 8, 387–412 (2007)

Representing Interventional Knowledge in Causal Belief Networks: Uncertain Conditional Distributions Per Cause

Oumaima Boussarsar, Imen Boukhris, and Zied Elouedi

LARODEC, Université de Tunis, Institut Supérieur de Gestion de Tunis, Tunisia
{oumaima.boussarsar}@hotmail.fr, zied.elouedi@gmx.fr,
imen.boukhris@hotmail.com

Abstract. Interventions are important for an efficient causal analysis. To represent and reason with interventions, the graphical structure is needed, the so-called causal networks are therefore used. This paper deals with the handling of uncertain causal information where uncertainty is represented with a belief function knowledge. To simplify knowledge acquisition and storage, we investigate the representational point of view of interventions when conditional distributions are defined per single parent. The mutilated and augmented causal belief networks are used in order to efficiently infer the effect of both observations and interventions.

1 Introduction

Causality is a crucial concept in Artificial Intelligence (AI) as well as in information systems. It comes to describe, interpret and analyze information and phenomena of our environment [3]. Besides, it enables to anticipate the dynamics of events when the system is evolving using interventions [8], [10] which are exterior manipulations that force target variables to have specific values. Upon acting on the cause, we are in a good position to say that the effect will also happen. However, an observation is seeing and monitoring phenomena happening by themselves without any manipulation on the system. Indeed, the "do" operator is used [8] to to deal with the effects of interventions.

Graphical models are compact and simple representations of uncertainty distributions. They are increasingly popular for reasoning under uncertainty due to their simplicity, their ability to easily express the human reasoning. A probability distribution, as good as it is, does not distinguish between equiprobability and ignorance situations. To tackle this problem, alternative networks have been proposed (e.g. possibility theory [1], [6], belief function theory [2], [11]). A causal network [8] is a graphical model which plays an important role for the achievement of a coherent causal analysis. In this network, arcs do not only represent dependencies, but also cause/effect relationships.

In this paper, we use the belief function theory [9] as a tool for knowledge representation and reasoning with uncertainty. Indeed, this theory has well understood connections to other frameworks such as probability, possibility and imprecise probability theories.

A. Laurent et al. (Eds.): IPMU 2014, Part III, CCIS 444, pp. 223–232, 2014.

Note that unlike Bayesian networks [7], conditional distributions in belief networks can be defined per single parent [11]. The main advantage of defining conditionals per edge is to simplify knowledge acquisition and storage. In fact, it first enables experts to express their beliefs in a more flexible way. Then, it allows to reduce the size of the conditional tables and therefore to decrease the storage complexity. Therefore, this paper focuses on the representation of uncertain knowledge and on the handling of standard interventions in causal belief networks where conditional distributions are defined per single parent.

The rest of the paper is organized as follows: in Section 2, we provide a brief background on the belief function theory. In Section 3, we recall the existing causal belief networks where beliefs are defined for all parents as for Bayesian networks. In Section 4, we propose a mutilated and an augmented based approaches to handle interventions on causal belief networks where beliefs are defined per edge. Section 5 concludes the paper.

2   Belief Function Theory

Let Θ be a finite set of elementary events, called the frame of discernment. Beliefs are expressed on subsets belonging to the powerset of Θ denoted 2^Θ. The basic belief assignment (bba), denoted by m^Θ or m, is a mapping from 2^Θ to $[0,1]$ such that: $\sum_{A \subseteq \Theta} m(A) = 1$. $m(A)$ is a basic belief mass (bbm) assigned to A. It represents the part of belief exactly committed to the event A of Θ. Subsets of Θ such that $m(A) > 0$ are called focal elements. A bba is said to be certain if the whole mass is allocated to a unique singleton of Θ. If the bba has Θ as a unique focal element, it is called vacuous and it represents the case of total ignorance. Two bbas provided by two distinct and independent sources m_1 and m_2 may be combined to give one resulting mass using the Dempster's rule of combination.

$$m_1 \oplus m_2(A) = K \cdot \sum_{B \cap C = A} m_1(B) m_2(C), \forall B, C \subseteq \Theta \qquad (1)$$

where $K^{-1} = 1 - \sum_{B \cap C = \emptyset} m_1(B) m_2(C)$.

Dempster's rule of conditioning allows us to update the knowledge of an expert in the light of a new information. $m(A|B)$ denotes the degree of belief of A in the context of B with A, B $\subseteq \Theta$.

$$m(A|B) = \begin{cases} K. \sum_{C \subseteq \bar{B}} m(A \cup C) & \text{if } A \subseteq B, A \neq \emptyset \\ 0 & \text{if } A \not\subseteq B \end{cases} \qquad (2)$$

where $K^{-1} = 1 - m(\emptyset)$. Knowing that experts are not fully reliable, a method of discounting seems imperative to update experts beliefs. The idea is to quantify the reliability of each expert. Let (1- α) be the degree of trust assigned to an expert. The corresponding bba can be weakened by the discounting method defined as:

$$m^\alpha(A) = \begin{cases} (1 - \alpha) \cdot m(A), \forall \ A \subset \Theta \\ \alpha + (1 - \alpha) \cdot m(A), & \text{if } A = \Theta \end{cases} \qquad (3)$$

The discounting operation is controlled by a *discount rate* α taking values between 0 and 1. If $\alpha = 0$, the source is fully reliable; whereas if $\alpha = 1$, the source of information is not reliable at all.

3   Causal Belief Networks

Causal belief networks [5] are graphical models under an uncertain environment where the uncertainty is represented by belief masses. This model represents an alternative and an extension to Bayesian causal networks. It allows the detection of causal relationships resulting from acting on some events. Moreover, it is no more necessary to define all a priori distributions. Indeed, it is possible to define vacuous *bbas* on some nodes. It is defined on two levels:

- Qualitative level: represented by a directed acyclic graph (DAG) named G where $G = (V, E)$ in which the nodes V represent variables, and arcs E describe the cause-effect relations embedded in the model. Parents of a given variable A_i denoted by $PA(A_i)$ are seen as its immediate causes. An instance from the set of parents of A_i is denoted $Pa(A_i)$. Each variable A_i is associated with its frame of discernment Θ_{A_i} representing all its possible instances.
- Quantitative level: is the set of normalized *bbas* associated to each node in the graph. Conditional distributions can be defined for each subset of each variable A_i ($sub_{ik} \subseteq \Theta_{A_i}$) in the context of its parents such that:
$$\sum_{sub_{ik} \subseteq \Theta_{A_i}} m^{A_i}(sub_{ik}|Pa(A_i)) = 1$$

4   Handling Interventions: Conditional *bbas* Defined Per Edge

In belief networks, conditional distributions can be defined as for Bayesian networks [2], [5] or per edge [11] in order to simplify knowledge acquisition and storage under uncertainty. In this latter, the knowledge about the relations between two nodes are assumed to be issued from different sources (called local conditional beliefs). Conditional distributions can be conjunctively aggregated into one distribution representing the relation between a node and all its parents. Thus, each node A_i is defined in the context of its parent. An instance from a single parent of A_i is denoted by $Pa_j(A_i)$.

Handling interventions and computing their effects on the system can be done by making changes on the structure of the causal belief network. The two methods developed are namely, belief graph mutilation and belief graph augmentation methods. These methods were proved to be equivalent when conditional distributions are defined for all parents [5]. Besides, they were used to handle non standard interventions [4]. In the following, we will propose and explain how to define the different conditional distributions on the altered causal belief graphs to handle interventions.

4.1   A Mutilated Based Approach

An external action will alter the system. Thus, an intervention is interpreted as a surgery by cutting off the edges pointing to the node concerned by the action. The rest of the network remains unchanged. Therefore, this action makes the direct causes (parents) of the variable concerned by the intervention not more responsible of its state. However, beliefs on these direct causes should not be modified.

Let $G = (V, E)$ be a causal belief network where conditional distributions are defined per edge and let A_i be a variable in G forced to take the value a_{ij} upon to the intervention $do(a_{ij})$. We define mutilation on two steps:

- Each arc linking A_i to each one of its parents $Pa_j(A_i)$ will be deleted. The resulting graph is denoted G_{mut}. In the mutilated graph, it corresponds to observing $A_i = a_{ij}$. Thus, it simply consists of conditioning the mutilated graph by the value a_{ij} and it is defined as follows : $m_{G_{mut}}(.|a_{ij}) = m_G(.|do(a_{ij}))$
- The external action $do(a_{ij})$ imposes the specific value a_{ij} to the variable A_i. The conditional distribution of the target variable becomes a certain bba which is defined as follows:

$$m_{G_{mut}}^{A_i}(sub_{ik}) = \begin{cases} 1 & \text{if } sub_{ik} = \{a_{ij}\} \\ 0 & \text{otherwise} \end{cases} \quad (4)$$

Example 1. *Let us consider the network presented in Fig. 1. They concern a description of knowledge regarding the link between smoking S ($\Theta_S = \{s_1, s_2\}$ where s1 is yes and s2 is no) and having yellow teeth T ($\Theta_T = \{t_1, t_2\}$ where t_1 is yellow and t_2 is otherwise). After acting on a variable by forcing T to take the value t_1, the state of T will become independent from the fact of smoking (S). Therefore, the link relating S to T will be deleted. This is represented by the network in Fig. 2. Note that initial beliefs about smoking remain the same.*

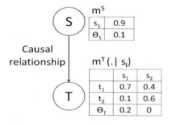

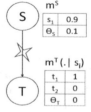

Fig. 1. A causal belief network

Fig. 2. Graph mutilation upon the intervention $do(t_1)$

4.2   An Augmented Based Approach

Another alternative to graphically represent interventional knowledge is to add a new fictive variable DO as a parent node of the variable A_i concerned by the manipulation. This added variable will totally control its state and its conditional distributions becomes a certain *bba*. Thus, the parents set of the variable A_i denoted PA is transformed to $PA' = PA \cup \{DO\}$. The DO node takes values in $do(x)$, $x \in \{\Theta_{A_i} \cup \{nothing\}\}$. do(*nothing*) represents the case when no interventions are made. $do(a_{ij})$ means that the variable A_i is forced to take the certain value a_{ij}. The resulting graph is called an augmented graph and denoted by G_{aug}. This method allows to represent the effect of observations and interventions.

Regarding the *bba* assigned to the added fictive node (i.e., DO), two cases are considered:

- If there is no interventions, then the causal belief network allows to model the effect of observations as on the initial causal graph. Hence, $m_{G_{aug}}^{DO}(do(x))$ is defined by:

$$m_{G_{aug}}^{DO}(do(x)) = \begin{cases} 1 & \text{if } x = nothing \\ 0 & otherwise \end{cases} \tag{5}$$

- If there is an intervention forcing the variable A_i to take the value a_{ij}, then $m_{G_{aug}}^{DO}(do(x))$ is defined by:

$$m_{G_{aug}}^{DO}(do(x)) = \begin{cases} 1 & \text{if } x = \{a_{ij}\} \\ 0 & otherwise \end{cases} \tag{6}$$

Remind that unlike [5], the conditional distributions of the initial causal belief network are defined per single parent. Doing this way is assuming that conditional distributions defined in the context of each cause can be issued by different local sources. Consequently, on the augmented causal belief graph, we will have a source given the DO node, i.e., the intervention and a source or multiple sources given the initial causes. In the following, we will explain how to define and modify the conditional distributions in context of all these causes.

(A) Defining conditionals given the DO node

The DO node can take values in $do(x)$, $x \in \{\Theta_{A_i} \cup nothing\}$. do(*nothing*) means that there are no actions on the variable concerned by the action. $do(a_{ij})$ means that the variable A_i is forced to take the value a_{ij}. In the following, we will detail how to define the conditional distribution given the DO node in both cases.

- Interventions not occurring:

In the case where no interventions are performed, the state of the variable concerned by the action will not depend on the intervention (i.e., from the DO node). The conditional *bba* given the DO node is not informative. Indeed, it only depends on the observations of the initial causes. This situation represents the state of total ignorance represented with the belief function theory by the vacuous *bba*. Therefore, the conditional distribution given the DO node becomes a vacuous *bba* by allocating one to the whole frame of discernment. Accordingly, the new distribution of the node DO is defined as:

- For each $sub_{ik} \subseteq \Theta_{A_i}$ and $x = \{nothing\}$

$$m_{G_{aug}}^{A_i}(sub_{ik}|do(nothing)) = \begin{cases} 1 & \text{if } sub_{ik} = \Theta_A \\ 0 & \text{otherwise} \end{cases} \tag{7}$$

Example 2. *Let us continue with the same events mentioned in Example 1. When there is no intervention, the DO node is taking the value nothing and the conditional distributions of Θ_T given the DO node is vacuous. It expresses the state of total ignorance (i.e., if we do not act on T, we ignore if the state of T will be t_1 or t_2). This is represented by the causal belief network depicted in Fig. 3.*

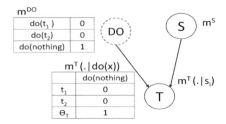

Fig. 3. Defining conditionals given the DO node when no interventions occurring

The main advantage of handling interventions by graph augmentation, is that it allows to represent the effect of observations (when the added node takes the value nothing).

Proposition 1. *Let G_{aug} be an augmented causal belief network where conditional beliefs are defined per single parent. In the case of not acting on the system, i.e., the DO node is set to the value nothing, the bba $m_{G_{aug}}(.|do(nothing))$ encodes the same joint distribution as the original causal belief network where $V' = DO \cup V$.*

$$m_{G_{aug}}^{V'}(.|do(nothing)) = m_G^V \tag{8}$$

– Interventions occurring:

Upon an intervention, the experimenter puts the target variable A_i into exactly one specific state $a_{ij} \in \Theta_{A_i}$. Accordingly, he is assumed to completely control his manipulation and he is in a good position to know the state of the variable concerned by the action. Thus, the *bba* of the target variable becomes a certain *bba*. Therefore, the conditional distribution given the DO node is defined as:

- For each $sub_{ik} \subseteq \Theta_{A_i}$ and $x = \{a_{ij}\}$

$$m_{G_{aug}}^{DO}(sub_{ik}|do(a_{ij})) = \begin{cases} 1 & \text{if } x = \{a_{ij}\} \\ 0 & \text{otherwise} \end{cases} \qquad (9)$$

Example 3. *Let us continue with the Fig. 4 which illustrates a causal belief augmented graph on which an intervention $do(t_1)$ forces the variable T to take the specific value t_1. The conditional bba given the intervention is transformed into a certain bba focused on t_1.*

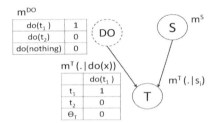

Fig. 4. Defining conditionals given the DO node upon $do(t_1)$

(B) Defining conditionals given the initial causes

In what follows, we will explain how to modify the conditional distribution of the target variable given the initial causes when no intervention is occurring and also in the case of an intervention.

– Interventions not occurring:

When there is no interventions, only the initial parents of each node affect its state. Therefore, conditional beliefs given the initial causes on the augmented graph in the case of no interventions are kept the same as for the initial graph where:

$$m_{G_{aug}}^{A_i}(sub_{ik}|Pa_j(A_i)) = m_G^{A_i}(sub_{ik}|Pa_j(A_i)) \qquad (10)$$

Example 4 *When there is no interventions, the DO node takes the value nothing. It does not control the state of the variable T concerned by the action. This is interpreted as having yellow teeth are dependent and effects of only one cause which is smoking. The conditional distributions of T given S in the initial graph represented in Example 1 are kept the same for the causal belief augmented network represented in Fig. 5.*

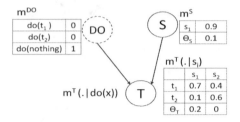

Fig. 5. Defining conditionals given the initial causes in the case of not acting

– Interventions occurring:

An intervention is an external action which completely controls the state of the target variable and succeeds to force the variable A_i to take a certain value a_{ij}. After acting on a variable, we assume that its initial causes are no more responsible of its state. In fact, the effect of the initial causes should no longer be considered. Thus, the source who predicted that the distribution of the target variable is defined by the initial distribution is considered as unreliable and should be weakened. The belief function theory allows to take into account the reliability of the sources of information using the discounting method. Thus, the difference between what was predicted and the actual value is considered as its discounting factor. Hence, the modified conditional *bba* of the target variable given the initial causes becomes as follows:

$$m_{G_{aug}}^{A_i, \alpha}(sub_{ik}|Pa_j(A_i)) = \begin{cases} \alpha & \text{if } sub_{ik} = \Theta_A \\ 1 - \alpha & \text{otherwise} \end{cases} \tag{11}$$

In the case of interventions, the source is considered as unreliable since by definition only the intervention controls the state of its target. Hence, the source predicting that the state of the target variable will depend on the initial causes is totally unreliable, i.e., $\alpha = 1$, the conditional *bba* becomes as follows:

$$m_{G_{aug}}^{A_i, \alpha(.|Pa_j(A_i))}(sub_{ik}|Pa_j(A_i)) = \begin{cases} 1 & \text{if } sub_{ik} = \Theta_A \\ 0 & \text{otherwise} \end{cases} \tag{12}$$

As for causal belief networks where conditional beliefs are defined for all parents, the augmented based approach and the mutilated based approach are equivalent to graphically represent interventions using causal belief networks where conditional beliefs are defined per single parent and lead to the same results.

Proposition 2. *Let G be a causal belief network where conditional distributions are defined per single parent and G_{mut} and G_{aug} its corresponding mutilated and augmented graphs after acting on the variable A_i by forcing it to take the value*

a_{ij}. *Computing the effects of interventions using the mutilation of the graph or its augmentation gives the same results.*

$$m^V_{G_{mut}}(.|a_{ij}) = m^{V'}_{G_{aug}}(.|do(a_{ij}))$$
(13)

If it is needed, one can combine the conditional distributions defined per edge using the Dempster's rule of combination. Note that after the fusion process, the causal belief network where conditional distributions are defined per single parents collapse into a causal belief network where conditional distributions are originally defined for all parents. This is true in the case where no interventions are performed and also when acting on the system.

To get the conditional *bba* given all the parent nodes, Dempster's rule of combination is used to aggregate the conditional distribution given the initial causes with the conditional *bba* given the DO parent.

Proposition 3. *Let G be a causal belief network where conditional distributions are defined per single parent, the new distribution of the target variable A_i defined in the context of all its initial parents $(m^{A_i}_{G_{aug}}(.|PA(A_i)))$ including the DO node $(m^{A_i}_{G_{aug}}(.|do(x)))$ is computed by combining the aggregated conditional bbas defined per single parent Pa_j, with the conditional distribution given the intervention as follows:*

$$m^{A_i}_{G_{aug}}(.|do(x), Pa(A_i)) = m^{A_i}_{G_{aug}}(.|do(x)) \oplus (\underset{Pa_j(A_i) \in PA(A_i)}{\oplus} m^{A_i}_{G_{aug}}(.|Pa_j(A_i)))$$
(14)

Example 5 *Let us continue with the two events smoking S and having yellow teeth T. The distribution of the target variable T is computed from the combined effect of the intervention with the original cause S. Let us consider $m^V_{G_{aug}}(.|do(t_1))$ and $m^V_{G_{aug}}(.|s_i)$ as depicted in Fig. 6. After combining them using Dempster's rule of combination, we find the same conditional table of the one of Fig. 7 where beliefs where defined from the beginning for all parents.*

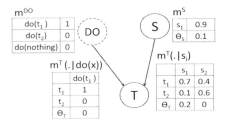
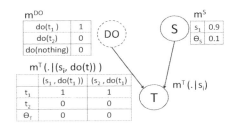

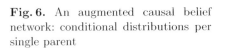

Fig. 6. An augmented causal belief network: conditional distributions per single parent

Fig. 7. An augmented causal belief network: conditional distributions for all parents

5   Conclusion

This paper provides a mutilated and an augmented based approach to handle interventions on a causal belief network where conditional distributions are defined per single parent. By reducing the size of the conditional tables, our model allows to simplify knowledge acquisition and storage. Doing this way is assuming that conditional distribution can be defined by different local sources. Consequently, we have a source given the DO node, i.e., the intervention and a source or multiple sources given the initial causes. For that, we have modified the distributions defined in the context of the initial causes when beliefs are defined per edge.

As future works, we intend to investigate on propagation algorithms to deal with interventions in causal belief networks where conditional beliefs are defined per single parent.

References

1. Ben Amor, N., Benferhat, S., Mellouli, K.: Anytime propagation algorithm for min-based possibilistic graphs. Soft Computing 8(2), 150–161 (2003)
2. Ben Yaghlane, B., Mellouli, K.: Inference in directed evidential networks based on the transferable belief model. International Journal of Approximate Reasoning 48, 399–418 (2008)
3. Benferhat, S.: Interventions and belief change in possibilistic graphical models. Artificial Intelligence 174(2), 177–189 (2010)
4. Boukhris, I., Elouedi, Z., Benferhat, S.: Handling interventions with uncertain consequences in belief causal networks. In: Greco, S., Bouchon-Meunier, B., Coletti, G., Fedrizzi, M., Matarazzo, B., Yager, R.R. (eds.) IPMU 2012, Part III. CCIS, vol. 299, pp. 585–595. Springer, Heidelberg (2012)
5. Boukhris, I., Elouedi, Z., Benferhat, S.: Dealing with external actions in causal belief networks. International Journal of Approximate Reasoning, 978–999 (2013)
6. Fonck, P.: Conditional independence in possibility theory. In: Uncertainty in Artificial Intelligence, pp. 221–226 (1994)
7. Pearl, J.: Graphical Models for probabilistic and causal reasonning. In: Gabbay, D.M., Smets, P. (eds.) The Handbook of Defeasible Resonning and Uncertainty Management Systems, pp. 367–389 (1998)
8. Pearl, J.: Causality: Models, Reasonning and Inference. Cambridge University Press (2000)
9. Shafer, G.: A Mathematical Theory of Evidence. Princeton Univ. Press, Princeton (1976)
10. Spirtes, P., Glymour, C., Scheines, R.: Causation, prediction, and search, vol. 1. MIT Press (2001)
11. Xu, H., Smets, P.: Ph. Smets. Reasoning in evidential networks with conditional belief functions. International Journal of Approximate Reasoning 14, 155–185 (1996)

An Extension of the Analytic Hierarchy Process Method under the Belief Function Framework

Amel Ennaceur[1,2], Zied Elouedi[1], and Eric Lefevre[2]

[1] LARODEC, University of Tunis, Institut Supérieur de Gestion, Tunisia
amel_naceur@yahoo.fr, zied.elouedi@gmx.fr
[2] Univ. Lille Nord of France, UArtois EA 3926 LGI2A, France
eric.lefevre@univ-artois.fr

Abstract. In this paper, an extension of the belief Analytic Hierarchy Process (AHP) method is proposed, based on the belief function framework. It takes into account the fact that the pair-wise comparison between criteria and alternatives may be uncertain and imprecise. Therefore, it introduces a new way to cope with expert judgments. Thus to express his preferences, the decision maker is allowed to use a belief assessment instead of exact ratios. The proposed extension also models the relationship between the alternative and criterion levels through conditional beliefs. Numerical examples explain in detail and illustrate the proposed approach.

1 Introduction

Analytic Hierarchy Process (AHP) method [5] is one of the widely preferred multi-criteria decision making (MCDM) methods and has successfully been applied to many practical problems. Using this approach, the decision maker models a problem as a hierarchy of criteria and alternatives. Then, the expert assesses the importance of each element at each level using a pair-wise comparison matrix, where elements are compared to each other.

Though its main purpose is to capture the expert's knowledge, the standard AHP still cannot reflect the human thinking style. It is often criticized for its use of an unbalanced scale of estimations and its inability to adequately handle the uncertainty and imprecision associated with the mapping of the decision maker's perception to a crisp number [4].

In order to model imperfect judgments, the AHP method was modified by many researchers. Under the belief functions framework, Beynon et al. have proposed a method called the DS/AHP method [1] comparing not only single alternatives but also groups of alternatives. Besides, several works has been defined by Utkin [10]. Also, Ennaceur et al. [2] [3] have developed the belief AHP approach that compares groups of criteria to subsets of alternatives. Then, they model the causality relationship between these groups of alternatives and criteria.

Taking into account the above, we propose an extension of the belief AHP method [3], a Multi-Criteria Decision Making (MCDM) method under the belief

A. Laurent et al. (Eds.): IPMU 2014, Part III, CCIS 444, pp. 233–242, 2014.

function framework. On the one hand, our proposed method takes into account the conditional relationships between alternatives and criteria. In fact, our aim is to more imitate the expert reasoning since he tries to express his preferences over the sets of alternatives regarding each criterion and not regardless of the criteria. Consequently, we try to represent the influences of the criteria on the evaluation of alternatives. On the other hand, our method takes into account the fact that the pair-wise comparison may be uncertain and imprecise. Therefore, it introduces a new way to cope with expert judgments. Thus to express his assessments, the decision maker is allowed to use subjective assessments instead of using numerical values. Then, a preference degree may be assigned to each expert's response. With our method, the expert does not require to complete all the comparison matrix. He can then derive priorities from incomplete set of judgments. Therefore, a new procedure is employed, he only selects the related linguistic variable to indicate whether a criterion or alternative was more or less important to its partner by "yes" or "no".

The proposed method uses the pair-wise comparisons with the minimal information. Therefore, using our proposed approach, we cannot get the best alternative but at least we can choose the most cautious one.

In what follows, we first present some definitions needed for belief function context. Next, we describe the belief AHP method in section 3. Then, section 4 details our new MCDM method, and gives an example to show its application. Finally, section 5 concludes the paper.

2   Belief Function Theory

2.1   Basic Concepts

The Transferable Belief Model (TBM) is a model to represent quantified belief functions [9]. Let Θ be the frame of discernment representing a finite set of elementary hypotheses related to a problem domain. We denote by 2^{Θ} the set of all the subsets of Θ [6].

The impact of a piece of evidence on the different subsets of the frame of discernment Θ is represented by the so-called basic belief assignment (bba) [6]. A bba is a function denoted by m that assigns a value in $[0, 1]$ to every subset A of Θ such that:

$$\sum_{A \subseteq \Theta} m(A) = 1 .$$
(1)

The value $m(A)$, named a basic belief mass (bbm), represents the portion of belief committed exactly to the event A.

2.2   Operations on the Product Space

Vacuous Extension. This operation is useful, when the referential is changed by adding new variables. Thus, a marginal mass function m^{Θ} defined on Θ will be expressed in the frame $\Theta \times \Omega$ as follows [7]:

$$m^{\Theta \uparrow \Theta \times \Omega}(C) = m^{\Theta}(A) \quad \text{if } C = A \times \Omega, A \subseteq \Theta .$$
(2)

Marginalization. Given a mass distribution defined on $\Theta \times \Omega$, marginalization corresponds to mapping over a subset of the product space by dropping the extra coordinates. The new belief defined on Θ is obtained by [7]:

$$m^{\Theta \times \Omega \downarrow \Theta}(A) = \sum_{\{B \subseteq \Theta \times \Omega | B^{\downarrow \Theta} = A)\}} m^{\Theta \times \Omega}(B), \forall A \subseteq \Theta . \quad (3)$$

$B^{\downarrow \Theta}$ denotes the projection of B onto Θ.

Ballooning Extension. Let $m^{\Theta}[\omega]$ represents your beliefs on Θ conditionnally on ω a subset of Ω. To get rid of conditioning, we have to compute its ballooning extension. The ballooning extension is defined as [7]:

$$m^{\Theta}[\omega]^{\Uparrow \Theta \times \Omega}(A \times \omega \cup \Theta \times \bar{\omega}) = m^{\Theta}[\omega](A), \forall A \subseteq \Theta . \quad (4)$$

3 Belief AHP Method

The belief AHP method is a MCDM method that combines the AHP approach with the belief function theory [3]. This method investigates some ways to define the influences of the criteria on the evaluation of alternatives.

3.1 Identification of the Candidate Alternatives and Criteria

Let $\Omega = \{c_1, \ldots, c_m\}$ be a set of criteria, and let C_k be the notation of a subset of Ω. The groups of criteria can be defined as [2]:

$$\forall \, k, j | C_k, C_j \in 2^{\Omega}, C_k \cap C_j = \emptyset \text{ and } \cup_j C_j = \Omega \text{ (with } C_j \text{ exclusive).} \quad (5)$$

This method suggests to allow the expert to express his opinions on groups of criteria instead of single one. So, he chooses these subsets by assuming that criteria having the same degree of preference are grouped together. On the other hand and similarly to the criterion level, the decision maker compares not only pairs of single alternatives but also sets of alternatives between each other ($\Theta = \{a_1, \ldots, a_n\}$ is a set of alternatives)[2].

3.2 Pair-Wise Comparisons and Preference Elicitation

After identifying the set of criteria and alternatives, the weights of each element have to be defined. The expert has to provide all the pair-wise comparisons matrices. In this study, Saaty's scale is chosen in order to evaluate the importance of pairs of grouped elements in terms of their contribution. Thus, the priority vectors are then generated using the eigenvector method.

3.3   Updating the Alternatives Priorities

Within this framework, we have $C_i \subseteq 2^\Omega$ and we have the criterion priority vector is regarded as a bba, denoted by m^Ω.

Furthermore, Belief AHP tries to model the influences of the criteria on the evaluation of alternatives by conditional belief. So, given a pair-wise comparison matrix which compares the sets of alternatives according to a specific criterion, a conditional bba can be represented by: $m^\Theta[c_j](A_k) = w_k$, $\forall A_k \subseteq 2^\Theta$ and $c_j \in \Omega$ where $m^\Theta[c_j](A_k)$ means that we know the belief about A_k regarding c_j.

Then, the aggregation procedure can be represented as follows. In fact, priorities concerning criteria and groups of criteria are defined on the frame of discernment Ω, whereas the sets of alternatives are defined on Θ. The idea was to standardize the frame of discernment. First, at the criterion level, the bba that represents criteria weights is extended from Ω to $\Theta \times \Omega$:

$$m^{\Omega \uparrow \Theta \times \Omega}(B) = m^\Omega(C_i) \quad B = \Theta \times C_i, C_i \subseteq \Omega . \tag{6}$$

Second, at the alternative level, the idea was to use the deconditionalization process in order to transform the conditional belief into a new belief function. In this case, the ballooning extension technique is applied:

$$m^\Theta[c_j]^{\Uparrow \Theta \times \Omega}(A_k \times c_j \cup \Theta \times \bar{c}_j) = m^\Theta[c_j](A_k), \forall A_k \subseteq \Theta . \tag{7}$$

Once the frame of discernment $\Theta \times \Omega$ is formalized, the belief AHP approach proposes to combine the obtained bba with the importance of their respective criteria to measure their contribution using the conjunctive rule of combination $\textcircled{n}$ and we get [8]:

$$m^{\Theta \times \Omega} = \left[\textcircled{n}_{j=1,\ldots,m} m^\Theta[c_j]^{\Uparrow \Theta \times \Omega} \right] \textcircled{n} m^{\Omega \uparrow \Theta \times \Omega} . \tag{8}$$

Finally, to choose the best alternatives, this method proposes to marginalize the obtained bba (in the previous step) on Θ (frame of alternatives) by transferring each mass $m^{\Theta \times \Omega}$ to its projection on Θ. Then, the pignistic probabilities [8] are used to make our choices:

$$BetP(a_j) = \sum_{A_i \subseteq \Theta} \frac{|a_j \cap A_i|}{|A_i|} \frac{m^{\Theta \times \Omega \downarrow \Theta}(A_i)}{(1 - m^{\Theta \times \Omega \downarrow \Theta}(\emptyset))}, \forall a_j \in \Theta . \tag{9}$$

3.4   Example

To describe this approach, we consider the problem of "purchasing a car" presented in [3]. Suppose that this problem involves four criteria: $\Omega = \{$Comfort (c_1), Style (c_2), Fuel (c_3), Quietness $(c_4)\}$, and three selected alternatives: $\Theta = \{$Peugeot(p), Renault(r), Ford$(f)\}$. For more details see [3].

At the criterion level, the criterion weights are expressed by a basic belief assessment (bba). We get: $m^\Omega(\{c_1\}) = 0.58$, $m^\Omega(\{c_4\}) = 0.32$ and $m^\Omega(\{c_2, c_3\}) = 0.1$.

Table 1. Priorities values

c_1	Priority	c_2	Priority	c_3	Priority	c_4	Priority
$\{p\}$	0.806	$\{p\}$	0.4	$\{r\}$	0.889	$\{f\}$	0.606
$\{p,r,f\}$	0.194	$\{r,f\}$	0.405	$\{p,r,f\}$	0.111	$\{p,r,f\}$	0.394
		$\{p,r,f\}$	0.195				

Next, we propose to model the alternative score by means of conditional bba (see Table 1).

According to the belief AHP approach, the next step is to standardize the criterion and the alternative frames of discernment. For the criterion level, the resulting bba's is summarized in Table 2.

Table 2. Vacuous extension of bba

bbm	Vacuous extension	Values
$m^{\Omega}(\{c_1\})$	$\{(p,c_1),(r,c_1),(f,c_1)\}$	0.58
$m^{\Omega}(\{c_4\})$	$\{(p,c_4),(r,c_4),(f,c_4)\}$	0.32
$m^{\Omega}(\{c_2,c_3\})$	$\{(p,c_2),(r,c_2),(f,c_2),(p,c_3),(r,c_3),(f,c_3)\}$	0.1

After normalizing the criteria's bba, the next step is to transform the conditional belief into joint distribution using Equation 7 (see Table 3).

Table 3. Ballooning extension of conditional bba

Conditional bbm	Ballooning extension	Values
$m^{\Theta}[c_1](\{p\})$	$\{(p,c_1),(p,c_2),(p,c_3),(p,c_4),(r,c_2),$ $(r,c_3),(r,c_4),(f,c_2),(f,c_3),(f,c_4)\}$	0.806
$m^{\Theta}[c_1](\{p,r,f\})$	$\{(p,c_1),(p,c_2),(p,c_3),(p,c_4),$ $(r,c_1),(r,c_2),(r,c_3),(r,c_4),(f,c_1),(f,c_2),(f,c_3),(f,c_4)\}$	0.194

As explained before, once the ballooning extensions are obtained, we can apply Equation 8, to combine the obtained bba with the criterion weights (bba).

Next, to choose the best alternatives, we must define the beliefs over the frame of alternatives Θ and the pignistic probabilities can be computed. We get: $BetP(p) = 0.567$, $BetP(r) = 0.213$ and $BetP(f) = 0.220$.

As a consequence, the alternative "Peugeot" is the recommended car since it has the highest values.

4   An Extension of the Belief AHP Method

The Belief AHP method is an interesting tool for solving multi-criteria decision problems. It provides the expert the possibility to select only some subsets of alternatives and groups of criteria.

However, this approach suffers from some weaknesses. In fact, in reality, the elicitation of preferences may be rather difficult since expert would not be able to efficiently express any kind of preference degree between the available alternatives and criteria. Therefore, the belief AHP method is extended to handle the presented problems.

4.1   Belief Pair-Wise Comparison

Under this approach, a new elicitation procedure is introduced. Thus to model his assessments, the decision maker has to express his opinions qualitatively. He indicated whether a criterion (or alternative) was more or less important to its partner by "yes" or "no". Moreover, we accept that the expert may define uncertain or even unknown assessments. Indeed, we assume that each subset of criteria is described by a basic belief assignment defined on the possible responses. For instance, in a problem of purchasing a car, the following type of subjective judgments was frequently used: "the comfort criterion is evaluated to be more important than style with a confidence degree of 0.8". In fact, the decision maker responses to the question "is comfort criterion important regarding the style criterion?". Thus, the answer was: comfort criterion is more preferable than style criterion and 0.8 is referred to the degree of belief. Then, to compute the criteria weight, we describe a new pair-wise comparison procedure where the following steps must be respected:

1. The first step is to model the pair-wise comparison matrix. Let d_{ij} is the entry from the ith column of pair-wise comparison matrix (d_{ij} represents the different bbm's of the identified bba).

$$\text{If } m_j^{\Omega_{C_i}}(.) = d_{ij}, \text{ then } m_i^{\Omega_{C_j}}(.) = \bar{m}_j^{\Omega_{C_i}}(.) = d_{ij} \tag{10}$$

where $m_j^{\Omega_{C_i}}$ represents the importance of C_i with respect to the subset of criteria C_j (for simplicity, we denote the subset of criteria by j instead of C_j), $i \neq j$, and $\bar{m}$ is the negation of m. The negation $\bar{m}$ of a bba m is defined as $\bar{m}(A) = m(\bar{A}), \forall A \subset \Omega$.

As regarding the previous example, if we have "the comfort criterion (C) is evaluated to be more important than style criterion (S) with a confidence degree of 0.8", that is $m_S^{\Omega_C}(\{yes\}) = 0.8$, then we can say that "the style criterion is evaluated to be less important than comfort criterion with a confidence degree of 0.8": $m_C^{\Omega_S}(\{no\}) = 0.8$.

2. Once the pair-wise comparison matrix is completed, our objective is then to obtain the priority of each subset of criteria. The idea is to combine the obtained bba using the conjunctive rule of combination [8] $((m_1 \textcircled{$\cap$} m_2)(A) = \sum_{B,C \subseteq \Theta, B \cap C = A} m_1(B) m_2(C))$.

Indeed, this function is chosen since we can regard each subset of criteria as a distinct source of information which provides distinct pieces of evidence. We will get the following bba:

$$m^{\Omega_{C_i}} = \textcircled{$\cap$} m_j^{\Omega_{C_i}}, \text{ where } j = \{1, \ldots, k\} \tag{11}$$

At this stage, we want to know which criterion is the most important. In fact, the obtained bba measures the confidence degree assigned to a specific criterion regarding the overall criteria. However, these obtained bba represents the belief over all possible answers (yes or no). The idea is then to standardize all the frames of discernment. Obviously, we propose to use the concept of refinement operations [6], which allows to establish relationships between different frames of discernment in order to express beliefs on any-one of them. The objective consists in obtaining one frame of discernment Ω from the set Ω_{C_k} by splitting some or all of its events:

$$m^{\Omega_{C_k} \uparrow \Omega}(\rho_k(\omega)) = m^{\Omega_{C_k}}(\omega) \qquad \forall \omega \subseteq \Omega_{C_k} \qquad (12)$$

where the mapping ρ_k from Ω_{C_k} to Ω is a refinement, and $\rho_k(\{yes\}) = \{C_k\}$ and $\rho_k(\{no\}) = \overline{\{C_k\}}$.

3. Finally, the obtained bba $m^{\Omega_{C_k} \uparrow \Omega}$ can be combined using the conjunctive rule of combination in order to get m^{Ω}.

The similar process is repeated to get the alternatives priorities $m^{\Theta}[c_k](A_i)$ representing the opinions-beliefs of the expert about his preferences regarding the set of alternatives.

Then, the vacuous extension is used at the criterion level and the ballooning extension is assumed at the alternative level in order to standardize the frame of discernment. So, the vacuous extension is used to extend the frame of criteria to the frame of alternatives and the ballooning is applied for the deconditioning process. After that, these obtained bba can be combined. Next, the marginalization technique is applied by transferring each mass to its projection on Θ. The final priority is then computed using the pignistic probabilities to make our choice.

4.2 Illustrative Example

Let us consider the previous example (see Section 3.5). After identifying the subsets of criteria and alternatives, the pair-wice comparison matrices should be constructed.

Computing the Criteria Weights. After collecting the expert beliefs, we have generated the following associated belief functions (see Table 4).

From Table 4, the expert may say that $\{c_1\}$ is evaluated to be more important than $\{c_4\}$ with a confidence degree of 0.4. That means, 0.4 of beliefs are exactly committed to the criterion $\{c_1\}$ is more important than $\{c_4\}$, whereas 0.6 is assigned to the whole frame of discernment (ignorance).

Then, the next step consists in combining the bba's corresponding to each criterion using the Equation 11. The obtained bba is reported in Table 5.

Subsequently, we proceed now with the standardization of our frame of discernment. By applying the Equation 12, we get for example: $m^{\Omega_{(c_1)} \uparrow \Omega}(\{c_1\}) =$

Table 4. The weights preferences assigned to the criteria according to the expert's opinion

	$\{c_1\}$	$\{c_4\}$	$\Omega_1 = \{c_2, c_3\}$
$\{c_1\}$	$m_{\{c_1\}}^{\Omega_{\{c_1\}}}(\Omega_{\{c_1\}}) = 1$	$m_{\{c_4\}}^{\Omega_{\{c_1\}}}(\{yes\}) = 0.4$ $m_{\{c_4\}}^{\Omega_{\{c_1\}}}(\Omega_{\{c_1\}}) = 0.6$	$m_{\Omega_1}^{\Omega_{\{c_1\}}}(\{yes\}) = 0.9$ $m_{\Omega_1}^{\Omega_{\{c_1\}}}(\Omega_{\{c_1\}}) = 0.1$
$\{c_4\}$	$m_{\{c_1\}}^{\Omega_{\{c_4\}}}(\{no\}) = 0.4$ $m_{\{c_1\}}^{\Omega_{\{c_4\}}}(\Omega_{\{c_4\}}) = 0.6$	$m_{\{c_4\}}^{\Omega_{\{c_4\}}}(\Omega_{\{c_4\}}) = 1$	$m_{\Omega_1}^{\Omega_{\{c_4\}}}(\{no\}) = 0.3$ $m_{\Omega_1}^{\Omega_{\{c_4\}}}(\Omega_{\{c_4\}}) = 0.7$
$\Omega_1 = \{c_2, c_3\}$	$m_{\{c_1\}}^{\Omega_{\Omega_1}}(\{no\}) = 0.9$ $m_{\{c_1\}}^{\Omega_{\Omega_1}}(\Omega_{\Omega_1}) = 0.1$	$m_{\{c_4\}}^{\Omega_{\Omega_1}}(\{yes\}) = 0.3$ $m_{\{c_4\}}^{\Omega_{\Omega_1}}(\Omega_{\Omega_1}) = 0.7$	$m_{\Omega_1}^{\Omega_{\Omega_1}}(\Omega_{\Omega_1}) = 1$

Table 5. Belief pair-wise matrix: Partial combination

	$\{c_1\}$	$\{c_4\}$	$\Omega_1 = \{c_2, c_3\}$
Weight	$m^{\Omega_{\{c_1\}}}(\{yes\}) = 0.94$ $m^{\Omega_{\{c_1\}}}(\Omega_{\{c_1\}}) = 0.06$	$m^{\Omega_{\{c_4\}}}(\{no\}) = 0.58$ $m^{\Omega_{\{c_4\}}}(\Omega_{\{c_4\}}) = 0.42$	$m^{\Omega_{\Omega_1}}(\{yes\}) = 0.03$ $m^{\Omega_{\Omega_1}}(\{no\}) = 0.63$ $m^{\Omega_{\Omega_1}}(\emptyset) = 0.27$ $m^{\Omega_{\Omega_1}}(\Omega_{\Omega_1}) = 0.07$

Table 6. Belief pair-wise matrix: Refinement

	$\{c_1\}$	$\{c_4\}$	$\Omega_1 = \{c_2, c_3\}$
Weight	$m_{\{c_1\}}^{\Omega}(\{c_1\}) = 0.94$ $m_{\{c_1\}}^{\Omega}(\Omega) = 0.06$	$m_{\{c_4\}}^{\Omega}(\{c_1, c_2, c_3\}) = 0.58$ $m_{\{c_4\}}^{\Omega}(\Omega) = 0.42$	$m_{\{c_2, c_3\}}^{\Omega}(\{c_2, c_3\}) = 0.03$ $m_{\{c_2, c_3\}}^{\Omega}(\{c_1, c_4\}) = 0.63$ $m_{\{c_2, c_3\}}^{\Omega}(\emptyset) = 0.27$ $m_{\{c_2, c_3\}}^{\Omega}(\Omega) = 0.07$

$m^{\Omega_{\{c_1\}}}(\{yes\})$. To simplify, we can note by $m_{\{c_1\}}^{\Omega}$ the bba $m^{\Omega_{\{c_1\}}\uparrow\Omega}$. These bba's are presented in Table 6.

At this stage, the obtained bba's can be combined using the conjunctive rule of combination. We get: $m^{\Omega}(\emptyset) = 0.2982$, $m^{\Omega}(\{c_1\}) = 0.6799$, $m^{\Omega}(\{c_2, c_3\}) = 0.0018$, $m^{\Omega}(\{c_1, c_2, c_3\}) = 0.0024$, $m^{\Omega}(\{c_1, c_4\}) = 0.0159$ and $m^{\Omega}(\Omega) = 0.0018$.

Computing the Alternatives Priorities. Like the criterion level, the judgments between decision alternatives over different criteria are dealt within an identical manner. For example, to evaluate the alternatives according to the criterion c_1 we get Table 7.

As in the criterion level, for the subset of alternatives $\{p\}$, we use Equation 11 in order to combine the obtained bba: $m^{\Theta_{\{p\}}}[c_1] = m_{\{p\}}^{\Theta_{\{p\}}}[c_1] \textcircled{\odot} m_{\{r,f\}}^{\Theta_{\{p\}}}[c_1]$ $(m^{\Theta_{\{p\}}}[c_1](\{yes\}) = 0.95$ and $m^{\Theta_{\{p\}}}[c_1](\{\Theta_{\{p\}}\}) = 0.05)$. Then, a similar process is repeated for the rest of alternatives, and we get $m^{\Theta_{\{r,f\}}}[c_1]$ $(m^{\Theta_{\{r,f\}}}[c_1](\{no\}) = 0.95$ and $m^{\Theta_{\{r,f\}}}[c_1](\Theta_{\{r,f\}}) = 0.05)$.

Subsequently, we proceed now with the standardization of our frame of discernment. By applying Equation 12, we get the following: $m^{\Theta_{\{p\}}\uparrow\Theta}[c_1](\{p\}) =$

Table 7. Belief pair-wise matrix regarding c_1 criterion

c_1	$\{p\}$	$\{r,f\}$
$\{p\}$	$m_{\{p\}}^{\Theta_{\{p\}}}[c_1](\Theta_{\{p\}}) = 1$	$m_{\{r,f\}}^{\Theta_{\{p\}}}[c_1](\{yes\}) = 0.95$ $m_{\{r,f\}}^{\Theta_{\{p\}}}[c_1](\Theta_{\{p\}}) = 0.05$
$\{r,f\}$	$m_{\{p\}}^{\Theta_{\{r,f\}}}[c_1](\{no\}) = 0.95$ $m_{\{p\}}^{\Theta_{\{r,f\}}}[c_1](\Theta_{\{r,f\}}) = 0.05$	$m_{\{r,f\}}^{\Theta_{\{r,f\}}}[c_1](\Theta_{\{r,f\}}) = 1$

0.95 and $m^{\Theta_{\{p\}}\uparrow\Theta}[c_1](\Theta) = 0.05$. Also, $m^{\Theta_{\{r,f\}}\uparrow\Theta}[c_1](\{p\}) = 0.95$ and $m^{\Theta_{\{r,f\}}\uparrow\Theta}[c_1](\Theta) = 0.05$.

Finally, the obtained bba's can be directly combined using the conjunctive rule of combination. For simplicity, we denote $m^{\Theta_{\{p\}}\uparrow\Theta}[c_1]$ by $m^{\Theta}[c_1]$, we get: $m^{\Theta}[c_1](\{p\}) = 0.9975$ and $m^{\Theta}[c_1](\{\Theta\}) = 0.0025$.

Then, as shown in the previous step, the computation procedure is repeated for the rest of comparison matrices.

Updating the Alternatives Priorities. As shown in the previous example, at the criterion level, the vacuous extension is used to standardize the frame of discernment $m^{\Omega\uparrow\Theta\times\Omega}$. At the alternative level, the ballooning extension is applied $m^{\Theta}[c_j]^{\uparrow\Theta\times\Omega}$. Then, the obtained bba can be directly combined by using Equation 8 as exposed in Table 8.

Table 8. The obtained bba: $m^{\Theta\times\Omega}$

$m^{\Theta\times\Omega}$	bbm	$m^{\Theta\times\Omega}$	bbm
$\{(p,c_1),(f,c_1),(r,c_1)\}$	0.28	$\{(p,c_1),(f,c_1)\}$	0.16
$\{(p,c_1)\}$	0.008	$\{(r,c_2),(r,c_3),(f,c_2),(p,c_2)\}$	0.03
$\{(f,c_4)\}$	0.0016	$\{(p,c_4),(f,c_4),(r,c_4)\}$	0.11
$\{(p,c_2),(f,c_2),(r,c_2),(p,c_3),(f,c_3)\}$	0.007	$\emptyset$	0.4034

To choose the best alternatives, we must define our beliefs over the frame of alternatives. As a result, the obtained bba is marginalized on Θ, we obtain the following distribution: $m^{\Theta\times\Omega\downarrow\Theta}(\{p,r,f\}) = 0.427$, $m^{\Theta\times\Omega\downarrow\Theta}(\{p\}) = 0.008$, $m^{\Theta\times\Omega\downarrow\Theta}(\{f\}) = 0.0016$, $m^{\Theta\times\Omega\downarrow\Theta}(\{p,f\}) = 0.16$ and $m^{\Theta\times\Omega\downarrow\Theta}(\emptyset) = 0.4034$.

We can now calculate the overall performance for each alternative and determine its corresponding ranking by computing the pignistic probabilities: $BetP(p) = 0.3863$, $BetP(r) = 0.3752$ and $BetP(f) = 0.2385$.

As a consequence, the alternative "Peugeot" is the recommended car since it has the highest values. The alternative r may also be chosen since it has a value close to p. For the sake of comparison, we have obtained the same best alternative as in the previous example. This would give the expert reasonable assurance in decision making. Our objective is then not to obtain the best alternative but to identify the most cautious one since it is defined with less information.

5   Conclusion

In this paper, the proposed method has extended the belief AHP model into more uncertain environment. Indeed, our approach develops a new pair-wise comparison technique in order to facilitate the elicitation process and to handle the problem of uncertainty. It leads to more simple comparison procedure without eliciting additional information. In fact, experts do not need to provide precise comparison judgments. They select only some subsets of alternatives in accordance with a certain criterion, and groups of criteria. Then, the proposed method models the imprecise judgments based on an appropriate mathematical framework of the belief function theory.

References

1. Beynon, M., Curry, B., Morgan, P.: The Dempster-Shafer theory of evidence: An alternative approach to multicriteria decision modelling. OMEGA 28(1), 37–50 (2000)
2. Ennaceur, A., Elouedi, Z., Lefevre, E.: Handling partial preferences in the belief AHP method: Application to life cycle assessment. In: Pirrone, R., Sorbello, F. (eds.) AI*IA 2011. LNCS (LNAI), vol. 6934, pp. 395–400. Springer, Heidelberg (2011)
3. Ennaceur, A., Elouedi, Z., Lefevre, E.: Reasoning under uncertainty in the AHP method using the belief function theory. In: Greco, S., Bouchon-Meunier, B., Coletti, G., Fedrizzi, M., Matarazzo, B., Yager, R.R. (eds.) IPMU 2012, Part IV. CCIS, vol. 300, pp. 373–382. Springer, Heidelberg (2012)
4. Holder, R.D.: Some comments on Saaty's AHP. Management Science 41, 1091–1095 (1995)
5. Saaty, T.: A scaling method for priorities in hierarchical structures. Journal of Mathematical Psychology 15, 234–281 (1977)
6. Shafer, G.: A Mathematical Theory of Evidence. Princeton University Press (1976)
7. Smets, P.: Belief functions: the disjunctive rule of combination and the generalized bayesian theorem. International Journal of Approximate Reasoning 99, 1–35 (1993)
8. Smets, P.: The combination of evidence in the Transferable Belief Model. IEEE Pattern Analysis and Machine Intelligence 12, 447–458 (1990)
9. Smets, P., Kennes, R.: The Transferable Belief Model. Artificial Intelligence 66, 191–234 (1994)
10. Utkin, L.V.: A new ranking procedure by incomplete pairwise comparisons using preference subsets. Intelligent Data Analysis 13(2), 229–241 (2009)

Pairwise and Global Dependence
in Trivariate Copula Models

Fabrizio Durante[1], Roger B. Nelsen[2],
José Juan Quesada-Molina[3], and Manuel Úbeda-Flores[4]

[1] School of Economics and Management
Free University of Bozen-Bolzano, Bolzano, Italy
fabrizio.durante@unibz.it
[2] Department of Mathematical Sciences
Lewis and Clark College, Portland, OR 97219, USA
nelsen@lclark.edu
[3] Department of Applied Mathematics
University of Granada, Granada, Spain
jquesada@ugr.es
[4] Department of Mathematics
University of Almería, Almería, Spain
e-mail: mubeda@ual.es

Abstract. We investigate whether pairwise dependence properties related to all the bivariate margins of a trivariate copula imply the corresponding trivariate dependence property. The main finding is that, in general, information about the pairwise dependence is not sufficient to infer some aspects of global dependence. In essence, dependence is a multi-facet property that cannot be easily reduced to simplest cases.

Keywords: Copula, Dependence, Stochastic model.

1 Introduction

Copula models have enjoyed a great popularity, especially in recent years, since they allow us to express the link (i.e. the dependence) among different random variables in a concise, yet powerful, way (see, for instance, [9,10,15] and references therein). However, while our state-of-the-art knowledge of copula theory seems quite established in the bivariate case, the higher–dimensional case still poses several challenging problems. In particular, the study of dependence properties of a random vector requires special care when we pass from dimension 2 to dimension 3 (or more), since possible extensions are not so obvious.

Here, we are interested whether it is possible that (some) trivariate models preserve dependence properties that are related to their bivariate margins. For instance, it is well known that pairwise independence among random variables does not correspond to global independence of all vector components. In the same way, we will show that positive pairwise dependence rarely corresponds to global positive dependence. This poses the natural question whether it is possible

A. Laurent et al. (Eds.): IPMU 2014, Part III, CCIS 444, pp. 243–251, 2014.

to infer some multivariate property of our random vector starting with a lower-dimensional and marginal information about the bivariate margins. This problem has important practical implications since it warns against the indiscriminate use of step-wise statistical procedures for fitting copulas to data.

In this paper, by using basic notions of positive dependence, we present some illustrations about the preservation of pairwise properties in the trivariate case. The main message is that stochastic dependence has so many facets that cannot be recovered from its lower–dimensional projections.

2   Pairwise and Global Independence and Comonotonicity

In this paper, we assume that the reader is familiar with basic properties of copulas and quasi–copulas, as presented, for instance, in [15]. Moreover, the results are mainly presented in the trivariate case for sake of illustration.

We start with the concept of independence for random variables. We recall that a d–dimensional continuous random vector $\mathbf{X} = (X_1, \ldots, X_d)$ is formed by independent components if, and only if, the copula of $\mathbf{X}$ is given by $\Pi_d(\mathbf{u}) = \prod_{i=1}^d u_i$.

As known, if C is a 3–copula whose bivariate margins are equal to Π_2, then C is not necessarily equal to Π_3. Consider, for instance, the copula

$$C(u_1, u_2, u_3) = u_1 u_2 u_3 (1 + \alpha(1 - u_1)(1 - u_2)(1 - u_3)) \tag{1}$$

for any $\alpha \in [-1, 1]$, with $\alpha \neq 0$.

In fact, if we consider the class $\mathcal{P}_{\Pi_2}^{3,2}$ of all 3–copulas whose bivariate margins are all equal to Π_2, the elements of $\mathcal{P}_{\Pi_2}^{3,2}$ actually may vary according to the following bounds (see [2,4,16,19]).

Proposition 1. *If $C \in \mathcal{P}_{\Pi_2}^{3,2}$, then, for every u_1, u_2 and u_3 in $[0,1]$,*

$$C_L(u_1, u_2, u_3) \leq C(u_1, u_2, u_3) \leq C_U(u_1, u_2, u_3),$$

where

$$C_L(u_1, u_2, u_3) = \max\{u_1 W_2(u_2, u_3), u_2 W_2(u_1, u_3), u_3 W_2(u_1, u_2)\} \tag{2}$$

$$C_U(u_1, u_2, u_3) = \min\{u_1 u_2, u_1 u_3, u_2 u_3, (1 - u_1)(1 - u_2)(1 - u_3) + u_1 u_2 u_3\}, \tag{3}$$

and $W_2(u, v) = \max\{u + v - 1, 0\}$ is the Fréchet–Hoeffding lower bound for 2–copulas.

Notice that both C_L and C_U are proper 3–quasi–copulas, as noted in [19, Example 10].

In order to give an idea about the way the bounds on special classes of (quasi)–copulas may improve our information with respect to the general case, we may consider a measure recently introduced in [16]. This (multivariate) measure is defined for any d–quasi-copula Q by

$$\mu_d(Q) = \frac{(d+1)! \int_{[0,1]^d} Q(\mathbf{u}) \, d\mathbf{u} - 1}{d! - 1}.$$

Basically, μ_d measures the distance between a d-quasi-copula Q and the d–dimensional Fréchet–Hoeffding lower bound $W_d(\mathbf{u}) = \max\{\sum_{i=1}^d u_i - d + 1, 0\}$. Note that $\mu_d(M_d) = 1$, where $M_d(\mathbf{u}) = \min\{u_1, \ldots, u_d\}$, and $\mu_d(W_d) = 0$. Thus, given two d-quasi-copulas Q_1 and Q_2 such that $Q_1(\mathbf{u}) \leq Q_2(\mathbf{u})$ for all $\mathbf{u}$ in $[0, 1]^d$, $\mu_d(Q_2) - \mu_d(Q_1)$ represents the normalized L_1-distance between Q_1 and Q_2. Note that if C_L and C_U are the 3–quasi-copulas given by (2) and (3), respectively, then $\mu_3(C_L) = 31/100$ and $\mu_3(C_U) = 49/100$. In this case the Fréchet-Hoeffding bounds have been narrowed by a factor of $\mu_3(C_U) - \mu_3(C_L) = 9/50$ (or 18%).

Together with the independence case, there are two other basic dependence properties of random vectors.

The first property is the comonotonicity. We say that a continuous random vector $\mathbf{X}$ is *comonotonic* if its copula is the Fréchet–Hoeffding upper bound M_d. The following result follows (see, for instance, [3, Theorem 3]).

Proposition 2. *Let C be a d–copula. Then $C = M_d$ if, and only if, all bivariate margins of C are equal to M_2.*

The second property is the counter-monotonicity. In two dimensions, we say that a continuous random vector $\mathbf{X}$ is *counter–comonotonic* if its copula is the Fréchet–Hoeffding lower bound W_2. However, this notion does not have a natural extension to the trivariate case, since a trivariate vector with all pairwise margins equal to W_2 does not exist (see also [13, section 2]).

3   Pairwise and Global Dependence

We start by considering the definition of orthant dependence (see [11]).

Definition 1. *Let $\mathbf{X} = (X_1, X_2, \cdots, X_d)$ be a d–dimensional continuous random vector $(d \geq 3)$. We say that:*

(a) $\mathbf{X}$ *is positively lower orthant dependent (PLOD) if*

$$P[\mathbf{X} \leq \mathbf{x}] \geq \prod_{i=1}^d P[X_i \leq x_i]$$

for all $\mathbf{x} = (x_1, x_2, \cdots, x_d) \in \mathbb{R}^d$.

(b) $\mathbf{X}$ *is is positively upper orthant dependent (PUOD) if*

$$P[\mathbf{X} \geq \mathbf{x}] \geq \prod_{i=1}^d P[X_i \geq x_i]$$

for all $\mathbf{x} = (x_1, x_2, \cdots, x_d) \in \mathbb{R}^d$.

Related notions of negative dependence like NLOD and NUOD can be given by changing the middle inequality sign in previous definitions.

If the random vector $\mathbf{X}$ has associated copula C, then we say that C is PLOD if $C(\mathbf{u}) \geq \Pi_d(\mathbf{u})$ for all $\mathbf{u} \in [0,1]^d$, and C is PUOD if $\overline{C}(\mathbf{u}) \geq \overline{\Pi}_d(\mathbf{u})$ for all $\mathbf{u} \in [0,1]^d$, where $\overline{C}$ denotes the *survival function* related to C (for formal definition, see [5]). We recall that the survival function of a 3–copula C is given, for every $\mathbf{u} \in [0,1]^3$, by

$$\overline{C}(\mathbf{u}) = 1 - u_1 - u_2 - u_3$$
$$+ C_{12}(u_1, u_2) + C_{13}(u_1, u_3) + C_{23}(u_2, u_3) - C(u_1, u_2, u_3), \qquad (4)$$

where C_{12}, C_{13} and C_{23} are the bivariate margins of C. Notice that, if $\widehat{C}$ is the survival copula related to C, then $\hat{C} = C$ (in fact, $\hat{C}(\mathbf{u}) = P[\mathbf{X} > \mathbf{1} - \mathbf{u}]$). Thus we have the following result.

Proposition 3. *Let C be a d–copula. Then, C is PLOD if and only if $\hat{C}$ is PUOD. Similarly, C is NLOD if and only if $\hat{C}$ is NUOD.*

Notice that, in the bivariate case, PLOD and PUOD coincide, and they are denoted by the term PQD (positively quadrant dependent). Analogously, NLOD and NUOD coincide and are denoted by NQD (negatively quadrant dependent).

We denote by $\mathcal{P}_{PQD}^{3,2}$ and $\mathcal{P}_{NQD}^{3,2}$ the class of all 3–copulas whose all bivariate margins are PQD and NQD, respectively. The class $\mathcal{P}_{PQD}^{3,2}$ (respectively, $\mathcal{P}_{NQD}^{3,2}$) includes the class of all trivariate copulas that are PLOD or PUOD (respectively, NLOD or NUOD): see, for instance, [14]. We denote the classes of 3–copulas that are PLOD, PUOD, NLOD and NUOD, respectively, by means of the symbols: $\mathcal{P}_{PLOD}^3$, $\mathcal{P}_{PUOD}^3$, $\mathcal{P}_{NLOD}^3$ and $\mathcal{P}_{NUOD}^3$.

Notice that $\mathcal{P}_{PLOD}^3 \setminus \mathcal{P}_{PUOD}^3 \neq \emptyset$ and $\mathcal{P}_{PUOD}^3 \setminus \mathcal{P}_{PLOD}^3 \neq \emptyset$. Analogous considerations hold for the related negative concepts. In fact, the following example holds.

Example 1. Let C be a member of the FGM family of 3-copulas given by

$$C(u_1, u_2, u_3) = u_1 u_2 u_3 [1 + \alpha_{12}\overline{u_1}\,\overline{u_2} + \alpha_{13}\,\overline{u_1}\,\overline{u_3} + \alpha_{23}\overline{u_2}\,\overline{u_3} + \alpha_{123}\overline{u_1}\,\overline{u_2}\,\overline{u_3}],$$

where, for every $t \in [0,1]$, $\bar{t} := 1 - t$, and the parameters satisfy the following inequality

$$1 + \sum_{1 \leq i < j \leq 3} \alpha_{ij}\xi_i\xi_j + \alpha_{123}\xi_1\xi_2\xi_3 \geq 0$$

for any $\xi_1, \xi_2, \xi_3 \in \{-1, 1\}$. For a copula C in this family, it holds that C is PUOD if, and only if,

$$\alpha_{12}u_1 u_2 + \alpha_{13}u_1 u_3 + \alpha_{23}u_2 u_3 - \alpha_{123}u_1 u_2 u_3 \geq 0$$

for every $(u_1, u_2, u_3) \in [0,1]^3$. Moreover, C is PLOD if, and only if,

$$\alpha_{12}u_1\,u_2 + \alpha_{13}u_1\,u_3 + \alpha_{23}u_2\,u_3 + \alpha_{123}u_1\,u_2\,u_3 \geq 0$$

for every $(u_1, u_2, u_3) \in [0, 1]^3$. So, for example, if

$$(\alpha_{12}, \alpha_{13}, \alpha_{23}, \alpha_{123}) = (1/9, 1/9, 1/9, -8/9),$$

then $C \in \mathcal{P}^{3,2}_{PQD} \cap \mathcal{P}^3_{PUOD}$, but C is not PLOD. Instead, if

$$(\alpha_{12}, \alpha_{13}, \alpha_{23}, \alpha_{123}) = (1/9, 1/9, 1/9, 8/9),$$

then $C \in \mathcal{P}^{3,2}_{PQD} \cap \mathcal{P}^3_{PLOD}$, but it is not PUOD.

However, under some additional properties $C \in \mathcal{P}^{3,2}_{PQD}$ may have a global dependence property.

Proposition 4. *Let $C \in \mathcal{P}^{3,2}_{PQD}$.*

(a) *If C is NLOD then C is PUOD.*
(b) *If C is NUOD then C is PLOD.*

Proof. Let C be a NLOD copula belonging to $\mathcal{P}^{3,2}_{PQD}$. Then, for all $\mathbf{u} \in [0, 1]^3$, we have

$$\overline{C}(\mathbf{u}) = 1 - u_1 - u_2 - u_3 + C_{12}(u_1, u_2) + C_{13}(u_1, u_3) + C_{23}(u_2, u_3) - C(u_1, u_2, u_3)$$
$$\geq 1 - u_1 - u_2 - u_3 + u_1 u_2 + u_1 u_3 + u_2 u_3 - u_1 u_2 u_3,$$

from which it follows that C is PUOD.
Analogously, let C be a NUOD copula belonging to $\mathcal{P}^{3,2}_{PQD}$. Then, for all $\mathbf{u} \in [0, 1]^3$, we have

$$\overline{C}(\mathbf{u}) = 1 - u_1 - u_2 - u_3 + C_{12}(u_1, u_2) + C_{13}(u_1, u_3) + C_{23}(u_2, u_3) - C(u_1, u_2, u_3)$$
$$\leq (1 - u_1)(1 - u_2)(1 - u_3),$$

which implies that

$$C(u_1, u_2, u_3) \geq C_{12}(u_1, u_2) - u_1 u_2 + C_{13}(u_1, u_3) - u_1 u_3 + C_{23}(u_2, u_3) - u_2 u_3$$
$$+ u_1 u_2 u_3$$

Therefore, C is PLOD.

Analogously, we can prove the following result.

Proposition 5. *Let $C \in \mathcal{P}^{3,2}_{NQD}$.*

(a) *If C is PLOD then C is NUOD.*
(b) *If C is PUOD then C is NLOD.*

In particular, in the case of pairwise independence, Propositions 4 and 5 imply the following result.

Corollary 1. *Let C be in $\mathcal{P}^{3,2}_{\Pi_2}$. The following statements hold:*

(a) C is PUOD if and only if $\overline{C}$ is NLOD;
(b) C is PLOD if and only if $\overline{C}$ is NUOD.

Example 2. Consider the 3-copula C of type (1), where $\alpha \in [-1, 1]$. If $\alpha \geq 0$, then C is PLOD and NUOD. If $\alpha \leq 0$, then C is NLOD and PUOD.

If we impose the 3–copula C to have a specific form, then the pairwise positive dependence may be globally preserved as the following proposition shows.

Proposition 6. *Let C be a 3–copula of type*

$$C(u_1, u_2, u_3) = C_1(C_2(u_1, u_2), u_3) \tag{5}$$

for every $(u_1, u_2, u_3) \in [0, 1]^3$ and suitable 2–copulas C_1 and C_2. The following statements hold:

(a) C_1, C_2 *are PQD if and only if C is PLOD;*
(b) C_1, C_2 *are NQD if and only if C is NLOD.*

Trivially, the previous result applies to Archimedean and nested Archimedean copulas [7,8,17]. Notice that copulas of type (5) have been investigated in [18].

Proposition 7. *Let C be a 3–copula of type*

$$C(u_1, u_2, u_3) = u_3 C_1\left(\frac{C_2(u_1, u_3)}{u_3}, \frac{C_3(u_2, u_3)}{u_3}\right) \tag{6}$$

for every $(u_1, u_2, u_3) \in\,]0, 1]^3$ and suitable 2–copulas C_1, C_2 and C_3. The following statements hold:

(a) C_1, C_2, C_3 *are PQD if and only if C is PLOD;*
(b) C_1, C_2, C_3 *are NQD if and only if C is NLOD.*

Notice that copulas of type (6) have been investigated in [1] (for an application, see [20]).

Finally, we present some bounds for the class of copulas with a given pairwise dependence.

If C is a trivariate copula with bivariate margins C_{12}, C_{13} and C_{23}, it was proved in [11, Theorem 3.11] that

$$F_L \leq C \leq F_U, \tag{7}$$

where

$$\begin{aligned} F_U(u_1, u_2, u_3) = \min\{&C_{12}(u_1, u_2), C_{13}(u_1, u_3), C_{23}(u_2, u_3), 1 - u_1 - u_2 - u_3 \\ &+ C_{12}(u_1, u_2) + C_{13}(u_1, u_3) + C_{23}(u_2, u_3)\} \end{aligned} \tag{8}$$

$$\begin{aligned} F_L(u_1, u_2, u_3) = \max\{&0, C_{12}(u_1, u_2) + C_{13}(u_1, u_3) - u_1, C_{12}(u_1, u_2) \\ &+ C_{23}(u_2, u_3) - u_2, C_{13}(u_1, u_3) + C_{23}(u_2, u_3) - u_3\}. \end{aligned} \tag{9}$$

Thus, we have:

Proposition 8. *For every $C \in \mathcal{P}^{3,2}_{\mathrm{PQD}}$ we have*

$$C_L(\mathbf{u}) \le C(\mathbf{u}) \le M_3(\mathbf{u}), \tag{10}$$

and for every $D \in \mathcal{P}^{3,2}_{\mathrm{NQD}}$ we have

$$W_3(\mathbf{u}) \le D(\mathbf{u}) \le C_U(\mathbf{u}), \tag{11}$$

where C_L and C_U are the 3–quasi-copulas given by (2) and (3), respectively. Moreover, the bounds are best-possible.

Proof. If $C \in \mathcal{P}^{3,2}_{\mathrm{PQD}}$, from (7) it follows that $C(\mathbf{u}) \ge C_L(\mathbf{u})$. Since this lower bound is the best-possible lower bound in the class of trivariate copulas whose bivariate margins are Π_2 (recall Proposition 1), it is the best-possible lower bound in $\mathcal{P}^{3,2}_{\mathrm{PQD}}$, and since M_3 is a 3-copula, from Proposition 2, (10) follows.

If $D \in \mathcal{P}^{3,2}_{\mathrm{NQD}}$, the proof of the upper bound of (11) is similar to the previous case. The lower bound follows by the general proof of sharpness of lower Fréchet–Hoeffding bounds for copulas provided, for instance, in [15, Theorem 2.10.13]. In fact, it is a consequence of the fact that the copula constructed in [15, Theorem 2.10.13] via a multilinear interpolation preserves the property of having the bivariate margins that are NQD, as shown in [6, Proposition 11, part b].

We compare the bounds in Proposition 8 to the Fréchet-Hoeffding bounds by using the measure μ_3. Observe that $\mu_3(M_3) - \mu_3(C_L) = 0.69$ and $\mu_3(C_U) = 0.49$.

4   Concluding Remarks

In this work, we have investigated the possible preservation of some bivariate properties of copulas to higher dimensions. In general, loosely speaking we have noticed that if the three 2-margins of a 3-copula C have some bivariate dependence property, then C rarely has the corresponding trivariate dependence property. For instance,

– Pairwise independence does not imply mutual independence;
– PQD of all bivariate margins does not imply PLOD or PUOD for the corresponding 3–copula.

The same considerations also apply to other interesting dependence properties of copulas. For instance:

– Pairwise exchangeability does not imply mutual exchangeability. Consider, e.g., the copula
$$C(u_1, u_2, u_3) = u_1 \min(u_2, u_3).$$

– Bivariate Gaussian margins do not imply that C is trivariate Gaussian. See [12].

Even some pairwise measure-theoretic properties of copulas do not extend to the multivariate case, as the following example shows.

Example 3. Let C_0 be a 3-copula with the following probability mass distribution: Assign probability mass $1/2$ uniformly to the triangle in $[0,1]^3$ with vertices $(1,0,0)$, $(0,1,0)$, and $(0,0,1)$; and a probability mass $1/2$ uniformly to the triangle with vertices $(1,1,0)$, $(1,0,1)$, and $(0,1,1)$ (see [14, Example 4]). The three 2-margins of C_0 are Π_2.

A probabilistic interpretation for uniform $(0,1)$ random variables U, V, and W with this copula C_0 as their joint distribution function is that

$$\mathbb{P}(\langle U + V + W \rangle = 0) = 1,$$

where $\langle x \rangle = x - \lfloor x \rfloor$ denotes the fractional part of the non-negative number x. Then these random variables (and this copula C_0) have the following properties:

(a) U, V, and W are pairwise independent but not mutually independent.
(b) All 2-margins of C_0 are PQD, but C_0 is neither PLOD nor PUOD.
(c) All 2-margins of C_0 have full support, but C_0 does not.
(d) All 2-margins of C_0 are absolutely continuous, but C_0 is singular.

Notice that the copula C_0 above is a member of the family $\{C_\theta\}_{\theta \in [0,1]}$ of 3-copulas which are distribution functions of U, V, W for which $\mathbb{P}(\langle U + V + W \rangle = \theta) = 1$.

To conclude, we can say that pairwise properties are only one aspect of the multi-facet nature of high–dimensional models. Following [13], we can say that the step from 2 to 3 dimensions is a giant step.

Acknowledgments. The first author acknowledges the support of the Free University of Bozen-Bolzano, School of Economics and Management, via the project "Multivariate Aggregation and Decision Analysis".

References

1. Chakak, A., Koehler, K.J.: A strategy for constructing multivariate distributions. Comm. Statist. Simulation Comput. 24(3), 537–550 (1995)
2. Deheuvels, P.: Indépendance multivariée partielle et inégalités de Fréchet. In: Studies in Probability and Related Topics, Nagard, Rome, pp. 145–155 (1983)
3. Dhaene, J., Denuit, M., Goovaerts, M.J., Kaas, R., Vyncke, D.: The concept of comonotonicity in actuarial science and finance: theory. Insurance Math. Econom. 31(1), 3–33 (2002)
4. Durante, F., Klement, E., Quesada-Molina, J.J.: Bounds for trivariate copulas with given bivariate marginals. J. Inequal. Appl. 2008, 1–9 (2008), article ID 161537
5. Durante, F., Sempi, C.: Copula theory: an introduction. In: Jaworski, P., Durante, F., Härdle, W., Rychlik, T. (eds.) Copula Theory and its Applications. Lecture Notes in Statistics - Proceedings, vol. 198, pp. 3–31. Springer, Heidelberg (2010)

6. Genest, C., Nešlehová, J.: A primer on copulas for count data. Astin Bull. 37(2), 475–515 (2007)
7. Hofert, M.: A stochastic representation and sampling algorithm for nested Archimedean copulas. J. Stat. Comput. Simul. 82(9), 1239–1255 (2012)
8. Hofert, M., Scherer, M.: CDO pricing with nested Archimedean copulas. Quant. Finance 11(5), 775–787 (2011)
9. Jaworski, P., Durante, F., Härdle, W.K. (eds.): Copulae in Mathematical and Quantitative Finance. Lecture Notes in Statistics - Proceedings, vol. 213. Springer, Heidelberg (2013)
10. Jaworski, P., Durante, F., Härdle, W.K., Rychlik, T. (eds.): Copula Theory and its Applications. Lecture Notes in Statistics - Proceedings, vol. 198. Springer, Heidelberg (2010)
11. Joe, H.: Multivariate models and dependence concepts. Monographs on Statistics and Applied Probability, vol. 73. Chapman & Hall, London (1997)
12. Loisel, S.: A trivariate non-Gaussian copula having 2-dimensional Gaussian copulas as margins. Tech. rep., Cahiers de recherche de l'Isfa (2009)
13. Mai, J.F., Scherer, M.: What makes dependence modeling challenging? Pitfalls and ways to circumvent them. Stat. Risk. Model. 30(4), 287–306 (2013)
14. Nelsen, R.B.: Nonparametric measures of multivariate association. In: Distributions with Fixed Marginals and Related Topics (Seattle, WA, 1993). IMS Lecture Notes Monogr. Ser., vol. 28, pp. 223–232, Inst. Math. Statist., Hayward, CA (1996)
15. Nelsen, R.B.: An Introduction to Copulas, 2nd edn. Springer Series in Statistics. Springer, New York (2006)
16. Nelsen, R.B., Úbeda-Flores, M.: How close are pairwise and mutual independence? Statist. Probab. Lett. 82(10), 1823–1828 (2012)
17. Okhrin, O., Okhrin, Y., Schmid, W.: On the structure and estimation of hierarchical Archimedean copulas. J. Econometrics 173(2), 189–204 (2013)
18. Quesada-Molina, J.J., Rodríguez-Lallena, J.A.: Some advances in the study of the compatibility of three bivariate copulas. J. Ital. Stat. Soc. 3(3), 397–417 (1994)
19. Rodríguez-Lallena, J.A., Úbeda-Flores, M.: Compatibility of three bivariate quasi-copulas: applications to copulas. In: Soft Methodology and Random Information Systems. Adv. Soft Comput., vol. 26, pp. 173–180. Springer, Berlin (2004)
20. Salvadori, G., De Michele, C.: Statistical characterization of temporal structure of storms. Adv. Water Resour. 29(6), 827–842 (2006)

On Additive Generators of Grouping Functions

Graçaliz Pereira Dimuro[1], Benjamín Bedregal[2],
Humberto Bustince[3], Radko Mesiar[4], and Maria José Asiain[3]

[1] Programa de Pós-Graduação em Computação, Centro de Ciência Computacionais
Universidade Federal do Rio Grande
Av. Itália km 08, Campus Carreiros, 96201-900 Rio Grande, Brazil
gracaliz@gmail.com
[2] Departamento de Informática e Matemática Aplicada
Universidade Federal do Rio Grande do Norte
Campus Universitário s/n, 59072-970 Natal, Brazil
bedregal@dimap.ufrn.br
[3] Departamento de Automática y Computación, Universidad Pública de Navarra
Campus Arrosadia s/n, P.O. Box 31006, Pamplona, Spain
bustince@unavarra.es
[4] Department of Mathematics and Descriptive Geometry, Faculty of Civil Engineering
Slovak University of Technology in Bratislava
Radlinského 11, 813 68 Bratislava, Slovak Republic
mesiar@math.sk

Abstract. This paper continues previous work on additive generators of overlap functions, introducing the construction of grouping functions by means of additive generators, defined as a pair of one-place functions satisfying appropriate properties. Some important results are discussed.

Keywords: Grouping functions, additive generators.

1 Introduction

Classification problems often faced in many applications, such as image processing [1], decision making and preference modeling [2,3], may involve the assignment of an object to one of the possible classes that are considered in the problem, according to the membership degree (e.g., a preference or support degree) of such object to each one of those classes. It is well known that t-norms and t-conorms have been widely used as the main aggregation operators in this context. However, in many situations the associativity property satisfied by t-norms and t-conorms is not required. This is the case when the classification problem consists of deciding between one of just two classes [2]. Based on this idea, Bustince et al. [3,4] introduced the concepts of *overlap* and *grouping* functions.

Overlap and grouping functions are used to measure, respectively, the degree in which an object is simultaneously supported by both classes and the degree in which the object is supported by the combination of the two classes [5]. In fuzzy preference modeling and decision making, overlap functions allow the definition of the concept of indifference, and Bustince et al. [3] developed an algorithm, which makes use of

A. Laurent et al. (Eds.): IPMU 2014, Part III, CCIS 444, pp. 252–261, 2014.

overlap functions, to elaborate on an alternative preference ranking that penalizes the alternatives for which the expert is not sure about his/her preference. On the other hand, the concept of grouping functions – the N-dual notion of overlap functions – can be used for evaluating the amount of evidence in favor of either of the two alternatives. Thus, its negation provides a measure of incomparability [3].

Notice that there exists a close relation between overlap and grouping functions and some particular classes of t-norms and t-conorms, respectively. A very important way for constructing t-norms and t-conorms is the use of additive generators [6] (see, also, [7] for the representation of aggregation functions, and [8,9] for a study of additive generators of t-norms and t-conorms considering an interval-valued fuzzy approach [10,11,12,13]), and also. This practice offers more flexibility and can improve the performance in applications [14].

Keeping the advantages of using additive generators in mind, in previous work we introduced the notion of additive generators of overlap functions [15]. Now, the present paper extends the previous work, by introducing the construction of grouping functions by means of additive generators, so that one can simplify the choice of an appropriate overlap/grouping function for a given problem, reducing also the computational complexity by only considering single-variable functions instead of bivariate ones.

The paper is organized as follows. Section 2 summarizes preliminary concepts. In Section 3, we present the main results related to additive generators of overlap functions introduced in previous work. Section 4 introduces the concept of additive generator pair of grouping functions, with some important results and examples. Section 5 is the Conclusion, pointing to future work.

2 Preliminary Concepts

A function $A : [0, 1]^n \to [0, 1]$ is said to be an n-ary aggregation function if it satisfies the following two conditions:

(A1) A is increasing[1] in each argument: for each $i \in \{1, \ldots, n\}$, if $x_i \leq y$, then $A(x_1, \ldots, x_n) \leq A(x_1, \ldots, x_{i-1}, y, x_{i+1}, \ldots, x_n)$;

(A2) Boundary conditions: $A(0, \ldots, 0) = 0$ and $A(1, \ldots, 1) = 1$.

In this paper, we are interested in *overlap* and *grouping* functions [3,4,5,16], important classes of aggregation functions that are related in some sense to *triangular norms* (t-norms) and *triangular conorms* (t-conorms), respectively.

Definition 1. *A bivariate aggregation function* $T : [0, 1]^2 \to [0, 1]$ *is said to be a t-norm if it satisfies the following properties: (T1) commutativity; (T2) associativity; and (T3) boundary condition:* $\forall x \in [0, 1] : T(x, 1) = x$.

Definition 2. *A bivariate aggregation function* $S : [0, 1]^2 \to [0, 1]$ *is a t-conorm if it satisfies the following properties: (S1) commutativity; (S2) associativity; and (S3) boundary condition:* $\forall x \in [0, 1] : S(x, 0) = x$.

A t-conorm is *positive* if and only if it has no non-trivial one divisor, that is, if $S(x, y) = 1$ then either $x = 1$ or $y = 1$.

---

[1] In this paper, an increasing (decreasing) function does not need to be strictly increasing (decreasing).

Definition 3. *A function* $N : [0, 1] \to [0, 1]$ *is said to be a fuzzy negation if the following conditions hold:* **(N1)** *boundary conditions:* $N(0) = 1$ *and* $N(1) = 0$; **(N2)** N *is decreasing: if* $x \le y$ *then* $N(y) \le N(x)$.

The function $N : [0, 1] \to [0, 1]$, defined by $N(x) = 1 - x$, is usually called *standard negation* or simply *fuzzy negation*. If a fuzzy negation N is a continuous strictly decreasing function then it is said to be a *strict* negation. For a strict negation N and an aggregation function $A : [0, 1]^2 \to [0, 1]$, its N-dual $A^N : [0, 1]^2 \to [0, 1]$ is given by

$$A^N(x, y) = N^{-1}(A(N(x), N(y))). \tag{1}$$

3   Additive Generators of Overlap Functions

The concept of *overlap functions* was firstly introduced by Bustince et al. [4]. Several recent studies have analysed important properties of overlap functions, such as migrativity, homogeneity, idempotency, Lipschitzianity and their additive generators. Observe that, contrary to the case of t-norms, the class of overlap functions is convex. [4,5,15,16]

Definition 4. *A bivariate function* $O : [0, 1]^2 \to [0, 1]$ *is said to be an overlap function if it satisfies the following conditions:* **(O1)** O *is commutative;* **(O2)** $O(x, y) = 0$ *if and only if* $xy = 0$; **(O3)** $O(x, y) = 1$ *if and only if* $xy = 1$; **(O4)** O *is increasing;* **(O5)** O *is continuous.*

Example 1. [15, Example 3] It is possible to find several examples of overlap functions, such as any continuous t-norm with no zero divisors (property **(O2)**). On the other hand, the function $O_{mM} : [0, 1]^2 \to [0, 1]$, given by $O_{mM}(x, y) = \min(x, y) \max(x^2, y^2)$, is a non associative overlap functions having 1 as neutral element, and, thus, it is not a t-norm. The overlap function $O_2(x, y) = x^2 y^2$ or, more generally, $O_p(x, y) = x^p y^p$, with $p > 0, p \ne 1$, is neither associative nor has 1 as neutral element.

The concept of additive generator pair of an overlap function was firstly introduced by Dimuro and Bedregal [15], allowing the definition of overlap functions (as two-place functions) by means of one-place functions (their additive generator pair). This concept of additive generator pair was inspired by Viceník's work [6] related to additive generators of (non-continuous) t-norms, based on a pair of functions composed by a (non-continuous) one-place function and its pseudo-inverse. In the definition introduced in [15], since overlap functions are continuous, the additive generator pair is composed by continuous functions, satisfying certain properties that allow them to play analogous role to the Viceník's additive generator (and its pseudo-inverse).

Proposition 1. *[15, Corollary 2] Let* $\theta : [0, 1] \to [0, \infty]$ *and* $\vartheta : [0, \infty] \to [0, 1]$ *be continuous and decreasing functions such that:* (1) $\theta(x) = \infty$ *if and only if* $x = 0$; (2) $\theta(x) = 0$ *if and only if* $x = 1$; (3) $\vartheta(x) = 1$ *if and only if* $x = 0$; (4) $\vartheta(x) = 0$ *if and only if* $x = \infty$. *Then, the function* $O_{\theta,\vartheta} : [0, 1]^2 \to [0, 1]$, *defined by* $O_{\theta,\vartheta}(x, y) = \vartheta(\theta(x) + \theta(y))$, *is an overlap function.*

Proposition 2. *[15, Proposition 2] Let* $\theta : [0, 1] \to [0, \infty]$ *and* $\vartheta : [0, \infty] \to [0, 1]$ *be continuous and decreasing functions such that* (1) $\vartheta(x) = 1$ *if and only if* $x = 0$; (2)

$\vartheta(x) = 0$ *if and only if* $x = \infty$; *(3)* $0 \in Ran(\theta)$; *(4)* $O_{\theta,\vartheta}(x,y) = \vartheta(\theta(x)+\theta(y))$ *is an overlap function. Then, the following conditions also hold: (5)* $\theta(x) = \infty$ *if and only if* $x = 0$; *(6)* $\theta(x) = 0$ *if and only if* $x = 1$;

(θ, ϑ) is called an *additive generator pair* of the overlap function $O_{\theta,\vartheta}$, and $O_{\theta,\vartheta}$ is said to be additively generated by the pair (θ, ϑ).

Example 2. [15, Example 4] Consider the functions $\theta : [0,1] \to [0,\infty]$ and $\vartheta : [0,\infty] \to [0,1]$, defined, respectively by:

$$\theta(x) = \begin{cases} -2\ln x & \text{if } x \neq 0 \\ \infty & \text{if } x = 0 \end{cases} \quad \text{and} \quad \vartheta(x) = \begin{cases} e^{-x} & \text{if } x \neq \infty \\ 0 & \text{if } x = \infty, \end{cases}$$

which are continuous and decreasing functions, satisfying the conditions 1-4 of Corollary 1. Then, whenever $x \neq 0$ and $y \neq 0$, one has that:

$$O_{\theta,\vartheta}(x,y) = \vartheta(\theta(x)+\theta(y)) = e^{-(-2\ln x - 2\ln y)} = e^{\ln x^2 y^2} = x^2 y^2.$$

Otherwise, if $x = 0$, it holds that $O_{\theta,\vartheta}(0,y) = \vartheta(\theta(0)+\theta(y)) = \vartheta(\infty + \theta(y)) = 0$, and, similarly, if $y = 0$, then $O_{\theta,\vartheta}(x,0) = 0$. It follows that $O_{\theta,\vartheta}(x,y) = x^2 y^2$, and so we recover the non associative overlap function O_2, given in Example 1.

4   Additive Generators of Grouping Functions

The notion of grouping function was firstly introduced by Bustince et al. [3]:

Definition 5. *A bivariate function* $G : [0,1]^2 \to [0,1]$ *is said to be a grouping function if it satisfies the following conditions:*

(G1) *G is symmetric;*
(G2) $G(x,y) = 0$ *if and only if* $x = y = 0$;
(G3) $G(x,y) = 1$ *if and only if* $x = 1$ *or* $y = 1$;
(G4) *G is increasing;*
(G5) *G is continuous.*

Grouping and overlap functions are N-dual concepts (Eq. (1)). It is immediate that:

(i) A bivariate function $O : [0,1]^2 \to [0,1]$ is an overlap function if and only if $G_O : [0,1]^2 \to [0,1]$, defined by

$$G_O(x,y) = O^N = 1 - O(1-x,1-y), \tag{2}$$

is a grouping function;
(ii) Dually, a function $G : [0,1]^2 \to [0,1]$ is a grouping function if and only if $O_G : [0,1]^2 \to [0,1]$, defined by

$$O_G(x,y) = G^N = 1 - G(1-x,1-y), \tag{3}$$

is an overlap function.

Example 3. Considering the overlap functions discussed in Example 1, then the N-duals of such functions are grouping functions. Thus, it is immediate that continuous t-conorms with no divisors of 1 (property **(G3)**) are grouping functions. Moreover, by Eq. (2), the function $G_2(x, y) = G_{O_2}(x, y) = 1 - (1 - x)^2(1 - y)^2$, or, more generally, $G_p(x, y) = G_{O_p}(x, y) = 1 - (1 - x)^p(1 - y)^p$, with $p > 1$, is a grouping function that is neither associative nor has 0 as neutral element (and, therefore, it is not a t-conorm). Other example of grouping function that is not a t-conorm is the non associative grouping function that has 0 as neutral element given by

$$G_{mM}(x, y) = G_{O_{mM}}(x, y) = 1 - \min\{1 - x, 1 - y\} \max\{(1 - x)^2, (1 - y)^2\}.$$

Lemma 1. *Let* $\sigma : [0, 1] \to [0, \infty]$ *be an increasing function such that*
1. $\sigma(x) + \sigma(y) \in Ran(\sigma)$, *for* $x, y \in [0, 1]$ *and*
2. *if* $\sigma(x) = \sigma(1)$ *then* $x = 1$.

Then $\sigma(x) + \sigma(y) \geq \sigma(1)$ *if and only if* $x = 1$ *or* $y = 1$.

Proof. ($\Rightarrow$) Since σ is increasing and $\sigma(x) + \sigma(y) \in Ran(\sigma)$, for each $x, y \in [0, 1]$, one has that $\sigma(x) + \sigma(y) \leq \sigma(1)$. Thus, if $\sigma(x) + \sigma(y) \geq \sigma(1)$, then it holds that $\sigma(x) + \sigma(y) = \sigma(1)$. Suppose that $\sigma(1) = 0$. Then, since σ is increasing, one has that $\sigma(x) = 0$, for each $x \in [0, 1]$, which is contradiction with condition 2, and, thus, it holds that $\sigma(1) > 0$. Now, suppose that $\sigma(1) \neq 0$ and $\sigma(1) \neq \infty$. Then, since $\sigma(1) \neq 0$, one has that $\sigma(1) + \sigma(1) > \sigma(1)$, which is also a contradiction. It follows that $\sigma(1) = \infty$ and, therefore, since $\sigma(x) + \sigma(y) = \sigma(1)$, we have that $\sigma(x) = \infty$ or $\sigma(y) = \infty$. Hence, by condition 2, one has that $x = 1$ or $y = 1$. ($\Leftarrow$) It is straightforward. $\square$

Lemma 2. [15, Lemma 3] *Consider functions* $\sigma : [0, 1] \to [0, \infty]$ *and* $\varsigma : [0, \infty] \to [0, 1]$ *such that:* $\forall x_0 \in [0, 1] : \varsigma(\sigma(x)) = x_0 \Leftrightarrow x = x_0$. *Then it holds that:* $\forall x_0 \in [0, 1] : \sigma(x) = \sigma(x_0) \Leftrightarrow x = x_0$.

Theorem 1. *Let* $\sigma : [0, 1] \to [0, \infty]$ *and* $\varsigma : [0, \infty] \to [0, 1]$ *be continuous and increasing functions satisfying the conditions of Lemmas 1 and 2 such that*
1. $\sigma(x) + \sigma(y) \in Ran(\sigma)$, *for* $x, y \in [0, 1]$ *;*
2. $\varsigma(\sigma(x)) = 0$ *if and only* $x = 0$*;*
3. $\varsigma(\sigma(x)) = 1$ *if and only if* $x = 1$*;*
4. $\sigma(x) + \sigma(y) = \sigma(0)$ *if and only if* $x = 0$ *and* $y = 0$.

Then, the function $G_{\sigma,\varsigma} : [0, 1]^2 \to [0, 1]$, *defined by*

$$G_{\sigma,\varsigma}(x, y) = \varsigma(\sigma(x) + \sigma(y)), \tag{4}$$

is a grouping function.

Proof. We show that the conditions of Definition 5 holds. The proofs of the commutativity (condition **(G1)**) and continuity (Condition **(G5)**) properties are immediate. Then:
$G_{\sigma,\varsigma}(x, y) = 0 \Leftrightarrow \varsigma(\sigma(x) + \sigma(y)) = 0$ by Eq. (4)
$\qquad\qquad \Leftrightarrow \varsigma(\sigma(z)) = 0$ for some $z \in [0, 1]$ by Condition 1
$\qquad\qquad \Leftrightarrow z = 0 \Leftrightarrow \sigma(z) = \sigma(0) = \sigma(x) + \sigma(y) = \sigma(0)$ by Condition 2, Lemma 2
$\qquad\qquad \Leftrightarrow x = 0$ and $y = 0$ by Condition 4

which proves the condition **(G2)**. Also, one has that:

$$G_{\sigma,\varsigma}(x,y) = 1 \Leftrightarrow \varsigma(\sigma(x) + \sigma(y)) = 1 \text{ by Eq. (4)}$$
$$\Leftrightarrow \varsigma(\sigma(z)) = 1 \text{ for some } z \in [0,1] \quad \text{by Condition 1;}$$
$$\Leftrightarrow z = 1 \Leftrightarrow \sigma(z) = \sigma(1) \quad \text{by Condition 3, Lemma 2}$$
$$\Leftrightarrow \sigma(x) + \sigma(y) = \sigma(1) \Leftrightarrow x = 1 \text{ or } y = 1 \quad \text{by Lemma 1,}$$

which proves the condition **(G3)**. Finally, to prove the condition **(G4)**, considering $z \in [0,1]$ with $y \leq z$, then $\sigma(y) \leq \sigma(z)$. It follows that $G_{\sigma,\varsigma}(x,y) = \varsigma(\sigma(x) + \sigma(y)) \leq \varsigma(\sigma(x) + \sigma(z)) = G_{\sigma,\varsigma}(x,z)$, since ς and σ are increasing. $\square$

Corollary 1. *Let* $\sigma : [0,1] \to [0,\infty]$ *and* $\varsigma : [0,\infty] \to [0,1]$ *be continuous and increasing functions such that*

1. $\sigma(x) = \infty$ *if and only if* $x = 1$;
2. $\sigma(x) = 0$ *if and only if* $x = 0$;
3. $\varsigma(x) = 1$ *if and only if* $x = \infty$;
4. $\varsigma(x) = 0$ *if and only if* $x = 0$.

Then, the function $G_{\sigma,\varsigma} : [0,1]^2 \to [0,1]$, *defined by* $G_{\sigma,\varsigma}(x,y) = \varsigma(\sigma(x) + \sigma(y))$, *is a grouping function.*

Proof. It follows from Theorem 1. $\square$

Proposition 3. *Let* $\sigma : [0,1] \to [0,\infty]$ *and* $\varsigma : [0,\infty] \to [0,1]$ *be continuous and increasing functions such that*

1. $\varsigma(x) = 0$ *if and only if* $x = 0$;
2. $\varsigma(x) = 1$ *if and only if* $x = \infty$;
3. $0 \in Ran(\sigma)$;
4. $G_{\sigma,\varsigma}(x,y) = \varsigma(\sigma(x) + \sigma(y))$ *is a grouping function.*
   Then, the following conditions also hold:
5. $\sigma(x) = \infty$ *if and only if* $x = 1$;
6. $\sigma(x) = 0$ *if and only if* $x = 0$;

Proof. If $G_{\sigma,\varsigma}$ is a grouping function, then it follows that:

5. ($\Rightarrow$) By Condition 2, it holds that: $\sigma(x) = \infty \Rightarrow \sigma(x) + \sigma(y) = \infty$ (for some $y \neq 1$) $\Rightarrow \varsigma(\sigma(x) + \sigma(y)) = 1 \Rightarrow G_{\sigma,\varsigma}(x,y) = 1 \Rightarrow x = 1$.
   ($\Leftarrow$) Consider $y \in [0,1]$ such that $\sigma(y) \neq \infty$. By Condition 2, one has that: $x = 1 \Rightarrow G_{\sigma,\varsigma}(x,y) = 1 \Rightarrow \varsigma(\sigma(x) + \sigma(y)) = 1 \Rightarrow \sigma(x) + \sigma(y) = \infty \Rightarrow \sigma(x) = \infty$.
6. ($\Rightarrow$) Consider $y \in [0,1]$ such that $\sigma(y) = 0$, which is possible due to Condition 3. Then, by Condition 1, one has that: $sigma(x) = 0 \Rightarrow \sigma(x) + \sigma(y) = 0 \Rightarrow \varsigma(\sigma(x) + \sigma(y)) = 0 \Rightarrow G_{\sigma,\varsigma}(x,y) = 0 \Rightarrow x = y = 0$.
   ($\Leftarrow$) By Condition 1, it follows that: $x = 0 \Rightarrow G_{\sigma,\varsigma}(x,0) = 0 \Rightarrow \varsigma(\sigma(x) + \sigma(0)) = 0 \Rightarrow \sigma(x) + \sigma(0) = 0 \Rightarrow \sigma(x) = \sigma(0) = 0$ $\square$

(σ, ς) is called an *additive generator pair* of the grouping function $G_{\sigma,\varsigma}$, and $G_{\sigma,\varsigma}$ is said to be additively generated by the pair (σ, ς).

Theorem 2. *Let $\theta : [0, 1] \to [0, \infty]$ and $\vartheta : [0, \infty] \to [0, 1]$ be continuous and decreasing functions, satisfying the conditions 1-4 of Proposition 1, and consider the functions $\sigma_\theta : [0, 1] \to [0, \infty]$ and $\varsigma_\vartheta : [0, \infty] \to [0, 1]$, defined, respectively, by:*

$$\sigma_\theta(x) = \theta(1 - x) \quad and \quad \varsigma_\vartheta(x) = 1 - \vartheta(x). \tag{5}$$

If (θ, ϑ) is an additive generator pair of an overlap function $O_{\theta,\vartheta} : [0, 1]^2 \to [0, 1]$, then the function $G_{\sigma_\theta, \varsigma_\vartheta} : [0, 1]^2 \to [0, 1]$, defined by

$$G_{\sigma_\theta, \varsigma_\vartheta}(x, y) = \varsigma_\vartheta(\sigma_\theta(x) + \sigma_\theta(y)) \tag{6}$$

is a grouping function.

Proof. It follows that:

$$G_{\sigma_\theta, \varsigma_\vartheta}(x, y) = \varsigma_\vartheta(\sigma_\theta(x) + \sigma_\theta(y)) = 1 - \vartheta(\theta(1 - x), \theta(1 - y)) \quad \text{by Eqs. (6), (5)}$$
$$= 1 - O_{\theta,\vartheta}(1 - x, 1 - y) = G_{O_{\theta,\vartheta}}(x, y), \quad \text{by Prop. (1), Eq. (2)}$$

and it is immediate that $G_{\sigma_\theta, \varsigma_\vartheta}$ is a grouping function. $\square$

Proposition 4. *Let $\theta : [0, 1] \to [0, \infty]$ and $\vartheta : [0, \infty] \to [0, 1]$ be continuous and decreasing functions, satisfying the conditions 1-4 of Proposition 1, and consider the functions $\sigma_\theta : [0, 1] \to [0, \infty]$ and $\varsigma_\vartheta : [0, \infty] \to [0, 1]$, defined by Equations (5). Then, σ_θ and ς_ϑ satisfy the conditions of Corollary 1.*

Proof. Since θ and ϑ are continuous and decreasing, then, for all $x, y \in [0, 1]$, whenever $x < y$ it follows that $\sigma_\theta(x) = \theta(1 - x) < \theta(1 - y) = \sigma_\theta(y)$, and, for all $x, y \in [0, \infty]$, whenever $x < y$, one has that $\varsigma_\vartheta(x) = 1 - \vartheta(x) < 1 - \vartheta(y) = \varsigma_\vartheta(y)$, and, thus σ_θ and ς_ϑ are both continuous and increasing. Since θ and ϑ satisfy the conditions 1-4 of Proposition 1, it follows that: (1) $\sigma_\theta(x) = \infty$ if and only if $\theta(1 - x) = \infty$ if and only if $x = 1$; (2) $\sigma_\theta(x) = 0$ if and only if $\theta(1 - x) = 0$ if and only if $x = 0$; (3) $\varsigma_\vartheta(x) = 1$ if and only if $1 - \vartheta(x) = 1$ if and only if $x = \infty$; (4) $\varsigma_\vartheta(x) = 0$ if and only if $1 - \vartheta(x) = 0$ if and only if $x = 0$. Therefore, σ_θ and ς_ϑ satisfy the conditions of Corollary 1. $\square$

Corollary 2. *Let $\theta : [0, 1] \to [0, \infty]$ and $\vartheta : [0, \infty] \to [0, 1]$ be continuous and decreasing functions, and $\sigma_\theta : [0, 1] \to [0, \infty]$ and $\varsigma_\vartheta : [0, \infty] \to [0, 1]$, defined, respectively, by Equations (5). Then, if (θ, ϑ) is an additive generator pair of an overlap function $O_{\theta,\vartheta} : [0, 1]^2 \to [0, 1]$, then $(\sigma_\theta, \varsigma_\vartheta)$ is an additive generator pair of the grouping function $G_{\sigma_\theta, \varsigma_\vartheta} = G_{O_{\theta,\vartheta}}$.*

Proof. It follows from Theorem 2, Proposition 4 and Corollary 1. $\square$

Conversely, it is immediate that:

Corollary 3. *Let $\sigma : [0, 1] \to [0, \infty]$ and $\varsigma : [0, \infty] \to [0, 1]$ be continuous and increasing functions such that (σ, ς) is an additive generator pair of a grouping function $G_{\sigma,\varsigma} : [0, 1]^2 \to [0, 1]$. Then the functions $\theta_\sigma : [0, 1] \to [0, \infty]$ and $\vartheta_\varsigma : [0, \infty] \to [0, 1]$, defined, respectively by $\theta_\sigma(x) = \sigma(1 - x)$ and $\vartheta_\varsigma(x) = 1 - \varsigma(x)$, constitute an additive generator pair of the overlap function $O_{G_{\sigma,\varsigma}}$.*

Example 4. Consider the functions $\theta : [0,1] \to [0,\infty]$ and $\vartheta : [0,\infty] \to [0,1]$, defined, respectively by:

$$\theta(x) = \begin{cases} -2 \ln x & \text{if } x \neq 0 \\ \infty & \text{if } x = 0 \end{cases} \quad \text{and} \quad \vartheta(x) = \begin{cases} e^{-x} & \text{if } x \neq \infty \\ 0 & \text{if } x = \infty, \end{cases}$$

which are continuous and decreasing functions, satisfying the conditions 1-4 of Proposition 1. In Example 2, we showed that (θ, ϑ) is an additive generator pair of the overlap function $O_2(x,y) = x^2 y^2$, given in Example 1. Now, consider the functions $\sigma_\theta : [0,1] \to [0,\infty]$ and $\varsigma_\vartheta : [0,\infty] \to [0,1]$, given in Equations (5). It follows that

$$\sigma_\theta(x) = \begin{cases} -2 \ln(1-x) & \text{if } x \neq 1 \\ \infty & \text{if } x = 1 \end{cases} \quad \text{and} \quad \varsigma_\vartheta(x) = \begin{cases} 1 - e^{-x} & \text{if } x \neq \infty \\ 1 & \text{if } x = \infty. \end{cases}$$

Then, whenever $x \neq 1$ and $y \neq 1$, one has that

$$G_{\sigma_\theta, \varsigma_\vartheta}(x,y) = \varsigma_\vartheta(\sigma_\theta(x) + \sigma_\theta(y)) = 1 - e^{-(-2\ln(1-x) - 2\ln(1-y))} = 1 - (1-x^2)(1-y^2).$$

Otherwise, if $x = 1$, then $\sigma_\theta(1) = \infty$, and, thus, $G_{\sigma_\theta, \varsigma_\vartheta}(1, y) = 1$. Similarly, if $x = 1$, one has that $G_{\sigma_\theta, \varsigma_\vartheta}(x, 1) = 1$. It follows that $G_{\sigma_\theta, \varsigma_\vartheta}(x, y) = 1 - (1 - x^2)(1 - y^2) = G_2(x, y)$, where G_2 is the grouping function given in Example 3, such that $G_2 = G_{O_2}$.

It is immediate that:

Corollary 4. *In the same conditions of Theorem 1, whenever $\varsigma = \sigma^{(-1)}$ then $G_{\sigma,\varsigma}$ is a positive t-conorm.*

Theorem 3. *Let $G : [0,1]^2 \to [0,1]$ be a grouping function having 0 as neutral element. Then, if G is additively generated by a pair (σ, ς), with $\sigma : [0,1] \to [0,\infty]$ and $\varsigma : [0,\infty] \to [0,1]$ satisfying the conditions of Theorem 1, then G is associative.*

Proof. If 0 is the neutral element of G, then, since $\sigma(0) = 0$, one has that:

$$y = G(0, y) = \varsigma(\sigma(0) + \sigma(y)) = \varsigma(0 + \sigma(y)) = \varsigma(\sigma(y)). \tag{7}$$

Since $\sigma(x) + \sigma(y) \in Ran(\sigma)$, for $x, y \in [0,1]$, then there exists $w \in [0,1]$ such that

$$\sigma(w) = \sigma(x) + \sigma(y). \tag{8}$$

It follows that

$$\begin{aligned} G(x, G(y, z)) &= \varsigma(\sigma(x) + \sigma(\varsigma(\sigma(y) + \sigma(z)))) && \text{by Eq. (4)} \\ &= \varsigma(\sigma(x) + \sigma(\varsigma(\sigma(w)))) = \varsigma(\sigma(x) + \sigma(w)) && \text{by Eqs. (8), (7)} \\ &= \varsigma(\sigma(x) + \sigma(y) + \sigma(z)) = && \text{by Eq. (8)} \\ &= \varsigma(\sigma(u) + \sigma(z)) = \varsigma(\sigma(\varsigma(\sigma(u))) + \sigma(z)) && \text{by Eqs. (8), (7)} \\ &= \varsigma(\sigma(\varsigma(\sigma(x) + \sigma(y))) + \sigma(z)) = G(G(x, y), z) && \text{by Eqs. (8), (4).} \quad \square \end{aligned}$$

Corollary 5. *Let $G : [0,1]^2 \to [0,1]$ be a grouping function additively generated by a pair (σ, ς). G is a t-conorm if and only if 0 is a neutral element of G.*

Notice that whenever S is a positive continuous t-conorm (that is, a grouping function) that is additively generated by a function $s : [0,1] \to [0,\infty]$ [6], then it is also additively generated by a pair (σ, ς) in the sense of Theorem 1, with $\sigma = s$ and $\varsigma = s^{(-1)}$, and vice-versa, where $s^{(-1)}$ is the pseudo-inverse of s. We have also the next results.

Proposition 5. *A grouping function $G : [0,1]^2 \to [0,1]$ is additively generated if and only if the function $f_G : [0,1] \to [0,1]$ given by $f_G(x) = G(x,0)$ is strictly increasing, and $f_G^{-1} \circ G : [0,1]^2 \to [0,1]$ is a positive t-conorm.*

Proposition 6. $G : [0,1]^2 \to [0,1]$ *is an additively generated grouping function if and only if there are two strict negations $N_1, N_2 : [0,1] \to [0,1]$ such that $G(x,y) = N_2(S_p(N_1(x), N_1(y)))$, where $S_p : [0,1]^2 \to [0,1]$ is the probabilistic sum given by $S_p(x,y) = x + y - xy$.*

5   Conclusion

In this paper, we have introduced the notion of additive generator pair of grouping functions, presenting results and examples, so giving some directions in order to reduce the computational complexity of algorithms that apply grouping functions, and to obtain a more systematic way for their choice in practical applications.

Future theoretical work is concerned with the study of the influence of important properties (e.g., the migrativity, homogeneity and idempotency properties) in additively generated grouping functions and their respective additive generator pairs. We also intend to extend the concepts of additive generator pairs of overlap and grouping functions to the interval-valued setting, following the approach adopted in [8,9].

Future applied work is related to the application in the context of hybrid BDI-fuzzy[2] agent models, where the evaluation of social values and exchanges are of a qualitative and subjective nature [18,19,20]. Overlap and grouping functions will be used for dealing with indifference and incomparability when reasoning on the agent's fuzzy belief base. Observe that the computational complexity in this kind of problems is crucial since, in general, we have a large belief base.

Acknowledgments.   Work supported by CNPq (Proc. 305131/10-9, 481283/2013-7, 306970/2013-9), NSF (ref. TIN2010-15055, TIN2011-29520), VEGA 1/0171/12 and APVV-0073-10.

References

1. Bustince, H., Barrenechea, E., Pagola, M.: Image thresholding using restricted equivalence functions and maximizing the measures of similarity. Fuzzy Sets and Systems 158(5), 496–516 (2007)
2. Fodor, J., Roubens, M.: Fuzzy preference modelling and multicriteria decision support. Kluwer, New York (1994)
3. Bustince, H., Pagola, M., Mesiar, R., Hüllermeier, E., Herrera, F.: Grouping, overlaps, and generalized bientropic functions for fuzzy modeling of pairwise comparisons. IEEE Transactions on Fuzzy Systems 20(3), 405–415 (2012)
4. Bustince, H., Fernandez, J., Mesiar, R., Montero, J., Orduna, R.: Overlap functions. Nonlinear Analysis 72(3-4), 1488–1499 (2010)

---

[2] BDI stands for "Beliefs, Desires, Intentions", a particular cognitive agent model [17].

5. Jurio, A., Bustince, H., Pagola, M., Pradera, A., Yager, R.: Some properties of overlap and grouping functions and their application to image thresholding. Fuzzy Sets and Systems 229, 69–90 (2013)
6. Viceník, P.: Additive generators of associative functions. Fuzzy Sets and Systems 153(2), 137–160 (2005)
7. Mayor, G., Trillas, E.: On the representation of some aggregation functions. In: Proceedings of the XVI International Symposium on Multiple-Valued Logic (ISMVL 1986), Blacksburg, pp. 110–114 (1986)
8. Dimuro, G.P., Bedregal, B.C., Santiago, R.H.N., Reiser, R.H.S.: Interval additive generators of interval t-norms and interval t-conorms. Information Sciences 181(18), 3898–3916 (2011)
9. Dimuro, G.P., Bedregal, B.R.C., Reiser, R.H.S., Santiago, R.H.N.: Interval additive generators of interval t-norms. In: Hodges, W., de Queiroz, R. (eds.) WoLLIC 2008. LNCS (LNAI), vol. 5110, pp. 123–135. Springer, Heidelberg (2008)
10. Bedregal, B.C., Dimuro, G.P., Santiago, R.H.N., Reiser, R.H.S.: On interval fuzzy S-implications. Information Sciences 180(8), 1373–1389 (2010)
11. Dimuro, G.: On interval fuzzy numbers. In: 2011 Workshop-School on Theoretical Computer Science, WEIT 2011, pp. 3–8. IEEE, Los Alamitos (2011)
12. Reiser, R.H.S., Dimuro, G.P., Bedregal, B.C., Santiago, R.H.N.: Interval valued QL-implications. In: Leivant, D., de Queiroz, R. (eds.) WoLLIC 2007. LNCS, vol. 4576, pp. 307–321. Springer, Heidelberg (2007)
13. Bedregal, B.C., Dimuro, G.P., Reiser, R.H.S.: An approach to interval-valued R-implications and automorphisms. In: Carvalho, J.P., Dubois, D., Kaymak, U., da Costa Sousa, J.M. (eds.) Proceedings of the Joint 2009 International Fuzzy Systems Association World Congress and 2009 European Society of Fuzzy Logic and Technology Conference, IFSA/EUSFLAT, pp. 1–6 (2009)
14. Mesiarová-Zemánková, A.: Ranks of additive generators. Fuzzy Sets and Systems 160(14), 2032–2048 (2009)
15. Dimuro, G.P., Bedregal, B.: Additive generators of overlap functions. In: Bustince, H., Fernandez, J., Mesiar, R., Calvo, T. (eds.) Aggregation Functions in Theory and in Practise. AISC, vol. 228, pp. 167–178. Springer, Heidelberg (2013)
16. Bedregal, B.C., Dimuro, G.P., Bustince, H., Barrenechea, E.: New results on overlap and grouping functions. Information Sciences 249, 148–170 (2013)
17. Rao, A.S., Georgeff, M.P.: Modeling rational agents within a BDI-architecture. In: Fikes, R., Sandewall, E. (eds.) Proceedings of the 2nd International Conference on Principles of Knowledge Representation and Reasoning, pp. 473–484. Morgan Kaufmann, San Mateo (1991)
18. Dimuro, G.P., da Rocha Costa, A.C., Gonçalves, L.V., Hübner, A.: Centralized regulation of social exchanges between personality-based agents. In: Noriega, P., Vázquez-Salceda, J., Boella, G., Boissier, O., Dignum, V., Fornara, N., Matson, E. (eds.) COIN 2006. LNCS (LNAI), vol. 4386, pp. 338–355. Springer, Heidelberg (2007)
19. Dimuro, G.P., da Rocha Costa, A.C., Gonçalves, L.V., Pereira, D.: Recognizing and learning models of social exchange strategies for the regulation of social interactions in open agent societies. Journal of the Brazilian Computer Society 17, 143–161 (2011)
20. Pereira, D.R., Gonçalves, L.V., Dimuro, G.P., da Rocha Costa, A.C.: Towards the self-regulation of personality-based social exchange processes in multiagent systems. In: Zaverucha, G., da Costa, A.L. (eds.) SBIA 2008. LNCS (LNAI), vol. 5249, pp. 113–123. Springer, Heidelberg (2008)

Fusion Functions and Directional Monotonicity

Humberto Bustince[1], Javier Fernandez[1], Anna Kolesárová[2],
and Radko Mesiar[3,4]

[1] Public University of Navarra
Campus Arrosadía s/n, 31006 Pamplona, Spain
[2] Slovak University of Technology, Fac. of Chemical and Food Technology,
Radlinského 9, 812 37 Bratislava, Slovakia
[3] Slovak University of Technology, Fac. of Civil Engineering,
Radlinského 11, 813 68 Bratislava, Slovakia
[4] University of Ostrava,
Institute for Research and Applications of Fuzzy Modelling,
30. dubna 22, Ostrava, Czech Republic
{bustince,fcojavier.fernandez}@unavarra.es,
{anna.kolesarova;radko.mesiar}@stuba.sk

Abstract. After introducing fusion functions, the directional monotonicity of fusion functions is introduced and studied. Moreover, in special cases the sets of all vectors with respect to which the studied fusion functions are increasing (decreasing) are completely characterized.

Keywords: Aggregation function, fusion function, directional monotonicity, piece-wise linear function.

1 Introduction

Aggregation of data into one representative value is a basic tool in many scientific domains such as multi-criteria decision support, fuzzy logic, economy, sociology, etc. The common framework for inputs and outputs of aggregation functions is usually the unit interval $[0,1]$. A function $A\colon [0,1]^n \to [0,1]$ is said to be an aggregation function if it is increasing, i.e., $A(x_1,\ldots,x_n) \leq A(y_1,\ldots,y_n)$ whenever $(x_1,\ldots,x_n)$, $(y_1,\ldots,y_n) \in [0,1]^n$, $x_i \leq y_i$, $i = 1,\ldots,n$, and if it satisfies the boundary conditions $A(0,\ldots,0) = 0$, $A(1,\ldots,1) = 0$. However, increasing monotonocity that is one of the basic properties of aggregation functions, is violated in the case of several other fusion techniques frequently applied in real data processing, e.g., implications which are characterized by hybrid monotonicity [1,7], or the mode function [2,12].

The aim of this paper is to generalize the theory of aggregation functions [6,2]. Note, that the unit interval can be replaced by any interval $[a,b] \subseteq [-\infty, \infty]$, but in this paper, we will only deal with the unit interval $[0,1]$, both for inputs and outputs of generalized aggregation functions.

The paper is organized as follows: In Section 2, fusion functions on the interval $[0,1]$ and their directional monotonicity are introduced and exemplified. Section

A. Laurent et al. (Eds.): IPMU 2014, Part III, CCIS 444, pp. 262–268, 2014.

3 is devoted to the study of some properties of r-monotone fusion functions. In Section 4, for some distinguished fusion functions we investigate the sets of all vectors r with respect to which they are r-increasing. Finally, some concluding remarks are added.

2   Fusion Functions

Let $\mathbf{I}$ be the unit interval $[0, 1]$ and $\mathbf{I}^n = \{\mathbf{x} = (x_1, \ldots, x_n) \mid x_i \in \mathbf{I}, i = 1, \ldots, n\}$.

Definition 2.1. *Let $n \in \mathbb{N}$, $n \geq 2$. A fusion function is an arbitrary function $F \colon \mathbf{I}^n \to \mathbf{I}$.*

Definition 2.2. *Let r be a real vector, $r \neq \mathbf{0}$. A fusion function $F \colon \mathbf{I}^n \to \mathbf{I}$ is r-increasing (r-decreasing) if for all points $\mathbf{x} \in \mathbf{I}^n$ and all $c > 0$ such that $\mathbf{x} + c r \in \mathbf{I}^n$, it holds*

$$F(\mathbf{x} + c r) \geq F(\mathbf{x}) \quad (F(\mathbf{x} + c r) \leq F(\mathbf{x})). \tag{1}$$

Note, that the common name for r-increasing and r-decreasing fusion functions is "r-monotone" fusion functions. From now on, the set of all real n-dimensional vectors $r \neq \mathbf{0}$ will be denoted by V_n. These vectors will be called directions, and monotonicity introduced in Definition 2.2 will be called the directional monotonicity of fusion functions.

Example 2.1. Consider the Gödel implication $I_G \colon \mathbf{I}^2 \to \mathbf{I}$, given by

$$I_G(x, y) = \begin{cases} 1 \text{ if } x \leq y, \\ y \text{ otherwise.} \end{cases}$$

It is clear that I_G is a $(1, 1)$-increasing fusion function. It is also $(0, 1)$-increasing, but it is not $(1, 0)$-increasing (in fact it is $(1, 0)$-decreasing).

However, the Reichenbach implication $I_R \colon \mathbf{I}^2 \to \mathbf{I}$, $I_R(x, y) = 1 - x + xy$, is neither $(1, 1)$-increasing nor $(1, 1)$-decreasing. If we consider $r = (1, 1)$ and take, for example, $\mathbf{x} = (0.1, 0.2)$ and $c = 0.2$, then $\mathbf{x} + c r = (0.3, 0.4)$, and we have $I_R(0.3, 0.4) = 0.82$ and $I_R(0.1, 0.2) = 0.92$, i.e., I_R is not $(1, 1)$-increasing. If we consider $\mathbf{x} = (0.4, 0.5)$ and $c = 0.2$, then $\mathbf{x} + c r = (0.6, 0.7)$ and because $I_R(0.6, 0.7) = 0.82 > I_R(0.4, 0.5) = 0.8$, I_R is not $(1, 1)$-decreasing.

Example 2.2. Consider the function $F \colon \mathbf{I}^2 \to \mathbf{I}$, $F(x, y) = x - (\max\{0, x - y\})^2$. F is a continuous fusion function. Though $F(0, 0) = 0$ and $F(1, 1) = 1$, F is not an aggregation function. However, this function can be used in decision making when a decision is made by a very skilled person (evaluation x) and some less skilled person (evaluation y). F is, e.g., $(1, 1)$-increasing, $(0, 1)$-increasing, but it is neither $(1, 0)$-increasing nor $(1, 0)$-decreasing.

Example 2.3. The weighted Lehmer mean $L_\lambda \colon \mathbf{I}^2 \to \mathbf{I}$, $L_\lambda(x, y) = \frac{\lambda x^2 + (1-\lambda)y^2}{\lambda x + (1-\lambda)y}$ (with convention $\frac{0}{0} = 0$), where $\lambda \in {]0, 1[}$, is $(1 - \lambda, \lambda)$-increasing.

For each $i \in \{1, \ldots, n\}$, let $\boldsymbol{e}_i = (\epsilon_1, \ldots, \epsilon_n)$ be the vector with $\epsilon_i = 1$ and $\epsilon_j = 0$ for each $j \neq i$. Functions which are $\boldsymbol{e}_i$-increasing ($\boldsymbol{e}_i$-decreasing) for each $i = 1, \ldots, n$, are in fact functions known as increasing (decreasing) functions. So, aggregation functions $A \colon \mathbf{I}^n \to \mathbf{I}$ are fusion functions which are $\boldsymbol{e}_i$-increasing for each $i = 1, \ldots, n$, and satisfy the boundary conditions $A(0, \ldots, 0) = 0$ and $A(1, \ldots, 1) = 1$, compare, e.g., [2,6]. Similarly, implications $I \colon \mathbf{I}^2 \to \mathbf{I}$ are $\boldsymbol{e}_1$-decreasing and $\boldsymbol{e}_2$-increasing fusion functions, satisfying the boundary conditions $I(0,0) = I(1,1) = 1$ and $I(1,0) = 0$.

3   Some Properties of r-monotone Fusion Functions

Let us summarize several important properties of r-monotone fusion functions.

Proposition 3.1. *A fusion function $F \colon \mathbf{I}^n \to \mathbf{I}$ is r-decreasing if and only if F is $(-r)$-increasing.*

Theorem 3.1. *Let a fusion function $F \colon \mathbf{I}^n \to \mathbf{I}$ have the first-order partial derivatives with respect to each variable. If F is r-increasing and s-increasing for some vectors r and s in V_n, then F is u increasing for each $u = ar + bs$, where $a \geq 0$, $b \geq 0$, $a + b > 0$.*

Proposition 3.2. *Let $F \colon \mathbf{I}^n \to \mathbf{I}$ be an r-increasing fusion function. If $\varphi \colon \mathbf{I} \to \mathbf{I}$ is an increasing (decreasing) function then $G = \varphi \circ F$ is an r-increasing (decreasing) fusion function.*

Proposition 3.3. *Let $F_1, \ldots, F_k \colon \mathbf{I}^n \to \mathbf{I}$ be r-increasing fusion functions and let $F \colon \mathbf{I}^k \to \mathbf{I}$ be an increasing function. Then the function $G = F(F_1, \ldots, F_k)$ is an r-increasing fusion function.*

Remark 3.1. Note, that in the previous proposition, the r-increasing monotonicity of F is not sufficient for the r-increasing monotonicity of G. For example, consider the functions F, F_1, $F_2 \colon [0,1]^2 \to [0,1]$, given by $F_1(x,y) = x - (\max\{0, x - y\})^2$, see Example 2.2, $F_2(x,y) = I_L(x,y)$, where I_L is the Łukasiewicz implication given by $I_L(x,y) = \min\{1, 1 - x + y\}$, and $F(x,y) = \frac{x^2 + y^2}{x + y}$, i.e., the Lehmer mean for $\lambda = 1/2$, see Example 2.3. All these functions are $(1,1)$-increasing. Put $G = F(F_1, F_2)$. As $G(x,x) = \frac{x^2 + 1}{x + 1}$, we get $G(0,0) = 1$, $G(1/2, 1/2) = 5/6$, and $G(1,1) = 1$, which shows that G is not $(1,1)$-increasing.

Proposition 3.4. *Let $F \colon \mathbf{I}^n \to \mathbf{I}$ be an increasing fusion function and $f_i \colon [0,1] \to [0,1]$, $i = 1, \ldots, n$, be monotone functions. If a function $G \colon \mathbf{I}^n \to \mathbf{I}$ is defined by*

$$G(x_1, \ldots, x_n) = F(f_1(x_1), \ldots, f_n(x_n)),$$

then G is r-increasing for each vector $r = (r_1, \ldots, r_n)$ with the property:

$$\forall\, i \in \{1, \ldots, n\}, \quad r_i \geq 0 \text{ if } f_i \text{ is increasing and } r_i \leq 0 \text{ if } f_i \text{ is decreasing.}$$

Proposition 3.5. *Let* $F\colon \mathbf{I}^n \to \mathbf{I}$ *be an* $\mathbf{r}$*-increasing fusion function and let functions* g *and* f_i, $i = 1,\dots,n$, *be in* $\{id_{\mathbf{I}}, 1 - id_{\mathbf{I}}\}$. *If a function* $H\colon \mathbf{I}^n \to \mathbf{I}$ *is defined by*

$$H(x_1,\dots,x_n) = g\left(F\left(f_1(x_1),\dots,f_n(x_n)\right)\right),$$

then H *is* $\mathbf{s}$*-increasing where*

$$\mathbf{s} = (-1)^{g(0)}\left((-1)^{f_1(0)}r_1,\dots,(-1)^{f_n(0)}r_n\right).$$

For a fusion function $F\colon [0,1]^n \to [0,1]$, let us define its dual as the fusion function $F^d\colon [0,1]^n \to [0,1]$, $F^d(x_1,\dots,x_n) = 1 - F(1 - x_1,\dots,1 - x_n)$.

Corollary 3.1. *If a fusion function* $F\colon \mathbf{I}^n \to \mathbf{I}$ *is* $\mathbf{r}$*-increasing then also its dual* F^d *is* $\mathbf{r}$*-increasing.*

Corollary 3.2. *Functions* F *and* F^d *are directionally increasing with respect to the same vectors.*

4 $\mathcal{D}^{\uparrow}$ -sets of Fusion Functions

For a fusion function $F\colon \mathbf{I}^n \to \mathbf{I}$, let $\mathcal{D}^{\uparrow}(F)$ be a set of all real vectors $\mathbf{r} = (r_1,\dots,r_n) \in V_n$ for which F is $\mathbf{r}$-increasing. Clearly, for each aggregation function $A\colon \mathbf{I}^n \to \mathbf{I}$, $\mathcal{D}^{\uparrow}(A)$ contains the set $V_n^+ = \{\mathbf{r} \in V_n \mid r_1 \geq 0,\dots,r_n \geq 0\}$. Similarly, for each implication $\colon \mathbf{I}^2 \to \mathbf{I}$, $\mathcal{D}^{\uparrow}(I) \supseteq \{\mathbf{r} = (r_1,r_2) \in V_2 \mid r_1 \leq 0, r_2 \geq 0\}$. Note, that by Corollary 3.2, for each fusion function F we have $\mathcal{D}^{\uparrow}(F) = \mathcal{D}^{\uparrow}(F^d)$.

Example 4.1. Consider the Lukasiewicz t-norm $T_L\colon \mathbf{I}^2 \to \mathbf{I}$, $T_L(x,y) = \max\{0, x + y - 1\}$. It holds $\mathcal{D}^{\uparrow}(T_L) = \{\mathbf{r} = (r_1,r_2) \mid r_2 \geq -r_1\}$. So, $\mathcal{D}^{\uparrow}(T_L)$ is a superset of the set V_2^+. Moreover, for the Lukasiewicz t-conorm S_L, as $S_L = T_D^d$, we have $\mathcal{D}^{\uparrow}(S_L) = \mathcal{D}^{\uparrow}(T_L)$, see Corollary 3.2.

For the Lukasiewicz implication $I_L\colon \mathbf{I}^2 \to \mathbf{I}$, $I_L(x,y) = \min\{1, 1 - x + y\}$, we have $\mathcal{D}^{\uparrow}(I_L) = \{\mathbf{r} = (r_1,r_2) \mid r_2 \geq r_1\}$. The result follows from the fact that $I_L(x,y) = 1 - T_L(x, 1 - y)$, see Proposition 3.5.

From Example 2.3, it follows that the Lehmer mean $L_{1/2}$ is $(1,1)$-increasing (i.e., weakly monotone in terminology used by Wilkin and Beliakov in [12]). Even a stronger result is valid.

Proposition 4.1. *Let* $L_{1/2}\colon \mathbf{I}^2 \to \mathbf{I}$ *be the Lehmer mean given by*

$$L_{1/2}(x,y) = \frac{x^2 + y^2}{x + y}.$$

The only vectors with respect to which the function $L_{1/2}$ *is* $\mathbf{r}$*-increasing are vectors* $\mathbf{r} = \alpha(1,1)$ *with* $\alpha > 0$.

Remark 4.1. It can be shown that, in general, for the weighted Lehmer mean L_λ, $\lambda \in]0,1[$, see Example 2.3, it holds $\mathcal{D}^{\uparrow}(L_\lambda) = \{\alpha(1 - \lambda, \lambda) \mid \alpha > 0\}$.

In several applications, linear/piece-wise linear fusion functions are considered. Recall that a continuous fusion function F is piece-wise linear if and only if it can be obtained by the patchwork technique with linear functions B_j, $j = 1, \ldots, k$.

Lemma 1. *The next claims are equivalent:*

(i) *A linear function $F \colon \mathbf{I}^n \to \mathbf{I}$ is $\boldsymbol{r}$-increasing.*
(i) *For a fixed connected set $E \subseteq \mathbf{I}^n$ with positive volume, $F|_E$ is $\boldsymbol{r}$-increasing.*

Lemma 2. *If $F \colon \mathbf{I}^n \to \mathbf{I}$ is a continuous piece-wise linear fusion function determined by linear functions $B_1, \ldots, B_k$, then $\mathcal{D}^\uparrow(F) = \bigcap\limits_{j=1}^{k} \mathcal{D}^\uparrow(B_j)$.*

In the next proposition, $\mathcal{D}^\uparrow$-sets of linear fusion functions are determined.

Proposition 4.2. *Let a fusion function $F \colon \mathbf{I}^n \to \mathbf{I}$ be linear, i.e., $F(\mathbf{x}) = b + \sum\limits_{i=1}^{n} a_i x_i$. Then*

$$\mathcal{D}^\uparrow(F) = \{(r_1, \ldots, r_n) \in V_n \mid \sum_{i=1}^{n} a_i r_i \geq 0\}.$$

Corollary 4.1. *Let $\boldsymbol{w} = (w_1, \ldots, w_n)$ be a weighting vector, i.e., for each $i \in \{1, \ldots, n\}$, $w_i \geq 0$ and $\sum\limits_{i=1}^{n} w_i = 1$. Let $W_{\boldsymbol{w}} \colon \mathbf{I}^n \to \mathbf{I}$ be the weighted arithmetic mean, $W_{\boldsymbol{w}}(x_1, \ldots, x_n) = \sum\limits_{i=1}^{n} w_i x_i$. Then*

$$\mathcal{D}^\uparrow(W_{\boldsymbol{w}}) = \left\{(r_1, \ldots, r_n) \in V_n \mid \sum_{i=1}^{n} w_i r_i \geq 0\right\}.$$

OWA operators [13] and Choquet integrals [3,5] are also special piece-wise linear fusion functions. From Proposition 4.2 and Lemma 2 one can deduce the following consequences for OWA operators and Choquet integrals.

Corollary 4.2. *Let $A_{\boldsymbol{w}} \colon \mathbf{I}^n \to \mathbf{I}$ be an OWA operator corresponding to a weighting vector $\boldsymbol{w}$. Then*

$$\mathcal{D}^\uparrow(A_{\boldsymbol{w}}) = \bigcap_{\sigma \in \Omega} \mathcal{D}^\uparrow\left(W_{\sigma(\boldsymbol{w})}\right)$$

$$= \bigcap_{\sigma \in \Omega} \left\{\boldsymbol{r} = (r_1, \ldots, r_n) \in V_n \mid \sum_{i=1}^{n} w_{\sigma(i)} r_i \geq 0\right\}, \tag{2}$$

where Ω is the set of all permutations $\sigma \colon \{1, \ldots, n\} \to \{1, \ldots, n\}$ and $\sigma(\boldsymbol{w}) = (w_{\sigma(1)}, \ldots, w_{\sigma(n)})$.

Note, that (2) can also be written as

$$\mathcal{D}^{\uparrow}(A_{\boldsymbol{w}}) = \{\boldsymbol{r} = (r_1, \ldots, r_n) \in V_n \mid \forall \sigma \in \Omega : \sum_{i=1}^{n} w_{\sigma(i)} r_i \geq 0\}.$$

Corollary 4.3. *Let* Ch_μ *be the Choquet integral corresponding to a capacity* μ. *Then*

$$\mathcal{D}^{\uparrow}(\mathrm{Ch}_\mu) = \bigcap_{\sigma \in \Omega} \mathcal{D}^{\uparrow}\left(W_{\sigma(\mu)}\right)$$

$$= \bigcap_{\sigma \in \Omega} \left\{\boldsymbol{r} = (r_1, \ldots, r_n) \in V_n \mid \sum_{i=1}^{n} w_{\sigma,i}^{\mu} r_i \geq 0\right\}, \qquad (3)$$

where $w_{\sigma,i}^{\mu} = \mu(\{\sigma(i), \ldots, \sigma(n)\}) - \mu(\{\sigma(i+1), \ldots, \sigma(n)\})$ *with convention* $w_{\sigma,n}^{\mu} = \mu(\{\sigma(n)\})$.

Again, (3) can be written as

$$\mathcal{D}^{\uparrow}(\mathrm{Ch}_\mu) = \left\{\boldsymbol{r} = (r_1, \ldots, r_n) \in V_n \mid \forall \sigma \in \Omega : \sum_{i=1}^{n} w_{\sigma,i}^{\mu} r_i \geq 0\right\}.$$

Example 4.2. Consider the OWA operator $A \colon \mathbf{I}^2 \to \mathbf{I}$, $A(x,y) = \frac{1}{3}\min\{x,y\} + \frac{2}{3}\max\{x,y\}$. Then

$$\mathcal{D}^{\uparrow}(A) = \mathcal{D}^{\uparrow}\left(W_{(1/3,2/3)}\right) \cap \mathcal{D}^{\uparrow}\left(W_{(2/3,1/3)}\right)$$
$$= \{(r_1, r_2) \in V_2 \mid r_1 + 2r_2 \geq 0,\ 2r_1 + r_2 \geq 0\},$$

see Fig. 4.1.

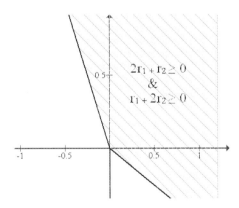

Fig. 4.1. Graphical illustration of the set $\mathcal{D}^{\uparrow}(A)$ for the OWA operator from Example 4.2

Note, that the Sugeno integral [10], Shilkret integral [9], hierarchical Choquet integral [4,8], twofold integral [11] are also continuous piece-wise linear functions, which allow to determine the set of all directions in which they are increasing.

5   Concluding Remarks

We have introduced and discussed fusion functions on the unit interval **I** and their directional monotonicity. This property is related to the directional derivative (if it exists). We have only considered the unit interval **I** for input and output values. All introduced notions can easily be rewritten for any subinterval of the extended real line. However, not all of our results have the same form when open or unbounded intervals are considered. The aim of this contribution is to open the theory of fusion functions and directional monotonicity. Observe that our results generalize the results of Wilkin and Beliakov [12] concerning so-called weak monotonicity.

Acknowledgment. The authors kindly acknowledge the support of the project TIN15055-2010, the grants VEGA 1/0171/12, 1/0419/13 and the project of Science and Technology Assistance Agency under the contract No. APVV–0073–10. Moreover, the work of R. Mesiar on this paper was supported by the European Regional Development Fund in the IT4Innovations Centre of Excellence project reg. no. CZ.1.05/1.1.00/ 02.0070.

References

 1. Baczynski, M., Jayaram, B.: Fuzzy Implications. STUDFUZZ, vol. 231. Springer, Heidelberg (2008)
 2. Beliakov, G., Pradera, A., Calvo, T.: Aggregation Functions: A Guide for Practitioners. Springer, Heidelberg (2007)
 3. Choquet, G.: Theory of capacities. Ann. Inst. Fourier 5, 131–295 (1953)
 4. Fujimoto, K., Sugeno, M., Murofushi, T.: Hierarchical decomposition of Choquet integrals models. Int. J. Uncertainty, Fuzziness and Knowledge-Based Systems 3, 1–15 (1995)
 5. Grabisch, M.: Fuzzy integral in multicriteria decision making. Fuzzy Sets and Systems 69, 279–298 (1995)
 6. Grabisch, M., Marichal, J.-L., Mesiar, R., Pap, E.: Aggregation Functions. Cambridge University Press, Cambridge (2009)
 7. Klement, E.P., Mesiar, R., Pap, E.: Triangular Norms. Kluwer Academic Publishers, Dordrecht (2000)
 8. Mesiar, R., Vivona, D.: Two-step integral with respect to a fuzzy measure. Tatra Mount. Math. Publ. 16, 358–368 (1999)
 9. Shilkret, N.: Maxitive measure and integration. Indag. Math. 33, 109–116 (1971)
10. Sugeno, M.: Theory of Fuzzy Integrals and Applications. PhD Thesis, Tokyo Inst. of Technology, Tokyo (1974)
11. Torra, V., Narukawa, Y.: Twofold integral and multistep Choquet integral. Kybernetika 40, 39–50 (2004)
12. Wilkin, T., Beliakov, G.: Weakly monotone aggregation functions. Fuzzy Sets and Systems (2013) (submitted)
13. Yager, R.: On ordered weighted averaging aggregation operators in multi-criteria decision making. IEEE Transactions on Systems, Man and Cybernetics 18, 183–190 (1988)

Solving Multi-criteria Decision Problems under Possibilistic Uncertainty Using Optimistic and Pessimistic Utilities

Nahla Ben Amor[1], Fatma Essghaier[1,2], and Hélène Fargier[2]

[1] LARODEC, Institut Supérieur de Gestion Tunis, Tunisie
[2] IRIT-CNRS, UMR 5505 Université de Toulouse, France
nahla.benamor@gmx.fr, essghaier.fatma@gmail.com , fargier@irit.fr

Abstract. This paper proposes a qualitative approach to solve multi-criteria decision making problems under possibilistic uncertainty. Depending on the decision maker attitude with respect to uncertainty (i.e. optimistic or pessimistic) and on her attitude with respect to criteria (i.e. conjunctive or disjunctive), four *ex-ante* and four *ex-post* decision rules are defined and investigated. In particular, their coherence w.r.t. the principle of monotonicity, that allows Dynamic Programming is studied.

1 Introduction

A popular criterion to compare decisions under risk is the expected utility model (*EU*) axiomatized by Von Neumann and Morgenstern [9]. This model relies on a probabilistic representation of uncertainty: an elementary decision is represented by a probabilistic lottery over the possible outcomes. The preferences of the decision maker are supposed to be captured by a utility function assigning a numerical value to each consequence. The evaluation of a lottery is then performed through the computation of its expected utility (the greater, the better). When several independent criteria are to be taken into account, the utility function is the result of the aggregation of several utility functions u_i (one for each criterion). The expected utility of the additive aggregation can then be used to evaluate the lottery, and it is easy to show that it is equal to the additive aggregation of the mono-criterion expected utilities.

These approaches presuppose that both numerical probability and additive utilities are available. When the information about uncertainty cannot be quantified in a probabilistic way the topic of possibilistic decision theory is often a natural one to consider. Giving up the probabilistic quantification of uncertainty yields to give up the EU criterion as well. The development of possibilistic decision theory has lead to the proposition and the characterization of (mono-criterion) possibilistic counterparts of expected utility: Dubois and Prade [3] propose two criteria based on possibility theory, an optimistic and a pessimistic one, whose definitions only require a finite ordinal scale for evaluating both utility and plausibility. Likewise, qualitative approaches of multi-criteria decision

A. Laurent et al. (Eds.): IPMU 2014, Part III, CCIS 444, pp. 269–279, 2014.

making have been advocated, leading to the use of Sugeno Integrals (see e.g. [1,8]) and especially weighted maximum and weighted minimum [2].

In this paper, we consider possibilistic decision problems in the presence of multiple criteria. The difficulty is here to make a double aggregation. Several attitudes are possible: shall we consider the pessimistic/optimistic utility value of a weighted min (or max)? or shall we rather aggregate with a weighted min (or max) the individual pessimistic (or optimistic) utilities provided by the criteria? In short, shall we proceed in an *ex-ante* or *ex-post* way?

The remainder of the paper is organized as follows: Section 2 presents a refresher on possibilistic decision making under uncertainty using Dubois and Prade's pessimistic and optimistic utilities, on one hand, and on the qualitative approaches of MCDM (mainly, weighted min and weighted max), on the other hand. Section 3 develops our proposition, defining four ex-ante and four ex-post aggregations, and shows that when the decision maker attitude is homogeneous, i.e. either fully min-oriented or fully max-oriented, the ex-ante and the ex-post possibilistic aggregations provide the same result. Section 4 studies the monotonicity of these decision rules, in order to determine the applicability of Dynamic Programming to sequential decision making problems.[1]

2   Background on One-Stage Decision Making in a Possibilistic Framework

2.1   Decision Making under Possibilistic Uncertainty (U^+ and U^-)

Following Dubois and Prade's possibilistic approach of decision making under qualitative uncertainty, a one stage decision can be seen as a possibility distribution over a finite set of outcomes also called a (simple) *possibilistic lottery* [3]. Since we consider a finite setting, we shall write $L = \langle \lambda_1/x_1, \ldots, \lambda_n/x_n \rangle$ s.t. $\lambda_i = \pi_f(x_i)$ is the possibility that decision f leads to outcome x_i; this possibility degree can also be denoted by $L[x_i]$. We denote $\mathcal{L}$ the set of all simple possibilistic lotteries.

In this framework, a decision problem is thus fully specified by a set Δ of possibilistic lotteries on a set of consequences X and a utility function $u : X \mapsto [0, 1]$. Under the assumption that the utility scale and the possibility scale are commensurate and purely ordinal, Dubois and Prade have proposed the following qualitative degrees for evaluating any simple lottery $L = \langle \lambda_1/x_1, \ldots, \lambda_n/x_n \rangle$:

Optimistic utility (U^+) [3,15,16]: $U^+(L) = \max_{x_i \in X} \min(\lambda_i, u(x_i))$ (1)

Pessimistic utility (U^-) [3,14]: $U^-(L) = \min_{x_i \in X} \max((1 - \lambda_i), u(x_i))$ (2)

---

[1] Proofs relative to this paper are omitted for lack of space; they are available on ftp://ftp.irit.fr/IRIT/ADRIA/PapersFargier/ipmu14.pdf

The value $U^-(L)$ is high only if L gives good consequences in every "rather plausible" state. This criterion generalizes the Wald criterion, which estimates the utility of an act by its worst possible consequence. $U^-(L)$ is thus "pessimistic" or "cautious". On the other hand, $U^+(L)$ is a mild version of the maximax criterion which is "optimistic", or "adventurous": act L is good as soon as it is totally plausible that it gives a good consequence.

2.2 Multi-criteria Decision Making (MCDM) Using Agg^+ and Agg^-

The previous setting assumes a clear ranking of X by a single preference criterion, hence the use of a single utility function u. When several criteria, say a set $C = \{c_1...c_p\}$ of p criteria, have to be taken into account, u shall be replaced by a vector $\boldsymbol{u} = \langle u_1, \ldots, u_p \rangle$ of utility functions $u_j : X \mapsto [0,1]$ and the global (qualitative) utility of each consequence $x \in X$ can be evaluated either in a conjunctive, cautious, way according to the Wald aggregation $(Agg^-(x) = \min_{c_j \in C} u_j(x))$, or in an disjunctive way according to its max-oriented counterpart $(Agg^+(x) = \max_{c_j \in C} u_j(x))$. When the criteria are not equally important, a weight $w_j \in [0,1]$ can be associated to each c_j. Hence the following definitions relative to multi-criteria utilities [2]:

$$Agg^+(x) = \max_{c_j \in C} \min(w_j, u_j(x)). \tag{3}$$

$$Agg^-(x) = \min_{c_j \in C} \max((1 - w_j), u_j(x)). \tag{4}$$

These utilities are particular cases of the Sugeno integral [1,8,13]:

$$Agg_{\gamma,u}(L) = \max_{\lambda \in [0,1]} \min(\lambda, \gamma(F_\lambda)) \tag{5}$$

where $F_\lambda = \{c_j \in C, u_j(x) \geq \lambda\}$, γ is a monotonic set-function that reflects the importance of criteria's set. Agg^+ is recovered when γ is the *possibility measure* based on the weight distribution $(\gamma(E) = \max_{c_j \in E} w_j)$, and Agg^- is recovered when γ corresponds to *necessity measure* $(\gamma(E) = \min_{c_j \notin E}(1 - w_j))$.

3   Multi-criteria Decision Making under Possibilistic Uncertainty

Let us now study possibilistic decision making in a multi-criteria context. Given a set X of consequences, a set C of independent criteria we define a multi-criteria decision problem under possibilistic uncertainty as triplet $\langle \Delta, \boldsymbol{w}, \boldsymbol{u} \rangle^2$ where:

---

[2] Classical problems of decision under possibilistic uncertainty are recovered when $|C| = 1$; Classical MCDM problems are recovered when all the lotteries in Δ associate possibility 1 to some x_i and possibility 0 to all the other elements of X: Δ is identified to X, i.e. is a set of "alternatives" for the MCDM decision problem.

- Δ is a set of possibilistic lotteries;
- $\boldsymbol{w} \in [0,1]^p$ is a weighting vector: w_j denotes the weight of criterion c_j;
- $\boldsymbol{u} = \langle u_1, \ldots, u_p \rangle$ is a vector of p utility functions on X: $u_j(x_i) \in [0,1]$ is the utility of x_i according to criterion c_j;

Our aim consists in comparing lotteries according to decision maker's preferences relative to their different consequences (captured by the utility functions) and the importance of the criteria (captured by the weighting vector). To do this, we can proceed in two different ways namely *ex-ante* or *ex-post*:

- The *ex-ante* aggregation consists in first determining the aggregated utilities (Agg^+ or Agg^-) relative to each possible consequence x_i of L and then combine them with the possibility degrees.
- The *ex-post* aggregation consists in computing the (optimistic or pessimistic) utilities relative to each criterion c_j, and then perform the aggregation (Agg^+ or Agg^-) using the criteria's weights.

We borrow this terminology from economics and social welfare economics, were agents play the role played by criteria in the present context (see e.g. [7,10]). Setting the problem in a probabilistic context, these works have shown that the two approaches can lead to different results (this is the so-called "timing effect") and that their coincide iff the collective utility is affine. As a matter of fact, it is easy to show that the expected utility of the weighted sum is the sum of the expected utilities.

Let us go back to possibilistic framework. The decision maker's attitude with respect to uncertainty can be either optimistic (U^+) or pessimistic (U^-) and her attitude with respect to criteria can be either conjunctive (Agg^-) or disjunctive (Agg^+), hence the definition of four approaches of MCDM under uncertainty, namely U^{++}, U^{+-}, U^{-+} and U^{--}; the first (resp. the second) indices denoting the attitude of the decision maker w.r.t. uncertainty (resp. criteria).

Each of these utilities can be computed either *ex-ante* or *ex-post*. Hence the definition of eight utilities:

Definition 1. *Given a possibilistic lottery L on X, a set of criteria C defining a vector of utility functions $\boldsymbol{u}$ and weighting vector $\boldsymbol{w}$, let:*

$$U^{++}_{ante}(L) = \max_{x_i \in X} \min(L[x_i], \max_{c_j \in C} \min(u_j(x_i), w_j)). \tag{6}$$

$$U^{--}_{ante}(L) = \min_{x_i \in X} \max((1 - L[x_i]), \min_{c_j \in C} \max(u_j(x_i), (1 - w_j))). \tag{7}$$

$$U^{+-}_{ante}(L) = \max_{x_i \in X} \min(L[x_i], \min_{c_j \in C} \max(u_j(x_i), (1 - w_j))). \tag{8}$$

$$U^{-+}_{ante}(L) = \min_{x_i \in X} \max((1 - L[x_i]), \max_{c_j \in C} \min(u_j(x_i), w_j)). \tag{9}$$

$$U^{++}_{post}(L) = \max_{c_j \in C} \min(w_j, \max_{x_i \in X} \min(u_j(x_i), L[x_i])). \tag{10}$$

$$U^{--}_{post}(L) = \min_{c_j \in C} \max((1 - w_j), \min_{x_i \in X} \max(u_j(x_i), (1 - L[x_i]))). \qquad (11)$$

$$U^{+-}_{post}(L) = \min_{c_j \in C} \max((1 - w_j), \max_{x_i \in X} \min(u_j(x_i), L[x_i])). \qquad (12)$$

$$U^{-+}_{post}(L) = \max_{c_j \in C} \min(w_j, \min_{x_i \in X} \max(u_j(x_i), (1 - L[x_i]))). \qquad (13)$$

Interestingly, the optimistic aggregations are related to their pessimistic counterparts by duality as stated by the following proposition.

Proposition 1. *Let $P = \langle \Delta, \boldsymbol{w}, \boldsymbol{u} \rangle$ be a qualitative decision problem, let $P^\tau = \langle \Delta, \boldsymbol{w}, \boldsymbol{u}^\tau \rangle$ be the inverse problem, i.e. the problem such that for any $x_i \in X, c_j \in C, u^\tau_j(x_i) = 1 - u_j(x_i)$. Then, for any $L \in \Delta$:*

$$\begin{array}{ll}
U^{++}_{ante}(L) = 1 - U^{\tau--}_{ante}(L) & U^{++}_{post}(L) = 1 - U^{\tau--}_{post}(L) \\
U^{--}_{ante}(L) = 1 - U^{\tau++}_{ante}(L) & U^{--}_{post}(L) = 1 - U^{\tau++}_{post}(L) \\
U^{+-}_{ante}(L) = 1 - U^{\tau-+}_{ante}(L) & U^{+-}_{post}(L) = 1 - U^{\tau-+}_{post}(L) \\
U^{-+}_{ante}(L) = 1 - U^{\tau+-}_{ante}(L) & U^{-+}_{post}(L) = 1 - U^{\tau+-}_{post}(L)
\end{array}$$

As previously said the *ex-ante* and the *ex-post* approaches coincide in the probabilistic case. Likewise, the following Proposition 2 shows that when the decision maker attitude is homogeneous, i.e. either fully min-oriented or fully max-
oriented, the *ex-ante* and the *ex-post* possibilistic aggregations provide the same
result.

Proposition 2. *For any $L \in \mathcal{L}$, $U^{++}_{ante}(L) = U^{++}_{post}(L)$ and $U^{--}_{ante}(L) = U^{--}_{post}(L)$.*

Hence, for any multi-criteria decision problem under possibilistic uncertainty,
U^{++}_{ante} (resp. U^{--}_{ante}) is equal to U^{++}_{post} (resp. U^{--}_{post}). Such an equivalence between
the *ex-ante* and *ex-post* does not hold for U^{+-} nor for U^{-+}, as shown by the
following counter-example.

Counter-example 1. *Consider a set C of two equally important criteria c_1
and c_2, and a lottery L (cf. Figure 1) leading to two equi-possible consequences
x_1 and x_2 such that x_1 is good for c_1 and bad for c_2, and x_2 is bad for c_1 and
good for c_2; i.e. $L[x_1] = L[x_2] = 1$, $w_1 = w_2 = 1$, $u_1(x_1) = u_2(x_2) = 1$ and
$u_2(x_1) = u_1(x_2) = 0$.*

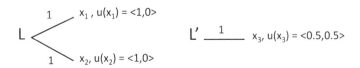

Fig. 1. Lotteries L and L' relative to counter-example 1

We can check that $U^{+-}_{ante}(L) = 0 \neq U^{+-}_{post}(L) = 1$:

$$U_{ante}^{+-}(L) = \max \ (\ \min(L[x_1], \min(\max(u_1(x_1), (1 - w_1)), \max(u_2(x_1), (1 - w_2)))),$$
$$\min(L[x_2], \min(\max(u_1(x_2), (1 - w_1)), \max(u_2(x_2), (1 - w_2)))))).$$
$$= \max \ (\ \min(1, \min(\max(1, (1 - 1)), \max(0, (1 - 1)))) \ ,$$
$$\min(1, \min(\max(0, (1 - 1)), \max(1, (1 - 1))))))$$
$$= 0.$$
$$U_{post}^{+-}(L) = \min \ (\ \max((1 - w_1), \max(\min(u_1(x_1), L[x_1]), \min(u_1(x_2), L[x_2]))),$$
$$\max((1 - w_2), \max(\min(u_2(x_1), L[x_1]), \min(u_2(x_2), L[x_2]))))).$$
$$= \min \ (\ \max((1 - 1), \max(\min(1, 1), \min(0, 1))) \ ,$$
$$\max((1 - 1), \max(\min(0, 1), \min(1, 1)))))$$
$$= 1 \ .$$

The ex-ante and ex-post approaches may lead to different rankings of lotteries. Consider for instance, a lottery L' leading to the consequence x_3 for sure, i.e. $L'[x_3] = 1$ and $L'[x_i] = 0, \forall i \neq 3$ (such a lottery is called a constant lottery), with $u_1(x_3) = u_2(x_3) = 0.5$. It is easy to check that $U_{ante}^{+-}(L') = U_{post}^{+-}(L') = 0.5$ i.e. $U_{ante}^{+-}(L) < U_{ante}^{+-}(L')$ while $U_{post}^{+-}(L) > U_{post}^{+-}(L')$.

Using the same lotteries L and L', we can show that:
$U_{ante}^{-+}(L) = 1 \neq U_{post}^{-+}(L) = 0$ *and that* $U_{ante}^{-+}(L') = U_{post}^{-+}(L') = 0.5$; *then* $U_{post}^{-+}(L') > U_{post}^{-+}(L)$ *while* $U_{ante}^{-+}(L') < U_{ante}^{-+}(L)$: *like* U^{+-}, U^{-+} *are subject to the timing effect.*

In summary, U^{-+} and U^{+-} suffer from the timing effect, but U^{--} and U^{++} do not.

4 Multi-criteria Sequential Decision Making under Possibilistic Uncertainty

Possibilistic sequential decision making relies on *possibilistic compound lotteries*[3], that is a possibility distributions over (simple or compound) lotteries. Compound lotteries indeed allow the representation of decision policies or "strategies", that associate a decision to each decision point: the execution of the decision may lead to several more or less possible situations, where new decisions are to be made, etc. For instance, in a two stages decision problem, a first decision is made and executed; then, depending on the observed situation, a new, one stage, decision is to be made, that lead to the final consequences. The decisions at the final stage are simple lotteries, and the decision made at the first stage branches on each of them. The global strategy thus defines to a compound lottery .

To evaluate a strategy by U^+, U^- or, in the case of MCDM under uncertainty by any of the eight aggregated utility proposed in Section 3, the idea is to "reduce" its compound lottery into an equivalent simple one. Consider a compound lottery $L = \langle \lambda_1/L_1, \ldots, \lambda_m/L_m \rangle$; the possibility of getting consequence $x_i \in X$ from one of its sub lotteries L_k is $\pi_{k,i} = \min(\lambda_k, L_k[x_i])$ (for the shake of simplicity, suppose that L'_k are simple lotteries; the principle trivially extends to the general case). Hence, the possibility of getting x_i from L is the *max*, over all the L_k's, of $\pi_{k,i}$. Thus, [3] have proposed to reduce a compound lottery $\langle \lambda_1/L_1, \ldots, \lambda_m/L_m \rangle$ into a simple lottery, denoted by $Reduction(\langle \lambda_1/L_1, \ldots, \lambda_m/L_m \rangle)$, that is considered as equivalent to the compound one: $Reduction(\langle \lambda_1/L_1, \ldots, \lambda_m/L_k \rangle)$ is the simple lottery that associate

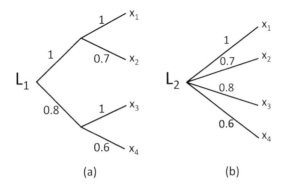

Fig. 2. A compound lottery L_1 (a) and its reduction L_2 (b)

to each x_i the possibility degree $\max_{k=1..m} \min(\lambda_k, L_k[x_i])$ (with $L[x_i] = 0$ when none of the L_k's give a positive possibility degree to consequence x_i). See Figure 2 for an example.

From a practical point of view, sequential decision problems are generally stated through the use of compact representation formalisms, such as possibilistic decision trees [4], possibilistic influence diagrams [5,6] or possibilistic Markov decision processes [11,12]. The set of potential strategies to compare, Δ, is generally exponential w.r.t. the input size. So, an explicit evaluation of each strategy in Δ is not realistic. Such problems can nevertheless be solved efficiently, without an explicit evaluation of the strategies, by Dynamic Programming algorithms as soon as the decision rule leads to transitive preferences and satisfies the principle of weak monotonicity. Formally, for any decision rule O (e.g. U^+, U^- or even any of the decision rules proposed in the previous Section) over possibilistic lotteries, $\succeq_O$ is said to be weakly monotonic iff whatever L, L' and L'' and whatever (α, β) such that $\max(\alpha, \beta) = 1$:

$$L \succeq_O L' \Rightarrow \langle \alpha/L, \beta/L'' \rangle \succeq_O \langle \alpha/L', \beta/L'' \rangle. \tag{14}$$

Such property ensures that each sub-strategy of an optimal strategy is optimal in its sub-problem. This allows *Dynamic Programming* algorithms to build optimal strategies in an incremental way (e.g. in decision tree, from the last decision to the root of the tree).

[5,4] have shown that U^+ and U^- are monotonic. Let us now study whether it is also the case for the *ex-ante* and *ex-post* rules proposed in the previous Section. The *ex-ante* approaches are the easiest to handle: once the vectors of utilities have been aggregated according to Agg^- (resp. Agg^+), these approaches collapse to the classical U^+ and U^- approaches. It is then easy to show that:

Proposition 3. $U^{++}_{ante}, U^{--}_{ante}, U^{+-}_{ante}$ and U^{-+}_{ante} satisfy the weak monotonicity.

Concerning U^{++}_{post} and U^{--}_{post}, recall that the full optimistic and full pessimistic *ex-post* utilities are equivalent to their *ex-ante* counterparts thanks to Proposition 2. This allows us to show that:

Proposition 4. U_{post}^{--} and U_{post}^{++} satisfy the weak monotonicity.

It follows from Propositions 3 and 4 that when the decision is based either on an ex-ante approach, or on U_{post}^{++} or U_{post}^{--}, the algorithms proposed by Sabbadin et al. [12,5] can be used on multi-criteria possibilistic decision trees and influence diagrams after their transformation into single-criterion problems: it is enough to aggregate the vectors of utilities leading to the consequences x into single utilities using Agg^+ (for $U_{ante}^{++}, U_{ante}^{-+}, U_{post}^{++}$) or Agg^- (for $U_{ante}^{--}, U_{ante}^{+-}, U_{post}^{--}$) to get an equivalent single criterion problem where the criterion to optimize is simply U^+ (for $U_{ante}^{++}, U_{ante}^{+-}, U_{post}^{++}$) or U^- (for $U_{ante}^{--}, U_{ante}^{-+}, U_{post}^{--}$).

Such approach cannot be applied when optimizing U_{post}^{-+} or U_{post}^{+-}. First because $U_{post}^{+-}(L) \neq U_{ante}^{+-}(L)$ and $U_{post}^{-+}(L) \neq U_{ante}^{-+}(L)$, i.e. the reduction of the problem to the optimization w.r.t. U^+ (resp. U^-) of a single criterion problem obtained by aggregating the utilities with Agg^- (resp. Agg^+) can lead to a wrong result. Worst, it is not even possible to apply Dynamic Programming, since U_{post}^{-+} and U_{post}^{+-} do not satisfy the weak monotonicity property, as shown by the following counter-example:

Counter-example 2. Let $X = \{x_1, x_2, x_3\}$ and consider two equally important criteria c_1 and c_2 ($w_1 = w_2 = 1$) with : $u_1(x_1) = 1$, $u_1(x_2) = 0.8$, $u_1(x_3) = 0.5$; $u_2(x_1) = 0.6$, $u_2(x_2) = 0.8$, $u_2(x_3) = 0.8$. Consider the lotteries $L = \langle 1/x_1, 0/x_2, 0/x_3 \rangle$, $L' = \langle 0/x_1, 1/x_2, 0/x_3 \rangle$ and $L'' = \langle 0/x_1, 0/x_2, 1/x_3 \rangle$: L gives consequence x_1 for sure, L' gives consequence x_2 for sure and L'' gives consequence x_3 for sure. It holds that:

$U_{post}^{-+}(L) = Agg^-(x_1) = \max(1, 0.6) = 1$
$U_{post}^{-+}(L') = Agg^-(x_2) = \max(0.8, 0.8) = 0.8$.

Hence $L >_{U_{post}^{-+}} L'$ with respect to the U_{post}^{-+} rule.
Consider now the compound lotteries $L_1 = \langle 1/L, 1/L'' \rangle$ and $L_2 = \langle 1/L', 1/L'' \rangle$. If the weak monotonicity principle were satisfied, we would get: $L_1 >_{U_{post}^{-+}} L_2$. Since: $Reduction(\langle 1/L, 1/L'' \rangle) = \langle 1/x_1, 0/x_2, 1/x_3 \rangle$ and $Reduction(\langle 1/L', 1/L'' \rangle) = \langle 0/x_1, 1/x_2, 1/x_3 \rangle$. We have:

$U_{post}^{-+}(L_1) = U_{post}^{-+}(Reduction(\langle 1/L, 1/L'' \rangle)) = 0.6.$
$U_{post}^{-+}(L_2) = U_{post}^{-+}(Reduction(\langle 1/L', 1/L'' \rangle)) = 0.8.$

Hence, $L_1 <_{U_{post}^{-+}} L_2$ while $L >_{U_{post}^{-+}} L'$, which contradicts weak monotonicity.

Using the fact that $U_{post}^{+-} = 1 - U_{post}^{\tau -+}$, this counter-example is modified to show that also U_{post}^{+-} does not satisfy the monotonicity principle. Consider two equally important criteria, c_1^τ and c_2^τ with $w_1 = w_2 = 1$ with: $u_1^\tau(x_1) = 0$, $u_1^\tau(x_2) = 0.2$, $u_1^\tau(x_3) = 0.5$; $u_2^\tau(x_1) = 0.4$, $u_2^\tau(x_2) = 0.2$, $u_2^\tau(x_3) = 0.2$. Consider now the same lotteries L, L' and L'' presented above. It holds that:

$U_{post}^{+-}(L) = Agg^-(x_1) = 0 < U_{post}^{+-}(L') = Agg^-(x_2) = 0.2$, while
$U_{post}^{+-}(Reduction(\langle 1/L, 1/L'' \rangle)) = 0.4 > U_{post}^{+-}(Reduction(\langle 1/L', 1/L'' \rangle)) = 0.2.$

The lack of monotonicity of U_{post}^{-+} is not as dramatic as it may seem. When optimizing $U_{post}^{-+}(L)$, the decision maker is looking for a strategy that is good w.r.t. U^- for at least one (important) criterion since we can write:

$$U_{post}^{-+}(L) = \max_{c_j \in C} \min(w_j, U_j^-(L))$$

where $U_j^-(L) = \min_{x_i \in X} \max((1 - L[x_i]), u_j(x_i))$ is the pessimistic utility of L according to the sole criterion c_j. This means that if it is possible to get for each criterion c_j a strategy that optimizes U_j^- (and this can be done by Dynamic Programming, since the pessimistic utility do satisfy the principle of monotonicity), the one with the higher U_{post}^{-+} is globally optimal. Formally:

Proposition 5. *Let $\mathcal{L}$ be the set of lotteries that can be built on X and let:*

- *$\Delta^* = \{L_1^*, \ldots, L_p^*\}$ s.t. $\forall L \in \mathcal{L}, c_j \in C, U_j^-(L_j^*) \geq U_j^-(L)$;*
- *$L^* \in \Delta^*$ s.t. $\forall L_j^* \in \Delta^*$: $\max_{c_j \in C} \min(w_j, U_j^-(L^*)) \geq \max_{c_j \in C} \min(w_j, U_j^-(L_i^*))$.*

It holds that, for any $L \in \mathcal{L}, U_{post}^{-+}(L^*) \geq U_{post}^{-+}(L)$.

Hence, it is enough to optimize w.r.t. each criterion c_j separately to get one optimal strategy for each of them ; a strategy optimal w.r.t. U_{post}^{-+} is then obtained by comparing the U_{post}^{-+} values of these p candidate strategies. .

Let us finally study U_{post}^{+-}; it is always possible to define, for each j, $U_j^+(L) = \max_{x_i \in X} \min(L[x_i], u_j(x_i))$ as the optimistic utility of L according to this sole criterion and to write that $U_{post}^{+-}(L) = \min_{c_j \in C} \max((1 - w_j), U_j^+(L))$. But this remark is helpless, since the lottery L maximizing this quantity is not necessarily among those maximizing the U_j^+'s: one lottery optimal for U_1^+ w.r.t. criterion c_1 can be very bad for U_2^+ and thus bad for U_{post}^{+-}. The polynomial approach proposed in the previous paragraph for optimizing U_{post}^{-+} does not hold for U_{post}^{+-}.

5 Conclusion

This paper has provided a first decision theoretical approach for evaluating multi-criteria decision problems under possibilistic uncertainty. The combination of the multi-criteria dimension, namely the conjunctive aggregation with a weighted min (Agg^-) or the disjunctive aggregation with a weighted max (Agg^-) and the decision maker's attitude with respect to uncertainty (i.e. optimistic utility U^+ or pessimistic utility U^-) leads to four approaches of MCDM under possibilistic uncertainty. Considering that each of these utilities can be computed either *ex-ante* or *ex-post*, we have proposed the definition of eight aggregations, that eventually reduce to six: U_{ante}^{++} (resp. U_{ante}^{--}) has been shown to coincide with U_{post}^{++} (resp. U_{post}^{--}); such a coincidence does not happen for U^{+-} and U^{-+}, that suffer from timing effect.

Then, in order to use these decision rules in sequential decision problems, we have proven that all *ex-ante* utilities (i.e. $U_{ante}^{++}, U_{ante}^{--}, U_{ante}^{+-}, U_{ante}^{-+}$) satisfy the weak monotonicity while for the *ex-post* utilities, only U_{post}^{++} and U_{post}^{--} satisfy this property. This result means that Dynamic Programming algorithms can be used to compute strategies that are optimal w.r.t. the rules. We have also shown that the optimization of U_{post}^{+-} can be handled thanks to a call of a series of optimization of pessimistic utilities (one for each criterion). The question of the optimization of U_{post}^{-+} still remains open.

This preliminary work call for several developments. From a theoretical point of view we have to propose a general axiomatic characterization of our six decision rules. Moreover, considering that the possibilistic aggregations used here are basically specializations of the Sugeno integral, we aim at generalizing the study of MCDM decision making under uncertainty through the development of double Sugeno Integrals. We shall also extend the approach, considering that the importance of the criteria is not absolute but may depend on the node considered. From a more practical point of view, we shall propose and test suitable algorithms to solve sequential qualitative multi-criteria decision problems, e.g. influence diagrams and decision trees.

References

1. Dubois, D., Marichal, J.L., Prade, H., Roubens, M., Sabbadin, R.: The use of the discrete sugeno integral in decision-making: A survey. International Journal of Uncertainty, Fuzziness and Knowledge-Based Systems 9(5), 539–561 (2001)
2. Dubois, D., Prade, H.: Weighted minimum and maximum operations in fuzzy set theory. Journal of Information Sciences 39, 205–210 (1986)
3. Dubois, D., Prade, H.: Possibility theory as a basis for qualitative decision theory. In: Proceedings of IJCAI, pp. 1924–1930 (1995)
4. Fargier, H., Ben Amor, N., Guezguez, W.: On the complexity of decision making in possibilistic decision trees. In: Conference on UAI, pp. 203–210 (2011)
5. Garcias, L., Sabbadin, R.: Possibilistic influence diagrams. In: Proceedings of ECAI, pp. 372–376 (2006)
6. Guezguez, W., Ben Amor, N., Mellouli, K.: Qualitative possibilistic influence diagrams based on qualitative possibilistic utilities. EJOR 195, 223–238 (2009)
7. Harsanyi, J.C.: Cardinal welfare, individualistic ethics, and interpersonal comparisons of utility. Journal of Political Economy 63(4), 309–321 (1955)
8. Marichal, J.L.: An axiomatic approach of the discrete sugeno integral as a tool to aggregate interacting criteria in a qualitative framework. IEEE T. on Fuzzy Systems 9(1), 164–172 (2001)
9. Morgenstern, O., Neumann, J.V.: Theory of Games and Economic Behavior, 2nd edn. Princeton University Press (1947)
10. Myerson, R.B.: Utilitarianism, egalitarianism, and the timing effect in social choice problems. Econometrica 49(4), 883–897 (1981)
11. Sabbadin, R.: Empirical comparison of probabilistic and possibilistic markov decision processes algorithms. In: Proceedings of European Conference on Artificial Intelligence, pp. 586–590 (2000)
12. Sabbadin, R.: Possibilistic markov decision processes. Engineering Applications of Artificial Intelligence 14, 287–300 (2001)

13. Sugeno, M.: Theory of fuzzy integrals and its applications. PhD thesis, Tokyo Institute of Technology (1974)
14. Whalen, T.: Decision making under uncertainty with various assumptions about available information. IEEE T. on SMC 14, 888–900 (1984)
15. Yager, R.: Possibilistic decision making. IEEE T. on SMC 9, 388–392 (1979)
16. Zadeh, L.A.: Fuzzy sets as a basis for a theory of possibility. Fuzzy Sets and Systems 1, 3–28 (1978)

Copula - Based Generalizations
of OWA Operators

Radko Mesiar and Andrea Stupňanová

Slovak University of Technology, Faculty of Civil Engineering
Radlinského 11,
813 68 Bratislava, Slovak Republic
{radko.mesiar,andrea.stupnanova}@stuba.sk

Abstract. We recall first graded classes of copula - based integrals and their specific form when a finite universe X is considered. Subsequently, copula - based generalizations of OWA operators are introduced, as copula - based integrals with respect to symmetric capacities. As a particular class of our new operators, recently introduced OMA operators are obtained. Several particular examples are introduced and discussed to clarify our approach.

Keywords: copula, decomposition integral, OMA operator, OWA operator, symmetric capacity, universal integral.

1 Introduction

Ordered weighted averages, in short OWA operators, were introduced in 1988 by Yager [20] in order to unify in one class arithmetic mean, maximum and minimum operators. Formally, OWA operators can be seen as convex combinations of order statistics. Namely, for a given weighting vector $\mathbf{w} \in [0,1]^n$, $\sum_{i=1}^{n} w_i = 1$, $\mathrm{OWA}_{\mathbf{w}} : [0,1]^n \to [0,1]$ is given by

$$\mathrm{OWA}_{\mathbf{w}}(\mathbf{x}) = \sum_{i=1}^{n} w_i \, x_{(i)}, \tag{1}$$

where $(\cdot)$ is a permutation of $\{1, \dots n\}$ such that $x_{(1)} \geq \cdots \geq x_{(n)}$, i.e., $x_{(1)} = \max\{x_1, \dots x_n\}$, $x_{(2)}$ is the second biggest value from the input n-tuple $\mathbf{x} = \{x_1, \dots, x_n\}$, $\dots$, $x_{(n)} = \min\{x_1, \dots, x_n\}$. OWA operators are popular both from theoretical and applied point of view, see e.g. edited volumes [22,23].
Note also that there are several generalizations of OWA operators, such as weighted OWA (WOWA) [19], induced OWA (IOWA) [21], generalized OWA (GOWA) [1], etc.

An important step in understanding the nature of OWA operators was their representation by means of Choquet integral due to Grabisch [5], compare also [13]. For a given weighting vector $\mathbf{w} \in [0,1]^n$, consider the universe $X = \{1, \dots, n\}$ and introduce a *capacity* $m_{\mathbf{w}} : 2^X \to [0,1]$ given by

A. Laurent et al. (Eds.): IPMU 2014, Part III, CCIS 444, pp. 280–288, 2014.

$$m_{\mathbf{w}}(A) = \sum_{i=1}^{|A|} w_i = v_{|A|},$$

where $|A|$ is the cardinality of the set A, and $v_0 = 0$ by convention. The capacity $m_{\mathbf{w}}$ is called a *symmetric capacity*, reflecting the fact that for any integral the corresponding functional is symmetric. Observe also that $v_n = 1 = m_{\mathbf{w}}(X)$, and that $v_i - v_{i-1} = w_i$, $i = 1, \ldots, n$. Formally, the vector $\mathbf{v} \in [0,1]^n$ can be seen as cumulative version of the weighting vector $\mathbf{w} \in [0,1]^n$. Denoting the Choquet integral [2] with respect to a capacity m as Ch_m, Grabisch's representation of OWA operators can be written as

$$\text{OWA}_{\mathbf{w}}(\mathbf{x}) = Ch_{m_{\mathbf{w}}}(\mathbf{x}). \tag{2}$$

Due to Schmeidler's axiomatic characterization of the Choquet integral [15], OWA operators can be seen as symmetric comonotone additive aggregation functions [6], i.e., non-decreasing mappings $A : [0,1]^n \to [0,1]$ satisfying the boundary conditions $A(\mathbf{0}) = 0, A(\mathbf{1}) = 1$, and such that :

$\diamond$ for each $\mathbf{x}, \mathbf{y} \in [0,1]^n$ satisfying $\mathbf{x} + \mathbf{y} \in [0,1]^n$ and $(x_i - x_j)(y_i - y_j) \geq 0$ for each $i, j \in \{1, \ldots, n\}$, it holds

$$A(\mathbf{x} + \mathbf{y}) = A(\mathbf{x}) + A(\mathbf{y}). \tag{3}$$

$\diamond$ for each permutation σ of $\{1, \ldots, n\}$, and each $\mathbf{x} \in [0,1]^n$, $A(\mathbf{x}) = A(x_{\sigma(1)}, \ldots, x_{\sigma(n)})$. Observe that then $A = \text{OWA}_{\mathbf{w}}$, where the weighting vector $\mathbf{w} \in [0,1]^n$ is linked to the cumulative vector $\mathbf{v} \in [0,1]^n$ determined by $v_i = A(\underbrace{1, \ldots, 1}_{i-\text{times}}, 0, \ldots, 0)$.

Recently, ordered modular averanges (OMA operators) were introduced in [11], characterized as symmetric comonotone modular idempotent aggregation functions, i.e., the comonotone additivity (3) was replaced by the idempotency

$$A(x, \ldots, x) = x \text{ for all } x \in [0,1],$$

and the comonotone modularity

$$A(\mathbf{x} \vee \mathbf{y}) + A(\mathbf{x} \wedge \mathbf{y}) = A(\mathbf{x}) + A(\mathbf{y})$$

valid for all comonotone $\mathbf{x}, \mathbf{y} \in [0,1]^n$. OMA operators were shown to coincide with copula - based integrals introduced by Klement et al. [9,8] with respect to symmetric capacities. In particular, considering the independence copula Π, OWA operators are rediscovered.

In the last months, new hierarchical families of copula - based integrals were introduced [12,10]. Here the ideas of universal integrals [8] and decomposition integrals [4] interfered first into the description of functionals being simultaneously universal and decomposition integrals, and afterwards resulted into the introduction of new copula - based families of integrals. The main idea of this contribution is to introduce and study the functionals obtained from the above mentioned copula - based integrals with respect to symmetric capacities. Then, as particular cases, OWA and OMA operators are also covered, and, moreover, several other operators known from the literature, such as ordered weighted maximum OWMax [3].

The paper is organized as follows. In the next section, the basics of copulas are given, and hierarchical families of copula - based integrals introduced in [10] are recalled. Section 3 brings copula - based generalizations of OWA operators and discusses some particular cases. In Section 4, product - based generalizations of OWA operators are studied. Finally, some concluding remarks are added.

2   Copulas and Copula - Based Integrals

A (binary) copula $C : [0, 1]^2 \to [0, 1]$ was introduced by Sklar [17] to characterize the link between marginal and joint distribution functions of random vector (U, V). Axiomatically, a copula is an aggregation function with a neutral element $e = 1$, i.e.,

$$C(x, 1) = C(1, x) = x \quad \text{for all } x \in [0, 1],$$

which is supermodular, i. e.,

$$C(\mathbf{x} \vee \mathbf{y}) + C(\mathbf{x} \wedge \mathbf{y}) \geq C(\mathbf{x}) + C(\mathbf{y}) \quad \text{for all } \mathbf{x}, \mathbf{y} \in [0, 1]^2.$$

For more details on copulas we recommend the monographs [7,14].

In [10], compare also [12], C-based integrals forming a hierarchical family

$$I_C^{(1)} \leq I_C^{(2)} \leq \cdots \leq I_C^{(n)} \leq \cdots \leq I_C \leq I^C \leq \cdots \leq I_{(n)}^C \leq \cdots \leq I_{(2)}^C \leq I_{(1)}^C,$$

were introduced. These integrals act on any measurable space $(X, \mathcal{A})$, however, we will consider them only on a fixed space $(X, 2^X)$, with $X = \{1, \ldots, n\}$. Then

$$I_C^{(n)} = I_C^{(n+1)} = \cdots = I_C = I^C = \cdots = I_{(n+1)}^C = I_{(n)}^C,$$

thus we restrict our considerations to integrals

$$I_C^{(1)} \leq \cdots \leq I_C^{(n-1)} \leq I_C^{(n)} = I_{(n)}^C \leq I_{(n-1)}^C \leq \cdots \leq I_{(1)}^C.$$

For a given capacity $m : 2^X \to [0, 1]$, and any input vector $\mathbf{x} \in [0, 1]^n$, they are respectively defined by, for $i = 1, \ldots, n$,

$$I_C^{(i)} = \max \left\{ \sum_{j=1}^{i} \left[C\left(x_{(k_j)}, m(\mathbf{x} \geq x_{(k_j)})\right) - C\left(x_{(k_j)}, m(\mathbf{x} \geq x_{(k_{j-1})})\right)\right] \mid \right.$$

$$\left. 1 \leq k_1 < \cdots < k_i \leq n \right\}, \quad (4)$$

and by

$$I_{(i)}^C = \min\left\{ \sum_{j=1}^{i} \left[C\left(x_{(k_j)}, m(\mathbf{x} \geq x_{(k_{j+1}-1)})\right) - C\left(x_{(k_j)}, m(\mathbf{x} \geq x_{(k_j-1)})\right)\right] \mid \right.$$

$$\left. 1 = k_1 < \cdots < k_i \leq n \right\}, \quad (5)$$

where $k_j \in \{1, \ldots, n\}$, $m(\mathbf{x} \geq t) = m(\{i \in X | x_i \geq t\})$, and $x_{(k_0)} = 2$ by convention (i.e., $m(\mathbf{x} \geq x_{(k_0)}) = 0$) and $k_{i+1} = \begin{cases} \min\{r \mid x_{(r)} = 0\} & \text{if } x_{(n)} = 0 \\ n+1 & \text{else.} \end{cases}$

Observe that the integrals $I_{(n)}^C = I_C^{(n)}$ can be represented as follows, compare [9,8]:

$$(I_{(n)}^C)_m(\mathbf{x}) = P_C\left(\{(x,y) \in [0,1]^2 | y \leq m(\mathbf{x} \geq x)\}\right) =$$
$$= P_C\left(\{(x,y) \in [0,1]^2 | y < m(\mathbf{x} \geq x)\}\right) = (I_C^{(n)})_m(\mathbf{x}),$$

where $P_C : \mathcal{B}([0,1]^2) \to [0,1]$ is the unique probability measure related to the copula C (for more details see [14]).

Note that for each copula C,

$$(I_{(1)}^C)_m(\mathbf{x}) = C(x_{(1)}, m(\mathbf{x} > 0))$$

and

$$(I_C^{(1)})_m(\mathbf{x}) = \sup\left\{C(x_i, m(\mathbf{x} \geq x_i)) \mid i \in \{1, \ldots, n\}\right\}.$$

For the product copula $\Pi : [0,1]^2 \to [0,1]$, $\Pi(x,y) = xy$, $I_{(n)}^{\Pi} = I_{\Pi}^{(n)}$ is the Choquet integral on X, while for the comonotonicity copula $M : [0,1]^2 \to [0,1]$, $M(x,y) = \min\{x,y\}$, $I_{(n)}^M = I_M^{(n)}$ is the Sugeno integral [18]. Observe also that $I_{\Pi}^{(1)}$ is the Shilkret integral [16] and

$$I_{\Pi}^{(1)} < I_{\Pi}^{(2)} < \cdots < I_{\Pi}^{(n)} = I_{(n)}^{\Pi} < I_{(n-1)}^{\Pi} < \cdots < I_{(1)}^{\Pi}.$$

On the other hand, $I_M^{(1)} = \cdots = I_{(2)}^M$ is the Sugeno integral, and $I_{(2)}^M < I_{(1)}^M$ (i.e., there is a capacity m and an input vector $\mathbf{x}$ so that $(I_{(2)}^M)_m(\mathbf{x}) < (I_{(1)}^M)_m(\mathbf{x})$).

3   Copula - Based Generalizations of OWA Operators

We propose to generalize OWA operators as copula - based integrals with respect to symmetric measures.

Definition 1. *Let $C : [0,1]^2 \to [0,1]$ be a fixed copula, let $\mathbf{w} \in [0,1]^n$ be a weighting vector and consider an integral parameter $i \in \{1, \ldots, 2n-1\}$. Then the operator $\mathrm{COWA_w}^{(i)} : [0,1]^n \to [0,1]$ given by*

$$\mathrm{COWA_w}^{(i)} = (I_C^{(i)})_{m_{\mathbf{w}}}(\mathbf{x}) \quad \text{if } i \leq n$$

and

$$\text{COWA}_{\mathbf{w}}^{(i)} = (I_{(2n-i)}^C)_{m_{\mathbf{w}}}(\mathbf{x}) \quad \text{if } i \geq n$$

is called a copula C - based generalized OWA operator of order i.

Theorem 1. *Let* $\mathbf{w} \in [0,1]^n$ *be a weighting vector and let* $\mathbf{v} \in [0,1]^n$ *be the corresponding cumulative vector. Then,*

$\diamondsuit$ *if* $i \leq n$

$$\text{COWA}_{\mathbf{w}}^{(i)}(\mathbf{x}) = \max \left\{ \sum_{j=1}^{i} \left[C(x_{(k_j)}, v_{k_j}) - C(x_{(k_j)}, v_{k_{j-1}}) \right] \Big| 1 \leq k_1 < \cdots < k_i \leq n \right\},$$

$\diamondsuit$ *if* $i \geq n$

$$\text{COWA}_{\mathbf{w}}^{(i)}(\mathbf{x}) = \min \left\{ \sum_{j=1}^{2n-i} \left[C(x_{(k_j)}, v_{k_{j+1}-1}) - C(x_{(k_j)}, v_{k_j-1}) \right] \Big| 1 = k_1 < \cdots < k_{2n-i} \leq n \right\}.$$

Proposition 1. *C-based generalized OWA operators of order n coincide with OMA operators introduced in [11]. In particular,* $\Pi\text{OWA}_{\mathbf{w}}^{(n)} = \text{OWA}_{\mathbf{w}}$ *and* $\text{MOWA}_{\mathbf{w}}^{(n)} = \text{OWMax}_{\mathbf{w}}$.

Proposition 2. *For a given weighting vector* $\mathbf{w} \in [0,1]^n$, *let* $\mathbf{v} \in [0,1]^n$ *be the corresponding cumulative vector. Then, for any given copula C,*

$$\text{COWA}_{\mathbf{w}}^{(1)}(\mathbf{x}) = \sup \left\{ C(x_{(j)}, v_j) \big| j \in \{1, \ldots, n\} \right\},$$

and

$$\text{COWA}_{\mathbf{w}}^{(2n-1)}(\mathbf{x}) = C(x_{(1)}, v_k), \quad \text{where } k = \left| \{j \in \{1, \ldots, n\} \big| x_j > 0\} \right|.$$

Observe that $\Pi\text{OWA}_{\mathbf{w}}^{(1)}(\mathbf{x}) = \sup\{x_{(j)} \cdot v_j \big| j \in \{1, \ldots, n\}\}$ can be seen as the biggest area of a rectangle $R = [0,u] \times [0,v]$ contained in the area $\{(x,y) \in [0,1]^2 \big| y \leq v_{|\{\mathbf{x} \geq x\}|}\}$.
The smallest cumulative vector $\mathbf{v}_* = (0, \ldots, 0, 1)$ is related to the weighting vector $\mathbf{w}_* = (0, \ldots, 0, 1)$ and to the smallest capacity $m_{w_*} = m_* : 2^X \to [0,1]$,

$$m_*(A) = \begin{cases} 1 & \text{if } A = X \\ 0 & \text{else.} \end{cases}$$

Similarly, the greatest cumulative vector $\mathbf{v}^* = (1, 1, \ldots, 1)$ is related to $\mathbf{w}^* = (1, 0, \ldots, 0)$ and to the greatest capacity $m^* : 2^X \to [0,1]$,

$$m^*(A) = \begin{cases} 0 & \text{if } A = \emptyset \\ 1 & \text{else.} \end{cases}$$

It is not difficult to check that, independently of copula C and order i,

$$\text{COWA}_{\mathbf{w}_*}(\mathbf{x}) = \min\{x_1, \ldots, x_n\} = x_{(n)}$$

and

$$\mathrm{COWA}_{\mathbf{w}^*}(\mathbf{x}) = \max\{x_1, \ldots, x_n\} = x_{(1)}.$$

Moreover, for any weighting vector $\mathbf{w}$, and any constant input vector $\mathbf{c} = (c, \ldots, c)$,

$$\mathrm{COWA}_{\mathbf{w}}^{(i)}(\mathbf{c}) = c.$$

The symmetry of all $\mathrm{COWA}_{\mathbf{w}}^{(i)}$ operators is obvious.

4   Product - Based Generalizations of OWA Operators

Based on Theorem 1, we see that if the order $i \leq n$, then

$$\Pi\mathrm{OWA}_{\mathbf{w}}^{(i)}(\mathbf{x}) = \max\left\{ \sum_{j=1}^{i} x_{(k_j)} \cdot (v_{k_j} - v_{k_{j-1}}) \Big| 1 \leq k_1 < \cdots < k_i \leq n \right\}$$

$$= \max\left\{ \sum_{j=1}^{i} \left(\sum_{r=k_{j-1}+1}^{k_j} w_r \right) \cdot x_{(k_j)} \Big| 1 \leq k_1 < \cdots < k_i \leq n \right\},$$

compare also [12], while if $i \geq n$, then

$$\Pi\mathrm{OWA}_{\mathbf{w}}^{(i)}(\mathbf{x}) = \min\left\{ \sum_{j=1}^{2n-i} x_{(k_j)} \cdot (v_{k_{j+1}-1} - v_{k_j-1}) \Big| 1 = k_1 < \cdots < k_{2n-i} \leq n \right\}$$

$$= \min\left\{ \sum_{j=1}^{2n-i} \left(\sum_{r=k_j}^{k_{j+1}-1} w_r \right) \cdot x_{(k_j)} \Big| 1 \leq k_1 < \cdots < k_{2n-i} \leq n \right\}.$$

Observe that each product - based generalization of OWA operators, i.e., each $\Pi\mathrm{OWA}_{\mathbf{w}}^{(i)}$, $i = 1, \ldots, 2n-1$, is positively homogeneous and shift invariant, i.e.,

$$\Pi\mathrm{OWA}_{\mathbf{w}}^{(i)}(c \cdot \mathbf{x}) = c \cdot \Pi\mathrm{OWA}_{\mathbf{w}}^{(i)}(\mathbf{x})$$

and

$$\Pi\mathrm{OWA}_{\mathbf{w}}^{(i)}(\mathbf{d} + \mathbf{x}) = d + \Pi\mathrm{OWA}_{\mathbf{w}}^{(i)}(\mathbf{x})$$

for all $\mathbf{x} \in [0,1]^n$ and $d, c > 0$ such that $c \cdot \mathbf{x}$ and $\mathbf{d} + \mathbf{x} \in [0,1]^n$ (here $\mathbf{d} = (d, \ldots, d)$).

Observe that due to the genuine extension of the product copula Π to the product on $[0, \infty[^2$, all operators $\Pi\mathrm{OWA}^{(i)}$ can be straightforwardly extended to act on $[0, \infty[$ (the same holds for the domain $]-\infty, \infty[$ of all real numbers).

Example 1. Consider the unique additive symmetric capacity $\mu : 2^X \to [0,1]$, $\mu(A) = \frac{|A|}{n}$, related to the constant weighting vector $\mathbf{w} = \left(\frac{1}{n}, \ldots, \frac{1}{n}\right)$, linked to the cumulative vector $\mathbf{v} = \left(\frac{1}{n}, \frac{2}{n}, \ldots, \frac{n}{n}\right)$, and consider the input vector $\mathbf{x} = \left(1, \frac{n-2}{n-1}, \ldots, \frac{1}{n-1}, 0\right)$. Then:

$$\text{ПOWA}_{\mathbf{w}}^{(1)}(\mathbf{x}) \quad = \max\left\{\frac{n-j}{n-1} \cdot \frac{j}{n} \,\Big|\, j \in \{1, \ldots, n\}\right\} = \begin{cases} \frac{n}{4(n-1)} & \text{if} \quad n \text{ is even} \\ \frac{n+1}{4n} & \text{if} \quad n \text{ is odd} \end{cases},$$

$$\text{ПOWA}_{\mathbf{w}}^{(2)}(\mathbf{x}) \quad = \begin{cases} \frac{n}{3(n-1)} & \text{if} \quad n = 3k \\ \frac{n+1}{3n} & \text{else} \end{cases},$$

$$\text{ПOWA}_{\mathbf{w}}^{(n)}(\mathbf{x}) \quad = \frac{1}{2},$$

$$\text{ПOWA}_{\mathbf{w}}^{(2n-2)}(\mathbf{x}) \quad = \begin{cases} \frac{3n-4}{4n-4} & \text{if} \quad n \text{ is even} \\ \frac{3n-1}{4n} & \text{if} \quad n \text{ is odd} \end{cases},$$

and

$$\text{ПOWA}_{\mathbf{w}}^{(2n-1)}(\mathbf{x}) \quad = \frac{n-1}{n}.$$

Observe that in all cases, if $n \to \infty$, then the corresponding $\text{ПOWA}_{\mathbf{w}}$ operators are approaching in limit the corresponding Π-based integral on $X = [0, 1]$, with respect to the standard Lebesque measure λ, and from the identity function $f : X \to [0, 1]$, $f(x) = x$. Thus:

$$\lim_{n\to\infty} \text{ПOWA}_{\mathbf{w}}^{(1)}(\mathbf{x}) \quad = (I_{\Pi}^{(1)})_\lambda(f) \quad = \frac{1}{4} \quad \text{(Shilkret integral)},$$

$$\lim_{n\to\infty} \text{ПOWA}_{\mathbf{w}}^{(2)}(\mathbf{x}) \quad = (I_{\Pi}^{(2)})_\lambda(f) \quad = \frac{1}{3},$$

$$\lim_{n\to\infty} \text{ПOWA}_{\mathbf{w}}^{(n)}(\mathbf{x}) \quad = (I_{\Pi}^{(\infty)})_\lambda(f) \quad = \frac{1}{2} \quad \text{(Choquet integral)},$$

$$\lim_{n\to\infty} \text{ПOWA}_{\mathbf{w}}^{(2n-2)}(\mathbf{x}) \quad = (I_{(2)}^{\Pi})_\lambda(f) \quad = \frac{3}{4},$$

and

$$\lim_{n\to\infty} \text{ПOWA}_{\mathbf{w}}^{(2n-1)}(\mathbf{x}) \quad = (I_{(1)}^{\Pi})_\lambda(f) \quad = 1.$$

5   Concluding Remarks

We have introduced a concept of copula - based generalizations of OWA operators, covering among others also OWMax and OMA operators. Formally, COWA operators can be seen as solutions of optimization problems related to their order i. For example, in the case of $\text{COWA}_{\mathbf{w}}^{(2)}$ ($\text{ПOWA}_{\mathbf{w}}^{(2)}$), we choose two indices $1 \le j < k \le n$, replace the input vector $\mathbf{x} = (x_i, \ldots, x_n)$ by

$$\mathbf{x}_{j,k} = (\underbrace{x_{(j)}, \ldots, x_{(j)}}_{j-\text{times}}, \underbrace{x_{(k)}, \ldots, x_{(k)}}_{(k-j)-\text{times}}, 0, \ldots, 0),$$

and apply to corresponding C-based OMA operator $\text{OMA}_{\mathbf{w}}$ to $\mathbf{x}_{j,k}$ (we compute $\text{OWA}_{\mathbf{w}}(\mathbf{x}_{j,k})$). Then we look for the maximal value $\text{OMA}_{\mathbf{w}}(\mathbf{x}_{j,k})$ ($\text{OWA}_{\mathbf{w}}(\mathbf{x}_{j,k})$), running over all pairs $(j,k) \in \{1,\ldots,n\}^2, j < k$. We expect applications of COWA operators in several branches based on multicriteria decision procedures.

Acknowledgments. The support of the grants VEGA 1/0171/12 and APVV–0073–10 is kindly announced.

References

1. Beliakov, G.: Learning weights in the generalized OWA operators. Fuzzy Optimization and Decision Making 4(2), 119–130 (2005)
2. Choquet, G.: Theory of capacities. Ann. Inst. Fourier, 131–295 (1953/1954)
3. Dubois, D., Prade, H.: A review of fuzzy set aggregation connectives. Inform. Sci. 36, 85–121 (1985)
4. Even, Y., Lehrer, E.: Decomposition-Integral: Unifying Choquet and the Concave Integrals. Economic Theory, 1–26 (2013) (article in Press)
5. Grabisch, M.: Fuzzy integral in multicriteria decision making. Fuzzy Sets and Systems 69(3), 279–298 (1995)
6. Grabisch, M., Marichal, J.L., Mesiar, R., Pap, E.: Aggregation Functions. Cambridge University Press, New York (2009)
7. Joe, H.: Multivariate Models and Dependence Concepts. Monographs on Statics and Applied Probability, vol. 73. Chapman & Hall, London (1997)
8. Klement, E.P., Mesiar, R., Pap, E.: A universal integral as common frame for Choquet and Sugeno integral. IEEE Transactions on Fuzzy Systems 18, 178–187 (2010)
9. Klement, E.P., Mesiar, R., Pap, E.: Measure-based aggregation operators. Fuzzy Sets and Systems 142(1), 3–14 (2004)
10. Klement, E.P., Mesiar, R., Spizzichino, F., Stupňanová, A.: Universal integrals based on copulas. Submitted to Fuzzy Optimization and Decision Making
11. Mesiar, R., Mesiarová-Zemánková, A.: The ordered modular averages. IEEE Transactions on Fuzzy Systems 19(1), 42–50 (2011)
12. Mesiar, R., Stupňanová, A.: Decomposition integrals. International Journal of Approximate Reasoning 54, 1252–1259 (2013)
13. Murofushi, T., Sugeno, M.: Some quantities represented by the Choquet integral. Fuzzy Sets and Systems 56(2), 229–235 (1993)
14. Nelsen, R.B.: An Introduction to Copulas, 2nd edn. Springer, New York (2006)
15. Schmeidler, D.: Integral representation without additivity. Proc. Amer. Math. 97(2), 255–270 (1986)
16. Shilkret, N.: Maxitive measure and integration. Indag. Math. 33, 109–116 (1971)
17. Sklar, A.: Fonctions de répartition à n dimensions et leurs marges. Publ. Inst. Statist. Univ. Paris 8, 229–231 (1959)
18. Sugeno, M.: Theory of fuzzy integrals and its applications. PhD thesis, Tokyo Institute of Technology (1974)
19. Torra, V.: The weighted OWA operator. International Journal of Intelligent Systems 12(2), 153–166 (1997)
20. Yager, R.R.: On ordered weighted averaging aggregation operators in multicriteria decision making. IEEE Trans. Systems Man Cybernet. 18(1), 183–190 (1988)

21. Yager, R.R., Filev, D.P.: Induced Ordered Weighted Averaging operators. IEEE Transactions on Systems, Man, and Cybernetics, Part B: Cybernetics 29(2), 141–150 (1999)
22. Yager, R.R., Kacprzyk, J. (eds.): The Ordered Weighted Averanging Operators. Kluwer Academic Publishers, USA (1999)
23. Yager, R.R., Kacprzyk, J., Beliakov, G. (eds.): Recent Developments in the Ordered Weighted Averaging Operators: Theory and Practice. STUDFUZZ. Springer, Berlin (2011)

Generalized Product

Salvatore Greco[1,2], Radko Mesiar[3,4], and Fabio Rindone[5,⋆]

[1] Department of Economics and Business
95029 Catania, Italy
[2] Portsmouth Business School, Operations & Systems Management University of Portsmouth
Portsmouth PO1 3DE, United Kingdom
salgreco@unict.it
[3] Department of Mathematics and Descriptive Geometry, Faculty of Civil Engineering
Slovak University of Technology
Bratislava, Slovakia
[4] Institute for Research and Applications of Fuzzy Modeling
Division of University Ostrava
NSC IT4Innovations
30. dubna 22, Ostrava 701 03
Czech Republic
radko.mesiar@stuba.sk
[5] Department of Economics and Business
95029 Catania, Italy
frindone@unict.it

Abstract. Aggregation functions on $[0,1]$ with annihilator 0 can be seen as a generalized product on $[0,1]$. We study the generalized product on the bipolar scale $[-1,1]$, stressing the axiomatic point of view. Based on newly introduced bipolar properties, such as the bipolar increasingness, bipolar unit element, bipolar idempotent element, several kinds of generalized bipolar product are introduced and studied. A special stress is put on bipolar semicopulas, bipolar quasi-copulas and bipolar copulas.

Keywords: Aggregation function, bipolar copula, bipolar scale, bipolar semicopula, symmetric minimum.

1 Introduction

Recall that an aggregation function $A : [0,1]^2 \to [0,1]$ is characterized by boundary conditions $A(0,0) = 0$ and $A(1,1) = 1$, and by the increasingness of A, i.e., the sections $A(x, \cdot)$ and $A(\cdot, y)$ are increasing for each $x, y \in [0,1]$. For more details see [2,5]. The product $\Pi : [0,1]^2 \to [0,1]$ has additionally 0 as its annihilator, and thus each aggregation function $A : [0,1]^2 \to [0,1]$ with annihilator 0 (i.e., $A(x,0) = A(0,y)$ for all $x, y \in [0,1]$) can be seen as a generalization of the product Π on the unipolar scale $[0,1]$. Observe that the class $\mathcal{P}$ of generalized products on the scale $[0,1]$ has the smallest aggregation function $A_* : [0,1]^2 \to [0,1]$ given by

$$A_*(x,y) = \begin{cases} 1 \text{ if } x = y = 1, \\ 0 \text{ else} \end{cases} \tag{1}$$

⋆ Corresponding author.

A. Laurent et al. (Eds.): IPMU 2014, Part III, CCIS 444, pp. 289–295, 2014.
© Springer International Publishing Switzerland 2014

as its minimal element, and its maximal element $A^* : [0,1]^2 \to [0,1]$ is given by

$$A^*(x,y) = \begin{cases} 0 \text{ if } xy = 0, \\ 1 \text{ else.} \end{cases} \tag{2}$$

Moreover $\mathcal{P}$ is a complete lattice (with respect to pointwise suprema and infima), and it contains, among others, geometric mean G, harmonic mean H, minimum M, etc. Convex sums of these extremal generalized products, give rise to a parametric family connecting them.

The most distinguished subclasses of $\mathcal{P}$ are

- the class $\mathcal{S}$ of semicopulas, i.e., aggregation functions from $\mathcal{P}$ having $e = 1$ as neutral element, $S(x,1) = S(1,x) = x$ for all $x \in [0,1]$, see [1,3];
- the class $\mathcal{T}$ of triangular norms, i.e., of associative and commutative semicopulas, [9,13];
- the class $\mathcal{Q}$ of quasi-copulas, i.e. $1-$Lipschitz aggregation functions from $\mathcal{P}$ (observe that $\mathcal{Q} \subsetneq \mathcal{S}$), [12];
- the class $\mathcal{C}$ of copulas, i.e. supermodular functions from $\mathcal{P}$ (observe that $\mathcal{C} \subsetneq \mathcal{Q}$), [12].

Observe that the product Π belongs to any of mentioned classes, similarly as M. Among several applications of the generalized product functions, recall their role as conjunctions in fuzzy logic [8], or their role of multiplications in the area of general integrals [10,14].

Integration on bipolar scale $[-1,1]$ requires a bipolar function $B : [-1,1]^2 \to [-1,1]$ related to the standard product $\Pi : [-1,1]^2 \to [-1,1]$ (we will use the same notation Π for the product independently of the actual scale). Up to standard product Π applied, e.g., in the case of Choquet integral on $[-1,1]$, or in the case of bipolar capacities based Choquet integral, Grabisch [4] has introduced a symmetric Sugeno integral on $[-1,1]$ based on the symmetric minimum $B_M : [-1,1]^2 \to [-1,1]$, $B_M(x,y) = sign(xy)\min(|x|,|y|)$. The aim of this paper is to generalize the bipolar product on $[-1,1]$ in a way similar to generalized product on $[0,1]$, and to study special classes of such generalizations. Clearly, the idea to study generalized products stems from bipolar integrals. In turn, bipolar integrals are inserted in the rich domain of bipolarity. A wide literature, developed in very recent years, has demonstrated how bipolarity must be considered in modeling human reasoning.

The paper is organized as follows. In the next section, several properties of bipolar functions are proposed, and the generalized bipolar product is introduced. Section 3 is devoted to bipolar semicopulas and bipolar triangular norms, while in Section 4 we study bipolar quasi-copulas and bipolar copulas. Finally, some concluding remarks are added.

2   Generalized Bipolar Product

Considering the function $F : [-1,1]^2 \to [-1,1]$, several algebraic and analytic properties can be considered in their standard form, such as the commutativity, associativity, annihilator 0, continuity, Lipschitzianity, supermodularity, etc. Note that the bipolar product $\Pi : [-1,1]^2 \to [-1,1]$ as well as the symmetric minimum $B_M : [-1,1]^2 \to [-1,1]$ satisfy all of them. However, there are some properties reflecting the bipolarity of the scale $[-1,1]$.

Recall that a mapping $S : [0,1]^2 \to [0,1]$ is a semicopula [1,3], whenever it is non-decreasing in both variables and 1 is the neutral element, i.e., $S(x,1) = S(1,x) = x$ for all $x \in [0,1]$. When considering the product $\Pi : [-1,1]^2 \to [-1,1]$, we see that 1 is its neutral element. More, it holds $\Pi(-1,x) = \Pi(x,-1) = -x$ for all $x \in [-1,1]$.

Definition 1. *Let $F : [-1,1]^2 \to [-1,1]$ be a mapping such that $F(x,1) = F(1,x) = x$ and $F(-1,x) = F(x,-1) = -x$ for all $x \in [-1,1]$. Then 1 is called a bipolar neutral element for F.*

Simple bipolar semicopulas B_S, introduced for bipolar universal integrals in [6], are fully determined by standard semicopulas $S : [0,1]^2 \to [0,1]$, by means of $B_S(x,y) = (sign(xy))S(|x|,|y|)$. Observe that 1 is a bipolar neutral element for any simple bipolar semicopula B_S. Concerning the monotonicity required for semicopulas, observe that considering the product Π, or any simple bipolar semicopula B_S (note that $B_\Pi = \Pi$, abusing the notation Π both for the product on $[-1,1]$ and on $[0,1]$), these mappings are non-decreasing in both coordinates when fixing an element from the positive part of the scale, while they are non-increasing when fixing an element from the negative part of the scale $[-1,1]$.

Definition 2. *Let $F : [-1,1]^2 \to [-1,1]$ be a mapping such that the partial mappings $F(x,\cdot)$ and $F(\cdot,y)$ are non-decreasing for any $x,y \in [0,1]$ and they are non-increasing for any $x,y \in [-1,0]$. Then F will be called a bipolar increasing mapping.*

Similarly, inspired by the symmetric minimum B_M, we introduce the notion of a bipolar idem potent element.

Definition 3. *Let $F : [-1,1]^2 \to [-1,1]$ be given. An element $x \in [0,1]$ is called a bipolar idempotent element of F whenever it satisfies $F(x,x) = F(-x,-x) = x$ and $F(-x,x) = F(x,-x) = -x$.*

Recall that the class $\mathcal{P}$ of generalized products on $[0,1]$ can be characterized as the class of all the increasing mappings $F : [0,1]^2 \to 0,1]$ such that $F|_{\{0,1\}^2} = \Pi|_{\{0,1\}^2}$. Inspired by this characterization, we introduce the class $\mathcal{BP}$ of all generalized bipolar products as follows.

Definition 4. *A function $B : [-1,1]^2 \to [-1,1]$ is a generalized bipolar product whenever it is bipolar increasing and $B|_{\{-1,0,1\}^2} = \Pi|_{\{-1,0,1\}^2}$.*

Theorem 1. $B \in \mathcal{BP}$ *if and only if there are $A_1, A_2, A_3, A_4 \in \mathcal{P}$ such that*

$$B(x,y) = \begin{cases} A_1(x,y) & \text{if } (x,y) \in [0,1]^2 \\ -A_2(-x,y) & \text{if } (x,y) \in [-1,0] \times [0,1] \\ A_3(-x,-y) & \text{if } (x,y) \in [-1,0]^2 \\ -A_4(x,-y) & \text{if } (x,y) \in [0,1] \times [-1,0]. \end{cases} \tag{3}$$

Due to Theorem 1, each $B \in \mathcal{BP}$ can be identified with a quadruple $(A_1, A_2, A_3, A_4) \in \mathcal{P}^4$.

Definition 5. *Let* $A \in \mathcal{P}$. *then* $B_A = (A, A, A, A) \in \mathcal{BP}$, *given by* $B_A(x, y) = sign(xy)A(|x|, |y|)$, *is called a simple generalized bipolar product (simple GBP, in short).*

Evidently, B_M is a simple GBP related to M, while $B_\Pi = \Pi$. Observe that

$$B_{A_*}(x, y) = \begin{cases} \Pi(x, y) & \text{if} \quad (x, y) \in \{-1, 1\} \\ 0 & \text{else,} \end{cases}$$

and

$$B_{A^*}(x, y) = sign(xy).$$

However B_{A_*} and B_{A^*} are not extremal elements of $\mathcal{BP}$. The class $\mathcal{BP}$ is a complete lattice (considering pointwise sup and inf) with top element $B^* = (A^*, A_*, A^*, A_*)$ and bottom element $B_* = (A_*, A^*, A_*, A^*)$, given by

$$B^*(x, y) = \begin{cases} 1 & \text{if} \quad xy > 0 \\ -1 & \text{if} \quad xy = -1 \\ 0 & \text{else,} \end{cases} \tag{4}$$

and

$$B_*(x, y) = \begin{cases} -1 & \text{if} \quad xy < 0 \\ 1 & \text{if} \quad xy = 1 \\ 0 & \text{else.} \end{cases} \tag{5}$$

As in the case of generalized products on the scale $[0, 1]$, to obtain a parametric family connecting B_* and B^* it is enough to consider their convex sum.

3   Bipolar Semicopulas and Bipolar t-norms

Based on the idea of a bipolar neutral element $e = 1$, we introduce now the bipolar semicopulas, compare also [7].

Definition 6. *A mapping* $B : [-1, 1]^2 \to [-1, 1]$ *is called a bipolar semicopula whenever it is bipolar increasing and* 1 *is a bipolar neutral element of* B.

Based on Theorem 1 we have the next result

Corollary 1. *A mapping* $B : [-1, 1]^2 \to [-1, 1]$ *is a bipolar semicopula if and only if there is a quadruple* (S_1, S_2, S_3, S_4) *of semicopulas so that*

$$B(x, y) = \begin{cases} S_1(x, y) & \text{if} \quad (x, y) \in [0, 1]^2 \\ -S_2(-x, y) & \text{if} \quad (x, y) \in [-1, 0] \times [0, 1] \\ S_3(-x, -y) & \text{if} \quad (x, y) \in [-1, 0]^2 \\ -S_4(x, -y) & \text{if} \quad (x, y) \in [0, 1] \times [-1, 0]. \end{cases} \tag{6}$$

It is not difficult to check that the extremal bipolar semicopulas are related to extremal semicopulas M (the greatest semicopula given by $M(x, y) = \min(x, y)$) and Z (the smallest semicopula, called also the drastic product, and given by $Z(x, y) = \min(x, y)$ if $1 \in \{x, y\}$ and $Z(x, y) = 0$ else).

We have also the next results.

Proposition 1. *Let $B \in \mathcal{B}$ be a bipolar semicopula such that each $x \in [0, 1]$ is its bipolar idempotent element. Then $B = B_M$ is the symmetric minimum introduced by Grabisch [4].*

Associativity of binary operations (binary functions) is a strong algebraic property, which, in the case of bipolar semicopulas characterizes a particular subclass of $\mathcal{B}$.

Theorem 2. *A bipolar semicopula $B \in \mathcal{B}$ is associative if and only if B is a simple bipolar semicopula, $B = B_S$, where $S \in \mathcal{S}$ is an associative semicopula.*

Typical examples of associative bipolar semicopulas are the product Π and the symmetric minimum B_M. Recall that a symmetric semicopula $S \in \mathcal{S}$, i.e., $S(x, y) = S(y, x)$ for all $x, y \in [0, 1]$, which is also associative is called a triangular norm [13,9].

Definition 7. *A symmetric associative bipolar semicopula $B \in \mathcal{B}$ is called a bipolar triangular norm.*

Due to Theorem 2 it is obvious that a bipolar semicopula $B \in \mathcal{B}$ is a bipolar triangular norm if and only if $B = B_T$, where $T : [0, 1]^2 \to [0, 1]$ is a triangular norm, i.e. if $B(x, y) = (sign(xy))T(|x|, |y|)$. Obviously, the product, Π, and the symmetric minimum, B_M, are bipolar triangular norms. The smallest semicopula Z is also a triangular norm and the corresponding bipolar triangular norm $B_Z : [-1, 1]^2 \to [-1, 1]$ is given by

$$B_Z(x, y) = \begin{cases} 0 & \text{if } (x, y) \in]-1, 1[^2, \\ xy & \text{else.} \end{cases} \tag{7}$$

Observe that the genuine n-ary extension $B_Z : [-1, 1]^n \to [-1, 1]$, $n > 2$, is given by

$$B_Z(x_1, \ldots, x_n) = \begin{cases} 0 & \text{if } \#\{i \mid x_i \in]-1, 1[\} \geq 2, \\ \Pi_{i=1}^n x_i & \text{else.} \end{cases} \tag{8}$$

Also $W : [-1, 1]^2 \to [-1, 1]$ given by $W(x, y) = \max(0, x + y - 1)$ is a triangular norm, and consequently also $B_W : [-1, 1]^2 \to [-1, 1]$ given by $B_W(x, y) = (sign(xy)) \max(0, x+y-1)$ is a bipolar triangular norm. Moreover, its n-ary extension $B_W : [-1, 1]^n \to [-1, 1]$, $n > 2$, is given by

$$B_W(x_1, \ldots, x_n) = (sign(\Pi_{i=1}^n x_i)) \max(0, \sum x_i - n + 1).$$

Note that several construction methods for bipolar semicopulas were proposed in [7].

4   Bipolar Quasi-copulas and Copulas

In this section, we extend the notion of quasi-copulas and copulas acting on the unipolar scale $[0, 1]$ to the bipolar scale $[-1, 1]$.

Definition 8. *Let $B \in \mathcal{BP}$ be $1-$Lipschitz, i.e.,*

$$|B(x_1, x_2) - B(x_2, y_2)| \leq |x_1 - x_2| + |y_1 - y_2|, \quad \text{for all } x_1, x_2, y_1, y_2 \in [-1, 1].$$

Then B is called a bipolar quasi-copula.

Based on theorem 1 we have the following result.

Corollary 2. $B \in \mathcal{GP}$ *is a bipolar quasi-copula if and only if $B = (Q_1, Q_2, Q_3, Q_4) \in \mathcal{Q}^4$.*

Evidently, each bipolar quasi-copula is also a bipolar semicopula.

Definition 9. *Let $B \in \mathcal{GP}$ has a bipolar neutral element $e = 1$ and let B be super-modular, i.e.,*

$$B(x_1 \vee x_2, y_1 \vee y_2) + B(x_1 \wedge x_2, y_1 \wedge y_2) \geq B(x_1, y_1) + B(x_2, y_2),$$

for all $x_1, x_2, y_1, y_2 \in [-1, 1]$. Then B is called a bipolar copula.

Corollary 3. $B \in \mathcal{GP}$ *is a bipolar copula if and only if $B = (C_1, C_2, C_3, C_4) \in \mathcal{C}^4$.*

Observe that each bipolar copula B is also a bipolar quasi-copula, and that the class of all bipolar quasi-copulas $\mathcal{BQ}$ is a $\sup - (\inf -)$ closure of the class $\mathcal{BC}$ of all bipolar copulas. Π and B_M are typical example of simple bipolar copulas.

As an example of a bipolar copula B which is not simple, we consider the function $B : [-1, 1]^2 \to [-1, 1]$ given by

$$B(x, y) = xy + |xy|(1 - |x|)(1 - |y|).$$

Then $B = (C_1, C_2, C_1, C_2)$ where $C_1, C_2 \in \mathcal{C}$ are Farlie-Gumbel-Morgenstern copulas [12] given by

$$C_1(x, y) = xy + xy(1 - x)(1 - y)$$

and

$$C_2(x, y) = xy - xy(1 - x)(1 - y).$$

5   Concluding Remarks

We have introduced and discussed bipolar generalizations of the product, including bipolar semicopulas, bipolar triangular norms, bipolar quasi-copulas and bipolar copulas. Observe that our approach to bipolar aggregation can be seen as a particular case of the multi-polar aggregation proposal as given in [11] for dimension n=2. Though there is a minor overlap with conjunctive aggregation functions in the area of Atanassov's intuitionistic framework, our concept is rather different (especially, the bipolar neutral element makes the difference). Observe that our approach brings a generalization of the product (considered as a multiplication for integrals) into the bipolar framework, where it is supposed to play the role of multiplication when constructing integrals. For the future research, it could be interesting to look on our generalized product as a conjunctive-like connective and the try to develop a dual concept of generalized bipolar sums, and related bipolar connectives. We expect application of our results in multicriteria decision support when considering bipolar scales, especially when dealing with bipolar capacities based integrals. Observe that simple bipolar semicopulas were already applied when introducing universal integrals on the bipolar scale $[-1, 1]$, see [6].

Acknowledgment. The work on this contribution was partially supported by the grants APVV-0073-10, VEGA 1/0171/12 and by the European Regional Development Fund in the IT4 Innovations Centre of Excellence Project (C.Z.1.05/1.1.0002.0070).

References

1. Bassan, B., Spizzichino, F.: Relations among univariate aging, bivariate aging and dependence for exchangeable lifetimes. Journal of Multivariate Analysis 93(2), 313–339 (2005)
2. Beliakov, G., Pradera, A., Calvo, T.: Aggregation functions: A guide for practitioners. STUDFUZZ, vol. 221. Springer, Heidelberg (2008)
3. Durante, F., Sempi, C.: Semicopulæ. Kybernetika 41(3), 315–328 (2005)
4. Grabisch, M.: The symmetric Sugeno integral. Fuzzy Sets and Systems 139(3), 473–490 (2003)
5. Grabisch, M., Marichal, J.L., Mesiar, R., Pap, E.: Aggregation Functions (Encyclopedia of Mathematics and its Applications). Cambridge University Press (2009)
6. Greco, S., Mesiar, R., Rindone, F.: The Bipolar Universal Integral. In: Greco, S., Bouchon-Meunier, B., Coletti, G., Fedrizzi, M., Matarazzo, B., Yager, R.R. (eds.) IPMU 2012, Part III. CCIS, vol. 299, pp. 360–369. Springer, Heidelberg (2012)
7. Greco, S., Mesiar, R., Rindone, F.: Bipolar Semicopulas. Submitted to FSS (2013)
8. Hájek, P.: Metamathematics of fuzzy logic, vol. 4. Springer (1998)
9. Klement, E.P., Mesiar, E., Pap, R.: Triangular norms. Kluwer, Dordrecht (2000)
10. Klement, E.P., Mesiar, R., Pap, E.: A universal integral as common frame for Choquet and Sugeno integral. IEEE Transactions on Fuzzy Systems 18(1), 178–187 (2010)
11. Mesiarová-Zemánková, A., Ahmad, K.: Multi-polar Choquet integral. Fuzzy Sets and Systems (2012)
12. Nelsen, R.B.: An introduction to copulas. Springer, New York (2006)
13. Schweizer, B., Sklar, A.: Statistical metric spaces. Pacific J. Math. 10, 313–334 (1960)
14. Wang, Z., Klir, G.J.: Generalized measure theory, vol. 25. Springer, New York (2009)

Improving the Performance of FARC-HD in Multi-class Classification Problems Using the One-Versus-One Strategy and an Adaptation of the Inference System

Mikel Elkano[1], Mikel Galar[1], José Sanz[1], Edurne Barrenechea[1],
Francisco Herrera[2], and Humberto Bustince[1]

[1] Dpto. de Automática y Computación, Universidad Publica de Navarra, Campus Arrosadia s/n,
31006 Pamplona, Spain
elkano.mikel@gmail.com, {mikel.galar,joseantonio.sanz,
edurne.barrenechea,bustince}@unavarra.es
[2] Department of Computer Science and Artificial Intelligence,
CITIC-UGR (Research Center on Information and Communications Technology)
University of Granada, 18071 Granada, Spain
herrera@decsai.ugr.es

Abstract. In this work we study the behavior of the FARC-HD method when addressing multi-class classification problems using the *One-vs-One* (OVO) decomposition strategy. We will show that the confidences provided by FARC-HD (due to the use of the product in the inference process) are not suitable for this strategy. This problem implies that robust strategies like the weighted vote obtain poor results. For this reason, we propose two improvements: 1) the replacement of the product by greater aggregations whose output is independent of the number of elements to be aggregated and 2) the definition of a new aggregation strategy for the OVO methodology, which is based on the weighted vote, in which we only take into account the confidence of the predicted class in each base classifier. The experimental results show that the two proposed modifications have a positive impact on the performance of the classifier.

Keywords: Classification, One-vs-One, Fuzzy Rule-Based Classification Systems, Aggregations.

1 Introduction

There are different techniques in data mining to solve classification problems. Among them, Fuzzy Rule-Based Classification Systems (FRBCSs) are widely used because they provide an interpretable model by using linguistic labels in the antecedents of their rules [8]. In this work we use the FARC-HD algorithm [2], which is currently one of the most accurate and interpretable FRBCSs.

We can distinguish two types of problems in classification: binary (two-class) and multi-class problems. Multi-class classification is usually more difficult because of the higher complexity in the definition of the decision boundaries. A solution to cope with multi-class classification problems is to use decomposition techniques [10], which try to divide the original multi-class problem into binary problems that are easier to solve.

A. Laurent et al. (Eds.): IPMU 2014, Part III, CCIS 444, pp. 296–306, 2014.

In the specialized literature a number of decomposition strategies have been proposed [10]. One of the most used is the *One-vs-One* (OVO) decomposition strategy. In OVO, the original multi-class problem is divided into as many new sub-problems as possible pairs of classes, where each one is addressed with an independent base classifier. When classifying a new instance, all base classifiers are queried and their outputs are combined to make the final decision. This technique usually works better than addressing the problem directly [5].

In this work we will try to improve the performance of the FARC-HD algorithm in multi-class classification problems using the OVO strategy. Obviously, the aggregation of the classifiers in the OVO strategy directly depends on the confidences provided by the base classifiers. In the case of FARC-HD, we consider the association degree obtained for each class of the problem as the confidence.

We will show that using aggregation strategies that usually have a robust and accurate performance (such as the weighted vote [7]), we do not obtain good results using FARC-HD as base classifier due to the confidences obtained. This is due the fact that using the product in the inference process of FARC-HD produces very small confidences for each pair of classes and it also implies that the rules with the largest number of antecedents are penalized. Another problem is that the confidences returned by each binary classifier are not related. In order to address these problems, we propose to adapt the confidences of FARC-HD to OVO, aiming to obtain a more accurate aggregation, which consequently can lead to improve the classification in the OVO model.

In our proposal, we apply in the inference process aggregation operators whose result is greater than that of the product and we impose them the condition that they cannot decrease their results when the number of antecedents of the rule increases. Furthermore, we propose an alternative to the usage of the weighted voting in the aggregation phase of OVO strategy. More specifically, we propose an aggregation strategy named WinWV in which we do not consider the confidences obtained by non-predicted classes, since its usage is not appropriate for the classification.

In order to assess the quality of the methods, we use twenty numerical datasets [1] and we contrast the results obtained using non-parametric statistical tests. In these experiments, we study which is the best aggregations in the inference process for FARC-HD that allows us to obtain a better overall OVO classification. To do so, we compare the results of this new hybridization against the original FARC-HD classifier when dealing with multi-class classification problems using different aggregations.

The structure of the paper is as follows. In Section 2, we briefly introduce FRBCSs. Section 3 contains an introduction to the decomposition of multi-class problems. In Section 4, we describe in detail our proposals to use FARC-HD with the OVO strategy. Section 5 contains the experimental framework description and the analysis of the achieved results. Finally, in section 6 we draw the conclusions.

2   Fuzzy Rule-Based Classification Systems

Any classification problem consists of P training examples $x_p = (x_{p1}, \ldots, x_{pn})$, $p = 1, 2, \ldots, P$ where x_{pi} is the value of the i-th attribute $(i = 1, 2, \ldots, n)$ of the p-th training example. Each example belongs to a class $y_p \in \mathbb{C}$ and $\mathbb{C} = \{1, 2, \ldots, m\}$, where m is the number of classes of the problem.

In this work we use the FARC-HD method [2], which is a FRBCSs that makes use of rules whose structure is as follows:

$$\text{Rule } R_j: \text{ If } x_1 \text{ is } A_{j1} \text{ and } \dots \text{ and } x_n \text{ is } A_{jn} \text{ then Clase} = C_j \text{ with } RW_j \tag{1}$$

where R_j is the label of the j-th rule, $x = (x_1, \dots, x_n)$ is an n-dimensional pattern vector that represents the example, A_{ji} is a fuzzy set, $C_j \in \mathbb{C}$ is the class label and RW_j is the rule weight, which is computed using the confidence of the rule [9].

The three steps of the fuzzy rule learning method of FARC-HD are the following:

1. *Fuzzy association rules extraction for classification*: A search tree is constructed for each class from which the fuzzy rules are obtained. The number of linguistic terms in the antecedents of the rules is limited by the maximum depth of the tree.
2. *Candidate rule prescreening*: In this phase the most interesting fuzzy rules among the fuzzy rules obtained in the previous stage are selected. To do so, a weighted scheme of examples is applied, which is based on the coverage of the fuzzy rules.
3. *Genetic rule selection and lateral tuning*: An evolutionary algorithm is used both to tune the lateral position of the membership function and to select the most precise rules from the rule base generated in the previous steps.

In order to classify a new example, FARC-HD uses a fuzzy reasoning method called additive combination, which predicts the class that obtains the largest confidence according to Equation 2.

$$conf_l(x_p) = \sum_{R_j \in BR; C_j = l} b_j(x_p), \qquad l = 1, 2, \dots, m \tag{2}$$

with $$b_j(x_p) = \mu_{A_j}(x_p) \cdot RW_j = T\left(\mu_{A_{j1}}(x_{p1}), \dots, \mu_{A_{jn_j}}(x_{pn_j})\right) \cdot RW_j \tag{3}$$

where $b_j(x_p)$ is the association degree of the example x_p with the rule j, $\mu_{A_j}(x_p)$ is the matching degree of the example x_p with the antecedent of the rule j, $\mu_{A_{ji}}(x_{pi})$ is the membership degree of the example with the i-th antecedent of the rule, T is a t-norm (in the case of FARC-HD the product t-norm), and n_j the number of antecedents of the rule j.

3   Decomposition of Multi-class Problems: One-vs-One Strategy

Decomposition strategies [10] face multi-class classification problems using binary classifiers, commonly known as base classifiers. These strategies are not only useful when working with classifiers that only distinguish between two classes but also with those that can directly handle multi-class problems. Even in this later case, the results achieved by means of the application of decomposition strategies are usually enhanced [5].

3.1   One-vs-One Strategy

The OVO strategy is one of the the most used decomposition strategies, where a problem of m classes is divided in $m(m-1)/2$ binary sub-problems (all possible pairs of classes) that are faced by independent base classifiers. A new instance is classified by combining the output of each base classifier. Each base classifier distinguishes a pair of classes $\{C_i, C_j\}$ and it returns a couple of confidence degrees $r_{ij}, r_{ji} \in [0, 1]$ in favor of classes C_i, C_j, respectively. These outputs are stored in the score-matrix as follows.

$$R = \begin{pmatrix} - & r_{12} & \cdots & r_{1m} \\ r_{21} & - & \cdots & r_{2m} \\ \vdots & & & \vdots \\ r_{m1} & r_{m2} & \cdots & - \end{pmatrix} \tag{4}$$

Finally, the outputs of the base classifiers are aggregated and the predicted class obtained. In the next section, we present the aggregation strategies of the score-matrices that we consider in this work.

3.2   Aggregation Methods

The method used to aggregate the outputs of the base classifiers is a key factor for the classification process [5]. In this work, we consider three well-known aggregation methods.

– *Voting strategy (VOTE) [4]*: Each base classifier votes for the predicted class and the class having the largest number of votes is given as output.

$$Class = arg\ \underset{i=1,\dots,m}{max} \sum_{1 \le j \ne i \le m} s_{ij} \tag{5}$$

where s_{ij} is 1 if $r_{ij} > r_{ji}$ and 0 otherwise.
– *Weighted Voting (WV) [7]*: Each base classifier votes for both classes based on the confidences provided for the pair of classes. The class having the largest value is given as output.

$$Class = arg\ \underset{i=1,\dots,m}{max} \sum_{1 \le j \ne i \le m} r_{ij}^n \tag{6}$$

where R^n is the normalized voting matrix.
– *Non-Dominance Criteria (ND) [3]*: The score-matrix is considered as a fuzzy preference relation. Then non-dominance degree is computed, being the winning class the one with the highest value.

$$Class = arg\ \underset{i=1,\dots,m}{max} \left\{ 1 - \underset{j \in \mathscr{C}}{sup}\ r_{ji}' \right\} \tag{7}$$

where R' is the normalized and strict score-matrix.

3.3   Confidence Provided by the Base Classifiers

According to the aforementioned aggregation methods, it is clear that one of the most important factors is the value used as confidence. In the case of FARC-HD, we consider the association degree of each class as the confidence, i.e., the obtained value in

Equation 2. However, as we will show when using the weighted voting strategy (considered one of the most robust ones [7]), these values do not provide the expected results. If FARC-HD is directly used to face multi-class classification problems, the obtained confidences are irrelevant once the example is classified. However, in the case of using FARC-HD as a base classifier, the confidences provided for each class are aggregated either when they are the predicted class or not. Therefore, it is very important to obtain good confidences since they are the values aggregated when making the final classification.

Another problem inherent to the OVO decomposition method is the non-competent classifiers [6]. The learning process of each base classifier is performed using only the examples belonging to the two classes that this classifiers will classify and consequently, it ignores the examples belonging to other classes. Therefore, the remainder classes are unknown for these classifiers and their outputs are irrelevant when classifying examples of those classes. However, these outputs are aggregated in the same way as the relevant ones. This problem can lead to incorrectly label the example in certain situations, which stresses the problem we have mentioned about the lack of suitability of the confidences provided by the FARC-HD classifier.

4   FARC-HD and OVO: Improving the Combination

Following Equation 3 shown in Section 2, it can be observed that the inference method used by FARC-HD diminishes the values of the association degrees of the fuzzy rules having more antecedents, which affects the combination in OVO as we have stated in the previous section. This is due to the usage of the product as the t-norm when computing the matching degree of the example with the fuzzy rules. Thus, fuzzy rules with more antecedents will provide lower confidences.

In order address this problem and to obtain more appropriate confidences for its subsequent treatment with the OVO strategy, we propose to modify both the first two stages of inference process of FARC-HD (Section 4.1) and the aggregation phase of the OVO method (Section 4.2).

4.1   Improving Confidence Estimations for OVO: Computing the Association Degree with Specific Aggregations

In this work, we propose the usage of t-norms greater than the product or even means with certain properties to compute the association degrees. In order to do so, we have considered the minimum as an alternative to the product, since it is the largest t-norm, and the geometric mean as a representative of means, since it has some properties similar to those of the product. Moreover, in both cases, the result obtained is not affected by the number of antecedents of the fuzzy rules (their results do not decrease as the product does).

Therefore, we propose to modify Equation 3, replacing the product t-norm by either the minimum t-norm or the geometric mean. This change affects the computation of the association degree $b_j(x_p)$, which in turn modifies the value of $conf_l(x_p)$, that is, the confidences used in the score-matrix of OVO. Thus, we intend to show that the usage of these aggregation methods can significantly improve the confidences of the base classifiers for their subsequent processing in the classification process. Next, the proposed modifications of Equation 3 are presented.

- *Minimum:* The obtained association degree is the minimum among the membership degrees of the example with the antecedents of the fuzzy rule and the rule weight.

$$b_j(x_p) = min(\min_{i=1.....n_j} \mu_{A_{ji}}(x_{pi}), RW_j) \tag{8}$$

- *Geometric Mean:* The matching degree is computed as the geometric mean of the membership degrees of the example with the antecedent of the fuzzy rules and then, the association degree is computed as the geometric mean among the matching degree and the rule weight.

$$b_j(x_p) = \left(\sqrt[n_j]{\prod_{i=1}^{n_j} \mu_{A_{ji}}(x_{pi})} \cdot RW_j \right)^{1/2} \tag{9}$$

4.2   Adapting the Weighted Voting to OVO: WinWV

Besides the modification of the aggregation methods in the inference process of the base classifiers, we propose a new aggregation method for OVO strategy named *WinWV*, which is a modification of the WV. Since all the obtained confidences by FARC-HD are not appropriate for the WV, we propose to consider only the confidence of the predicted class, whereas that of the non-predicted class is not taken into account:

$$Class = arg \max_{i=1,...,m} \sum_{1 \le j \ne i \le m} s_{ij} \tag{10}$$

where s_{ij} is r_{ij}^n if $r_{ij} > r_{ji}$ and 0 otherwise.

5   Experimental Study

The experimental study is carried out with three main objectives:

1. To observe the improvement obtained when we change the product in the inference method of FARC-HD by either the minimum t-norm or the geometric mean, which return greater values, in the OVO model (Section 5.3).
2. To show the necessity of a new aggregation strategy (WinWV) for FARC-HD using OVO, instead of the commonly used WV (Section 5.4).
3. To study the improvement obtained when combining the OVO decomposition technique using FARC-HD as base classifier versus the direct usage of FARC-HD in multi-class classification problems (Section 5.5).

In the remainder of this section we present the experimental framework (Section 5.1), the achieved results (Section 5.2) and their correponding analysis (Sections 5.3, 5.4 and 5.5).

Table 1. Summary of the features of the datasets used in the experimental study

Dataset	#Ex.	#Atr.	#Clas.	Dataset	#Ex.	#Atr.	#Clas.
autos	159	25	6	penbased	1100	16	10
balance	625	4	3	satimage	643	36	7
cleveland	297	13	5	segment	2310	19	7
contraceptive	1473	9	3	shuttle	2175	9	5
ecoli	336	7	8	tae	151	5	3
glass	214	9	7	thyroid	720	21	3
hayes-roth	132	4	3	vehicle	846	18	4
iris	150	4	3	vowel	990	13	11
newthyroid	215	5	3	wine	178	13	3
pageblocks	548	10	5	yeast	1484	8	10

5.1 Experimental Framework

In order to carry out the experimental study, we use twenty numerical datasets selected from the KEEL dataset repository [1], whose main features are introduced in Table 1.

The configuration used for the FARC-HD method is the one recommended by the authors: 5 labels for each fuzzy partition, the maximum tree depth is 3, minimum support of 0.05, minimum confidence of 0.8, populations formed by 50 individuals, 30 bits per gene for the Gray codification and a maximum of 20000 evaluations.

We have used a 5 folder cross-validation scheme so as to obtain the results for each method. In order to support the quality of the methods we apply non-parametric statistical tests [11]. More specifically, we use the Wilcoxon's rank test to compare two methods, the Friedman aligned ranks test to check whether there are statistical differences among a group of methods and the Holm *post-hoc* test to find the algorithms that reject the null hypothesis of equivalence against the selected control method.

5.2 Experimental Results

Table 2 shows the accuracy obtained in testing in every dataset by all the methods studied. We observe 5 groups of results, where the first one corresponds to the original FARC-HD algorithm using the three aggregation methods to compute the association degree in the inference process. The four remainder groups correspond to the different aggregation strategies considered in OVO. The best result of each group is highlighted in **bold-face** while the best overall result in each dataset is underlined.

5.3 Analysis of Aggregations in the Association Degree

In this section, we analyze the effect produced by the modification of the aggregation method in the computation of association degree in the inference process of FARC-HD. In Table 2 we can observe that in the case of the original FARC-HD, the replacement of the product by a greater aggregation does not produce significant changes in the results. However, in the case of the aggregation strategies considered in OVO, the geometric mean obtains the best result in all of them, generally with large differences. Anyway, we can not base our conclusions on the overall results, so we carry out a multiple comparison in order to compare the two proposed aggregation methods and the product t-norm when using the original FARC-HD method and for each combination method used in the OVO strategy.

Table 2. Accuracy obtained in testing by each method

| | FARC | | | OVO | | | | | | | | | | | |
| | | | | VOTE | | | ND | | | WV | | | WinWV | | |
Dataset	PROD	MIN	GM	PROD	MIN	GM	PROD	MIN	GM	PROD	MIN	GM	PROD	MIN	GM
autos	**83,26**	81,44	77,49	78,85	80,66	**80,84**	78,85	79,32	**80,89**	75,76	**79,89**	79,62	76,30	**81,26**	78,35
balance	86,08	85,59	**88,16**	84,79	84,77	**85,90**	84,31	85,10	**86,22**	**80,95**	79,82	**80,95**	83,99	83,81	**85,26**
cleveland	**56,93**	56,30	54,89	57,97	55,24	**59,29**	56,92	54,90	**59,99**	55,93	57,24	**60,66**	56,94	56,58	**59,66**
contraceptive	**53,91**	53,57	53,56	**54,31**	54,04	53,23	**53,97**	53,70	53,37	53,97	**54,58**	54,25	**53,97**	53,84	53,57
ecoli	**83,50**	82,60	81,37	**83,77**	83,68	83,18	83,72	**84,26**	82,66	84,96	**85,27**	81,97	85,24	84,62	84,62
glass	66,25	66,86	**71,59**	68,22	**72,08**	72,00	67,27	69,68	**71,93**	68,23	**72,55**	68,44	67,68	**72,00**	70,66
hayes-roth	**81,26**	79,73	78,19	81,92	81,92	**83,41**	81,15	81,15	**82,64**	81,92	81,92	**83,41**	81,92	81,92	**83,41**
iris	**96,00**	95,33	95,33	**94,67**	**94,67**	**94,67**	94,67	**94,67**	94,00	**95,33**	95,33	94,67	**94,67**	**94,67**	**94,67**
newthyroid	94,88	**97,67**	94,42	96,28	**96,74**	96,28	96,28	**96,74**	96,28	**95,81**	94,42	94,88	96,28	**96,74**	96,28
pageblocks	94,01	93,63	**94,72**	93,45	**94,89**	94,36	93,99	**94,69**	94,53	93,63	**93,97**	93,27	93,63	**95,07**	94,35
penbased	93,20	91,65	**94,01**	94,02	93,92	**94,37**	**94,83**	94,37	93,55	**93,83**	93,64	91,93	94,19	**94,29**	93,83
satimage	80,60	**80,75**	80,10	**83,07**	82,92	82,44	**81,52**	81,22	80,27	78,09	75,45	**80,10**	82,44	82,00	**83,54**
segment	93,03	92,77	**93,12**	92,77	94,42	**95,46**	93,33	94,33	**95,33**	92,25	**93,25**	90,74	93,33	**94,37**	94,07
shuttle	94,39	98,44	**99,54**	93,80	98,71	**99,77**	93,66	98,75	**99,77**	**93,34**	91,77	92,78	93,75	97,65	**98,62**
tae	55,48	**58,78**	54,88	54,70	58,75	**60,77**	55,35	58,75	**61,42**	53,44	55,42	**58,77**	56,73	59,42	**61,42**
thyroid	**93,62**	92,51	92,51	**93,47**	92,65	92,51	**93,47**	92,65	92,37	**93,06**	92,65	92,51	**93,47**	92,65	92,51
vehicle	68,80	**73,04**	70,68	73,41	71,85	**73,62**	73,29	72,92	**73,74**	72,82	72,33	**72,91**	74,00	72,09	73,86
vowel	75,35	72,63	66,77	89,80	89,09	**91,92**	**90,00**	89,49	**90,00**	82,53	81,82	81,82	86,46	86,77	**87,37**
wine	**95,03**	94,38	94,43	95,01	96,09	**97,78**	95,55	96,09	**97,19**	95,01	96,04	**97,17**	94,96	95,50	**96,63**
yeast	58,97	58,77	**59,17**	60,72	**61,00**	59,10	61,33	**61,46**	59,84	60,04	**60,65**	57,69	61,19	**61,67**	59,57
Total	80,23	**80,32**	79,75	81,25	81,90	**82,54**	81,17	81,71	**82,30**	80,05	80,40	**80,43**	81,06	81,85	**82,11**

Table 3 shows the results obtained by both the Friedman's aligned ranks test and the Holm's test. These results are grouped by the method used to perform the comparison (FARC-HD or each OVO aggregation) and by the aggregation methods used to compute the association degree (which is the subject of the study). The first column corresponds to the different aggregations over the original FARC-HD, while in the other columns the different OVO strategies considered in this work (ND, VOTE, WV and WinWV) are shown. In each row, we show the aggregation used to compute the association degree (PROD, MIN and GM). The value of each cell corresponds to the rank obtained with the Friedman's aligned-rank test that compares the three ways of computing the association degree in each column. The value that is shown in brackets corresponds to the adjusted p-value obtained by the Holm's test using as control method the one having the smallest rank in the same column, which is shown in bold-face. The adjusted p-value is underlined when there are statistical differences ($\alpha = 0.05$).

Table 3. Friedman's Aligned Ranks Test and Holm's test

	FARC	OVO^{VOTE}	OVOND	OVO^{WV}	OVO^{WinWV}
PROD	**28,00**	39,05 (0,003)	36,90 (0,057)	31,67 (1,489)	39,55 (0,006)
MIN	28,75 (0,891)	30,72 (0,103)	29,80 (0,365)	**29,87**	28,70 (0,324)
GM	34,75 (0,443)	**21,72**	**24,8**	29,95 (1,489)	**23,25**

As it can be observed in the first column of Table 3, the best aggregation method is the product when we use the original FARC-HD, although there are no statistical differences between the three aggregations. However, in the case of the OVO aggregations, we can observe that the best method to compute the association degree is the geometric mean in all cases. In addition, the geometric mean is statistically better than the product and it is better than the minimum in almost all cases despite not being significant differences. The exception to this situation is when we use the WV, since it is severely affected by the poorer quality of the confidences of the non-predicted classes (studied in the next subsection). Thus, the aggregation to compute the association degree that shows the best results is that obtaining the highest aggregated values , that is, the geometric mean.

5.4 Analysis of the WV: WinWV vs. WV

Looking at the results presented in Table 2, we can observe how those obtained by WV are quite different from those obtained by the remainder OVO aggregations. In this subsection, we will show that despite the fact that apparently the difference between WV and WinWV are small, when using FARC-HD it allows us to obtain significant statistical differences.

In order to do so, we have carried out a number of pair-wise comparisons using the Wilcoxon's test, where we confront the original $\overline{WV}$ method against the proposed modification, with each of the aggregations that we have considered to compute the association degree in FARC-HD. Table 4 shows the results of these comparisons, where we can see how the usage of the new aggregation strategy provides significantly better results than the original WV method.

Table 4. Wilcoxon's test to compare the weighted vote and WinWV

Comparison		R+	R-	Hypothesis	p-value
OVO_{PROD}^{WV}	vs. OVO_{PROD}^{WinWV}	56,5	153,5	Rejected OVO_{GM}^{WinWV} 90%	0,059
OVO_{MIN}^{WV}	vs. OVO_{MIN}^{WinWV}	46,5	163,5	Rejected OVO_{GM}^{WinWV} 95%	0,018
OVO_{GM}^{WV}	vs. OVO_{GM}^{WinWV}	28,0	182,0	Rejected OVO_{GM}^{WinWV} 95%	0,003

5.5 Best Multi-class Model for FARC-HD: FARC-HD vs. FARC-HD_OVO

Finally, we want to check which is the best model to address multi-class classification problems with FARC-HD. We have shown that the geometric mean is better for any of the OVO aggregations, whereas the product provides better results in the case of FARC-HD. Therefore, we will carry out different pair-wise comparisons to check whether the usage of OVO (with the geometric mean) statistically improves the behavior of the original FARC-HD classifier.

Table 5. Wilcoxon's test to compare the original FARC-HD and the considered combinations for FARC-HD_OVO

Comparison		R+	R-	Hypothesis	p-value
$FARC_{PROD}$ vs.	OVO_{GM}^{ND}	164,0	46,0	Rejected by OVO_{GM}^{ND} 95%	0,03
	OVO_{GM}^{VOTE}	173,0	37,0	Rejected OVO_{GM}^{VOTE} 95%	0,01
	OVO_{GM}^{WinWV}	171,0	39,0	Rejected OVO_{GM}^{WinWV} 95%	0,01
	OVO_{GM}^{WV}	108,5	101,5	Not rejected	0,84

The results of the statistical tests are shown in the Table 5. According to them, we can confirm that using OVO model is statistically better than the usage of FARC-HD in all cases except for the one using the original WV (which is solved with WinWV).

6   Conclusions

In this work, we have faced multi-class classification problems with FARC-HD. In order to improve the obtained results, we have used the OVO strategy modifying both the inference process of FARC-HD (substituting the product t-norm with greater aggregations) and the WV aggregation of OVO (not using the confidences that have been obtained by the non-predicted classes).

These adaptations have allowed us to show the importance of the inference process when the OVO model is considered, since the confidence values are used beyond the FARC-HD classification. In addition, we found that the usage of OVO is suitable for the FARC-HD classifier, but this synergy is better when the inference process is adapted appropriately.

Acknowledgments. This work was supported in part by the Spanish Ministry of Science and Technology under projects TIN2011-28488 and the Andalusian Research Plan P10-TIC-6858 and P11-TIC-7765.

References

1. Alcalá-Fdez, J., Fernandez, A., Luengo, J., Derrac, J., García, S., Sánchez, L., Herrera, F.: KEEL data-mining software tool: Data set repository, integration of algorithms and experimental analysis framework. Journal of Multiple-Valued Logic and Soft Computing 17(2-3), 255–287 (2011)
2. Alcala-Fdez, J., Alcala, R., Herrera, F.: A fuzzy association rule-based classification model for high-dimensional problems with genetic rule selection and lateral tuning. IEEE Transactions on Fuzzy Systems 19(5), 857–872 (2011)
3. Fernández, A., Calderón, M., Barrenechea, E., Bustince, H., Herrera, F.: Solving mult-class problems with linguistic fuzzy rule based classification systems based on pairwise learning and preference relations. Fuzzy Sets and Systems 161(23), 3064–3080 (2010)
4. Friedman, J.H.: Another approach to polychotomous classification. Tech. rep., Department of Statistics, Stanford University (1996),
http://www-stat.stanford.edu/~jhf/ftp/poly.ps.Z
5. Galar, M., Fernández, A., Barrenechea, E., Bustince, H., Herrera, F.: An overview of ensemble methods for binary classifiers in multi-class problems: Experimental study on one-vs-one and one-vs-all schemes. Pattern Recognition 44(8), 1761–1776 (2011)
6. Galar, M., Fernández, A., Barrenechea, E., Bustince, H., Herrera, F.: Dynamic classifier selection for One-vs-One strategy: Avoiding non-competent classifiers. Pattern Recognition 46(12), 3412–3424 (2013)
7. Hüllermeier, E., Vanderlooy, S.: Combining predictions in pairwise classification: An optimal adaptive voting strategy and its relation to weighted voting. Pattern Recognition 43(1), 128–142 (2010)

8. Ishibuchi, H., Nakashima, T., Nii, M.: Classification and modeling with linguistic information granules: Advanced approaches to linguistic Data Mining. Springer, Berlin (2004)
9. Ishibuchi, H., Yamamoto, T.: Rule weight specification in fuzzy rule-based classification systems. IEEE Transactions on Fuzzy Systems 13(4), 428–435 (2005)
10. Lorena, A.C., Carvalho, A.C., Gama, J.M.: A review on the combination of binary classifiers in multiclass problems. Artificial Intelligence Review 30(1-4), 19–37 (2008)
11. Sheskin, D.: Handbook of parametric and nonparametric statistical procedures. Chapman & Hall/CRC (2006)

Pre-orders and Orders Generated by Conjunctive Uninorms

Dana Hliněná[1], Martin Kalina[2], and Pavol Kráľ[3]

[1] Dept. of Mathematics FEEC Brno Uni. of Technology
Technická 8, Cz-616 00 Brno, Czech Republic
hlinena@feec.vutbr.cz
[2] Slovak University of Technology in Bratislava
Faculty of Civil Engineering, Department of Mathematics
Radlinského 11, Sk-813 68 Bratislava, Slovakia
kalina@math.sk
[3] Dept. of Quantitative Methods and Information Systems,
Faculty of Economics, Matej Bel University
Tajovského 10, Sk-975 90 Banská Bystrica, Slovakia
pavol.kral@umb.sk

Abstract. This paper is devoted to studying of (pre-)orders of the unit interval generated by uninorms. We present properties of such generated pre-orders. Further we give a condition under which the generated relation is just a pre-order, i.e., under which it is not anti-symmetric. We present also a new type of uninorms, which is interesting from the point of view of generated pre-orders.

Keywords: Uninorm, non-representable uninorm, generated pre-order, twisted minimum, twisted maximum.

1 Introduction and Preliminaries

In this paper we study pre-orders generated by uninorms. The main idea is based on that of Karaçal and Kesicioğlu [8]. Further we introduce a new type of uninorms containing twisted minimum and/or twisted maximum. This type of uninorms is interesting from the point of view of generated pre-orders. As we will see later in the text, they provide arbitrarily many classes of equivalence containing infinitely many elements.

1.1 Uninorms

In 1996 Yager and Rybalov [15] proposed uninorms as a natural generalisation of both t-norms and t-conorms (for details on t-norms and their duals, t-conorms, see, e.g., [10]). Since that time researchers study properties of several distinguished families of uninorms. A complete characterization of representable uninorms can be found in [4].

A. Laurent et al. (Eds.): IPMU 2014, Part III, CCIS 444, pp. 307–316, 2014.
© Springer International Publishing Switzerland 2014

Definition 1. *A* uninorm *U is a function $U : [0,1]^2 \to [0,1]$ that is increasing, commutative, associative and has a neutral element $e \in [0,1]$.*

Some special classes of uninorms were studied, e.g., in papers [2–5, 7, 12, 13]. An overview of basic properties of uninorms is in [1]. Because of lack of space we provide only a very brief introduction to uninorms.

A uninorm U is said to be *conjunctive* if $U(x,0) = 0$, and U is said to be disjunctive if $U(1,x) = 1$, for all $x \in [0,1]$.

In the theory of fuzzy measures and integrals with respect to fuzzy measures, uninorms play the role of pseudo-multiplication [11].

A uninorm U is called *representable* if it can be written in the form

$$U(x,y) = g^{-1}(g(x) + g(y)),$$

where $g : [0,1] \to [-\infty, \infty]$ is a continuous strictly increasing function with $g(0) = -\infty$ and $g(1) = \infty$. Note yet that for each generator g there exist two different uninorms depending on convention we take: $\infty - \infty = \infty$, or $\infty - \infty = -\infty$. In the former case we get a disjunctive uninorm, in the latter case a conjunctive uninorm.

Representable uninorms are "almost continuous", i.e., they are continuous everywhere on $[0,1]^2$ except of points $(0,1)$ and $(1,0)$.

Conjunctive and disjunctive uninorms are dual in the following way

$$U_d(x,y) = 1 - U_c(1 - x, 1 - y),$$

where U_c is an arbitrary conjunctive uninorm and U_d its dual disjunctive uninorm. Assuming U_c has a neutral element e, the neutral element of U_d is $1 - e$.

For an arbitrary uninorm U and arbitrary $(x,y) \in]0, e[\times]e, 1] \cup]e, 1] \times]0, e[$ we have

$$\min\{x,y\} \leq U(x,y) \leq \max\{x,y\}. \tag{1}$$

We say that a uninorm U contains a "*zoomed-out*" representable uninorm in a square $]a, b[^2$ for $0 \leq a < e < b \leq 1$ (where $a \neq 0$ and/or $b \neq 1$), if there exists a continuous function $\tilde{g} : [a,b] \to [-\infty, \infty]$ such that $\tilde{g}(a) = -\infty$, $\tilde{g}(b) = \infty$, $\tilde{g}(e) = 0$ and

$$U(x,y) = \tilde{g}^{-1}(\tilde{g}(x) + \tilde{g}(y)) \quad \text{for } x,y \in]a,b[. \tag{2}$$

1.2   Orders Generated by T-norms

In [9] t-norms on bounded lattices were introduced.

Definition 2. *Let L be a bounded lattice. A function $T : L^2 \to L$ is said to be a t-norm if T is commutative, associative, monotone and $\mathbf{1}_L$ is its neutral element.*

Each uninorm U with a neutral element $0 < e < 1$, when restricted to the square $[0,e]^2$, is a t-norm (on the lattice $L = [0,e]$ equipped with meet and join) and when restricted to the square $[e, 1]^2$, is a t-conorm (on the lattice $L = [e, 1]$ equipped with meet and join). We will denote this t-norm by T_U and the t-conorm by S_U.

In [8], for a given t-norm T on a bounded lattice L a relation $\preceq_T$ generated by T was introduced. The definition is as follows

Definition 3. *Let $T : L^2 \to L$ be a given t-norm. For arbitrary $x, y \in L$ we denote $x \preceq_T y$ if there exists $\ell \in L$ such that $T(y, \ell) = x$.*

Proposition 1. *([8]) Let T be an arbitrary t-norm. The relation $\preceq_T$ is a partial order. Moreover, if $x \preceq_T y$ holds for $x, y \in L$ then $x \leq y$, where $\leq$ is the order generated by lattice operations.*

Dually, we can introduce a partial order $\preceq_S$ for arbitrary t-conorm S by

$$x \preceq_S y \quad \text{if there exists } \ell \in [0, 1] \text{ such that } S(y, \ell) = x.$$

However, in this case we have

$$x \preceq_S y \quad \Rightarrow \quad x \geq y.$$

2 Examples of Uninorms and Consequences for Generated (Pre-)orders

In this section we provide basic definitions necessary for our considerations, some examples of uninorms illustrating what problems we are facing and finally properties of the (pre-)order $\preceq_U$.

2.1 Basic Definitions

We start our considerations on uninorms by introducing a relation $\preceq_U$.

Definition 4. *Let U be arbitrary uninorm. By $\preceq_U$ we denote the following relation*

$$x \preceq_U y \quad \text{if there exists } \ell \in [0, 1] \text{ such that } U(y, \ell) = x.$$

Associativity of U implies transitivity of $\preceq_U$. Existence of a neutral element e implies reflexivity of $\preceq_U$. However, anti-symmetry of $\preceq_U$ is rather problematic.

Since for representable uninorm U and for arbitrary $x \in]0, 1[$ and $y \in [0, 1]$ there exists ℓ_y such that $U(x, \ell_y) = y$, the relation $\preceq_U$ is not necessarily anti-symmetric.

Lemma 1. *Let U be arbitrary uninorm. The relation $\preceq_U$ is a pre-order.*

We introduce a relation $\sim_U$.

Definition 5. *Let U be arbitrary uninorm. We say that $x, y \in [0, 1]$ are U-indifferent if*

$$x \preceq_U y \quad \text{and} \quad y \preceq_U x.$$

If x, y are U-indifferent, we write $x \sim_U y$.

Lemma 2. *For arbitrary uninorm U the relation $\sim_U$ is an equivalence relation.*

We can factorize $[0,1]$ by $\sim_U$. On the equivalence classes we introduce a relation $\trianglelefteq_U$.

Definition 6. *Let U be arbitrary uninorm. Further, let $\tilde{x}$, $\tilde{y}$ be equivalence classes of $\sim_U$. We denote $\tilde{x} \trianglelefteq_U \tilde{y}$ if for all $x \in \tilde{x}$ and all $y \in \tilde{y}$ we have $x \preceq_U y$.*

Lemma 3. *For arbitrary uninorm U the relation $\trianglelefteq_U$ is an order on $[0,1]$ factorized by $\sim_U$.*

2.2 Examples of Uninorms

In [6] some examples of non-representable uninorms were presented. We pick up at least two of them, namely the uninorm constructed in Example 2.1.(b) and in Example 2.4.(a). The latter one was published in [6] with a mistake. This means that formula (4) is an erratum to Example 2.4.(a) from [6].

Example 1. **(a)**

$$U_1(x,y) = \begin{cases} \max\{x,y\}, & \text{if } \min\{x,y\} \geq \frac{1}{2}, \\ \min\{x,y\}, & \text{if } \max\{x,y\} = \frac{1}{2}, \\ 0, & \text{if } \min\{x,y\} \leq \frac{1}{4} \text{ and } \max\{x,y\} < \frac{1}{2}, \\ & \text{or if } \min\{x,y\} = 0 \text{ and } \max\{x,y\} > \frac{1}{2}, \\ \frac{2^i-1}{2^{i+1}}, & \text{for } i \in \{1,2,3,\ldots\}, \\ & \text{if } \frac{2^{i-1}-1}{2^i} < \min\{x,y\} \leq \frac{2^i-1}{2^{i+1}} \text{ and } \max\{x,y\} > \frac{1}{2}, \\ & \text{or if } \frac{2^i-1}{2^{i+1}} < \min\{x,y\} \leq \frac{2^{i+1}-1}{2^{i+2}} \\ & \text{and } \max\{x,y\} < \frac{1}{2}. \end{cases} \tag{3}$$

(b)

$$U_2(x,y) = \begin{cases} \max\{x,y\}, & \text{if } \min\{x,y\} \geq 0.5, \\ 0, & \text{if } \max\{x,y\} < 0.5 \text{ and } \min\{x,y\} \leq 0.4, \\ & \text{or if } \min\{x,y\} = 0, \\ 0.4y - 0.1, & \text{if } y \geq 0.5 \text{ and } 0.1 \leq x \leq 0.4y - 0.1, \\ 0.4x - 0.1, & \text{if } x \geq 0.5 \text{ and } 0.1 \leq y \leq 0.4x - 0.1, \\ \min\{x,y\}, & \text{otherwise.} \end{cases} \tag{4}$$

Now we present new types of uninorms which will play an important role in describing of properties of the pre-order $\preceq_U$. In Example 2**(b)** we use a t-norm, constructed in [14].

Example 2. We show two uninorms, both of them can be further generalized. For simplicity we will set $e = \frac{1}{2}$. In the square $]\frac{1}{4}, \frac{3}{4}[^2$ a representable uninorm will be zoomed-out. Particularly, we will consider its generator

$$g(x) = \begin{cases} \ln(4x - 1), & \text{for } x \in]\frac{1}{4}, \frac{1}{2}], \\ -\ln(3 - 4x), & \text{for } x \in]\frac{1}{2}, \frac{3}{4}[. \end{cases} \tag{5}$$

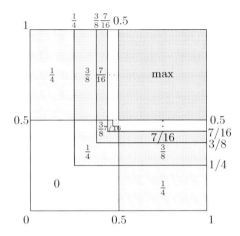

Fig. 1. Uninorm U_1

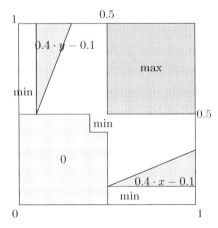

Fig. 2. Uninorm U_2

We will denote $U_r(x, y) = g^{-1}(g(x) + g(y))$ for $(x, y) \in]\frac{1}{4}, \frac{3}{4}[$, though U_r is not a uninorm because of its domain.

(a) We are going to construct a uninorm U_3. To do this, we choose function $f(z) = \frac{z}{2} - \frac{1}{8}$ that will represent values $U_3(\frac{1}{8}, z)$ for $z \in [\frac{1}{4}, \frac{3}{4}]$. Using this function and U_r we will compute all other values in the rectangle $]0, \frac{1}{4}[\times]\frac{1}{4}, \frac{3}{4}[$, and by commutativity we get also values in the rectangle $]\frac{1}{4}, \frac{3}{4}[\times]0, \frac{1}{4}[$. In general, $f : [\frac{1}{4}, \frac{3}{4}] \to [0, \frac{1}{4}]$ is a function that is continuous, strictly increasing and fulfilling $f(\frac{1}{4}) = 0$, $f(\frac{1}{2}) = \frac{1}{8}$ and $f(\frac{3}{4}) = \frac{1}{4}$. These properties (and the way of construction) guarantee that for arbitrary $x \in]0, \frac{1}{4}[$ and $z \in]0, \frac{1}{4}[$ there exists unique $y \in]\frac{1}{4}, \frac{3}{4}[$ such that $U(x, y) = z$ (meaning that all elements of $]0, \frac{1}{4}[$ will be U_3-indifferent).

Let us now construct the values of U_3 in the rectangle $A =]0, \frac{1}{4}[\times]\frac{1}{4}, \frac{3}{4}[$. Using function f, for arbitrary $x \in]0, \frac{1}{4}[$ there exists unique $\bar{y} \in]\frac{1}{4}, \frac{3}{4}[$ such that $x = U(\frac{1}{8}, \bar{y})$. Namely, $\bar{y} = f^{-1}(x) = 2x + \frac{1}{4}$. For arbitrary $(x, y) \in A$ we get

$$U_3(x, y) = U_3\left(U_3(\frac{1}{8}, \bar{y}), y\right) = U_3\left(\frac{1}{8}, U_3(\bar{y}, y)\right), \qquad (6)$$

and $U_3(\bar{y}, y) = U_r(\bar{y}, y)$. Since $U_3(\frac{1}{8}, z) = f(z)$ for $z \in [\frac{1}{4}, \frac{3}{4}]$, this implies $U_3(x, y) = U_3\left(\frac{1}{8}, U_r(\bar{y}, y)\right) = f(U_r(\bar{y}, y))$. Finally, using the definition of f and the formula for $\bar{y}$, we have that

$$U_3(x, y) = \frac{U_r(2x + \frac{1}{4}, y)}{2} - \frac{1}{8}. \qquad (7)$$

For $(x, y) \notin]0, \frac{1}{4}[\times]\frac{1}{4}, \frac{3}{4}[\cup]\frac{1}{4}, \frac{3}{4}[\times]0, \frac{1}{4}[$, we define

$$U_3(x, y) = \begin{cases} 0, & \text{if } \min\{x, y\} = 0, \text{ or if } \max\{x, y\} \leq \frac{1}{4}, \\ 1, & \text{if } \min\{x, y\} \geq \frac{3}{4}, \\ \frac{1}{4}, & \text{if } 0 < \min\{x, y\} \leq \frac{1}{4} \text{ and } \max\{x, y\} \geq \frac{3}{4}, \\ & \text{or if } \min\{x, y\} = \frac{1}{4} \text{ and } \max\{x, y\} > \frac{1}{4}, \\ U_r(x, y), & \text{if } (x, y) \in]\frac{1}{4}, \frac{3}{4}[^2, \\ \max\{x, y\}, & \text{if } \frac{1}{4} < \min\{x, y\} < \frac{3}{4} \text{ and } \max\{x, y\} \geq \frac{3}{4}. \end{cases} \qquad (8)$$

In Figure 3 we have sketched level-functions of U_3 for level $\frac{1}{16}, \frac{1}{8}, \frac{3}{16}$ in the rectangles $]0, \frac{1}{4}[\times]\frac{1}{4}, \frac{3}{4}[$ and $]\frac{1}{4}, \frac{3}{4}[\times]0, \frac{1}{4}[$. Yet we show associativity of U_3. Of course, only the case $x \in]0, \frac{1}{4}[$ and $y_1, y_2 \in]\frac{1}{4}, \frac{3}{4}[$ is interesting. For all other cases associativity of U_3 is trivial. Formula (6) implies

$$U_3(x, U_3(y_1, y_2)) = U_3(x, U_r(y_1, y_2)) = \frac{U_r(2x + \frac{1}{4}, U_r(y_1, y_2))}{2} - \frac{1}{8},$$

$$U_3(U_3(x, y_1), y_2) = U_3\left(\frac{U_r(2x + \frac{1}{4}, y_1)}{2} - \frac{1}{8}, y_2\right) = \frac{U_r(U_r(2x + \frac{1}{4}, y_1), y_2)}{2} - \frac{1}{8},$$

and by associativity of U_r we get equality of the above two formulas and the proof is finished. In the last formula we have used twice equality (6).

(b) We use similar considerations as by constructing of the uninorm U_3. Just in this case we have two different function, $g_1(z) = \frac{z}{4} - \frac{1}{16} = U_4(\frac{1}{16}, z)$ and

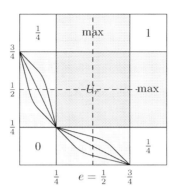

Fig. 3. Uninorm U_3

$g_2(z) = \frac{z}{4} + \frac{1}{16} = U_4(\frac{3}{16}, z)$. I.e., g_1 will act in the rectangle $]0, \frac{1}{8}[\times]\frac{1}{4}, \frac{3}{4}[$, and g_2 in $]\frac{1}{8}, \frac{1}{4}[\times]\frac{1}{4}, \frac{3}{4}[$.

For $x \in]0, \frac{1}{8}[$ we have uniquely given $\bar{y}$ such that $x = g_1(\bar{y})$, implying $\bar{y} = 4x + \frac{1}{4}$. Followingly, for $x \in]0, \frac{1}{8}[$ and $y \in]\frac{1}{4}, \frac{3}{4}[$ we get formula

$$U_4(x, y) = U_4\left(U_4(\frac{1}{16}, 4x + \frac{1}{4}), y\right) = \frac{U_4(y, 4x + \frac{1}{4})}{4} - \frac{1}{16}. \tag{9}$$

For $x \in]\frac{1}{8}, \frac{1}{4}[$ we have uniquely given $\bar{y}$ such that $x = g_2(\bar{y})$, implying $\bar{y} = 4x - \frac{1}{4}$. Hence, for $x \in]\frac{1}{8}, \frac{1}{4}[$ and $y \in]\frac{1}{4}, \frac{3}{4}[$ we get formula

$$U_4(x, y) = U_4\left(U_4(\frac{1}{16}, 4x + \frac{1}{4}), y\right) = \frac{U_r(y, 4x - \frac{1}{4})}{4} + \frac{1}{16}. \tag{10}$$

For $(x, y) \notin]0, \frac{1}{4}[\times]\frac{1}{4}, \frac{3}{4}[\cup]\frac{1}{4}, \frac{3}{4}[\times]0, \frac{1}{4}[$, we define

$$U_4(x, y) = \begin{cases} 0, & \text{if } \min\{x, y\} = 0, \\ & \text{or if } \max\{x, y\} \leq \frac{1}{4} \text{ and } \min\{x, y\} \leq \frac{1}{8}, \\ \frac{1}{8}, & \text{if } \min\{x, y\} > \frac{1}{8} \text{ and } \max\{x, y\} \leq \frac{1}{4}, \\ & \text{or if } 0 < \min\{x, y\} \leq \frac{1}{8} \text{ and } \max\{x, y\} \geq \frac{3}{4}, \\ & \text{or if } \min\{x, y\} = \frac{1}{8} \text{ and } \max\{x, y\} \in]\frac{1}{4}, \frac{3}{4}[, \\ \frac{1}{4}, & \text{if } \frac{1}{8} < \min\{x, y\} \leq \frac{1}{4} \text{ and } \max\{x, y\} \geq \frac{3}{4}, \\ & \text{if } \frac{1}{8} < \min\{x, y\} \leq \frac{1}{4} \text{ and } \max\{x, y\} \geq \frac{3}{4}, \\ & \text{or if } \min\{x, y\} = \frac{1}{4} \text{ and } \max\{x, y\} > \frac{1}{4}, \\ 1, & \text{if } \min\{x, y\} \geq \frac{3}{4}, \\ U_r(x, y), & \text{if } (x, y) \in]\frac{1}{4}, \frac{3}{4}[^2, \\ \max\{x, y\}, & \text{if } \frac{1}{4} < \min\{x, y\} < \frac{3}{4} \text{ and } \max\{x, y\} \geq \frac{3}{4}. \end{cases} \tag{11}$$

In Figure 4 we have sketched level-functions of U_4 for level $\frac{1}{32}, \frac{1}{16}, \frac{3}{32}, \frac{5}{32}, \frac{3}{16}, \frac{7}{32}$ in the rectangles $]0, \frac{1}{4}[\times]\frac{1}{4}, \frac{3}{4}[$ and $]\frac{1}{4}, \frac{3}{4}[\times]0, \frac{1}{4}[$. We skip proving associativity of U_4 since this proof follows the same idea as that one used for U_3.

The construction method used in uninorms U_3, U_4 in the rectangles $]0, \frac{1}{4}[\times]\frac{1}{4}, \frac{3}{4}[$ and $]\frac{1}{4}, \frac{3}{4}[\times]0, \frac{1}{4}[$, is called *twisting of minimum*.

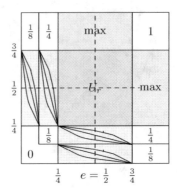

Fig. 4. Uninorm U_4

Remark 1.

- In the same way as we have twisted minimum in uninorms U_3, U_4, we could twist also maximum.
- In general, if U is a uninorm with a zoomed-out representable uninorm U_r in a square $]a, b[$, and an L-shaped area $]c_1, a[^2 \setminus]c_2, a[^2$ is an area of constantness, we can twist minimum in the rectangle $]c_1, c_2[\times]a, b[$, in case $c_2 < e$ and we can twist maximum in the same rectangle in case $e < c_1$.
- Particularly, we could modify U_3 (and also U_4) in such a way that in any rectangle $]c_1, c_2[\times]a, b[$ with $c_2 < \frac{1}{4}$ we could twist minimum, and in any rectangle $]c_1, c_2[\times]a, b[$ with $c_1 > \frac{3}{4}$ we could twist maximum. In this way we could construct uninorms U having arbitrarily many open intervals such that elements of each of these intervals are U-indifferent.

2.3   Properties of $\preceq_U$

A natural question concerning the pre-order $\preceq_U$ is, under which conditions it is an order and under which conditions this order $\preceq_U$ restricted to $[0, e]$ coincides with $\preceq_{T_U}$, and restricted to $[e, 1]$ coincides with $\preceq_{S_U}$.

Proposition 2. *Let U be a uninorm. The following statements are equivalent*

(a) *For arbitrary $x \in [0, e]$ and $y \in [e, 1]$ we have $U(x, y) \in \{x, y\}$.*
(b) *Arbitrary $x_1, x_2 \in [0, e]$ yield $x_1 \preceq_U x_2$ if and only if $x_1 \preceq_{T_U} x_2$, arbitrary $y_1, y_2 \in [e, 1]$ yield $y_1 \preceq_U y_2$ if and only if $y_1 \preceq_{S_U} y_2$.*

Remark 2. There are many non-representable uninorms which do not fulfil Proposition 2(a) (cf. uninorms $U_1 - U_4$ from Examples 1 and 2).
Though if U is a uninorm fulfilling Proposition 2(a), i.e., we have an order $\preceq_U$ that, restricted to $[0, e]$ coincides with $\preceq_{T_U}$, and restricted to $[e, 1]$ coincides with $\preceq_{S_U}$, not all properties of the structures $([0, e], \preceq_{T_U})$ and $([e, 1], \preceq_{S_U})$ are

inherited in the structure $([0, 1], \preceq_U)$. Particularly, if

$$U(x, y) = \begin{cases} \min\{x, y\}, \text{ if } \min\{x, y\} = 0, \\ \qquad \text{ or if } \min\{x, y\} < e \text{ and } \max\{x, y\} < 1, \\ \max\{x, y\}, \text{ otherwise,} \end{cases}$$

then the interval $]0, e[$ has no join and the interval $]e, 1[$ has no meet in the order $\preceq_U$, but we have

$$\bigvee_{\preceq_{T_U}}]0, e[= e, \qquad \bigwedge_{\preceq_{S_U}}]e, 1[= 1 \, .$$

Proposition 3. *Let U is either a representable uninorm or it has a zoomed-out representable uninorm in a square $]a, b[^2$.*
(a) The relation $\preceq_U$ is not anti-symmetric.
(b) The relation $\trianglelefteq_U$ is an order on the classes of equivalence. The class containing e is the greatest in $\trianglelefteq_U$. All other classes of equivalence containing infinitely many elements are mutually incomparable.

3 Open Problem and Conclusions

In Proposition 3 we have formulated a sufficient condition for a uninorm U under which $\preceq_U$ is not anti-symmetric. Whether this condition is also sufficient for violating the anti-symmetry of $\preceq_U$, remains an open problem.

In this paper we have started studying the relation $\preceq_U$. Because of lack of space we have presented just two basic properties formulated in Propositions 2 and 3. Further we have introduced a new type of uninorms constructed using twisting of minimum and/or twisting of maximum.

Acknowledgments. Martin Kalina has been supported from the Science and Technology Assistance Agency under contract No. APVV-0073-10, and from the VEGA grant agency, grant number 1/0143/11. Pavol Kráľ has been supported from the VEGA grant agency, grant number 1/0647/14.

This paper has been supported by the Project: Mobility - enhancing research, science and education at the Matej Bel University, ITMS code: 26110230082, under the Operational Program Education co-financed by the European Social Fund.

References

1. Calvo, T., Kolesárová, A., Komorníková, M., Mesiar, R.: Aggregation operators: Properties, classes and construction methods. In: Calvo, T., Mayor, G., Mesiar, R. (eds.) Aggregation Operators, pp. 3–104. Physica-Verlag, Heidelberg (2002)
2. Drewniak, J., Drygaś, P.: Characterization of uninorms locally internal on A(e). Fuzzy Set and Systems (submitted)

3. Drygaś, P.: On monotonic operations which are locally internal on some subset of their domain. In: Štepnička, et al. (eds.) New Dimensions in Fuzzy Logic and Related Technologies, Proceedings of the 5th EUSFLAT Conference 2007, vol. II, pp. 359–364. Universitas Ostraviensis, Ostrava (2007)

4. Fodor, J., De Baets, B.: A single-point characterization of representable uninorms. Fuzzy Sets and Systems 202, 89–99 (2012)

5. Fodor, J., Yager, R.R., Rybalov, A.: Structure of uninorms. International Journal of Uncertainty, Fuzziness and Knowledge-based Systems 5, 411–422 (1997)

6. Hliněná, D., Kalina, M., Kráľ, P.: Non-representable uninorms. In: EUROFUSE 2013, Uncertainty and Imprecision Modelling in Decision Making, pp. 131–138, Servicio de Publicaciones de la Universidad de Oviedo, Oviedo (2013)

7. Hu, S., Li, Z.: The structure of continuous uninorms. Fuzzy Sets and Systems 124, 43–52 (2001)

8. Karaçal, F., Kesicioğlu, M.N.: A t-partial order obtained from t-norms. Kybernetika 47(2), 300–314 (2011)

9. Karaçal, F., Khadijev, D.: $\bigvee$-distributive and infinitely $\bigvee$-distributive t-norms on complete lattices. Fuzzy Sets and Systems 151, 341–352 (2005)

10. Klement, E.P., Mesiar, R., Pap, E.: Triangular Norms. Springer, Heidelberg (2000)

11. Mesiar, R.: Choquet-like integrals. J. Math. Anal. Appl. 194, 477–488 (1995)

12. Petrík, M., Mesiar, R.: On the structure of special classes of uninorms. Fuzzy Sets and Systems 240, 22–38 (2014)

13. Ruiz-Aguilera, D., Torrens, J., De Baets, B., Fodor, J.: Some remarks on the characterization of idempotent uninorms. In: Hüllermeier, E., Kruse, R., Hoffmann, F. (eds.) IPMU 2010. LNCS (LNAI), vol. 6178, pp. 425–434. Springer, Heidelberg (2010)

14. Smutná, D.: Limit t-norms as a basis for the construction of a new t-norms. International Journal of Uncertainty, Fuzziness and Knowledge-based Systems 9(2), 239–247 (2001)

15. Yager, R.R., Rybalov, A.: Uninorm aggregation operators. Fuzzy Sets and Systems 80, 111–120 (1996)

On Extensional Fuzzy Sets
Generated by Factoraggregation

Pavels Orlovs and Svetlana Asmuss

Institute of Mathematics and Computer Science
University of Latvia
Raina blvd. 29, Riga, LV-1459, Latvia
pavels.orlovs@gmail.com, svetlana.asmuss@lu.lv

Abstract. We develop the concept of a general factoraggregation operator introduced by the authors on the basis of an equivalence relation and applied in two recent papers for analysis of bilevel linear programming solving parameters. In the paper this concept is generalized by using a fuzzy equivalence relation instead of the crisp one. We show how the generalized factoraggregation can be used for construction of extensional fuzzy sets and consider approximations of arbitrary fuzzy sets by extensional ones.

Keywords: Aggregation operator, general aggregation operator, fuzzy equivalence relation, extensional fuzzy set.

1 Introduction

The paper deals with some generalization of the concept of factoraggregation introduced, studied and applied by the authors in [7] and [8]. Factoraggregation is a special construction of a general aggregation operator based on an equivalence relation. The idea of factoraggregation is based on factorization, which allows to aggregate fuzzy subsets taking into account the classes of equivalence, i.e. the partition generated by an equivalence relation. The factoraggregation operator was specially designed in the context of bilevel linear programming in order to analyse the satisfaction degree of objectives on each level and to choose solving parameters values.

In this paper we develop the concept of a factoraggregation operator by using a t-norm T and a T-fuzzy equivalence relation E instead of the crisp one. We define generalized T-fuzzy factoraggregation with respect to E and consider its properties. Taking into account that fuzzy equivalence relations represent the fuzzification of equivalence relations and extensional fuzzy subsets play the role of fuzzy equivalence classes we consider the generalized fuzzy factoraggregation in the context of extensional fuzzy sets.

Within the theory of fuzzy logic the first researcher, who pointed the relevance of extensional fuzzy sets, was L.A. Zadeh [10]. Extensional fuzzy sets are a key concept in the comprehension of the universe set under the effect of an T-fuzzy equivalence relation as they correspond with the observable sets or granules of

A. Laurent et al. (Eds.): IPMU 2014, Part III, CCIS 444, pp. 317–326, 2014.

the universe set [5]. In our paper we show how the generalized factoraggregation operator can be used for construction of extensional fuzzy sets.

The paper is structured as follows. In the second section we recall the definitions of an ordinary aggregation operator and of a general aggregation operator acting on fuzzy structures. Then we give the definition of a factoraggregation operator corresponding to an equivalence relation, which is a case of the general aggregation operator.

The third section is devoted to the concept of a generalized T-fuzzy factoraggregation with respect to a T-fuzzy equivalence relation. We recall the definition of a T-fuzzy equivalence relation E and modify the construction of factoraggregation by involving E. We show that all properties of the definition of a general aggregation operator such as the boundary conditions and the monotonicity hold for the generalized T-fuzzy factoraggregation operator.

The fourth section contains the main result on generalized fuzzy factoraggregation in the context of extensional fuzzy sets. We show that in the case of lower semi-continuous t-norm T the result of generalized T-fuzzy factoraggregation corresponding to E is extensional with respect to T-fuzzy equivalence E. Finally the generalized T-fuzzy factoraggregation approach is applied for approximation of arbitrary fuzzy sets by extensional ones. The proposed method of approximation is tested by considering numerical examples from [5].

2   Factoraggregation

In this section we recall the definition of a factoraggregation operator, which is based on an equivalence relation. This concept was introduced and studied in [7, 8]. Let us start with the classical notion of an aggregation operator (see e.g. [1–3]).

Definition 1. *A mapping $A\colon [0,1]^n \to [0,1]$ is called an aggregation operator if and only if the following conditions hold:*

(A1) $A(0,\ldots,0) = 0$;
(A2) $A(1,\ldots,1) = 1$;
(A3)   *for all $x_1,\ldots,x_n, y_1,\ldots,y_n \in [0,1]$:*
        if $x_1 \le y_1, \ldots, x_n \le y_n$, then $A(x_1,\ldots,x_n) \le A(y_1,\ldots,y_n)$.

The integer n represents the arity of the aggregation operator, that is, the number of its variables. Conditions (A1) and (A2) are called the boundary conditions of A, but (A3) means the monotonicity of A.

The general aggregation operator $\tilde{A}$ acting on $[0,1]^D$, where $[0,1]^D$ is the set of all fuzzy subsets of a set D, was introduced in 2003 by A. Takaci [9]. We denote by $\preceq$ a partial order on $[0,1]^D$ with the least and the greatest elements $\tilde{0}$ and $\tilde{1}$ respectively.

Definition 2. *A mapping $\tilde{A}\colon ([0,1]^D)^n \to [0,1]^D$ is called a general aggregation operator if and only if the following conditions hold:*

($\tilde{A}$1) $\tilde{A}(\tilde{0}, \ldots, \tilde{0}) = \tilde{0}$;
($\tilde{A}$2) $\tilde{A}(\tilde{1}, \ldots, \tilde{1}) = \tilde{1}$;
($\tilde{A}$3)   for all $\mu_1, \ldots, \mu_n, \eta_1, \ldots, \eta_n \in [0,1]^D$:
          if $\mu_1 \preceq \eta_1, \ldots, \mu_n \preceq \eta_n$, then $\tilde{A}(\mu_1, \ldots, \mu_n) \preceq \tilde{A}(\eta_1, \ldots, \eta_n)$.

We consider the case:

$$\mu \preceq \eta \text{ if and only if } \mu(x) \leq \eta(x) \text{ for all } x \in D,$$

for $\mu, \eta \in [0,1]^D$. The least and the greatest elements are indicators of $\varnothing$ and D respectively, i.e.

$$\tilde{0}(x) = 0 \text{ and } \tilde{1}(x) = 1 \text{ for all } x \in D.$$

There exist several approaches to construct a general aggregation operator $\tilde{A}$ based on an ordinary aggregation operator A. The most simplest one is the pointwise extension of an aggregation operator A:

$$\tilde{A}(\mu_1, \mu_2, \ldots, \mu_n)(x) = A(\mu_1(x), \mu_2(x), \ldots, \mu_n(x)),$$

where $\mu_1, \mu_2, \ldots, \mu_n \in [0,1]^D$ are fuzzy sets and $x \in D$.

A widely used approach to constructing a general aggregation operator $\tilde{A}$ is the T - extension [9], whose idea comes from the classical extension principle and uses a t-norm T (see e.g. [4]):

$$\tilde{A}(\mu_1, \mu_2, \ldots, \mu_n)(x) = \sup_{x = A(u_1, u_2, \ldots, u_n)} T(\mu_1(u_1), \mu_2(u_2), \ldots, \mu_n(u_n)).$$

Here $\mu_1, \mu_2, \ldots, \mu_n \in [0,1]^D$ and $x, u_1, u_2, \ldots, u_n \in D$, where $D = [0,1]$.

Another method of constructing a general aggregation operator is factoraggregation [7, 8]. This method is based on an equivalence relation ρ defined on a set D and it allows to aggregate fuzzy subsets of D taking into account the classes of equivalence ρ, i.e. the corresponding partition of D.

Definition 3. *Let $A \colon [0,1]^n \to [0,1]$ be an ordinary aggregation operator and ρ be an equivalence relation defined on a set D. An operator*

$$\tilde{A}_\rho \colon ([0,1]^D)^n \to [0,1]^D$$

such as

$$\tilde{A}_\rho(\mu_1, \mu_2, \ldots, \mu_n)(x) = \sup_{u \in D : (u,x) \in \rho} A(\mu_1(u), \mu_2(u), \ldots, \mu_n(u)), \qquad (1)$$

where $x \in D$ and $\mu_1, \mu_2, \ldots, \mu_n \in [0,1]^D$, is called a factoraggregation operator corresponding to ρ.

Relation ρ factorizes D into the classes of equivalence. Operator $\tilde{A}_\rho$ aggregates fuzzy sets $\mu_1, \mu_2, \ldots, \mu_n$ in accordance with these classes of equivalence. In this construction for evaluation of $\tilde{A}_\rho(\mu_1, \mu_2, \ldots, \mu_n)(x)$ we take the supremum of aggregation A of values $\mu_1(u), \mu_2(u), \ldots, \mu_n(u)$ on the set of all points u, which are equivalent to x with respect to ρ, i.e. we consider all elements $u \in D$ such that $(u, x) \in \rho$.

In our previous papers [7, 8] this construction was used for the analysis of solving parameters for bilevel linear programming problems with one objective on the upper level P^U with the higher priority in optimization than multiple objectives on the lower level $P^L = (P_1^L, P_2^L, ..., P_n^L)$:

$$P^U : \quad y_0(x) = c_{01}x_1 + c_{02}x_2 + ... + c_{0k}x_k \longrightarrow \min$$
$$P_i^L : \quad y_i(x) = c_{i1}x_1 + c_{i2}x_2 + ... + c_{ik}x_k \longrightarrow \min, \quad i = \overline{1, n}$$

$$D : \begin{cases} a_{j1}x_1 + a_{j2}x_2 + ... + a_{jk}x_k \leq b_j, \ j = \overline{1, m}, \\ x_l \geq 0, \ l = \overline{1, k}, \end{cases}$$

where $k, l, m, n \in \mathbb{N}$, $a_{jl}, b_j, c_{il} \in \mathbb{R}$, $j = \overline{1, m}$, $l = \overline{1, k}$, $i = \overline{0, n}$, and $x = (x_1, ..., x_k) \in \mathbb{R}^k$, $D \subset \mathbb{R}^k$ is non-empty and bounded.

The factoraggregation was applied to the membership functions of the objectives, which characterise how the corresponding objective function is close to its optimal value (see [11]):

$$\mu_i(x) = \begin{cases} 1, & y_i(x) < y_i^{min}, \\ \dfrac{y_i(x) - y_i^{max}}{y_i^{min} - y_i^{max}}, & y_i^{min} \leq y_i(x) \leq y_i^{max}, \\ 0, & y_i(x) > y_i^{max}, \end{cases}$$

where y_i^{min} and y_i^{max} are the individual minimum and the individual maximum of the objective y_i subject to the given constraints, $i = \overline{0, n}$. The introduced operator aggregates the membership functions on the lower level considering the classes of equivalence generated by the membership function on the upper level:

$$\tilde{A}(\mu_1, \mu_2, ..., \mu_n)(x) = \max_{\mu_0(x) = \mu_0(u)} A(\mu_1(u), \mu_2(u), ..., \mu_n(u)), \quad x \in D.$$

In this case μ_0 generates the equivalence relation ρ_{μ_0}:

$$(u, v) \in \rho_{\mu_0} \iff \mu_0(u) = \mu_0(v).$$

The role of this equivalence in the construction of factoraggregation follows from the hierarchy between the objectives, it was explained in details in [7, 8].

The factoraggregation operator (1) is a general aggregation operator. In the next section we generalize this construction by using a t-norm T and a T-fuzzy equivalence relation E instead of the crisp one.

3   Generalization of Factoraggregation

In order to generalize the concept of factoraggregation we recall the definition of a T-fuzzy equivalence relation. Fuzzy equivalence relations were introduced in 1971 by L.A. Zadeh [10] for the strongest t-norm T_M and later were developed and applied by several authors in more general cases.

Definition 4. *Let T be a t-norm and E be a fuzzy relation on a set D, i.e. E is a fuzzy subset of $D \times D$. A fuzzy relation E is a T-fuzzy equivalence relation if and only if for all $x, y, z \in D$ it holds*

(E1) $E(x, x) = 1$ *(reflexivity)*;
(E2) $E(x, y) = E(y, x)$ *(symmetry)*;
(E3) $T(E(x, y), E(y, z)) \leq E(x, z)$ *(T-transitivity)*.

We modify the construction (1) by using a T-fuzzy equivalence relation in order to obtain a T-fuzzy generalization of factoraggregation.

Definition 5. *Let $A \colon [0, 1]^n \to [0, 1]$ be an ordinary aggregation operator, T be a t-norm and E be a T-fuzzy equivalence relation defined on D. An operator*

$$\tilde{A}_{E,T} \colon ([0, 1]^D)^n \to [0, 1]^D$$

such as

$$\tilde{A}_{E,T}(\mu_1, \mu_2, \ldots, \mu_n)(x) = \sup_{u \in D} T(E(x, u), A(\mu_1(u), \mu_2(u), \ldots, \mu_n(u))), \qquad (2)$$

where $x \in D$ and $\mu_1, \mu_2, \ldots, \mu_n \in [0, 1]^D$, is called a generalized T-fuzzy factor-aggregation corresponding to E.

Let us note that in the case of crisp equivalence relations, i.e. when $E = E_\rho$ for an equivalence relation ρ, where

$$E_\rho(x, y) = \begin{cases} 1, & (x, y) \in \rho, \\ 0, & (x, y) \notin \rho, \end{cases}$$

we obtain $\tilde{A}_{E_\rho, T} = \tilde{A}_\rho$ (see (1)):

$$\tilde{A}_{E_\rho, T}(\mu_1, \mu_2, \ldots, \mu_n)(x) = \sup_{u \in D} T(E_\rho(x, u), A(\mu_1(u), \mu_2(u), \ldots, \mu_n(u)))$$

$$= \sup_{u \in D : (u, x) \in \rho} T(1, A(\mu_1(u), \mu_2(u), \ldots, \mu_n(u)))$$

$$= \sup_{u \in D : (u, x) \in \rho} A(\mu_1(u), \mu_2(u), \ldots, \mu_n(u)) = \tilde{A}_\rho(\mu_1, \mu_2, \ldots, \mu_n)(x).$$

Let us prove that the construction (2) gives us a general aggregation operator. We must show that conditions $(\tilde{A}1)$, $(\tilde{A}2)$ and $(\tilde{A}3)$ are satisfied.

Proposition 1. *Let $A \colon [0, 1]^n \to [0, 1]$ be an ordinary aggregation operator, T be a t-norm and E be a T-fuzzy equivalence relation defined on D. Operator $\tilde{A}_{E,T}$ given by (2) is a general aggregation operator.*

Proof. First we prove the boundary conditions:

1)
$$\tilde{A}_{E,T}(\tilde{0},\ldots,\tilde{0})(x) = \sup_{u\in D} T(E(x,u), A(\tilde{0}(u),\ldots,\tilde{0}(u)))$$

$$= \sup_{u\in D} T(E(x,u), A(0,\ldots,0)) = \sup_{u\in D} T(E(x,u),0) = \tilde{0}(x);$$

2)
$$\tilde{A}_{E,T}(\tilde{1},\ldots,\tilde{1})(x) = \sup_{u\in D} T(E(x,u), A(\tilde{1}(u),\ldots,\tilde{1}(u)))$$

$$= \sup_{u\in D} T(E(x,u), A(1,\ldots,1)) = \sup_{u\in D} T(E(x,u),1) = \sup_{u\in D} E(x,u) = \tilde{1}(x).$$

To prove the monotonicity of $\tilde{A}_{E,T}$ we use the monotonicity of A and T:

$$\mu_i \preceq \eta_i, \quad i = 1, 2, \ldots, n$$

$$\implies A(\mu_1(u),\ldots,\mu_n(u)) \le A(\eta_1(u),\ldots,\eta_n(u)) \text{ for all } u \in D$$

$$\implies T(E(x,u), A(\mu_1(u),\ldots,\mu_n(u))) \le T(E(x,u), A(\eta_1(u),\ldots,\eta_n(u)))$$

$$\text{for all } x \in D \text{ and for all } u \in D$$

$$\implies \sup_{u\in D} T(E(x,u), A(\mu_1(u),\ldots,\mu_n(u)))$$

$$\le \sup_{u\in D} T(E(x,u), A(\eta_1(u),\ldots,\eta_n(u))) \text{ for all } x \in D$$

$$\implies \tilde{A}_{E,T}(\mu_1,\ldots,\mu_n) \preceq \tilde{A}_{E,T}(\eta_1,\ldots,\eta_n).$$

Now we illustrate the generalized T-fuzzy factoraggregation on some particular numerical example. Here and throughout the paper the numerical inputs are taken from [5].

Let us consider the discrete universe $D = \{x_1, x_2, x_3, x_4, x_5\}$ and the following T_M-fuzzy equivalence relation E in matrix form, where T_M is the minimum t-norm:

$$E = \begin{pmatrix} 1 & 0.9 & 0.7 & 0.4 & 0.2 \\ 0.9 & 1 & 0.7 & 0.4 & 0.2 \\ 0.7 & 0.7 & 1 & 0.4 & 0.2 \\ 0.4 & 0.4 & 0.4 & 1 & 0.2 \\ 0.2 & 0.2 & 0.2 & 0.2 & 1 \end{pmatrix}.$$

This equivalence relation is also T_L-transitive and T_P-transitive, i.e. transitive with respect to the Lukasiewicz t-norm T_L and product t-norm T_P respectively.

Let us take the following fuzzy subsets of D:

$$\mu_1 = \begin{pmatrix} 0.9 \\ 0.5 \\ 0.6 \\ 0.8 \\ 0.3 \end{pmatrix}, \quad \mu_2 = \begin{pmatrix} 0.2 \\ 0 \\ 0.2 \\ 0.6 \\ 0.9 \end{pmatrix}, \quad \mu_3 = \begin{pmatrix} 0.7 \\ 0.5 \\ 0.1 \\ 0.8 \\ 0.6 \end{pmatrix}, \quad \mu_4 = \begin{pmatrix} 0.1 \\ 0.9 \\ 0.2 \\ 0.8 \\ 0.5 \end{pmatrix}.$$

Now we consider the minimum aggregation operator $A = MIN$ and obtain the following generalized T-fuzzy factoraggregation:

$$\tilde{A}_{E,T}(\mu_1, \mu_2, \mu_3, \mu_4)(x) = \max_{u \in D} T(E(x, u), \min(\mu_1(u), \mu_2(u), \mu_3(u), \mu_4(u))).$$

Taking $T = T_L$, $T = T_M$ and $T = T_P$ as a result we obtain the fuzzy subsets μ_{T_L}, μ_{T_M} and μ_{T_P} respectively:

$$\mu_{T_L} = \begin{pmatrix} 0.1 \\ 0 \\ 0.1 \\ 0.6 \\ 0.3 \end{pmatrix}, \; \mu_{T_M} = \begin{pmatrix} 0.4 \\ 0.4 \\ 0.4 \\ 0.6 \\ 0.3 \end{pmatrix}, \; \mu_{T_P} = \begin{pmatrix} 0.24 \\ 0.24 \\ 0.24 \\ 0.6 \\ 0.3 \end{pmatrix}.$$

Similarly, taking $A = AVG$, $T = T_L$, $T = T_M$ and $T = T_P$ we obtain the following fuzzy subsets:

$$\mu_{T_L} = \begin{pmatrix} 0.5 \\ 0.5 \\ 0.3 \\ 0.8 \\ 0.6 \end{pmatrix}, \; \mu_{T_M} = \begin{pmatrix} 0.5 \\ 0.5 \\ 0.5 \\ 0.8 \\ 0.6 \end{pmatrix}, \; \mu_{T_P} = \begin{pmatrix} 0.5 \\ 0.5 \\ 0.3 \\ 0.8 \\ 0.6 \end{pmatrix}.$$

4   Construction of Extensional Fuzzy Sets

Dealing with fuzzy equivalence relations usually extensional fuzzy sets attracts an additional attention. These sets correspond to the fuzzification of classical classes of equivalence, they play the role of fuzzy equivalence classes altogether with their intersections and unions.

Definition 6. *Let T be a t-norm and E be a T-fuzzy equivalence relation on a set D. A fuzzy subset $\mu \in [0,1]^D$ is called extensional with respect to E if and only if:*
$$T(E(x, y), \mu(y)) \leq \mu(x) \text{ for all } x, y \in D.$$

Let us show that we can obtain extensional fuzzy sets as a result of generalized T-fuzzy factoraggregation. It is important that the result of generalized T-fuzzy factoraggregation corresponding to E is extensional with respect to E even in the case, when aggregated fuzzy sets are not extensional.

Proposition 2. *Let T be a lower semi-continuous t-norm and E be a T-fuzzy equivalence relation on a set D. Then fuzzy set $\tilde{A}_{E,T}(\mu_1, \mu_2, \ldots, \mu_n)$ is extensional with respect to E for each $n \in \mathbb{N}$ and for all fuzzy sets $\mu_1, \ldots, \mu_n \in [0,1]^D$.*

Proof. We prove the following inequality

$$T(E(x, y), \tilde{A}_{E,T}(\mu_1, \mu_2, \ldots, \mu_n)(y)) \leq \tilde{A}_{E,T}(\mu_1, \mu_2, \ldots, \mu_n)(x)$$

for all $x, y \in D$:

$$T(E(x, y), \tilde{A}_{E,T}(\mu_1, \mu_2, \ldots, \mu_n)(y))$$

$$= T(E(x, y), \sup_{u \in D} T(E(y, u), A(\mu_1(u), \mu_2(u), \ldots, \mu_n(u))))$$

$$= \sup_{u \in D} T(E(x, y), T(E(y, u), A(\mu_1(u), \mu_2(u), \ldots, \mu_n(u))))$$

$$= \sup_{u \in D} T(T(E(x, y), E(y, u)), A(\mu_1(u), \mu_2(u), \ldots, \mu_n(u)))$$

$$\leq \sup_{u \in D} T(E(x, u), A(\mu_1(u), \mu_2(u), \ldots, \mu_n(u)))$$

$$= \tilde{A}_{E,T}(\mu_1, \mu_2, \ldots, \mu_n)(x).$$

And now we will propose a method to approximate an arbitrary fuzzy set by an extensional one. We recall two approximation operators ϕ_E and ψ_E, which appear in a natural way in the theory of fuzzy rough sets (see e.g. [6]). Fuzzy sets $\phi_E(\mu)$ and $\psi_E(\mu)$ were introduced to provide upper and lower approximation of a fuzzy set μ by extensional fuzzy sets with respect to fuzzy equivalence relation E.

Definition 7. *Let T be a lower semi-continuous t-norm and E be a T-fuzzy equivalence relation on a set D.*
The maps $\phi_E \colon [0,1]^D \to [0,1]^D$ and $\psi_E \colon [0,1]^D \to [0,1]^D$ are defined by:

$$\phi_E(\mu)(x) = \sup_{y \in D} T(E(x, y), \mu(y)), \qquad \psi_E(\mu)(x) = \inf_{y \in D} \overrightarrow{T}(E(x, y)|\mu(y))$$

for all $x \in D$ and for all $\mu \in [0,1]^D$, where $\overrightarrow{T}$ is the residuation of T defined for all $x, y \in [0,1]$ by

$$\overrightarrow{T}(x|y) = \sup\{\alpha \in [0,1] \mid T(\alpha, x) \leq y\}.$$

The approximations $\phi_E(\mu)$ and $\psi_E(\mu)$ in general are not the best approximations of μ by extensional fuzzy subsets. It is important to understand how $\phi_E(\mu)$ and $\psi_E(\mu)$ should be aggregated to obtain an aggregation with good approximative properties. In [5] the authors provided and compared two methods for Archimedean t-norms: one is by taking the weighted quasi-arithmetic mean of $\phi_E(\mu)$ and $\psi_E(\mu)$ and another by taking powers with respect to a t-norm of lower approximation $\psi(\mu)$. Unfortunately, the proposed methods could not be applied in the case when t-norm T does not satisfy the required restrictions.

It is easy to show that by taking an arbitrary aggregation of $\phi_E(\mu)$ and $\psi_E(\mu)$ we will not necessary obtain an extensional fuzzy set. If we take the following T_M-fuzzy equivalence relation E' and fuzzy set μ' to approximate:

$$E' = \begin{pmatrix} 1 & 0.7 \\ 0.7 & 1 \end{pmatrix}, \quad \mu' = \begin{pmatrix} 0.2 \\ 0.8 \end{pmatrix},$$

and apply the arithmetic mean aggregation operator AVG to upper and lower approximations $\phi_{E'}(\mu')$ and $\psi_{E'}(\mu')$, then the result

$$\tilde{\mu}' = \begin{pmatrix} 0.45 \\ 0.5 \end{pmatrix}$$

is not extensional with respect to E'.

Our approach is to apply to $\phi_E(\mu)$ and $\psi_E(\mu)$ the generalized T-fuzzy factoraggregation corresponding to E, thus obtaining an approximation of μ by extensional fuzzy subset

$$\tilde{A}_{E,T}(\phi_E(\mu), \psi_E(\mu)).$$

As ordinary aggregation operator A one should take an aggregation operator, which satisfies the property of internality (or compensation) (see e.g. [3]). In the case when $A = MAX$ we obtain the upper approximation:

$$\tilde{A}_{E,T_M}(\phi_E(\mu), \psi_E(\mu))(x) = \phi_E(\mu).$$

By analogy, in the case when $A = MIN$ we obtain the lower approximation:

$$\tilde{A}_{E,T_M}(\phi_E(\mu), \psi_E(\mu))(x) = \psi_E(\mu).$$

Now we will illustrate our approach with several examples. Let us consider E and fuzzy sets $\mu_1, \mu_2, \mu_3, \mu_4$ from the previous section. In the case of finite D to evaluate the error of approximation we use the Euclidean distance between an original fuzzy set $\mu \in [0,1]^D$ and the result of factoraggregation:

$$d_T(A, \mu) = \|\mu - \tilde{A}_{E,T}(\phi_E(\mu), \psi_E(\mu))\|.$$

For example, by taking $T = T_M$ and $A = AVG$ and applying the corresponding factoraggregation $\tilde{A}_{E,T_M}(\phi_E(\mu), \psi_E(\mu))$, we obtain the approximations of fuzzy sets $\mu_1, \mu_2, \mu_3, \mu_4$, and then evaluate $d_{T_M}(AVG, \mu_i)$ and compare with $d_{T_M}(MAX, \mu_i)$, $d_{T_M}(MIN, \mu_i)$ for different i:

$$d_{T_M}(MAX, \mu_1) = 0.4123, \quad d_{T_M}(MIN, \mu_1) = 0.4123, \quad d_{T_M}(AVG, \mu_1) = 0.3000,$$

$$d_{T_M}(MAX, \mu_2) = 0.4899, \quad d_{T_M}(MIN, \mu_2) = 1.1180, \quad d_{T_M}(AVG, \mu_2) = 0.6344,$$

$$d_{T_M}(MAX, \mu_3) = 0.6325, \quad d_{T_M}(MIN, \mu_3) = 1.1225, \quad d_{T_M}(AVG, \mu_3) = 0.6124,$$

$$d_{T_M}(MAX, \mu_4) = 0.9434, \quad d_{T_M}(MIN, \mu_4) = 1.1402, \quad d_{T_M}(AVG, \mu_4) = 0.7566.$$

The approximation by factoraggregation of $\phi_E(\mu)$ and $\psi_E(\mu)$ for some initial fuzzy sets provides better results than upper and lower approximations and could be improved by involving weights into averaging operator. The problem of choosing optimal weights remains beyond the frames of this paper.

Next we take $T = T_L$ and $A = MAX$, $A = MIN$, $A = AVG$, and evaluate the approximation errors:

$$d_{T_L}(MAX, \mu_1) = 0.3000, \quad d_{T_L}(MIN, \mu_1) = 0.3000, \quad d_{T_L}(AVG, \mu_1) = 0.2121,$$

$$d_{T_L}(MAX, \mu_2) = 0.1000, \quad d_{T_L}(MIN, \mu_2) = 0.1414, \quad d_{T_L}(AVG, \mu_2) = 0.0866,$$

$$d_{T_L}(MAX, \mu_3) = 0.3162, \quad d_{T_L}(MIN, \mu_3) = 0.3317, \quad d_{T_L}(AVG, \mu_3) = 0.2179,$$

$$d_{T_L}(MAX, \mu_4) = 0.8062, \quad d_{T_L}(MIN, \mu_4) = 0.7071, \quad d_{T_L}(AVG, \mu_4) = 0.5362.$$

As one can see, the smallest errors for all the given sets are obtained for the aggregation $A = AVG$, and these results are comparable with the approximations obtained in [5].

Our approach has been tested for one particular fuzzy equivalence relation E and for four fuzzy subsets of the discrete universe. A deeper analysis should be performed to make a conclusion on approximative properties of factoraggregation depending on the choice of ordinary aggregation operator A. Our future work will focus on the problem of choosing optimal weights considering approximations by generalized factoraggregation based on weighted averaging operators.

Acknowledgments. This work has been supported by the European Social Fund within the project 2013/0024/1DP/1.1.1.2.0/13/APIA/VIAA/045.

References

1. Calvo, T., Mayor, G., Mesiar, R.: Aggregation Operators. Physical-Verlag, Heidelberg (2002)
2. Detyniecki, M.: Fundamentals on Aggregation Operators, Berkeley (2001)
3. Grabisch, M., Marichal, J.-L., Mesiar, R., Pap, E.: Aggregation Functions. Cambridge University Press (2009)
4. Klement, E.P., Mesiar, R., Pap, E.: Triangular Norms. Kluwer Academic Publishers, Dordrecht (2000)
5. Mattioli, G., Recasens, J.: Comparison of different algorithms of approximation by extensional fuzzy subsets. In: Bustince, H., Fernandez, J., Mesiar, R., Calvo, T. (eds.) Aggregation Functions in Theory and in Practise. AISC, vol. 228, pp. 307–317. Springer, Heidelberg (2013)
6. Morsi, N.N., Yakout, M.M.: Axiomatics for fuzzy rough sets. Fuzzy Sets and Systems 100, 327–342 (1998)
7. Orlovs, P., Montvida, O., Asmuss, S.: A choice of bilevel linear programming solving parameters: factoraggregation approach. In: Proceedings of the 8th Conference of the European Society for Fuzzy Logic and Technology. Advances in Intelligent Systems Research, vol. 32, pp. 489–496 (2013)
8. Orlovs, P., Montvida, O., Asmuss, S.: An analysis of bilevel linear programming solving parameters based on factoraggregation approach. In: Bustince, H., Fernandez, J., Mesiar, R., Calvo, T. (eds.) Aggregation Functions in Theory and in Practise. AISC, vol. 228, pp. 345–354. Springer, Heidelberg (2013)
9. Takaci, A.: General aggregation operators acting on fuzzy numbers induced by ordinary aggregation operators. Novi Sad J. Math. 33(2), 67–76 (2003)
10. Zadeh, L.: Similarity relations and fuzzy ordering. Inform. Sci. 3(2), 177–200 (1971)
11. Zimmermann, H.J.: Fuzzy programming and linear programming with several objective functions. Fuzzy Sets and Systems 1, 45–55 (1978)

Preassociative Aggregation Functions

Jean-Luc Marichal and Bruno Teheux

Mathematics Research Unit, FSTC, University of Luxembourg
6, rue Coudenhove-Kalergi, L-1359 Luxembourg, Luxembourg
{jean-luc.marichal,bruno.teheux}@uni.lu

Abstract. We investigate the associativity property for varying-arity aggregation functions and introduce the more general property of preassociativity, a natural extension of associativity. We discuss this new property and describe certain classes of preassociative functions.

Keywords: associativity, preassociativity, aggregation, axiomatizations.

1 Introduction

Let X, Y be nonempty sets (e.g., nontrivial real intervals) and let $F \colon X^* \to Y$ be a varying-arity function, where $X^* = \cup_{n \geqslant 0} X^n$. The *n-th component* F_n of F is the restriction of F to X^n, i.e., $F_n = F|_{X^n}$. We convey that $X^0 = \{\varepsilon\}$ and that $F_0(\varepsilon) = \varepsilon$, where ε denotes the 0-tuple. For tuples $\mathbf{x} = (x_1, \ldots, x_n)$ and $\mathbf{y} = (y_1, \ldots, y_m)$ in X^*, the notation $F(\mathbf{x}, \mathbf{y})$ stands for $F(x_1, \ldots, x_n, y_1, \ldots, y_m)$, and similarly for more than two tuples. The *length* $|\mathbf{x}|$ of a tuple $\mathbf{x} \in X^*$ is a nonnegative integer defined in the usual way: we have $|\mathbf{x}| = n$ if and only if $\mathbf{x} \in X^n$.

In this note we are first interested in the associativity property for varying-arity functions. Actually, there are different equivalent definitions of this property (see, e.g., [6, 7, 11, 13, 15]). Here we use the one introduced in [15, p. 24].

Definition 1 ([15]). *A function* $F \colon X^* \to X$ *is said to be* associative *if for every* $\mathbf{x}, \mathbf{y}, \mathbf{z} \in X^*$ *we have* $F(\mathbf{x}, \mathbf{y}, \mathbf{z}) = F(\mathbf{x}, F(\mathbf{y}), \mathbf{z})$.

As an example, the real-valued function $F \colon \mathbb{R}^* \to \mathbb{R}$ defined by $F_n(\mathbf{x}) = \sum_{i=1}^{n} x_i$ is associative.

Associative varying-arity functions are closely related to associative binary functions $G \colon X^2 \to X$, which are defined as the solutions of the functional equation

$$G(G(x, y), z) = G(x, G(y, z)), \qquad x, y, z \in X.$$

In fact, we show (Corollary 6) that a binary function $G \colon X^2 \to X$ is associative if and only if there exists an associative function $F \colon X^* \to X$ such that $G = F_2$.

Based on a recent investigation of associativity (see [7, 8]), we show that an associative function $F \colon X^* \to X$ is completely determined by its first two components F_1 and F_2. We also provide necessary and sufficient conditions on the

A. Laurent et al. (Eds.): IPMU 2014, Part III, CCIS 444, pp. 327–334, 2014.

components F_1 and F_2 for a function $F\colon X^* \to X$ to be associative (Theorem 7). These results are gathered in Section 3.

The main aim of this note is to introduce and investigate the following generalization of associativity, called *preassociativity*.

Definition 2. *We say that a function* $F\colon X^* \to Y$ *is* preassociative *if for every* $\mathbf{x}, \mathbf{y}, \mathbf{y}', \mathbf{z} \in X^*$ *we have*

$$F(\mathbf{y}) \;=\; F(\mathbf{y}') \quad \Rightarrow \quad F(\mathbf{x}, \mathbf{y}, \mathbf{z}) \;=\; F(\mathbf{x}, \mathbf{y}', \mathbf{z}).$$

For instance, any real-valued function $F\colon \mathbb{R}^* \to \mathbb{R}$ defined as $F_n(\mathbf{x}) = f(\sum_{i=1}^n x_i)$ for every $n \in \mathbb{N}$, where $f\colon \mathbb{R} \to \mathbb{R}$ is a continuous and strictly increasing function, is preassociative.

It is immediate to see that any associative function $F\colon X^* \to X$ necessarily satisfies the equation $F_1 \circ F = F$ (take $\mathbf{x} = \varepsilon$ and $\mathbf{z} = \varepsilon$ in Definition 1). Actually, we show (Proposition 8) that a function $F\colon X^* \to X$ is associative if and only if it is preassociative and satisfies $F_1 \circ F = F$.

It is noteworthy that, contrary to associativity, preassociativity does not involve any composition of functions and hence allows us to consider a codomain Y that may differ from the domain X. For instance, the length function $F\colon X^* \to \mathbb{R}$ defined as $F(\mathbf{x}) = |\mathbf{x}|$ is preassociative.

In this note we mainly focus on those preassociative functions $F\colon X^* \to Y$ for which F_1 and F have the same range. (When $Y = X$, the latter condition is an immediate consequence of the condition $F_1 \circ F = F$ and hence those preassociative functions include the associative ones). Similarly to associative functions, we show that those functions are completely determined by their first two components (Proposition 12) and we provide necessary and sufficient conditions on the components F_1 and F_2 for a function $F\colon X^* \to Y$ to be preassociative and have the same range as F_1 (Theorem 15). We also give a description of these functions as compositions of the form $F = f \circ H$, where $H\colon X^* \to X$ is associative and $f\colon \operatorname{ran}(H) \to Y$ is one-to-one (Theorem 13). This is done in Section 4. Finally, in Section 5 we focus on some noteworthy axiomatized classes of associative functions and show how they can be extended to classes of preassociative functions.

The terminology used throughout this paper is the following. We denote by $\mathbb{N}$ the set $\{1, 2, 3, \dots\}$ of strictly positive integers. The domain and range of any function f are denoted by $\operatorname{dom}(f)$ and $\operatorname{ran}(f)$, respectively. The identity function is the function $\operatorname{id}\colon X \to X$ defined by $\operatorname{id}(x) = x$.

The proofs of our results have intentionally been omitted due to space limitation but will be available in an extended version of this note.

2   Preliminaries

Recall that a function $F\colon X^n \to X$ is said to be *idempotent* (see, e.g., [11]) if $F(x, \dots, x) = x$ for every $x \in X$. A function $F\colon X^* \to X$ is said to be *idempotent* if F_n is idempotent for every $n \in \mathbb{N}$.

We now introduce the following definitions. We say that $F\colon X^* \to X$ is *unarily idempotent* if $F_1(x) = x$ for every $x \in X$, i.e., $F_1 = \mathrm{id}$. We say that $F\colon X^* \to X$ is *unarily range-idempotent* if $F(x) = x$ for every $x \in \mathrm{ran}(F)$, or equivalently, $F_1 \circ F = F$. We say that $F\colon X^* \to Y$ is *unarily quasi-range-idempotent* if $\mathrm{ran}(F_1) = \mathrm{ran}(F)$. Since this property is a consequence of the condition $F_1 \circ F = F$, we see that unarily range-idempotent functions are necessarily unarily quasi-range-idempotent.

We now show that any unarily quasi-range-idempotent function $F\colon X^* \to Y$ can always be factorized as $F = F_1 \circ H$, where $H\colon X^* \to X$ is a unarily range-idempotent function. First recall that a function g is a *quasi-inverse* [17, Sect. 2.1] of a function f if

$$f \circ g|_{\mathrm{ran}(f)} = \mathrm{id}|_{\mathrm{ran}(f)},$$
$$\mathrm{ran}(g|_{\mathrm{ran}(f)}) = \mathrm{ran}(g).$$

For any function f, denote by $Q(f)$ the set of its quasi-inverses. This set is nonempty whenever we assume the Axiom of Choice (AC), which is actually just another form of the statement "every function has a quasi-inverse."

Proposition 3. *Assume AC and let $F\colon X^* \to Y$ be a unarily quasi-range-idempotent function. For any $g \in Q(F_1)$, the function $H\colon X^* \to X$ defined as $H = g \circ F$ is a unarily range-idempotent solution of the equation $F = F_1 \circ H$.*

3   Associative Functions

As observed in [15, p. 25] (see also [5, p. 15] and [11, p. 33]), associative functions $F\colon X^* \to X$ are completely determined by their unary and binary components. Indeed, by associativity we have

$$F_n(x_1, \ldots, x_n) \;=\; F_2(F_{n-1}(x_1, \ldots, x_{n-1}), x_n), \qquad n \geqslant 3, \qquad (1)$$

or equivalently,

$$F_n(x_1, \ldots, x_n) \;=\; F_2(F_2(\ldots F_2(F_2(x_1, x_2), x_3) \ldots), x_n), \qquad n \geqslant 3. \qquad (2)$$

We state this immediate result as follows.

Proposition 4. *Let $F\colon X^* \to X$ and $G\colon X^* \to X$ be two associative functions such that $F_1 = G_1$ and $F_2 = G_2$. Then $F = G$.*

A natural and important question now arises: Find necessary and sufficient conditions on the components F_1 and F_2 for a function $F\colon X^* \to X$ to be associative. To answer this question we first yield the following characterization of associative functions.

Theorem 5. *A function $F\colon X^* \to X$ is associative if and only if*

(i) $F_1 \circ F_1 = F_1$ and $F_1 \circ F_2 = F_2$,

(ii) $F_2(x_1, x_2) = F_2(F_1(x_1), x_2) = F_2(x_1, F_1(x_2))$,
(iii) F_2 *is associative, and*
(iv) *condition (1) or (2) holds.*

Corollary 6. *A binary function* $F: X^2 \to X$ *is associative if and only if there exists an associative function* $G: X^* \to X$ *such that* $F = G_2$.

Theorem 5 enables us to answer the question raised above. We state the result in the following theorem.

Theorem 7. *Let* $F_1: X \to X$ *and* $F_2: X^2 \to X$ *be two functions. Then there exists an associative function* $G: X^* \to X$ *such that* $G_1 = F_1$ *and* $G_2 = F_2$ *if and only if conditions (i)–(iii) of Theorem 5 hold. Such a function* G *is then uniquely determined by* $G_n(x_1, \ldots, x_n) = G_2(G_{n-1}(x_1, \ldots, x_{n-1}), x_n)$ *for* $n \geqslant 3$.

Thus, two functions $F_1: X \to X$ and $F_2: X^2 \to X$ are the unary and binary components of an associative function $F: X^* \to X$ if and only if these functions satisfy conditions (i)–(iii) of Theorem 5. In the case when only a binary function F_2 is given, any unary function F_1 satisfying conditions (i) and (ii) can be considered, for instance the identity function. Note that it may happen that the identity function is the sole possibility for F_1, for instance when we consider the binary function $F_2: \mathbb{R}^2 \to \mathbb{R}$ defined by $F_2(x_1, x_2) = x_1 + x_2$. However, there are examples where F_1 may differ from the identity function. For instance, for any real number $p \geqslant 1$, the p-norm $F: \mathbb{R}^* \to \mathbb{R}$ defined by $F_n(\mathbf{x}) = (\sum_{i=1}^n |x_i|^p)^{1/p}$ is associative but not unarily idempotent (here $|x|$ denotes the absolute value of x). Of course an associative function F that is not unarily idempotent can be made unarily idempotent simply by setting $F_1 = \mathrm{id}$. By Theorem 5 the resulting function is still associative.

4   Preassociative Functions

In this section we investigate the preassociativity property (see Definition 2) and describe certain classes of preassociative functions.

As mentioned in the introduction, any associative function $F: X^* \to X$ is preassociative. More precisely, we have the following result.

Proposition 8. *A function* $F: X^* \to X$ *is associative if and only if it is preassociative and unarily range-idempotent (i.e.,* $F_1 \circ F = F$*).*

Remark 9. The function $F: \mathbb{R}^* \to \mathbb{R}$ defined as $F_n(\mathbf{x}) = 2 \sum_{i=1}^n x_i$ is an instance of preassociative function which is not associative.

Let us now see how new preassociative functions can be generated from given preassociative functions by left and right compositions with unary maps.

Proposition 10 (Right composition). *If* $F: X^* \to Y$ *is preassociative then, for every function* $g: X \to X$, *the function* $H: X^* \to Y$, *defined as* $H_n = F_n \circ (g, \ldots, g)$ *for every* $n \in \mathbb{N}$, *is preassociative. For instance, the squared distance function* $F: \mathbb{R}^* \to \mathbb{R}$ *defined as* $F_n(\mathbf{x}) = \sum_{i=1}^n x_i^2$ *is preassociative.*

Proposition 11 (Left composition). *Let $F\colon X^* \to Y$ be a preassociative function and let $g\colon Y \to Y$ be a function. If $g|_{\mathrm{ran}(F)}$ is constant or one-to-one, then the function $H\colon X^* \to Y$ defined as $H = g \circ F$ is preassociative. For instance, the function $F\colon \mathbb{R}^* \to \mathbb{R}$ defined as $F_n(\mathbf{x}) = \exp(\sum_{i=1}^n x_i)$ is preassociative.*

We now focus on those preassociative functions which are unarily quasi-range-idempotent, that is, such that $\mathrm{ran}(F_1) = \mathrm{ran}(F)$. As we will now show, this special class of functions has interesting properties. First of all, just as for associative functions, preassociative and unarily quasi-range-idempotent functions are completely determined by their unary and binary components.

Proposition 12. *Let $F\colon X^* \to Y$ and $G\colon X^* \to Y$ be preassociative and unarily quasi-range-idempotent functions such that $F_1 = G_1$ and $F_2 = G_2$, then $F = G$.*

We now give a description of the preassociative and unarily quasi-range-idempotent functions as compositions of associative functions with unary maps.

Theorem 13. *Assume AC and let $F\colon X^* \to Y$ be a function. The following assertions are equivalent.*

(i) F is preassociative and unarily quasi-range-idempotent.
(ii) There exists an associative function $H\colon X^* \to X$ and a one-to-one function $f\colon \mathrm{ran}(H) \to Y$ such that $F = f \circ H$. In this case we have $F = F_1 \circ H$, $f = F_1|_{\mathrm{ran}(H)}$, $f^{-1} \in Q(F_1)$, and we may choose $H = g \circ F$ for any $g \in Q(F_1)$.

Remark 14. If condition (ii) of Theorem 13 holds, then by (1) we see that F can be computed recursively by

$$F_n(x_1, \ldots, x_n) = F_2((f^{-1} \circ F_{n-1})(x_1, \ldots, x_{n-1}), x_n), \qquad n \geqslant 3.$$

A similar observation was already made in a more particular setting for the so-called quasi-associative functions, see [18].

We now provide necessary and sufficient conditions on the unary and binary components for a function $F\colon X^* \to X$ to be preassociative and unarily quasi-range-idempotent. We have the following two results.

Theorem 15. *Assume AC. A function $F\colon X^* \to Y$ is preassociative and unarily quasi-range-idempotent if and only if $\mathrm{ran}(F_2) \subseteq \mathrm{ran}(F_1)$ and there exists $g \in Q(F_1)$ such that*

(i) $H_2(x_1, x_2) = H_2(H_1(x_1), x_2) = H_2(x_1, H_1(x_2))$,
(ii) H_2 is associative, and
(iii) the following holds

$$F_n(x_1, \ldots, x_n) = F_2((g \circ F_{n-1})(x_1, \ldots, x_{n-1}), x_n), \qquad n \geqslant 3,$$

or equivalently,

$$F_n(x_1, \ldots, x_n) = F_2(H_2(\ldots H_2(H_2(x_1, x_2), x_3) \ldots), x_n), \qquad n \geqslant 3,$$

where $H_1 = g \circ F_1$ and $H_2 = g \circ F_2$.

Theorem 16. *Assume AC and let $F_1 \colon X \to Y$ and $F_2 \colon X^2 \to Y$ be two functions. Then there exists a preassociative and unarily quasi-range-idempotent function $G \colon X^* \to Y$ such that $G_1 = F_1$ and $G_2 = F_2$ if and only if $\mathrm{ran}(F_2) \subseteq \mathrm{ran}(F_1)$ and there exists $g \in Q(F_1)$ such that conditions (i) and (ii) of Theorem 15 hold, where $H_1 = g \circ F_1$ and $H_2 = g \circ F_2$. Such a function G is then uniquely determined by $G_n(x_1, \ldots, x_n) = G_2((g \circ G_{n-1})(x_1, \ldots, x_{n-1}), x_n)$ for $n \geqslant 3$.*

5   Axiomatizations of Some Function Classes

In this section we derive axiomatizations of classes of preassociative functions from certain existing axiomatizations of classes of associative functions. We restrict ourselves to a small number of classes. Further axiomatizations can be derived from known classes of associative functions.

5.1   Preassociative Functions Built from Aczélian Semigroups

Let us recall an axiomatization of the Aczélian semigroups due to Aczél [1] (see also [2, 8, 9]).

Proposition 17 ([1]). *Let I be a nontrivial real interval (i.e., nonempty and not a singleton). A function $H \colon I^2 \to I$ is continuous, one-to-one in each argument, and associative if and only if there exists a continuous and strictly monotonic function $\varphi \colon I \to J$ such that*

$$H(xy) = \varphi^{-1}\left(\varphi(x) + \varphi(y)\right),$$

where J is a real interval of one of the forms $]-\infty, b[$, $]-\infty, b]$, $]a, \infty[$, $[a, \infty[$ or $\mathbb{R} =]-\infty, \infty[$ ($b \leqslant 0 \leqslant a$). For such a function H, the interval I is necessarily open at least on one end. Moreover, φ can be chosen to be strictly increasing.

Proposition 17 can be extended to preassociative functions as follows.

Theorem 18. *Let I be a nontrivial real interval (i.e., nonempty and not a singleton). A function $F \colon I^* \to \mathbb{R}$ is preassociative and unarily quasi-range-idempotent, and F_1 and F_2 are continuous and one-to-one in each argument if and only if there exist continuous and strictly monotonic functions $\varphi \colon I \to J$ and $\psi \colon J \to \mathbb{R}$ such that*

$$F_n(\mathbf{x}) = \psi(\varphi(x_1) + \cdots + \varphi(x_n)), \qquad n \in \mathbb{N},$$

where J is a real interval of one of the forms $]-\infty, b[$, $]-\infty, b]$, $]a, \infty[$, $[a, \infty[$ or $\mathbb{R} =]-\infty, \infty[$ ($b \leqslant 0 \leqslant a$). For such a function F, we have $\psi = F_1 \circ \varphi^{-1}$ and I is necessarily open at least on one end. Moreover, φ can be chosen to be strictly increasing.

5.2 Preassociative Functions Built from t-norms and Related Functions

Recall that a *t-norm* (resp. *t-conorm*) is a function $H\colon [0,1]^2 \to [0,1]$ which is nondecreasing in each argument, symmetric, associative, and such that $H(1,x) = x$ (resp. $H(0,x) = x$) for every $x \in [0,1]$. Also, a *uninorm* is a function $H\colon [0,1]^2 \to [0,1]$ which is nondecreasing in each argument, symmetric, associative, and such that there exists $e \in \,]0,1[$ for which $H(e,x) = x$ for every $x \in [0,1]$. For general background see, e.g., [3, 10–13, 17].

T-norms can be extended to preassociative functions as follows.

Theorem 19. *Let $F\colon [0,1]^* \to \mathbb{R}$ be a function such that F_1 is strictly increasing (resp. strictly decreasing). Then F is preassociative and unarily quasi-range-idempotent, and F_2 is symmetric, nondecreasing (resp. nonincreasing) in each argument, and satisfies $F_2(1,x) = F_1(x)$ for every $x \in [0,1]$ if and only if there exists a strictly increasing (resp. strictly decreasing) function $f\colon [0,1] \to \mathbb{R}$ and a t-norm $H\colon [0,1]^* \to [0,1]$ such that $F = f \circ H$. In this case we have $f = F_1$.*

If we replace the condition "$F_2(1,x) = F_1(x)$" in Theorem 19 with "$F_2(0,x) = F_1(x)$" (resp. "$F_2(e,x) = F_1(x)$ for some $e \in \,]0,1[$"), then the result still holds provided that the t-norm is replaced with a t-conorm (resp. a uninorm).

5.3 Preassociative Functions Built from Ling's Axiomatizations

Recall an axiomatization due to Ling [14]; see also [4, 16].

Proposition 20 ([14]). *Let $[a,b]$ be a real closed interval, with $a < b$. A function $H\colon [a,b]^2 \to [a,b]$ is continuous, nondecreasing in each argument, associative, and such that $H(b,x) = x$ for all $x \in [a,b]$ and $H(x,x) < x$ for all $x \in \,]a,b[$, if and only if there exists a continuous and strictly decreasing function $\varphi\colon [a,b] \to [0,\infty[$, with $\varphi(b) = 0$, such that*

$$H(xy) = \varphi^{-1}(\min\{\varphi(x) + \varphi(y), \varphi(a)\}).$$

Proposition 20 can be extended to preassociative functions as follows.

Theorem 21. *Let $[a,b]$ be a real closed interval and let $F\colon [a,b]^* \to \mathbb{R}$ be a function such that F_1 is strictly increasing (resp. strictly decreasing). Then F is unarily quasi-range idempotent and preassociative, and F_2 is continuous and nondecreasing (resp. nonincreasing) in each argument, $F_2(b,x) = F_1(x)$ for every $x \in [a,b]$, $F_2(x,x) < F_1(x)$ (resp. $F_2(x,x) > F_1(x)$) for every $x \in \,]a,b[$ if and only if there exists a continuous and strictly decreasing function $\varphi\colon [a,b] \to [0,\infty[$, with $\varphi(b) = 0$, and a strictly decreasing (resp. strictly increasing) function $\psi\colon [0,\varphi(a)] \to \mathbb{R}$ such that*

$$F_n(\mathbf{x}) = \psi(\min\{\varphi(x_1) + \cdots + \varphi(x_n), \varphi(a)\}).$$

For such a function, we have $\psi = F_1 \circ \varphi^{-1}$.

References

1. Aczél, J.: Sur les opérations définies pour nombres réels. Bull. Soc. Math. France 76, 59–64 (1949)
2. Aczél, J.: The associativity equation re-revisited. In: Erikson, G., Zhai, Y. (eds.) Bayesian Inference and Maximum Entropy Methods in Science and Engineering, pp. 195–203, American Institute of Physics, Melville-New York (2004)
3. Alsina, C., Frank, M.J., Schweizer, B.: Associative functions: triangular norms and copulas. World Scientific, London (2006)
4. Bacchelli, B.: Representation of continuous associative functions. Stochastica 10, 13–28 (1986)
5. Beliakov, G., Pradera, A., Calvo, T.: Aggregation functions: a guide for practitioners. STUDFUZZ, vol. 221. Springer, Heidelberg (2007)
6. Calvo, T., Kolesárová, A., Komorníková, M., Mesiar, R.: Aggregation operators: properties, classes and construction methods. In: Aggregation Operators. New Trends and Applications. STUDFUZZ, vol. 97, pp. 3–104. Physica-Verlag, Heidelberg (2002)
7. Couceiro, M., Marichal, J.-L.: Associative polynomial functions over bounded distributive lattices. Order 28, 1–8 (2011)
8. Couceiro, M., Marichal, J.-L.: Aczélian n-ary semigroups. Semigroup Forum 85, 81–90 (2012)
9. Craigen, R., Páles, Z.: The associativity equation revisited. Aeq. Math. 37, 306–312 (1989)
10. Fodor, J., Roubens, M.: Fuzzy preference modelling and multicriteria decision support. In: Theory and Decision Library. Series D: System Theory, Knowledge Engineering and Problem Solving. Kluwer Academic Publisher, Dordrecht (1994)
11. Grabisch, M., Marichal, J.-L., Mesiar, R., Pap, E.: Aggregation functions. In: Encyclopedia of Mathematics and its Applications, vol. 127, Cambridge University Press, Cambridge (2009)
12. Klement, E.P., Mesiar, R. (eds.): Logical, algebraic, analytic, and probabilistic aspects of triangular norms. Elsevier Science, Amsterdam (2005)
13. Klement, E.P., Mesiar, R., Pap, E.: Triangular norms. In: Trends in Logic - Studia Logica Library, vol. 8. Kluwer Academic, Dordrecht (2000)
14. Ling, C.H.: Representation of associative functions. Publ. Math. Debrecen 12, 189–212 (1965)
15. Marichal, J.-L.: Aggregation operators for multicriteria decision aid. PhD thesis, Department of Mathematics, University of Liège, Liège, Belgium (1998)
16. Marichal, J.-L.: On the associativity functional equation. Fuzzy Sets and Systems 114, 381–389 (2000)
17. Schweizer, B., Sklar, A.: Probabilistic metric spaces. North-Holland Series in Probability and Applied Mathematics. North-Holland Publishing Co., New York (1983) (New edition in: Dover Publications, New York, 2005)
18. Yager, R.R.: Quasi-associative operations in the combination of evidence. Kybernetes 16, 37–41 (1987)

An Extended Event Reasoning Framework for Decision Support under Uncertainty

Wenjun Ma[1], Weiru Liu[1], Jianbing Ma[2], and Paul Miller[1]

[1] EEECS, Queen's University Belfast, Belfast, UK, BT7 1NN
[2] DEC, Bournemouth University, Bournemouth, UK, BH12 5BB

Abstract. To provide in-time reactions to a large volume of surveillance data, uncertainty-enabled event reasoning frameworks for closed-circuit television and sensor based intelligent surveillance system have been integrated to model and infer events of interest. However, most of the existing works do not consider decision making under uncertainty which is important for surveillance operators. In this paper, we extend an event reasoning framework for decision support, which enables our framework to predict, rank and alarm threats from multiple heterogeneous sources.

1   Introduction

In recent years, intelligent surveillance systems have received significant attentions for public safety due to the increasing threat of terrorist attack, anti-social and criminal behaviors in the present world. In order to analyze a large volume of surveillance data, in the literature, there are a couple of event modeling and reasoning systems. For example, Finite State Machines [5], Bayesian Networks [3], and Event composition with imperfect information [10,11], etc.

However, the decision support issue has not been properly addressed, especially on how to rank the potential threats of multiple suspects and then focus on some of the suspects for further appropriate actions (taking immediate actions or reenforced monitoring, etc.) based on imperfect and conflicting information from different sources. This problem is extremely important in the sense that in real-world situations, a security operator is likely to make decisions under a condition that the security resources are limited whilst several malicious behaviors happen simultaneously. Consider an airport scenario, a surveillance system detects that there is a very high chance that two young people are fighting in the shopping area, and at the same time, there are a medium chance that a person may leave a bomb in airport terminal 1. Now suppose there is only one security team available at that moment, which security problem should be first presented to the security team?

In order to address this problem, in this paper, we extend the event modeling framework [10,11] with a decision support model for distributed intelligent surveillance systems, using a multi-criteria fusion architecture. More specifically, based on Dempster-Shafer (D-S) theory [14], we first improve the event modeling framework proposed in [10,11] to handle the multi-criteria event modeling.

A. Laurent et al. (Eds.): IPMU 2014, Part III, CCIS 444, pp. 335–344, 2014.
© Springer International Publishing Switzerland 2014

Then we use a normalized version of the Hurwicz's criterion [4] to obtain the degree of potential threat of each suspect (or suspects if they work as a team) with respect to each criterion. Finally, according to some background knowledge in surveillance, we apply a weighted aggregation operation to obtain the overall degree of potential threat for each suspect after considering all related criteria, from which we can set the priority for each subject.

This paper advances the state of the art on information analysis for intelligent surveillance systems in the following aspects. (i) We identify two factors that influence the potential threats in surveillance system: belief and utility. (ii) We propose an event modeling and reasoning framework to estimate the potential threats based on heterogeneous information from multiple sources. (iii) We introduce a weighted aggregation operator to combine the degrees of potential threats of each criterion and give an overall estimation to each subject.

The rest of this paper is organized as follows. Section 2 recaps D-S theory and the event modeling framework in [10,11]. Section 3 extends the event modeling framework in [10,11] to handle the multi-criteria issue. Section 4 develops a decision support model with an aggregation operator to handle the problem of judging the degrees of potential threats for multiple suspects. Sections 5 provides a case study to illustrate the usefulness of our model. Finally, Section 6 discusses the related work and concludes the paper with future work.

2   Preliminaries

This section recaps some basic concepts in D-S theory [14].

Definition 1. *Let Θ be a set of exhaustive and mutually exclusive elements, called a frame of discernment (or simple a frame). Function $m : 2^{\Theta} \to [0,1]$ is a mass function if $m(\emptyset) = 0$ and $\sum_{A \subseteq \Theta} m(A) = 1$.*

One advantage of D-S theory is that it provides a method to accumulate and combine evidence from multiple sources by using *Dempster combination rule*:

Definition 2 (Dempster combination rule). *Let m_1 and m_2 be two mass functions over a frame of discernment Θ. Then Dempster combination rule $m_{12} = m_1 \bigoplus m_2$ is given by:*

$$
m_{12}(x) = \begin{cases} 0 & \text{if } x = \emptyset \\ \dfrac{\sum\limits_{A_i \cap B_j = x} m_1(A_i) m_2(B_j)}{1 - \sum\limits_{A_i \cap B_j = \emptyset} m_1(A_i) m_2(B_j)} & \text{if } x \neq \emptyset \end{cases} \tag{1}
$$

In order to reflect the reliability of evidence, a *Discount rate* was introduced by which a mass function can be discounted [8]:

Definition 3. *Let m be a mass function over frame Θ and τ $(0 \leq \tau \leq 1)$ be a discount rate, then the discounted mass function m^{τ} is defined as:*

$$
m^{\tau}(A) = \begin{cases} (1 - \tau) m(A) & \text{if } A \subset \Theta \\ \tau + (1 - \tau) m(\Theta) & \text{if } A = \Theta \end{cases} \tag{2}
$$

Finally, in order to reflect the belief distributions from preconditions to the conclusion in an inference rule, in [9], a modeling and propagation approach was proposed based on the notion of *evidential mapping Γ^**.

Definition 4. $\Gamma^* : 2^{\Theta_E} \to 2^{2^{\Theta_H} \times [0,1]}$ *is an evidential mapping, which establishes relationships between two frames of discernment Θ_E, Θ_H, if Γ^* assigns a subset $E_i \subseteq \Theta_E$ to a set of subset-mass pairs in the following way:*

$$\Gamma^*(E_i) = ((H_{i1}, \ f(E_i \to H_{i1})), \ \ldots, \ (H_{it}, \ f(E_i \to H_{it}))) \tag{3}$$

where $H_{ij} \subseteq \Theta_H$, $i = 1, \ \ldots, \ n$, $j = 1, \ \ldots, \ t$, and $f : 2^{\Theta_E} \times 2^{\Theta_H} \to [0, 1]$ satisfying: (i) $H_{ij} \neq \emptyset$, $j = 1, \ \ldots, \ t$; (ii) $f(E_i \to H_{ij}) \geq 0$, $j = 1, \ \ldots, \ t$; (iii) $\sum_{j=1}^{t} f(E_i \to H_{ij}) = 1$; (iv) $\Gamma^*(\Theta_E) = ((\Theta_H, 1))$.

So a piece of evidence on Θ_E can be propagated to Θ_H through evidential mapping Γ^* as follows:

$$m_{\Theta_H}(H_j) = \sum_i m_{\Theta_E}(E_i) f(E_i \to H_{ij}). \tag{4}$$

3   Multiple Criteria Event Modeling Framework

In this section, we extend the event reasoning framework introduced in [10] to include multi-criteria, in order to allow for better decision making.

Definition 5. *In a multi-criteria event modeling framework, an elementary event e for detecting the potential threats is a tuple $(EType, occT, ID_s, rb, sig, Criterion, Weight, ID_p, s_1, \ldots, s_n)$, where: (i) EType: describes the event type; (ii) occT: a time point (or duration) for the observed event; (iii) ID_s: the source ID for a detected event; (iv) rb: the degree of reliability of a source; (v) sig: the degree of significance of a given event based on domain knowledge; (vi) Criterion: describes one of the attributes that can reveal some level of potential threat for a target, such as age, gender, and so on;[1] (vii) Weight: the degree of a criterion's importance for detecting a potential threat; (viii) ID_p: person ID for a detected event; (ix) s_i: additional attributes required to define event e.*

We can associate an event with a mass value and a utility function for a given criterion. For example: $e_{42}^{g,1}$=(FCE, 9:01pm − 9:05pm, 42, 0.9, 0.6, gender, 0.2, 13, $m_{42}^{g}(\{male\})$=0.3, U^g) means that for an event type FCE at 9:01pm to 9:05pm, the *gender* classification program used by camera 42, whose degree of reliability is 0.9, detects that at Foreign Currency Exchange office (FCE), person with $ID = 13$ is recognized as *male* with a certainty of 30%. The significance of this event is 0.6, the weight of *gender* criterion for detecting a potential threat is

---

[1] We will use the word "criterion" in this paper to define an attribute that can reveal some level of potential threat of an observed subject. Therefore, we can distinguish "criterion" from other attributes, such as person ID, location, etc.

0.2, and U^g is the utility function that shows the level of potential threat for the *gender* criterion. Related events are grouped together to form *event clusters* where events in the same cluster share the same *event type, occT, Criterion, ID_s, ID_p*, but may assign different mass values to subsets of the same frame. For example, two events for the person with ID 13 that are detected by camera 42 at 9:01pm to 9:05pm at FCE with the mass values $m_{42}^g(\{male\}) = 0.3$ and $m_{42}^g(\{female, male\}) = 0.7$ respectively are from the same cluster. Moreover, The mass items (i.e., the subsets and the mass values on the subsets) of all the events in an event cluster exactly form a mass function.

Compared with the model in [10], we retain the components $EType$, sID and rb, whilst the differences are: (i) we represent events with a time duration, which is more realistic in real-life applications; (ii) we keep the person ID as a common attribute, since it is important for surveillance applications; (iii) the degree of significance in our definition indicates the relative importance of a given event based on the background information. For example, an event detected in an area with high crime statistics at midnight is more significant than that in an area of low-crime in the morning; (iv) in our model, an elementary event can only have one criterion attribute. For example, a *man* boards a bus at 8:00am is an elementary event, but a *young man* boards a bus at 8:00am is not an elementary event since both age (yonng) and gender (man) are criterion attributes. In fact, since a classification algorithm only focuses on one criterion attribute, this semantics of *elementary* is natural; (v) we introduce the attribute *Weight* to reflect the importance of a criterion when determining a potential threat. For example, *age* and *gender* usually are not the critical evidence to detect the potential threat, while *behaviors*, such as holding a knife, fighting, etc., are more important in determining the dangerous level of a subject; (iv) we introduce the Utility function to distinguish different levels of threats for the outcomes of each criterion. For example, the threat level of a young person should be higher than the threat level of an old person.

There might be a set of event clusters that have the same criterion and event type but with different source IDs or observation times. For example, a person broads a bus with its back facing camera 4 at 9:15pm and then sits down with its face partially detected by camera 6 at 9:20pm. Suppose the gender classification algorithm shows that $m_4^g(\{male\}) = 0.5$ and $m_4^g(\{female, male\}) = 0.5$ by camera 4 and $m_6^g(\{male\}) = 0.7$ and $m_6^g(\{female, male\}) = 0.3$ by camera 6. Since these two classification results (in two clusters) refer to the same criterion about the same subject from different sources, mass functions (after possible discounting) defined in the two clusters are combined using Dempster's rule. When an event is described with a duration, this event can be instantiated at any time point within this duration, That is, we can replace the duration with any time point within the duration. This is particularly useful when combining mass functions from different clusters, since events in these clusters may not share exactly the same time points or durations, but as long as their durations overlap, they can be combined. This is an improvement over [10], which cannot handle such situation. Finally, a combined mass function must assign mass values to events in a derived event cluster. In this cluster, every derived event shares the same $EType$, sig, $Criterion$, $Weight$, ID_p, $location$, U_c as the original events, but $occT$, ID_s, and rb are the union of those of the original events, respectively.

Now, we consider event inference in our framework. Different from the rule definition in [10], an inference rule in our framework is defined as a tuple ($EType$, $Condition$, m^{IET}, U^{IET}), where: (i) m^{IET} in our method is a mass function for possible intentions of a subject. It means the prediction of the subject's intention based on historic data or experts' judgement when the rule is triggered. For example, m^{IET} for the event inference rule about loitering in a ticket counter is a mass function over a frame of discernment $\{Rob, Wait For Some Friends\}$. (ii) We only consider the behavior of the subjects to infer their intentions. Here, we divide behavior into different categories, such as movements (obtained by trajectory tracking), relations with objects, relations with peoples (obtained by the binary spatial relations of objects or people [12]), hand actions to detect a fight (obtained by 2D locations of the hands), etc.

The reasons of these changes are: (i) It is not reasonable to ask experts to directly assign the degree of a potential threat without aggregating the factors that contribute to the threat, such as the significance of an event, the weight of different attributes, the outcomes of each criterion, etc.. Thus, defining m_{IET} over the frame of discernment about possible intentions of a subject is more reasonable than the fixed frame of discernment: $\{Theart, Not Theart\}$. (ii) It can reduce the amount of inference rules since we only consider the behaviors of the subjects to infer their intentions. (iii) It follows the well-known social psychology study result that humans can infer the intentions of others through observations of their behaviors [6].

Finally, since events appeared in the condition of inference rules are themselves uncertain, we also apply the notion *evidential mapping* to obtain the mass functions of inferred events as [11]. Here is an example for the event inference rule in our model about the possible intentions of a subject in the shopping area.

Example 1. *The rule describing that* a person loitering in the Foreign Currency Exchange office (FCE) could be suspicious *can be defined as* ($EType$, $Conditions$, m^{IPL}, U^{IPL}) *where EType is the Intention of Person loitering in FCE; Conditions is* $m_i^m(\{loitering\}) > 0.5$ AND $e.location = FCE$ AND $t_n - t_0 > 10\,min$; m^{IPL} *can be* $m^{IPL}(\{Rob\}) = 0.5$, $m^{IPL}(\{Waiting\ Friends\}) = 0.3$, $m^{IPL}(\{Rob, Waiting\ Friends\}) = 0.2$; *and* U^{IPL} *can be* $U^{IPL} = \{u(Rob) = 9, u(Waiting\ Friends) = 3\}$.

4   A Multi-criteria System for Threat Ranking

In this section, we will construct a decision support system that can automatically rank the potential threat degree of different subjects in a multiple criteria surveillance environment under uncertainty.

First, we calculate the degrees of potential threat for each criterion by extending the approach in [15]:

Definition 6. *For a subject with ID x w.r.t. a given criterion c specified by mass function $m_{c,x}$ over $\Theta = \{h_1, ..., h_n\}$, where h_i is a positive value indicating the utility (level of potential threat) of each possible outcome for criterion c, its expected utility interval (interval degree of potential threat) is $EUI_c(x) = [\underline{E}_c(x), \overline{E}_c(x)]$, where*

$$\underline{E}_c(x) = \sum_{A \subseteq \Theta} m_{x,c}(A) \min\{h_i \mid h_i \in A\}, \overline{E}_c(x) = \sum_{A \subseteq \Theta} m_{x,c}(A) \max\{h_i \mid h_i \in A\}.$$

Second, we apply the transformational form of the Hurwicz's criterion [4], to find the point-valued degree of potential threat w.r.t. each criterion:

Definition 7. *Let $EUI_c(x) = [\underline{E}_c(x), \overline{E}_c(x)]$ be an interval-valued expected level of potential threat of criterion c for subject with ID x, $\delta_c(x) = sig$ be the degree of significance for the events, then the point-valued degree of potential threat for subject with ID x w.r.t. criterion c is given by:*

$$\nu_c(x) = (1 - \delta_c(x))\underline{E}_c(x) + \delta_c(x)\overline{E}_c(x). \tag{5}$$

Finally, we combine the potential threats w.r.t. each criterion by the following aggregation operator and then obtain the overall degree of potential threat of each subject.

Definition 8. *Let C be the whole set of related criteria, $nu_c(x)$ be the point-valued degree of potential threat for subject with ID x w.r.t. criterion c, w_c be the weight of each criterion c, and k be the highest utility value for the outcomes of all criteria, then the overall degree of potential threat for subject x, denoted as O_x, is given by*

$$O_x = \frac{2\sum_{c \subseteq C} w_c nu_c(x)}{\sum_{c \subseteq C}(k+1)w_c} \tag{6}$$

In fact, Equation (6) is a form of weighting average, where $\sum_{c \subseteq C} w_c nu_c(x)$ is the overall value that considers the weighting effect of each criterion for the overall evaluation of potential threat, $\sum_{c \subseteq C}(k+1)w_c/2$ is the averaging operator designed to avoid the situation that the more criteria the surveillance system detects, the higher value the potential threat is, and $(k+1)/2$ can be consider as an intermediate value to distinguish low threat levels from high threat levels.

Now, we reveal some properties of our model by the following Theorems:

Theorem 1. *Let $EUI_c(x) = [\underline{E}_c(x), \overline{E}_c(x)]$ be an interval-valued expected utility of criterion c for subject x, $\delta_c(x)$ and $\delta'_c(x)$ be the degrees of significance for two different surveillance scenarios, and $\delta_c(x) \geq \delta'_c(x)$, then $\nu_c(x) \geq \nu'_c(x)$.*

Proof. By Equation (5), $\delta_c(x) \geq \delta'_c(x)$ and $\overline{E} \geq \underline{E}$, we have:

$$\nu_c(x) - \nu'_c(x) = (\delta_c(x) - \delta'_c(x))(\overline{E}_c(x) - \underline{E}_c(x)) > 0. \qquad \square$$

From Theorem 1 with $\delta_c(x)$ representing *significance*, we can see that the point-valued degree of potential threats w.r.t. each criterion for the subjects would be higher if the set of events happen in an area with high crime statistics than that if the set of events happen in a lower crime area.

Theorem 2. *Let $EUI_c(i) = [\underline{E}_c(i), \overline{E}_c(i)]$ be an interval-valued expected level of potential threat of the criterion c for subject with ID i ($i \in \{x, y\}$), and $\delta_c(i)$ be the degrees of significance for the event of subject with ID i, then the point-valued degree of potential threat for these two subjects satisfies:*

(i) if $\underline{E}_c(x) > \overline{E}_c(y)$, then $\nu_c(x) > \nu_c(y)$;

(ii) if $\underline{E}_c(x) > \underline{E}_c(y)$, $\overline{E}_c(x) > \overline{E}_c(y)$, and $\delta_c(x) \geq \delta_c(y)$, then $\nu_c(x) > \nu_c(y)$.

Proof. (i) By $\delta_c(k) \in [0, 1]$ ($k \in \{x, y\}$) and Definition 7, we have $\underline{E}_c(k) \leq \nu_c(k) \leq \overline{E}_c(k)$. As a result, by $\underline{E}_c(x) > \overline{E}_c(y)$, we have

$$\nu_c(x) - \nu_c(y) \geq \underline{E}_c(k) - \nu_c(y) \geq \underline{E}_c(k) - \overline{E}_c(y) > 0.$$

So, item (i) holds.

(ii) When $\underline{E}_c(x) > \underline{E}_c(y)$, $\overline{E}_c(x) > \overline{E}_c(y)$, and $0 \leq \delta_c(y) \leq \delta_c(x) \leq 1$, we have

$$
\begin{aligned}
\nu_c(x) - \nu_c(y) &\geq (\underline{E}_c(x) - \underline{E}_c(y)) + \delta_c(x)(\overline{E}_c(x) - \underline{E}_c(x)) - \delta_c(x)(\overline{E}_c(y) - \underline{E}_c(y)) \\
&= (1 - \delta_c(x))(\underline{E}_c(x) - \underline{E}_c(y)) + \delta_c(x)(\overline{E}_c(x) - (\overline{E}_c(y)) \\
&> 0.
\end{aligned}
$$

So, item (ii) holds. □

In fact, Theorem 2 states two intuitions when considering the point-valued degree of potential threat of any two suspects: (i) for a given criterion, if the lowest expected level of potential threat for a suspect is higher than the highest expected level of potential threat of another suspect, the point-valued degree of potential threat of the first one should be higher; and (ii) if the degree of significance for the events of a suspect is not less than that of another, and the lowest and highest expected levels of potential threat of this suspect are higher than those of another respectively, the point-valued degree of potential threat of the first one should be higher.

Theorem 3. *Let O_x and O_y be the overall degrees of potential threat for subject x and y, $C' = C \cup s$ and $C'' = C \cup r$ be the whole sets of related criteria for subjects x and y, k be the highest utility value for the outcomes of all criteria, and for any criterion $c \in C$, we have $nu_c(x) = nu_c(y)$. Suppose $\nu_s(x) > \nu_r(y)$ and $w_s = w_r$, then $O_x > O_y$.*

Proof. By Definition 8, $nu_c(x) = nu_c(y)$, $\nu_s(x) > \nu_r(y)$, and $w_s = w_r$ we have:

$$
\begin{aligned}
O_x - O_y &= \frac{2(w_s nu_s(x) + \sum_{c \subseteq C} w_c nu_c(x))}{(k+1)(w_s + \sum_{c \subseteq C} w_c)} - \frac{2(w_r nu_r(y) + \sum_{c \subseteq C} w_c nu_c(x))}{(k+1)(w_r + \sum_{c \subseteq C} w_c)} \\
&= \frac{2w_s(nu_s(x) - \nu_r(y))}{(k+1)(w_s + \sum_{c \subseteq C} w_c)} \\
&> 0.
\end{aligned}
$$

Actually, Theorem 3 means that the increase of the point-valued degree of potential threat about a given criterion for a subject will cause the increase of the overall degree of potential threat for this subject, *ceteris paribus*.

Table 1. Event modeling for the airport security surveillance scenario

event	Etype	occT	ID_s	rb	sig	Criterion	Weight	ID_p	Location	mass value	utility
$e_{42}^{a,1}$	SA	9:01pm	42	0.9	0.7	age	0.3	13	FCE	$\{young\}, 0.3$	U^a
$e_{42}^{a,2}$	SA	9:01pm	42	0.9	0.7	age	0.3	13	FCE	$\{young, old\}, 0.7$	U^a
$e_{45}^{a,1}$	SA	9:03-9:15pm	45	0.9	0.7	age	0.3	13	FCE	$\{young\}, 0.6$	U^a
$e_{45}^{a,2}$	SA	9:03-9:15pm	45	0.9	0.7	age	0.3	13	FCE	$\{young, old\}, 0.4$	U^a
$e_{42}^{g,1}$	SA	9:01pm	42	0.9	0.7	gender	0.3	13	FCE	$\{female\}, 0.4$	U^g
$e_{42}^{g,2}$	SA	9:01pm	42	0.9	0.7	gender	0.3	13	FCE	$\{female, male\}, 0.6$	U^g
$e_{45}^{g,1}$	SA	9:03-9:15pm	45	0.9	0.7	gender	0.3	13	FCE	$\{male\}, 0.7$	U^g
$e_{45}^{g,2}$	SA	9:03-9:15pm	45	0.9	0.7	gender	0.3	13	FCE	$\{female, male\}, 0.3$	U^g
$e_{42}^{m,1}$	SA	9:01pm	42	0.9	0.7	move	0.8	13	FCE	$\{to\,east, loitering\}, 0.8$	
$e_{42}^{m,2}$	SA	9:01pm	42	0.9	0.7	move	0.8	13	FCE	$\Theta_m, 0.2$	
$e_{45}^{m,1}$	SA	9:03-9:15pm	45	0.9	0.7	move	0.8	13	FCE	$\{loiter\}, 0.9$	
$e_{45}^{m,2}$	SA	9:03-9:15pm	45	0.9	0.7	move	0.8	13	FCE	$\Theta_m, 0.1$	
$e_{29}^{a,1}$	CC	9:03pm	29	1	0.9	age	0.3	19	MoC	$\{young\}, 0.7$	U^a
$e_{29}^{a,2}$	CC	9:03pm	29	1	0.9	age	0.3	19	MoC	$\{young, old\}, 0.3$	U^a
$e_{29}^{g,1}$	CC	9:03pm	29	1	0.9	age	0.3	19	MoC	$\{male\}, 0.7$	U^g
$e_{29}^{g,2}$	CC	9:03pm	29	1	0.9	age	0.3	19	MoC	$\{male, female\}, 0.3$	U^g
$e_{29}^{sr,1}$	CC	9:03pm	29	1	0.9	sr	0.8	19	MoC	$\{unmatch\}, 0.8$	U^{sr}
$e_{29}^{sr,2}$	CC	9:03pm	29	1	0.9	sr	0.8	19	MoC	$\{unmatch, match\}, 0.2$	U^{sr}

Where $\Theta_m = \{to\,east, \ldots, to\,north, stay, loitering\}$; $U^a : \{u^a(young) = 6, u^a(old) = 2\}$; $U^g : \{u^g(male) = 6, u^g(female) = 4\}$; $U^{sr} = \{u^{sr}(unmatch) = 8, u^{sr}(match) = 4\}$; and the scale of measurement for the level of potential threat is $H = \{1, \ldots, 9\}$.

5   Case Study

Let us consider a scenario in an airport between at 9:00pm to 9:15pm, which covers the following two areas: Shopping Area (SA) and Control Center (CC).

- in the Shopping Area (SA), a person (id: 13) loiters near a Foreign Currency Exchange office (FCE) for a long time. Also, camera 42 catches its back image at the entrance of the shopping area at 9:01pm and camera 45 catches its side face image at FCE from 9:03pm to 9:15pm;
- in the Control Center (CC), the face of a person (id: 19) appears in the camera 29 in the middle of the corridor (MoC) to the control center at 9:03pm. However, the person's face does not appear in camera 23 monitoring the entrance to the corridor.

We assume that video classification algorithms can detect age, gender, behavior, and then re-acquire subjects (sr) when needed. We also assume that there is only one security team available. What should the system do at this moment?

First, the surveillance system detects the elementary events for each person as shown in Table 1 based on the information of multiple sensors.

For example, the first row in Table 1 means that for an event type SA in FCE at 9:01pm, the *age* classification program used by camera 42, whose degree of reliability is 0.9, detects a person with $ID = 13$ as *male* with a certainty of 30%. The significance of this event is 0.6, the weight of *age* criterion for detecting a potential threat is 0.3, and U^a is the utility function that shows the level of potential threat for the *age* criterion. Moreover, some events in the table are within the same event cluster, such as $e_{42}^{a,1}$ and $e_{42}^{a,2}$ which share the same *event type, occT, Criterion, ID_s, ID_p*, but assign different mass values to different subsets of the same frame and the sum of these mass values is 1.

Second, since some sensors are not completely reliable in our example, we obtain the discounted mass functions by Definition 2. For example, consider the *age* criterion for the person in FCE, we have :

$$m_{42}^a(\{young\}) = 0.3 \times 0.9 = 0.27, m_{42}^a(\{young, old\}) = 0.1 + 0.7 \times 0.9 = 0.73;$$
$$m_{45}^a(\{young\}) = 0.54, m_{45}^a(\{young, old\}) = 0.46.$$

Third, we consider the combination of mass functions associated with events in different clusters (from different sources) where these events are all about a common criterion, using Dempster's rule in Definition 1. For example, consider the *age* criterion for the person in FCE, we have:

$$m_{42\&45}^a(\{young\}) = (0.27 \times 0.54 + 0.27 \times 0.46 + 0.73 \times 0.54)/1 = 0.664,$$
$$m_{42\&45}^a(\{young, old\}) = (0.73 \times 0.46)/1 = 0.336.$$

Note that each mass value is associated with a derived event, such as, for the person in FCE, $m_{42\&45}^a(\{young\}) = 0.664$ is associated with $e_{42\&45}^{a,1} = $(SA, 9:01 − 9:15 pm, 42&45, 0.9, 0.7, age, 0.3, 13, FCE, $m_{42\&45}^a(\{young\})$=0.664, U^a).

Forth, we consider event inference. For the person in FCE, by the inference rule in Example 1, $m_{42\&45}^m(\{to\ east, lotering\}) = 0.137$, $m_{42\&45}^m(\{lotering\}) = 0.81$, $m_{42\&45}^m(\Theta_m) = 0.053$, and Equation (4), we have $m^{IPL}(\{Rob\}) = 0.41, m^{IPL}(\{Waiting\ Friends\}) = 0.24, m^{IPL}(\{Rob, Waiting\ Friends\}) = 0.35$.

Fifth, we obtain the expected utility interval for each criterion of each person by Definition 6. For example, for the person (id:13) in FCE, we have

$$\underline{E}_{13,a} = 4.656, \overline{E}_{13,a} = 6; \underline{E}_{13,g} = 5.044, \overline{E}_{13,g} = 5.656; \underline{E}_{13,IPL} = 5.43, \overline{E}_{13,IPL} = 7.542.$$

Sixth, we obtain the point-valued degree of potential threat for each criterion of each person by Definition 7. For example, for the person (id:13) in FCE:

$$\nu_a(13) = (1 - 0.7) \times 4.656 + 0.7 \times 6 = 5.6; \nu_g(13) = 5.47; \nu_{IPL}(13) = 6.91.$$

Seventh, we get the overall degree of potential threat of each target after considering all relative criteria at 9:15pm by Definition 8:

$$O_{13} = \frac{2(0.3 \times 5.6 + 0.3 \times 5.47 + 0.8 \times 6.91)}{(9 + 1)(0.3 + 0.3 + 0.8)} = 1.26; O_{19} = 1.41.$$

Hence, in this example, we derive that *id* 19 is more dangerous than *id* 13. Thus, If we have only one security team available at that moment, the surveillance system will suggest to prevent the further action of the person (id: 19) in the control center first.

6   Related Work and Summary

Ahmed and Shirmohammadi in [1] designed a probabilistic decision support engine to prioritizes multiple events in different cameras. In this model, they incorporated the feedbacks of operators, event correlation and decision modulation to rank the importance of events. Jousselme *et al.* [7] presented the concept of a decision support tool together with the underlying multi-objective optimization algorithm for a ground air traffic control application. However, none of these models provides a method to handle multiple criteria information under uncertainty as our model does. Moreover, the problem of information fusion has become a key challenge in the realm of intelligent systems. A common method to handle this challenge is to introduce aggregation operators. Albusac *et al.* in [2] analyzed different aggregation operators and proposed a new aggregation method based on the Sugeno integral for multiple criteria in the domain of intelligent surveillance. Also, Rudas *et al.* in [13] offered a comprehensive study of information aggregation in intelligence systems from different application fields such as robotics, vision, knowledge based systems and data mining, etc. However, to the best of our knowledge, there is no research considering the decision making problem under uncertainty in surveillance systems.

In this paper, we introduced our extended event reasoning framework, integrating a multi-criteria decision making element in sensor network based surveillance systems. We also discussed some properties of our framework. Our next step of work is to experiment the decision making element with surveillance data.

References

1. Ahmed, D.T., Shirmohammadi, S.: A decision support engine for video surveillance systems. In: ICME, pp. 1–6 (2011)
2. Albusac, J., Vallejo, D., Jimenez, L., Castro-Schez, J.J., Glez-Morcillo, C.: Combining degrees of normality analysis in intelligent surveillance systems. In: FUSION 2012, pp. 2436–2443 (2012)
3. Cheng, H.Y., Weng, C.C., Chen, Y.Y.: Vehicle detection in aerial surveillance using dynamic Bayesian networks. IEEE Transactions on Image Processing 21(4), 2152–2159 (2012)
4. Etner, J., Jeleva, M., Tallon, J.M.: Decision theory under ambiguity. Journal of Economic Surveys 26, 324–370 (2012)
5. Fernández-Caballero, A., Castillo, J.C., Rodríguez-Sánchez, J.M.: Human activity monitoring by local and global finite state machines. Expert Systems with Applications 39(8), 6982–6993 (2012)
6. Gallese, V., Goldman, A.: Mirror-neurons and the simulation theory of mind reading. Trends Cogn. Sci. 2, 493–501 (1998)
7. Jousselme, A.L., Maupin, P., Debaque, B., Prevost, D.: A decision support tool for a Ground Air Traffic Control application. In: FUSION 2012, pp. 60–67 (2012)
8. Lowrance, J.D., Garvey, T.D., Strat, T.M.: A framework for evidential reasoning systems. In: Proc. of 5th AAAI, pp. 896–903 (1986)
9. Liu, W., Hughes, J.G., McTear, M.F.: Representating heuristic knowledge in D-S theory. In: Proc. of UAI, pp. 182–190 (1992)
10. Ma, J., Liu, W., Miller, P., Yan, W.: Event composition with imperfect information for bus surveillance. In: AVSS 2009, pp. 382–387 (2009)
11. Ma, J., Liu, W., Miller, P.: Event modelling and reasoning with uncertain information for distributed sensor networks. In: Deshpande, A., Hunter, A. (eds.) SUM 2010. LNCS, vol. 6379, pp. 236–249. Springer, Heidelberg (2010)
12. Pei, M., Jia, Y., Zhu, S.C.: Parsing video events with goal inference and intent prediction. In: ICCV 2011, pp. 487–494 (2011)
13. Rudas, I.J., Pap, E., Fodor, J.: Information aggregation in intelligent systems: An application oriented approach. Knowledge-Based Systems 38, 3–13
14. Shafer, G.: A Mathematical Theory of Evidence. Princeton University Press, Princeton (1976)
15. Strat, T.M.: Decision Analysis Using Belief Functions. International Journal of Approximate Reasoning 4(5-6), 391–418 (1990)

Adjoint Triples and Residuated Aggregators[*]

María Eugenia Cornejo[1], Jesús Medina[2], and Eloisa Ramírez-Poussa[1]

[1] Department of Statistic and O.R., University of Cádiz, Spain
{mariaeugenia.cornejo,eloisa.ramirez}@uca.es
[2] Department of Mathematics, University of Cádiz, Spain
jesus.medina@uca.es

Abstract. Several domains, such as fuzzy logic programming, formal concept analysis and fuzzy relation equations, consider basic operators which need to have associated residuated implications. Adjoint triples are formed by operators satisfying weak properties, usefully used in these domains. This paper presents the comparison of these triples with other general operators considered in these frameworks.

Keywords: Adjoint triples, implication triples, u-norms, uninorms.

1 Introduction

Many domains in mathematics and information sciences are formed by different types of algebraic structures such as many-valued logics, generalized measure and integral theory, quantum logics and quantum computing, etc.

T-norms [16,26,27], which are defined in the unit interval $[0, 1]$, are the operators used most often in applicative examples. However, these operators are very restrictive and several applications need more general structures.

Adjoint triples are a general structure which has been developed to increase the flexibility of the framework, since conjunctors are neither required to be commutative nor associative. These operators are useful in fuzzy logic programming [22,23], fuzzy formal concept analysis [20] and fuzzy relation equations [10].

Other important generalizations of t-norms and residuated implications exist, such as implication triples [25], sup-preserving aggregations [3], unorms [17], uninorms [28,14] and extended-order algebras [15]. These operators are defined following the same motivation of adjoint triples in order to reduce the mathematical requirements of the basic operators used for computation in the considered framework.

This paper recalls the settings in which these operators are defined, shows a comparison among them and investigates when the general considered operators have adjoint implications and so, they can be used to define concept-forming operators, rules in a residuated logic program or in a general fuzzy relation equation.

---

[*] Partially supported by the Spanish Science Ministry projects TIN2009-14562-C05-03 and TIN2012-39353-C04-04, and by Junta de Andalucía project P09-FQM-5233.

A. Laurent et al. (Eds.): IPMU 2014, Part III, CCIS 444, pp. 345–354, 2014.

2   Adjoint Triples and Duals

The operators forming these triples arise as a generalization of a t-norm and its residuated implication. Since the general conjunctor assumed in an adjoint triple need not be commutative, we have two different ways of generalizing the well-known adjoint property between a t-norm and its residuated implication, depending on which argument is fixed.

Definition 1. *Let $(P_1, \leq_1)$, $(P_2, \leq_2)$, $(P_3, \leq_3)$ be posets and $\&: P_1 \times P_2 \to P_3$, $\swarrow: P_3 \times P_2 \to P_1$, $\nwarrow: P_3 \times P_1 \to P_2$ be mappings, then $(\&, \swarrow, \nwarrow)$ is an adjoint triple with respect to P_1, P_2, P_3 if $\&$, $\swarrow, \nwarrow$ satisfy the adjoint property, for all $x \in P_1$, $y \in P_2$ and $z \in P_3$:*

$$x \leq_1 z \swarrow y \quad \text{iff} \quad x \& y \leq_3 z \quad \text{iff} \quad y \leq_2 z \nwarrow x$$

Note that this property generalizes the residuation condition [13] for no commutative operators. Straightforward consequences can be obtained from the adjoint property, such as the monotonicity properties.

Proposition 1. *If $(\&, \swarrow, \nwarrow)$ is an adjoint triple, then*

1. *$\&$ is order-preserving on both arguments, i.e. if $x_1, x_2, x \in P_1$, $y_1, y_2, y \in P_2$ and $x_1 \leq_1 x_2$, $y_1 \leq_2 y_2$, then $(x_1 \& y) \leq_3 (x_2 \& y)$ and $(x \& y_1) \leq_3 (x \& y_2)$; and*

2. *$\swarrow, \nwarrow$ are order-preserving on the first argument and order-reversing on the second argument, i.e., if $x_1, x_2, x \in P_1$, $y_1, y_2, y \in P_2$, $z_1, z_2, z \in P_3$ and $x_1 \leq_1 x_2$, $y_1 \leq_2 y_2$, $z_1 \leq_3 z_2$, then $(z_1 \swarrow y) \leq_1 (z_2 \swarrow y)$, $(z \swarrow y_2) \leq_1 (z \swarrow y_1)$, $(z_1 \nwarrow x) \leq_2 (z_2 \nwarrow x)$ and $(z \nwarrow x_2) \leq_2 (z \nwarrow x_1)$;*

Moreover, the adjoint implications are unique.

Proposition 2. *Given a conjunctor $\&$, its residuated implications are unique.*

Example of adjoint triples are the Gödel, product and Łukasiewicz t-norms together with their residuated implications. Note that, the Gödel, product and Łukasiewicz t-norms are commutative, then the residuated implications satisfy that $\swarrow^G = \nwarrow_G$, $\swarrow^P = \nwarrow_P$ and $\swarrow^L = \nwarrow_L$. These adjoint triples are defined on $[0,1]$ as:

$$\&_G(x, y) = \min(x, y) \qquad z \nwarrow_G x = \begin{cases} 1 & \text{if } x \leq z \\ z & \text{otherwise} \end{cases}$$

$$\&_P(x, y) = x \cdot y \qquad z \nwarrow_P x = \min(1, z/x)$$

$$\&_L(x, y) = \max(0, x + y - 1) \qquad z \nwarrow_L x = \min(1, 1 - x + z)$$

Example 1. Given $m \in \mathbb{N}$, the set $[0, 1]_m$ is a regular partition of $[0, 1]$ in m pieces, for example $[0, 1]_2 = \{0, 0.5, 1\}$ divide the unit interval in two pieces.

A discretization of the product t-norm is the operator $\&_P^* \colon [0,1]_{20} \times [0,1]_8 \to [0,1]_{100}$ defined, for each $x \in [0,1]_{20}$ and $y \in [0,1]_8$ as:

$$x \,\&_P^*\, y = \frac{\lceil 100 \cdot x \cdot y \rceil}{100}$$

whose residuated implications $\swarrow_P^* \colon [0,1]_{100} \times [0,1]_8 \to [0,1]_{20}$, $\nwarrow_P^* \colon [0,1]_{100} \times [0,1]_{20} \to [0,1]_8$ are defined as:

$$b \swarrow_P^* a = \frac{\lfloor 20 \cdot \min\{1, b/a\} \rfloor}{20} \qquad\qquad b \nwarrow_P^* c = \frac{\lfloor 8 \cdot \min\{1, b/c\} \rfloor}{8}$$

Hence, the triple $(\&_P^*, \swarrow_P^*, \nwarrow_P^*)$ is an adjoint triple and the operator $\&_P^*$ is straightforward neither commutative nor associative. Similar adjoint triples can be obtained from the Gödel and Łukasiewicz t-norms.

"Dual" adjoint triples. In several settings is useful considering the duals of the assumed operators. For instance, this relation can be used to relate different frameworks [3,18], in which the assumed operators are used to compute in one setting and the duals are considered to obtain another "dual" setting. However, in general, they cannot be used in the same framework as we show next.

Given an adjoint triple $(\&, \swarrow, \nwarrow)$, we may consider two extra operators $\swarrow^d \colon P_3 \times P_2 \to P_1$, $\nwarrow_d \colon P_3 \times P_1 \to P_2$ satisfying the equivalence

$$z \swarrow^d y \leq_1 x \quad \text{iff} \quad z \leq_3 x \& y \quad \text{iff} \quad z \nwarrow_d x \leq_2 y$$

which can be called *dual adjoint property*.

However these operators cannot be associated with the conjunctor $\&$ of an adjoint triple, when, for instance, P_1 and P_3 have a maximum element, $\top_1, \top_3$ and P_2 and P_3 have a minimum element, $\bot_2, \bot_3$, where $\bot_3 \neq \top_3$. This is because, from the adjoint property of $\&$ and $\nwarrow$, we obtain the condition $\top_1 \& y = \top_3$, for all $y \in P_2$, and from the dual adjoint property of $\&$ and $\swarrow^d$ the equality $x \& \bot_2 = \bot_3$ holds, for all $x \in P_1$. Therefore, from both equalities, we obtain the chain $\bot_3 = \top_1 \& \bot_2 = \top_3$, which leads us to a contradiction.

3   Comparison with Other Operators

This section presents four important kinds of operators used in several framework and the comparison with the adjoint triples. We will show that these operators are particular cases of the conjunctor of an adjoint triple, when the residuated implications are considered.

3.1   Implication Triples

Implication triples and adjointness algebras were introduced in [25] and a lot of properties were studied in several papers.

Definition 2. Let $(L, \leq_L)$ $(P, \leq_P)$ be two posets with a top element $\top_P$ in $(P, \leq_P)$. An adjointness algebra is an 8-tuple $(L, \leq_L, P, \leq_P, \top_P, A, K, H)$, satisfying the following four conditions:

1. The operation $A \colon P \times L \to L$ is antitone in the left argument and isotone in the right argument, and it has $\top_P$ as a left identity element, that is $A(\top_P, \gamma) = \gamma$, for all $\gamma \in L$. We call A an implication on (L, P).

2. The operation $K \colon P \times L \to L$ is isotone in each argument and has $\top_P$ as a left identity element, that is $K(\top_P, \beta) = \beta$, for all $\beta \in L$. We call K a conjunction on (L, P).

3. The operation $H \colon L \times L \to P$ is antitone in the left argument and isotone in the right argument, and it satisfies, for all $\beta, \gamma \in L$, that

$$H(\beta, \gamma) = \top_P \quad \text{if and only if} \quad \beta \leq_L \gamma$$

   We call H a forcing-implication on L.

4. The three operations A, K and H, are mutually related by the following condition, for all $\alpha \in P$ and $\beta, \gamma \in L$:

$$\beta \leq_L A(\alpha, \gamma) \quad \text{iff} \quad K(\alpha, \beta) \leq_L \gamma \quad \text{iff} \quad \alpha \leq_P H(\beta, \gamma)$$

   which is called the adjointness condition.

We call the ordered triple (A, K, H) an implication triple on (L, P).

These operators were initially related in [7]. In order to relate adjoint triples to implication triples, we must consider that the antecedent and consequent elements are evaluated in the right and left arguments of the implications of the adjoint triples, respectively, and that this evaluation is the opposite for the implications in the implication triples. Therefore, we define $\swarrow^{\mathrm{op}} \colon P_2 \times P_3 \to P_1$, $\nwarrow_{\mathrm{op}} \colon P_1 \times P_3 \to P_2$, as $y \swarrow^{\mathrm{op}} z = z \swarrow y$ and $x \nwarrow_{\mathrm{op}} z = z \nwarrow x$, for all $x \in P_1$, $y \in P_2$ and $z \in P_3$.

The following result relates the adjoint implication to implication triples and it was proven in [6].

Theorem 1 ([6]). Given an adjoint triple $(\&, \swarrow, \nwarrow)$ with respect to the posets $(P_1, \leq_1)$, $(P_2, \leq_2)$, $(P_3, \leq_3)$. If $P_2 = P_3$ and P_1 has a maximum $\top_1$ as a left identity element for $\&$, then $(\&, \swarrow^{\mathrm{op}}, \nwarrow_{\mathrm{op}})$ is an implication triple.

Example 2. The adjoint triple $(\&_P^*, \swarrow_P^*, \nwarrow_P^*)$ given in Example 1 is not an implication triple. This operator is non-commutative and non-associative at the same time, but it does not satisfy that 1 is a left identity for $\&_P^*$, that is, there exists an element $y \in [0, 1]_8$ such that $1 \&_P^* y \neq y$. For instance, if $y = 0.625$, then $1 \&_P^* 0.625 = 0.63$. In addition, $[0, 1]_8 \neq [0, 1]_{100}$.

Therefore, adjoint triple are more general than implication triples.

3.2   Sup-preserving Aggregation Operators

In [2] an interesting problem in concept lattice theory is solved using an inf-residuum relation equation. These results have been extended in [11]. Specifically, in this last paper the property-oriented concept lattices are considered to solve sup-t-norm and inf-residuum relational equations and, moreover, several results in order to obtain the whole set of solutions of these equations, using fuzzy formal concept analysis, have been presented.

With respect to the flexibility in the definition of the fuzzy relation equations, to the best of our knowledge, one of the more general frameworks is given in [3], in which general operators are considered and an unification of the sup-t-norm and inf-residuum products of relations is given.

Next, we will compare these operators with the adjoint triples and show that the latter are more general and so, the equations introduced in [3] are a particular case of the multi-adjoint relation equations.

Definition 3. *A sup-preserving aggregation structure (aggregation structure, for short) is a quadruple* $(\mathbf{L}_1, \mathbf{L}_2, \mathbf{L}_3, \square)$, *where* $\mathbf{L}_i = (L_i, \preceq_i)$, *with* $i \in \{1, 2, 3\}$, *are complete lattices and* $\square \colon L_1 \times L_2 \to L_3$ *is a function which commutes with suprema in both arguments, that is,*

$$\left(\bigvee_{j \in J} a_j \right) \square\, b = \bigvee_{j \in J} (a_j \square\, b) \qquad a \,\square \left(\bigvee_{j' \in J'} b_{j'} \right) = \bigvee_{j' \in J'} (a \,\square\, b_{j'})$$

for all $a, a_j \in L_1$ $(j \in J)$, $b, b_{j'} \in L_2$ $(j' \in J')$.

From the results given in [3] the following theorem is obtained.

Theorem 2 ([3]). *Given an aggregation structure* $\square \colon L_1 \times L_2 \to L_3$, *there exist two mappings* $_\square \!\circ \colon L_3 \times L_2 \to L_1$ *and* $\circ_\square \colon L_1 \times L_3 \to L_2$, *satisfying that*

$$a_1 \,\square\, a_2 \leq_3 a_3 \quad \textit{iff} \quad a_2 \leq_2 a_1 \circ_\square a_3 \quad \textit{iff} \quad a_1 \leq_1 a_3 \,_\square\!\circ\, a_2$$

for all $a_1 \in L_1$, $a_2 \in L_2$, $a_3 \in L_3$.

This is the adjoint property, hence the following result is obtained.

Corollary 1. *The triple* $(\square, {}_\square\!\circ, \circ_\square{}^d)$, *where* $a_3 \circ_\square{}^d a_1 = a_1 \circ_\square a_3$, *for all* $a_1 \in L_1$, $a_3 \in L_3$, *is an adjoint triple.*

However the inverse is not true, since only posets are needed in the definition of an adjoint triple. Although in multi-adjoint relation equations we need to consider adjoint triples related to two complete lattices $\mathbf{L}_1$, $\mathbf{L}_2$ and a poset $(P, \leq)$. This last one need not be a complete lattice. Therefore, in the setting of fuzzy formal concept analysis both triples are not equivalent either.

The generality of $(P, \leq)$ is very important since, for instance, it can be a multilattice [4,5,19,24]. For example, an interesting fuzzy relation defined in a multilattice is given in [24]. Furthermore, in order to solve the proposed problem in multi-adjoint logic programming, several adjoint triples are needed, which permits the multi-adjoint relation equations.

3.3 U-norms

Another general operators are the *u-norms* [17]. U-norms have been presented as a generalization of the following three functions: arithmetic mean, continuous Archimedean t-norm and "fuzzy and" with $\gamma < 1$.

Definition 4. *A* u-norm *is a function* $U\colon [0,1]\times[0,1] \to [0,1]$, *such that* $U(0,0) = 0$, $U(1,1) = 1$, *and* U *is strictly increasing on the set* $D = \{(x,y) \in [0,1] \times [0,1] \mid 0 < U(x,y)\}$.

Clearly, the Gödel t-norm is not a u-norm and so, the conjunctor of an adjoint triple could not be a u-norm. On the other hand, an arbitrary u-norm may not form an adjoint triple:

Example 3. The weighted power mean $M_{w_1,w_2,r}$ with $w_1, w_2 \in [0,1]$, $w_1 + w_2 = 1$ and $r \in \mathbb{R}$ is defined as

$$M_{w_1,w_2,r}(x,y) = \begin{cases} (w_1 x^r + w_2 y^r)^{1/r} & \text{if } r \neq 0 \\ x^{w_1} y^{w_2} & \text{if } r = 0 \end{cases}$$

for all $x, y \in \mathbb{R}$. This operator is a u-norm if $w_1, w_2 > 0$ and $0 < r < \infty$ [17]. If we consider $w_1 = w_2 = 0.5$ and $r = 1$, then we have that

$$M_{0.5,0.5,1}(x,y) = 0.5x + 0.5y, \quad \text{for all } x, y \in \mathbb{R}$$

and this u-norm has not got adjoint implications, since the conjunctor of an adjoint triple satisfies that $x \,\&\, \bot = \bot$ and, in this case, $M_{0.5,0.5,1}(x,0) = \frac{x}{2} \neq 0$, for all $x \in \mathbb{R} \setminus \{0\}$. Therefore, $M_{0.5,0.5,1}(x,y)$ cannot be in an adjoint triple.

However, in order to solve fuzzy relation equations with u-norms introduced in [17], the authors need to consider continuous u-norms and these operators have two adjoint implications [21], which straightforwardly shows that the considered u-norms are a particular case of a conjunctor in an adjoint triple. As a consequence, the setting given in [17] is a particular case of multi-adjoint relation equations as well.

3.4 Uninorms

Uninorms are a generalisation of t-norms and t-conorms [28]. These operators are characterized by having a neutral element which is not necessarily equal to 0 (as for t-norms) or 1 (as for t-conorms).

Definition 5. *A* uninorm *on* $([0,1], \leq)$ *is a commutative, associative, increasing mapping* $U\colon [0,1] \times [0,1] \to [0,1]$ *for which there exists an element* $e \in [0,1]$ *such that* $U(e,x) = x$, *for all* $x \in [0,1]$. *The element* e *is called the neutral element of* U.

Given a uninorm U on the unit interval, there exists a t-norm T_U, a t-conorm S_U on $([0,1], \leq)$ and two increasing bijective mappings $\phi_e\colon [0,e] \to [0,1]$ and $\psi_e\colon [e,1] \to [0,1]$ with increasing inverse, such that [14]

1. For all $x, y \in [0, e]$, $U(x, y) = \phi_e^{-1}(T_U(\phi_e(x), \phi_e(y)))$
2. For all $x, y \in [e, 1]$, $U(x, y) = \psi_e^{-1}(S_U(\psi_e(x), \psi_e(y)))$

The last property shows that the behavior of a uninorm is similar to a t-norm for elements less or equal than e and as a t-conorm for elements greater or equal than e. Therefore, a uninorm might not necessarily be the conjunctor of an adjoint triple, which will be shown next.

Note that, if $(\&, \swarrow, \nwarrow)$ is an adjoint triple then it satisfies the adjoint property. This property can be adequately interpreted in terms of multiple-valued inference as asserting both that the truth-value of $z \swarrow y$ is the maximal x satisfying $x \& z \leq y$.

In the following, we will consider a uninorm on $([0, 1], \leq)$ which does not satisfy this last property.

Example 4. Given $e \in]0, 1[$, the operator U_e on $([0, 1], \leq)$ defined, for each $x, y \in [0, 1]$, by

$$U_e(x, y) = \begin{cases} max(x, y) & \text{if } x \geq e \text{ and } y \geq e \\ min(x, y) & \text{if } otherwise \end{cases}$$

is a uninorm. If we assume that U_e is the conjunctor of an adjoint triple, then its adjoint implication $\swarrow$ must satisfy that:

$$z \swarrow y = max\{x \in [0, 1] \mid U_e(x, y) \leq z\}$$

If we consider $e = 0.5$, $y = 0.7$ and $z = 0.6$, it can be proven that the maximum does not exist.

Therefore, in general, a uninorm may not have adjoint implications and so, be part of an adjoint triple. A particular case of uninorms, widely studied and used [1,8], are the left-continuous uninorms. As these operators have adjoint implications, the left-continuous uninorms are a particular case of adjoint conjunctors. For example, the conjunctor of the adjoint triple defined in Example 1 is not a left-continuous uninorm since $\&_P^*$ is neither commutative nor associative.

3.5   Extended-Order Algebras

In [9] simple implications are considered in order to introduce different algebraic structures. This section only recalls several algebraic structures given in [15] and compares the considered operators with the conjunctor of an adjoint triple.

Definition 6 ([15]). *A w-eo algebra is a triple $(P, \rightarrow, \top)$ where P is a non-empty set, $\rightarrow : P \times P \rightarrow P$ is a binary operation and $\top$ a fixed element of P, satisfying for all $a, b, c \in P$ the following conditions*[1]

(o_1) $a \rightarrow \top = \top$ *(upper bound condition)*
(o_2) $a \rightarrow a = \top$ *(reflexivity condition)*

---

[1] Note that the names of the properties are those in [15].

(o_3) $a \to b = \top$ and $b \to a = \top$ then $a = b$ (antisymmetry condition)
(o_4) $a \to b = \top$ and $b \to c = \top$ then $a \to c = \top$ (weak transitivity condition)

Given a w-eo algebra $(P, \to, \top)$, the relation $\leq$ determined by the operation $\to$, by means of the equivalence

$$a \leq b \quad \text{if and only if} \quad a \to b = \top, \quad \text{for all } a, b \in P$$

is an order relation in P. Moreover, $\top$ is the greatest element in $(P, \leq)$. This order relation was called *the natural order in P*, in [15].

Note that the previous equivalence is the property that satisfies a forcing-implication [25]. Hence, given a poset $(P, \leq)$ with a greatest element $\top$ and $\to \colon P \times P \to P$ is a forcing-implication on P, then $(P, \to, \top)$ is a w-eo algebra.

When the natural order in a w-eo algebra $(P, \to, \top)$ provides a complete lattice, we say that $(P, \to, \top)$ is a *complete w-eo algebra* $(P, \to, \top)$, in short, a *w-ceo algebra*. In this case, we will write L and $\preceq$ instead of P and $\leq$, respectively.

Definition 7. *Let L be a non-empty set, $\to \colon L \times L \to L$ a binary operation and $\top$ a fixed element of L. The triple $(L, \to, \top)$ is a right-distributive w-ceo algebra, if it is a w-ceo algebra and satisfies the following condition, for any $a \in L$ and $B \subseteq L$*

$$(d'_r) \quad a \to \bigwedge_{b \in B} b = \bigwedge_{b \in B} (a \to b)$$

Next results relate the notions presented above to the implication triples introduced in [25].

Proposition 3. *Let $(L, \preceq)$ be a complete lattice and (A, K, H) an implication triple on (L, L). Then, the triple $(L, H, \top)$ is a right-distributive w-ceo algebra.*

Proposition 4. *Let $(L, \preceq)$ be a complete lattice and (A, K, H) an implication triple on (L, L). If $K(x, \top) = x$, for all $x \in L$, then $(L, A, \top)$ is a right-distributive w-ceo algebra.*

From now on, we only consider the operation H since less hypotheses need to be assumed. Analogous results for A can be obtained considering the boundary condition noted in Proposition 4.

Note that, if $(L, \preceq)$ is a complete lattice and (A, K, H) is an implication triple on (L, L), by Proposition 3, we have that $(L, H, \top)$ is a right-distributive w-ceo algebra. In this environment, the adjoint product $\otimes \colon L \times L \to L$ introduced in [9,15] and defined as

$$a \otimes x = \bigwedge \{t \in L \mid x \preceq a \to t\}$$

for all $a, x \in L$, is not the operator K given in Definition 2, whenever $\otimes$ or K are not commutative. Therefore, the properties shown in Proposition 4.1 of [9] cannot be related to properties of implication triples presented in [25], in general. Although, these can be related if $\otimes$ or K are commutative.

In [9] is introduced an additional binary operation $\rightsquigarrow \colon L \times L \to L$ satisfying

$$a \preceq b \rightsquigarrow c \quad \text{iff} \quad a \otimes b \preceq c \quad \text{iff} \quad b \preceq a \to c$$

for all $a, b, c \in L$. Then the triple $(\rightsquigarrow, \otimes, \to)$ is not an implication triple on (L, L) since K is not defined in the same way that $\otimes$. Moreover, we cannot identify $\to$ with H because $(\otimes, \to)$ does not satisfy the same adjoint property that (K, H). Moreover, $\to$ cannot be A because A is not a forcing-implicaton.

However, we can define the operation $\otimes' \colon L \times L \to L$ by

$$a \otimes' x = \bigwedge \{t \in L \mid a \preceq x \to t\}$$

for all $a, x \in L$. This operator coincides with the operator K, given in Definition 2, and satisfies the adjointness condition given in the same definition. As a consequence, the triple $(\rightsquigarrow, \otimes', \to)$ is an implication triple on (L, L).

Proposition 5. *Given a complete lattice $(L, \preceq)$ and the mappings $\rightsquigarrow$, $\otimes$ and $\to$ defined above, the triple $(\rightsquigarrow, \otimes, \to)$ is an adjoint triple with respect to L.*

4 Conclusions and Future Work

From the comparison of adjoint triples with u-norm, uninorms, sup-preserving operators, extended-order algebras, we have proven that these operators are a particular case of adjoint triples when the residuated implications are demanded. From this comparison, the properties introduced in different papers can be considered to provide extra properties among these operators.

In the future more operators and properties will be studied. Moreover, a generalization of the transitivity property will be introduced in which these operators will be involved.

References

1. Aguiló, I., Suñer, J., Torrens, J.: A characterization of residual implications derived from left-continuous uninorms. Information Sciences 180(20), 3992–4005 (2010)
2. Bělohlávek, R.: Concept equations. Journal of Logic and Computation 14(3), 395–403 (2004)
3. Bělohlávek, R.: Sup-t-norm and inf-residuum are one type of relational product: Unifying framework and consequences. Fuzzy Sets and Systems 197, 45–58 (2012)
4. Benado, M.: Les ensembles partiellement ordonnés et le théorème de raffinement de Schreier, II. Théorie des multistructures. Czechoslovak Mathematical Journal 5(80), 308–344 (1955)
5. Cordero, P., Gutiérrez, G., Martínez, J., de Guzmán, I.P.: A new algebraic tool for automatic theorem provers. Annals of Mathematics and Artificial Intelligence 42(4), 369–398 (2004)
6. Cornejo, M., Medina, J., Ramírez, E.: A comparative study of adjoint triples. Fuzzy Sets and Systems 211, 1–14 (2013)

7. Cornejo, M.E., Medina, J., Ramírez, E.: Implication triples versus adjoint triples. In: Cabestany, J., Rojas, I., Joya, G. (eds.) IWANN 2011, Part II. LNCS, vol. 6692, pp. 453–460. Springer, Heidelberg (2011)
8. De Baets, B., Fodor, J.: Residual operators of uninorms. Soft Computing 3(2), 89–100 (1999)
9. Della Stella, M.E., Guido, C.: Associativity, commutativity and symmetry in residuated structures, vol. 30(2), pp. 363–401. Springer Science+Business Media (2013)
10. Díaz, J.C., Medina, J.: Multi-adjoint relation equations: Definition, properties and solutions using concept lattices. Information Sciences 253, 100–109 (2013)
11. Díaz, J.C., Medina, J.: Solving systems of fuzzy relation equations by fuzzy property-oriented concepts. Information Sciences 222, 405–412 (2013)
12. Díaz-Moreno, J., Medina, J., Ojeda-Aciego, M.: On basic conditions to generate multi-adjoint concept lattices via galois connections. International Journal of General Systems 43(2), 149–161 (2014)
13. Dilworth, R.P., Ward, M.: Residuated lattices. Transactions of the American Mathematical Society 45, 335–354 (1939)
14. Fodor, J.C., Yager, R.R., Rybalov, A.: Structure of uninorms. Int. J. Uncertain. Fuzziness Knowl.-Based Syst. 5(4), 411–427 (1997)
15. Guido, C., Toto, P.: Extended-order algebras. Journal of Applied Logic 6(4), 609–626 (2008)
16. Klement, E., Mesiar, R., Pap, E.: Triangular norms. Kluwer Academic (2000)
17. Lin, J.-L., Wu, Y.-K., Guu, S.-M.: On fuzzy relational equations and the covering problem. Information Sciences 181(14), 2951–2963 (2011)
18. Medina, J.: Multi-adjoint property-oriented and object-oriented concept lattices. Information Sciences 190, 95–106 (2012)
19. Medina, J., Ojeda-Aciego, M., Ruiz-Calviño, J.: Fuzzy logic programming via multilattices. Fuzzy Sets and Systems 158, 674–688 (2007)
20. Medina, J., Ojeda-Aciego, M., Ruiz-Calviño, J.: Formal concept analysis via multi-adjoint concept lattices. Fuzzy Sets and Systems 160(2), 130–144 (2009)
21. Medina, J., Ojeda-Aciego, M., Valverde, A., Vojtáš, P.: Towards biresiduated multi-adjoint logic programming. In: Conejo, R., Urretavizcaya, M., Pérez-de-la-Cruz, J.-L. (eds.) CAEPIA-TTIA 2003. LNCS (LNAI), vol. 3040, pp. 608–617. Springer, Heidelberg (2004)
22. Medina, J., Ojeda-Aciego, M., Vojtáš, P.: Multi-adjoint logic programming with continuous semantics. In: Eiter, T., Faber, W., Truszczyński, M. (eds.) LPNMR 2001. LNCS (LNAI), vol. 2173, pp. 351–364. Springer, Heidelberg (2001)
23. Medina, J., Ojeda-Aciego, M., Vojtáš, P.: Similarity-based unification: a multi-adjoint approach. Fuzzy Sets and Systems 146, 43–62 (2004)
24. Medina, J., Ruiz-Calviño, J.: Fuzzy formal concept analysis via multilattices: first prospects and results. In: The 9th International Conference on Concept Lattices and Their Applications (CLA 2012), pp. 69–79 (2012)
25. Morsi, N.N.: Propositional calculus under adjointness. Fuzzy Sets and Systems 132(1), 91–106 (2002)
26. Nguyen, H.T., Walker, E.: A First Course in Fuzzy Logic, 3rd edn. Chapman & Hall, Boca Ratón (2006)
27. Schweizer, B., Sklar, A.: Associative functions and abstract semigroups. Publ. Math. Debrecen 10, 69–81 (1963)
28. Yager, R.R., Rybalov, A.: Uninorm aggregation operators. Fuzzy Sets and Systems 80(1), 111–120 (1996)

First Approach of Type-2
Fuzzy Sets via Fusion Operators

María Jesús Campión[1], Juan Carlos Candeal[2], Laura De Miguel[3],
Esteban Induráin[1], and Daniel Paternain[3,*]

[1] Departamento de Matemáticas, Universidad Pública de Navarra,
Campus Arrosadía s/n 31006, Pamplona, Spain
{mjesus.campion,steiner}@unavarra.es
[2] Departamento de Análisis Económico, Facultad de Ciencias Económicas y
Empresariales, Universidad de Zaragoza,
Gran Vía 2-4 50005, Zaragoza, Spain
candeal@unizar.es
[3] Departamento de Automática y Computación, Universidad Pública de Navarra,
Campus Arrosadía s/n 31006, Pamplona, Spain
{demiguel.81238,daniel.paternain}@e.unavarra.es

Abstract. In this work we introduce the concept of a fusion operator
for type-2 fuzzy sets as a mapping that takes m functions from $[0,1]$
to $[0,1]$ and brings back a new function of the same type. We study
in depth the properties of pointwise fusion operators and representable
fusion operators. Finally, we study the union and intersection of type-
2 fuzzy sets and we analyze when these functions are pointwise and
representable fusion operators.

Keywords: Type-2 fuzzy sets, Fusion operators, Pointwise fusion operator, Representable fusion operator.

1 Introduction

Since Zadeh introduced the concept of a fuzzy set in 1965 [13], the interest
for the different possible extensions of such sets has been increasingly growing,
both from a theoretical and from an applied point of view [2,7,10]. Among the
extensions of fuzzy sets some of the most relevant ones are the interval-valued
fuzzy sets [11], the Atanassov's intuitionistic fuzzy sets [1] and the type-2 fuzzy
sets [14], which encompass the two previous ones. In this work we focus on the
latter.

Although many works devoted to such sets can be found in the literature,
we believe that it is necessary to expand the theoretical framework for them,
with an eye kept in the application in fields such as decision making or image
processing.

* We acknowledge financial support from the Ministry of Economy and Competitive-
ness of Spain under grant MTM2012-37894-C02-02 and TIN2011-29520.

A. Laurent et al. (Eds.): IPMU 2014, Part III, CCIS 444, pp. 355–363, 2014.
© Springer International Publishing Switzerland 2014

As an extension of the concept of a fusion operator relative to fuzzy sets (functions that take m values (membership degrees) in $[0,1]$ and give back a new value in $[0,1]$), and taking into account that the membership values of the elements in a type-2 fuzzy set are given in terms of new fuzzy sets over the referential $[0,1]$ (that is, by means of functions defined over $[0,1]$) we define fusion operators for type-2 fuzzy sets as mappings that take m functions from $[0,1]$ to $[0,1]$ into a new function in the same domain. That is, functions of the type $F : \left([0,1]^{[0,1]}\right)^m \to [0,1]^{[0,1]}$. Our goal is to study these functions in a way as general as possible because no restriction is imposed to the membership values of a type-2 fuzzy set. So we do not require a priori any property such as continuity, monotonicity, symmetry, etc.

In the theoretical study of fusion operators for type-2 fuzzy sets we pay special atention to some basic notions, namely the concept of pointwise fusion and the concept of representable fusion. Both concepts are related to the fact that, in order to know the value of the fusion operator F at some point x of its domain, we only need to know the values at that point (and not elsewhere) of the functions we are going to fuse.

The structure of this work is the following: in the next section we recall the notion of type-2 fuzzy set and we introduce the concept of a fusion operator. In section 3 we study the characterization of representable fusion operators. Then, we consider union and intersection between type-2 fuzzy sets as a special case of fusion operators. We finish the paper with some concluding remarks, as well as pointing out some future research lines.

2   Fusion of Type-2 Fuzzy Sets

In this section we recall the concepts of a fuzzy set and a type-2 fuzzy set. Throughout this paper we will denote by X a non-empty set that will represent the universe of discourse.

Definition 1. *A fuzzy set A on the universe X is defined as*

$$A = \{(x, \mu_A(x)) | x \in X\}$$

where $\mu_A : X \longrightarrow [0,1]$ is the membership degree of the element x to the set A.

In [14], Zadeh proposed type-2 fuzzy sets (T2FS) as a generalization of fuzzy sets (also called type-1 fuzzy sets). In type-2 fuzzy sets, the membership degree of an element to the set considered is given by a fuzzy set whose referential set is $[0,1]$. That is, the membership degree of an element to a type-2 fuzzy set is a function in $[0,1]^{[0,1]}$, the set of all possible functions from $[0,1]$ to $[0,1]$. Throughout the paper, we will adopt the mathematical formalization of the notion of type-2 fuzzy set introduced in [3,9].

Definition 2. *A type-2 fuzzy set A on the universe X is defined as*

$$A = \{(x, \mu_A(x)) | x \in X\}$$

where $\mu_A : X \to [0,1]^{[0,1]}$.

In almost every application of fuzzy logic (e.g. decision making problems, approximate reasoning, machine learning or image processing, among others), it is very usual to fuse or aggregate several membership degrees into a single value which preserves as much information as possible from the inputs. This kind of functions take m degrees of membership and gives back a new membership degree. That is, those fusing functions are mappings

$$F : [0, 1]^m \to [0, 1].$$

If we want to fuse m fuzzy sets over the same referential X, we use a function F that acts on the m-tuple of the membership degrees of the element $x \in X$ in each of the considered fuzzy sets.

A particular case of fusion operators are the well-known aggregation functions. Such functions are fusion operators to which some additional properties (monotonicity and boundary conditions) are demanded.

Our idea in this work is to study the fusion of type-2 fuzzy sets. Notice that for these sets it does not exist a natural order, so the definition of aggregation functions may be difficult. On the other hand, as the membership of each element is given by a function in $[0, 1]^{[0,1]}$, in order to fuse several membership functions we just need to define and study mappings

$$F : \left([0, 1]^{[0,1]}\right)^m \to [0, 1]^{[0,1]}.$$

In the next section we formally define the concept of a fusion operator and we study its former properties.

3   Definition of a Fusion Operator, and Related Concepts

In this section we define and study the main properties of fusion operators $F : ([0, 1]^{[0,1]})^m \to [0, 1]^{[0,1]}$.

Definition 3. *Let $(f_1, \ldots, f_m) \in ([0, 1]^{[0,1]})^m$ stand for a m-tuple of maps from $[0, 1]$ into $[0, 1]$. A map $f_{m+1} \in [0, 1]^{[0,1]}$ is said to be a fusion of $(f_1, \ldots, f_m)$ if there exists a map*

$$F : ([0, 1]^{[0,1]})^m \to [0, 1]^{[0,1]}$$

such that

$$f_{m+1} = F(f_1, \ldots, f_m) \tag{1}$$

In this case the map F is said to be an m-dimensional fusion operator.

First of all, we should observe that we do not require any specific property to fusion operators, since such study does not lie in the scope of the present work. Our general purpose is to make an approach to fusion operators as general as possible. Of course, in the future we will study those fusion operators which are continuous, monotone, etc.

3.1   Pointwise Fusion Operators

From the definition of a fusion operator, the first question that arises is the following: To what extent is necessary to know the values taken by the functions f_i? Depending on the situation it may happen that given a point $x \in [0,1]$, in order to calculate the value $f_{m+1}(x)$ we compulsorily need to have at hand the values of the functions f_i at points different from x. However, sometimes it suffices to know, just, the values of the functions f_i at the point $x \in [0,1]$. This idea leads to the notion of a pointwise fusion operator.

Definition 4. *A map $f_{m+1} \in [0,1]^{[0,1]}$ is said to be a* pointwise fusion *of* $(f_1, \ldots, f_m)$ *if there exists a map $W : [0,1]^m \to [0,1]$ such that:*

$$f_{m+1}(x) = W(f_1(x), \ldots, f_m(x)) \tag{2}$$

for every $x \in [0,1]$. In this case, the map W is said to be a pointwise m-*dimensional fusion operator, whereas the functional equation $f_{m+1}(x) = W(f_1(x), \ldots, f_m(x))$ is said to be the* structural functional equation *of pointwise fusion operators.*

Example 1. A prototypical example of a pointwise fusion operator is the arithmetic mean of the m functions, where $f_{m+1}(x) = \frac{f_1(x) + \ldots + f_m(x)}{m}$ for all $x \in [0,1]$.

3.2   Representable Fusion Operators

Next, we study a class of fusion operators which are closely related to pointwise fusion operators, namely, the so-called representable fusion operators.

Definition 5. *Let $F : ([0,1]^{[0,1]})^m \to [0,1]^{[0,1]}$ denote an m-dimensional fusion operator of maps from $[0,1]$ into $[0,1]$. Then F is said to be* representable *if there is a map $W : [0,1]^m \to [0,1]$ such that:*

$$F(f_1, \ldots, f_m)(x) = W(f_1(x), \ldots, f_m(x)) \tag{3}$$

holds for every $x \in [0,1]$ and $(f_1, \ldots, f_m) \in ([0,1]^{[0,1]})^m$.

Remark. Notice that in Definition 4 the maps $f_1, \ldots, f_n$ are *fixed* a priori, so W depends on $f_1, \ldots, f_n$. However, in Definition 5, $f_1, \ldots, f_n$ may vary in $[0,1]^{[0,1]}$, so that W will depend, directly, on F.

For instance, if $f_1 = |\sin(x)|$, $f_2(x) = \sin^2(x)$, $f_3(x) = \sin^4(x)$ it is true that $f_3(x) \le f_2(x) \le f_1(x)$ holds for every $x \in [0,1]$.

If $f_4(x) = median\{f_1(x), f_2(x), f_3(x)\}$ it is clear that f_4 is the projection over the second component, namely f_2. Obviously, this will not happen when the tuple (f_1, f_2, f_3) takes arbitrary values in $([0,1]^{[0,1]})^3$.

Given a set of m functions $f_1, \ldots, f_m$, the function W satisfies the properties of the definition of a pointwise fusion operator.

Proposition 1. *Let $f_1, \ldots, f_m \in ([0,1]^{[0,1]})^m$ be a fixed m-tuple of functions. If $F : ([0,1]^{[0,1]})^m \to [0,1]^{[0,1]}$ is a representable fusion operator, then F is a pointwise fusion operator.*

Proof: If we call $f_{m+1} = F(f_1, \ldots, f_m)$, then it follows that $f_{m+1}(x) = W(f_1(x), \ldots, f_m(x))$ for all $x \in [0, 1]$.

Observe that, although there exists a relation between the properties of pointwise and representability, they are conceptually different. For a pointwise fusion operator, the functions $f_1, \ldots, f_m$ are fixed beforehand. Therefore, the problem consists in solving the functional equation:
$f_{m+1}(x) = W(f_1(x), \ldots, f_m)(x)$. On the other hand, for representable fusion operators, we start from a fusion operator F that satisfies the functional equation $F(f_1, \ldots, f_m)(x) = W(f_1(x), \ldots, f_m(x))$ for every $x \in [0, 1]$ and every tuple $(f_1, \ldots, f_m) \in ([0, 1]^{[0,1]})^m$. Therefore, the idea is to find the map W depending only on F and not on the tuple of functions $f_1, \ldots, f_m$. The difference between both concepts can also be seen discussed in Section 5 through a case study.

4   Characterization of Representable Fusion Operators

In this section we characterize representable fusion operators. To do so, we start by introducing three properties that are going to help us to achieve the desired characterization.

Definition 6. *Let* $F : ([0, 1]^{[0,1]})^m \to [0, 1]^{[0,1]}$ *be an m-dimensional fusion operator. Then F is said to be:*

- *fully independent if it holds that:*

$$(f_1(x), \ldots, f_m(x)) = (g_1(t), \ldots g_m(t)) \Rightarrow F(f_1, \ldots, f_m)(x) = F(g_1, \ldots, g_m)(t) \tag{4}$$

  for every $x, t \in [0, 1]$ *and* $(f_1, \ldots, f_m), (g_1, \ldots, g_m) \in ([0, 1]^{[0,1]})^m$.
- *independent as regards maps if it holds that:*

$$(f_1(x), \ldots, f_m(x)) = (g_1(x), \ldots g_m(x)) \Rightarrow F(f_1, \ldots, f_m)(x) = F(g_1, \ldots, g_m)(x) \tag{5}$$

  for every $x \in [0, 1]$ *and* $(f_1, \ldots, f_m), (g_1, \ldots, g_m) \in ([0, 1]^{[0,1]})^m$
- *pointwise independent if it holds that:*

$$(f_1(x), \ldots, f_m(x)) = (f_1(t), \ldots f_m(t)) \Rightarrow F(f_1, \ldots, f_m)(x) = F(f_1, \ldots, f_m)(t) \tag{6}$$

  for every $x, t \in [0, 1]$ *and* $(f_1, \ldots, f_m) \in ([0, 1]^{[0,1]})^m$

Theorem 1. *Let* $F : ([0, 1]^{[0,1]})^n \to [0, 1]^{[0,1]}$ *be an m-dimensional fusion operator. The following statements are equivalent:*

(i) F is representable,
(ii) F is fully independent,
(iii) F is independent as regards maps, and pointwise independent.

Proof. The implications (i) $\Rightarrow$ (ii) $\Rightarrow$ (iii) are straightforward. To prove (iii) $\Rightarrow$ (i), let $(y_1, \ldots, y_m) \in [0, 1]^m$. Let $c_{y_i} : [0, 1] \to [0, 1]$ be the constant map defined by $c_{y_i}(x) = y_i$ for every $x \in [0, 1]$ ($i = 1, \ldots, m$). Fix an element $x_0 \in [0, 1]$. Define now $W : [0, 1]^m \to [0, 1]$ as $W(y_1, \ldots, y_m) = F(c_{y_1}, \ldots, c_{y_m})(x_0)$. Observe that the choice of x_0 is irrelevant here, since F is pointwise independent. In order to see that W represents F, fix $x \in [0, 1]$ and $(f_1, \ldots, f_m) \in ([0, 1]^{[0,1]})^m$. Let $c_i : [0, 1] \to [0, 1]$ be the constant map given by $c_i(t) = f_i(x)$ for every $t \in [0, 1]$ $(i = 1, \ldots, m)$. Since F is independent as regards maps, it follows that $F(f_1, \ldots, f_m)(x) = F(c_1, \ldots, c_m)(x)$. But, by definition of W, we also have that $F(c_1, \ldots, c_m)(x) = W(f_1(x), \ldots, f_m(x))$. Therefore $F(f_1, \ldots, f_m)(x) = W(f_1(x), \ldots, f_m(x))$ and we are done. $\qquad\square$

Example 2. The following m-dimensional fusion operators from $[0, 1]$ into $[0, 1]$ are obviously representable:

(i) each *projection* $\pi_i : ([0, 1]^{[0,1]})^m \to [0, 1]^{[0,1]}$, where $\pi_i(f_1, \ldots, f_m) = f_i$ for every $(f_1, \ldots, f_m) \in ([0, 1]^{[0,1]})^m$ $(i = 1, \ldots, m)$,
(ii) each *constant operator* mapping any m-tuple $(f_1, \ldots, f_m) \in ([0, 1]^{[0,1]})^m$ to a (fixed a priori) map $g : [0, 1] \to [0, 1]$.

Moreover, any m-ary operation in $[0, 1]$ immediately gives rise to a representable m-dimensional fusion operator. Indeed, given a map $H : [0, 1]^m \to [0, 1]$, it is clear that the m-dimensional fusion operator $F_H : ([0, 1]^{[0,1]})^m \to [0, 1]^{[0,1]}$ given by $F_H(f_1, \ldots, f_m)(x) = H(f_1(x), \ldots, f_m(x))$ for every $x \in [0, 1]$ and $(f_1, \ldots, f_m) \in ([0, 1]^{[0,1]})^m$ is representable through H.

In the spirit of Definition 6 and Theorem 1 we finish this section by considering that an m-tuple of functions from a set $[0, 1]$ into another set $[0, 1]$ has been *fixed*. We furnish a result concerning the structural functional equation of pointwise aggregation.

Theorem 2. *Let* $(f_1, \ldots, f_m) \in ([0, 1]^{[0,1]})^m$ *denote a fixed m-tuple of maps from $[0, 1]$ into $[0, 1]$. Let $f_{m+1} : [0, 1] \to [0, 1]$ be a map. The following statements are equivalent:*

(i) *There exists a solution $W : [0, 1]^m \to [0, 1]$ of the structural functional equation of pointwise aggregation so that $f_{m+1}(x) = W(f_1(x), \ldots, f_m(x))$ holds for every $x \in [0, 1]$.*
(ii) *The implication $(f_1(x), \ldots, f_m(x)) = (f_1(t), \ldots f_m(t)) \Rightarrow f_{m+1}(x) = f_{m+1}(t)$ holds true for all $x, t \in [0, 1]$.*

Proof. The implication (i) $\Rightarrow$ (ii) is obvious. To prove that (ii) $\Rightarrow$ (i), we choose an element $y_0 \in [0, 1]$, and define W as follows: given $(y_1, \ldots y_m) \in [0, 1]^m$ we declare that $W(y_1, \ldots y_m) = f_{m+1}(x)$ if there exists $x \in [0, 1]$ such that $(y_1, \ldots y_m) = (f_1(x), \ldots f_m(x))$; otherwise, $W(y_1, \ldots y_m) = y_0$. $\qquad\square$

5   A Case Study: Union and Intersection of Type-2 Fuzzy Sets as Fusion Operators

In this section we focus on two key concepts to deal with type-2 fuzzy sets: the union and intersection. Recall that the union and intersection of two type-2 fuzzy sets is a new type-2 fuzzy set. Therefore, we can interpret the union and intersection of type-2 fuzzy sets as a special case of fusion of type-2 fuzzy sets.

It is important to say that it does not exist a unique definition for union and intersection of type-2 fuzzy sets. However, the operations considered in this work cover several cases since they act in the same way. For each element in the referential set, we use a function that fuses the membership functions of that element to each set. So these operations can be seen as fusion operators $F : \left([0,1]^{[0,1]}\right)^2 \to [0,1]^{[0,1]}$.

5.1   Union and Intersection Based on Minimum and Maximum

Considering type-2 fuzzy sets as a special case of L-fuzzy sets launched by Goguen [5], the union and intersection of type-2 fuzzy sets is stated leaning on the union and intersection of fuzzy sets, as follows [6].

Definition 7. *Let $f_1, f_2 \in [0,1]^{[0,1]}$ be two maps. The operations (respectively called union and intersection)* $\cup, \cap : \left([0,1]^{[0,1]}\right)^2 \to [0,1]^{[0,1]}$ *are defined as*

$$(f_1 \cup f_2)(x) = \max(f_1(x), f_2(x)) \text{ and} \tag{7}$$

$$(f_1 \cap f_2)(x) = \min(f_1(x), f_2(x)). \tag{8}$$

Proposition 2. *The mappings* $\cup, \cap : \left([0,1]^{[0,1]}\right)^2 \to [0,1]^{[0,1]}$ *are representable fusion operators for all $f_1, f_2 \in [0,1]^{[0,1]}$.*

5.2   Union and Intersection of Type-2 Fuzzy Sets Based on Zadeh's Extension Principle

The problem with the previous definition of union and intersection of type-2 fuzzy sets is that these concepts do not retrieve the usual union and intersection of fuzzy sets [4]. In order to avoid this trouble, another definition of union and intersection of type-2 fuzzy sets was given based on Zadeh's extension principle [9,4,8,12].

Definition 8. *Let $f_1, f_2 \in [0,1]^{[0,1]}$ be two maps. The operations (again, respectively called union and intersection)* $\sqcup, \sqcap : \left([0,1]^{[0,1]}\right)^2 \to [0,1]^{[0,1]}$ *are defined as*

$$(f_1 \sqcup f_2)(x) = \sup\{(f_1(y) \wedge f_2(z)) : y \vee z = x\} \text{ and} \tag{9}$$

$$(f_1 \sqcap f_2)(x) = \sup\{(f_1(y) \wedge f_2(z)) : y \wedge z = x\}. \tag{10}$$

Observe that the fusion operators $\sqcup$ and $\sqcap$ are completely different from $\cup$ and $\cap$. In general, for any $f_1, f_2 \in [0,1]^{[0,1]}$ it is not possible to know the value $(f_1 \sqcup f_2)(x)$ knowing only the values $f_1(x)$ and $f_2(x)$.

Proposition 3. *The mappings* $\sqcup, \sqcap : \left([0,1]^{[0,1]}\right)^2 \to [0,1]^{[0,1]}$ *fail to be representable fusion operators.*

Another result in this direction may be seen in [12] and states that, under certain conditions, if we fix the functions f_1, f_2, we can see union and intersection of fuzzy sets as pointwise fusion operators.

Proposition 4. *The following statements hold true:*

(i) *If* $f_1, f_2 \in [0,1]^{[0,1]}$ *are increasing mappings then the operation* $\sqcup$ *is a pointwise fusion operator.*

(ii) *If* $f_1, f_2 \in [0,1]^{[0,1]}$ *are decreasing mappings then the operation* $\sqcap$ *is a pointwise fusion operator.*

Proof. This follows from Corollary 5 in [12] but we include here an alternative proof, for the sake of completeness.

(i) For the case $\sqcup$, let us see that $\{(y,z)|y \vee z = x\} = \{(x,z)|z \leq x\} \cup \{(y,x)|y \leq x\}$.

In the first situation, namely for $\{(x,z)|z \leq x\}$, since the funtion f_2 is increasing, we get $f_2(z) \leq f_2(x)$ for all $z \leq x$. So, $f_1(x) \wedge f_2(z) \leq f_1(x) \wedge f_2(x)$ for all $z \leq x$. In particular, $\sup_{z \leq x}\{f_1(x) \wedge f_2(z)\} \leq f_1(x) \wedge f_2(x)$. Moreover, since the point (x,x) lies in the considered set, we have that $\sup_{z \leq x}\{f_1(x) \wedge f_2(z)\} = f_1(x) \wedge f_2(x)$.

In a similar way, in the second situation, for $\{(y,x)|y \leq x\}$, it holds that $\sup_{y \leq x}\{f_1(y) \wedge f_2(x)\} = f_1(x) \wedge f_2(x)$.

Therefore $(f_1 \sqcup f_2)(x) = \sup\{(f_1(y) \wedge f_2(z) : y \vee z = x\} =$
$\vee(\sup_{z \leq x}\{f_1(x) \wedge f_2(z)\}, \sup_{y \leq x}\{f_1(y) \wedge f_2(x)\}) = f_1(x) \wedge f_2(x)$.

Hence the union is a pointwise fusion operator.

(ii) This case, for the intersection $\sqcap$, is handled in an entirely analogous way to the case $\sqcup$ just discussed. $\square$

6   Conclusions

In this work we have studied a first approach to the fusion of type-2 fuzzy sets. We have defined the concepts of a pointwise fusion operator and a representable fusion operator. Their properties are very interesting if we have in mind possible future applications in problems of decision making or image processing.

We have focused on two operations on type-2 fuzzy sets, namely the union and the intersection. Because these operations are particular cases of fusion operators in our sense, we have accordingly analyzed their properties of pointwise aggregationship and representability.

In future works, our idea is to study fusion operators that satisfy additional properties as continuity, boundary conditions or monotonicity. This study could be very useful to establish a link between our fusion operators and aggregation functions on type-2 fuzzy sets.

References

1. Atanassov, K.T.: Intuitionistic fuzzy sets. Fuzzy Sets Syst. 20, 87–96 (1986)
2. Bustince, H., Herrera, F., Montero, J.: Fuzzy Sets and their extensions. Springer, Berlin (2007)
3. Dubois, D., Prade, H.: Operations in a fuzzy-valued logic. Inform. Control 43, 224–254 (1979)
4. Dubois, D., Prade, H.: Fuzzy sets and systems: Theory and applications. Academic Press (1980)
5. Goguen, J.A.: L-fuzzy sets. J. Math. Anal. Appl. 18, 145–174 (1967)
6. Hardig, J., Walker, C., Walker, E.: The variety generated by the truth value algebra of type-2 fuzzy sets. Fuzzy Sets Syst. 161, 735–749 (2010)
7. Hernández, P., Cubillo, S., Torres-Blanc, C.: Negations on type-2 fuzzy sets. Fuzzy Sets Syst. (Available online December 11, 2013), http://dx.doi.org/10.1016/j.fss.2013.12.004, ISSN 0165-0114
8. Mendel, J.M., John, R.I.: Type-2 fuzzy sets made simple. IEEE Trans. Fuzzy Syst. 10, 117–127 (2002)
9. Mizumoto, M., Tanaka, K.: Some properties of fuzzy sets of type 2. Infor. Control 31, 312–340 (1976)
10. Pagola, M., Lopez-Molina, C., Fernandez, J., Barrenechea, E., Bustince, H.: Interval Type-2 Fuzzy Sets Constructed From Several Membership Functions: Application to the Fuzzy Thresholding Algorithm. IEEE Trans. Fuzzy Syst. 21(2), 230–244 (2013)
11. Sambuc, R.: Functions Φ-Flous, Application à l'aide au Diagnostique en Pathologie Thyroidenne, Thèse Doctorat en Medicine, University of Marseille (1975)
12. Walker, C.L., Walker, E.A.: The algebra of fuzzy truth values. Fuzzy Sets Syst. 149, 309–347 (2005)
13. Zadeh, L.A.: Fuzzy Sets. Inform. Control, 338–353 (1965)
14. Zadeh, L.A.: Quantitative fuzzy semantics. Inform. Sci. 3, 159–176 (1971)

Weakly Monotone Averaging Functions

Tim Wilkin[1], Gleb Beliakov[1], and Tomasa Calvo[2]

[1] School of Information Technology, Deakin University, 221 Burwood Hwy,
Burwood 3125, Australia
{gleb,tim.wilkin}@deakin.edu.au
[2] Universidad de Alcalá, Alcala de Henares, Spain
tomasa.calvo@uah.es

Abstract. Averaging behaviour of aggregation functions depends on the
fundamental property of monotonicity with respect to all arguments. Unfortunately this is a limiting property that ensures that many important
averaging functions are excluded from the theoretical framework. We propose a definition for weakly monotone averaging functions to encompass
the averaging aggregation functions in a framework with many commonly
used non-monotonic means. Weakly monotonic averages are robust to
outliers and noise, making them extremely important in practical applications. We show that several robust estimators of location are actually
weakly monotone and we provide sufficient conditions for weak monotonicity of the Lehmer and Gini means and some mixture functions. In
particular we show that mixture functions with Gaussian kernels, which
arise frequently in image and signal processing applications, are actually
weakly monotonic averages. Our concept of weak monotonicity provides
a sound theoretical and practical basis for understanding both monotone
and non-monotone averaging functions within the same framework. This
allows us to effectively relate these previously disparate areas of research
and gain a deeper understanding of averaging aggregation methods.

Keywords: aggregation functions, monotonicity, means, penalty-based
functions, non-monotonic functions.

1 Introduction

The aggregation of a set of inputs into a single representative quantity arises naturally in many application domains. Growing interest in aggregation problems
has led to the development of many new mathematical techniques to support
both the application and understanding of aggregation methods. A wide range of
aggregation functions now appear within the literature, including the classes of
weighted quasi-arithmetic means, ordered weighted averages, triangular norms
and co-norms, Choquet and Sugeno integrals and many more. Recent books [2,6]
provide a comprehensive overview of this field.

The class of averaging functions, commonly known as *means*, are frequently
applied in problems in statistical analysis, automated decision support and in
image and signal processing. Classical means (such as the median and arithmetic mean) share a fundamental property with the broader class of aggregation

A. Laurent et al. (Eds.): IPMU 2014, Part III, CCIS 444, pp. 364–373, 2014.

functions; that of monotonicity with respect to all arguments [2,6]. In decision making, monotonicity of the aggregation plays an important role: an increase in any input or criterion should not lead to a decrease of the overall score or utility. In many application areas though, where data may contain outliers or be corrupted by noise, monotonic averages generally perform poorly as aggregation operators. This is due specifically to monotonicity which (apart from the median) makes them sensitive to translation of individual input values. In such domains non-monotonic averages are often applied. For example, in statistical analysis, the robust estimators of location are used to measure central tendancy of a data set, with many estimators having a breakdown point of 50% [11].

There are many other non-monotonic means appearing in the literature: the mode (an average possibly known to the Greeks), Gini means, Lehmer means, Bajraktarevic means [2,3] and mixture functions [7,10] being particularly well known cases. These means have significant practical and theoretical importance and are well studied, however being non-monotonic they are not classified as aggregation functions.

Ideally we want a formal framework for averaging functions that encompasses the non-monotonic means and places them within in context with existing (monotonic) averaging aggregation. This new framework would enable a better understanding of the relationships within this broader class of functions and lead to a deeper understanding of non-monotonic averaging as aggregation.

We achieve these aims in this article by relaxing the monotonicity requirement for averaging aggregation and propose a new definition that encompasses many non-monotonic averaging functions. This definition rests on the property of directional monotonicity in the direction of the vector $(1, 1, \ldots, 1)$, which is obviously implied by shift-invariance as well as by the standard definition of monotonicity. We call this property *weak monotonicity* within the context of aggregation functions and we investigate it herein in the context of means.

The remainder of this article is structured as follows. In Section 2 we provide the necessary mathematical foundations that underpin the subsequent material. Section 3 contains our fundamental contribution, the definition of weakly monotonic averaging functions, along with some relevant examples. We examine several classes of non-monotonic means derived from the mean of Bajraktarevic in Section 4 and provide sufficient conditions for weak monotonicity. In Section 5 we consider $\phi-$transforms of weakly monotonic averages and determine conditions for the preservation of weak monotonicity under transformations. Our conclusions are presented in Section 6.

2   Preliminaries

2.1   Aggregation Functions

The following notations and assumptions are used throughout this article. Consider any closed, non-empty interval $\mathbb{I} = [a, b] \subseteq \bar{\mathbb{R}} = [-\infty, \infty]$. Without loss of generality we assume that the domain of interest is quantified by $\mathbb{I}^n$, with n implicit within the context of use. Tuples in $\mathbb{I}^n$ are defined as $\mathbf{x} = (x_{i,n} | n \in$

$\mathbb{N}, i \in \{1, ..., n\}$), though we write x_i as the shorthand for $x_{i,n}$ such that it is implicit that $i \in \{1, ..., n\}$. $\mathbb{I}^n$ is assumed ordered such that for $\mathbf{x}, \mathbf{y} \in \mathbb{I}^n$, $\mathbf{x} \leq \mathbf{y}$ implies that each component of $\mathbf{x}$ is no greater than the corresponding component of $\mathbf{y}$. Unless otherwise stated, a constant vector given as $\mathbf{a}$ is taken to mean $\mathbf{a} = a\underbrace{(1, 1, ..., 1)}_{n-times} = a\mathbf{1}$, $a \in \mathbb{R}$.

Consider now the following definitions:

Definition 1. A function $F : \mathbb{I}^n \to \bar{\mathbb{R}}$ is **monotonic** (non decreasing) if and only if, $\forall \mathbf{x}, \mathbf{y} \in \mathbb{I}^n, \mathbf{x} \leq \mathbf{y}$ then $F(\mathbf{x}) \leq F(\mathbf{y})$.

Definition 2. A function $F : \mathbb{I}^n \to \mathbb{I}$ is an **aggregation function** in $\mathbb{I}^n$ if and only if F is monotonic non-decreasing in $\mathbb{I}$ and $F(\mathbf{a}) = a$, $F(\mathbf{b}) = b$.

The two fundamental properties defining an aggregation function are monotonicity with respect to all arguments and bounds preservation.

Definition 3. A function F is called **idempotent** if for every input $\mathbf{x} = (t, t, ..., t)$, $t \in \mathbb{I}$ the output is $F(\mathbf{x}) = t$.

Definition 4. A function F has **averaging behaviour** (or is averaging) if for every $\mathbf{x}$ it is bounded by $\min(\mathbf{x}) \leq F(\mathbf{x}) \leq \max(\mathbf{x})$.

Definition 5. A function $F : \mathbb{I}^n \to \mathbb{I}$ is **shift-invariant** (stable for translations) if $F(\mathbf{x} + a\mathbf{1}) = F(\mathbf{x}) + a$ *whenever* $\mathbf{x}, \mathbf{x} + a\mathbf{1} \in \mathbb{I}^n$ *and* $F(\mathbf{x}) + a \in \mathbb{I}$.

Technically the definition of shift-invariance expresses stability of aggregation functions with respect to translation. Because the term shift-invariance is much in use, e.g. [2, 5], we adopt it for the remainder of this paper.

Definition 6. Let $\phi : \mathbb{I} \to \mathbb{I}$ be a bijection. The ϕ-**transform** of a function F is the function $F_\phi(\mathbf{x}) = \phi^{-1}(F(\phi(x_1), \phi(x_2), ..., \phi(x_n)))$.

An important example of ϕ-transforms are the negations, with the standard negation being $\phi(t) = 1 - t$.

2.2   Means

For monotonic functions averaging behaviour and idempotency are equivalent, however without monotonicity, idempotence does not imply averaging. We will follow many recent authors (e.g., [3]) and take the definition of a *mean* - a term used synonymously with averaging aggregation - to be any function (not necessarily monotonic) that has averaging behaviour.

Definition 7. *A function* $M : \mathbb{I}^n \to \mathbb{I}$ *is called a* **mean** *if and only if it has averaging behaviour.*

Examples of well known monotonic means found within the literature include the weighted arithmetic mean, weighted quasi-arithmetic mean, ordered weighted average (OWA) and the median.

An important class of means that are not necessarily monotonic are those expressed by the mean of Bajraktarevic (a generalisation of the weighted quasi-arithmetic means).

Definition 8. *Let* $\mathbf{w} = (w_1, ..., w_n)$ *be a vector of weight functions* $w_i : \mathbb{I} \to [0, \infty)$*, and let* $g : \mathbb{I} \to \bar{\mathbb{R}}$ *be a strictly monotonic function. The **mean of Bajraktarevic** is the function*

$$M_{\mathbf{w}}^g(\mathbf{x}) = g^{-1}\left(\frac{\sum_{i=1}^{n} w_i(x_i)g(x_i)}{\sum_{i=1}^{n} w_i(x_i)}\right). \tag{2.1}$$

In the case that $g(x_i) = x_i$ and all w_i are distinct functions, the Bajraktarevic mean is a *generalised mixture function* (or *generalised mixture operator*). If all w_i are equivalent then we obtain the *mixture functions*

$$M_w(\mathbf{x}) = \frac{\sum_{i=1}^{n} w(x_i)x_i}{\sum_{i=1}^{n} w(x_i)}. \tag{2.2}$$

A particularly interesting sub-class of Bajraktarevic means are Gini means $G^{p,q}$, obtained by setting $w(x_i) = x_i^q$ and $g(x_i) = x_i^{p-q}$ when $p \neq q$, or $g(x_i) = \log(x_i)$ if $p = q$,

$$G^{p,q}(\mathbf{x}) = \left(\frac{\sum_{i=1}^{n} x_i^p}{\sum_{i=1}^{n} x_i^q}\right)^{\frac{1}{p-q}}. \tag{2.3}$$

Gini means include as a special case ($q = 0$) the power means and hence include the minimum, maximum and the arithmetic mean as specific cases. Another special case of the Gini mean is the Lehmer mean, obtained when $p = q+1$. The contra-harmonic mean is the Lehmer mean with $q = 1$. We investigate these means further in Section 4.

3   Weak Monotonicity

3.1   Main Definition

The definition of weak monotonicity provided herein is prompted by applications and intuition, which suggests that an aggregate value does not decrease if all inputs increase equally, as the relative positions of the inputs are unchanged. Additionally, in many situations it is reasonable to expect that a small increase in one or a few values will also cause an increase in the aggregate value, however a larger increase may then cause a decrease in the aggregate. Such a circumstance would occur when a data value is moved so that it is then considered an outlier from the significant cluster present in the inputs. These properties lead us to the concept of *weak monotonicity* and thus the following definition.

Definition 9. A function f is called **weakly monotonic** non-decreasing (or directionally monotonic) if $F(\mathbf{x} + a\mathbf{1}) \geq F(\mathbf{x})$ for any $a > 0$, $\mathbf{1} = \underbrace{(1, ..., 1)}_{n-\text{times}}$, such that $\mathbf{x}, \mathbf{x} + a\mathbf{1} \in \mathbb{I}^n$.

Remark 1. *Monotonicity implies weak monotonicity and hence all aggregation functions are weakly monotonic. By Definition 5 all shift-invariant functions are also weakly monotonic.*

Remark 2. *If F is directionally differentiable in $\mathbb{I}^n$ then weak monotonicity is equivalent to non-negativity of the directional derivative $D_{\mathbf{1}}(F)(\mathbf{x}) \geq 0$.*

3.2   Examples of Weakly Monotonic Means

Herein we consider several examples of means generally considered to be non-monotonic and we show that they are actually weakly monotonic. We begin by considering several of the robust estimators of location, then investigate mixture functions and some interesting means from the literature.

Example 1. **Mode:** The mode is perhaps the most widely applied estimator of location, depicting the most frequent input. It may be expressed in penalty form [4] as the minimiser of the function

$$\mathcal{P}(\mathbf{x}, y) = \sum_{i=1}^{n} p(x_i, y) \qquad \text{where} \quad p(x_i, y) = \begin{cases} 0 & x_i = y \\ 1 & \text{otherwise} \end{cases}.$$

Since $F(\mathbf{x} + a\mathbf{1}) = \arg\min_y \mathcal{P}(\mathbf{x} + a\mathbf{1}, y) = \arg\min_y \sum_{i=1}^{n} p(x_i + a, y)$ is minimised by the value $y = F(\mathbf{x}) + a$ then the mode is shift-invariant and thus weakly monotonic.

Remark 3. *The penalty $\mathcal{P}$ associated with the mode is not quasi-convex and as such it may have several minimisers. Thus the mode may not be uniquely defined, in which case we must apply a reasonable convention. Suitable conventions include choosing the the smallest or the largest value from the set of minimisers. All non-monotonic means expressed by non-quasi-convex penalties also require such conventions.*

Robust estimators of location are calculated using the shortest contiguous sub-sample of $\mathbf{x}$ containing at least half of the values. The candidate sub-samples are the sets $X_k = \{x_j : j \in \{k, k+1, ..., k + \lfloor \frac{n}{2} \rfloor\}, \ k = 1, ..., \lfloor \frac{n+1}{2} \rfloor$. The length of each contiguous set is taken as $\|X_k\| = \left| x_{k+\lfloor \frac{n}{2} \rfloor} - x_k \right|$ and thus the index of the shortest sub-sample is

$$ k^* = \arg\min_i \|X_i\|, \ i = 1, ..., \left\lfloor \frac{n+1}{2} \right\rfloor. $$

Under the translation $\bar{\mathbf{x}} = \mathbf{x} + a\mathbf{1}$ the length of each sub-sample is unaltered since
$$ \|\bar{X}_k\| = \left| \bar{x}_{k+\lfloor \frac{n}{2} \rfloor} - \bar{x}_k \right| = \left| (x_{k+\lfloor \frac{n}{2} \rfloor} + a) - (x_k + a) \right| = \left| x_{k+\lfloor \frac{n}{2} \rfloor} - x_k \right| = \|X_k\| $$
and thus k^* remains the same. Thus, to establish the weak monotonicity of estimators based on the shortest half of the data we need only consider the effect of translation on the function applied to $X_{k^*}^*$.

Example 2. **Least Median of Squares (LMS):** The LMS is the midpoint of the set X_k^* and is given by the minimiser μ of the penalty function

$$ \mathcal{P}(\mathbf{x}, y) = median\left\{ (x_i - y)^2 \, | y \in \mathbb{I}, \ x_i \in X_{k^*} \right\}. $$

It is evident that the value y minimising the penalty $\mathcal{P}(\mathbf{x} + a\mathbf{1}, y)$ is $y = \mu + a$. Hence $F(\mathbf{x} + a\mathbf{1}) = F(\mathbf{x}) + a$ and the LMS is shift-invariant and thus weakly monotonic.

Example 3. The **Shorth** [1] is the arithmetic mean of X_{k^*} and is given by

$$ F(\mathbf{x}) = \frac{1}{h} \sum_{i=1}^{h} x_i, \ x_i \in X_{k^*}, \ h = \left\lfloor \frac{n}{2} \right\rfloor + 1. $$

Since membership of the set X_{k^*} is unaltered by translation and the arithmetic mean is shift-invariant, then the shorth is shift-invariant and hence weakly monotonic.

In the next section we present conditions for the weak monotonicity of several classes of means derived from the mean of Bajraktarevic.

4   Weakly Monotonic Mixture Functions

Monotonicity of mixture functions given by Eqn. (2.2) was investigated in [8, 9] where some sufficient conditions have been established in cases where the weighting function w is monotonic.

A particularly interesting case for which w is not monotonic is that in which the weighting function is Gaussian; i.e., $w(x) = \exp(-(x-a)^2/b^2)$. Such mixture functions arise in the implementation of convolution-based filters in image and signal processing and thus represent an extremely important example of non-monotonic means. With some calculation it can be shown that the directional derivative of a mixture function A with Gaussian weight function is given by

$$D_u(A) \propto \sum_{i=1}^{n} w_i^2 + \frac{1}{b^2} \sum_{i=1}^{n} \sum_{j \neq i}^{n} (b^2 - (x_i - x_j)^2) w_i w_j,$$

with $\mathbf{u} = \frac{1}{\sqrt{n}}(1, 1, ..., 1)$ and $w_i = w(x_i)$. This expression is non-negative for all $x_i, x_j \in [0, 1], b \geq 1$, independent of a. By varying parameters a and b we generate either monotone increasing, decreasing (both convex and concave) or unimodal quasi-convex weight functions, which produce weakly monotonic mixture functions.

Proposition 1. *Let A be a mixture function defined by eqn. (2.2) with generator $w(x) = e^{-(\frac{x-a}{b})^2}$. Then A is weakly monotonic for all $a, b \in \mathbb{R}$ and $\mathbf{x} \in [0, 1]^n$.*

Mesiar [9] showed that mixture functions may be expressed in penalty form by

$$P(\mathbf{x}, y) = \sum_{i=1}^{n} w(x_i)(x_i - y)^2.$$

It follows that if w is shift-invariant then so too is the mixture function and hence it is weakly monotonic. We can extend this result to weight functions of the form $w = w(\|x_i - f(\mathbf{x})\|_p)$ such that if f is shift-invariant then the mixture function is weakly monotonic.

For the Gini means, given by Eqn. (2.3), we present the following results without proof, which are beyond the scope of this paper and which will appear in a subsequent publication.

Theorem 1. *The Gini mean $G^{p,q}$ for $p, q \in \mathbb{R}$ defined in (2.3) is weakly monotonic on $[0, \infty)^n$ if and only if*

$$(n-1) \left(\left(\frac{q}{p-1} \right)^{p-1} \left(\frac{q-1}{p} \right)^{q-1} \right)^{\frac{1}{p-q}} \leq 1$$

and $p > q$ with $q \notin (0, 1)$ or $p < q$ with $p \notin (0, 1)$.

Taking the case $p = q + 1$, we obtain the Lehmer means, L^q, which are also not generally monotonic. Lehmer means are also mixture functions with weight function $w(t) = t^q$, which is neither increasing for all $q \in \mathbb{R}$ nor shift-invariant. For $q < 0$ the value of the Lehmer mean at $\mathbf{x}$ with at least one component $x_i = 0$ is defined as the limit when $x_i \to 0^+$, so that L^q is continuous on $[0, \infty)^n$.

Theorem 2. *The Lehmer mean of n arguments,*

$$L^q(\mathbf{x}) = \frac{\sum_{i=1}^{n} x_i^{q+1}}{\sum_{i=1}^{n} x_i^q}, \quad q \in \mathbb{R} \setminus (0, 1),$$

is weakly monotonic on $[0, \infty)^n$ *if* $n \leq 1 + \left(\frac{q+1}{q-1} \right)^{q-1}$.

Remark 4. *Interestingly, as*

$$\left(\frac{q+1}{q-1} \right)^{q-1} = \left(\left(1 + \frac{2}{q-1} \right)^{\frac{q-1}{2}} \right)^2,$$

and the right hand side is increasing (with q) and approaches e^2 as $q \to \infty$, we have a restriction that for all $q > 1$ weak monotonicity holds for at most $n < 9$ arguments. This restricts the use of Lehmer means for positive q in applications requiring some aspect of monotonicity.

Corollary 1 *The contra-harmonic mean ($L^q, q = 1$) is weakly monotonic only for two arguments.*

5   Duality and ϕ-transforms

Herein we consider the ϕ−transform of a weakly monotonic function. We know that standard monotonicity is preserved under ϕ−transform, when ϕ is a strictly monotonic function. Consider now functions of the form

$$\varphi(\mathbf{x}) = (\varphi(x_1), \varphi(x_2), ..., \varphi(x_n))$$

with φ any twice differentiable and invertible function.

Proposition 2. *If $A : \mathbb{I}^n \to \mathbb{I}^n$ is weakly monotonic and $\varphi : \mathbb{I} \to \mathbb{I}$ is a linear function then the ϕ−transform $A_\varphi(\mathbf{x}) = F(\mathbf{x}) = \varphi^{-1}(A(\varphi(\mathbf{x})))$ is weakly monotonic.*

Corollary 2 *The dual A^d of a weakly monotonic function A is also weakly monotonic under standard negation.*

Weak monotonicity is not preserved under all nonlinear transforms, although of course for specific functions this property will be preserved for some specific choices of ϕ.

Proposition 3. *The only functions ϕ which preserve weak monotonicity of **all** weakly monotonic functions are linear functions.*

Proof. Consider the example of a shift-invariant and hence weakly monotonic function whose ϕ−transform is not weakly monotonic in general. The function $F(x, y) = D(y - x) + \frac{x+y}{2}$, where D is the Dirichlet function (which takes values 0 at rational numbers and 1 at irrational numbers) is one such example. Taking generalised derivatives of F we obtain

$$F_x = -D'(y - x) + \frac{1}{2} \quad \text{and} \quad F_y = D'(y - x) + \frac{1}{2},$$

from which it follows that $F_x + F_y = 1 \geq 0$.

Consider the ϕ-transform of F given by $F_\phi(x, y) = \phi^{-1}(F(\phi(x), \phi(y)))$ and the sum of its partial derivatives. We need only consider the function $F(\phi(x), \phi(y))$ in which case we find that

$$F(\phi(x), \phi(y)))_x + F(\phi(x), \phi(y)))_y = D'(\phi(y) - \phi(x))(\phi'(y) - \phi'(x))$$
$$+ \frac{\phi'(x) + \phi'(y)}{2}.$$

The first term is zero only when $\phi'(x) = \phi'(y)$ for all x, y. That is, $\phi'(x)$ is constant and hence ϕ is a linear function. In all other cases the generalised derivative of D takes values $+\infty$ and $-\infty$, and hence the sum of partial derivatives can be negative.

The next result shows that for some shift-invariant functions, such as the Shorth and LMS estimators, weak monotonicity is preserved under φ-transforms with increasing invertible functions ϕ such that $\ln(\varphi')$ is concave.

Lemma 1. *Let φ be any twice differentiable and invertible function. Then the function $\varphi(\varphi^{-1}(\mathbf{x}) + c)$ is concave if and only if: $\ln \varphi'$ is concave when $\varphi' > 0$; or, $\ln|\varphi'|$ is convex for $\varphi' < 0$, for every $c \in \mathbb{R}, c \geq 0$ and $\mathbf{x} \in \mathbb{I}^n$.*

Lemma 2. *Let $\mathbf{x}$ be ordered such that $x_i \leq x_j$ for $i < j$ and let X_i denote the subset $\{x_i, \ldots, x_{i+h}\}$ for some fixed h. Let $\Delta X_i = \|X_i\| = |x_{i+h} - x_i|$ denote the length of the interval containing X_i. If φ is a concave increasing function then for $i < j$, $\Delta X_j \leq \Delta X_i$ implies that $\Delta \varphi(X_i) \geq \Delta \varphi(X_j)$, where $\Delta \varphi(X_i) = |\varphi(x_{i+h}) - \varphi(x_i)|$.*

Proposition 4. *Consider a robust estimator of location A based on the shortest contiguous half of the data. Let φ be a twice differentiable strictly increasing and thus invertible function, such that $\ln \phi'$ is concave. Then $A(\mathbf{x} + a\mathbf{1}) \geq A(\mathbf{x})$ implies $\varphi^{-1}(A(\varphi(\mathbf{x} + a\mathbf{1}))) \geq \varphi^{-1}(A(\varphi(\mathbf{x})))$.*

Proof. Denote by $\mathbf{y} = \varphi(\mathbf{x})$ and $\mathbf{x} = \varphi^{-1}(\mathbf{y})$ (with functions applied componentwise) and we have $\varphi' > 0$. We must show that

$$A(\psi_c(\mathbf{y})) = A(\varphi(\varphi^{-1}(\mathbf{y}) + c)) \geq A(\mathbf{y}).$$

By Lemma 1 function ψ_c is concave. By Lemma 2 we have $\Delta X_i \leq \Delta X_j \Rightarrow \Delta \psi_c(X_i) \leq \Delta \psi_c(X_j)$ for $i \geq j$. Hence the starting index of the shortest half cannot decrease after the transformation ψ_c and the result follows directly.

Corollary 3 *Let A be a robust estimator of location based on the shortest contiguous half of the data. Let φ be twice differentiable strictly increasing function such that $\ln|\phi'|$ is convex. Then the φ-dual of A is weakly monotonic.*

Proposition 4 serves as a simple test to determine which φ-transforms preserve weak monotonicity of averages such as the Shorth and LMS estimator. For example $\varphi(x) = e^x$ does preserve weak monotonicity and $\varphi(x) = \ln(x)$ does not.

6   Conclusion

We have presented a novel definition of weak monotonicity that enables us to encompass within the same framework the monotonic averaging aggregation functions and many means thought to be purely non-monotonic. We have demonstrated that several of the commonly used robust estimators of location are weakly monotonic and provided sufficient conditions for the weak monotonicity of the Lehmer means, Gini means and certain mixture functions. In particular we have shown that the nonlinear convolution-based filters with Gaussian kernels, used commonly in signal and image processing, are actually weakly monotonic averaging functions. We demonstrated that weak monotonicity is not preserved under all nonlinear transforms, and we found a class of transforms that preserve weak monotonicity of such functions as the shorth and LMS estimator.

This study was prompted by two significant issues that have arisen within our work on aggregation functions. The first is that many means reported in the literature lie outside the current definition of aggregation functions, which requires monotonicity in all arguments. The second issue is that of practical application. Monotonic means are generally sensitive to noise and outliers, common features of data collected in real world contexts. In averaging these data we employ non-monotonic averages. Improving our understanding of these functions and how they fit within the existing framework of aggregation functions is an important step in the development of this field of study.

References

1. Andrews, D., Bickel, P., Hampel, F., Huber, P., Rogers, W., Tukey, J.: Robust Estimates of Location: Surveys and Advances. Princeton University Press, Princeton (1972)
2. Beliakov, G., Pradera, A., Calvo, T.: Aggregation Functions: A Guide for Practitioners. STUDFUZZ, vol. 221. Springer, Heidelberg (2007)
3. Bullen, P.S.: Handbook of Means and Their Inequalities. Kluwer, Dordrecht (2003)
4. Calvo, T., Beliakov, G.: Aggregation functions based on penalties. Fuzzy Sets and Systems 161(10), 1420–1436 (2010)
5. Calvo, T., Mayor, G., Mesiar, R. (eds.): Aggregation Operators. New Trends and Applications. Physica-Verlag, Heidelberg (2002)
6. Grabisch, M., Marichal, J.-L., Mesiar, R., Pap, E.: Aggregation Functions. Encyclopedia of Mathematics and Its Foundations. Cambridge University Press (2009)
7. Marques Pereira, R.A., Ribeiro, R.A.: Aggregation with generalized mixture operators using weighting functions. Fuzzy Sets and Systems 137, 43–58 (2003)
8. Mesiar, R., Spirkova, J.: Weighted means and weighting functions. Kybernetika 42(2), 151–160 (2006)
9. Mesiar, R., Spirkova, J., Vavrikova, L.: Weighted aggregation operators based on minimization. Inform. Sci. 178, 1113–1140 (2008)
10. Ribeiro, R., Marques Pereira, R.A.: Generalized mixture operators using weighting functions. Europ. J. Oper. Research 145, 329–342 (2003)
11. Rousseeuw, P., Leroy, A.: Robust Regression and Outlier Detection. John Wiley and Sons (1987)

F-transform and Its Extension as Tool for Big Data Processing

Petra Hodáková, Irina Perfilieva, and Petr Hurtík

University of Ostrava,
Institute for Research and Applications of Fuzzy Modeling,
Centre of Excellence IT4Innovations,
30. dubna 22, 701 03 Ostrava 1,
Czech Republic
{Petra.Hodakova,Irina.Perfilieva,Petr.Hurtik}@osu.cz

Abstract. In this contribution, the extension of F-transform to F^s-transform for functions of two variables is introduced. The F^s-transform components are characterized as orthogonal projections, and some of their properties are discussed. The aim of this study is to present the possibility of using the technique of F^1-transform in big data processing and to suggest a good searching mechanism for large databases.

Keywords: F-transform, F^s-transform, big data.

1 Introduction

In this paper, we focus on the F-transform technique and its extension to the F-transform of a higher degree (F^s-transform). The aim is to describe the formal conception of this technique, demonstrate some of its properties and introduce its application to big data processing.

"Big data" refers to data that exceeds the processing capacity of conventional database systems, e.g., geographical data, medical data, social networks, and banking transactions. The data are generally of high volume, high velocity or high variety. The benefit gained from the ability to process large amounts of data is that we can create effective models and use them for databases of a custom size.

It is not feasible to handle large databases with classical analytic tools. We are not able to process every item of data in a reasonable amount of time. Therefore, we are searching for an alternative way to obtain the desired information from these data.

The F-transform is a technique that was developed as tool for a fuzzy modeling [1]. Similar to conventional integral transforms (the Fourier and Laplace transforms, for example), the F-transform performs a transformation of an original universe of functions into a universe of their "skeleton models". Each component of the resulting skeleton model is a weighted local mean of the original function over an area covered by a corresponding basic function. The F-transform is a

A. Laurent et al. (Eds.): IPMU 2014, Part III, CCIS 444, pp. 374–383, 2014.

simplified representation of the original function, and it can be used instead of the original function to make further computations easier.

Initially, the F-transform was introduced for functions of one or two variables. This method proved to be very general and powerful in many applications. The F-transform of functions of two variables shows great potential in applications involving image processing, particularly, image compression [2], image fusion [3], and edge detection [4], [5].

Generalization of the ordinary F-transform to the F-transform of a higher degree in the case of functions of one variable was introduced in [6]. An extension of the F-transform of the first degree (F^1-transform) to functions of many variables was introduced in [7]. Many interesting properties and results have been proven in those studies .

The aim of this contribution is to introduce the F-transform of a higher degree (F^s-transform) for functions of two variables and to show how this technique can be successfully applied for searching for patterns in big data records.

The paper is organized as follows: Section 2 recalls the basic tenets of the fuzzy partition and introduces a particular Hilbert space. In Section 3, the F^s-transform of functions of two variables is introduced. The inverse F^s-transform is established in Section 4. In Section 5, an illustrative application of F^1-transform to big data is presented. Finally, conclusions and comments are provided in Section 6.

2 Preliminaries

In this section, we briefly recall the basic tenets of the fuzzy partition and introduce a particular Hilbert space.

2.1 Generalized Fuzzy Partition

Let us recall the concept of a generalized fuzzy partition [8].

Definition 1. *Let $x_0, x_1, \ldots, x_n, x_{n+1} \in [a, b]$ be fixed nodes such that $a = x_0 \leq x_1 < \ldots < x_n \leq x_{n+1} = b$, $n \geq 2$. We say that the fuzzy sets $A_1, \ldots, A_n : [a, b] \to [0, 1]$ constitute a generalized fuzzy partition of $[a, b]$ if for every $k = 1, \ldots, n$ there exists $h'_k, h''_k \geq 0$ such that $h'_k + h''_k > 0$, $[x_k - h'_k, x_k + h''_k] \subseteq [a, b]$ and the following conditions are fulfilled:*

1. *(locality) – $A_k(x) > 0$ if $x \in (x_k - h'_k, x_k + h''_k)$ and $A_k(x) = 0$ if $x \in [a, b] \setminus [x_k - h'_k, x_k + h''_k]$;*
2. *(continuity) – A_k is continuous on $[x_k - h'_k, x_k + h''_k]$;*
3. *(covering) – for $x \in [a, b]$, $\sum_{k=1}^{n} A_k(x) > 0$.*

By the *locality* and *continuity*, it follows that $\int_a^b A_k(x)dx > 0$.

If the nodes $x_0, x_1, \ldots, x_n, x_{n+1}$ are equidistant, i.e., for all $k = 1, \ldots, n+1$, $x_k = x_{k-1} + h$, where $h = (b-a)/(n+1)$, $h' > h/2$ and two additional properties are satisfied,

6. $h_1' = h_1'' = h_2' = \ldots = h_{n-1}'' = h_n' = h_n'' = h'$, and $A_k(x_k - x) = A_k(x_k + x)$
   for all $x \in [0, h']$, $k = 1, \ldots, n$;
7. $A_k(x) = A_{k-1}(x - h)$ and $A_{k+1}(x) = A_k(x - h)$ for all $x \in [x_k, x_{k+1}]$,
   $k = 2, \ldots, n-1$;

then the fuzzy partition is called an (h, h')-*uniform generalized fuzzy partition*.

Remark 1. A fuzzy partition is called a Ruspini partition if

$$\sum_{k=1}^{n} A_k(x) = 1, \quad x \in [a, b].$$

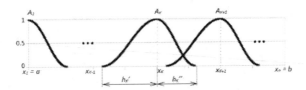

Fig. 1. Example of a generalized fuzzy partition of $[a, b]$

The concept of generalized fuzzy partition can be easily extended to the universe $D = [a, b] \times [c, d]$. We assume that $[a, b]$ is partitioned by $A_1, \ldots, A_n$ and that $[c, d]$ is partitioned by $B_1, \ldots, B_m$, according to **Definition 1**. Then, the Cartesian product $[a, b] \times [c, d]$ is partitioned by the Cartesian product of corresponding partitions where a basic function $A_k \times B_l$ is equal to the product $A_k \cdot B_l$, $k = 1, \ldots, n$, $l = 1, \ldots, m$.

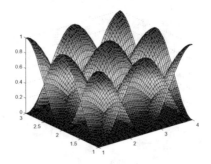

Fig. 2. Example of a cosine-shaped fuzzy partition of $[a, b] \times [c, d]$

For the remainder of this paper, we fix the notation related to fuzzy partitions of the universe $D = [a, b] \times [c, d]$.

2.2   Spaces $L_2(A_k)$, $L_2(A_k) \times L_2(B_l)$

Throughout the following subsections, we fix integers k, l from $\{1, \ldots, n\}$, $\{1, \ldots, m\}$, respectively.

Let $L_2(A_k)$ be a Hilbert space of square-integrable functions $f : [x_{k-1}, x_{k+1}] \to \mathbb{R}$ with the inner product $\langle f, g \rangle_k$ given by

$$\langle f, g \rangle_k = \int_{x_{k-1}}^{x_{k+1}} f(x) g(x) A_k(x) dx. \tag{1}$$

Analogously, the same holds for the space $L_2(B_l)$.

Then, the Hilbert space $L_2(A_k) \times L_2(B_l)$ of functions of two variables $f : [x_{k-1}, x_{k+1}] \times [y_{l-1}, y_{l+1}] \to \mathbb{R}$ is given by the Cartesian product of the respective spaces $L_2(A_k)$ and $L_2(B_l)$. The inner product is defined analogously to (1).

Remark 2. The functions $f, g \in L_2(A_k) \times L_2(B_l)$ are *orthogonal* in $L_2(A_k) \times L_2(B_l)$ if

$$\langle f, g \rangle_{kl} = 0.$$

In the sequel, by $L_2([a, b] \times [c, d])$, we denote a set of functions $f : [a, b] \times [c, d] \to \mathbb{R}$ such that for all $k = 1, \ldots, n$, $l = 1, \ldots, m$, $f|_{[x_{k-1}, x_{k+1}] \times [y_{l-1}, y_{l+1}]} \in L_2(A_k) \times L_2(B_l)$, where $f|_{[x_{k-1}, x_{k+1}] \times [y_{l-1}, y_{l+1}]}$ is the restriction of f on $[x_{k-1}, x_{k+1}] \times [y_{l-1}, y_{l+1}]$.

2.3   Subspaces $L_2^p(A_k)$, $L_2^s(A_k \times B_l)$

Let space $L_2^p(A_k)$, $p \geq 0$, $(L_2^r(B_l)$, $r \geq 0)$ be a closed linear subspace of $L_2(A_k)$ $(L_2(B_l))$ with the orthogonal basis given by polynomials

$$\{P_k^i(x)\}_{i=0,\ldots,p}, \ (\{Q_l^j(y)\}_{j=0,\ldots,r}),$$

where p (r) denotes a degree of polynomials and orthogonality is considered in the sense of (2).

Then, we can introduce space $L_2^s(A_k \times B_l)$, $s \geq 0$ as a closed linear subspace of $L_2(A_k) \times L_2(B_l)$ with the basis given by orthogonal polynomials

$$\{S_{kl}^{ij}(x, y)\}_{i=0,\ldots,p; \ j=0,\ldots,r; \ i+j \leq s} = \{P_k^i(x) \cdot Q_l^j(y)\}_{i=0,\ldots,p; \ j=0,\ldots,r; \ i+j \leq s}. \tag{2}$$

Remark 3. Let us remark that the space $L_2^s(A_k \times B_l)$ is not the same as the Cartesian product $L_2^p(A_k) \times L_2^r(B_l)$; the difference is in using fewer of the possible combinations of orthogonal basis polynomials. Therefore, $s \leq (p+1)(r+1)$. In the case where $s = (p+1)(r+1)$, the space $L_2^s(A_k \times B_l)$ coincides with $L_2^p(A_k) \times L_2^r(B_l)$.

In point of fact, s is the maximal degree of products $P_k^i(x) Q_l^j(y)$ such that $i + j \leq s$. For example, the basis of the space $L_2^1(A_k \times B_l)$ is established by the following polynomials

$$\underbrace{P_k^0(x) Q_l^0(y)}_{S_{kl}^{00}(x,y)}, \ \underbrace{P_k^1(x) Q_l^0(y)}_{S_{kl}^{10}(x,y)}, \ \underbrace{P_k^0(x) Q_l^1(y)}_{S_{kl}^{01}(x,y)}. \tag{3}$$

The following lemma characterizes the orthogonal projection of a function $f \in L_2([a,b] \times [c,d])$ or the best approximation of f in the space $L_2^s(A_k \times B_l)$.

Lemma 1. *Let $f \in L_2([a,b] \times [c,d])$ and let $L_2^s(A_k \times B_l)$ be a closed linear subspace of $L_2(A_k) \times L_2(B_l)$, as specified above. Then, the orthogonal projection F_{kl}^s of f on $L_2^s(A_k \times B_l)$, $s \geq 0$, is equal to*

$$F_{kl}^s = \sum_{0 \leq i+j \leq s} c_{kl}^{ij} S_{kl}^{ij} \qquad (4)$$

where

$$c_{kl}^{ij} = \frac{\langle f, S_{kl}^{ij} \rangle_{kl}}{\langle S_{kl}^{ij}, S_{kl}^{ij} \rangle_{kl}} = \frac{\int_{y_{l-1}}^{y_{l+1}} \int_{x_{k-1}}^{x_{k+1}} f(x,y) S_{kl}^{ij}(x,y) A_k(x) B_l(y) dx \, dy}{\int_{y_{l-1}}^{y_{l+1}} \int_{x_{k-1}}^{x_{k+1}} (S_{kl}^{ij}(x,y))^2 A_k(x) B_l(y) dx \, dy}. \qquad (5)$$

3   Direct F^s-transform

Now let $f \in L_2([a,b] \times [c,d])$ and let $L_2^s(A_k \times B_l)$, $s \geq 0$ be a space with the basis given by

$$\{S_{kl}^{ij}(x,y)\}_{i=0,\ldots,p; \; j=0,\ldots,r; \; i+j \leq s}.$$

In the following, we define the direct F^s-transform of the given function f.

Definition 2. *Let $f \in L_2([a,b] \times [c,d])$. Let F_{kl}^s, $s \geq 0$ be the orthogonal projection of $f|_{[x_{k-1},x_{k+1}] \times [y_{l-1},y_{l+1}]}$ on $L_2^s(A_k \times B_l)$, $k = 1,\ldots,n$, $l = 1,\ldots,m$. We say that $(n \times m)$ matrix $\mathbf{F_{nm}^s}[f]$ is the direct F^s-transform of f with respect to $A_1,\ldots,A_n$, $B_1,\ldots,B_m$, where*

$$\mathbf{F_{nm}^s}[f] = \begin{pmatrix} F_{11}^s & \cdots & F_{1m}^s \\ \vdots & \vdots & \vdots \\ F_{n1}^s & \cdots & F_{nm}^s \end{pmatrix}. \qquad (6)$$

F_{kl}^s, $k = 1,\ldots,n$, $l = 1,\ldots,m$ *is called the F^s-transform component.*

By **Lemma 1**, the F^s-transform components have the representation given by (4).

We will briefly recall the main properties of the F^s-transform, $s \geq 0$.

(A) The F^s-transform of f, $s \geq 0$, is an image of a linear mapping from $L_2([a,b] \times [c,d])$ to $L_2^s(A_1 \times B_1) \times \ldots \times L_2^s(A_n \times B_m)$ where, for all functions $f, g, h \in L_2([a,b] \times [c,d])$ such that $f = \alpha g + \beta h$, where $\alpha, \beta \in \mathbb{R}$, the following holds:

$$\mathbf{F_{nm}^s}[f] = \alpha \mathbf{F_{nm}^s}[g] + \beta \mathbf{F_{nm}^s}[h]. \qquad (7)$$

(B) Let $f \in L_2([a,b] \times [c,d])$. The kl-th component of the F^s-transform, $s \geq 0$, of the given function f gives the minimum to the function

$$c_{kl}^{00},\ldots,c_{kl}^{ij} = \int_a^b \int_c^d (f(x,y) - \sum_{i+j \leq s} c_{kl}^{ij} S_{kl}^{ij})^2 A_k(x) B_l(y) dx \, dy, \qquad (8)$$

Therefore, F_{kl}^s is the best approximation of f in $L_2^s(A_k \times B_l)$, $k = 1,\ldots,n, l = 1,\ldots,m$.

(C) Let f be a polynomial of degree $t \leq s$. Then, any F^s-transform component F^s_{kl}, $s \geq 0$, $k = 1, \ldots, n$, $l = 1, \ldots, m$ coincides with f on $[x_{k-1}, x_{k+1}] \times [y_{l-1}, y_{l+1}]$.

(D) Every F^s-transform component F^s_{kl}, $s \geq 1$, $k = 1, \ldots, n$, $l = 1, \ldots, m$, fulfills the following recurrent equation:

$$F^s_{kl} = F^{s-1}_{kl} + \sum_{i+j=s} c^{ij}_{kl} S^{ij}_{kl}. \tag{9}$$

The following lemma describes the relationship between the F^0-transform and F^s-transform components.

Lemma 2. *Let* $\mathbf{F^s_{nm}}[f] = (F^s_{11}, \ldots, F^s_{nm})$, *where* F^s_{kl}, $k = 1, \ldots, n$, $l = 1, \ldots, m$ $s \geq 0$ *is given by (4), be the* F^s-transform of f with respect to the given partition $\{A_k \times B_l\}$, $k = 1, \ldots, n$, $l = 1, \ldots, m$. *Then,* $(c^{00}_{11}, \ldots, c^{00}_{nm})$ *is the* F^0-transform *of* f *with respect to* $A_k \times B_l$, $k = 1, \ldots, n$, $l = 1, \ldots, m$.

The proof is analogous to that of the case of functions of one variable given in [6].

Any F^s-transform component $F^0_{kl}, F^1_{kl}, \ldots, F^s_{kl}$, $k = 1, \ldots, n$, $l = 1, \ldots, m$ $s \geq 0$, can approximate the original function $f \in L_2([a, b] \times [c, d])$ restricted to $[x_{k-1}, x_{k+1}] \times [y_{l-1}, y_{l+1}]$. The following lemma says that the quality of approximation increases with the degree of the polynomial.

Lemma 3. *Let the polynomials* F^s_{kl}, F^{s+1}_{kl}, $k = 1, \ldots, n$, $l = 1, \ldots, m$ $s \geq 0$, *be the orthogonal projections of* $f|_{[x_{k-1}, x_{k+1}] \times [y_{l-1}, y_{l+1}]}$ *on the subspaces* $L^s_2(A_k \times B_l)$ *and* $L^{s+1}_2(A_k \times B_l)$, *respectively. Then,*

$$\| f|_{[x_{k-1}, x_{k+1}] \times [y_{l-1}, y_{l+1}]} - F^{s+1}_{kl} \|_{kl} \leq \| f|_{[x_{k-1}, x_{k+1}] \times [y_{l-1}, y_{l+1}]} - F^s_{kl} \|_{kl}. \tag{10}$$

The proof is analogous to that of the case of functions of one variable given in [6].

3.1   Direct F^1-transform

In this section, we assume that $s = 1$ and give more details to the F^1-transform and its components.

The F^1-transform components F^1_{kl}, $k = 1, \ldots, n, l = 1, \ldots, m$, are in the form of linear polynomials

$$F^1_{kl} = c^{00}_{kl} + c^{10}_{kl}(x - x_k) + c^{01}_{kl}(y - y_l), \tag{11}$$

where the coefficients are given by

$$c^{00}_{kl} = \frac{\int_{y_{l-1}}^{y_{l+1}} \int_{x_{k-1}}^{x_{k+1}} f(x, y) A_k(x) B_l(y) dx \, dy}{(\int_{x_{k-1}}^{x_{k+1}} A_k(x) dx)(\int_{y_{l-1}}^{y_{l+1}} B_l(y) dy)}, \tag{12}$$

$$c_{kl}^{10} = \frac{\int_{y_{l-1}}^{y_{l+1}} \int_{x_{k-1}}^{x_{k+1}} f(x,y)(x-x_k)A_k(x)B_l(y)dx\,dy}{(\int_{x_{k-1}}^{x_{k+1}} (x-x_k)^2 A_k(x)dx)(\int_{y_{l-1}}^{y_{l+1}} B_l(y)dy)}, \tag{13}$$

$$c_{kl}^{01} = \frac{\int_{y_{l-1}}^{y_{l+1}} \int_{x_{k-1}}^{x_{k+1}} f(x,y)(y-y_l)A_k(x)B_l(y)dx\,dy}{(\int_{x_{k-1}}^{x_{k+1}} A_k(x)dx)(\int_{y_{l-1}}^{y_{l+1}} (y-y_l)^2 B_l(y)dy)}. \tag{14}$$

Lemma 4. *Let* $f \in L_2([a,b] \times [c,d])$ *and* $\{A_k \times B_l\}$, $k = 1, \ldots, n$, $l = 1, \ldots, m$ *be an* (h, h')-*uniform generalized fuzzy partition of* $[a,b] \times [c,d]$. *Moreover, let functions* f, A_k, B_l *be four times continuously differentiable on* $[a,b] \times [c,d]$. *Then, for every* k, l, *the following holds:*

$$c_{kl}^{00} = f(x_k, y_l) + O(h^2), \quad c_{kl}^{10} = \frac{\partial f}{\partial x}(x_k, y_l) + O(h), \quad c_{kl}^{01} = \frac{\partial f}{\partial y}(x_k, y_l) + O(h).$$

The proof can be found in [7].

4   Inverse F^s-transform

The inverse F^s-transform of the original function f is defined as a linear combination of basic functions and F^s-transform components.

Definition 3. *Let* $\mathbf{F}_{nm}^s[f] = (F_{kl}^s)$, $k = 1, \ldots, n$, $l = 1, \ldots, m$ *be the* F^s-*transform of given function* $f \in L_2([a,b] \times [c,d])$. *We say that the function* $\hat{f}^s : [a,b] \times [c,d] \to \mathbb{R}$ *represented by*

$$\hat{f}^s(x,y) = \frac{\sum_{k=1}^{n} \sum_{l=1}^{m} F_{kl}^s A_k(x)B_l(y)}{\sum_{k=1}^{n} \sum_{l=1}^{m} A_k(x)B_l(y)}, \quad x \in [a,b], \; y \in [c,d], \tag{15}$$

is the inverse F^s-*transform of the function* f.

Remark 4. From **Definition 3** and property (9) of the F^s-transform components, the recurrent formula below easily follows:

$$\hat{f}^s(x,y) = \hat{f}^{s-1}(x,y) + \frac{\sum_{k=1}^{n} \sum_{l=1}^{m} \sum_{i+j=s} c_{kl}^{ij} S_{kl}^{ij} A_k(x)B_l(y)}{\sum_{k=1}^{n} \sum_{l=1}^{m} A_k(x)B_l(y)}. \tag{16}$$

In the following theorem, we show that the inverse F^s-transform approximates an original function, and we estimate the quality of the approximation. Based on **Lemma 3**, the quality of the approximation increases with the increase of s.

Theorem 1. *Let* $\{A_k \times B_l\}$, $k = 1, \ldots, n$, $l = 1, \ldots, m$ *be an* h-*uniform fuzzy partition (with the Ruspini condition) of* $[a,b] \times [c,d]$, *and let* $\hat{f}^s$ *be the inverse* F^s-*transform of* f *with respect to the given partition. Moreover, let functions* f, A_k, *and* B_l *be four times continuously differentiable on* $[a,b] \times [c,d]$. *Then, for all* $(x,y) \in [a,b] \times [c,d]$, *the following estimation holds true:*

$$\int_a^b \int_c^d |f(x,y) - \hat{f}^s(x,y)|dxdy \le O(h). \tag{17}$$

5 Illustrative Application

In this section, we present an illustrative application of F^1-transform to big data. The application shows how an effective searching mechanism in large databases can be constructed on the basis of F^1-transform. The detailed characterization is as follows.

Let o be a sound (voice) signal represented by the function $o : T \rightarrow V$, where $T = \{0, ..., t_{max}\}$ is a discrete set of regular time moments and V is a certain range.

Assume that the signal o is an example of a very large record (big data), i.e., $t_{max} = 220 \times 60$ sec, sampled at every second. We are given a small sound pattern o^P that is represented by $o^P : T^P \rightarrow V$, where T^P is a set of time moments measured in seconds such that $T^P = \{1, ..., 6\}$. The goal is to find occurrences of the pattern o^P in the given sound signal o in a reasonable amount of time. See the illustrative example on Fig. 3.

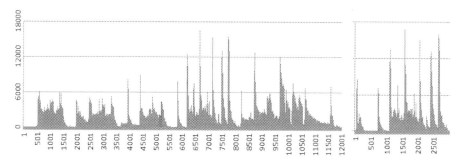

Fig. 3. An extraction of a big sound signal o (*Left*) and a small sound pattern o^P (*Right*). The red mark indicates the first occurrence of the recognized pattern in the given signal.

The naive approach is to make a sliding comparison between the values of the pattern o^P and the values in o. For this comparison, the following measure of closeness can be used:

$$\sum_{j=0}^{|T^P|} |o(t+j) - o^P(j)|, \ t \in T. \tag{18}$$

This method is very computationally complex and time consuming.

Our approach is as follows. We apply the direct F^1-transform to the record of the sound signal o and to the pattern o^P, and we obtain their vectors of components $\mathbf{F}_n^1[o]$, $\mathbf{F}_m^1[o^P]$, respectively. The dimensions of $\mathbf{F}_n^1[o]$, $\mathbf{F}_m^1[o^P]$ are significantly less than the dimensions of the original data o, o^P. Instead of comparing all values of o^P and o, we suggest to compare the components of F^1-transform $\mathbf{F}_n^1[o]$ and $\mathbf{F}_m^1[o^P]$.

Finally, the algorithm is realized by the following steps:

S 1: Read data o and compute $\mathbf{F_n^1}[o] = (F_1^1, \ldots, F_n^1)_o$ w.r.t. the (h, h')-uniform generalized fuzzy partition. This step is realized just once and can be stored independently.

S 2: Read data o^P and compute $\mathbf{F_m^1}[o^P] = (F_1^1, \ldots, F_m^1)_{o^P}$ w.r.t. the same fuzzy partition.

S 3: Compute the measure of closeness (18) between components of $\mathbf{F_n^1}[o]$ and $\mathbf{F_m^1}[o^P]$. The pattern is recognized if the closeness is less than a predefined threshold of tolerance.

Experiment

For our experiment, we took a record of a sound signal, o with $t_{max} = 220 \times 60$ sec, and a record of a small sound pattern, o^P with $t_{max} \approx 6.4$ sec. The records were taken unprofessionally. In fact, the sounds were both part of a piece of music recorded by a microphone integrated in a notebook computer. Therefore, the records are full of noise (because of the microphone, surroundings, etc.) and they may, for example, differ in volume of the sounds.

We applied the naive approach to these records and obtained the following results:

- $5.463 \cdot 10^{12}$ computations,
- run time ≈ 11 h.

Then, we applied the approach based on F^1-transform and obtained the following results:

- $1.081 \cdot 10^7$ computations,
- run time ≈ 0.008 s.

In this experiment, we used a fuzzy partition with triangular shaped fuzzy sets and tested the fuzzy partition for $h = 1 * 10^n$, $n = 2, 4, 6, 8$. The optimal length for the experiment demonstrated above is for $n = 4$. For the larger $n = 6, 8$, multiple false occurrences of the searched pattern were found. A possible solution is to make the algorithm hierarchical, i.e., use the larger h at the beginning and then use the smaller h for the detected results. This approach can achieve extremely fast recognition by making the algorithm sequentially more accurate.

Remark 5. The task discussed here is usually solved as speech recognition, where words are individually separated and then each of them is compared with words in a database. The most similar words are then found as results. The speech recognition is mostly solved by neural networks. Comparing different speech recognition approaches will be our future task.

From a different point of view, this task can be discussed as an example of reducing the dimension of big data. The sub-sampling algorithm is often used in this task. We tried to apply this algorithm to the task above, and, for comparison, we used the same reduction of dimensions as in the F^1-transform algorithm. The sub-sampling algorithm failed; it did not find the searched pattern correctly.

The example demonstrates the effectiveness of the technique of F^1-transform in big data processing. A good searching mechanism for large databases can be developed on the basis of this technique. Similar applications can be developed in the area of image processing, where searching of patterns is a very popular problem. The F^s-transform, $s > 1$, technique can be efficiently applied as well. This will be the focus of our future research.

6 Conclusion

In this paper, we presented the technique of F-transform and our vision of its application in big data processing. We discussed the extension to the F^s-transform, $s \geq 1$, for functions of two variables. We characterized the components of the F^s-transform as orthogonal projections and demonstrated some of their properties. Finally, we introduced an illustrative application of using the F^1-transform in searching for a pattern in a large record of sound signals.

Acknowledgments. The research was supported by the European Regional Development Fund in the IT4Innovations Centre of Excellence project (CZ.1.05/1.1.00/02.0070) and SGS18/PRF/2014 (Research of the F-transform method and applications in image processing based on soft-computing).

References

1. Perfilieva, I.: Fuzzy transforms: Theory and applications. Fuzzy Sets and Systems 157, 993–1023 (2006)
2. Di Martino, F., Loia, V., Perfilieva, I., Sessa, S.: An image coding/decoding method based on direct and inverse fuzzy transforms. International Journal of Appr. Reasoning 48, 110–131 (2008)
3. Vajgl, M., Perfilieva, I., Hodáková, P.: Advanced F-transform-based image fusion. Advances in Fuzzy Systems (2012)
4. Daňková, M., Hodáková, P., Perfilieva, I., Vajgl, M.: Edge detection using F-transform. In: Proc. of the ISDA 2011, Spain, pp. 672–677 (2011)
5. Perfilieva, I., Hodáková, P., Hurtík, P.: F^1-transform edge detector inspired by canny's algorithm. In: Greco, S., Bouchon-Meunier, B., Coletti, G., Fedrizzi, M., Matarazzo, B., Yager, R.R. (eds.) IPMU 2012, Part I. CCIS, vol. 297, pp. 230–239. Springer, Heidelberg (2012)
6. Perfilieva, I., Daňková, M., Bede, B.: Towards a higher degree F-transform. Fuzzy Sets and Systems 180, 3–19 (2011)
7. Perfilieva, I., Hodáková, P.: F^1-transform of functions of two variables. In: Proc. EUSFLAT 2013, Advances in Intelligent Systems Research, Milano, Italy, pp. 547–553 (2013)
8. Stefanini, L.: F-transform with parametric generalized fuzzy partitions. Fuzzy Sets and Systems 180, 98–120 (2011)

Fuzzy Queries over NoSQL Graph Databases: Perspectives for Extending the Cypher Language

Arnaud Castelltort and Anne Laurent

LIRMM, Montpellier, France
{castelltort,laurent}@lirmm.fr

Abstract. When querying databases, users often wish to express vague concepts, as for instance asking for the *cheap* hotels. This has been extensively studied in the case of relational databases. In this paper, we propose to study how such useful techniques can be adapted to NoSQL graph databases where the role of fuzziness is crucial. Such databases are indeed among the fastest-growing models for dealing with big data, especially when dealing with network data (e.g., social networks). We consider the Cypher declarative query language proposed for Neo4j which is the current leader on this market, and we present how to express fuzzy queries.

Keywords: Fuzzy Queries, NoSQL Graph Databases, Neo4j, Cypher, Cypherf.

1 Introduction

Graph databases have attracted much attention in the last years, especially because of the collaborative concepts of the Web 2.0 (social and media networks etc.) and the arriving Web 3.0 concepts.

Specific databases have been designed to handle such data relying on big dense network structures, especially within the NoSQL world. These databases are built to remain robust against huge volumes of data, against their heterogeneous nature and the high speed of the treatments applied to them, thus coping with the so-called Big Data paradigm.

They are currently gaining more and more interest and are applied in many real world applications, demonstrating their power compared to other approaches. NoSQL graph databases are known to offer great scalability [1].

Among these NoSQL graph databases, Neo4j appears to be one of the most mature and deployed [2]. In such databases, as for graphs, nodes and relationships between nodes are considered. Neo4j includes nodes and relationships labeling with the so-called types. Moreover, properties are attached to nodes and relationships. These properties are managed in Neo4j using the *key:value* paradigm.

Fig. 1 shows an example of hotels and customers database. The database contains hotels located in some cities and visited by some customers. Links are represented by the :LOCATED and :VISIT relationships. The hotels and people and relationships are described by properties: id, price, size (number of rooms) for hotels; id, name, age for people. One specificity is that relationships in Neo4j are provided with types (e.g., type "hotel" or "people" in the example) and can also have properties as for nodes. This

A. Laurent et al. (Eds.): IPMU 2014, Part III, CCIS 444, pp. 384–395, 2014.

allows to represent in a very intuitive and efficient manner many data from the real world. For instance, :LOCATED has property *distance*, standing for the distance to city center.

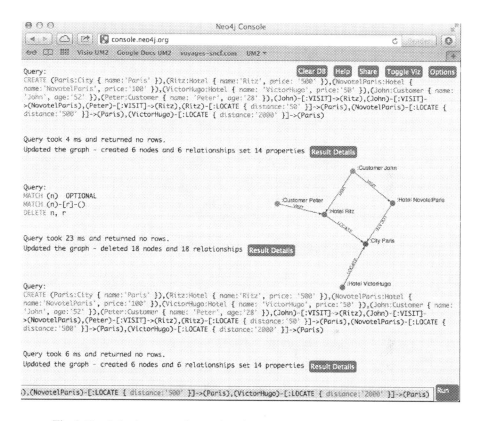

Fig. 1. Neo4j database console user interface: Example for Hotels and Customers

All NoSQL graph databases require the developers and users to use graph concepts to query data. As for any other repository, when querying such NoSQL graph databases, users either require specific focused knowledge (e.g., retrieving Peter's friends) or ask for trend detection (e.g., detecting trends and behaviours within social networks).

Queries are called *traversals*. A graph traversal refers to visiting elements, *i.e.* nodes and relations. There are three main ways to traverse a graph:

- programmaticaly, by the use of an API that helps developers to operate on the graph;
- by functional traversal, a traversal based on a sequence of functions applied to a graph;
- by declarative traversal, a way to explicit what we want to do and not how we want to do it. Then, the database engine defines the best way to achieve the goal.

In this paper, we focus on declarative queries over a NoSQL graph database. The Neo4j language is called Cypher.

For instance on Fig. 1, one query is displayed to return the customers who have visited the "Ritz" hotel.They are both displayed in the list and circled in red in the graph.

We consider in this paper the manipulating queries in READ mode.

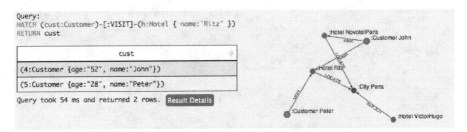

Fig. 2. Displaying the Result of a Cypher Query

However, none of the query languages embeds a way for dealing with flexible queries, for instance to get *cheap* hotels or *popular* ones, where *cheap* and *popular* are fuzzy sets.

This need has nevertheless been intensively studied when dealing with other database paradigms, especially with relational databases.

In this paper, we thus focus on the declarative way of querying the Neo4j system with the Cypher query language and we extend it for dealing with vague queries.

The rest of the paper is organised as follows. Section 2 reports existing work from the literature regarding fuzzy queries and presents the Cypher language. Section 3 introduces the extension of the Cypher language to Cypherf and Section 4 shows how such an extension can be implemented. Section 5 concludes the paper and provides some ideas for future work.

2   Related Work

2.1   Neo4j Cypher Language

Queries in Cypher have the following syntax[1]:

```
[START]
[MATCH]
[OPTIONAL MATCH WHERE]
[WITH [ORDER BY] [SKIP] [LIMIT]]
RETURN [ORDER BY] [SKIP] [LIMIT]
```

---

[1] http://docs.neo4j.org/refcard/2.0/
  http://docs.neo4j.org/chunked/milestone/cypher-query-lang.html

As shown above, Cypher is comprised of several distinct clauses:

- START: Starting points in the graph, obtained via index lookups or by element IDs.
- MATCH: The graph pattern to match, bound to the starting points in START.
- WHERE: Filtering criteria.
- RETURN: What to return.
- CREATE: Creates nodes and relationships.
- DELETE: Removes nodes, relationships and properties.
- SET: Set values to properties.
- FOREACH: Performs updating actions once per element in a list.
- WITH: Divides a query into multiple, distinct parts.

2.2 Fuzzy Queries

Many works have been proposed for dealing with fuzzy data and queries. All cannot be reported here. [3] proposes a survey of these proposals.

[4, 5] consider querying regular databases by both extending the SQL language and studying aggregating subresults. The FSQL/SQLf and FQL languages have been proposed to extend queries over relational databases in order to incorporate fuzzy descriptions of the information being searched for.

Some works have been implemented as fuzzy database engines and systems have incorporated such fuzzy querying features [6, 7].

In such systems, fuzziness in the queries is basically associated to fuzzy labels, fuzzy comparators (e.g., *fuzzy greater than*) and aggregation over clauses. Thresholds can be defined for the expected fulfillment of fuzzy clauses.

For instance, on a crisp database describing hotels, users can ask for *cheap* hotels that are *close_to_city_center*, *cheap* and *close_to_city_center* being fuzzy labels described by fuzzy sets and their membership functions respectively defined on the universe of prices and distance to the city center.

Many works have been proposed to investigate how such fuzzy clauses can be defined by users and computed by the database engine, especially when several clauses must be merged (e.g., *cheap* **AND** *close_to_city_center*).

Such aggregation can consider preferences, for instance for queries where price is prefered to distance to city center using weighted t-norms.

Thresholds can be added for working with α−cuts, such as searching for hotels where the degree *cheap* is greater than 0.7.

As we consider graph data, the works on fuzzy ontology querying are very close and relevant for us [8, 9].

[8] proposes the f-SPARQL query language that supports fuzzy querying over ontologies by extending the SPARQL language. This extension is based on threshold query (e.g., asking for people who are *tall* at a degree greater than 0.7) or general fuzzy queries based on semantic functions.

It should be noted that many works have dealt with fuzzy databases for representing and storing imperfect information in databases: fuzzy ER models, fuzzy object databases, fuzzy relational databases, fuzzy ontologies-OWL [10], etc. Fuzziness can then impact many levels, from metadata (attributes) to data (tuples), and cover many semantics (uncertainty, imprecision, inconsistency, etc.) as recalled in [3]. These works are not reported here as we consider fuzzy queries over crisp data.

3 Fuzzy Queries over NoSQL Graph databases: Towards the Cypherf Language

In this paper, we address fuzzy READ queries over regular NoSQL Neo4j graph databases. We claim that fuzziness can be handled at the following three levels:

- over properties,
- over nodes,
- over relationships.

3.1 Cypherf over Properties

Dealing with fuzzy queries over properties is similar to the queries from the literature on relational databases and ontologies. Such queries are defined by using linguistic labels (fuzzy sets) and/or fuzzy comparators.

Such fuzzy queries impact the $START$, $MATCH$, $WHERE$ and $RETURN$ clauses from Cypher.

In the $WHERE$ clause, it is then possible to search for *cheap* hotels in some databases, or for hotels located $close\_to\_city\_center$[2]. Note that these queries are different as the properties being addressed are respectively linked to a node and a relationship.

Listing 1.1. Cheap Hotels

```
1  MATCH (h:Hotel)
2  WHERE CHEAP(price) > 0
3  RETURN h
4  ORDER BY CHEAP(h) DESC
```

Listing 1.2. Hotels Close to City Center

```
1  MATCH (c:City)<-[:LOCATED]-(h:Hotel)
2  WHERE CLOSE(c,h) > 0
3  RETURN h
4  ORDER BY CLOSE(c,h) DESC
```

In the $START$ clause, it is possible to define which nodes and relationships to start from by using fuzzy labels, as for instance:

Listing 1.3. Starting from Cheap Hotels

```
1  START h:Hotel(CHEAP(price) > 0)
2  RETURN h
3  ORDER BY CHEAP(h) DESC
```

---

[2] For the sake of simplicity, the fuzzy labels and membership functions are hereafter denoted by the same words.

Listing 1.4. Starting from location links close to city center

```
1   START l=relationship:LOCATED(CLOSE(distance)>0)
2   MATCH (h:Hotel)-[:LOCATED]->(c:City)
3   RETURN h
4   ORDER BY CLOSE(h,c) DESC
```

In the $MATCH$ clause, integrating fuzzy labels is also possible:

Listing 1.5. Matching Hotels Close to City Center

```
1   MATCH (h:Hotel)-[:LOCATED {CLOSE(distance)>0}]->(c:City)
2   RETURN h
3   ORDER BY CLOSE(h,c) DESC
```

In the $RETURN$ clause, no selection will be operated, but fuzzy labels can be added in order to show the users the degree to which some values match fuzzy sets, as for instance:

Listing 1.6. Fuzziness in the Return Clause

```
1   MATCH (h:Hotel)-[:LOCATED]->(c:City)
2   RETURN h, CLOSE(h,c) AS 'ClosenessToCityCenter'
3   ORDER BY ClosenessToCityCenter DESC
```

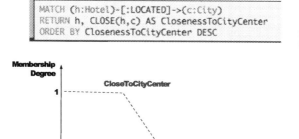

Fig. 3. Fuzzy Cypher Queries: an Example

When considering fuzzy queries over relational databases, the results are listed and can be ranked according to some degrees. When considering graph data, graphical representations are of great interest for the user comprehension and interaction on the data. For instance, Fig 2 shows how a result containing two items (the two customers who went to Ritz hotel) is displayed in the Cypher console, demonstrating the interest of the graphic display.

It would thus be interesting to investigate how fuzzy queries over graph may be displayed, showing the graduality of membership of the objects to the result. For this purpose, we propose to use the work from the literature on fuzzy graph representation and distored projection as done in anamorphic maps [11].

3.2 Cypherf over Nodes

Dealing with fuzzy queries over nodes allows to retrieve similar nodes. It is set at a higher level from queries over properties although it may use the above-defined queries.

For instance, it is possible to retrieve similar hotels:

Listing 1.7. Getting Similar Hotel Nodes

```
1  MATCH (h1:Hotel),(h2:Hotel)
2  WITH h1 AS hot1, h2 AS hot2, SimilarTo(hot1,hot2) AS sim
3  WHERE sim > 0.7
4  RETURN hot1,hot2,sim
```

In this framework, the link between nodes is based on the definition of measures between the descriptions. Such measures integrate aggregators to deal with the several properties they embed. Similarity measures may for instance be used and hotels may all the more be considered as their prices and size are similar.

It should be noted that such link could be materialized by relationships, either for performance concerns, or because it was designed this way. In the latter case, such query amounts to a query as defined above.

3.3 Cypherf over Relationships

As for nodes, such queries may be based on properties. But it can also be based on the graph structure in order to better exploit and benefit from it.

In Cypher, the structure of the pattern being searched is mostly defined in the $MATCH$ clause.

The first attempt to extend pattern matching to fuzzy pattern matching is to consider chains and depth matching. Chains are defined in Cypher in the $MATCH$ clause with consecutive links between objects. If a node a is linked to an object b at depth 2, the pattern is writen as $(a) - [*2] - > (b)$. If a link between a and b without regarding the depth in-between is searched, then it is writen $(a) - () - > (b)$. The mechanism also applies for searching objects linked trough a range of nodes (e.g., between 3 and 5): $(a) - [*3..5] - > (b)$.

We propose here to introduce fuzzy descriptors to define extended patterns where the depth is imprecisely described. It will then for instance be possible to search for customers linked through *almost 3* hops. The syntax $**$ is proposed to indicate a fuzzy linker.

Listing 1.8. Fuzzy Patterns

```
1  MATCH (c1:customer)-[:KNOWS**almost3]->(c2:customer)
2  RETURN c1,c2
```

It is related to fuzzy tree and graph mining [12] where some patterns emerge from several graphs even they do not occur exactly the same way everywhere regarding the structure.

Another possibility is not to consider chains but patterns where several links from and to nodes.

In our running example, *popular* hotels may for instance be chosen when they are chosen by many people. This is similar as the way famous people are detected if they are followed by many people on social networks.

In this example, a hotel is popular if a large proportion of customers visited it.

In Cypher, such queries are defined by using aggregators. For instance, the following query retrieves hotels visited by at least 2 customers:

Listing 1.9. Aggregation

```
1  MATCH (c:Customer)-[:VISIT]->(h:Hotel)
2  WITH c AS cust, count(*) AS cpt
3  WHERE cpt>1
4  RETURN cust
```

Such crisp queries can be extended to consider fuzziness:

Listing 1.10. Aggregation

```
1  MATCH (c:Customer)-[:VISIT]->(h:Hotel)
2  WITH c AS cust, count(*) AS cpt
3  WHERE POPULAR(cpt) > 0
4  RETURN cust
```

All fuzzy clauses described in this section can be combined. The question then risen is to implement them in the existing Neo4j engine.

4 Implementation Challenges

4.1 Architecture

There are several ways to implement fuzzy Cypher queries:

1. Creating an overlay language on top of the Cypher language that will produce as ouput Cypher well formatted queries to do fuzzy work;
2. Extending the Cypher queries and using the existing low level API behind;
3. Extending the low level API with optimized functions, offering the possibility only to developpers to use it;
4. Combining the last two possibilities: using an extended cypher query language over an enhanced low level API.

Every possibility is debated in this section. The reader will find at the end of this section a summary of the debates.

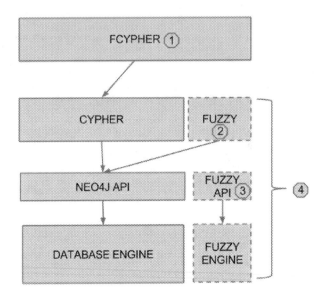

Fig. 4. Implementation Ways

4.2 Creating an Overlay Language

Concept. The concept is to create a high-level fuzzy DSL language that will be used to generate Cypher well-formed queries. The generated Cypher queries will be executed by the existing Neo4j engine.

A grammar must be defined for this external DSL which can rely on the existing Cypher syntax and only enhance it with new fuzzy features. The output of the generation process is pure Cypher code. In this scenario, Cypher is used as a low level language to achieve fuzzy queries.

Discussion. This solution is a cheap and non intrusive solution but has several huge drawbacks:

- Features missing, indeed every fuzzy query shown in Section 3 cannot be expressed by the current cypher language (e.g., listing 1.4);
- Performance issue, Cypher is not designed for fuzzy queries neither for being used as an algorithmic language. All the fuzzy queries will produce Cypher query codes that are not optimized for fuzzy tasks;
- Lack of user-friendliness, Each query cannot be executed directly against the Neo4j environnement, it needs a two-step process: (i) write a fuzzy query, then compile it to get the cypher query; (ii) use the cypher generated queries on the Neo4j database

4.3 Extending the Cypher Queries

Concept. The idea is to extend the Cypher language to add new features. Cypher offers various types of functions: scalar functions, collection functions, predicate functions, mathematical functions, etc. To enhance this language with fuzzy features, we propose to add a new type of functions: fuzzy functions. Fuzzy functions are used in the same way as other functions of Cypher (or SQL) as shown in section 3.

Cypher is an external DSL. Therefore, somewhere it needs to be parsed. The query correctness must be checked and then it should be executed. In the Cypher case, retrieving the results we asked for.

In order to write Cypher, the Neo4j's team had defined its grammar, which gives the guidelines of how the language is supposed to be structured and what is and isnt valid. In order to express this definition, we can use some variation of EBNF syntax [13], which provides a clear way to expose the language definition. To parse this syntax, Cypher uses Scala language Parser Combinator library.

Then, to extend the Cypher engine, the Cypher grammar must be extended regarding the current grammar parser. Once the cypher query is parsed, the code has to be bound on the current programmatic API to achieve the desired result.

Discussion. This work needs a deeper comprehension of the Neo4j engine and more skills on Java/Scala programming language (used to write the Neo4j engine and API) than the previous solutions. The main advantage of this is to offer an easy and user-friendly way to use the fuzzy feature. The disavantages of this solution are:

– Performance issue. This solution should have better performance than the previous one but it stills built on the current Neo4j engine API that is not optimized for fuzzy queries (e.g., degree computing);
– Cost of maintenance. Until Neo4j accepts to inlude this contribution to the Neo4j project, it will be needed to upgrade each new version of Neo4j with these enhancements. If this feature is built in a plugin, it will be necessary to check that the API has not been broken by the new version (if so an upgrade of the fuzzy plugin will be required).

4.4   Extending Low Level API

Concept The scenario is to enhance the core database engine with a framework to handle efficiently the fuzzy queries and to extend the programming API built on it to provide to developpers access to this new functionnality.

Discussion. This solution offers a high performance improvment but needs high Neo4j skills, possibly high maintenance costs, a poor user friendly experience (only developpers can use it) and a costly development process.

4.5   Extending Cypher over an Enhanced Low Level API

Concept. The last and not the least possibility is to combine the solutions from Sections 4.3 and 4.4: adding to the database engine the feactures to handle the fuzzy queries, extending the API and extending the Cypher language.

Discussion. This solution is user-friendly, provides optimized performance but has a heavy development cost (skills, tasks, etc.) and a high cost of maintenance.

4.6   Summary and Prototype

The first solution is a non intrusive solution with limited perspectives. It is more a hack than a real long termes solution. The best, but most costly, solution still the last one: extend cypher query language and build a low level API framework to extend the Neo4j database engine to support such kind of queries.

A prototype based on the extension of cypher over an enhanced API is under developement, fuzzy queries can be run, as shown in Fig. 5.

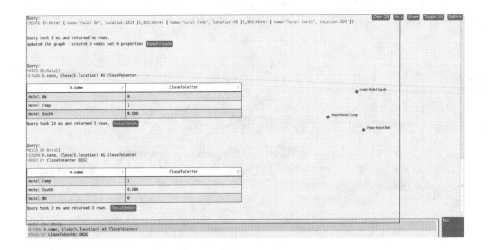

Fig. 5. Protoype Developed

5   Conclusion

In this paper, we propose an extension of the declarative NoSQL Neo4j graph database query language (Cypher). This language is applied on large graph data which represent one of the challenges for dealing with big data when considering social networks for instance. A protoype has been developed and is currently being enhanced.

As we consider the existing Neo4j system which is efficient, performance is guaranteed. The main property of NoSQL graph databases, i.e. the optimized $O(1)$ low complexity for retrieving nodes connected to a given one, and the efficient index structures ensure that performances are optimized.

Future works include the extension of our work to the many concepts possible with fuzziness (e.g., handling fuzzy modifiers), the study of fuzzy queries over historical NoSQL graph databases as introduced in [14] and the study of definition fuzzy structures: Fuzzy Cypher queries for Data Definition or in WRITE mode (e.g., inserting imperfect data). The implementation of the full solution relying on our work, currently in progress, will be completed by these important extensions.

References

1. Rodriguez, M.A., Neubauer, P.: The graph traversal pattern. CoRR abs/1004.1001 (2010)
2. Board, T.T.A.: Technology radar (May 2013),
   http://thoughtworks.fileburst.com/assets/
   technology-radar-may-2013.pdf
3. Meng, X., Ma, Z., Zhu, X.: A knowledge-based fuzzy query and results ranking approach for relational databases. Journal of Computational Information Systems 6(6) (2010)
4. Bosc, P., Pivert, O.: Sqlf: a relational database language for fuzzy querying. IEEE Transactions on Fuzzy Systems 3(1), 1–17 (1995)
5. Takahashi, Y.: A fuzzy query language for relational databases. IEEE Transactions on Systems, Man, and Cybernetics 21(6), 1576–1579 (1991)
6. Zadrożny, S., Kacprzyk, J.: Implementing fuzzy querying via the internet/WWW: Java applets, activeX controls and cookies. In: Andreasen, T., Christiansen, H., Larsen, H.L. (eds.) FQAS 1998. LNCS (LNAI), vol. 1495, pp. 382–392. Springer, Heidelberg (1998)
7. Galindo, J., Medina, J.M., Pons, O., Cubero, J.C.: A server for fuzzy SQL queries. In: Andreasen, T., Christiansen, H., Larsen, H.L. (eds.) FQAS 1998. LNCS (LNAI), vol. 1495, pp. 164–174. Springer, Heidelberg (1998)
8. Pan, J.Z., Stamou, G.B., Stoilos, G., Taylor, S., Thomas, E.: Scalable querying services over fuzzy ontologies. In: Huai, J., Chen, R., Hon, H.W., Liu, Y., Ma, W.Y., Tomkins, A., Zhang, X. (eds.) WWW, pp. 575–584. ACM (2008)
9. Cheng, J., Ma, Z.M., Yan, L.: f-SPARQL: A flexible extension of SPARQL. In: Bringas, P.G., Hameurlain, A., Quirchmayr, G. (eds.) DEXA 2010, Part I. LNCS, vol. 6261, pp. 487–494. Springer, Heidelberg (2010)
10. Stoilos, G., Stamou, G.B., Tzouvaras, V., Pan, J.Z., Horrocks, I.: Fuzzy owl: Uncertainty and the semantic web. In: Grau, B.C., Horrocks, I., Parsia, B., Patel-Schneider, P.F. (eds.) OWLED. CEUR Workshop Proceedings, vol. 188. CEUR-WS.org (2005)
11. Griffin, T.: Cartographic transformations of the thematic map base. Cartography 11(3), 163–174 (1980)
12. López, F.D.R., Laurent, A., Poncelet, P., Teisseire, M.: Ftmnodes: Fuzzy tree mining based on partial inclusion. Fuzzy Sets and Systems 160(15), 2224–2240 (2009)
13. Pattis, R.: Teaching ebnf first in cs 1. In: Beck, R., Goelman, D. (eds.) SIGCSE, pp. 300–303. ACM (1994)
14. Castelltort, A., Laurent, A.: Representing history in graph-oriented nosql databases: A versioning system. In: Proc. of the Int. Conf. on Digital Information Management (2013)

Learning Categories
from Linked Open Data

Jesse Xi Chen and Marek Z. Reformat

Electrical and Computer Engineering,
University of Alberta, T6G 2V4, Canada
{jesse.chen,marek.reformat}@ualberta.ca

Abstract. The growing presence of Resource Description Framework
(RDF) as a data representation format on the web brings opportunity
to develop new approaches to data analysis. One of important tasks is
learning categories of data. Although RDF-based data is equipped with
properties indicating its type and subject, building categories based on
similarity of entities contained in the data provides a number of benefits.
It mimics an experience-based learning process, leads to construction
of an extensional-based hierarchy of categories, and allows to determine
degrees of membership of entities to the identified categories. Such a
process is addressed in the paper.

Keywords: RDF triples, similarity, clustering, fuzziness.

1 Introduction

Any software agent that wants to make decisions as well as reason about things
related to its environment should be able to collect data and build a model of
things it experienced. In order to do this, it requires abilities to process data and
put it in some kind of a structure.

If we think about the web not only as a repository of data but also as an
environment where agents reside, it becomes important to represent data and
information in a way that is suitable for processing. However, textual format is
the most common on the web. Multiple processing tools and methods – from
language processing to expert systems – are required to analyze textual documents.

One of the most important contributions of the Semantic Web concept [1] is
the Resource Description Framework (RDF) [13]. This framework is a recommended format for representing data. Its fundamental idea is to represent each
piece of data as a triple: <subject-property-object>, where the subject
is an entity being described, object is an entity describing the subject, and
property is a "connection" between subject and object. In other words, the
property-object is a description of the subject. For example, *London is city*
is a triple with *London* as its subject, *is* the property, and *city* its object.
In general a subject of one triple can be an object of another triple, and
vice versa. This results in a network of interconnected triples, and constitutes

A. Laurent et al. (Eds.): IPMU 2014, Part III, CCIS 444, pp. 396–405, 2014.

an environment suitable for constructing new processes for analyzing data, and converting it to useful and more structured information.

With a growing number of RDF triples on the web – more than 62 billions right now (http://stats.lod2.eu) – processing data represented as RDF triples is gaining attention. There are multiple works focusing on RDF data storage and querying strategies using a specialized query language SPARQL [10] [9]. More and more publications look at handling RDF triples directly.

The work described in [4] looks at the efficient processing of information in RDF data sources for detecting communities. A process of identifying relevant datasets for a specific task or topic is addressed in [8]. A hierarchical clustering algorithm is used for inferring structural summaries to support querying LD sources in [2]. Linked data classification is a subject of the work presented in [3], while an approach for formalizing the hierarchy of concepts from linked data is described in [12].

In this paper we focus on building categories as a learning process in which agents model experienced environment via hierarchy of categories. It is a data-driven process that depends on a set of collected data. We explore an idea of treating RDF triples as feature-based descriptions of entities. We describe a method for determining similarity between them. Further, a methodology for building and updating categories of entities, as well as naming them is introduced. We also incorporate aspects of fuzziness in calculating degrees of conformance of entities to categories.

2   RDF Data and Category Learning: Overview

2.1   RDF-Triples as Definitions of Entities

A single RDF-triple <subject-property-object> can be perceived as a feature of an entity identified by the subject. In other words, each single triple is a feature of its subject. Multiple triples with the same subject constitute a definition of a given entity. A simple illustration of this is shown in Fig. 1(a). It is a definition of London. If we think "graphically" about it, a definition of entity resembles a star, we will call it an RDF-star.

Quite often a subject and object of one triple can be involved in multiple other triples, i..e, they can be objects or subjects of other triples. In such a case, multiple definitions – RDF-stars – can share features, or some of the features can be centres of another RDF-stars. Such interconnected triples constitute a network of interleaving definitions of entities, Fig. 1(b).

Due to the fact that everything is connected to everything, we can state that numerous entities share features among themselves. In such a case, comparison of entities is equivalent to comparison of RDF-stars. This idea is a pivotal aspect of the learning approach described here. It enables categorization, incremental updates, as well as establishing degrees of belonging of entities to categories.

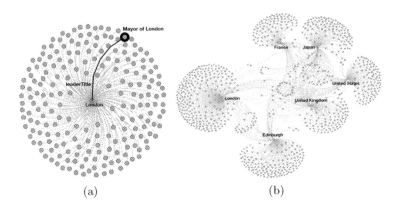

(a)   (b)

Fig. 1. RDF-stars: a definition of London with one of its features enhanced (a), interconnected RDF-stars representing: London, Edinburgh, France, Japan, United Kingdom and United States (b)

2.2   Learning: Construction and Augmenting Processes

In a nutshell, the proposed process of building categories contains a number of activities that can be divided into phases: identification of clusters of RDF-stars; augmenting the clusters with names and memberships degrees; and incremental updating of the clusters and their hierarchy.

Identification of clusters starts with constructing a similarity matrix. Once a set of triples (RDF-stars) is obtained, for example as the result of a walk done by an agent, similarity values are determined for all pairs of RDF-stars. The matrix is an input to an aggregative clustering algorithm. The result is a hierarchy of clusters (groups of RDF-stars) with the most specific clusters at the bottom, and the most abstract one (one that contains everything) at the top (see Section 3).

The next phase is augmenting clusters. Each cluster is labeled with a set of common features of RDF-stars that belong to the same cluster. Elements of the similarity matrix are used to determine the most characteristic – representative – RDF-star for each cluster. Degrees of membership of each RDF-star to its cluster are also calculated based on the similarity matrix. The process is described in Section 4.

A very important aspect of the proposed learning process is handling of incoming, new RDF-stars and their placement in the hierarchy of categories. An approach presented here, Section 5, is an adaptation of [5] to our clustering algorithm and RDF environment. Any time a new RDF-star is encountered by an agent, its similarity values to all other RDF-stars are determined. They are used to put the new RDF-star in a proper place in the hierarchy.

3   Building Clusters

All interconnected RDF-stars constitute a graph, and it seemed graph segmentation could be used to identify groups of highly interconnected – similar – nodes. However, all nodes (entities) of this graph are not equally important. Some of them are the centres of RDF-stars, i.e., they are `subjects` of multiple triples, and we will call them *defined entities*, while some are just `objects` of RDF triples, we will call them *defining entities*. All nodes which play only the role of *defining entities* should not be involved in the categorization process. They do not have any features for the comparison purposes. Therefore, instead of graph segmentation methods we use an agglomerative hierarchical clustering method.

3.1   Similarity of RDF Entities

The agglomerative clustering requires a single similarity matrix. The similarity matrix is built with RDF-stars as rows and columns. The similarity between RDF-stars is calculated using a feature-based similarity measure that resembles the Jaccard's index [7]

In the proposed approach, a similarity value between two *defined entities* (RDF-stars) is determined by a number of common features. In the case of RDF-stars, it nicely converts into checking how many *defining entities* they shared. The idea is presented in Fig. 2. The *defined entities* Edinburgh and London share a number of *defining entities*, and some of these entities are connected to the *defined entities* with the same `property` (black circles in Fig. 2).

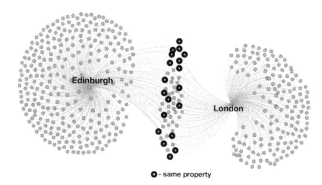

Fig. 2. Similarity of RDF-stars: based on shared objects connected to *the defined entities* with the same properties

In general, a number of different comparison scenarios can be identified. It depends on interpretation of the term "entities they share". The possible scenarios are:

- identical `properties` and identical `objects`;
- identical `properties` and similar `objects`;
- similar `properties` and identical `objects`;
- similar `properties` and similar `objects`;

For details, please see [7]. The similarity assessment process used in the paper follows the first scenario.

3.2 Similarity Matrix and Clustering

A similarity matrix for a set of RDF-stars constructed using the similarity evaluation technique presented in the previous subsection is used for hierarchical clustering. The clusters are created via an aggregation process in a bottom-up approach. Two clusters of a lower level are merged to create a cluster at a higher level.

At the beginning each RDF-star is considered as a one-element cluster. All aggregation decisions are made based on a distance between clusters calculated using an extended Wards minimum variance measure [11]. This measure takes into account heterogeneity between clusters and homogeneity within clusters. The distances are calculated based on entries from the modified similarity matrix. The modified similarity matrix is de facto a distance matrix created from subtracting the similarity values from a constant equal to the highest similarity value plus epsilon. The two clusters with the smallest distance are merged to become a new cluster. Distances (Ward's measures) between the new cluster and the remaining clusters are calculated. This agglomeration process is repeated until only a single cluster is left. The pseudocode for the agglomerative hierarchical clustering used here is presented below.

```
WardClustering(RDFstars)
begin:
  create_distance_matrix(RDFstars)
  clusterList   ⇐  create_intial_clusters(RDFstars)
  pairList   ⇐  create_pairs(clusterList)
  clusterPairDistances  = calculate_distances(pairList)
  while (length(clusterList) > 1)
     find_pair_with_minimum_distance(clusterPairDistances)
     newCluster = aggregate_clusters_with_min_distance()
     remove_clusters_with_min_distance(clusterList)
     for (each cluster from clusterList)
           calculate_Ward_measure(cluster, newCluster)
           update(pairList)
           update(clusterPairDistances)
     update(clusterList, newCluster)
end.
```

3.3 Running Example: Clustering

The described approach to construct categories is illustrated with a simple running example. The data used here is presented in Fig 3. The part (a) is a visualization of six entities – RDF-stars – that constitute an input to the algorithm. They are: London, Edinburgh, United Kingdom, France, Japan, and United States. The part (b) of the figure, shows the clustering results in the form of the dendogram.

4   From Clusters to Categories

4.1   RDF Properties and Concept Naming

The clustering algorithm operates on *defined entities* (centres of RDF-starts). Once the clusters are defined, we retrieve all entities – the whole network of

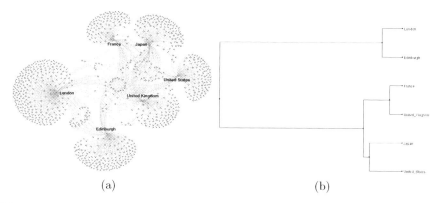

(a)   (b)

Fig. 3. Six entities used for the running example: RDF-stars (a), the dendogram of clustering results (b)

interconnected nodes. As the result, all clusters contain full RDF-stars with both *defined entities* and *defining entities*.

We "transform" each cluster into a category, i.e., we label each cluster by a name that represents the meaning of its members. This is accomplished via taking into account two properties and inspect all triples in a given cluster:

- <subject-**dcterm:subject**-object>,
- <subject-**rdf:type**-object>.

We look for all `objects` that are the same for all *defined entities* in a single cluster. These `objects` are labels of the cluster. We repeat this process for all clusters. We start at the top – the most general concept, and go further (deeper) into the hierarchy adding more labels. The categories at the bottom are the most specific, they have the largest number of labels.

4.2   Fuzziness: Degree of Belonging to Categories

So far, we have treated categories as crisp sets – all RDF-stars fully belong to their categories. However, when we look closer we find out that similarity between members of the same category has different values. This would imply that an RDF-star belongs to its category to a degree. Let us propose a method for determining that degree.

This task starts with identification of the centres of categories. We extract entries from the similarity matrix that are associated with RDF-stars from a given category, and identify a single RDF-star which has the largest number of commonalities with other RDF-stars. Let C_i is a category of N RDF-stars, and let RS_k represents k-th star. Its conformance to all other RDF-stars from this category is:

$$conf_{RS_k} = \sum_{m=1, m \neq k}^{N-1} sim(RS_k, RS_m)$$

where $sim(\cdot, \cdot)$ is an appropriate entry from the similarity matrix. Then the centre is:

$$centerID = \underset{m=1...N}{\arg\max}(conf_{RS_m})$$

We treat this RDF-star as the most representative entity of the category, and make it its centre. Once the centre is determined, we use its conformance level as a reference and compare it with conformance values of other RDF-stars in the same category:

$$\mu(RS_k) = \frac{conf_{RS_k}}{conf_{RS_{centerID}}}$$

In such a way we are able to determine degrees of membership of RDF-stars to the categories.

4.3   Running Example: Naming and Membership Degrees

Now, we name the clusters identified in our running example, Section 3.3, and assign membership values to RDF-stars (entities). The results are shown in Table 1. It contains labels associated with each category. Please note that the cluster **C1** is labeled with all labels of its predecessors in the hierarchy, i.e., labels of clusters **C5** and **C4**. The values of membership of entities to clusters are given besides entities' names.

Table 1. Running example: naming and membership values for identified clusters

C5:
Thing, Feature, Place, Populated_Place, Administrative_District, Physical_Entity Region, YagoGeoEntity, Location_Underspecified
France(0.95), **UK**(1.00), **Japan**(0.88), **US**(0.86), **London**(0.69), **Edinburgh**(0.67)
C4:
Member_states_of_the_United_Nations, G20_nations Liberal_democracies, G8_nations
France(1.00), **UK**(1.00), **Japan**(0.91), **US**(0.86)

C1	C3	C2
C1: Member_states_of_the_EU Countries_in_Europe Western_Europe Member_states_of_NATO Countries_bordering_the_Atlantic	**C3:** Countries Bordering ThePacific Ocean	**C2:** British_capitals Capitals_in_Europe Settlement, City
France(1.00), **UK**(1.00)	**Japan**(1.00), **US**(1.00)	**London**(1.00), **Edinburgh**(1.00)

5   Incremental Concept Building

5.1   Process Overview

The process of learning is never complete. Therefore, a procedure suitable for updating the already existing hierarchy of categories is proposed. Any incoming new data – the result of agent's walk on the web – can be added to the hierarchy.

We have adopted an incremental procedure proposed in [5]. We use Ward's measure to decide about a node for placing a new RDF-star. The flowchart of the algorithm can be seen below:

```
incrementalClustering(currentNode, newEntity)
begin:
distance  ⇐  calculate_Ward_measure(currentNode, newEntity)
height  ⇐  obtain_Node_Distance(currentNode)
if (height < distance)
    newNode  ⇐  createLeafNode(newEntity)
    attachNode(newNode, currentNode)
else
    LChild  ⇐  getChildL(currentNode)
    RChild  ⇐  getChildR(currentNode)
    distL = calculate_Ward_measure(LChild, newEntity)
    distR = calculate_Ward_measure(RChild, newEntity)
    if (distL < distR)
            incrementalClustering(LChild, newEntity)
    else
            incrementalClustering(RChild, newEntity)
end.
```

The process starts at the top, and as long as the distance of a new entity to the category at a given node is smaller than the node's height (the value of Ward's measure calculated for sibling nodes) we go deeper into the hierarchy. A decision is made, based on Ward's values, which way to go: should it be a left branch or a right one. The Ward's measures are used to ensure the final hierarchy is the same as the hierarchy that would be constructed if all data were used from the beginning.

5.2   Running Example: Incremental Building

In order to illustrate an incremental updating of the hierarchy, let us go back to our running example. We added a number of entities on one-by-one basis: Basel, Paul Gauguin, Buster Keaton, Henri Matisse, Pablo Picasso, and Charlie Chaplin. The dendogram of a new hierarchy is shown in Fig 4.

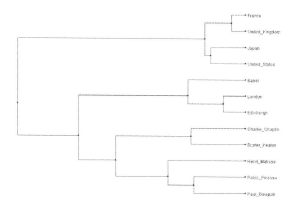

Fig. 4. The dendogram of all entities used for the running example

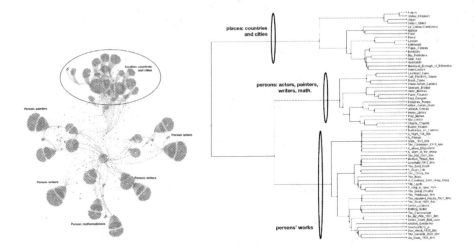

Fig. 5. Case Study: original RDF data and the dendogram with the results

6   Case Study

The application of the proposed learning method with a larger data set is presented here. A number of triples are loaded from dbpedia.org. The data contains: four groups of individuals – painters, actors, writers and mathematicians; and a number of geographical locations – countries and cities, Fig. 5(a). The dendogram representing the hierarchy is shown in Fig. 5(b). It contains not only entities that have been explicitly loaded, but also entities – the cluster "persons' works" – that are parts of the originally loaded RDF-stars.

7   Discussion and Conclusion

One of the most important aspects of the proposed method is the fact that it is an experience- and data-driven process. In the case of both running example and case study, we can observe changes in membership values across categories at different hierarchy levels. Also, the centres of categories are changing. For example, two individuals: Picasso and Gauguin belong to the same low level category. Their membership values are 0.79 and 0.86, respectively. If we look at Picasso at higher level categories its membership values changes to 0.90 and 0.61. These changes in Picasso's membership are the result of different content of categories and changes in their representative entities – centres. Our experiments indicate that the composition of categories and membership values depend on collected information, especially what it is and how detailed it is, but not when it is absorbed. The method mimics a learning process, i.e., clusters, their centres and members, together with their degrees of belonging, are changing and improving when more information is collected.

The paper introduce a methodology for a gradual learning categories of RDF data based on data that are being collected via agents and software systems. The hierarchy of categories can be updated via constant inflow of new data. Additionally, categories are labeled, and all entities are associated with values representing degrees of membership of these entities to categories at different levels of the constructed hierarchy. It would be interesting to further investigate how selection of a similarity measure from possible ones (Section 3.1) influences the clusterization process.

References

1. Berners-Lee, T., Hendler, J., Lassila, O.: The Semantic Web. Scientific American, 29–37 (2001)
2. Christodoulou, K., Paton, N.W., Fernandes, A.A.A.: Structure Inference for Linked Data Sources using Clustering. In: EDBT/ICDT Workshops, pp. 60–67 (2013)
3. Ferrara, A., Genta, L., Montanelli, S.: Linked Data Classification: a Feature-based Approach. In: EDBT/ICDT Workshops, pp. 75–82 (2013)
4. Giannini, S.: RDF Data Clustering. In: Abramowicz, W. (ed.) BIS 2013 Workshops. LNBIP, vol. 160, pp. 220–231. Springer, Heidelberg (2013)
5. Gurrutxaga, I., Arbelaitz, O., Marin, J.I., Muguerza, J., Perez, J.M., Perona, I.: SIHC: A Stable Incremental Hierarchical Clustering Algorithm. In: ICEIS, pp. 300–304 (2009)
6. Hossein Zadeh, P.D., Reformat, M.Z.: Semantic Similarity Assessment of Concepts Defined in Ontology. Information Sciences (2013)
7. Hossein Zadeh, P.D., Reformat, M.Z.: Context-aware Similarity Assessment within Semantic Space Formed in Linked Data. Journal of Ambient Intelligence and Humanized Computing (2012)
8. Lalithsena, S., Hitzler, P., Sheth, A., Jain, P.: Automatic Domain Identification for Linked Open Data. In: IEEE/WIC/ACM Inter. Conf. on Web Intelligence and Intelligent Agent Technology, pp. 205–212 (2013)
9. Levandoski, J.J., Mokbel, M.F.: RDF Data-Centric Storage. In: IEEE International Conference on Web Services, ICWS, pp. 911–918 (2009)
10. Schmidt, M., Hornung, T., Küchlin, N., Lausen, G., Pinkel, C.: An Experimental Comparison of RDF Data Management Approaches in a SPARQL Benchmark Scenario. In: Sheth, A.P., Staab, S., Dean, M., Paolucci, M., Maynard, D., Finin, T., Thirunarayan, K. (eds.) ISWC 2008. LNCS, vol. 5318, pp. 82–97. Springer, Heidelberg (2008)
11. Szekely, G.J., Rizzo, M.L.: Hierarchical Clustering via Joint Between-Within Distances: Extending Wards Minimum Variance Method. Journal of Classification 22, 151–183 (2005)
12. Zong, N., Im, D.-H., Yang, S., Namgoon, H., Kim, H.-G.: Dynamic Generation of Concepts Hierarchies for Knowledge Discovering in Bio-medical Linked Data Sets. In: 6th Inter. Conf. on Ubiquitous Inf. Management and Commun., vol. 12 (2012)
13. http://www.w3.org/RDF/ (accessed December 30, 2013)

The X-μ Approach: In Theory and Practice

Daniel J. Lewis and Trevor P. Martin

Intelligent Systems Laboratory, University of Bristol,
Woodland Road, Bristol, BS8 1UB, United Kingdom
{Daniel.Lewis,Trevor.Martin}@bristol.ac.uk
http://intelligentsystems.bristol.ac.uk/

Abstract. Criticisms exist for fuzzy set theory which do not reside in classical (or "crisp") set theory, some such issues exhibited by fuzzy set theory regard the law of excluded middle and law of contradiction. There are also additional complexities in fuzzy set theory for monotonicity, order and cardinality. Fuzzy set applications either avoid these issues, or use them to their advantage. The X-μ approach, however, attempts to solve some of these issues through analysis of inverse fuzzy membership functions. Through the inverse fuzzy membership function it is possible to computationally calculate classical set operations over an entire fuzzy membership. This paper firstly explores how the X-μ approach compares to both classical/crisp set theory and conventional fuzzy set theory, and explores how the problems regarding the laws of excluded middle and contradiction might be solved using X-μ. Finally the approach is implemented and applied to an area of big data over the world-wide-web, using movie ratings data.

1 Introduction

Uncertainty plays a very large role in data analysis and data mining, and therefore probabilistic methods have permeated almost every aspect of data analysis and data mining. It has, however, been argued [1] that probability is good for modelling some uncertainties, and fuzzy set theories are suitable for other uncertainties and vagueness. There has been some work on translation of fuzzy sets into probabilistic models, including initial work in the use of Kleene Logic by Lawry and Martin [2] and of multisets by Goertzel [3] as stepping stones. It should be noted that fuzzy set theory and probability are not competitors, and have been associated, with the earliest work in the form of fuzzy probabilities [4]. The research presented in this paper uses fuzzy set theory to model vagueness and statistical measures of cardinality. Big data, particularly data that is put together by a wide community, also provides its own advantages and disadvantages, where some aspects of uncertainties and vagueness are resolved and others become increasingly challenging.

There have, however, been many critiques of fuzzy set theory (e.g. [5], [6]) which highlight potential flaws in conventional fuzzy set theory. These are usually in respect to either the methods of fuzzification and defuzzification of sets, or, conventional fuzzy set operations not adhering to the same laws of classical set

A. Laurent et al. (Eds.): IPMU 2014, Part III, CCIS 444, pp. 406–415, 2014.

theory, such as the law of contradiction and the law of excluded middle. There have been a number of methods proposed in existing research which attempt to solve many of these issues [7],[8],[9]. The research presented in this paper extends the existing development of the X-μ approach [9],[10],[11]. Section 2 discusses the mathematics behind the X-μ approach and how it differs from conventional fuzzy set theory, particularly by detailing how the existing problems with the laws of contradiction and excluded middle are resolved. It then discusses the precise methods of X-μ set operations from an analytic/symbolic perspective. Section 3 first describes one particular implementation of the X-μ approach using a symbolic computation engine within a popular programming language, then describes the application of the theory and its implementation within the domain of association rule mining over a big data set found on the world wide web - in the form of the MovieLens data set[1].

2   Theory

Fuzzy set theory is an extension to classical set theory where each element within a set (finite or infinite) has a membership value μ. Fuzzy membership values are either stored within a lookup table, or have some algebraic membership function. A membership function is defined as:

$$\mu : U \rightarrow [0, 1]$$

Where μ is a membership function applicable to a particular fuzzy set, it is often simplified to $A(x)$ where A is the label for the fuzzy set, and x is some element of the universe.

Conventional fuzzy set theory states that there are the following fuzzy set operations:

- Fuzzy Negation: $\neg A(x) = 1 - A(x)$
- Fuzzy Union: $(A \cup B)(x) = \max(A(x), B(x))$
- Fuzzy Intersection: $(A \cap B)(x) = \min(A(x), B(x))$

2.1   Laws of Set Theory: Conventional Fuzzy Set Theory

Some laws of classic set theory do not necessarily hold for fuzzy sets. For example, the classical law of contradiction states:

$$A \cap \neg A = \emptyset$$

Whereas, when applied to a fuzzy set; the membership function returns values between 0 and 1, therefore creating a non-empty set, for each item x:

$$(A \cap \neg A)(x) = \min(A(x), 1 - A(x)) >= 0$$

---

[1] MovieLens Dataset is publicly available from the world-wide-web http://grouplens.org/datasets/movielens/ (December 2013).

The result of fuzzy intersection can have elements which are greater than 0, and the law of contradiction requires all members to have elements which are exactly equal to 0 (i.e. an empty set).

The law of excluded middle can also be a problem in conventional fuzzy set theory, for example in classical/crisp set theory:

$$A \cup \neg A = U$$

Whereas, when applied to a fuzzy set, for each item x:

$$(A \cup \neg A)(x) = \max(A(x), 1 - A(x)) <= 1$$

The result of a fuzzy union may have elements which are greater than 0 and less than 1, and the law of excluded middle requires all members to have elements which are exactly equal to 1 (i.e. in this case all members of the universal set).

α-cuts allow us to slice a fuzzy set at a defined point (called α). An α-cut selects elements of a fuzzy set which have a membership value greater than or equal to α. The definition of a traditional α-cut is:

$$^{\alpha}A = \{x \in X \mid A(x) >= \alpha\}$$

The result is a classical (or "crisp") set, the membership property is eliminated from the resultant set with the assumption that all members are now full members. The α-cut function is anti-monotonic (decreasing or staying the same in size with each alpha-cut).

Let $\alpha1 < \alpha2$, then

$$^{\alpha1}A \supseteq {}^{\alpha2}A$$

2.2 Laws of Set Theory: X-μ Approach

The X-μ approach [9][10][11] uses an inverse fuzzy membership function:

$$\mu^{-1} : [0,1] \to P(U)$$

Which takes a value between 0 and 1 (inclusive), and returns a subset from the (crisp) power set P of the universe U. This could be calculated using the following set builder notation:

$$\mu^{-1}(\alpha) = \{x \mid x \in U, \mu(x) >= \alpha\}$$

Here α is used as in a conventional fuzzy α-cut, the difference being that it is applied to the universal set rather than a particular subset of the universal set. We also use X as an alternative label to μ^{-1}. The result can be visualised graphically as a rotated chart. Figure 1 would be where:

$$\mu_{medium}(x) = \begin{cases} 0.5x - 0.5 & \text{for } x > 1.0 \wedge x < 3.0 \\ 1.0 & \text{for } x \geq 3.0 \wedge x \leq 4.0 \\ -0.5x + 3.0 & \text{for } x > 4.0 \wedge x < 6.0 \\ 0.0 & \text{otherwise} \end{cases}$$

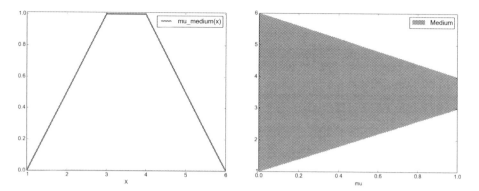

Fig. 1. An μ graph (Left) and X-μ graph (Right) for linguistic term 'Medium'

Or in X-μ form:

$$X_{medium} = \begin{cases} [2.0\alpha + 1.0, -2.0\alpha + 6.0] & \text{for } \alpha \geq 0.0 \wedge \alpha \leq 1.0 \\ \emptyset & \text{otherwise} \end{cases}$$

This technique differs from both traditional techniques, and techniques by Sánchez [8], in that it is also capable of working as an analytical or symbolic system. For example, when two fuzzy sets are combined through union or intersection, it is possible, to combine their X-μ functions into a new combined X-μ function. An analysis on the differences between this technique and the Sánchez technique is available in existing literature [10] [11].

2.3 X-μ Fuzzy Set Operations

X-μ set operations are classical set operations performed upon subsets returned by the inverse membership function of a fuzzy set. When a membership function is algebraic (rather than a lookup table), these set operations (union, intersection etc.) can be performed via algebraic modification. The clearest difference between traditional fuzzy set operations and X-μ operations is in the set difference operator. For a formal definition of X-μ operators please see Martin[10].

For the following set theory processes, the following two height-based examples are used, and are bounded between 0.0 and 7.0 (i.e. the universal set). For some membership function representing the fuzzy category "Tall":

$$X_{Tall}(\alpha) = \{x \mid x \in U \wedge (0.2 * \alpha + 5.8 <= x)\}$$
$$= [0.2\alpha + 5.8, 7.0]$$

For some membership function representing the fuzzy category "Small":

$$X_{Small}(\alpha) = \{x \mid x \in U \wedge (x <= 1.0 * \alpha + 6.0)\}$$
$$= [0.0, -1.0\alpha + 6.0]$$

X-μ Union

The result from an X-μ union is identical to a conventional fuzzy set union, but rotated. At each point along X there exists a lower and upper μ value. The result is also equivalent to standard set theory for any value of α. The membership functions can be merged thus to represent two membership functions under union:

$$X_{Small \cup Tall} = X_{Small}(\alpha) \cup X_{Tall}(\alpha) = [0.0, 6.0 - 1.0\alpha] \cup [5.8 + 0.2\alpha, 7.0]$$

X-μ Intersection

As with X-μ union, the result from an X-μ intersection are identical to a conventional fuzzy set intersection, but rotated. They are also equivalent to standard set theory for any value of α. The membership functions can be merged thus to represent two fuzzy memberships intersected:

$$X_{Small \cap Tall} = X_{Small}(\alpha) \cap X_{Tall}(\alpha) = [0.0, 6.0 - 1.0\alpha] \cap [5.8 + 0.2\alpha, 7.0]$$

X-μ Set Difference

Set difference is where the most clear difference lies in comparison to conventional fuzzy set difference. In conventional fuzzy set difference the difference of the μ value is found against full membership (i.e. 1.0). In the X-μ the interval difference is found using interval set difference thus:

$$X_{Small - Tall} = [0.0, 6.0 - 1.0\alpha] \cap [0.0, 5.8 + 0.2\alpha]$$

2.4   X-μ Numbers

Fuzzy Numbers, first developed by [12], fuzzify traditional numbers to allow for linguistic terms such as 'around 5', 'almost 42' or 'much larger than 333'. These are defined using the domain in which the fuzzy number exists, and the level of vagueness required for that particular context. X-μ numbers are single-valued quantities related to the gradual elements introduced by Dubois and Prade[7], in that the inverse is found from their fuzzy membership. If the membership is calculated analytically, then fuzzy arithmetic operations can also be performed analytically. The focus of X-μ numbers is on modelling single values that vary with membership, whereas other techniques often model imprecise values. An analysis of X-μ numbers shall appear in a future publication.

3   Practice

3.1   Implementation

An object-orientated class library has been developed to represent generic X-μ functions in addition to more specialised classes for functions in common shapes.

Common shapes include: upward gradients, downward gradients, triangular functions and trapezoidal functions. The python programming language[2] was used to develop this library. This language was chosen for its familiarity, in addition to its object-orientated and functional programming syntax, and its add-on libraries for symbolic computation and graphing. However, almost any other programming language could be chosen for such a library. The abstract X-μ function allows for the following operations:

- Set Operations: X-μ Set Union, X-μ Set Intersection, X-μ Set Difference and Negation
- Number Operations: X-μ Number Addition, X-μ Number Subtraction, X-μ Number Multiplication and Power, X-μ Number Division

Wherever possible the implementation of the above operations are performed symbolically using the SymPy framework[3], which is an industry and academic supported framework for Python providing classes and functions for symbolic computing. Our X-μ library also provides a simplified function for generating graphs, which utilises the matplotlib framework[4].

3.2 Example - MovieLens

MovieLens[5] is a web-based system by the University of Minnesota which allows its users to rate movies in a star-based framework, where one star indicates a dislike of a movie and 5 stars indicates a strong like of a movie. Their initial purpose for developing the system was for the data to be used in experimental recommender systems, and its dataset is provided freely and openly on the GroupLens Research Group[6] website under a bespoke usage license. The whole dataset is a little higher than 10 million ratings records, however two subsets of sizes 1 million and 100 thousand have also been provided on the GroupLens website in order to test scalability of experimental algorithms.

The MovieLens dataset has been used as a data source in our X-μ framework, for association rule calculations. Association Rule Mining was developed in the early 1990s[13][14], and an X-μ variant was developed by Lewis and Martin[11], and Martin[9]. The purpose for building a fuzzy model and performing association analysis on this dataset is to discover 'high' and 'low' rated films with

---

[2] The official documentation for the Python programming language is available on the world wide web: http://www.python.org/ (December 2013).

[3] SymPy is a symbolic computation framework for python, and is free and open source software: http://sympy.org/ (December 2013).

[4] Matplotlib is a plotting library based on the plotting functionality of other well known mathematics software, it is available over the world wide web: http://matplotlib.org/ (December 2013).

[5] MovieLens as a user-based system is available on the world wide web: http://movielens.umn.edu/ (December 2013).

[6] GroupLens content for the MovieLens database is available freely on the world wide web under a bespoke attribution license: http://grouplens.org/datasets/movielens/ (December 2013).

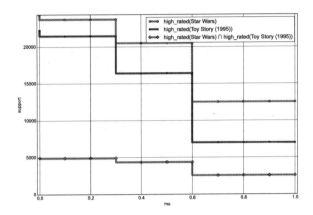

Fig. 2. X-μ support calculations for Star Wars, Toy Story and their intersection

a more human-like understanding, 'high' and 'low' both being naturally fuzzy. Two X-μ functions have been developed to represent such terms:

$$X_{low\_rated} = [1.0, -3.0\alpha + 4.0]$$
$$X_{high\_rated} = [3.0\alpha + 2.0, 5.0]$$

With the conjoint supports being symbolically created, for example:

$$high\_rating(\alpha) \cap low\_rating(\alpha) = [1.0, -3.0\alpha + 4.0] \cap [3.0\alpha + 2.0, 5.0]$$

The association confidence of each low and high rating for both films in both directions are then calculated, and the results are plotted with a predetermined resolution (or 'granularity'), as found in Figure 4, which shows the combination of the confidences in various aspects of 'Star Wars' and 'Toy Story', for both low and high rating memberships. Along the X axis is the membership value, and the Y axis holds the confidence value. These two particular films share a viewer-base which thinks highly of both films, as can be seen in the two association rules: $high\_rated(StarWars) \rightarrow high\_rated(ToyStory)$ and $high\_rated(ToyStory) \rightarrow high\_rated(StarWars)$, a movie critic might analyse this as being due to both movies being accessible and enjoyable by the whole family. If however, the high rating confidences are analysed for the original Star Wars and Star Trek movies, then our movie critic description may be different.

Figure 5 finds that the rule $high\_rated(StarWars) \rightarrow high\_rated(StarTrek)$ decreases as μ increases, whereas the association rule $high\_rated(StarTrek) \rightarrow high\_rated(StarWars)$ increases rapidly. This indicates that those fans of the original Star Trek movie are likely going to be impressed by the original Star Wars movie. A movie critics analysis may be that as Star Trek is a niche film in

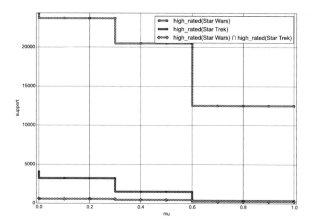

Fig. 3. X-μ support calculations for Star Trek, Star Wars and their intersection

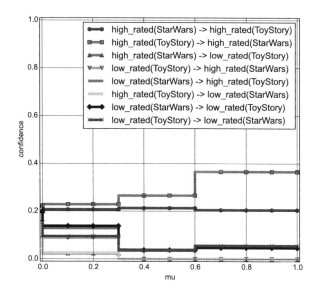

Fig. 4. X-μ association confidence calculations on the memberships of $\{low\_rating, high\_rating\} \times \{StarWars, ToyStory\}$

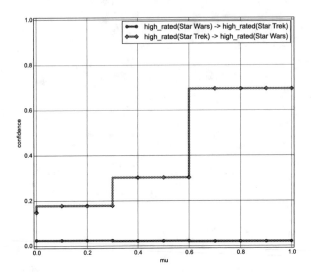

Fig. 5. X-μ association confidence calculations on high ratings for Star Wars and Star Trek original movies

the science fiction genre, and Star Wars also being in the science fiction genre then there is likely to be a common fan-base. However, Star Wars is a wider reaching and is more accessible to a range of movie tastes, and will therefore have a wider fan-base than Star Trek, meaning the confidence is likely to be lower than its inverse.

4   Conclusion

In this paper the essentials of the X-μ approach have been presented, and placed in context with both classical (crisp) set theory and conventional fuzzy set theory. The laws of contradiction and excluded middle have been discussed, and X-μ set operations provided which adhere to those laws. An implementation and an example application based on a form of data mining known as attribute-based association rule mining was described. This example used real-world data in the form of the MovieLens dataset, a large dataset (or 'big data') available for free on the world wide web. The X-μ approach to fuzzy numbers and fuzzy arithmetic has also been introduced, but due to space limitations and the theoretical nature these aspects were unable to be described in-depth in this paper. Future work will include in-depth research in the automatic discovery of rules and patterns, particularly utilising spatial and temporal fuzzy datasets.

References

1. Zadeh, L.A.: Discussion: Probability theory and fuzzy logic are complementary rather than competitive. Technometrics 37(3), 271–276 (1995)
2. Lawry, J., Martin, T.: Conditional beliefs in a bipolar framework. In: van der Gaag, L.C. (ed.) ECSQARU 2013. LNCS, vol. 7958, pp. 364–375. Springer, Heidelberg (2013)
3. Goertzel, I.F., Ikl, M., Heljakka, A.: Probabilistic logic networks: A comprehensive framework for uncertain inference. Springer (2008)
4. Zadeh, L.A.: Probability measures of fuzzy events. Journal of Mathematical Analysis and Applications 23(2), 421–427 (1968)
5. Elkan, C., Berenji, H., Chandrasekaran, B., de Silva, C., Attikiouzel, Y., Dubois, D., Prade, H., Smets, P., Freksa, C., Garcia, O., Klir, G., Yuan, B., Mamdani, E., Pelletier, F., Ruspini, E., Turksen, B., Vadiee, N., Jamshidi, M., Wang, P.Z., Tan, S.K., Tan, S., Yager, R., Zadeh, L.: The paradoxical success of fuzzy logic. IEEE Expert 9(4), 3–49 (1994)
6. Laviolette, M., Seaman Jr., J.W.: The efficacy of fuzzy representations of uncertainty. IEEE Transactions on Fuzzy Systems 2(1), 4–15 (1994)
7. Dubois, D., Prade, H.: Gradual elements in a fuzzy set. Soft Computing 12(2), 165–175 (2008)
8. Sánchez, D., Delgado, M., Vila, M.A., Chamorro-Martínez, J.: On a non-nested level-based representation of fuzziness. Fuzzy Sets and Systems 192, 159–175 (2012)
9. Martin, T., Azvine, B.: The x-mu approach: Fuzzy quantities, fuzzy arithmetic and fuzzy association rules. In: 2013 IEEE Symposium on Foundations of Computational Intelligence (FOCI), pp. 24–29 (2013)
10. Martin, T.: The x-mu representation of fuzzy sets - regaining the excluded middle. In: 2013 13th UK Workshop on Computational Intelligence (UKCI), pp. 67–73 (2013)
11. Lewis, D., Martin, T.: X-mu; fuzzy association rule method. In: 2013 13th UK Workshop on Computational Intelligence (UKCI), pp. 144–150 (2013)
12. Zadeh, L.A.: Calculus of fuzzy restrictions. Electronics Research Laboratory, University of California (1975)
13. Agrawal, R., Imieliski, T., Swami, A.: Mining association rules between sets of items in large databases. SIGMOD Rec. 22(2), 207–216 (1993)
14. Agrawal, R., Srikant, R. et al.: Fast algorithms for mining association rules. In: Proc. 20th Int. Conf. Very Large Data Bases, VLDB, vol. 1215, pp. 487–499 (1994)

Kolmogorov-Smirnov Test for Interval Data

Sébastien Destercke[1] and Olivier Strauss[2]

[1] Heudiasyc, UMR 7253, Rue Roger Couttolenc, 60203 Compiegne, France
sebastien.destercke@hds.utc.fr
[2] LIRMM (CNRS & Univ. Montpellier II), 161 Rue Ada, F-34392 Montpellier Cedex 5, France
olivier.strauss@lirmm.fr

Abstract. In this paper, we are interested in extending the classical Kolmogorov-Smirnov homogeneity test to compare two samples of interval-valued observed measurements. In such a case, the test result is interval-valued, and one major difficulty is to find the bounds of this set. We propose a very efficient computational method for approximating these bounds by using a p-box (pairs of upper and lower cumulative distributions) representation of the samples.

Keywords: Interval data, homogeneity test, approximation, p-box.

1 Introduction

In many applications, the precise value of data may only be known up to some precision, that is it may be interval-valued. Common examples are censored data (e.g., censor limitations) or digital data. When performing statistical tests, ignoring this imprecision may lead to unreliable decisions. For instance, in the case of digital data, quantization can hide the information contained in the data and provide unstable decision.

It is therefore advisable to acknowledge this imprecision in statistical tests, if only to provide results robust to this imprecision. By robust, we understand tests that will remain cautious (i.e., will abstain to say something about the null hypothesis) if not enough information is available. However, treating this imprecision usually leads to an increased computational costs, as shown by various authors in the past [6,7,3]. This means that developing efficient methods to compute statistics with interval data is a critical issue.

In this paper, we explore the extension of the Kolmogorov-Smirnov (KS) homogeneity test to interval data, and more precisely its computational aspects. To our knowledge, this aspect has not been considered in the past, even if some but not much works on the KS test with interval or fuzzy data exist [4,5]. Approximate and exact bounds that are straightforward to compute are provided in Section 3, while notations and reminders are given in Section 2.

In Section 4, we illustrate our results on a image based medical diagnosis problem. Indeed, in such problems a common task is to detect whether two regions of a quantized image have similar pixel distributions.

2 Preliminary Material

Komogorov-Smirnov (KS) homogeneity test [1] is commonly used to compare two samples $A = \{a_i | i = 1, \ldots, n, a_i \in \mathbb{R}\}$ and $B = \{b_i | i = 1, \ldots, m, b_i \in \mathbb{R}\}$ of measurements

A. Laurent et al. (Eds.): IPMU 2014, Part III, CCIS 444, pp. 416–425, 2014.

to determine whether or not they follow the same probability distribution. Those samples are supposed to be independently drawn from a continuous one-dimensional real-valued probability distributions.

If F_A (F_B) denote the empirical cumulative distributions built from A (B), that is if

$$F_A(x) = \frac{\#\{a \in A | a \leq x\}}{n} \tag{1}$$

with $\#E$ the cardinal of a set E, then the KS test statistic KS is defined by:

$$KS(A,B) = \sup_{x \in \mathbb{R}} |F_A(x) - F_B(x)|$$

Under the null hypothesis H_0 that the two-samples are drawn from the same distribution, the statistic $\beta(n,m)KS(A,B)$ converges to the Kolmogorov distribution, with $\beta(n,m) = \sqrt{\frac{1}{n} + \frac{1}{m}}$. Using the critical values of the Kolmogorov distribution, the null hypothesis can be rejected at level α if $KS(A,B) > \beta(n,m)\kappa_\alpha$. One common value of this rejection threshold is $\kappa_{0.05} = 1.36$.

As this test makes very few assumptions about the samples (i.e., it is non-parametric) and aims at testing a complex hypothesis (with respect to, e.g., comparing two means), it requires in practice relatively large samples to properly reject the null hypothesis.

In this paper, we explore the case where observations are interval-valued, i.e., they correspond to two sets $[A] = \{[\underline{a}_i, \overline{a}_i] | i = 1, \ldots, n\}$ and $[B] = \{[\underline{b}_i, \overline{b}_i] | i = 1, \ldots, m\}$ of real-valued intervals. As recalled in the introduction and further explored in Section 4, such imprecision may be the result of some quantization process.

In the next section, we study the interval-valued statistic resulting from such data, and in particular provide efficient approximative (and sometimes exact) bounds for it, using the notion of p-box.

3   Kolmogorov-Smirnov Test with Interval-Valued Data

Let us first introduce some notations. We will call *selection* of $[A]$ a set $S_{[A]}$ of values $S_{[A]} := \{a_i | i = 1, \ldots, n, a_i \in [\underline{a}_i, \overline{a}_i]\}$ where each a_i is picked inside the interval $[\underline{a}_i, \overline{a}_i]$, $i = 1, \ldots, n$. We will denote by $\mathscr{S}([A])$ the set of all selections of $[A]$. To a selection $S_{[A]}$ corresponds an empirical cumulative distribution $F_{S_{[A]}}$ obtained by Eq. (1), and we denote by $\mathscr{F}([A])$ the (non-convex) set of such empirical cumulative distributions.

Given this, the imprecise Kolmogorov-Smirnov Test

$$[KS]([A],[B]) = [\underline{KS}([A],[B]), \overline{KS}([A],[B])]$$

is an interval such that

$$\underline{KS}([A],[B]) = \inf_{\substack{S_{[A]} \in \mathscr{S}([A]), \\ S_{[B]} \in \mathscr{S}([B])}} \sup_{x \in \mathbb{R}} |F_{S_{[A]}}(x) - F_{S_{[B]}}(x)|, \tag{2}$$

$$\overline{KS}([A],[B]) = \sup_{\substack{S_{[A]} \in \mathscr{S}([A]), \\ S_{[B]} \in \mathscr{S}([B])}} \sup_{x \in \mathbb{R}} |F_{S_{[A]}}(x) - F_{S_{[B]}}(x)|. \tag{3}$$

Computing such values is not, a priori, a trivial task since the number of possible selections for both sets of intervals $[A]$ and $[B]$ are usually infinite. It should however be noted that, as the empirical cumulative distributions can only take a finite number of values (i.e., $\{0, 1/n, 2/n, \ldots, 1\}$ for $[A]$), so does the test. Yet, we are only interested in the extreme values it can take.

In the sequel, we propose to use the formalism of p-boxes to approximate those bounds $\underline{KS}([A], [B])$ and $\overline{KS}([A], [B])$

3.1 Approximating p-box

A p-box [2] $[\underline{F}, \overline{F}]$ is a pair of cumulative distributions such that $\underline{F}(x) \leq \overline{F}(x)$ for any $x \in \mathbb{R}$. The usual notion of cumulative distribution is retrieved when $\underline{F} = \overline{F}$, and a p-box usually describes an ill-known cumulative distribution that is known to lie between $\underline{F}$ and $\overline{F}$. That is, to a p-box $[\underline{F}, \overline{F}]$ we can associate a set $\Phi([\underline{F}, \overline{F}])$ of cumulative distributions such that

$$\Phi([\underline{F}, \overline{F}]) = \{F \mid \forall x \in \mathbb{R}, \underline{F}(x) \leq F(x) \leq \overline{F}(x)\}.$$

Here, we will use it as an approximating tool.

For a set of intervals $[A]$, let us denote by $S_{\underline{a}}$ and $S_{\overline{a}}$ the particular selections $S_{\underline{a}} = \{\underline{a}_i \mid i = 1, \ldots, n\}$ and $S_{\overline{a}} = \{\overline{a}_i \mid i = 1, \ldots, n\}$. Then, we define the p-box $[\underline{F}_{[A]}, \overline{F}_{[A]}]$ approximating $[A]$ as

$$\underline{F}_{[A]} := F_{S_{\overline{a}}} \quad \text{and} \quad \overline{F}_{[A]} := F_{S_{\underline{a}}}.$$

We have the following property

Proposition 1. *Given a set of intervals $[A]$, we have $\mathscr{F}([A]) \subseteq \Phi([\underline{F}_{[A]}, \overline{F}_{[A]}])$*

Proof. Consider a given selection S_A. For every a_i in this selection, we have

$$\underline{a}_i \leq a_i \leq \overline{a}_i.$$

Since this is true for every $i = 1, \ldots, n$, this means that F_{S_A} is stochastically dominated[1] by $F_{S_{\overline{A}}}$ and stochastically dominates $F_{S_{\underline{A}}}$, i.e.

$$F_{S_{\overline{a}}}(x) \leq F_{S_A}(x) \leq F_{S_{\underline{a}}}(x), \forall x \in \mathbb{R}$$

and as this is true for every selection S_A, we have $\mathscr{F}([A]) \subseteq \Phi([\underline{F}_{[A]}, \overline{F}_{[A]}])$. To see that the inclusion is strict, simply note that F_{S_A} can only take a finite number of values, while cumulative distributions in $\Phi([\underline{F}_{[A]}, \overline{F}_{[A]}])$ can be strictly monotonous.

This shows that the associated (convex) set $\Phi([\underline{F}_{[A]}, \overline{F}_{[A]}])$ is actually a conservative approximation of $\mathscr{F}([A])$. The next example illustrates both the p-box $[\underline{F}_{[A]}, \overline{F}_{[A]}]$ and Proposition 1.

---

[1] Recall that F_1 stochastically dominates F_2 if $F_1 \leq F_2$.

Example 1. Consider the case where we have 3 sampled intervals, with the three following intervals:

$$[\underline{a}_1, \overline{a}_1] = [2, 7]$$
$$[\underline{a}_2, \overline{a}_2] = [6, 12]$$
$$[\underline{a}_3, \overline{a}_3] = [10, 16]$$

Figure 1 illustrates the obtained p-box and one cumulative distribution ($\hat{F}$) included in $\Phi([\underline{F}_{[A]}, \overline{F}_{[A]}])$. However, $\hat{F}$ is not in $\mathscr{F}([A])$, since any empirical cumulative distribution obtained from a selection on 3 intervals can only takes its values in the set $\{0, 1/3, 2/3, 1\}$.

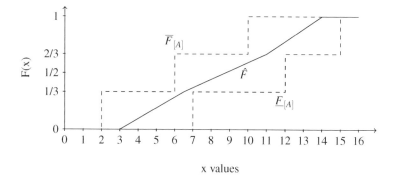

Fig. 1. P-box of Example 1

3.2   Approximating $\overline{KS}([A], [B])$ and $\underline{KS}([A], [B])$

Consider two samples $[A]$ and $[B]$ and the associated p-boxes $[\underline{F}_{[A]}, \overline{F}_{[A]}]$ and $[\underline{F}_{[B]}, \overline{F}_{[B]}]$. We can now introduce the approximated imprecise KS Test $[\widetilde{KS}] = [\underline{KS}, \widetilde{KS}]$ such that:

$$\widetilde{KS}([A], [B]) = \sup_{x \in \mathbb{R}} \max\{|\overline{F}_{[A]}(x) - \underline{F}_{[B]}(x)|, |\underline{F}_{[A]}(x) - \overline{F}_{[B]}(x)|\}, \tag{4}$$

$$\underset{\sim}{KS}([A], [B]) = \sup_{x \in \mathbb{R}} D_{[A], [B]}(x), \tag{5}$$

with

$$D_{[A], [B]}(x) = \begin{cases} 0 \text{ if } [\underline{F}_{[A]}(x), \overline{F}_{[A]}(x)] \cap [\underline{F}_{[B]}(x), \overline{F}_{[B]}(x)] \neq \emptyset \\ \min\{|\overline{F}_{[A]}(x) - \underline{F}_{[B]}(x)|, |\underline{F}_{[A]}(x) - \overline{F}_{[B]}(x)|\} \text{ otherwise} \end{cases}$$

These approximations are straightforward to compute (if $n + m$ intervals are observed, at worst they require $2n + 2m$ computations once the p-boxes are built). We also have the following properties:

Proposition 2. *Given a set of intervals* $[A]$ *and* $[B]$, *we have* $\widetilde{KS}([A],[B]) = \overline{KS}([A],[B])$

Proof. The value $\widetilde{KS}([A],[B])$ is reached on x either for a pair $\{\overline{F}_{[A]}, \underline{F}_{[B]}\}$ or $\{\underline{F}_{[A]}, \overline{F}_{[B]}\}$. As any pair F_1, F_2 with $F_1 \in \Phi([\underline{F}_{[A]}, \overline{F}_{[A]}])$ and $F_2 \in \Phi([\underline{F}_{[B]}, \overline{F}_{[B]}])$ would have a KS statistic lower than $\widetilde{KS}([A],[B])$, and given the inclusion of Proposition 1, this means that $\overline{KS}([A],[B]) \leq \widetilde{KS}([A],[B])$. To show that they coincide, it is sufficient to note that all distributions $\underline{F}_{[A]}, \overline{F}_{[A]}, \underline{F}_{[B]}, \overline{F}_{[B]}$ can be obtained by specific selections (i.e., the one used to build the p-boxes). ∎

This shows that the upper bound is exact. Concerning the lower bound, we only have the following inequality:

Proposition 3. *Given a set of intervals* $[A]$ *and* $[B]$, *we have* $\underline{KS}([A],[B]) \leq \widetilde{\underline{KS}}([A],[B])$

Proof. Immediate, given the inclusion of Proposition 1 and the fact that $\widetilde{\underline{KS}}([A],[B])$ is the minimal KS statistics reached by a couple of cumulative distributions respectively in $[\underline{F}_{[A]}, \overline{F}_{[A]}]$ and $[\underline{F}_{[B]}, \overline{F}_{[B]}]$ ∎

And unfortunately this inequality will usually be strict, as shows the next example.

Example 2. Consider the case where $n = 2$, $m = 3$ and where $\bigcap_{i=1}^{n}[\underline{a}_i, \overline{a}_i] = \emptyset$, $\bigcap_{i=1}^{m}[\underline{b}_i, \overline{b}_i] = \emptyset$. This means that, for every selection $S_{[A]} \in \mathscr{S}([A])$ and $S_{[B]} \in \mathscr{S}([B])$, we have that the empirical cumulative distributions $F_{S_{[A]}}$ and $F_{S_{[B]}}$ respectively takes at least one value in $\{1/2\}$ and in $\{1/3, 2/3\}$. This means that $\underline{KS}([A],[B]) \neq 0$ (as every cumulative distributions coming from selections will assume different values), while it is possible in such a situation to have $\widetilde{\underline{KS}}([A],[B]) = 0$.

Consider the following example:

$$[\underline{a}_1, \overline{a}_1] = [1,8] \qquad\qquad [\underline{b}_1, \overline{b}_1] = [2,7]$$
$$[\underline{a}_2, \overline{a}_2] = [9,15] \qquad\qquad [\underline{b}_2, \overline{b}_2] = [6,12]$$
$$\qquad\qquad\qquad\qquad\qquad [\underline{b}_3, \overline{b}_3] = [10,16]$$

A simple look at Figure 2 allows us to see that $\widetilde{\underline{KS}}([A],[B]) = 0$ in this case.

The inequality between $\underline{KS}([A],[B])$ and $\widetilde{\underline{KS}}([A],[B])$ can also be strict when $\widetilde{\underline{KS}}([A],[B]) \neq 0$. It should be noted that the discrepancy between $\underline{KS}([A],[B])$ and $\widetilde{\underline{KS}}([A],[B])$ will decrease as the number of sampled intervals increases. Finally, a noticeable situation where $\underline{KS}([A],[B])$ will be an exact bound ($\underline{KS}([A],[B]) = \widetilde{\underline{KS}}([A],[B])$) is when $[\underline{F}_{[A]}, \overline{F}_{[A]}]$ and $[\underline{F}_{[B]}, \overline{F}_{[B]}]$ are disjoint, that is either $\underline{F}([A]) > \overline{F}([B])$ or $\overline{F}([A]) < \underline{F}([B])$.

3.3 Decision Making Using an Imprecise-Valued Test

One of the main features of this extension is that it provides a pair of (conservative) bounds $\underline{KS}([A],[B])$ and $\widetilde{KS}([A],[B])$ rather than a precise value $KS(A,B)$. In contrast

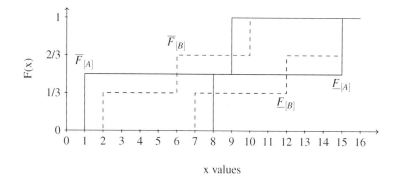

Fig. 2. P-boxes of Example 2

with usual tests that either reject or do not reject an hypothesis, this leads to three possible decisions: the answer to the test can be *yes*, *no* or *unknown*, the last one occurring when available information is insufficient.

In fact, interpreting this test is straightforward. Let $\gamma = \beta(n,m)\kappa_\alpha$ be the significance level.

- If $\underline{KS}([A],[B]) > \gamma$ then we can conclude that there is no possible selections $S_{[A]}$ of $[A]$ and $S_{[B]}$ of $[B]$ such that $KS(S_{[A]}, S_{[B]}) \leq \gamma$ and thus the hypothesis that the two-samples are drawn from the same distribution can be rejected at a level α.
- On the contrary, if $\widetilde{KS}([A],[B]) < \gamma$ then there is no possible selections $S_{[A]}$ of $[A]$ and $S_{[B]}$ of $[B]$ such that $KS(S_{[A]}, S_{[B]}) \geq \gamma$ and thus the hypothesis that the two-samples are drawn from the same distribution cannot be rejected at a level α.
- Otherwise, we will conclude that our information is too imprecise to lead to a clear decision about rejection.

This new test will therefore point out those cases where the data imprecision is too important to lead to a clear decision. As we shall see in the next section, it allows one to deal with quantization in a new way, namely it can detect when the disturbance or loss of information induced by the quantization makes the test inconclusive. It should be noted that, as $\widetilde{KS}([A],[B])$ is an approximated lower bound, indecision may also be due to this approximation, yet experiments of the next section indicate that this approximation is reasonable.

4   Experimentation

The experimentation we propose is based on a set of medical images acquired by a gamma camera. In such applications, statistical hypothesis testing is often used to determine whether pixels distribution in two different regions of an image are similar or not. Physicians usually try to control the probability of making a decision leading to harmful consequences, and make it as low as possible (usually 0.05).

The advantage of using a KS test in this case is that it makes very few assumption about the distribution. However, in such applications, it is quite common to deal with quantized information, i.e., real-valued information constrained to belong to a small subset of (integer) values. Since the KS test is designed to compare pairs of continuous distributions, it is necessary to ensure that the statistical test is robust with respect to the data model. Indeed, the value of the statistic computed from quantized data may differ markedly from the calculation based on original (non-quantized) but unavailable data.

Physicians would usually try to avoid a wrong decision, and prefer to acquire additional data when the actual data are not fully reliable. Thus, knowing that no decision can be taken based on the current set of data is a valuable piece of information.

We illustrate this weakness of the usual KS test with a set of medical images acquired by a gamma camera (nuclear medicine images) whose values are quantified on a restricted number of values. This experiment also highlights the ability of the extended KS test to avoid wrong decisions induced by quantization. It aims at mimicking real medical situations where the nuclear physician has to compare the distribution of values in two regions of interest in order to decide whether or not a patient has a specific disease.

The set of images is made of 1000 planar acquisitions of a Hoffman 2-D brain phantom (acquisition time: 1 second; average count per image 1.5 kcounts, 128×128 images to satisfy the Shannon condition), representing 1000 measures of a random $2D$ image (see Figure (3)). Due to the fact that nuclear images are obtained by counting the photons that have been emitted in a particular direction, pixel values in a nuclear image can be supposed to be contaminated by Poisson distributed noise. Due to the very short acquisition time, the images were very noisy, i.e. the signal to noise ratio was very low. More precisely, the average pixel value in the brain corresponded to a 69% coefficient of variation of the Poisson noise. Moreover, the number of different possible values to be assigned to a pixel was low and thus, within those images, the impact of quantization was high: pixel possible values were $\{0, 256, 512, 768, 1024, 1280, 1536, 1792, 2048\}$.

To obtain less noisy and less quantized images, we summed the raw images (see e.g. Figure (4)). The higher the number of summed images, the higher the average pixel value, and thus the higher the signal to noise ratio and the higher is the number of possible values for each pixel. When summing the 1000 raw images, we obtained the high dynamic resolution and high signal to noise ratio image depicted in Figure (5).a.

We use the KS test to decide whether the two regions depicted in Figures (5).b and (5).c can be considered as being similar or not (the null hypothesis). Considering the number of pixels in each region ($n = 183, m = 226$), the significance level for a p-value $\alpha = 0.05$ is $\gamma \approx 0.1910$. Testing the two regions with the reference image (Figure (5).a) provides the following values: $KS(A,B) \approx 0.2549$, $\underline{KS}([A],[B]) \approx 0.2505$ $\widetilde{KS}([A],[B]) \approx 0.2549$, leading to conclude that the similarity of regions A and B should be rejected at a level 0.05, which can be considered as our ground truth.

We use the KS test for comparing the same regions but with 300 pairs of images that have been randomly selected in the set of 1000 original images. In that case, the classical test accepts the similarity of the two regions, while the imprecise test is inconclusive for each pairs: $\underline{KS}([A],[B]) < \gamma < \widetilde{KS}([A],[B])$. We now do the same test with images having a higher dynamic obtained by summing $p = 2, 3, \ldots, 40$ images. For each value of p, we count the number of times the classical test provides the right answer, i.e. reject the null hypothesis at level 0.05 ($\gamma \leq KS(A,B)$). We then compute the ratio of

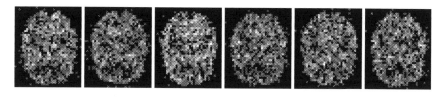

Fig. 3. 6 acquisitions of the Hoffman 2-D brain phantom

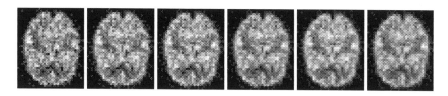

Fig. 4. 6 images obtained by summing up 10 raw acquisitions of the Hoffman 2-D brain phantom

Fig. 5. Reference image obtained by summing the 1000 raw images (a), region A (b) and region B (c) selected on the reference image

this count over 300. For the extended KS test, we compute two ratios: the ratio of times when $\gamma \leq \underline{KS}([A],[B])$, i.e. we **must** reject the null hypothesis at level 0.05, and the ratio of times when $\gamma \leq \widetilde{KS}([A],[B])$, i.e. we **can** reject the null hypothesis at level 0.05. We also compute the number of times where the test is inconclusive, i.e. $\underline{KS}([A],[B]) < \gamma < \widetilde{KS}([A],[B])$.

Figure (6) plots these ratio versus p, the number of summed images. On one hand, concerning the classical KS test, it can be noticed that depending on the quantification level, the answer to the test differs. In fact, when the number of pixel's possible values is low, the test concludes that H_0 cannot be rejected most of the time, leading to a decision that the two distributions are similar even though they are not. When p increases, so increases pixel's possible values and increases the ratio of correct answer. Thus, quantization has a high impact on the conclusions of a classical KS test.

On the other hand, concerning the extended KS test, it can be noticed that the null hypothesis **can** always be rejected. The impact of the quantization only affects the ratio of times when the null hypothesis **must** be rejected. Thus the impact of quantization here is much more sensible, in the sense that when quantization is too severe (information is too poor), the test abstains to make a decision. Also, in all cases, the test is either

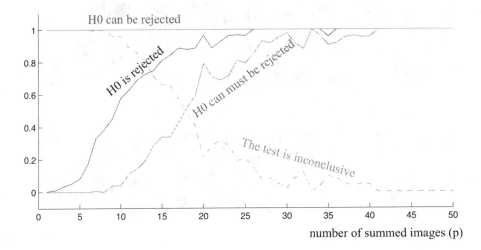

Fig. 6. Correct decision ratio with the classical test (black) and with the extended test (blue for *can be rejected*, red for *must be rejected*), superimposed with the proportion of times extended test is inconclusive (green)

inconclusive or provides the right answer, and is therefore never wrong, which is what we could expect from a robust test.

5   Conclusions

In this paper, we have introduced efficient methods to approximate the bounds of the KS test with interval-valued data. We have demonstrated that the upper bound is exact while the lower bound is, in general, only a lower approximation. However, the experiments have shown that this is not too conservative approximation and still allows to take decision when enough information is available.

The obvious advantages of this paper proposal is its efficiency (computational time is almost linear in the number of sampled intervals), however we may search in the future for exact rather than approximated lower bounds. Since KS test result only depends on the ordering (i.e., ranking) of sampled elements between them, a solution would be to explore the number of possible orderings among elements of $[A]$ and $[B]$, or to identify the orderings for which the lower bound is obtained (the number of such orderings, while finite, may be huge).

Finally, it would also be interesting to investigate other non-parametric homogeneous tests, such as the Cramer-Von Mises one.

Acknowledgements. This work was partially carried out in the framework of the Labex MS2T, which was funded by the French Government, through the program " Investments for the future" managed by the National Agency for Research (Reference ANR-11-IDEX-0004-02)

References

1. Conover, W.J.: Practical non-parametric statistic, 3rd edn. Wiley, New York (1999)
2. Ferson, S., Ginzburg, L., Kreinovich, V., Myers, D.M., Sentz, K.: Constructing probability boxes and dempster-shafer structures. Technical report, Sandia National Laboratories (2003)
3. Hébert, P.-A., Masson, M.-H., Denoeux, T.: Fuzzy multidimensional scaling. Computational Statistics & Data Analysis 51(1), 335–359 (2006)
4. Hesamian, G., Chachi, J.: Two-sample Kolmogorov–Smirnov fuzzy test for fuzzy random variables. Statistical Papers (to appear, 2014)
5. Mballo, C., Diday, E.: Decision trees on interval valued variables. The Electronic Journal of Symbolic Data Analysis 3, 8–18 (2005)
6. Nguyen, H.T., Kreinovich, V., Wu, B., Xiang, G.: Computing Statistics under Interval and Fuzzy Uncertainty. SCI, vol. 393. Springer, Heidelberg (2012)
7. Otero, J., Sánchez, L., Couso, I., Palacios, A.M.: Bootstrap analysis of multiple repetitions of experiments using an interval-valued multiple comparison procedure. J. Comput. Syst. Sci. 80(1), 88–100 (2014)

A Continuous Updating Rule
for Imprecise Probabilities

Marco E.G.V. Cattaneo

Department of Statistics, LMU Munich
Ludwigstraße 33, 80539 München, Germany
cattaneo@stat.uni-muenchen.de
www.statistik.lmu.de/~cattaneo

Abstract. The paper studies the continuity of rules for updating imprecise probability models when new data are observed. Discontinuities can lead to robustness issues: this is the case for the usual updating rules of the theory of imprecise probabilities. An alternative, continuous updating rule is introduced.

Keywords: coherent lower and upper previsions, natural extension, regular extension, α-cut, robustness, Hausdorff distance.

1 Introduction

Imprecise probability models must be updated when new information is gained. In particular, prior (coherent) lower previsions must be updated to posterior ones when new data are observed. Unfortunately, the usual updating rules of the theory of imprecise probabilities have some discontinuities. These can lead to robustness problems, because an arbitrarily small change in the prior lower previsions can induce a substantial change in the posterior ones.

In the next section, the discontinuity of the usual updating rules is illustrated by examples and formally studied in the framework of functional analysis. Then, in Sect. 3, an alternative, continuous updating rule is introduced and discussed. The final section gives directions for further research.

2 Discontinuous Updating Rules and Robustness Issues

Let Ω be a nonempty set of possible states of the world. A (bounded) uncertain payoff depending on the true state of the world $\omega \in \Omega$ can be represented by an element of $\mathcal{L}$, the set of all bounded real-valued functions on Ω. In the Bayesian theory, the uncertain belief or information about the true state of the world $\omega \in \Omega$ is described by a (finitely additive) probability measure P on Ω [1,2]. The expectation $P(X)$ of an uncertain payoff $X \in \mathcal{L}$ is its integral with respect to this probability measure [3, Chap. 4]. Hence, P denotes the probability as well as the expectation: $P(A) = P(I_A)$, where I_A denotes the indicator function of the event $A \subseteq \Omega$.

A. Laurent et al. (Eds.): IPMU 2014, Part III, CCIS 444, pp. 426–435, 2014.

Let $\mathcal{P}$ be the set of all expectation functionals $P : \mathcal{L} \to \mathbb{R}$ corresponding to integrals with respect to a (finitely additive) probability measure on Ω. In the theory of imprecise probabilities, the elements of $\mathcal{P}$ are called linear previsions and are a special case of coherent lower and upper previsions. A (coherent) lower prevision $\underline{P} : \mathcal{L} \to \mathbb{R}$ is the (pointwise) infimum of a nonempty set $\mathcal{M} \subseteq \mathcal{P}$ of linear previsions [4, Sect. 3.3]. Hence, a lower prevision $\underline{P}$ is determined by the set $\mathcal{M}(\underline{P}) = \{P \in \mathcal{P} : P \geq \underline{P}\}$ of all linear previsions (pointwise) dominating it.

Let $\underline{\mathcal{P}}$ be the set of all (coherent) lower previsions $\underline{P} : \mathcal{L} \to \mathbb{R}$. The upper prevision $\overline{P} : \mathcal{L} \to \mathbb{R}$ conjugate to a lower prevision $\underline{P}$ is the (pointwise) supremum of $\mathcal{M}(\underline{P})$. Hence, $\overline{P}(X) = -\underline{P}(-X)$ for all $X \in \mathcal{L}$, and $\underline{P}$ is linear (i.e., $\underline{P} \in \mathcal{P} \subseteq \underline{\mathcal{P}}$) if and only if $\overline{P} = \underline{P}$. As in the case of linear previsions, $\underline{P}$ and $\overline{P}$ denote also the corresponding lower and upper probabilities: $\underline{P}(A) = \underline{P}(I_A)$ and $\overline{P}(A) = \overline{P}(I_A) = 1 - \underline{P}(A^c)$ for all $A \subseteq \Omega$. However, contrary to the case of linear previsions, in general these lower and upper probability values do not completely determine the corresponding lower and upper previsions.

Linear previsions and lower (or upper) previsions are quantitative descriptions of uncertain belief or information. When an event $B \subseteq \Omega$ is observed, these descriptions must be updated. The updating is trivial when $B = \Omega$ or when B is a singleton. Hence, in order to avoid trivial results, it is assumed that B is a proper subset of Ω with at least two elements.

In the Bayesian theory, a linear prevision $P \in \mathcal{P}$ with $P(B) > 0$ is updated to $P(\cdot \mid B)$, the conditional linear prevision given B. That is, the integral with respect to the probability measure P conditioned on B. When $P(B) = 0$, the Bayesian updating of the linear prevision P is not defined.

In the theory of imprecise probabilities, there are two main updating rules: natural extension and regular extension [4, Appendix J]. According to both rules, a lower prevision $\underline{P} \in \underline{\mathcal{P}}$ with $\underline{P}(B) > 0$ is updated to the infimum $\underline{P}(\cdot \mid B)$ of all conditional linear previsions $P(\cdot \mid B)$ with $P \in \mathcal{M}(\underline{P})$. The natural extension updates each lower prevision $\underline{P} \in \underline{\mathcal{P}}$ such that $\underline{P}(B) = 0$ to the vacuous conditional lower prevision given B. That is, the lower prevision $\underline{V}(\cdot \mid B)$ such that $\underline{V}(X \mid B) = \inf_{\omega \in B} X(\omega)$ for all $X \in \mathcal{L}$, describing the complete ignorance about $\omega \in B$. By contrast, the regular extension updates a lower prevision $\underline{P} \in \underline{\mathcal{P}}$ such that $\overline{P}(B) > 0$ to the infimum $\underline{P}(\cdot \mid B)$ of all conditional linear previsions $P(\cdot \mid B)$ with $P \in \mathcal{M}(\underline{P})$ and $P(B) > 0$, and updates only the lower previsions $\underline{P} \in \underline{\mathcal{P}}$ such that $\overline{P}(B) = 0$ to the vacuous conditional lower prevision given B. Hence, the natural and regular extensions are always defined, and they agree for all lower previsions $\underline{P} \in \underline{\mathcal{P}}$ such that either $\underline{P}(B) > 0$ or $\overline{P}(B) = 0$.

Both natural and regular extensions generalize the Bayesian updating: if a lower prevision $\underline{P}$ is linear and $\overline{P}(B) = \underline{P}(B) > 0$, then $\underline{P}(\cdot \mid B)$ is the conditional linear prevision given B. If a lower prevision $\underline{P}$ is linear and $\overline{P}(B) = \underline{P}(B) = 0$, then the conditional linear prevision given B does not exist, and both natural and regular extensions update $\underline{P}$ to $\underline{V}(\cdot \mid B)$, which is not linear. But with lower previsions, besides the cases with $\underline{P}(B)$ and $\overline{P}(B)$ both positive or both zero, there is also the case with $\overline{P}(B) > \underline{P}(B) = 0$. The fact that there are two different updating rules for this case shows that it is more challenging than the others.

Example 1. A simple instance of discontinuity in the updating of lower and upper previsions is the following [5, Example 2]. Let X be a function on Ω with image $\{1, 2, 3\}$, and let $\underline{P}$ be the lower prevision determined by the set $\mathcal{M}(\underline{P}) = \{P \in \mathcal{P} : P(X) \geq x\}$, where $x \in [1, 3]$. That is, $\underline{P}$ is the lower prevision based on the unique assessment $\underline{P}(X) = x$.

Assume that the event $B = \{X \neq 2\}$ is observed and the lower prevision $\underline{P}$ must be updated. If $x > 2$, then $\underline{P}(B) = x - 2 > 0$, and both natural and regular extensions update $\underline{P}$ to $\underline{P}(\,\cdot\,|\,B)$, with in particular $\underline{P}(X\,|\,B) = x$. On the other hand, if $x \leq 2$, then $\overline{P}(B) = 1 > \underline{P}(B) = 0$, and the regular extension updates $\underline{P}$ to $\underline{P}(\,\cdot\,|\,B)$, while the natural extension updates it to $\underline{V}(\,\cdot\,|\,B)$. However, $\underline{P}(\,\cdot\,|\,B)$ and $\underline{V}(\,\cdot\,|\,B)$ are equal when $x < 2$, while they are different when $x = 2$, with in particular $\underline{P}(X\,|\,B) = 2$ and $\underline{V}(X\,|\,B) = 1$.

Therefore, according to both rules, the updated lower previsions of X are discontinuous functions of $x \in [1, 3]$ at $x = 2$. These functions are plotted as a solid line in Fig. 1 (the other functions plotted in Fig. 1 will be discussed in Sect. 3). Hence, inferences and decisions based on imprecise probability models can depend in a discontinuous way from the prior assessments, and this can lead to robustness issues [5].

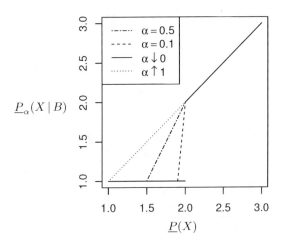

Fig. 1. Updated lower prevision $\underline{P}_\alpha(X\,|\,B)$ from Example 1 as a function of $\underline{P}(X) = x$, for some values of $\alpha \in (0, 1)$

Example 2. A more important instance of discontinuity in the updating of lower and upper previsions is the following [6, Example 9]. Let $X_1, \ldots, X_{10}, Y_1, \ldots, Y_{10}$ be 20 functions on Ω with image $\{0, 1\}$, and for each $t \in (0, 1)$, let $P_t \in \mathcal{P}$ be a linear prevision such that

$$P_t(X_1 = x_1, \ldots, X_{10} = x_{10}, Y_1 = y_1, \ldots, Y_{10} = y_{10})$$
$$= \int_0^1 \frac{\theta^{t-1+\sum_{i=1}^{10} x_i} (1-\theta)^{10-t-\sum_{i=1}^{10} x_i}}{\Gamma(t)\,\Gamma(1-t)}\, d\theta\, \varepsilon^{\sum_{i=1}^{10} |x_i - y_i|} (1-\varepsilon)^{10-\sum_{i=1}^{10} |x_i - y_i|}$$

for all $x_1, \ldots, x_{10}, y_1, \ldots, y_{10} \in \{0, 1\}$, where $\varepsilon \in [0, 1/2]$. That is, P_t describes a Bayesian model in which $X_1, \ldots, X_{10}$ follow the beta-Bernoulli distribution with (prior) expectation t and variance $t(1-t)/2$, while each Y_i is independent of the other 18 variables given $X_i = x_i$, with conditional probability ε that $Y_i \neq x_i$.

Let the lower prevision $\underline{P} \in \underline{\mathcal{P}}$ be the infimum of all linear previsions P_t with $t \in (0, 1)$, and assume that the event $B = \{\sum_{i=1}^9 Y_i = 7\}$ is observed. That is, $\underline{P}$ describes an imprecise beta-Bernoulli model for $X_1, \ldots, X_{10}$ with hyperparameter $s = 1$ [7], but instead of the (latent) variables X_i, only the proxy variables Y_i can be observed, which can be wrong with a (small) probability ε. In the first 9 binary experiments, 7 successes are (possibly incorrectly) reported, and the updated lower prevision of a success in the last experiment is the quantity of interest.

If $\varepsilon > 0$, then $\underline{P}(B) \geq \varepsilon^{10} > 0$, and both natural and regular extensions update $\underline{P}$ to $\underline{P}(\cdot \mid B)$, with in particular $\underline{P}(X_{10} \mid B) = 0$ [6]. On the other hand, if $\varepsilon = 0$, then $\overline{P}(B) > \underline{P}(B) = 0$, and the regular extension updates $\underline{P}$ to $\underline{P}(\cdot \mid B)$, while the natural extension updates it to $\underline{V}(\cdot \mid B)$, with in particular $\underline{P}(X_{10} \mid B) = 0.7$ and $\underline{V}(X_{10} \mid B) = 0$ [7].

Hence, according to the natural extension, the updated lower prevision of X_{10} is a continuous function of $\varepsilon \in [0, 1/2]$. However, since vacuous conditional previsions are not very useful in statistical applications, the usual updating rule for the imprecise beta-Bernoulli model is regular extension, according to which the updated lower prevision of X_{10} is a discontinuous function of $\varepsilon \in [0, 1/2]$ at $\varepsilon = 0$. That is, the assumption that the experimental results can be incorrectly reported with an arbitrarily small positive probability leads to very different conclusions than the assumption that the results are always correctly reported. This is an important robustness problem of the imprecise beta-Bernoulli and other similar models [6].

In order to investigate more carefully such discontinuities in the updating of imprecise probability models, a metric on the set $\underline{\mathcal{P}}$ of all (coherent) lower previsions can be introduced. Lower previsions are functionals on the set $\mathcal{L}$ of all bounded real-valued functions on Ω. The set $\mathcal{L}$ has a natural norm, which makes it a Banach space: the supremum norm $\|\cdot\|_\infty$, defined by $\|X\|_\infty = \sup_{\omega \in \Omega} |X(\omega)|$ for all $X \in \mathcal{L}$ [8, Sect. 4.5].

The set $\mathcal{P}$ of all linear previsions is a subset of the dual space $\mathcal{L}^*$ of $\mathcal{L}$, consisting of all continuous linear functionals on $\mathcal{L}$ [3, Sect. 4.7]. The dual space of $\mathcal{L}$ is a Banach space when endowed with the dual norm $\|\cdot\|$, defined by $\|F\| = \sup_{X \in \mathcal{L}:\, \|X\|_\infty \leq 1} |F(X)|$ for all $F \in \mathcal{L}^*$ [8, Sect. 2.3]. The dual norm induces a metric d on $\mathcal{P}$ such that the distance between two linear previsions $P, P' \in \mathcal{P}$ is 2 times the total variation distance between the corresponding (finitely additive) probability measures [8, Sect. 4.5]:

$$d(P, P') = \sup_{X \in \mathcal{L}: \|X\|_\infty \leq 1} |P(X) - P'(X)| = 2 \sup_{A \subseteq \Omega} |P(A) - P'(A)|.$$

A lower prevision is determined by a nonempty set of linear previsions. The Hausdorff distance between two nonempty sets $\mathcal{M}, \mathcal{M}' \subseteq \mathcal{P}$ of linear previsions is defined as

$$d_H(\mathcal{M}, \mathcal{M}') = \max\left\{\sup_{P \in \mathcal{M}} \inf_{P' \in \mathcal{M}'} d(P, P'), \sup_{P' \in \mathcal{M}'} \inf_{P \in \mathcal{M}} d(P, P')\right\}.$$

The next theorem shows that the Hausdorff distance can be used to extend d to a metric on $\underline{\mathcal{P}}$, which can also be interpreted as a direct generalization of the metric induced by the the dual norm. However, the connection with the total variation distance is lost, because in general the lower and upper probability measures do not completely determine the corresponding lower and upper previsions.

Theorem 1. *The metric d on $\mathcal{P}$ can be extended to a metric on $\underline{\mathcal{P}}$ by defining, for all $\underline{P}, \underline{P}' \in \underline{\mathcal{P}}$,*

$$d(\underline{P}, \underline{P}') = \sup_{X \in \mathcal{L}: \|X\|_\infty \leq 1} |\underline{P}(X) - \underline{P}'(X)| = d_H(\mathcal{M}(\underline{P}), \mathcal{M}(\underline{P'})).$$

Proof. The first equality clearly defines a metric on $\underline{\mathcal{P}}$ that extends the metric d on $\mathcal{P}$. The second equality can be shown by generalizing the proof of [9, Theorem 2] to the case of a possibly infinite set Ω. Let $\underline{P}, \underline{P}' \in \underline{\mathcal{P}}$, and define $f : (P, X) \mapsto P(X) - P'(X)$ on $\mathcal{M}(\underline{P}) \times \mathcal{L}$, where $P' \in \mathcal{P}$. In the weak* topology, $\mathcal{M}(\underline{P})$ is compact and Hausdorff [4, Appendix D], while f is continuous and convex in P, and concave in X. Therefore, the minimax theorem [10, Theorem 2] implies

$$\inf_{P \in \mathcal{M}(\underline{P})} \sup_{X \in \mathcal{L}: \|X\|_\infty \leq 1} |P(X) - P'(X)| = \sup_{X \in \mathcal{L}: \|X\|_\infty \leq 1} \inf_{P \in \mathcal{M}(\underline{P})} (P(X) - P'(X)),$$

since the absolute value on the left-hand side has no effect, because P, P' are linear and X can be replaced by $-X$. Hence,

$$\sup_{P' \in \mathcal{M}(\underline{P}')} \inf_{P \in \mathcal{M}(\underline{P})} d(P, P') = \sup_{X \in \mathcal{L}: \|X\|_\infty \leq 1} \sup_{P' \in \mathcal{M}(\underline{P}')} \inf_{P \in \mathcal{M}(\underline{P})} (P(X) - P'(X))$$

$$= \sup_{X \in \mathcal{L}: \|X\|_\infty \leq 1} (\underline{P}(X) - \overline{P}'(X)),$$

and analogously, by exchanging the roles of $\underline{P}$ and $\underline{P}'$,

$$\sup_{P \in \mathcal{M}(\underline{P})} \inf_{P' \in \mathcal{M}(\underline{P}')} d(P, P') = \sup_{X \in \mathcal{L}: \|X\|_\infty \leq 1} (\underline{P}'(X) - \overline{P}(X)),$$

from which the desired result follows. □

The continuity of the updating rules can now be studied with respect to the metric d on $\underline{\mathcal{P}}$. The next theorem shows that, contrary to Bayesian updating, the usual updating rules of the theory of imprecise probabilities are not continuous on the whole domain on which they are defined.

Theorem 2. *The Bayesian updating rule $P \mapsto P(\cdot \mid B)$ on $\{P \in \mathcal{P} : P(B) > 0\}$ is continuous.*

The regular extension $\underline{P} \mapsto \underline{P}(\cdot \mid B)$ is continuous on $\{\underline{P} \in \underline{\mathcal{P}} : \underline{P}(B) > 0\}$, but not on $\{\underline{P} \in \underline{\mathcal{P}} : \overline{P}(B) > 0\}$. The same holds for the natural extension.

Proof. Let $g : \mathcal{P} \to \mathcal{L}^*$ and $h : \mathcal{P} \to \mathbb{R}$ be the two functions that are defined by $g(P) = P(\cdot I_B)$ and $h(P) = P(B)$, respectively, for all $P \in \mathcal{P}$. The functions g, h are uniformly continuous with respect to the dual and Euclidean norms, respectively, since $\|P(\cdot I_B) - P'(\cdot I_B)\| \leq d(P, P')$ and $|P(B) - P'(B)| \leq 1/2\, d(P, P')$ for all $P, P' \in \mathcal{P}$. Hence, their ratio $g/h : P \mapsto P(\cdot \mid B)$ is continuous on the set $\{P \in \mathcal{P} : P(B) > 0\}$, and uniformly continuous on the set $\{P \in \mathcal{P} : P(B) > \alpha\}$, for all $\alpha \in (0, 1)$.

For each $\alpha \in (0, 1)$, let $\mathcal{S}_\alpha$ be the power set of $\{P \in \mathcal{P} : P(B) > \alpha\}$. The uniform continuity of $P \mapsto P(\cdot \mid B)$ on $\{P \in \mathcal{P} : P(B) > \alpha\}$ implies the uniform continuity of $\mathcal{M} \mapsto \{P(\cdot \mid B) : P \in \mathcal{M}\}$ on $\mathcal{S}_\alpha \setminus \{\varnothing\}$ with respect to the Hausdorff pseudometric d_H, for all $\alpha \in (0, 1)$.

Therefore, in order to complete the proof of the continuity of $\underline{P} \mapsto \underline{P}(\cdot \mid B)$ on $\{\underline{P} \in \underline{\mathcal{P}} : \underline{P}(B) > 0\}$, it suffices to show that the sets $\{P(\cdot \mid B) : P \in \mathcal{M}(\underline{P})\}$ and $\mathcal{M}(\underline{P}(\cdot \mid B))$ are equal for all $\underline{P} \in \underline{\mathcal{P}}$ such that $\underline{P}(B) > 0$. That is, it suffices to show that $\{P(\cdot \mid B) : P \in \mathcal{M}(\underline{P})\}$ is nonempty, weak*-compact, and convex, for all $\underline{P} \in \underline{\mathcal{P}}$ such that $\underline{P}(B) > 0$ [4, Sect. 3.6], but this is implied by the fact that the function $P \mapsto P(\cdot \mid B)$ on $\{P \in \mathcal{P} : P(B) > 0\}$ is weak*-continuous and maintains segments [11, Theorem 2].

Finally, let X be a function on Ω with image $\{1, 2, 3\}$ such that $B = \{X \neq 2\}$, and for each $x \in [1, 3]$, let $\underline{P}_x$ be the lower prevision based on the unique assessment $\underline{P}_x(X) = x$. Example 1 implies that both regular and natural extensions updating of $\underline{P}_x$ are discontinuous functions of $x \in [1, 3]$ at $x = 2$. Hence, in order to prove that both regular and natural extensions are not continuous on $\{\underline{P} \in \underline{\mathcal{P}} : \overline{P}(B) > 0\}$, it suffices to show that the function $x \mapsto \underline{P}_x$ on $[1, 3]$ is continuous.

Let $x, x' \in [1, 3]$ with $x < x'$. For each $P \in \mathcal{P}$ such that $x \leq P(X) < x'$, another $P' \in \mathcal{P}$ with $P'(X) = x'$ and $d(P, P') \leq 2(x' - x)$ can be obtained by moving some probability mass from $\{X \neq 3\}$ to $\{X = 3\}$. Therefore, $d_H(\mathcal{M}(\underline{P}_x), \mathcal{M}(\underline{P}_{x'})) \leq 2(x' - x)$, and the function $x \mapsto \underline{P}_x$ on $[1, 3]$ is thus continuous. $\qquad \square$

Hence, the case with $\overline{P}(B) > \underline{P}(B) = 0$ is a source of discontinuity in the usual updating rules of the theory of imprecise probabilities. Unfortunately, this case is very common and leads to robustness issues in practical applications of the theory, such as the ones based on the imprecise beta-Bernoulli model discussed in Example 2. In the next section, an alternative updating rule avoiding these difficulties is introduced.

3   A Continuous Updating Rule

The discontinuities of the regular extension when $\overline{P}(B) > \underline{P}(B) = 0$ are due to the fact that in this case $\mathcal{M}(\underline{P})$ contains linear previsions P with arbitrarily small $P(B)$, so that their updating $P \mapsto P(\cdot \,|\, B)$ is arbitrarily sensitive to small changes in P. However, these linear previsions are almost refuted by the observed data B. In fact, in a hierarchical Bayesian model with a second-order probability measure on $\mathcal{M}(\underline{P})$, the linear previsions $P \in \mathcal{M}(\underline{P})$ with arbitrarily small $P(B)$ would not pose a problem for the updating, since they would be downweighted by the (second-order) likelihood function $P \mapsto P(B)$. Theorem 2 implies that the updating of such a hierarchical Bayesian model would be continuous, if possible (i.e., if the second-order probability measure is not concentrated on the linear previsions $P \in \mathcal{M}(\underline{P})$ with $P(B) = 0$).

The regular extension updates a lower prevision $\underline{P} \in \mathcal{P}$ such that $\overline{P}(B) > 0$ by conditioning on B all linear previsions $P \in \mathcal{M}(\underline{P})$ with positive likelihood $P(B)$. A simple way to define an alternative updating rule avoiding the discontinuities of the regular extension is to discard not only the linear previsions with zero likelihood, but also the too unlikely ones. The α-cut rule updates each lower prevision $\underline{P} \in \mathcal{P}$ such that $\overline{P}(B) > 0$ to the infimum $\underline{P}_\alpha(\cdot \,|\, B)$ of all conditional linear previsions $P(\cdot \,|\, B)$ with $P \in \mathcal{M}(\underline{P})$ and $P(B) \geq \alpha\,\overline{P}(B)$, where $\alpha \in (0,1)$. That is, the linear previsions whose (relative) likelihood is below the threshold α are discarded from the set $\mathcal{M}(\underline{P})$, before conditioning on B all linear previsions in this set. Hence, the α-cut updating corresponds to the natural and regular extensions when $\underline{P}(B) \geq \alpha\,\overline{P}(B) > 0$, and thus in particular it generalizes the Bayesian updating.

From the standpoint of the theory of imprecise probabilities, the α-cut updating rule consists in replacing the lower prevision $\underline{P}$ by $\underline{P}_\alpha$ before updating it by regular extension (or natural extension, since they agree in this case), where the lower prevision $\underline{P}_\alpha$ is obtained from $\underline{P}$ by including the unique additional assessment $\underline{P}_\alpha(B) \geq \alpha\,\overline{P}(B)$. That is, $\mathcal{M}(\underline{P}_\alpha) = \mathcal{M}(\underline{P}) \cap \{P \in \mathcal{P} : P(B) \geq \alpha\,\overline{P}(B)\}$. Updating rules similar to the α-cut have been suggested in the literature on imprecise probabilities [12,13,14,15]. When $\overline{P}(B) > 0$, the regular extension $\underline{P}(\cdot \,|\, B)$ is the limit of the α-cut updated lower prevision $\underline{P}_\alpha(\cdot \,|\, B)$ as α tends to 0. Furthermore, when Ω is finite and $\underline{P}$ corresponds to a belief function on Ω, the limit of the α-cut updated lower prevision $\underline{P}_\alpha(\cdot \,|\, B)$ as α tends to 1 corresponds to the result of Dempster's rule of conditioning [16,12].

Example 1 (continued). Since $\overline{P}(B) = 1$, the α-cut updating of $\underline{P}$ is well-defined, and $\mathcal{M}(\underline{P}_\alpha) = \{P \in \mathcal{P} : P(X) \geq x$ and $P(B) \geq \alpha\}$. Therefore, $\underline{P}_\alpha$ and $\underline{P}$ are equal if and only if $x \geq 2 + \alpha$. However, $\underline{P}_\alpha(\cdot \,|\, B)$ and $\underline{P}(\cdot \,|\, B)$ can be equal also when $\underline{P}_\alpha$ and $\underline{P}$ are not, and in fact, $\underline{P}_\alpha(\cdot \,|\, B)$ and $\underline{P}(\cdot \,|\, B)$ differ if and only if $2 - \alpha < x < 2$. That is, the α-cut updating of $\underline{P}$ corresponds to the regular extension updating when $x \notin (2 - \alpha, 2)$ (and thus also to the natural extension updating when $x \notin (2 - \alpha, 2]$), with in particular $\underline{P}_\alpha(X \,|\, B) = 0$ when $x \leq 2 - \alpha$, and $\underline{P}_\alpha(X \,|\, B) = x$ when $x \geq 2$. If $2 - \alpha < x < 2$, then the α-cut updating of $\underline{P}$ differs from the natural or regular extensions updating, with in particular $\underline{P}_\alpha(X \,|\, B) = x/\alpha + 2 - 2/\alpha$.

Hence, the updated lower prevision $\underline{P}_\alpha(X \,|\, B)$ is a continuous, piecewise linear function of $x \in [1, 3]$, which is plotted in Fig. 1 for some values of $\alpha \in (0, 1)$. Since the slope of the central line segment is $1/\alpha$, the limit as α tends to 0 leads to the discontinuity of the regular extension $\underline{P}(X \,|\, B)$ at $x = 2$.

Example 2 (continued). Since $\overline{P}(B) > 0$, the α-cut updating of $\underline{P}$ is well-defined. The explicit consideration of $\underline{P}_\alpha$ is not necessary for calculating $\underline{P}_\alpha(X_{10} \mid B)$. In fact, the (relative) profile likelihood function

$$\lambda : p \mapsto \sup_{P \in \mathcal{M}(\underline{P}) \,:\, P(B) > 0,\, P(X_{10} \mid B) = p} \frac{P(B)}{\overline{P}(B)}$$

on $[0, 1]$ can be easily obtained [11, Theorem 3], and $\underline{P}_\alpha(X_{10} \mid B) = \inf\{\lambda \geq \alpha\}$. That is, $\underline{P}_\alpha(X_{10} \mid B)$ is the infimum of the α-cut of the profile likelihood λ, consisting of all $p \in [0, 1]$ such that $\lambda(p) \geq \alpha$.

The (relative) profile likelihood function λ is plotted in Fig. 2 for some values of $\varepsilon \in [0, 1/2]$, together with the threshold $\alpha = 0.01$. In particular, $\underline{P}_{0.01}(X_{10} \mid B)$ has approximately the values 0.701, 0.704, 0.718, and 0.727, when ε is 0, 0.01, 0.05, and 0.1, respectively. Hence, there is no hint of any discontinuity of the updated lower prevision $\underline{P}_\alpha(X_{10} \mid B)$ as a function of $\varepsilon \in [0, 1/2]$.

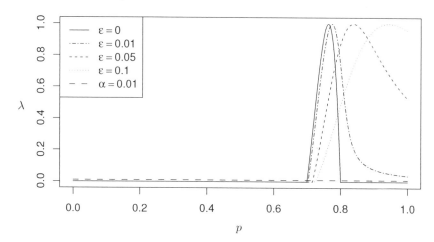

Fig. 2. Profile likelihood function of $P(X_{10} \mid B)$ from Example 2, for some values of $\varepsilon \in [0, 1/2]$

The next theorem shows that, contrary to the usual updating rules of the theory of imprecise probabilities, the α-cut updating rule is continuous on the set of all lower previsions $\underline{P}$ with $\overline{P}(B) > 0$, and thus avoids the robustness issues related to discontinuity of updating. Moreover, the α-cut rule has the huge advantage of allowing to use the vacuous prevision as prior imprecise prevision in statistical analyses, and still get non-trivial conclusions, while this is not possible with the natural or regular extensions. For these reasons, the α-cut updating rule has been successfully used in several practical applications of the theory of imprecise probabilities [17,18,19].

Theorem 3. *The α-cut updating rule $\underline{P} \mapsto \underline{P}_\alpha(\cdot \mid B)$ on $\{\underline{P} \in \mathcal{P} : \overline{P}(B) > 0\}$ is continuous for all $\alpha \in (0, 1)$.*

Proof. The α-cut updating rule is the composition of the two functions $\underline{P} \mapsto \underline{P}_\alpha$ on $\{\underline{P} \in \mathcal{P} : \overline{P}(B) > 0\}$ and $\underline{P} \mapsto \underline{P}(\cdot \mid B)$ on $\{\underline{P} \in \mathcal{P} : \underline{P}(B) > 0\}$. Theorem 2 implies that the latter is continuous, so only the continuity of the former remains to be proved.

Let $\delta \in \mathbb{R}_{>0}$ and let $\underline{P}, \underline{P}' \in \{\underline{P}'' \in \mathcal{P} : \overline{P}''(B) > 0\}$ such that $d(\underline{P}, \underline{P}') < \delta$. If $P \in \mathcal{M}(\underline{P}_\alpha)$, then $P \in \mathcal{M}(\underline{P})$, and thus there is a $P' \in \mathcal{M}(\underline{P}')$ with $d(P, P') < \delta$. Moreover, there is also a $P'' \in \mathcal{M}(\underline{P}')$ such that $P''(B) = \overline{P}'(B)$ [4, Sect. 3.6]. Finally, define $P_\delta = \gamma\, P' + (1 - \gamma)\, P''$, where

$$\gamma = \frac{2\,(1 - \alpha)\,\overline{P}'(B)}{2\,(1 - \alpha)\,\overline{P}'(B) + (\alpha + 1)\,\delta}.$$

Then $P_\delta \in \mathcal{M}(\underline{P}'_\alpha)$, since $\mathcal{M}(\underline{P}')$ is convex, and $P_\delta(B) > \alpha\, \overline{P}'(B)$ is implied by $P'(B) > P(B) - \delta/2 \geq \alpha\, \overline{P}(B) - \delta/2 > \alpha\,\big(\overline{P}'(B) - \delta/2\big) - \delta/2$. That is, for each $P \in \mathcal{M}(\underline{P}_\alpha)$, there is a $P_\delta \in \mathcal{M}(\underline{P}'_\alpha)$ such that $d(P, P_\delta) < \gamma\,\delta + (1 - \gamma)\,2$.

Since the roles of $\underline{P}$ and $\underline{P}'$ can be exchanged, $d(\underline{P}_\alpha, \underline{P}'_\alpha) \leq \gamma\,\delta + (1 - \gamma)\,2$, and therefore the function $\underline{P} \mapsto \underline{P}_\alpha$ on $\{\underline{P} \in \mathcal{P} : \overline{P}(B) > 0\}$ is continuous, because $\lim_{\delta \downarrow 0} (\gamma\,\delta + (1 - \gamma)\,2) = 0$. □

4   Conclusion

In the present paper, the α-cut updating rule for imprecise probability models has been introduced. It is continuous and allows to use the vacuous prevision as prior imprecise prevision in statistical analyses. Therefore, it avoids the robustness problems of the usual updating rules of the theory of imprecise probabilities, and the difficulties related to the choice of near-ignorance priors.

As a consequence, the α-cut rule cannot always satisfy the property of coherence between conditional and unconditional previsions [20], since this property is incompatible with both the continuity of the updating and the successful use of vacuous priors in statistical analyses. The relative importance of these properties depends on the application field and on the exact interpretation of imprecise probabilities, as will be discussed in future work.

Anyway, the α-cut updating rule is not iteratively consistent. That is, when the updating is done in several steps, the order in which the data come in can influence the result. This iterative inconsistency can be resolved if the imprecise probability models are generalized by including also the second-order likelihood functions as part of the models [21,11]. The corresponding generalization of the α-cut updating rule remains continuous, as will also be discussed in future work.

Acknowledgments. The author wishes to thank the anonymous referees and the participants of WPMSIIP 2013 for their valuable comments and suggestions.

References

1. de Finetti, B.: Theory of Probability, 2 vols. Wiley (1974-1975)
2. Savage, L.J.: The Foundations of Statistics, 2nd edn. Dover (1972)
3. Bhaskara Rao, K.P.S., Bhaskara Rao, M.: Theory of Charges. Academic Press (1983)
4. Walley, P.: Statistical Reasoning with Imprecise Probabilities. Chapman and Hall (1991)
5. Cattaneo, M.: On the robustness of imprecise probability methods. In: Cozman, F., Denœux, T., Destercke, S., Seidenfeld, T. (eds.) ISIPTA 2013, pp. 33–41. SIPTA (2013)
6. Piatti, A., Zaffalon, M., Trojani, F., Hutter, M.: Limits of learning about a categorical latent variable under prior near-ignorance. Int. J. Approx. Reasoning 50, 597–611 (2009)
7. Walley, P.: Inferences from multinomial data: learning about a bag of marbles. J. R. Stat. Soc., Ser. B 58, 3–57 (1996)
8. Dunford, N., Schwartz, J.T.: Linear Operators, 3 vols. Wiley (1958–1971)
9. Škulj, D., Hable, R.: Coefficients of ergodicity for Markov chains with uncertain parameters. Metrika 76, 107–133 (2013)
10. Fan, K.: Minimax theorems. Proc. Natl. Acad. Sci. USA 39, 42–47 (1953)
11. Cattaneo, M.: A generalization of credal networks. In: Augustin, T., Coolen, F.P.A., Moral, S., Troffaes, M.C.M. (eds.) ISIPTA 2009, pp. 79–88, SIPTA (2009)
12. Moral, S., de Campos, L.M.: Updating uncertain information. In: Bouchon-Meunier, B., Yager, R.R., Zadeh, L.A. (eds.) IPMU 1990. LNCS, vol. 521, pp. 58–67. Springer, Heidelberg (1991)
13. Moral, S.: Calculating uncertainty intervals from conditional convex sets of probabilities. In: Dubois, D., Wellman, M.P., D'Ambrosio, B., Smets, P. (eds.) UAI 1992, pp. 199–206. Morgan Kaufmann (1992)
14. Gilboa, I., Schmeidler, D.: Updating ambiguous beliefs. J. Econ. Theory 59, 33–49 (1993)
15. Moral, S., Wilson, N.: Revision rules for convex sets of probabilities. In: Coletti, G., Dubois, D., Scozzafava, R. (eds.) Mathematical Models for Handling Partial Knowledge in Artificial Intelligence, pp. 113–127. Springer (1995)
16. Shafer, G.: A Mathematical Theory of Evidence. Princeton University Press (1976)
17. Antonucci, A., Cattaneo, M.E.G.V., Corani, G.: Likelihood-based robust classification with Bayesian networks. In: Greco, S., Bouchon-Meunier, B., Coletti, G., Fedrizzi, M., Matarazzo, B., Yager, R.R. (eds.) IPMU 2012, Part III. CCIS, vol. 299, pp. 491–500. Springer, Heidelberg (2012)
18. Cattaneo, M., Wiencierz, A.: Likelihood-based Imprecise Regression. Int. J. Approx. Reasoning 53, 1137–1154 (2012)
19. Destercke, S.: A pairwise label ranking method with imprecise scores and partial predictions. In: Blockeel, H., Kersting, K., Nijssen, S., Železný, F. (eds.) ECML PKDD 2013, Part II. LNCS (LNAI), vol. 8189, pp. 112–127. Springer, Heidelberg (2013)
20. Miranda, E.: Updating coherent previsions on finite spaces. Fuzzy Sets Syst. 160, 1286–1307 (2009)
21. Cattaneo, M.: Fuzzy probabilities based on the likelihood function. In: Dubois, D., Lubiano, M.A., Prade, H., Gil, M.A., Grzegorzewski, P., Hryniewicz, O. (eds.) Soft Methods for Handling Variability and Imprecision, pp. 43–50. Springer (2008)

Stable Non-standard Imprecise Probabilities

Hykel Hosni[1,*] and Franco Montagna[2]

[1] London School of Economics
h.hosni@lse.ac.uk
[2] Uninversità di Siena
montagna@unisi.it

Abstract. Stability arises as the consistency criterion in a betting interpretation for hyperreal imprecise previsions, that is imprecise previsions
(and probabilities) which may take infinitesimal values. The purpose of
this work is to extend the notion of *stable coherence* introduced in [8]
to conditional hyperreal imprecise probabilities. Our investigation extends the de Finetti-Walley operational characterisation of (imprecise)
prevision to conditioning on events which are considered "practically
impossible" but not "logically impossible".

1 Introduction and Motivation

This paper combines, within a logico-algebraic setting, several extensions of the
imprecise probability framework which we aim to generalise so as to represent
infinitesimal imprecise probabilities on fuzzy events.

Imprecise conditional probabilities, as well as imprecise conditional previsions,
have been investigated in details by Walley [12] in the case where the conditioning
event ψ is boolean and has non-zero lower probability. There, a de Finetti-style
interpretation of upper and lower probability and of upper and lower prevision in
terms of bets is proposed. In Walley's approach, the conditional upper prevision,
$U(x|\psi)$ of the gamble x given the event ψ is defined to be a number α such that
$U(x_\psi \cdot (x - \alpha)) = 0$, where $x_\psi = 1$ if ψ is true and $x_\psi = 0$ if ψ is false.
When the lower probability of ψ is non-zero, there is exactly one α satisfying
the above condition, and hence, the upper conditional prevision is well-defined.
Likewise, the lower conditional prevision of x given ψ is the unique β such that
$L(x_\psi \cdot (x - \beta)) = 0$. However, the uniqueness of α and β is only guaranteed if
the lower probability of ψ is not zero, otherwise, there might be infinitely many
solutions of the above equations.

In terms of bets, the rationality of an assessment of upper probabilities (or
of upper previsions) corresponds to the absence of inadmissible bets, that is, of
bets for which there is an alternative strategy for the gambler which ensures
to him a strictly better payoff whatever the outcome of the experiment will be.
In the case of conditional upper and lower previsions, however, the absence of
inadmissible bets might be due to the fact that the conditioning event has lower

---

* Supported by the European Commission under the Marie Curie IEF-GA-2012-
327630 project *Rethinking Uncertainty: A Choice-based approach.*

A. Laurent et al. (Eds.): IPMU 2014, Part III, CCIS 444, pp. 436–445, 2014.

probability zero (remind that when the conditioning event is false the bet is invalidated, and hence the payoff of *any* bet on the conditional event is zero). So, the presence of events with lower probability zero might force the absence of inadmissible bets even in non-rational assessments. For instance, if we chose a point at random on the surface of Earth, and ϕ denotes the event: *the point belongs to the western hemisphere* and ψ denotes the event *the point belongs to the equator*, the assessment $\phi|\psi \mapsto 0$, $\neg\phi|\psi \mapsto 0$ avoids inadmissible bets (in case the point does not belong to the equator), but is not rational.

The goals of this paper are the following:

(1) To provide for a treatment of conditional upper and lower previsions when the conditioning event is many-valued and when the conditioning event has probability zero. For probabilities in the usual sense, this goal has been pursued in [8].

(2) To model conditional or unconditional bets in which truth values and betting odds may be non-standard. In particular, to every non-zero event with probability zero we assign an infinitesimal non-zero probability, so that we may avoid conditioning events with probability zero. Then taking standard parts we obtain a probability (or, in the case of many-valued events, a state) in the usual sense.

(3) The basic idea is the following: we replace the concept of coherence (absence of inadmissible bets) by a stronger concept, namely, *stable coherence*. Not only inadmissible bets are ruled out, but, in addition, the absence of inadmissible bets is preserved if we modify the assessment by an infinitesimal in such a way that no lower probability assessments equal to zero are allowed for events which are not impossible. The main result (Theorem 5) will be that stable coherence for an assessment of conditional upper probability corresponds to the existence of a non-standard upper probability which extends the assessment modulo an infinitesimal and assigns a non-zero lower probability to all non-zero events.

Our main result has important foundational consequences. For stable coherence allows us to distinguish between events which are regarded as practically impossible and events which are indeed logically impossible. It is well-known that this subtle but crucial difference can only be captured by so-called *regular* probability functions which are characterised by Shimony's notion of *strict coherence*. A companion paper will address this point in full detail.

For reasons of space all proofs are omitted from this version of the paper.

2   Algebraic Structures for Non-standard Probability

We build on [8], which can be consulted for further background on MV-algebra and related structures.[1] Specifically, we work in the framework of unital lattice ordered abelian groups, which can be represented as algebras[2] of bounded

---

[1] [2] provides a basic introduction and [11] a more advanced treatment, including states and their relationship with coherence. The basic notions of universal algebras we use are provided by [1].

[2] Algebras will be usually denoted by boldface capital letters (with the exception of the standard MV-algebra on $[0,1]$ and the standard PMV$^+$-algebra on $[0,1]$ which are denoted by $[0,1]_{MV}$ and by $[0,1]_{PMV}$, respectively) and their domains will be denoted by the corresponding lightface capital letters.

functions from a set X into a non-standard extension, $\mathbf{R}^*$, of $\mathbf{R}$. Hence, their elements may be interpreted as bounded random variables (also called *gambles* in [12]). The set of bounded random variables is closed under sum, subtraction and under the lattice operations. Since we are interested in extending the framework of [8] to *imprecise conditional* prevision and probability, we also need a product operation. It turns out that appropriate structure is constituted by the c-s-u-f integral domains.

Definition 1. *A commutative strongly unital function integral domain (c-s-u-f integral domain for short) is an algebra* $\mathbf{R} = (R, +, -, \vee, \wedge, \cdot, 0, 1)$ *such that:*

(i) $(R, +, -, \vee, \wedge, 0)$ *is a lattice ordered abelian group and* 1 *is a strong unit of this lattice ordered group.*

(ii) $(R, +, -, \cdot, 0, 1)$ *is a commutative ring with neutral element* 1.

(iii) The identity $x^+ \cdot (y \vee z) = (x^+ \cdot y) \vee (x^+ \cdot z)$ *holds, where* $x^+ = x \vee 0$.

(iv) The quasi identity: $x^2 = 0$ *implies that* $x = 0$ *holds.*

Remark 1. In [7] it is shown that every c-s-u-f integral domain embeds into an algebra of the form $(\mathbf{R}^*_{fin})^H$, where $\mathbf{R}^*$ is an ultrapower of the real field, $\mathbf{R}_{fin}$ is the c-s-u-f domain consisting of all finite elements of $\mathbf{R}^*$, and H is an index set[3]. In particular, every c-s-u-f integral domain embeds into the product of totally ordered integral domains, and this fact justifies the name of these structures.

Let $(\mathbf{G}, u)$ be a unital lattice ordered group. We define $\Gamma(\mathbf{G}, u)$ to be the algebra whose domain is the interval $[0, u]$, with the constant 0 and with the operations $\sim x = u - x$ and $x \oplus y = (x + y) \wedge u$. Moreover if h is a homomorphism of unital lattice ordered abelian groups $(\mathbf{G}, u)$ and $(\mathbf{G}', u')$ (i.e., a homomorphism of lattice ordered groups such that $h(u) = u'$), we define $\Gamma(h)$ to be the restriction of h to $\Gamma(\mathbf{G}, u)$. Likewise, given a c-s-u-f integral domain $(\mathbf{F}, u)$ and denoting by $\mathbf{F}^-$ the underlying lattice ordered abelian group, we define $\Gamma_R(\mathbf{F}, u)$ to be $\Gamma(\mathbf{F}^-, u)$ equipped with the restriction of $\cdot$ to $\Gamma(\mathbf{F}^-, u)$. Moreover given a homomorphism h of c-s-u-f integral domains from $(\mathbf{F}, u)$ into $(\mathbf{F}', u')$ we denote by $\Gamma_R(h)$ its restriction to $\Gamma_R(\mathbf{F}, u)$.

Theorem 1. *(1) (see [10]).* Γ *is a functor from the category of unital lattice ordered abelian groups into the category of MV-algebras. Moreover* Γ *has an adjoint* Γ^{-1} *such that the pair* (Γ, Γ^{-1}) *is an equivalence of categories.*

(2) (see [7]). Γ_R *is a functor from the category of c-s-u-f integral domains into the category of PMV$^+$-algebras. Moreover* Γ_R *has an adjoint* Γ_R^{-1} *such that the pair* $(\Gamma_R, \Gamma_R^{-1})$ *is an equivalence of categories.*

Remark 2. Theorem 1 tells us that the algebra of gambles, represented by a unital lattice ordered abelian group or of a c-s-u-f integral domain, is completely determined by the algebra of $[0, u]$-valued gambles, whose elements may be regarded as many-valued events. MV and PMV$^+$-algebras provide rich semantics for the logic of many-valued events.

---

[3] Of course, the embedding sends u, the neutral element for product, in to 1.

Let $\mathbf{A}$ be an MV-algebra, and let $(\mathbf{G}, u)$ be a lattice ordered unital abelian group such that $\Gamma(\mathbf{G}, u) = \mathbf{A}$. Let $\mathbf{R}^*$ be an ultrapower of the real field, and let $\mathbf{R}^*_{fin}$ be the set of all finite elements of $\mathbf{R}^*$ and let $[0, 1]^* = \Gamma(\mathbf{R}^*_{fin}, 1)$.

We say that $\mathbf{R}^*$ is $(\mathbf{G}, u)$-*amenable* if for all $g \in G$, if $g \neq 0$, then there is a homomorphism h from $\mathbf{G}$ into $\mathbf{R}^*_{fin}$, considered as a lattice ordered abelian group, such that $h(u) = 1$ and $h(g) \neq 0$. We say that $[0, 1]^*$ is $\mathbf{A}$-*amenable* if for all $a \in A$, if $a \neq 0$, then there is a homomorphism h from $\mathbf{A}$ into $[0, 1]^*$, such that $h(u) = 1$ and $h(g) \neq 0$.

Lemma 1. *Let $(\mathbf{G}, u)$ be a unital lattice ordered abelian group, let $\mathbf{R}^*$ be an ultrapower of $\mathbf{R}$ and $[0, 1]^* = \Gamma(\mathbf{R}^*, 1) = \Gamma(\mathbf{R}^*_{fin}, 1)$. Then $[0, 1]^*$ is $\Gamma(\mathbf{G}, u)$-amenable iff $\mathbf{R}^*$ is $(\mathbf{G}, u)$-amenable.*

In [8], the following result is shown:

Proposition 1. *For every MV-algebra $\mathbf{A}$, an $\mathbf{A}$-amenable ultrapower $[0, 1]^*$, of $[0, 1]$ exists.*

It follows:

Corollary 1. *For each unital lattice ordered abelian group $(\mathbf{G}, u)$, a $(\mathbf{G}, u)$-amenable ultrapower $\mathbf{R}^*$ of $\mathbf{R}$ exists.*

In the sequel, we will need MV-algebras with product or c-s-u-f domains in order to treat conditional probability in an algebraic setting. Moreover we need to treat probabilities in terms of bets in such a way that zero probabilities will be replaced by infinitesimal probabilities. We would like to have a richer structure in which not only MV-operations or lattice ordered group operations, but also product and hyperreal numbers are present. The construction presented in the next lines provides for such structures.

Definition 2. *Let $(\mathbf{R}^*_{fin}, 1)$ be $(\mathbf{G}, u)$-amenable, let H be the set of all homomorphisms from $(\mathbf{G}, u)$ into $(\mathbf{R}^*_{fin}, 1)$, and let Φ be defined, for all $g \in \mathbf{G}$, by $\Phi(g) = (h(g) : h \in H)$. By $\Pi(\mathbf{G}, u, \mathbf{R}^*_{fin})$, we denote the subalgebra of $(\mathbf{R}^*_{fin})^H$ (with respect to the lattice ordered group operations and to product in $\mathbf{R}^*_{fin}$) generated by $\Phi(\mathbf{G})$ and by the elements of $\mathbf{R}^*_{fin}$, thought of as constant maps from H into $\mathbf{R}^*$ (in the sequel, by abuse of language, we denote by α the function from H into $\mathbf{R}^*_{fin}$ which is constantly equal to α).*

Likewise, if $\mathbf{A}$ is an MV-algebra and $[0, 1]^*$ is an ultrapower of $[0, 1]_{MV}$ which is $\mathbf{A}$-amenable, if H is the set of all homomorphisms from $\mathbf{A}$ into $[0, 1]^*$ and Φ is defined, for all $a \in A$ by $\Phi(a) = (h(a) : h \in H)$, then $\Pi(\mathbf{A}, [0, 1]^*)$ denotes the subalgebra of $([0, 1]^*)^H$ generated by $[0, 1]^*$ and by $\Phi(\mathbf{A})$.

It can be proved that both $(\mathbf{G}, u)$ and $\mathbf{R}^*_{fin}$ are embeddable into $\Pi(\mathbf{G}, u, \mathbf{R}^*_{fin})$ and both $\mathbf{A}$ and $[0, 1]^*$ are embeddable in $\Pi(\mathbf{A}, [0, 1]^*)$, see [8].

Lemma 2. $\Pi(\mathbf{G}, u, \mathbf{R}^*_{fin}) = \Gamma^{-1}(\Pi(\Gamma(\mathbf{G}, u), [0, 1]^*))$.

3   Fuzzy Imprecise Probabilities over the Hyperreals

We are now in a position to introduce the notion of fuzzy imprecise hyperprevisions, which, as usual, lead in a special case, to probabilities. We will focus on *upper* hyper previsions and probabilities (*lower* notions will be obtained as usual). Naturally enough, our first step requires us to extend propositional valuations to the hyperreals.

Definition 3. *Let* $(\mathbf{G}, u)$ *be a unital lattice ordered abelian group and suppose that* $\mathbf{R}^*$ *is* $(\mathbf{G}, u)$ *amenable. Let* $\mathbf{A} = \Gamma(\mathbf{G}, u)$ *and* $[0,1]^* = \Gamma_R(\mathbf{R}^*_{fin}, 1)$, $\mathbf{A}^* = \Pi(\mathbf{A}, [0,1]^*)$, $\mathbf{G}^* = \Pi(\mathbf{G}, u, \mathbf{R}^*_{fin})$. *A hypervaluation on* $\mathbf{G}^*$ *(resp., on* $\mathbf{A}^*$*) is a homomorphism* v^* *from* $\mathbf{G}^*$ *into* $\mathbf{R}^*_{fin}$ *(resp., from* $\mathbf{A}^*$ *into* $[0,1]^*$*) such that for every* $\alpha \in \mathbf{R}^*_{fin}$ *(resp., in* $[0,1]^*$*),* $v^*(\alpha^*) = \alpha$. *A hyperprevision on* $(\mathbf{G}, u)$ *is a function* P^* *from* $\mathbf{G}^*$ *into* $\mathbf{R}^*_{fin}$ *such that for all* $\alpha, \in \mathbf{R}^*_{fin}$ *and* $x, y \in \mathbf{G}^*$, *the following conditions hold:*

(1) $P^*(\alpha^* x) = \alpha P^*(x)$.
(2) if $x \geq y$, *then* $P^*(x) \geq P^*(y)$.
(3) $P^*(x + y) = P^*(x) + P^*(y)$.
(4) $P^*(u) = 1$.
(5) There are hypervaluations v, w *such that* $v(x) \leq P^*(x) \leq w(x)$.

A hyperstate *on* $\mathbf{A}^*$ *is a map* S^* *from* $\mathbf{A}^*$*) into* $[0,1]^*$ *such that, for all* $\alpha \in [0,1]^*$ *and for all* $x, y \in \mathbf{A}^*$, *the following conditions hold:*

(a) $S^*(u) = 1$

(b) $S^*(\alpha^* \cdot x) = \alpha \cdot S^*(x)$

(c) if $x \odot y = 0$, *then* $S^*(x \oplus y) = S^*(x) + S^*(y)$

(d) there are hypervaluations v, w *such that* $v(x) \leq S^*(x) \leq w(x)$.

Definition 4. *An* upper hyperprevision *is a function* U^* *on* $\mathbf{G}^*$ *which satisfies (2), (4), (5), with* P^* *replaced by* U^*, *and*

(1)' $U^*(\alpha^* x) = \alpha U^*(x)$, *provided that* $\alpha \geq 0$.
(3)' $U^*(x + y) \leq U^*(x) + U^*(y)$.
(6) $U^*(x + \alpha) = U^*(x) + \alpha$.

An upper hyperstate *on* $\mathbf{A}^*$ *is a function* U_0^* *from* $\mathbf{A}^*$ *into* $[0,1]^*$ *such that, for all* $x, y \in \mathbf{A}^*$ *and for all* $\alpha \in [0,1]^*$, *the following conditions hold:*

(i) $U_0^*(u) = 1$.

(ii) If $x \leq y$, *then* $U_0^*(x) \leq U_0^*(y)$.

(iii) $U_0^*(\alpha \cdot x) = \alpha \cdot U_0^*(x)$ *and if* $\alpha \odot x = 0$, *then* $U_0^*(x \oplus \alpha) = U_0^*(x) + \alpha$.

(iv) $U_0^*(x \oplus y) \leq U_0^*(x) \oplus U_0^*(y)$.

(v) $U_0^*(x \oplus \alpha) = U_0^*(a) + \alpha$ *whenever* $x \odot \alpha = 0$.
(vi) There are hypervaluations v, w *such that* $v(x) \leq U_0^*(x) \leq w(x)$.

Remark 3. (1) A valuation on a lattice ordered abelian group $\mathbf{G}^*$ (resp., on $\mathbf{A}^*$) is a homomorphism v from $\mathbf{G}^*$ into $\mathbf{R}$ (resp., from $\mathbf{A}$ into $[0,1]_{PMV}$) such that $v(\alpha) = \alpha$ for every standard real α. Moreover, a prevision on $\mathbf{G}^*$ is a map into $\mathbf{R}$ which satisfies (2), (3) and (4), as well as (1) for all standard α and (5) with hypervaluations replaced by valuations. Likewise, a state on $\mathbf{A}^*$ is a map into $[0,1]$ which satisfies (a) and (c), as well as (b) for all standard α and (d) with hypervaluations replaced by valuations.

Moreover, an upper prevision on $\mathbf{G}^*$ is a map into $\mathbf{R}$ which satisfies (2), (4) and (3)', as well as (1)' and (6) for all standard α and (5) with hypervaluations replaced by valuations. Finally, an upper hyperstate on $\mathbf{A}^*$ is a map into $[0,1]$ which satisfies (i), (ii) and (iv), as well as (iii) and (v) for all standard α and (vi) with hypervaluations replaced by valuations.

Hence, hypervaluations, hyperprevisions, hyperstates, upper hypeprevisions and upper hyperstates are natural non-standard generalizations of valuations, previsions, states, upper previsions and upper states, respectively.

(2) The restriction to $\mathbf{A}^*$ of a hyperprevision P^* (resp., an upper hyperprevision U^*) on $\mathbf{G}^*$, is a hyperstate (resp., an upper hyperstate). Moreover, a hyperstate S^* (resp, a hyper upper state U_0^*) on $\mathbf{A}^*$ has a unique extension P^* (resp., U^*) to a hyper prevision (resp., to a hyper upper prevision) on $\mathbf{G}^*$. Indeed, given $a \in \mathbf{G}^*$, there are positive integers M, N such that $0 \leq \frac{a+N}{M} \leq u$. So, $\frac{a+N}{M} \in \mathbf{A}^*$, and it suffices to define $P^*(a) = M \cdot S^*(\frac{a+N}{M}) - N$, and $U^*(a) = M \cdot U_0^*(\frac{a+N}{M}) - N$. Note that in [5] it is shown that the definition does not depend on the choice of the integers M and N such that $\frac{a+N}{M} \in \mathbf{A}$.

(3) Let U_0^* be an upper hyperstate on $\mathbf{A}^*$ and U^* be the unique upper hyperprevision on $\mathbf{G}^*$ extending U_0^*. Then for all $x \in \mathbf{A}^*$, the upper hyperprobability, $U_0^*(x)$, of x, is a number α such that the upper hyperprevision, $U^*(x - \alpha)$ of the gamble $x - \alpha$, is 0. Indeed, $U^*(x - \alpha) = U^*(x) - \alpha$ (because α is a constant), and hence, $U_0^*(x) = U^*(x) = \alpha$ iff $U^*(x - \alpha) = 0$. This means that the upper hyperprevision of a gamble x is a number α such that the upper hyperprevision of the payoff of the gambler when he bets 1 with betting odd α, namely $x - \alpha$, is 0.

(4) Given an upper hyperprevision U^*, its corresponding lower hyperprevision is $L^*(x) = -U^*(-x)$. Likewise, if U_0^* is an upper hyperstate, its corresponding lower hyperstate is given by $L_0^*(x) = 1 - U_0^*(\neg x)$.

(5) If U^* is an upper hypeprevision, then $U^*(x) = U^*(x + y - y) \leq U^*(x + y) + U^*(-y)$, and hence, $U^*(x) + L^*(y) \leq U^*(x + y) \leq U^*(x) + U^*(y)$. Likewise, if U_0^* is an upper hyperstate and L_0^* is its corresponding lower hyperstate and if $x \odot y = 0$, then $U_0^*(x) + L_0^*(y) \leq U_0^*(x \oplus y) \leq U_0^*(x) + U_0^*(y)$.

We now present a betting interpretation of hyper upper previsions, which will lead to the appropriate notion of coherence. We begin by recalling a characterisation of coherence as *avoiding inadmissible bets* given in [6] (the terminology "bad bet" was used there.)

Definition 5. *Let* $(\mathbf{G}, u)$ *be a unital lattice ordered abelian group, and let* $\Lambda = x_1 \mapsto \alpha_1, \ldots, x_n \mapsto \alpha_n$ *be an assessment of upper previsions on the bounded* *random variables* $x_1, \ldots, x_n \in \mathbf{G}$. *The associated betting game is as follows: the gambler can bet only non-negative numbers* $\lambda_1, \ldots, \lambda_n$ *on* $x_1, \ldots, x_n$, *and the payoff for the bookmaker corresponding to the valuation* v *on* $(\mathbf{G}, u)$ *will be* $\sum_{i=1}^{n} \lambda_i \cdot (\alpha_i - v(x_i))$.

Let W *be a set of valuations on* $(\mathbf{G}, u)$. *An* inadmissible W *bet is a bet* $\mu_i \geq 0$ *on* x_i *(for some* $i \leq n$*) such that there is a system of non-negative bets* $\lambda_1, \ldots, \lambda_n$ *which guarantees a better payoff to the gambler, independently of the valuation* $v \in W$, *that is, for every valuation* $v \in W$, $\sum_{j=1}^{n} \lambda_j \cdot (v(x_j) - \alpha_j) > \mu_i \cdot (v(x_i) - \alpha_i)$. *An* inadmissible bet *is an inadmissible* W *bet, where* W *is the set of all valuations on* $(\mathbf{G}, u)$.

The assessment Λ *is said to be* W coherent *if it excludes inadmissible* W-*bets, and* coherent *if it excludes inadmissible bets.*

In [6] it is shown that an assessment of upper probability avoids inadmissible bets iff it can be extended to an upper prevision. The result was shown first, although in a different setting, by Walley in [12].

In [6] it is also shown that given gambles $x_1, \ldots, x_m$ and given an upper prevision U, for $i = 1, \ldots, m$ there is a prevision P_i such that $P_i(x_i) = U(x_i)$ and $P_i(x) \leq U(x)$ for every gamble x. Moreover, as shown in [11], there are valuations $v_{i,j}$ and non-negative reals $\lambda_{j,i}$, $i = 1, \ldots, m+1$, $j = 1, \ldots, m$ such that for $j = 1, \ldots, m$, $\sum_{i=1}^{m+1} \lambda_{i,j} = 1$ and for $h, j = 1, \ldots, m$, $s_j(x_h) = \sum_{i=1}^{m+1} \lambda_{i,j} v_{i,j}(x_h)$. In other words, we can assume that each P_i is a convex combination of valuations.

Hence, coherence for upper previsions is equivalent to the following condition:

Theorem 2. *Let* $\Lambda = x_1 \mapsto \alpha_1, \ldots, x_m \mapsto \alpha_m$ *be an assessment as in Definition 5. Then* Λ *is coherent (i.e., avoids inadmissible bets) iff there are valuations* $v_{i,j}$: $j = 1, \ldots, m$, $i = 1, \ldots, m+1$ *and non-negative real numbers* $\lambda_{i,j}$: $j = 1, \ldots, m$, $i = 1, \ldots, m+1$, *such that, letting for* $j = 1, \ldots, m$, $P_j(x) = \sum_{i=1}^{m+1} \lambda_{i,j} v_{i,j}(x)$, *the following conditions hold:*
(i) For $j = 1, \ldots, m$, $\sum_{i=1}^{m+1} \lambda_{i,j} = 1$.
(ii) For $j = 1, \ldots, m$, $P_i(x_j) \leq \alpha_j$.
(iii) $P_i(x_i) = \alpha_i$.

In words, Λ *avoids inadmissible bets iff there are* m *convex combinations,* $P_1, \ldots, P_m$, *of valuations, such that for* $j = 1, \ldots, m$, $\alpha_j = \max\{P_h(x_j) : h = 1, \ldots, m\}$.

The result above[4] may be extended to non-standard assessments, to hypervaluations and to upper hyper previsions. First of all, we consider a $(\mathbf{G}, u)$-amenable ultrapower, $\mathbf{R}^*$, of $\mathbf{R}$, and we set $\mathbf{G}^* = \Pi(\mathbf{G}, u, \mathbf{R}^*_{fin})$. Then we consider a hyperassessment $\Lambda := x_1 \mapsto \alpha_1, \ldots, x_n \mapsto \alpha_n$ with $x_1, \ldots, x_n \in \mathbf{G}^*$. Let W be a set of hypervaluations. We say that Λ is W-*coherent* if it rules out inadmissible W-bets, that is, for $i = 1, \ldots, n$ and for every $\lambda, \lambda_1, \ldots, \lambda_n \geq 0$, there is a hypervaluation $v^* \in W$ such that $\lambda \cdot (v^*(x_i) - \alpha_i) \geq \sum_{j=1}^{n+1} \lambda_j \cdot (v^*(x_j) - \alpha_j)$. We

---

[4] Our coherence criterion resembles very closely a number of similarly-minded generalisations of de Finetti's own notion of coherence, among others that of [3].

say that Λ is $\mathbf{R}^*$-*coherent* if it is W-coherent, where W is the set of all hyperval-uations on $\mathbf{G}^*$, and that Λ is *coherent* if it is $\mathbf{R}^\circ$-coherent for some ultrapower, $\mathbf{R}^\circ$, of $\mathbf{R}^*$. Similar definitions can be given for assessments of upper hyperprob-ability on algebras of the form $\mathbf{A}^* = \Pi(\mathbf{A}, [0,1]^*)$ (in this case, hypervaluations are homomorphisms from $\mathbf{A}^*$ into $[0,1]^*$ which preserve the elements of $[0,1]^*$, and (upper) hyperprevisions must be replaced by (upper) hyperstates).

Recall that there is a bijection between upper hyperprevisions on $\mathbf{G}^* = \Pi(\mathbf{G}, u, \mathbf{R}^*_{fin})$ and upper hyperstates on $\mathbf{A}^* \Gamma_R(\Pi(\mathbf{G}, u, \mathbf{R}^*_{fin}))$: the restriction to $\mathbf{A}^*$ of an upper hyperprevision is an upper hyperstate, and every upper hy-perstate on $\mathbf{A}^*$ has a unique extension to an upper hyperprevision U^* on $\mathbf{G}^*$.

By a similar argument, there is a bijection between coherent assessments on $\mathbf{G}^*$ and coherent assessments on $\mathbf{A}^*$. Indeed, clearly, a coherent assessment on $\mathbf{A}^*$ is also a coherent assessment on $\mathbf{G}^*$. Conversely, given any assessment $\Lambda =: x_1 \mapsto \alpha_1, \ldots, x_k \mapsto \alpha_k$ on $\mathbf{G}^*$, there are integers M_i, N_i, with $M_i > 0$, such that $0 \leq \frac{x_i + N_i}{M_i} \leq u$. Now let $a_i = \frac{x_i + N_i}{M_i}$, and let Λ_0 be the assessment: $\Lambda_0 =: a_1 \mapsto \frac{\alpha_1 + N_1}{M_1}, \ldots, a_k \mapsto \frac{\alpha_k + N_k}{M_k}$ on $\mathbf{A}^*$. Then Λ avoids inadmissible bets iff Λ_0 avoids inadmissible bets, and Λ can be extended to an upper hyperprevision U^* iff Λ_0 extends to its restriction U_0^* to $\mathbf{A}^*$, which is an upper hyperstate on $\mathbf{A}^*$. Hence, *in the sequel we will often identify upper hyperprevisions on $\mathbf{G}^*$ with their restriction to $\mathbf{A}^*$, and the assessment Λ on $\mathbf{G}^*$ with its corresponding assessment Λ_0 on $\mathbf{A}^*$.*

Theorem 3. *Let $\Lambda = x_1 \mapsto \alpha_1, \ldots, x_m \mapsto \alpha_m$ be a hyperassessment on an algebra of the form $\mathbf{G}^*$ (hence, $\alpha_i \in \mathbf{R}^*_{fin}$). Then the following are equivalent:*

(1) Λ is coherent.

(2) There is an upper hyperprevision U^* s.t. for $i = 1, \ldots, m$, $U^*(x_i) = \alpha_i$.

(3) There are hypervaluations $v_{i,j}^* : j = 1, \ldots, m$, $i = 1, \ldots, m+1$ and non-negative hyperreal numbers (possibly, in an ultrapower of $\mathbf{R}^*$) $\lambda_{i,j} : j = 1, \ldots, m$, $i = 1, \ldots, m+1$, such that, letting for $j = 1, \ldots, m$, $P_j^*(x) = \sum_{i=1}^{m+1} \lambda_{i,j} v_{i,j}^*(x)$, the following conditions hold:
  (i) For $j = 1, \ldots, m$, $\sum_{i=1}^{m+1} \lambda_{i,j} = 1$.
  (ii) For $j = 1, \ldots, m$, $P_i^*(x_j) \leq \alpha_j$.
  (iii) $P_i^*(x_i) = \alpha_i$.

We express this fact saying that Λ is the supremum of m convex combinations of $m + 1$ hypervaluations

We conclude this section with a result that, up to infinitesimals, a coherent assessment can be extended by a faithful upper hyperprevision (an upper hyper-prevision U^* is faithful if $U^*(x) = 1$ implies $x = 1$, or equivalently if $L^*(x) = 0$ implies $x = 0$, where $L^*(x) = 1 - U^*(\neg x)$ is the lower prevision associated to U^*).

Theorem 4. *For every coherent assessment $a_1 \mapsto \alpha_1, \ldots, a_n \mapsto \alpha_n$ of upper previsions on $(\mathbf{G}, u)$ there is a faithful upper hyperprevision U^* on $\mathbf{G}^*$ such that for $i = 1, \ldots, n$, $U^*(a_i) - \alpha_i$ is infinitesimal.*

4   Conditional Imprecise Non-standard Probabilities

[9] gives the following betting interpretation of conditional probability: when the conditioning event ψ is many-valued, we assume that betting on $\phi|\psi$ is like betting on ϕ with the proviso that only a part of the bet proportional to the truth value $v(\psi)$ of ψ will be valid. Hence, the gambler's payoff corresponding to the bet λ in a conditional bet on $\phi|\psi$ is $\lambda v(\psi)(v(\phi) - \alpha)$, where α is the betting odd. If $\lambda = 1$, and if we identify any formula with its truth value, the payoff is expressed by $\psi(\phi - \alpha)$. Hence, the upper conditional probability of ϕ given ψ is obtained by imposing the upper prevision of the payoff to be zero. That is, the upper conditional probability $U_0^*(\phi|\psi)$ of ϕ given ψ must be a number α such that $U^*(\psi(\phi-\alpha)) = 0$, where U^* is the unique upper hyperprevision which extends U_0^*.

A desirable condition is that for a given hyper upper probability U_0^* there is a unique α such that $U_0^*(\psi(\phi - \alpha)) = 0$. Clearly one cannot expect this condition to hold when for instance $U_0^*(\psi) = 0$. Indeed, the random variable $\phi - \alpha$ is bounded (it takes values in $[-1, 1]$), and hence for any choice of $\alpha \in [0, 1]$,

$-\psi \leq \psi(\phi - \alpha) \leq \psi$, and

$0 = -U^*(\psi) \leq U^*(-\psi) \leq U^*(\psi(\phi - \alpha)) \leq U^*(\psi) = 0$.

This equality holds independently of α, and hence we are in the bad situation where any α might serve as an upper probability of $\phi|\psi$. We will see that such an inadmissible situation is avoided when $U_0^*(\neg\psi) < 1$, or equivalently, when the lower prevision $L_0^*(\psi)$ is strictly positive.

Lemma 3. *Suppose $L_0^*(\psi) > 0$. Then there is at most one α such that $U^*(\psi(\phi - \alpha)) = 0$.*

The argument used to prove the lemma shows that if $L_0^*(\psi) = L^*(\psi) > 0$, then the upper conditional probability $U_0^*(\phi|\psi)$, if it exists, can be uniquely recovered from the (unconditional) upper hyperstate U_0^*. In the standard case, such a conditional upper probability is shown to exist by a continuity argument, while it is not clear whether it exists in the non-standard case (we have shown uniqueness, not existence). However, for a given finite assessment we will prove that such a conditional upper probability exists, in a sense which will be made precise in Theorem 5 below. Before discussing it, we introduce completeness.

Definition 6. *An assessment of conditional upper probability is said to be complete if for any betting odd $\phi|\psi \mapsto \alpha$ on a conditional event $\phi|\psi$, it also contains a betting odd $\neg\psi_i \mapsto \beta_i$ on the negation of the conditioning event ψ.*

A complete assessment $\Lambda : \phi_i|\psi_i \mapsto \alpha_i, \neg\psi_i \mapsto \beta_i, i = 1, \ldots, n$ of conditional upper probability is said to be stably coherent if there is a hyperassessment

$$\Lambda' : \phi_i|\psi_i \mapsto \alpha_i', \neg\psi_i \mapsto \beta_i', i = 1, \ldots, n$$

which avoids inadmissible bets, differs from Λ by an infinitesimal and such that, for every i, $\beta_i' < 1$.

The requirement of a betting odd β on the negation of the conditioning event and not on the conditioning event itself may look strange. However, imposing a betting odd for the *upper* hyperprevision of the negation of ψ is the same as imposing the betting odd $1 - \beta$ for the *lower* hyperprevision of ψ. So, in a complete assessment we really impose conditions on the lower prevision of the conditioning event.

Stable coherence is the consistency criterion for conditional hyper upper probabilities: when the probability assigned to the conditioning events is 0, it is quite possible that any assignment to conditional events avoids inadmissible bets. Now stable coherence says that inadmissible bets are also avoided if the bookmaker changes the lower probabilities of the conditioning events by an infinitesimal so that a positive number is assigned to them. Our main theorem shows that stable coherence corresponds to faithful upper hyperprevisions.

Theorem 5. *Let* $\Lambda : \phi_i|\psi_i \mapsto \alpha_i, \neg\psi_i \mapsto \beta_i, i = 1,\ldots,n$ *be an assessment of conditional upper probability. Then* Λ *is stably coherent if there is a* faithful *hyper upper prevision* U^* *s.t. for* $i = 1,\ldots,n$, $U^*(\neg\psi_i)) - \beta_i$ *is an infinitesimal, and* $U^*(\psi_i(\phi_i - \alpha_i')) = 0$ *for some* α_i' *such that* $\alpha_i - \alpha_i'$ *is infinitesimal.*

References

1. Burris, S., Sankappanavar, H.P.: A course in Universal Algebra. Springer (1981)
2. Cignoli, R., D'Ottaviano, I., Mundici, D.: Algebraic Foundations of Many-valued Reasoning. Kluwer, Dordrecht (2000)
3. Coletti, G., Scozzafava, R.: Stochastic independence in a coherent setting. Annals of Mathematics and Artificial Intelligence 35, 151–176 (2002)
4. Di Nola, A.: Representation and reticulation by quotients of MV-algebras. Ricerche di Matematica (Naples) 40, 291–297
5. Fedel, M., Hosni, H., Montagna, F.: A logical characterization of coherence for imprecise probabilities. International Journal of Approximate Reasoning 52(8), 1147–1170 (2011)
6. Fedel, M., Keimel, K., Montagna, F., Roth, W.D.: Imprecise probabilities, bets and functional analytic methods in Lukasiewicz logic. Forum Mathematicum 25(2), 405–441 (2013)
7. Montagna, F.: Subreducts of MV algebras with product and product residuation. Algebra Universalis 53, 109–137 (2005)
8. Montagna, F., Fedel, M., Scianna, G.: Non-standard probability, coherence and conditional probability on many-valued events. Int. J. Approx. Reasoning 54(5), 573–589 (2013)
9. Montagna, F.: A notion of coherence for books on conditional events in many-valued logic. Journal of Logic and Computation 21(5), 829–850 (2011)
10. Mundici, D.: Interpretations of $AF\ C^*$ algebras in Lukasiewicz sentential calculus. J. Funct. Analysis 65, 15–63 (1986)
11. Mundici, D.: Advanced Lukasiewicz calculus and MV-algebras. Trends in Logic, vol. 35. Springer (2011)
12. Walley, P.: Statistical Reasoning with Imprecise Probabilities. Monographs on Statistics and Applied Probability, vol. 42. Chapman and Hall, London (1991)

Coherent T-conditional Possibility Envelopes and Nonmonotonic Reasoning

Giulianella Coletti[1], Davide Petturiti[2], and Barbara Vantaggi[2]

[1] Dip. Matematica e Informatica, Università di Perugia, Italy
coletti@dmi.unipg.it
[2] Dip. S.B.A.I., Università di Roma "La Sapienza", Italy
{barbara.vantaggi,davide.petturiti}@sbai.uniroma1.it

Abstract. Envelopes of coherent T-conditional possibilities and coherent T-conditional necessities are studied and an analysis of some inference rules which play an important role in nonmonotonic reasoning is carried out.

Keywords: Envelopes, Coherence, T-conditional possibility, Possibilistic logic, Nonmonotonic reasoning.

1 Introduction

Starting from the seminal work by Adams [1,2], nonmonotonic reasoning has been developed under the interpretations of an uncertainty measure, in particular probability and possibility (see for instance [17,27,18,28,29,7,8,11,4,6,5,9]), although other measures have been considered [10,19,20].

In the coherent probabilistic setting (see [27,12]), the inference rules of System P have been analysed in the framework of g-coherent imprecise probabilistic assessments: a knowledge base formed by a set of constraints $\alpha_i \leq P(E_i|H_i) \leq \beta_i$, consistent with at least a coherent conditional probability, is considered and the propagation of the intervals $[\alpha_i, \beta_i]$ to other events is studied.

In this paper, relying on the general concepts of T-conditional possibility and necessity introduced in [13,21] and the relevant notions of coherent assessment and coherent extensions [13,21,22,3], we cope with a similar problem, but related to intervals $[\alpha_i, \beta_i]$ expressing lower and upper bounds of coherent extensions of a coherent T-conditional possibility (or necessity) or, in general, of a class of coherent T-conditional possibilities (or necessities).

In other words, our coherence-based approach is finalized to study how the lower and upper possibility (or necessity) bounds propagate in a generalized inference process: the concept of "entailment" is formalized as a possibilistic inferential problem and inference rules derived from it are investigated. In particular, we show that under a necessity knowledge base by taking the degenerate interval $[1, 1]$ as a numerical assessment to model the *weak implication*, the classical Adam's rules are obtained.

For this aim we need to study and characterize the envelopes of a class of coherent T-conditional possibilities or of coherent T-conditional necessities.

A. Laurent et al. (Eds.): IPMU 2014, Part III, CCIS 444, pp. 446–455, 2014.

2 T-Conditional Possibility

Different definitions of T-conditional possibility have been introduced as a derived concept of a possibility measure based on t-norms [31] and their residuum [23,25]. In this paper we consider T-conditional possibility as a primitive concept, that is as a function defined on conditional events satisfying a set of axioms [13,22] (for the relationship among this one and the other definitons see also [15]):

Definition 1. *Let $\mathcal{A}$ be a Boolean algebra, $\mathcal{H} \subseteq \mathcal{A}^0$ an additive set (i.e., closed under finite disjunctions), with $\mathcal{A}^0 = \mathcal{A} \setminus \{\emptyset\}$, and T a t-norm. A function $\Pi : \mathcal{A} \times \mathcal{H} \to [0,1]$ is a T-**conditional possibility** if it satisfies:*

(CP1) $\Pi(E|H) = \Pi(E \wedge H|H)$, *for every* $E \in \mathcal{A}$ *and* $H \in \mathcal{H}$;
(CP2) $\Pi(\cdot|H)$ *is a finitely maxitive possibility on* $\mathcal{A}$, *for every* $H \in \mathcal{H}$;
(CP3) $\Pi(E \wedge F|H) = T(\Pi(E|H), \Pi(F|E \wedge H))$, *for every* $H, E \wedge H \in \mathcal{H}$ *and* $E, F \in \mathcal{A}$.

The conditional *dual* function $N : \mathcal{A} \times \mathcal{H} \to [0,1]$ of a T-conditional possibility Π on $\mathcal{A} \times \mathcal{H}$ is defined for every $E|H \in \mathcal{A} \times \mathcal{H}$ as $N(E|H) = 1 - \Pi(E^c|H)$ and is called T-*conditional necessity*. By axiom **(CP2)** it follows that, for every $H \in \mathcal{H}$, $N(\cdot|H)$ is a finitely minitive necessity on $\mathcal{A}$. Due to duality we can limit to study the properties of T-conditional possibilities.

When $\mathcal{H} = \mathcal{A}^0$, Π $[N]$ is said to be a *full T-conditional possibility [full T-conditional necessity]* on $\mathcal{A}$. In Definition 1 no particular property is required to the t-norm T, nevertheless, since (as shown in [26,22,32]) the continuity of T is fundamental in order to guarantee the extendability of a T-conditional possibility to a full T-conditional possibility, in what follows we assume T is a continuous t-norm even when not explicitly stated.

Given an arbitrary Boolean algebra $\mathcal{A}$ and any continuous t-norm T, we denote by T-**CPoss**$(\mathcal{A})$ the class of all the full T-conditional possibilities on $\mathcal{A}$.

Theorem 1. *The set T-**CPoss**$(\mathcal{A})$ is a compact subset of $[0,1]^{\mathcal{A} \times \mathcal{A}^0}$ endowed with the usual product topology of pointwise convergence.*

Proof. By Thychonoff's theorem $[0,1]^{\mathcal{A} \times \mathcal{A}^0}$ is a compact space endowed with the usual product topology of pointwise convergence. Then to prove the assertion it is sufficient to show that T-**CPoss**$(\mathcal{A})$ is a closed subset of $[0,1]^{\mathcal{A} \times \mathcal{A}^0}$, i.e., that the limit Π of any pointwise convergent sequence $(\Pi_i)_{i \in \mathbb{N}}$ of elements of T-**CPoss**$(\mathcal{A})$ is a full T-conditional possibility on $\mathcal{A} \times \mathcal{A}^0$. This immediately follows from continuity on $[0,1]^2$ of both operators max and T. $\square$

Proposition 1. *The function $\Pi^* : \mathcal{A} \times \mathcal{A}^0 \to [0,1]$ defined for $E|H \in \mathcal{A} \times \mathcal{A}^0$ as*

$$\Pi^*(E|H) = \begin{cases} 0 \text{ if } E \wedge H = \emptyset, \\ 1 \text{ otherwise,} \end{cases} \tag{1}$$

*belongs to T-**CPoss**$(\mathcal{A})$, for every continuous t-norm T. Moreover, for any $E|H \in \mathcal{A} \times \mathcal{A}^0$ it holds $\Pi^*(E|H) = \max\{\Pi(E|H) : \Pi \in T$-**CPoss**$(\mathcal{A})\}$.*

Proof. It is immediate to see that Π^* is a full T-conditional possibility on $\mathcal{A}$, for every T, and that $\Pi \leq \Pi^*$ for every $\Pi \in T\text{-}\mathbf{CPoss}(\mathcal{A})$, where the inequality is intended pointwise on the elements of $\mathcal{A} \times \mathcal{A}^0$. $\square$

A full T-conditional possibility $\Pi(\cdot|\cdot)$ on $\mathcal{A}$ is not necessarily "represented" by means of a single finitely maxitive possibility measure (see [22,32,14]), but in general an *agreeing T-nested class* of possibility measures is needed.

3   Coherent T-conditional Possibilities

Boolean conditions on the domain of Π in Definition 1 are essential, since otherwise axioms **(CP1)**–**(CP3)** can result only vacuously satisfied.

In order to deal with an assessment Π on an *arbitrary* set of conditional events $\mathcal{G} = \{E_i|H_i\}_{i \in I}$ we need to resort to the concept of *coherence* [22], originally introduced by de Finetti (trough the concept of coherent bet) in the context of finitely additive probabilities [24].

Definition 2. *Let $\mathcal{G} = \{E_i|H_i\}_{i \in I}$ be an arbitrary family of conditional events and T a continuous t-norm. A function $\Pi : \mathcal{G} \to [0,1]$ is a **coherent T-conditional possibility (assessment)** on $\mathcal{G}$ if there exists a T-conditional possibility $\Pi' : \mathcal{A} \times \mathcal{H} \to [0,1]$, where $\mathcal{A} = \langle \{E_i, H_i\}_{i \in I} \rangle$ and $\mathcal{H}$ is the additive set generated by $\{H_i\}_{i \in I}$, such that $\Pi'_{|\mathcal{G}} = \Pi$.*

As already pointed out in Section 2, an equivalent definition can be given by requiring the existence of a full T-conditional possibility on $\mathcal{A}$ extending the assessment Π given on $\mathcal{G}$. In [32] the coherence of an assessment on $\mathcal{G}$ has been characterized in terms of the coherence on every finite subfamily $\mathcal{F} \subseteq \mathcal{G}$ (the latter studied in [22]), for every continuous t-norm. Moreover, for the minimum and strict t-norms, the following possibilistic version of the de Finetti's fundamental theorem has been proved [32,22].

Theorem 2. *Let $\mathcal{G} = \{E_i|H_i\}_{i \in I}$ be an arbitrary family of conditional events and $T = \min$ or a strict t-norm. A function $\Pi : \mathcal{G} \to [0,1]$ can be extended as a coherent T-conditional possibility Π' on $\mathcal{G}' \supset \mathcal{G}$ if and only if Π is a coherent T-conditional possibility.*

Let us stress that coherence does not impose any constraint of uniqueness, indeed, in general there can be infinitely many extensions of a coherent assessment. For this, denote

$$T\text{-}\mathbf{CExt}(\Pi, \mathcal{G}, \mathcal{A}) = \{\Pi' \in T\text{-}\mathbf{CPoss}(\mathcal{A}) \ : \ \Pi'_{|\mathcal{G}} = \Pi\}. \tag{2}$$

The following compactness result holds for every continuous t-norm.

Theorem 3. *The set $T\text{-}\mathbf{CExt}(\Pi, \mathcal{G}, \mathcal{A})$ is a compact subset of $[0,1]^{\mathcal{A} \times \mathcal{A}^0}$ endowed with the usual product topology of pointwise convergence.*

Proof. Theorem 2 implies $T\text{-}\mathbf{CExt}(\Pi, \mathcal{G}, \mathcal{A}) \neq \emptyset$, moreover, by Theorem 1, $T\text{-}\mathbf{CPoss}(\mathcal{A})$ is a compact space endowed with the relative topology inherited from $[0,1]^{\mathcal{A} \times \mathcal{A}^0}$. Hence, it is sufficient to prove that $T\text{-}\mathbf{CExt}(\Pi, \mathcal{G}, \mathcal{A})$ is a closed subset of $T\text{-}\mathbf{CPoss}(\mathcal{A})$. For that we need to show that the limit Π' of every pointwise convergent sequence $(\Pi'_i)_{i \in \mathbb{N}}$ of elements of $T\text{-}\mathbf{CExt}(\Pi, \mathcal{G}, \mathcal{A})$ extends Π, as Π' is a full T-conditional possibility on $\mathcal{A}$ by Theorem 1. Then the conclusion follows since, for every $E|H \in \mathcal{G}$, $\Pi'(E|H) = \lim_{i \to \infty} \Pi'_i(E|H) = \lim_{i \to \infty} \Pi(E|H) = \Pi(E|H)$. $\qquad \square$

In the particular case T is the minimum or a strict t-norm [26,22,32] the set $T\text{-}\mathbf{CExt}(\Pi, \mathcal{G}, \mathcal{A})$ can be written as the Cartesian product of (possibly degenerate) intervals

$$T\text{-}\mathbf{CExt}(\Pi, \mathcal{G}, \mathcal{A}) = \underset{E|H \in \mathcal{A} \times \mathcal{A}^0}{\times} K_{E|H}, \tag{3}$$

with $K_{E|H} = [\underline{\Pi}(E|H), \overline{\Pi}(E|H)] \subseteq [0,1]$, for every $E|H \in \mathcal{A} \times \mathcal{A}^0$. If $K_{E|H}$ is non-degenerate, by selecting an extension Π' of Π on $\mathcal{G}' = \mathcal{G} \cup \{E|H\}$ s.t. $\Pi'(E|H) \in K_{E|H}$ it follows $T\text{-}\mathbf{CExt}(\Pi', \mathcal{G}', \mathcal{A}) \subset T\text{-}\mathbf{CExt}(\Pi, \mathcal{G}, \mathcal{A})$.

4   Coherent T-conditional Possibility Envelopes

Let $\mathcal{G} = \{E_i|H_i\}_{i \in I}$ be an arbitrary family of conditional events and $\mathcal{A} = \langle \{E_i, H_i\}_{i \in I} \rangle$, we denote by $T\text{-}\mathbf{CCohe}(\mathcal{G})$ the set of coherent T-conditional possibilities on $\mathcal{G}$, that is:

$$T\text{-}\mathbf{CCohe}(\mathcal{G}) = \{\Pi_{|\mathcal{G}} \ : \ \Pi \in T\text{-}\mathbf{CPoss}(\mathcal{A})\}. \tag{4}$$

By Theorem 1 and Proposition 1 it trivially follows that $T\text{-}\mathbf{CCohe}(\mathcal{G})$ is a compact subset of $[0,1]^{\mathcal{G}}$ endowed with the usual product topology of pointwise convergence, moreover, the function $\Pi^* : \mathcal{G} \to [0,1]$ defined as in (1) dominates every coherent T-conditional possibility in $T\text{-}\mathbf{CCohe}(\mathcal{G})$ and belongs to this set.

Let us consider now the Goodman-Nguyen's relation $\subseteq_{GN}$ [30] between conditional events, which generalizes the usual implication among events:

$$E_i|H_i \subseteq_{GN} E_j|K_j \quad \Leftrightarrow \quad E_i \wedge H_i \subseteq E_j \wedge H_j \text{ and } E_j^c \wedge H_j \subseteq E_i^c \wedge H_i. \tag{5}$$

In case $T = \min$ or a strict t-norm, the following result can be proven (which has an analogue for conditional probability [16,28]).

Theorem 4. *For all $E_i|H_i, E_j|H_j \in \mathcal{G}$ the following statements are equivalent:*

(i) $E_i|H_i \subseteq_{GN} E_j|H_j$ or $E_i \wedge H_i = \emptyset$ or $E_j^c \wedge H_j = \emptyset$;
(ii) $\Pi(E_i|H_i) \leq \Pi(E_j|H_j)$, for every $\Pi \in T\text{-}\mathbf{CCohe}(\mathcal{G})$.

Proof. The implication *(i)* $\Rightarrow$ *(ii)* has been proven in [15]. To prove *(ii)* $\Rightarrow$ *(i)* suppose by absurd that condition *(i)* does not hold, i.e., $E_i|H_i \not\subseteq_{GN} E_j|H_j$ and $E_i \wedge H_i \neq \emptyset$ and $E_j^c \wedge H_j \neq \emptyset$, which implies $(E_i \wedge H_i \wedge E_j^c \wedge H_j) \neq \emptyset$.

It is easy to prove that the assessment $\Pi(E_i|H_i) = 1$ and $\Pi(E_j|H_j) = 0$ is coherent, thus by virtue of Theorem 2, Π can be extended as a coherent T-conditional possibility Π' on $\mathcal{G}$. Hence, there exists $\Pi' \in T\text{-}\mathbf{CCohe}(\mathcal{G})$ such that $\Pi'(E_i|H_i) > \Pi'(E_j|H_j)$ and we get a contradiction with condition *(ii)*. $\square$

A result analogous to Theorem 4 holds for T-conditional necessities.

The following theorem, which is the analogue of Theorem 8 in [27] given in the framework of conditional probability, follows by Theorem 1.

Theorem 5. *Let* $\mathcal{G} = \{E_i|H_i\}_{i \in I}$ *be an arbitrary set of conditional events. Then the following statements are equivalent:*

(i) *the assessment* $\Pi^*(E_i|H_i) = 1$ *for all* $i \in I$ *belongs to* $T\text{-}\mathbf{CCohe}(\mathcal{G})$;
(ii) *for every* $\epsilon > 0$, *there exists an assessment* $\Pi^\epsilon(E_i|H_i) \geq 1 - \epsilon$ *for all* $i \in I$, *belonging to* $T\text{-}\mathbf{CCohe}(\mathcal{G})$.

Now we consider a class $\mathbf{P} \subseteq T\text{-}\mathbf{CCohe}(\mathcal{G})$, whose information can be summarized by means of its lower and upper envelopes.

Definition 3. *Let* $\mathcal{G} = \{E_i|H_i\}_{i \in I}$ *be an arbitrary family of conditional events. A pair of functions* $(\underline{\Pi}, \overline{\Pi})$ *on* $\mathcal{G}$ *are* **coherent** T**-conditional possibility envelopes** *if there exists a class* $\mathbf{P} \subseteq T\text{-}\mathbf{CCohe}(\mathcal{G})$ *of coherent* T-conditional possibilities such that $\underline{\Pi} = \inf \mathbf{P}$ and $\overline{\Pi} = \sup \mathbf{P}$.

In general, $\underline{\Pi}, \overline{\Pi}$ are not elements of $T\text{-}\mathbf{CCohe}(\mathcal{G})$, since this space is not closed under pointwise minimum and maximum.

Example 1. Let $\mathcal{G} = \{E \wedge F|H, E|H, F|E \wedge H, E|H^c, E^c|H^c\}$ with E, F, H logically independent events, and $\Pi^1, \Pi^2 \in T\text{-}\mathbf{CCohe}(\mathcal{G})$ s.t. $\Pi^1(E \wedge F|H) = \Pi^1(E|H) = \Pi^1(E|H^c) = \Pi^2(E \wedge F|H) = \Pi^2(F|E \wedge H) = \Pi^2(E^c|H^c) = 0.2$, $\Pi^1(F|E \wedge H) = \Pi^1(E^c|H^c) = \Pi^2(E|H) = \Pi^2(E|H^c) = 1$. Consider $\Pi_* = \min\{\Pi^1, \Pi^2\}$ and $\Pi^* = \max\{\Pi^1, \Pi^2\}$. It holds $\max\{\Pi_*(E|H^c), \Pi_*(E^c|H^c)\} = 0.2 \neq 1$ and $\Pi^*(E \wedge F|H) = 0.2 \neq 1 = T(\Pi^*(E|H), \Pi^*(F|E \wedge H))$, thus $\Pi_*, \Pi^* \notin T\text{-}\mathbf{CCohe}(\mathcal{G})$.

By duality, a pair of functions $(\underline{N}, \overline{N})$ on $\mathcal{G}' = \{E_i^c|H_i\}_{i \in I}$ are coherent T-conditional necessity envelopes if there exists a pair of coherent T-conditional possibility envelopes $(\underline{\Pi}, \overline{\Pi})$ on $\mathcal{G}$, such that $\underline{N}(E_i^c|H_i) = 1 - \overline{\Pi}(E_i|H_i)$ and $\overline{N}(E_i^c|H_i) = 1 - \underline{\Pi}(E_i|H_i)$, for every $E_i^c|H_i \in \mathcal{G}'$.

Notice that no assumption is made about the class $\mathbf{P} \subseteq T\text{-}\mathbf{CCohe}(\mathcal{G})$, which could be a non-closed subset of $T\text{-}\mathbf{CCohe}(\mathcal{G})$: this implies that, in general, the pointwise infimum and supremum of $\mathbf{P}$ could not be attained by an element of $\mathbf{P}$. Nevertheless, since $T\text{-}\mathbf{CCohe}(\mathcal{G})$ is compact, the closure $\text{cl}(\mathbf{P})$ is a subset of $T\text{-}\mathbf{CCohe}(\mathcal{G})$ and it holds $\inf \mathbf{P} = \min \text{cl}(\mathbf{P})$ and $\sup \mathbf{P} = \max \text{cl}(\mathbf{P})$, thus we can always consider a closed set of coherent T-conditional possibilities on $\mathcal{G}$.

In case of a finite set $\mathcal{G} = \{E_1|H_1, \dots, E_n|H_n\}$, the coherence of an assessment $(\underline{\Pi}, \overline{\Pi})$ on $\mathcal{G}$ can be characterized in terms of the existence of a finite class $\mathbf{P}^{2n} \subseteq T\text{-}\mathbf{CCohe}(\mathcal{G})$, which is therefore a closed subset of $T\text{-}\mathbf{CCohe}(\mathcal{G})$. The class $\mathbf{P}^{2n}$ is composed at most by $2n$ coherent T-conditional possibilities on $\mathcal{G}$, that can be determined solving for each event $E_i|H_i$ two suitable sequences of systems.

Theorem 6. *Let $\mathcal{G} = \{E_1|H_1, \ldots, E_n|H_n\}$ be a finite set of conditional events and $(\underline{\Pi}, \overline{\Pi})$ a pair of assessments on $\mathcal{G}$. Let $\mathcal{A} = \langle\{E_i, H_i\}_{i=1,\ldots,n}\rangle$ and $\mathcal{C}_{\mathcal{A}}$ be the corresponding set of atoms. The following statements are equivalent:*

(i) $(\underline{\Pi}, \overline{\Pi})$ are coherent T-conditional possibility envelopes on $\mathcal{G}$;

(ii) for every $i \in \{1, \ldots, n\}$ and $u \in \{0, 1\}$ there exist a natural number $k \leq n$ and a sequence $\mathcal{S}_\alpha^{i,u}$ $(\alpha = 0, \ldots, k)$, with unknowns $x_r^\alpha \geq 0$ for $C_r \in \mathcal{C}_\alpha$,

$$
\mathcal{S}_\alpha^{i,u} : \begin{cases}
\max_{C_r \subseteq E_i \wedge H_i} x_r^\alpha = T\left(\theta_i, \max_{C_r \subseteq H_i} x_r^\alpha\right) & \text{if } \max_{C_r \subseteq H_i} \mathbf{x}_r^{\alpha-1} < 1 \\[4pt]
\left[\theta_i = \begin{cases} \underline{\Pi}(E_i|H_i) \text{ if } u = 0 \\ \overline{\Pi}(E_i|H_i) \text{ if } u = 1 \end{cases}\right] \\[10pt]
\max_{C_r \subseteq E_j \wedge H_j} x_r^\alpha = T\left(\theta_j, \max_{C_r \subseteq H_j} x_r^\alpha\right) & \text{if } j \neq i \text{ and } \max_{C_r \subseteq H_j} \mathbf{x}_r^{\alpha-1} < 1 \\[4pt]
\left[\underline{\Pi}(E_j|H_j) \leq \theta_j \leq \overline{\Pi}(E_j|H_j)\right] \\[6pt]
x_r^\alpha \geq \mathbf{x}_r^{\alpha-1} & \text{if } C_r \in \mathcal{C}_\alpha \\[6pt]
\mathbf{x}_r^{\alpha-1} = T\left(x_r^\alpha, \max_{C_j \in \mathcal{C}_\alpha} \mathbf{x}_j^{\alpha-1}\right) & \text{if } C_r \in \mathcal{C}_\alpha \\[6pt]
\max_{C_r \in \mathcal{C}_\alpha} x_r^\alpha = 1
\end{cases}
$$

admitting solution $\mathbf{x}^\alpha$ (whose r-th component is $\mathbf{x}_r^\alpha$) with $\mathbf{x}^{-1} = \mathbf{0}$, $\mathcal{C}_0 = \mathcal{C}_{\mathcal{A}}$, $\mathcal{C}_\alpha = \{C_r \in \mathcal{C}_{\alpha-1} : \mathbf{x}_r^{\alpha-1} < 1\}$ for $\alpha \geq 1$, and $\mathcal{C}_{k+1} = \emptyset$.

Proof. $(ii) \Rightarrow (i)$. For $i = 1, \ldots, n$, the solutions of systems $\mathcal{S}_\alpha^{i,0}$ and $\mathcal{S}_\alpha^{i,1}$ give rise to two full T-conditional possibilities on $\mathcal{A}$ whose restrictions $\Pi^{i,0}$ and $\Pi^{i,1}$ on $\mathcal{G}$ are such that $\Pi^{i,0}(E_i|H_i) = \underline{\Pi}(E_i|H_i)$ and $\underline{\Pi}(E_j|H_j) \leq \Pi^{i,0}(E_j|H_j) \leq \overline{\Pi}(E_j|H_j)$ for $i \neq j$, $\Pi^{i,1}(E_i|H_i) = \overline{\Pi}(E_i|H_i)$ and $\underline{\Pi}(E_j|H_j) \leq \Pi^{i,1}(E_j|H_j) \leq \overline{\Pi}(E_j|H_j)$ for $i \neq j$. Hence the class $\mathbf{P}^{2n} = \{\Pi^{1,0}, \Pi^{1,1}, \ldots, \Pi^{n,0}, \Pi^{n,1}\}$ is a subset of T-$\mathbf{CCohe}(\mathcal{G})$ and is such that $\underline{\Pi} = \min \mathbf{P}^{2n}$ and $\overline{\Pi} = \max \mathbf{P}^{2n}$, where the minimum and maximum are intended pointwise on the elements of $\mathcal{G}$.

$(i) \Rightarrow (ii)$. If $(\underline{\Pi}, \overline{\Pi})$ are coherent T-conditional possibility envelopes then there exists a class $\mathbf{P} \subseteq T$-$\mathbf{CPoss}(\mathcal{G})$ such that $\underline{\Pi} = \inf \mathbf{P}$ and $\overline{\Pi} = \sup \mathbf{P}$. By the compactness of T-$\mathbf{CPoss}(\mathcal{G})$, the closure $\mathrm{cl}(\mathbf{P})$ is a subset of T-$\mathbf{CPoss}(\mathcal{G})$ and it holds $\underline{\Pi} = \min \mathrm{cl}(\mathbf{P})$ and $\overline{\Pi} = \max \mathrm{cl}(\mathbf{P})$. In turn, this implies for $i = 1, \ldots, n$ the existence of two coherent T-conditional possibilities $\Pi^{i,0}, \Pi^{i,1} \in \mathrm{cl}(\mathbf{P})$ such that $\Pi^{i,0}(E_i|H_i) = \underline{\Pi}(E_i|H_i)$ and $\underline{\Pi}(E_j|H_j) \leq \Pi^{i,0}(E_j|H_j) \leq \overline{\Pi}(E_j|H_j)$ for $i \neq j$, $\Pi^{i,1}(E_i|H_i) = \overline{\Pi}(E_i|H_i)$ and $\underline{\Pi}(E_j|H_j) \leq \Pi^{i,1}(E_j|H_j) \leq \overline{\Pi}(E_j|H_j)$ for $i \neq j$. The coherence of $\Pi^{i,0}$ and $\Pi^{i,1}$ implies their extendability as full T-conditional possibilities $\Pi^{i,0'}$ and $\Pi^{i,1'}$ on $\mathcal{A}$: the distributions of the possibility measures in the T-nested classes representing $\Pi^{i,0'}$ and $\Pi^{i,1'}$ are exactly the solutions of sequences $\mathcal{S}_\alpha^{i,0}$ and $\mathcal{S}_\alpha^{i,1}$. $\qquad\square$

Example 2. Consider the events A, B, C such that $A \subseteq B$, which generate the atoms $C_1 = A^c \wedge B^c \wedge C^c$, $C_2 = A^c \wedge B \wedge C^c$, $C_3 = A \wedge B \wedge C^c$, $C_4 = A \wedge B \wedge C$, $C_5 = A^c \wedge B \wedge C$, $C_6 = A^c \wedge B^c \wedge C$. The following assessment is given:

$$\left(\underline{\mathit{\Pi}}(A|C) = 0.2, \overline{\mathit{\Pi}}(A|C) = 0.3\right), \left(\underline{\mathit{\Pi}}(B|C) = 0.5, \overline{\mathit{\Pi}}(B|C) = 1\right)$$
$$\left(\underline{\mathit{\Pi}}(A|B \vee C) = 0.3, \overline{\mathit{\Pi}}(A|B \vee C) = 0.5\right).$$

We prove that $(\underline{\mathit{\Pi}}, \overline{\mathit{\Pi}})$ are coherent min-conditional possibility envelopes. For $A|C$, the relevant system with unknowns $x_r^0 \geq 0$, $r = 1, \ldots, 6$, is

$$\mathcal{S}_0^{1,0} : \begin{cases} x_4^0 = \min\left\{0.2, \max\{x_4^0, x_5^0, x_6^0\}\right\} \\ \max\{x_4^0, x_5^0\} = \min\left\{\theta_2, \max\{x_4^0, x_5^0, x_6^0\}\right\} & [0.5 \leq \theta_2 \leq 1] \\ \max\{x_3^0, x_4^0\} = \min\left\{\theta_3, \max\{x_2^0, x_3^0, x_4^0, x_5^0, x_6^0\}\right\} & [0.3 \leq \theta_3 \leq 0.5] \\ \max\{x_1^0, x_2^0, x_3^0, x_4^0, x_5^0, x_6^0\} = 1 \end{cases}$$

and a solution is $\mathbf{x}_1^0 = \mathbf{x}_2^0 = \mathbf{x}_6^0 = 1$, $\mathbf{x}_3^0 = \mathbf{x}_5^0 = 0.5$, $\mathbf{x}_4^0 = 0.2$. Next system $\mathcal{S}_1^{1,0}$ has solution $\mathbf{x}_3^1 = \mathbf{x}_5^1 = 1$, $\mathbf{x}_4^1 = 0.2$, while the system $\mathcal{S}_2^{1,0}$ has trivial solution $\mathbf{x}_4^2 = 1$.

By analogous computations, $\mathbf{y}_1^0 = \mathbf{y}_2^0 = \mathbf{y}_5^0 = \mathbf{y}_6^0 = 1$, $\mathbf{y}_3^0 = \mathbf{y}_4^0 = 0.3$ is a solution for the systems $\mathcal{S}_0^{1,1}$, and $\mathbf{y}_3^1 = \mathbf{y}_4^1 = 1$ is a solution for the system $\mathcal{S}_1^{1,1}$.

The solutions of sequences $\mathcal{S}_\alpha^{1,0}$ and $\mathcal{S}_\alpha^{1,1}$ induce two full min-conditional possibilities on $\mathcal{A}$ whose restrictions on $\mathcal{G}$ are denoted as $\mathit{\Pi}^{1,0}$ and $\mathit{\Pi}^{1,1}$. The solutions $\mathbf{x}^0, \mathbf{x}^1, \mathbf{x}^2$ are still valid for the sequences $\mathcal{S}_\alpha^{2,0}$, related to $B|C$, and $\mathcal{S}_\alpha^{3,1}$, related to $A|B \vee C$. Analogously, the solutions $\mathbf{y}^0, \mathbf{y}^1$ are still valid for the sequences $\mathcal{S}_\alpha^{2,1}$ and $\mathcal{S}_\alpha^{3,0}$. Hence, $\mathbf{P}^{2n} = \{\mathit{\Pi}^{1,0}, \mathit{\Pi}^{1,1}\}$ and $\underline{\mathit{\Pi}} = \min \mathbf{P}^{2n}$ and $\overline{\mathit{\Pi}} = \max \mathbf{P}^{2n}$.

Given a pair of coherent T-conditional possibility envelopes $(\underline{\mathit{\Pi}}, \overline{\mathit{\Pi}})$ on $\mathcal{G}$, our aim is to enlarge the assessment on a new event $E|H \in (\mathcal{A} \times \mathcal{A}^0) \setminus \mathcal{G}$ in such a way that $(\underline{\mathit{\Pi}}', \overline{\mathit{\Pi}}')$ are coherent T-conditional possibility envelopes on $\mathcal{G} \cup \{E|H\}$ and $\underline{\mathit{\Pi}}'_{|\mathcal{G}} = \underline{\mathit{\Pi}}$, $\overline{\mathit{\Pi}}'_{|\mathcal{G}} = \overline{\mathit{\Pi}}$. Thus we set

$$\underline{\mathit{\Pi}}'(E|H) = \min\left\{\mathit{\Pi}'(E|H) : \mathit{\Pi}' \in T\text{-}\mathbf{CPoss}(\mathcal{A}), \underline{\mathit{\Pi}} \leq \mathit{\Pi}'_{|\mathcal{G}} \leq \overline{\mathit{\Pi}}\right\}, \quad (6)$$

$$\overline{\mathit{\Pi}}'(E|H) = \max\left\{\mathit{\Pi}'(E|H) : \mathit{\Pi}' \in T\text{-}\mathbf{CPoss}(\mathcal{A}), \underline{\mathit{\Pi}} \leq \mathit{\Pi}'_{|\mathcal{G}} \leq \overline{\mathit{\Pi}}\right\}. \quad (7)$$

As a consequence of Theorem 6, the bounds $\underline{\mathit{\Pi}}'(E|H)$ and $\overline{\mathit{\Pi}}'(E|H)$ can be computed by solving the two optimization problems over the sequence of system $\tilde{\mathcal{S}}_\alpha$ ($\alpha = 0, \ldots, k$) with unknowns $x_r^\alpha \geq 0$ for $C_r \in \mathcal{C}_\alpha$

$$\text{minimize} / \text{maximize } \theta$$

$$\tilde{\mathcal{S}}_\alpha : \begin{cases} \max_{C_r \subseteq E \wedge H} x_r^\alpha = T\left(\theta, \max_{C_r \subseteq H} x_r^\alpha\right) & \text{if } \max_{C_r \subseteq H} \mathbf{x}_r^{\alpha-1} < 1 \\ \max_{C_r \subseteq E_i \wedge H_i} x_r^\alpha = T\left(\theta_i, \max_{C_r \subseteq H_i} x_r^\alpha\right) & \text{if } \max_{C_r \subseteq H_i} \mathbf{x}_r^{\alpha-1} < 1 \\ \left[\underline{\mathit{\Pi}}(E_i|H_i) \leq \theta_i \leq \overline{\mathit{\Pi}}(E_i|H_i)\right] \\ x_r^\alpha \geq \mathbf{x}_r^{\alpha-1} & \text{if } C_r \in \mathcal{C}_\alpha \\ \mathbf{x}_r^{\alpha-1} = T\left(x_r^\alpha, \max_{C_j \in \mathcal{C}_\alpha} \mathbf{x}_j^{\alpha-1}\right) & \text{if } C_r \in \mathcal{C}_\alpha \\ \max_{C_r \in \mathcal{C}_\alpha} x_r^\alpha = 1 \end{cases}$$

Example 3. Given the coherent min-conditional possibility envelopes $(\underline{\Pi}, \overline{\Pi})$ of Example 2, in order to compute the extension on $A|\Omega$ we need to solve the following optimization problems involving the sequence $\tilde{S}_\alpha$, whose first system with unknowns $x_r^0 \geq 0$, $r = 1, \ldots, 6$, results

$$\text{minimize}/\text{maximize } \theta$$

$$\tilde{S}_0 : \begin{cases} \max\{x_3^0, x_4^0\} = \min\{\theta, \max\{x_1^0, x_2^0, x_3^0, x_4^0, x_5^0, x_6^0\}\} \\ x_4^0 = \min\{\theta_1, \max\{x_4^0, x_5^0, x_6^0\}\} & [0.2 \leq \theta_1 \leq 0.3] \\ \max\{x_4^0, x_5^0\} = \min\{\theta_2, \max\{x_4^0, x_5^0, x_6^0\}\} & [0.5 \leq \theta_2 \leq 1] \\ \max\{x_3^0, x_4^0\} = \min\{\theta_3, \max\{x_2^0, x_3^0, x_4^0, x_5^0, x_6^0\}\} & [0.3 \leq \theta_3 \leq 0.5] \\ \max\{x_1^0, x_2^0, x_3^0, x_4^0, x_5^0, x_6^0\} = 1 \end{cases}$$

A solution of $\tilde{S}_0$ minimizing θ is $\mathbf{x}_1^0 = 1$, $\mathbf{x}_r^0 = 0$ for $r = 2, \ldots, 6$. Next systems do not contain the equation involving θ, and for them a sequence of solutions is $\mathbf{x}_2^1 = \mathbf{x}_5^1 = \mathbf{x}_6^1 = 1$, $\mathbf{x}_3^1 = \mathbf{x}_4^1 = 0.3$ for the system $\tilde{S}_1$, and $\mathbf{x}_3^2 = \mathbf{x}_4^2 = 1$ for the system $\tilde{S}_2$. In analogy, a solution of $\tilde{S}_0$ maximizing θ is $\mathbf{y}_1^0 = \mathbf{y}_2^0 = \mathbf{y}_5^0 = \mathbf{y}_6^0 = 1$, $\mathbf{y}_3^0 = 0.5$, $\mathbf{y}_4^0 = 0.3$. Next systems do not contain the equation involving θ, and for them a sequence of solutions is $\mathbf{y}_3^1 = 1$, $\mathbf{y}_4^1 = 0.3$ for the system $\tilde{S}_1$, and $\mathbf{y}_4^2 = 1$ for the system $\tilde{S}_2$. This implies $\underline{\Pi}'(A|\Omega) = 0$ and $\overline{\Pi}'(A|\Omega) = 0.5$.

5 Nonmonotonic Inference Rules under Possibility and Necessity

Now we deal with extensions of coherent T-conditional possibility and necessity envelopes, in a way to study some relevant inferential rules. Given $\mathcal{G} = \{E_1|H_1, \ldots, E_n|H_n\}$ we consider a *possibility or necessity knowledge base (KB)*

$$\Delta = \{(E_i|H_i)[\underline{\pi}_i, \overline{\pi}_i] : i = 1, \ldots, n\}, \tag{8}$$

$$\Delta^* = \{(E_i|H_i)[\underline{n}_i, \overline{n}_i] : i = 1, \ldots, n\}, \tag{9}$$

where the assessments $\{(\underline{\Pi}(E_i|H_i) = \underline{\pi}_i, \overline{\Pi}(E_i|H_i) = \overline{\pi}_i) : i = 1, \ldots, n\}$ and $\{(\underline{N}(E_i|H_i) = \underline{n}_i, \overline{N}(E_i|H_i) = \overline{n}_i) : i = 1, \ldots, n\}$ are, respectively, a pair of coherent T-conditional possibility envelopes and of coherent T-conditional necessity envelopes on $\mathcal{G}$. We are interested in characterizing the intervals entailed on new conditional events, starting from a possibility or necessity KB.

The proofs of next theorems follow from the results presented in previous sections and are omitted for lack of space, moreover, it is easy to check that the following rules are entailed under both possibility and necessity frameworks:

(Reflexivity) $(A|A)[1, 1]$;
(Left Logical Equivalence) $A = B, (C|A)[\underline{\varphi}_1, \overline{\varphi}_1]$ entail $(C|B)[\underline{\varphi}_1, \overline{\varphi}_1]$;
(Right Weakening) $B \subseteq C, (B|A)[\underline{\varphi}_1, \overline{\varphi}_1]$ entail $(C|A)[\underline{\varphi}_1, 1]$.

In the following results other inferential rules are studied:

Theorem 7. *Under a possibility KB the following inference rules hold:*

(And) $(B|A)[\underline{\pi}_1, \overline{\pi}_1], (C|A)[\underline{\pi}_2, \overline{\pi}_2]$ *entail* $(B \wedge C|A)[0, \min\{\overline{\pi}_2, \overline{\pi}_2\}]$;

(Cautious Monotonicity) $(C|A)[\underline{\pi}_1, \overline{\pi}_1], (B|A)[\underline{\pi}_2, \overline{\pi}_2]$ *entail*

$(C|A \wedge B)[0, R^T(\overline{\pi}_2, \underline{\pi}_1)]$, *where* R^T *is the residuum of the t-norm* T;

(Or) $(C|A)[\underline{\pi}_1, \overline{\pi}_1], (C|B)[\underline{\pi}_2, \overline{\pi}_2]$ *entail* $(C|A \vee B)[\min\{\underline{\pi}_1, \underline{\pi}_2\}, \max\{\overline{\pi}_1, \overline{\pi}_2\}]$;

(Cut) $(C|A \wedge B)[\underline{\pi}_1, \overline{\pi}_1], (B|A)[\underline{\pi}_2, \overline{\pi}_2]$ *entail* $(C|A)[T(\underline{\pi}_1, \underline{\pi}_2), 1]$;

(Equivalence) $(B|A)[\underline{\pi}_1, \overline{\pi}_1], (A|B)[\underline{\pi}_2, \overline{\pi}_2], (C|A)[\underline{\pi}_3, \overline{\pi}_3]$ *entail* $(C|B)[0, 1]$.

Theorem 8. *Under a necessity KB the following inference rules hold:*

(And) $(B|A)[\underline{n}_1, \overline{n}_1], (C|A)[\underline{n}_2, \overline{n}_2]$ *entail* $(B \wedge C|A)[\min\{\underline{n}_1, \underline{n}_2\}, \min\{\overline{n}_1, \overline{n}_2\}]$;

(Cautious Monotonicity) $(C|A)[\underline{n}_1, \overline{n}_1], (B|A)[\underline{n}_2, \overline{n}_2]$ *entail* $(C|A \wedge B)[\alpha, \beta]$,

where $\alpha = 0$ *if* $\min\{\underline{n}_1, \underline{n}_2\} = 0$ *and* $\alpha = \underline{n}_1$ *otherwise,* $\beta = 1$ *if* $\overline{n}_1 \geq \underline{n}_2$ *and*

$\beta = \overline{n}_1$ *otherwise;*

(Or) $(C|A)[\underline{n}_1, \overline{n}_1], (C|B)[\underline{n}_2, \overline{n}_2]$ *entail* $(C|A \vee B)[\min\{\underline{n}_1, \underline{n}_2\}, \max\{\overline{n}_1, \overline{n}_2\}]$;

(Cut) $(C|A \wedge B)[\underline{n}_1, \overline{n}_1], (B|A)[\underline{n}_2, \overline{n}_2]$ *entail* $(C|A)[\min\{\underline{n}_1, \underline{n}_2\}, \beta]$, *where* $\beta = 1$ *if* $\overline{n}_1 = 1$ *or* $\underline{n}_2 = 0$ *and* $\beta = \underline{n}_1$ *otherwise;*

(Equivalence) $(B|A)[\underline{n}_1, \overline{n}_1], (A|B)[\underline{n}_2, \overline{n}_2], (C|A)[\underline{n}_3, \overline{n}_3]$ *entail* $(C|B)[\alpha, \beta]$, *w-here* $\alpha = 0$ *if* $\min\{\underline{n}_1, \underline{n}_2, \underline{n}_3\} = 0$ *and* $\alpha = \min\{\underline{n}_2, \underline{n}_3\}$ *otherwise,* $\beta = 1$ *if* $\overline{n}_3 \geq \underline{n}_1$ *or* $\underline{n}_2 = 0$ *and* $\beta = \overline{n}_3$ *otherwise.*

Restricting a possibility KB to the degenerate interval $[1, 1]$ for the premises, we get the degenerate interval $[1, 1]$ also for the consequence in case of Reflexivity, Left Logical Equivalence, Right Weakening, Or and Cut rules. A similar situation occurs for all the inference rules in case of a necessity KB. This results are in line with those in [20], where the case of a precise possibility or necessity KB is investigated. We emphasize that, starting from a possibility KB, we obtain weaker conclusions than under a necessity KB: this is due to the fact that, since possibility dominates necessity, the values in the intervals assessed for the necessity are coherent also for the possibility.

References

1. Adams, E.W.: The Logic of Conditionals. Reidel, Dordrecht (1975)
2. Adams, E.W.: On the logic of high probability. J. Phil. Log. 15, 255–279 (1986)
3. Baioletti, M., Coletti, G., Petturiti, D., Vantaggi, B.: Inferential models and relevant algorithms in a possibilistic framework. Int. J. of App. Reas. 52(5), 580–598 (2011)
4. Benferhat, S., Bonnefon, J., Da Silva Neves, R.: An overview of possibilistic handling of default reasoning: an experimental study. Synthese 146(1), 53–70 (2005)
5. Benferhat, S., Dubois, D., Prade, H.: Representing Default Rules in Possibilistic Logic. In: Proc. 3rd Int. Conf. on Princ. of Knowl. Rep. Reas., pp. 673–684 (1992)
6. Benferhat, S., Dubois, D., Prade, H.: Beyond Counter-Examples to Nonmonotonic Formalisms: A Possibility-Theoretic Analysis. Proc. In: 12th Europ. Conf. on A.I., pp. 652–656 (1996)
7. Benferhat, S., Dubois, D., Prade, H.: Nonmonotonic reasoning, conditional objects and possibility theory. Art. Int. J. 92, 259–276 (1997)
8. Benferhat, S., Dubois, D., Prade, H.: The Possibilistic Handling of Irrelevance in exception-Tolerant Reasoning. Ann. of Math. and Art. Int. 35(1), 29–61 (2002)
9. Benferhat, S., Lagrue, S., Papini, O.: Reasoning with partially ordered information in a possibilistic framework. Fuzzy Sets and Sys. 144, 25–41 (2004)

10. Benferhat, S., Smets, P., Saffiotti, A.: Belief functions and default reasoning. Art. Int. J. 122, 1–69 (2000)
11. Benferhat, S., Tabia, K., Sedki, K.: Jeffrey's rule of conditioning in a possibilistic framework. Ann. of Math. and Art. Int. 66(3), 185–202 (2011)
12. Biazzo, V., Gilio, A., Lukasiewicz, T., Sanfilippo, G.: Probabilistic logic under coherence: complexity and algorithms. Ann. of Math. and A. I. 45, 35–81 (2005)
13. Bouchon-Meunier, B., Coletti, G., Marsala, C.: Independence and Possibilistic Conditioning. Ann. of Math. and Art. Int. 35, 107–123 (2002)
14. Coletti, G., Petturiti, D.: Bayesian-like inference, complete disintegrability and complete conglomerability in coherent conditional possibility theory. In: Proc. 8th Int. Symp. on Imp. Prob.: Theor. and App., pp. 55–66 (2013)
15. Coletti, G., Petturiti, D., Vantaggi, B.: Possibilistic and probabilistic likelihood functions and their extensions: Common features and specific characteristics. Fuzzy Sets and Sys. (2014) (in press online since 25 September 2013), http://dx.doi.org/10.1016/j.fss.2013.09.010
16. Coletti, G., Scozzafava, R.: Probabilistic Logic in a Coherent Setting. Trends in Logic, vol. 15. Kluwer Academic Publisher, Dordrecht (2002)
17. Coletti, G., Scozzafava, R., Vantaggi, B.: Probabilistic Reasoning as a General Unifying Tool. In: Benferhat, S., Besnard, P. (eds.) ECSQARU 2001. LNCS (LNAI), vol. 2143, pp. 120–131. Springer, Heidelberg (2001)
18. Coletti, G., Scozzafava, R., Vantaggi, B.: Coherent conditional probability as a tool for default reasoning. In: Bouchon-Meunier, Foulloy, Yager (eds.) Intell. Systems for Inform. Processing: from Repres. to Appl., pp. 191–202. Elsevier (2003)
19. Coletti, G., Scozzafava, R., Vantaggi, B.: Weak Implication in Terms of Conditional Uncertainty Measures. In: Mellouli, K. (ed.) ECSQARU 2007. LNCS (LNAI), vol. 4724, pp. 139–150. Springer, Heidelberg (2007)
20. Coletti, G., Scozzafava, R., Vantaggi, B.: Possibilistic and probabilistic logic under coherence: default reasoning and System P. Math. Slovaca (in press, 2014)
21. Coletti, G., Vantaggi, B.: Possibility theory: conditional independence. Fuzzy Sets and Sys. 157(11), 1491–1513 (2006)
22. Coletti, G., Vantaggi, B.: T-conditional possibilities: Coherence and inference. Fuzzy Sets and Sys. 160(3), 306–324 (2009)
23. de Cooman, G.: Possibility theory II: Conditional possibility. Int. J. Gen. Sys. 25(4), 325–351 (1997)
24. de Finetti, B.: Teoria della probabilità, vol. I-II. Einaudi, Torino (1970)
25. Dubois, D., Prade, H.: Possibility Theory: An Approach to Computerized Processing of Uncertainty. Plenum Press, New York (1988)
26. Ferracuti, L., Vantaggi, B.: Independence and conditional possibility for strictly monotone triangular norms. Int. J. of Intell. Sys. 21(3), 299–323 (2006)
27. Gilio, A.: Probabilistic reasoning under coherence in System P. Ann. of Math. and Art. Int. 34, 5–34 (2002)
28. Gilio, A., Sanfilippo, G.: Quasi conjunction, quasi disjunction, t-norms and t-conorms: Probabilistic aspects. Inf. Sci. 245, 146–167 (2013)
29. Godo, L., Marchioni, E.: Reasoning about coherent conditional probability in a fuzzy logic setting. Logic J. of the IGPL 14(3), 457–481 (2006)
30. Goodman, I.R., Nguyen, H.T.: Conditional objects and the modeling of uncertainties. In: Gupta, Yamakawa (eds.) Fuzzy Computing, pp. 119–138. N.Holland (1988)
31. Klement, E.P., Mesiar, R., Pap, E.: Triangular Norms. Trends in Logic, vol. 8. Kluwer Academic Publishers, Dordrecht (2000)
32. Petturiti, D.: Coherent Conditional Possibility Theory and Possibilistic Graphical Modeling in a Coherent Setting. PhD thesis, Univ. degli Studi di Perugia (2013)

Decision Making with Hierarchical Credal Sets[*]

Alessandro Antonucci[1], Alexander Karlsson[2], and David Sundgren[3]

[1] Istituto Dalle Molle di Studi sull'Intelligenza Artificiale
Manno-Lugano, Switzerland
alessandro@idsia.ch
[2] Informatics Research Center
University of Skövde, Sweden
alexander.karlsson@his.se
[3] Department of Computer and Systems Sciences
Stockholm University, Sweden
dsn@dsv.su.se

Abstract. We elaborate on hierarchical credal sets, which are sets of probability mass functions paired with second-order distributions. A new criterion to make decisions based on these models is proposed. This is achieved by sampling from the set of mass functions and considering the Kullback-Leibler divergence from the weighted center of mass of the set. We evaluate this criterion in a simple classification scenario: the results show performance improvements when compared to a credal classifier where the second-order distribution is not taken into account.

Keywords: Credal sets, second-order models, hierarchical credal sets, shifted Dirichlet distribution, credal classification, decision making.

1 Introduction

Many different frameworks exist for modeling and perform reasoning with *uncertainty*, e.g., *Bayesian theory* [1], *Dempster-Shafer theory* [8] or *coherent lower previsions* [11]. *Imprecise probability* is a general term referred to theories where a sharp specification of the probabilities is not required. These approaches are often considered to be more realistic and robust, while a precise assessment of the parameters can be hard to motivate. One common way to model imprecision is by closed convex sets of probability functions, which are also called *credal sets*.

Even though credal sets are attractive from several viewpoints, one problem that one can encounter is that the posterior can be highly imprecise and thus uninformative for a decision maker [5,6]. This is also one of the strengths of imprecise probability: if there is a serious lack of information, a single decision cannot be taken unless more information is provided.

However, if one models second-order probability over a credal set, it has been shown that the distribution can be remarkably concentrated within the set [4].

---

[*] This work was partly supported by the Information Fusion Research Program (University of Skövde, Sweden), in partnership with the Swedish Knowledge Foundation under grant 2010-0320 (http://www.infofusion.se, UMIF project).

A. Laurent et al. (Eds.): IPMU 2014, Part III, CCIS 444, pp. 456–465, 2014.

We aim to further explore whether such a concentration could be exploited to reduce imprecision and at the same time maintain a high degree of accuracy.

The paper is organized as follows: In Sect. 2, we provide the preliminaries for the theory of credal sets. In Sects. 3–4, we clarify the relation between credal sets and second-order models and present the concept of hierarchical credal sets. In Sects. 5–6, we introduce a new decision criterion that takes second-order probability into account. In Sect. 7, we evaluate this procedure on a simple classification scenario, and lastly, in Sect. 8, we provide a summary and conclusions.

2   Credal Sets

Let X be a variable taking its values in $\mathcal{X} := \{x_1, \ldots, x_n\}$. Uncertainty about X can be modeled by a single *probability mass function* (PMF) $P(X)$.[1] Given a function of X, say $f : \mathcal{X} \to \mathbb{R}$, the corresponding expected value of f according to $P(X)$ is:[2]

$$E_P[f] := \sum_{i=1}^{n} P(x_i) \cdot f(x_i). \tag{1}$$

Yet, there are situations where a single PMF cannot be regarded as a realistic model of uncertainy, e.g., when information is scarce or incomplete [11]. A possible generalization consists in coping with sets of (instead of single) PMFs. Such a generalized uncertainty model is called *credal set* (CS) and notation $K(X)$ is used here for CSs over X. Expectations based on a CS cannot be computed precisely as in Eq. (1). Only lower and upper bounds w.r.t. the different PMFs belonging to the CS can be evaluated, i.e.,

$$\underline{E}_K[f] := \min_{P(X) \in K(X)} E_P[f], \tag{2}$$

$$\overline{E}_K[f] := \max_{P(X) \in K(X)} E_P[f]. \tag{3}$$

Optima in Eqs. (2) and (3) can be equivalently evaluated over the convex closure of $K(X)$. Without lack of generality we therefore assume CSs to be closed and convex. Furthermore, if the CS is generated by a finite number of linear constraints, the two above optimization tasks are linear programs, the CS being the feasible region and the precise expectation in Eq. (1) the objective function. The solution of such a linear program can be found in an extreme point of the CS. We denote the extreme points of $K(X)$ by $\text{ext}[K(X)]$. Under these assumptions, the CS has a finite number of extreme points, i.e., $\text{ext}[K(X)] = \{P_j(X)\}_{j=1}^{v}$. Accordingly, the two optimization tasks can be equivalently solved by computing the precise expectation only on the extreme points.

---

[1] I.e., a map $P : \mathcal{X} \to \mathbb{R}$, such that $P(x_i) \geq 0 \; \forall i = 1, \ldots, n$, and $\sum_{i=1}^{n} P(x_i) = 0$.

[2] Following the behavioural interpretation of probability, f is regarded as a *gamble* (i.e., an uncertain reward), the expectation being the *fair price* an agent is willing to pay to buy the gamble on the basis of his/her subjective knowledge about X.

As an example, the so-called *vacuous* CS $K_0(X)$ is obtained by considering all the PMFs over X. The extreme points of this CS are degenerate PMFs assigning all the mass to a single state of X. The expectations as in Eqs. (2) and (3) based on the vacuous CS are therefore the minimum and the maximum value of f. The vacuous CS is the least informative uncertainty model, modeling a condition of *ignorance* about X. More informative CSs can be induced by a set of *probability intervals* (PIs) $I := \{(l_i, u_i)\}_{i=1}^n$, yielding the following CS:

$$K_I(X) := \left\{ P(X) \left| \begin{array}{l} \max\{0, l_i\} \leq P(x_i) \leq u_i \quad \forall i = 1, \ldots, n \\ \sum_{i=1}^n P(x_i) = 1 \end{array} \right. \right\}. \qquad (4)$$

As an example, if $l_i = 0$ and $u_i = 1$ for each $i = 1, \ldots, n$, Eq. (4) returns the vacuous CS. To guarantee the CS in Eq. (4) to be non-empty, it is sufficient (and necessary) to require $\sum_{i=1}^n l_i \leq 1$ and $\sum_{i=1}^n u_i \geq 1$. To have so called *reachable* PIs, i.e., such that for each $p_i \in [l_i, u_i]$ there is at least a $P(X) \in K_I(X)$ for which $P(x_i) = p_i$, the additional condition $\sum_{j \neq i} l_j + u_i \leq 1$ and $\sum_{j \neq i} u_j + l_i \geq 1$ should be met. A non-reachable set of PIs leading to a non-empty CS can be always made reachable [3].

3   Credal Sets Are (Not) Second-Order Models

Consider an auxiliary variable T whose set of possible values $\mathcal{T}$ is in one-to-one correspondence with $\text{ext}[K(X)]$. For each $t \in \mathcal{T}$, the (conditional) PMF $P(X|t)$ is the extreme point of $K(X)$ associated to t. This defines a conditional model $P(X|T)$ for X given T. The following result holds.

Proposition 1 (Cano Cano Moral transformation). *Consider a vacuous CS $K_0(T)$ and combine it with $P(X|T)$ as follows:*[3]

$$K'(X, T) := \left\{ P'(X, T) \left| \begin{array}{l} P'(x, t) = P(x|t) \cdot P(t) \quad \forall (x, t) \in \mathcal{X} \times \mathcal{T} \\ P(T) \in K_0(T) \end{array} \right. \right\}. \qquad (5)$$

Then marginalize X by summing out T as follows:

$$K'(X) := \left\{ P'(X) \left| \begin{array}{l} P'(x) = \sum_{t \in \mathcal{T}} P'(x, t) \\ \forall P'(X, T) \in K'(X, T) \end{array} \right. \right\}. \qquad (6)$$

The resulting CS coincides with the original one, i.e., $K(X) = K'(X)$.

This result, originally derived for credal nets in [2], clarifies why CSs should not be considered hierarchical models: the elements of the CS can be parametrized by an auxiliary variable, but the imprecision is just moved to the second order, where we should assume a complete lack of knowledge (modeled as a vacuous CS). To see this, the above proposition, which only considers the extreme points, can be extended to the whole CS. We therefore replace the categorical variable

---

[3] This operation is called *marginal extension* in [11]. Both Eqs. (5) and (6) can be computed by coping only with the extreme points and then taking the convex hull.

T with a continuous Θ indexing the elements of $K(X)$. For each $P(X) \in K(X)$, a conditional $P(X|\Theta = \theta) := P(X)$ is defined, i.e., Θ takes values in $K(X)$. We denote by $\pi(\Theta)$ a probability density over $K(X)$, i.e., $\pi(\Theta) \geq 0$ for each $\Theta \in K(X)$ and $\int_{\Theta \in K(X)} \pi(\Theta) \cdot d\Theta = 1$. The vacuous CS $K_0(\Theta)$ includes all the probability densities over Θ and its convex closure is considered with respect to the weak topology [11, App. D].

Proposition 2. *Combine the unconditional CS $K_0(\Theta)$ with the conditional model $P(X|\Theta)$ to obtain the following joint CS:*

$$K'(X) := \left\{ P'(X) \middle| \begin{array}{l} P'(x) := \int_{\Theta \in K(X)} P(x|\Theta) \cdot \pi(\Theta) \cdot d\Theta \\ \pi(\Theta) \in K_0(\Theta) \end{array} \right\}. \tag{7}$$

Then $K'(X) = K(X)$.

4   Hierarchical Credal Sets

We define a *hierarchical credal set* (HCS) as a pair (K, π), with π density over K [5]. Expectations based on these models can be therefore precisely computed, being the weighted average of the expectations associated to the different PMFs:

$$E_{K,\pi}[f] := \int_{\Theta \in K(X)} E_\Theta[f] \cdot \pi(\Theta) \cdot d\Theta. \tag{8}$$

The following result shows how HCS-based expectations can be regarded as precise expectations.

Proposition 3. *The computation of Eq. (8) can be obtained as follows:*

$$E_{K,\pi}[f] = E_{P_{K,\pi}}[f] \tag{9}$$

with

$$P_{K,\pi}(x_i) := \int_{\Theta \in K(X)} \Theta_i \cdot \pi(\Theta) \cdot d\Theta, \tag{10}$$

where Θ_i is the value of $P(X = x_i)$ when $P(X)$ is the PMF associated to Θ.

An obvious corollary of this result is the compatibility of the expectations based on HCSs with the lower and upper expectation based on CS.

Proposition 4. *Given a CS $K(X)$ and a HCS (K, π) with the same CS, then:*

$$\underline{E}_K[f] \leq E_{K,\pi}[f] \leq \overline{E}_K[f]. \tag{11}$$

Consider for instance a HCS with a uniform density, i.e., $\pi(\Theta) \propto 1$ for each $\Theta \in K(X)$. With a CS $K(X)$ with only three vertices, $P_{K,\pi(X)}$ is the center of mass of the CS and the model can be *equivalently* formalized as a *discrete* HCS, where the density over the whole set of elements of $K(X)$ is replaced by a PMF assigning probability $\frac{1}{3}$ to the three extreme points. This result generalizes to any HCS as stated by the following proposition.

Proposition 5. *Expectations based on a HCS* $[K(X), \pi(\Theta)]$ *can be equivalently computed as expectations of a discrete HCS* $[K(X), P(T)]$. *The discrete variable* T *indexes the elements of* $\text{ext}[K(X)]$, *and the values of* $P(T)$ *are a solution of the following linear system, which always admits a solution:*

$$\sum_{t \in \mathcal{T}} P(t) \cdot P_t(x_i) = P_{K,\pi}(x_i),\tag{12}$$

for each $x_i \in \mathcal{X}$, *where* $P_t(X)$ *is the extreme point of* $K(X)$ *associated to* t.

5   The Shifted Dirichlet Distribution

We restrict HCSs to simplicial forms which means that the set of PIs $\{(l_i, u_i)\}_{i=1}^n$ strictly satisfies the sufficient inequality conditions of reachability, i.e., $u_i = 1 - \sum_{j \neq i} l_j$, for each $i = 1, \dots, n$. The lower bounds $\{l_i\}_{i=1}^n$ are therefore sufficient to specify the PIs. Given a CS $K_I(X)$ of this kind, a (continuous) HCS is obtained by pairing the CS with the so-called *shifted Dirichlet distribution* [5] (SDD), which is parametrized by an array of nonnegative weights $\boldsymbol{\alpha} := (\alpha_1, \dots, \alpha_n)$ and lower bounds $\boldsymbol{l} := (l_1, \dots, l_n)$:

$$\pi_{\boldsymbol{\alpha}}(\Theta) \propto \prod_{i=1}^n [\Theta_i - l_i]^{\alpha_i - 1},\tag{13}$$

for each Θ associated to a $P(X) \in K(X)$ and with $\Theta_i := P(X = x_i)$, with the proportionality constant obtained by normalization. The SDD generalizes the standard Dirichlet distribution where the latter is obtained if the lower bounds are zero, i.e., the underlying CS is vacuous. Even in this generalized setup the weights $\boldsymbol{\alpha}$ are associated to the relative strengths of different states. This is made explicit by the following result about the expectations of HCSs based on the SDD.

Proposition 6. *The weighted center of mass, as in Eq. (10), of HCS associated to a SDD, say* $[K_I(X), \pi_{\boldsymbol{\alpha}}(\Theta)]$, *is, for each* $i = 1, \dots, n$:[4]

$$P_{K,\pi}(x_i) = l_i + \frac{\alpha_i(1 - \sum_{i=1}^n l_i)}{\sum_{j=1}^n \alpha_j}.\tag{14}$$

This allows one to compute expectations as in Eq. (9). The above considered continuous HCSs can be therefore equivalently expressed in terms of discrete HCS with PMF obtained by solving the linear system in Prop. 5 (the equivalence relation being intended with respect to expectancy).

---

[4] The right-hand side of Eq. (14) rewrites as $l_i + t_i(u_i - l_i)$, where $t_i := \alpha_i / (\sum_i \alpha_i)$.

6   Decision Making with Hierarchical Credal Sets

Let us discuss the problem of making decisions based on a HCS. Consider a single $P(X)$ and $0/1$ losses. The decision corresponds to identify the most probable state of $\mathcal{X}$, i.e., $x_P^* := \arg\max_{x \in \mathcal{X}} P(x)$. Moving to CSs, multiple generalizations are possible. A popular approach is the *maximality* criterion [11], which is based on the notion of *credal dominance*. Given $x, x' \in \mathcal{X}$, x dominates x' if $P(x) > P(x')$ for each $P(X) \in K(X)$. After testing this dominance for each $x, x' \in \mathcal{X}$, the set of undominated states, to be denoted as $\mathcal{X}_K^* \subseteq \mathcal{X}$, is returned. It is straightforward to see that $P(X) \in K(X)$ implies $x_P^* \in \mathcal{X}_K^*$. According to Prop. 3, expectations based on a HCS (K, π) can be equivalently computed with the precise model $P_{K,\pi}(X)$. The decision is therefore $x_{P_{K,\pi}}^*$, which belongs to the maximal set $\mathcal{X}_K^*$.

Apart from special cases like in Prop. 6, the weighted center of mass $P_{K,\pi}$ can be computed only by Monte Carlo integration. This can be done by uniformly sampling M PMFs from the CS $K(X)$. The corresponding approximation converges to the exact value as follows:

$$\left\| P_{K,\pi}(X) - \frac{\sum_{j=1}^{M} w_j \cdot P^{(j)}(X)}{\sum_{j=1}^{M} w_j} \right\|_{M \to +\infty} = O\left(\frac{1}{\sqrt{M}}\right), \tag{15}$$

where, for each $j = 1, \ldots, M$, $P^{(j)}(X)$ is the j-th PMF sampled from $K(X)$ and $w_j := \pi(P^j(X))$. Uniform sampling from a polytope can be efficiently achieved by a MCMC-schema, called the "Hit-And-Run" (HAR) algorithm [7]. HAR has recently been utilized in multi-criteria decision making to sample weights [10], and here we use of the algorithm for a similar purpose, i.e., sample weights with respect to a second-order distribution. Note that the second term in the left-hand side of Eq. (15) is a convex combination of elements of $K(X)$, and hence belongs to $K(X)$. For sufficiently large M, this returns $x_{P_{K,\pi}}^*$. We propose a new criterion, described in Alg. 1 and called HCS-KL, also based on sampling, but allowing for multiple decisions. The decision based on (the approximation of) $P_{K,\pi}(X)$ is replaced by a set of maximal decisions $\mathcal{X}_{K'}^*$, based on CS $K'(X) \subseteq K(X)$, obtained by removing from the sampled PMFs those at high weighted KL distance from the weighted center of mass. The idea is that the imprecision of a CS can be significant whilst the second-order distribution can be quite concentrated [5], but not so concentrated to always return a single option [4].

7   Application to Classification

The ideas outlined in the previous section are extended here to the multivariate case and tested on a classification problem. Let us therefore consider a collection of variables $(X_0, X_1, \ldots, X_n)$. Regard X_0 as the variable of interest (i.e., the class variable) and the remaining ones as those to be observed (i.e., the features). To assess a joint model over these variables, the so-called *naive assumption*

Algorithm 1. The HCS-KL algorithm. The input is a set of PMFs with their weights, i.e., $\{P^{(j)}(X), w_j\}_{j=1}^M$. This can be obtained from a HCS (K, π) by uniformly sampling the PMFs from $K(X)$ (using the HAR algorithm) and computing the weights $w_j := \pi(P^{(j)}(X))$. Given a value of the parameter $0 \leq \beta \leq 1$, the algorithm returns a set of optimal states $\mathcal{X}_{K'}^* \subseteq \mathcal{X}_K^*$.

1: Compute the weighted center of mass $\tilde{P}(X) := (\sum_{j=1}^M w_j)^{-1} \sum_{j=1}^M w_j P^{(j)}(X)$
2: $\mathcal{P} \leftarrow \{P^{(k)}(X)\}_{k=1}^M$
3: **for** $j = 1, \ldots, M$ **do**
4:    **if** $w_j^{-1} \text{KL}(\tilde{P}, P^{(j)}) > \beta \cdot \max_{k=1}^M \left[w_k^{-1} \text{KL}(\tilde{P}, P^{(k)}) \right]$ **then**
5:        $\mathcal{P} \leftarrow \mathcal{P} \setminus \{P^{(j)}\}$
6:    **end if**
7: **end for**
8: **return** maximal states $K'(X)$, i.e. $\mathcal{X}_{K'}^*$, with $K'(X)$ convex closure of $\mathcal{P}$

---

assumes conditional independence between the observable variables given X_0. This corresponds to the following factorization:

$$P(x_0, x_1, \ldots, x_n) = P(x_0) \cdot \prod_{k=1}^n P(x_k | x_0). \tag{16}$$

Given an observation $\tilde{\mathbf{x}} := (\tilde{x}_1, \ldots, \tilde{x}_n)$, the most probable value of X_0 is $x_0^* := \arg\max_{x_0 \in \mathcal{X}_0} P(x_0, \tilde{\mathbf{x}})$, which can be solved by Eq. (16) in terms of the local models: $P(X_0)$ and $P(X_k | x_0)$ for each $k = 1, \ldots, n$ and $x_0 \in \mathcal{X}_0$. In the imprecise framework, these local models are replaced by CSs. The maximal states are obtained by testing credal dominance test for $x_0', x_0'' \in \mathcal{X}_0$, i.e., checking whether of not the left-hand side of the following equation is greater than one.

$$\min_{\substack{P(X_0) \in K(X_0), \\ P(X_j | x_0) \in K(X_j | x_0) \\ \forall j \forall x_0}} \frac{P(x_0' | \tilde{\mathbf{x}})}{P(x_0'' | \tilde{\mathbf{x}})} = \min_{P(X_0) \in K(X_0)} \frac{P(x_0') \prod_{k=1}^n \underline{P}(\tilde{x}_k | x_0')}{P(x_0'') \prod_{k=1}^n \overline{P}(\tilde{x}_k | x_0'')}. \tag{17}$$

The right-hand side is obtained by exploiting the factorization in Eq. (16): the optimizatiom reduces to a trivial linear-fractional task over $P(x_0')$ and $P(x_0'')$.

The *imprecise Dirichlet model* [12] (IDM) can be used to learn CSs from data. This induces the following PIs parametrized by the lower bounds only:

$$P(x_0) \geq \frac{n(x_0)}{N + s} \tag{18}$$

where $n(x_0)$ are the data such that $X_0 = x_0$, N is the total amount of data, and s is the equivalent sample size. The conditional CSs $K(X_j | x_0)$ are obtained likewise. For the precise case, a single Dirichlet prior with sample size s and uniform weights, leading to $P(x_0) := (N + s)^{-1}(n(x_0) + s/|\mathcal{X}_0|)$, is considered.

In the hierarchical (credal) case, we pair these CSs with an equal number of SDDs. If no expert information is available, the SDD parameters can be specified by the *relative independence* assumption [9]. This assumption models a lack of dependence relations, apart from the necessary normalization constraint, among the values of Θ. This corresponds to set uniform weights with sum $n/(n-1)$, where n is the cardinality of the variable. This is the equivalent sample size of the SDD: it seems therefore reasonable to use this value for the parameter s in the IDM (this being also consistent with Walley's recommendation $1 \leq s \leq 2$) and also in the Bayesian case.

To perform classification based on this model, we extend the decision making criterion HCS-KL described in Alg. 1 to the multivariate case by the procedure described in Alg. 2. For each local model we sample a PMF and evaluate its weight. We then compute the posterior PMF, whose weight is just the product of the weights of the local models. This yields a collection of PMFs with the corresponding weights, to be processed by HCS-KL.

Algorithm 2. Hierarchical credal classification. A HCS is provided for each X_j given each value of x_0 and a HCS over X_0 are provided. Given an observation of the attributes $\tilde{\mathbf{x}}$, a set of possible classes $\mathcal{X}_0^* \subseteq \mathcal{X}_0$ is returned.

1: **for** $j = 1, \ldots, M$ **do**
2:     Uniformly sample $P^j(X_0)$ from $K(X_0)$
3:     Uniformly sample $P^j(X_k|x_0)$ from $K(X_k|x_0)$, $\forall k \; \forall x_0$
4:     Compute $P^j(X_0|\tilde{\mathbf{x}})$ [see Eq. (16)].
5:     $w_j = \pi_{X_0}(P^j(X_0)) \cdot \prod_{k,x_0} \pi_{X_k,x_0}(P^j(X_k|x_0))$
6: **end for**
7: **return** $\mathcal{X}_0^* :=$HCS-KL($\{P^j(X_0|\tilde{\mathbf{x}}), w_j\}_{j=1}^M$) (see Alg. 1)

7.1 Numerical Results

We validate classification based on Alg. 2 against the traditional NBC (see Eq. (16)) and its credal extension as in Eq. (17). These three approaches are called *hierarchical*, *Bayesian*, and *credal*. We use four datasets from the UCI repository with twofold cross validation. The accuracy of the Bayesian is compared with the utility-based u_{80} performance descriptor for other approaches. This descriptor, proposed in [13], is the state of the art for compare credal models with traditional ones under the assumption of (high) risk aversion to variability in the previsions. Regarding the choice of β and M in Alg. 2, $\beta = .25$ appears a reasonable choice to obtain results that clearly differs from the Bayesian case (corresponding to $\beta \simeq 0$) and the credal (corresponding to $\beta \simeq 1$), and $M = 200$ was sufficient to always observe convergence in the outputs.[5] We see that the hierarchical approach always outperforms the credal one (see Table 1).

---

[5] A R implementation is freely available at http://ipg.idsia.ch/software.

Table 1. Numerical evaluation. For each dataset, size, number of classes, accuracy of the Bayesian and u_{80}-accuracy of the credal and hierarchical approaches are reported.

Dataset	Size	Classes	Bayesian	Credal	Hierarchical
Contact Lenses	24	3	77.2	53.7	72.2
Labor	51	2	87.0	92.7	93.7
Hayes	160	4	59.5	51.1	72.4
Monk	556	2	64.1	70.6	72.9

8   Summary and Conclusion

We have extended CSs to a hierarchical uncertainty structure where beliefs can be expressed over the imprecision. We have introduced a simple decision criterion, based on KL divergence, that take second-order information into consideration. Preliminary tests on a classification benchmark are promising: the second-order information leads as expected to more accurate decisions. In our future research, we will explore more ways of modeling second-order information for decision making, including how one can express second-order information over a CS that are not simplicial and the determination of some reasonable shape of a credbility region that contains a certain degree of second-order probability mass.

A   Proofs

Proof of Proposition 1. *Given a $\tilde{P}(X) \in K'(X)$, let $\tilde{P}(X,T) \in K'(X,T)$ be the joint PMF whose marginalization produced $\tilde{P}(X)$. Similarly let $\tilde{P}(T) \in K_0(T)$ denote the PMF whose combination with $P(X|T)$ produced $\tilde{P}(X,T)$. We have $\tilde{P}(X) = \sum_t P(X|t)\tilde{P}(t)$. This means that $\tilde{P}(X)$ is a convex combination of the extreme points of $K(X)$. Thus, it belongs to $K(X)$. This proves $K'(X) \subseteq K(X)$. To prove the opposite inclusion consider a $\hat{P}(X) \in K(X)$. By definition, this is a convex combination of the extreme points of $K(X)$, i.e., $\hat{P}(X) = \sum_j \alpha_j P_j(X)$. Thus, by simply setting $P(t) = \alpha_j$, where t is the element of $\mathcal{T}$ associated to the j-th vertex of $K(X)$ we prove the result.* □

Proof of Proposition 2. *The proof is a simplified version of that of Prop. 1. Given a $P'(X) \in K'(X)$, $P'(X)$ also belongs to $K(X)$ because it is a convex combination of elements of $K(X)$, which is a closed and convex set. This proves $K'(X) \subseteq K(X)$, while the opposite inclusion follows from the fact that any $\hat{P}(X) \in K(X)$ also belongs to $K'(X)$ and this can be seen by choosing a degenerate distribution $\pi(\Theta)$ assigning all the probability density to $\hat{P}(X)$.* □

Proof of Proposition 3. *Let us rewrite put the expression of a precise expectation as in Eq. (1) in Eq. (8):*

$$E_{K,\pi}[f] = \int_{\Theta \in K(X)} \sum_{i=1}^{n} \Theta_i \cdot f(x_i) \cdot \pi(\Theta) d\Theta \qquad (19)$$

The result in Eq. (9) follows by moving the sum and the value of the function out of the integral.

Proof of Proposition 4. *The proof is straightforward.*

Proof of Proposition 5. *It is sufficient to note that the left-hand side of Eq. (12) is the weighted center of mass of the discrete HCS. Thus, the two HCSs have the same weighted center of mass and hence they return the same expectations. Moreover, the matrix of the coefficient has full rank because of the definition of extreme point of a convex set, and the linear system therefore always admits a solution.*

Proof of Proposition 6. *The mean of variable θ_i in a Dirichlet distribution with parameters $\vec{\alpha} = (\alpha_1, \ldots, \alpha_n)$ is $\frac{\alpha_i}{\sum_{j=1}^{n} \alpha_j}$. In the SDD the variables θ_i are linearly transformed so that $0 \mapsto l_i$ and $1 \mapsto 1 - \sum_{j \neq i} l_i$. The mean is by this transformation equal to*

$$l_i + \frac{\alpha_i(1 - \sum_{i=1}^{n} l_i)}{\sum_{j=1}^{n} \alpha_j}.$$

References

1. Bernardo, J.M., Smith, F.M.: Bayesian Theory. John Wiley and Sons (2000)
2. Cano, A., Cano, J., Moral, S.: Convex sets of probabilities propagation by simulated annealing on a tree of cliques. In: Bouchon-Meunier, B., Yager, R.R., Zadeh, L.A. (eds.) IPMU 1994. LNCS, vol. 945, pp. 978–983. Springer, Heidelberg (1995)
3. de Campos, L., Huete, J., Moral, S.: Probability intervals: A tool for uncertain reasoning. Int. Journ. of Unc., Fuzz. and Knowledge-Based Syst. 2, 167–196 (1994)
4. Gärdenfors, P., Sahlin, N.: Unreliable probabilities, risk taking, and decision making. Synthese 53, 361–386 (1982)
5. Karlsson, A., Sundgren, D.: Evaluation of evidential combination operators. In: Proceedings of ISIPTA 2013 (2013)
6. Karlsson, A., Sundgren, D.: Second-order credal combination of evidence. In: Proceedings of ISIPTA 2013 (2013)
7. Lovász, L.: Hit-and-run mixes fast. Math. Programming 86(3), 443–461 (1999)
8. Shafer, G.: A Mathematical Theory of Evidence. Princeton University Press (1976)
9. Sundgren, D., Karlsson, A.: On dependence in second-order probability. In: Hüllermeier, E., Link, S., Fober, T., Seeger, B. (eds.) SUM 2012. LNCS, vol. 7520, pp. 379–391. Springer, Heidelberg (2012)
10. Tervonen, T., van Valkenhoef, G., Basturk, N., Postmus, D.: Hit-and-run enables efficient weight generation for simulation-based multiple criteria decision analysis. European Journal of Operational Research 224(3), 552–559 (2013)
11. Walley, P.: Statistical Reasoning with Imprecise Probabilities. Chapman and Hall (1991)
12. Walley, P.: Inferences from multinomial data: learning about a bag of marbles. Journal of the Royal Statistical Society. Series B (Methodological) 58, 3–57 (1996)
13. Zaffalon, M., Corani, G., Mauá, D.D.: Evaluating credal classifiers by utility-discounted predictive accuracy. International Journal of Approximate Reasoning 53(8), 1282–1301 (2012)

A Propositional CONEstrip Algorithm

Erik Quaeghebeur*

Centrum Wiskunde & Informatica
Science Park 123, 1098 XG Amsterdam, Netherlands

Abstract. We present a variant of the CONEstrip algorithm for checking whether the origin lies in a finitely generated convex cone that can be
open, closed, or neither. This variant is designed to deal efficiently with
problems where the rays defining the cone are specified as linear combinations of propositional sentences. The variant differs from the original
algorithm in that we apply row generation techniques. The generator
problem is WPMaxSAT, an optimization variant of SAT; both can be
solved with specialized solvers or integer linear programming techniques.
We additionally show how optimization problems over the cone can be
solved by using our propositional CONEstrip algorithm as a prepro-
cessor. The algorithm is designed to support consistency and inference
computations within the theory of sets of desirable gambles. We also
make a link to similar computations in probabilistic logic, conditional
probability assessments, and imprecise probability theory.

Keywords: sets of desirable gambles, linear programming, row generation, satisfiability, SAT, PSAT, WPMaxSAT, consistency, coherence,
inference, natural extension.

1 Introduction

The CONEstrip algorithm [12] determines whether a finitely generated general
convex cone contains the origin. A general convex cone can be open, closed, or
ajar, i.e., neither open nor closed. This linear programming-based algorithm is
designed for working with uncertainty models based on (non-simple) sets of desirable gambles [15,16,13] and their generalizations [14]. In particular, it can be used
for checking the consistency criteria such models have to satisfy—specifically,
coherence and avoiding partial loss—and for drawing deductive inferences from
such models—typically, performing natural extension.

In the CONEstrip algorithm, the so-called gambles defining the cone had to
be specified as vectors on some explicitly given, finite possibility space. The specification of the gambles as linear combinations of indicator functions of events

---

* This research was started while the author was a member of the SYSTeMS Research
Group of Ghent University and was at that time a visiting scholar at the Department
of Information and Computing Sciences of Utrecht University. It was finished during
the tenure of an ERCIM "Alain Bensoussan" Fellowship Programme. The research
leading to these results has received funding from the *European Union* Seventh
Framework Programme (FP7/2007-2013) under *grant agreement* n° 246016.

A. Laurent et al. (Eds.): IPMU 2014, Part III, CCIS 444, pp. 466–475, 2014.
© Springer International Publishing Switzerland 2014

belonging to some finite set often provides a more natural and economical formulation of the uncertainty model.

Events can be formalized as logical propositions; an elementary event of the underlying possibility space corresponds to a conjunction of all these propositions or their negation. The cardinality of the underlying possibility space is therefore exponential in the number of events. So even though it is possible to write down the gambles as vectors on the underlying possibility space and apply the CONEstrip algorithm, this would be very inefficient: the size of the linear programs involved is linear in the cardinality of the possibility space.

Therefore we need a variant of the algorithm that works efficiently in the propositional context sketched. We present such a variant in this paper. It preserves the structure of CONEstrip as an iteration of linear programs, each one 'stripping away' a superfluous part of the general cone. But now these linear programs are solved using a row generation technique—each row corresponding to an elementary event—as was already suggested by Walley et al. [17, Sec. 4, comment (d)] for a CONEstrip predecessor. Only the rows necessary for solving the linear programs are generated, and those already generated are carried over from one linear program to the next.

So instead of solving one big problem, we solve multiple smaller ones. The sub-problem to be solved to generate a row is WPMaxSAT, weighted partial maximum satisfiability. This was already discovered by Georgakopoulos et al. [6, Sec. 3] for the related dual PSAT, probabilistic satisfiability. WPMaxSAT is an optimization variant of SAT, propositional satisfiability [4]. Both can be tackled using binary linear programming techniques or specific algorithms [9,7,5].

The original impulse to work on this problem came from Cozman and di Ianni's recent contribution to the field [5], where many relevant references are listed, among which Hansen et al.'s classic review [8] cannot remain unmentioned here. The goal was to design a 'direct algorithm' [17] for sets of desirable gambles in their full generality. To present the result, we first get (re)acquainted with the standard CONEstrip algorithm in Section 2 and then build up towards the propositional variant in Section 3. In Section 4, we make a link to established problems that can be seen as special cases.

2   Preliminaries

A Representation of General Convex Cones. General convex cones can be open, closed, or ajar, i.e., neither open nor closed. Any finitely generated general convex cone can by definition be generated as the convex hull of a finite number of finitely generated open cones.

Let us formalize this definition: The possibility space is denoted by Ω. The set of all gambles is then $\mathcal{G} = \Omega \to \mathbb{R}$. Now consider a finite set $\mathcal{R}_0$ of finite subsets of $\mathcal{G}$, each containing the generators of an open cone, then a gamble f in $\mathcal{G}$ belongs to the general cone $\underline{\mathcal{R}_0}$ if and only if the following feasibility problem has a solution:

$$\text{find} \quad \lambda_{\mathcal{D}} \in [0,1] \quad \text{and} \quad \nu_{\mathcal{D}} \in (\mathbb{R}_{>0})^{\mathcal{D}} \quad \text{for all } \mathcal{D} \text{ in } \mathcal{R}_0$$
$$\text{such that} \quad \sum_{\mathcal{D} \in \mathcal{R}_0} \lambda_{\mathcal{D}} = 1 \quad \text{and} \quad \sum_{\mathcal{D} \in \mathcal{R}_0} \lambda_{\mathcal{D}} \sum_{g \in \mathcal{D}} \nu_{\mathcal{D},g} g \gtrless f. \tag{1}$$

Here $h := \sum_{\mathcal{D} \in \mathcal{R}_0} \lambda_{\mathcal{D}} \sum_{g \in \mathcal{D}} \nu_{\mathcal{D},g} g \gtrless f$ represents $h(\omega) \leq f(\omega)$ for all ω in Ω_Γ and $h(\omega) \geq f(\omega)$ for all ω in Ω_Δ, with Ω_Γ and Ω_Δ problem-specific sets that satisfy $\Omega_\Gamma \cup \Omega_\Delta = \Omega$. Using inequality instead of equality constraints allows us to omit (up to $|\Omega|$) indicator functions of singletons or their negation that may be present in the representation $\mathcal{R}_0$.

The formulation of this problem includes strict inequalities and bilinear constraints. The former issue puts this problem outside of the class of standard mathematical programming problems. The latter issue makes it a non-linear problem, and therefore one with non-polynomial worst-case computational complexity. Below, we are going to go over the solution to both issues [12].

The subscript 0 in $\mathcal{R}_0$ indicates that it is the representation of the original cone. In the algorithms we discuss, smaller cones will iteratively be derived from it; their representation $\mathcal{R}_i$ gets the iteration number i as a subscript.

The CONEstrip Algorithm. To eliminate the strict inequalities $\nu_{\mathcal{D}} \in (\mathbb{R}_{>0})^{\mathcal{D}}$ from Problem (1) we are going to replace the $\nu_{\mathcal{D}}$ using $\tau_{\mathcal{D}} \in (\mathbb{R}_{\geq 1})^{\mathcal{D}}$ and $\sigma \in \mathbb{R}_{\geq 1}$ such that $\tau_{\mathcal{D}} = \sigma \nu_{\mathcal{D}}$ for all $\mathcal{D}$ in $\mathcal{R}_0$:

$$\text{find} \quad \lambda_{\mathcal{D}} \in [0,1] \quad \text{and} \quad \tau_{\mathcal{D}} \in (\mathbb{R}_{\geq 1})^{\mathcal{D}} \quad \text{for all } \mathcal{D} \text{ in } \mathcal{R}_0 \quad \text{and} \quad \sigma \in \mathbb{R}_{\geq 1}$$
$$\text{s.t.} \quad \sum_{\mathcal{D} \in \mathcal{R}_0} \lambda_{\mathcal{D}} \geq 1 \quad \text{and} \quad \sum_{\mathcal{D} \in \mathcal{R}_0} \lambda_{\mathcal{D}} \sum_{g \in \mathcal{D}} \tau_{\mathcal{D},g} g \gtrless \sigma f. \tag{2}$$

Note that we have also relaxed the convex coefficient constraint, because it gives us the flexibility we need next, without changing the problem.

To get rid of the non-linearity, we first replace $\lambda_{\mathcal{D}} \tau_{\mathcal{D}}$ with new variables $\mu_{\mathcal{D}}$ for all $\mathcal{D}$ in $\mathcal{R}_0$ and add a constraint that forces $\mu_{\mathcal{D}}$ to behave as $\lambda_{\mathcal{D}} \tau_{\mathcal{D}}$:

$$\text{find} \quad \lambda_{\mathcal{D}} \in [0,1] \quad \text{and} \quad \mu_{\mathcal{D}} \in (\mathbb{R}_{\geq 0})^{\mathcal{D}} \quad \text{for all } \mathcal{D} \text{ in } \mathcal{R}_0 \quad \text{and} \quad \sigma \in \mathbb{R}_{\geq 1}$$
$$\text{s.t.} \quad \sum_{\mathcal{D} \in \mathcal{R}_0} \lambda_{\mathcal{D}} \geq 1 \quad \text{and} \quad \sum_{\mathcal{D} \in \mathcal{R}_0} \sum_{g \in \mathcal{D}} \mu_{\mathcal{D},g} g \gtrless \sigma f \tag{3}$$
$$\lambda_{\mathcal{D}} \leq \mu_{\mathcal{D}} \leq \lambda_{\mathcal{D}} \mu_{\mathcal{D}} \quad \text{for all } \mathcal{D} \text{ in } \mathcal{R}_0.$$

Notice that now $\lambda_{\mathcal{D}} \in \{0,1\}$ for any solution, functioning as a switch between $\mu_{\mathcal{D}} = 0\tau_{\mathcal{D}} = 0$ and $\mu_{\mathcal{D}} = 1\tau_{\mathcal{D}} \in (\mathbb{R}_{\geq 1})^{\mathcal{D}}$, so that $\mu_{\mathcal{D}}/\sigma$ effectively behaves as $\lambda_{\mathcal{D}} \nu_{\mathcal{D}}$.

We could replace the non-linear constraints $\mu_{\mathcal{D}} \leq \lambda_{\mathcal{D}} \mu_{\mathcal{D}}$ by $\mu_{\mathcal{D}} \leq \lambda_{\mathcal{D}} M_{\mathcal{D}}$, where $M_{\mathcal{D}}$ is a positive real number 'guaranteed to be' larger than $\max \mu_{\mathcal{D}}$, but this may be numerically problematic. Another approach is to remove the non-linear constraint, causing $\lambda_{\mathcal{D}} = 0$ to not force $\mu_{\mathcal{D}} = 0$ anymore—$\lambda_{\mathcal{D}} > 0$ still forces $\mu_{\mathcal{D}} \in (\mathbb{R}_{>0})^{\mathcal{D}}$. However, by maximizing $\sum_{\mathcal{D} \in \mathcal{R}_0} \lambda_{\mathcal{D}}$, the components of λ will function as a witness: if $\lambda_{\mathcal{D}} = 0$, then we *should* have $\mu_{\mathcal{D}} = 0$. This is the basis of the iterative CONEstrip algorithm:

Initialization. Set $i := 0$.

Iterand. Does the linear programming problem below have a solution $(\bar{\lambda}, \bar{\mu}, \bar{\sigma})$?

$$
\begin{aligned}
\text{maximize} \quad & \sum_{\mathcal{D} \in \mathcal{R}_i} \lambda_{\mathcal{D}} \\
\text{subject to} \quad & \lambda_{\mathcal{D}} \in [0,1] \quad \text{and} \quad \mu_{\mathcal{D}} \in (\mathbb{R}_{\geq 0})^{\mathcal{D}} \quad \text{for all } \mathcal{D} \text{ in } \mathcal{R}_i \quad \text{and} \quad \sigma \in \mathbb{R}_{\geq 1} \\
& \sum_{\mathcal{D} \in \mathcal{R}_i} \lambda_{\mathcal{D}} \geq 1 \quad \text{and} \quad \sum_{\mathcal{D} \in \mathcal{R}_i} \sum_{g \in \mathcal{D}} \mu_{\mathcal{D},g} g \gtreqless \sigma f \\
& \lambda_{\mathcal{D}} \leq \mu_{\mathcal{D}} \text{ for all } \mathcal{D} \text{ in } \mathcal{R}_i.
\end{aligned}
\tag{4}
$$

> **No.** $f \notin \underline{\mathcal{R}_0}$. **Stop.**
> **Yes.** Let $\mathcal{Q} := \{\mathcal{D} \in \mathcal{R}_i : \bar{\lambda}_{\mathcal{D}} = 0\}$ and set $\mathcal{R}_{i+1} := \mathcal{R}_i \setminus \mathcal{Q}$.
>> Is $\{\mathcal{D} \in \mathcal{Q} : \bar{\mu}_{\mathcal{D}} = 0\} = \mathcal{Q}$?
>> **Yes.** Set $t := i+1$; $f \in \underline{\mathcal{R}_t} \subseteq \underline{\mathcal{R}_0}$. **Stop.**
>> **No.** Increase i's value by 1. **Reiterate.**

This algorithm terminates after at most $|\mathcal{R}_0| - 1$ iterations. The 'raw' complexity of the ith linear programming problem is polynomial in $|\mathcal{R}_i|$, $\sum_{\mathcal{D} \in \mathcal{R}_i} |\mathcal{D}|$, and $|\Omega|$. The terminal cone $\underline{\mathcal{R}_t}$ is the largest 'subcone' of $\underline{\mathcal{R}_0}$ that contains f in its relative interior; so $\underline{\mathcal{R}_t}$ is included in a face of $\underline{\mathcal{R}_0}$.

Optimization Problems. We can solve optimization problems with continuous objective functions over the general cone: First ignore the objective and run the CONEstrip algorithm on the representation $\mathcal{R}_0$ to obtain a terminal representation $\mathcal{R}_t$. Then we can optimize over the *topological closure* of the terminal general cone $\underline{\mathcal{R}_t}$, due to the continuity of the objective function:

$$
\begin{aligned}
\text{optimize} \quad & \text{a continuous function of } \mu \\
\text{subject to} \quad & \mu \in (\mathbb{R}_{\geq 0})^{\cup \mathcal{R}_t} \quad \text{and} \quad \sum_{g \in \cup \mathcal{R}_t} \mu_g g \gtreqless f.
\end{aligned}
\tag{5}
$$

When the objective function is linear in μ, we get a linear programming problem.

3 An Algorithm for Proposition-Based Gambles

We saw that Problem (4) was polynomial in the size $|\Omega|$ of the possibility space. When, as is often done, the gambles involved in the representation of the general cone are specified as linear combinations of indicator functions of Boolean propositions, $|\Omega|$ becomes exponential in the number of propositions. In this section we present an approach that avoids having to deal directly with an exponential number of constraints in such a propositional context.

How Structured Gambles Generate Structured Problems. We are not going to dive into the propositional framework directly, but first we are going to show how gambles that are linear combinations of a number of basic functions generate a specific exploitable structure in the problems we consider. Assume that any gamble g in $\{f\} \cup \bigcup \mathcal{R}_0$ can be written as a linear combination

$\sum_{\phi \in \Phi} g_\phi \phi$ with a finite set of basic functions $\Phi \subset \mathcal{G}$ and coefficients g_ϕ in $\mathbb{R}$.[1] Then $(\sum_{D \in \mathcal{R}_i} \sum_{g \in D} \mu_{D,g} g) - \sigma f = \sum_{\phi \in \Phi} \kappa_\phi \phi$ if for all ϕ in Φ we define κ_ϕ as $(\sum_{D \in \mathcal{R}_i} \sum_{g \in D} \mu_{D,g} g_\phi) - \sigma f_\phi$. So in Problem 4 we can rewrite the constraints on the variables μ and σ in terms of gambles as constraints on the variables κ in terms of basic functions by adding constraints linking the κ with the μ and σ.

Increasing the number of constraints in this way is productive only when we can deal with them more efficiently. How this can be done is discussed below.

Row Generation. A standard technique for dealing with large numbers of constraints in mathematical programming is row generation: The original problem is first relaxed by removing most or all constraints; in our problem, the constraints $\sum_{\phi \in \Phi} \kappa_\phi \phi \gtrless 0$ are removed. Then, in an iterative procedure, constraints are added back. Each such constraint or 'row' corresponds to some elementary event ω in Ω, i.e., is of the form $\sum_{\phi \in \Phi} \kappa_\phi \phi(\omega) \gtrless 0$. Each iteration the problem is solved under the present constraints, resulting in a solution vector $\bar{\kappa}$.

So which constraints should be added back? Constraints that are satisfied by the present solution $\bar{\kappa}$ will have no discernable impact, as $\bar{\kappa}$ will remain feasible. Therefore, constraints that are *violated* by $\bar{\kappa}$ must be generated. There may be many violated constraints and one would want to generate deep 'cuts', those that constrain κ most, as less are needed than when generating shallow cuts. However, generating deep cuts may be computationally more complex than generating shallow cuts, so a trade-off needs to be made between the number of iterations and the complexity of the constraint generation process. For our problem, $\mathrm{argmax}_{\omega \in \Omega} |\sum_{\phi \in \Phi} \bar{\kappa}_\phi \phi(\omega)|$ would generate a deep cut.

So when does the procedure stop? The original problem is infeasible if constraint generation is infeasible, or when an intermediate problem turns out to be infeasible, given that it is a relaxation of the original problem. When no violated constraint can be generated given a solution $\bar{\kappa}$, then the problem is feasible and—in case of an optimization problem—this solution is optimal. That the problem stops eventually is guaranteed, because we could in principle add all constraints back. But actually, the number of iterations needed is polynomial in the number of variables involved [6, Lemma 2]; so, for us, polynomial in $|\Phi|$.

The Propositional Context. Now we take the basic functions ϕ in Φ to be expressed as propositional sentences. The operations permitted in such sentences are the binary disjunction $\vee$—'or'—and conjunction $\wedge$—'and'—, and the unary negation $\neg$. The $\{0,1\}$-valued binary variables appearing in the sentence are the so-called literals β_ℓ, where ℓ belongs to a given, finite index set $\mathcal{L}_\Phi := \bigcup_{\phi \in \Phi} \mathcal{L}_\phi$. For example, $\varphi(\beta) := (\beta_\spadesuit \vee \neg \beta_\clubsuit) \wedge \beta_\heartsuit$ is a sentence with $\mathcal{L}_\varphi := \{\spadesuit, \clubsuit, \heartsuit\}$.

The possibility space can be expressed in terms of the literals in the following way: $\Omega := \{\beta \in \{0,1\}^{\mathcal{L}_\Phi \cup \mathcal{L}_\psi} : \psi(\beta) = 1\}$, where ψ is a given sentence restricting

---

[1] The set Φ and objects that depend on it will in general also depend on the iteration number i: as gambles are effectively removed from $\mathcal{R}_0$ to obtain $\mathcal{R}_i$, basic functions appearing only in the expressions for removed gambles can be removed from Φ. We do not make this explicit in this paper to limit the notational burden.

the possible truth assignments—instantiations of β—that are valid elementary events. For example, there is a one-to-one relationship between $\{1, 2, 3\}$ and $\{\beta \in \{0, 1\}^2 : \beta_1 \vee \beta_2 = 1\}$ with element vectors viewed as bit strings. Similarly, $\Omega_\Gamma := \{\beta \in \Omega : \psi_\Gamma(\beta) = 1\}$ and $\Omega_\Delta := \{\beta \in \Omega : \psi_\Delta(\beta) = 1\}$. In a propositional context, we use sentences instead of the corresponding sets.

Two important special cases [1,11, PSAT vs. CPA] are (i) no restrictions, i.e., ψ identically one, and (ii) $\mathcal{L}_\psi = \mathcal{L}_\Phi := \Phi$ with $\phi(\beta) := \beta_\phi$ for all ϕ in Φ. It is actually always possible to have $\phi(\beta) = \beta_\phi$ by using extensions $\mathcal{L} := \mathcal{L}_\psi \cup \mathcal{L}_\Phi \cup \Phi$ and $\chi(\beta) := \psi(\beta) \wedge \bigwedge_{\phi \in \Phi}((\beta_\phi \wedge \phi(\beta)) \vee (\neg\beta_\phi \wedge \neg\phi(\beta)))$. We will do so.

The propositional sentences we consider can in principle take any form. However, it is useful for algorithm design to write such sentences in some canonical 'normal' form. Given the connections of this work with PSAT, probabilistic satisfiability, we use the form standard in that field, CNF, conjunctive— or clausal—normal form. In this form, a sentence is written as a conjunction of clauses; a clause is a disjunction of literals and negated literals. Formally, $\chi(\beta) = \bigwedge_{m=1}^k (\bigvee_{\ell \in \mathcal{P}_m} \beta_\ell \vee \bigvee_{\ell \in \mathcal{N}_m} \neg\beta_\ell)$, where k is the number of conjuncts in the CNF of $\chi(\beta)$ and $\mathcal{P}_m \subseteq \mathcal{L}$ and $\mathcal{N}_m \subseteq \mathcal{L}$ with $\mathcal{P}_m \cap \mathcal{N}_m = \varnothing$ are the index sets of the mth conjunct's plain and negated disjuncts, respectively. The transformation of any sentence φ into CNF with a number of clauses linear in the number of operations in φ is an operation with polynomial complexity [10].

Row Generation in the Propositional Context.

In the propositional context, we must in each iteration generate constraints of the form $\sum_{\phi \in \Phi} \kappa_\phi \bar{\beta}_\phi \gtrless 0$, by generating some truth assignment $\bar{\beta}$ in $\{0, 1\}^{\mathcal{L}}$. To generate deep cuts, the assignment $\bar{\beta}$ must be such that $|\sum_{\phi \in \Phi} \bar{\kappa}_\phi \bar{\beta}_\phi|$ is relatively large, where $\bar{\kappa}$ is the linear-program solution vector generated earlier in the iteration.

Generating valid truth assignments corresponds to solving a SAT problem: determining (whether there is) a β in $\{0, 1\}^{\mathcal{L}}$ such that $\chi(\beta) = 1$. General SAT is NP-complete. There exist many specialized SAT solvers. Also, any SAT problem can be formulated as a binary linear programming problem:

$$\begin{aligned} &\text{find} \quad \beta \in \{0, 1\}^{\mathcal{L}} \\ &\text{such that} \quad \sum_{\ell \in \mathcal{P}_m} \beta_\ell + \sum_{\ell \in \mathcal{N}_m} (1 - \beta_\ell) \geq 1 \quad \text{for all } 1 \leq m \leq k. \end{aligned} \tag{6}$$

Here, each constraint corresponds to a conjunct in the CNF of χ.

A SAT solver blindly generates instances, which will typically not result in deep cuts. To generate deep cuts, we need to take the constraint expression into account. A binary linear programming problem that does just this presents itself:

$$\begin{aligned} &\text{optimize} \quad \sum_{\phi \in \Phi} \bar{\kappa}_\phi \beta_\phi \\ &\text{subject to} \quad \beta \in \{0, 1\}^{\mathcal{L}} \\ &\qquad\qquad \sum_{\ell \in \mathcal{P}_m} \beta_\ell + \sum_{\ell \in \mathcal{N}_m} (1 - \beta_\ell) \geq 1 \quad \text{for all } 1 \leq m \leq k. \end{aligned} \tag{7}$$

In case of minimization, the β-instance generated is denoted $\bar{\bar{\delta}}$; in case of maximization it is denoted $\bar{\gamma}$.

In essence, this Problem (7) is WPMaxSAT, weighted partial maximum SAT: now, part of the clauses are hard—those of χ—and part of them are weighted soft clauses—the $\bar{\kappa}_\phi \beta_\phi$, essentially—, whose total weight is maximized. General WPMaxSAT is NP-hard. Specialized WPMaxSAT solvers generally accept only positive integers as weights. The integral nature can be ensured by rescaling and rounding, so we can assume $\bar{\kappa}$ has integer components. For positivity, the weights are replaced by their absolute value; their sign is expressed through the corresponding soft clause: For maximization, one must use $\bar{\kappa}_\phi \beta_\phi$ if $\bar{\kappa}_\phi > 0$ and $|\bar{\kappa}_\phi|(\neg\beta_\phi)$ if $\bar{\kappa}_\phi < 0$. For minimization, one must use $\bar{\kappa}_\phi(\neg\beta_\phi)$ if $\bar{\kappa}_\phi > 0$ and $|\bar{\kappa}_\phi|\beta_\phi$ if $\bar{\kappa}_\phi < 0$. A big advantage of WPMaxSAT-based row generation is that $\bar{\kappa}$ satisfies all the constraints generated in earlier iterations, so these will not be generated again.

A Propositional CONEstrip Algorithm. We combine the original CONEstrip algorithm as described at the end of Section 2 with as constraint generators SAT for bootstrapping and WPMaxSAT subsequently:

Initialization. Set $i := 0$.

   Are $\chi \wedge \psi_\Gamma$ and $\chi \wedge \psi_\Delta$ satisfiable?
   Neither. The original problem is not well-posed. ***Stop.***
   Either or both.
   - If $\chi \wedge \psi_\Gamma$ is satisfied by some $\bar{\gamma}$, set $\Gamma_0 := \{\bar{\gamma}\}$; otherwise set $\Gamma_0 := \varnothing$.
   - If $\chi \wedge \psi_\Delta$ is satisfied by some $\bar{\delta}$, set $\Delta_0 := \{\bar{\delta}\}$; otherwise set $\Delta_0 := \varnothing$.

Iterand.
   1. Does the linear programming problem below have a solution $(\bar{\lambda}, \bar{\mu}, \bar{\sigma}, \bar{\kappa})$?

$$\text{maximize} \quad \sum_{\mathcal{D} \in \mathcal{R}_i} \lambda_{\mathcal{D}}$$

$$\text{subject to} \quad \lambda_{\mathcal{D}} \in [0,1] \quad \text{and} \quad \mu \in (\mathbb{R}_{\geq 0})^{\mathcal{D}} \quad \text{for all } \mathcal{D} \text{ in } \mathcal{R}_i \quad \text{and} \quad \sigma \in \mathbb{R}_{\geq 1}$$

$$\sum_{\mathcal{D} \in \mathcal{R}_i} \lambda_{\mathcal{D}} \geq 1 \quad \text{and} \quad \begin{cases} \sum_{\phi \in \Phi} \kappa_\phi \bar{\beta}_\phi \leq 0 \text{ for all } \bar{\beta} \text{ in } \Gamma_i \\ \sum_{\phi \in \Phi} \kappa_\phi \bar{\beta}_\phi \geq 0 \text{ for all } \bar{\beta} \text{ in } \Delta_i \end{cases} \quad (8)$$

$$\lambda_{\mathcal{D}} \leq \mu_{\mathcal{D}} \text{ for all } \mathcal{D} \text{ in } \mathcal{R}_i$$

   where $\kappa_\phi := \left(\sum_{\mathcal{D} \in \mathcal{R}_0} \sum_{g \in \mathcal{D}} \mu_{\mathcal{D},g} g_\phi \right) - \sigma f_\phi \in \mathbb{R}$ for all ϕ in Φ.

   No. $f \notin \underline{\mathcal{R}_0}$. ***Stop.***
   Yes. Let $\mathcal{Q} := \{\mathcal{D} \in \mathcal{R}_i : \bar{\lambda}_{\mathcal{D}} = 0\}$ and set $\mathcal{R}_{i+1} := \mathcal{R}_i \setminus \mathcal{Q}$.

   2. - If $\Gamma_i \neq \varnothing$, let $\bar{\gamma}$ be the solution of the WPMaxSAT for maximizing $\sum_{\phi \in \Phi} \bar{\kappa}_\phi \beta_\phi$ under the hard clauses $\chi \wedge \psi_\Gamma$; set $\Gamma_{i+1} := \Gamma_i \cup \{\bar{\gamma}\}$. Otherwise, set $\bar{\gamma}$ identically zero.
      - If $\Delta_i \neq \varnothing$, let $\bar{\delta}$ be the solution of the WPMaxSAT for minimizing $\sum_{\phi \in \Phi} \bar{\kappa}_\phi \beta_\phi$ under the hard clauses $\chi \wedge \psi_\Delta$; set $\Delta_{i+1} := \Delta_i \cup \{\bar{\delta}\}$. Otherwise, set $\bar{\delta}$ identically zero.

   Is $\sum_{\phi \in \Phi} \bar{\kappa}_\phi \bar{\gamma}_\phi \leq 0 \leq \sum_{\phi \in \Phi} \bar{\kappa}_\phi \bar{\delta}_\phi$ and $\{\mathcal{D} \in \mathcal{Q} : \bar{\mu}_{\mathcal{D}} = 0\} = \mathcal{Q}$?
   Yes. Set $t := i + 1$; $f \in \underline{\mathcal{R}_t} \subseteq \underline{\mathcal{R}_0}$. ***Stop.***
   No. Increase i's value by 1. ***Reiterate.***

In this algorithm, the cone stripping and constraint generation iterations are merged. This can be done because if $\lambda_{\mathcal{D}} = 0$ for some $\mathcal{D}$ in $\mathcal{R}_i$, then certainly $\lambda_{\mathcal{D}} = 0$ when additional constraints are added.

The complexity of the algorithm is determined by (i) the number of iterations: polynomial in $|\Phi|$ [6,11] and linear in $|\mathcal{R}_0|$; (ii) the complexity of the 'master' linear program: polynomial in $|\Phi|$, $|\mathcal{R}_0|$, and $\sum_{\mathcal{D} \in \mathcal{R}_0} |\mathcal{D}|$; and (iii) the complexity of constraint generation—in the worst case exponential in k and polynomial in $|\mathcal{L}|$ [2]. So we have replaced a procedure with guaranteed exponential complexity due to an exponential number of constraints by a procedure that often has decent practical complexity. Because of the reduction to standard problems—SAT and WPMaxSAT, or binary linear programming—advances in solvers for those problems can directly be taken advantage of.

Before an implementation can be called mature, it must support 'restarting' of the SAT and WPMaxSAT solvers, meaning that preprocessing—such as variable elimination—needs to be done only once, before the first iteration. This could provide efficiency gains similar to those obtained by algorithms for probabilistic reasoning from the Italian school [1,3].

Optimization Problems. Again, we can solve optimization problems with continuous objective functions over the general cone: First ignore the objective and run the propositional CONEstrip algorithm on the representation $\mathcal{R}_0$ to obtain a terminal representation $\mathcal{R}_t$ and terminal instance sets Δ_t and Γ_t. Then we optimize over the topological closure of the terminal general cone $\underline{\mathcal{R}_t}$:

Initialization. Set $i := t$.
Iterand.

   1. Solve the following optimization problem to obtain the solution $(\bar{\mu}, \bar{\kappa})$:

      optimize   a continuous function of μ

$$\text{subject to} \quad \mu \in (\mathbb{R}_{\geq 0})^{\cup \mathcal{R}_t} \quad \text{and} \quad \begin{cases} \sum_{\phi \in \Phi} \kappa_\phi \bar{\beta}_\phi \leq 0 \text{ for all } \bar{\beta} \text{ in } \Gamma_i \\ \sum_{\phi \in \Phi} \kappa_\phi \bar{\beta}_\phi \geq 0 \text{ for all } \bar{\beta} \text{ in } \Delta_i \end{cases} \quad (9)$$

      where $\kappa_\phi := (\sum_{g \in \cup \mathcal{R}_t} \mu_g g_\phi) - f_\phi \in \mathbb{R}$ for all ϕ in Φ.

   2.   • If $\Gamma_i \neq \varnothing$, let $\bar{\gamma}$ be the solution of the WPMaxSAT for maximizing $\sum_{\phi \in \Phi} \bar{\kappa}_\phi \beta_\phi$ under the hard clauses $\chi \wedge \psi_\Gamma$; set $\Gamma_{i+1} := \Gamma_i \cup \{\bar{\gamma}\}$. Otherwise, set $\bar{\gamma}$ identically zero.
       • If $\Delta_i \neq \varnothing$, let $\bar{\delta}$ be the solution of the WPMaxSAT for minimizing $\sum_{\phi \in \Phi} \bar{\kappa}_\phi \beta_\phi$ under the hard clauses $\chi \wedge \psi_\Delta$; set $\Delta_{i+1} := \Delta_i \cup \{\bar{\delta}\}$. Otherwise, set $\bar{\delta}$ identically zero.

      Is $\sum_{\phi \in \Phi} \bar{\kappa}_\phi \bar{\gamma}_\phi \leq 0 \leq \sum_{\phi \in \Phi} \bar{\kappa}_\phi \bar{\delta}_\phi$?
      Yes. $\bar{\mu}$ is optimal. **Stop.**
      No. Increase i's value by 1. **Reiterate.**

When the objective function is linear in μ, we now get an iteration of linear programming problems.

4   Some Special Cases

The propositional CONEstrip algorithm presented in the preceding section can be directly applied to problems in desirability theory. We here make a link with other, more established theories by describing how the standard problems in those theories can be encoded as problems in desirability theory. In all of the cases discussed, ψ_Γ is identically one and ψ_Δ is identically zero.

A probability assessment $P(\phi) = p_\phi$ for some event ϕ corresponds to an open cone with representation $\mathcal{D}_{\phi,p_\phi} := \{\phi - p_\phi 1, p_\phi 1 - \phi, 1\}$, where 1 denotes the constant gamble 1. For the classical PSAT problem, assessments are given for a set Φ of events and the questions is asked whether a probability mass function on the possibility space exists that satisfies these assessments. This problem can be solved applying the propositional CONEstrip algorithm to $\mathcal{R}_0 := \{\mathcal{D}_{\phi,p_\phi} : \phi \in \Phi\}$ and f identically zero: there is a satisfying probability mass function if and only if the problem is infeasible; no such mass function is explicitly constructed.

This setup can be generalized to conditional probability assessments $P(\phi|\varphi) = p_{\phi|\varphi}$; these correspond to $\mathcal{D}_{(\phi|\varphi),p_{\phi|\varphi}} := \{\phi \wedge \varphi - p_{\phi|\varphi}\varphi, p_{\phi|\varphi}\varphi - \phi \wedge \varphi, \varphi\}$. Given assessments for a set Φ_C of conditional events, the propositional algorithm can be applied to $\mathcal{R}_0 := \{\mathcal{D}_{(\phi|\varphi),p_{\phi|\varphi}} : (\phi|\varphi) \in \Phi_C\}$ and f identically zero to determine whether there exists a satisfying full conditional probability mass function.

A further generalization is to lower (and upper) conditional expectations of gambles g given as linear combinations of sentences: $\underline{P}(g|\varphi) = p_{g|\varphi}$ corresponds to $\mathcal{D}_{(g|\varphi),p_{g|\varphi}} := \{g \wedge \varphi - p_{g|\varphi}\varphi, \varphi\}$. When given a set of such assessments, the propositional CONEstrip algorithm can be applied to check whether they incur partial loss. Also, in this context, to calculate the lower expectation via natural extension for a gamble h conditional on an event ξ, we need to include the set $\mathcal{D}_\xi := \{-\xi, 0, \xi\}$ in the representation $\mathcal{R}_0$, use $f := h \wedge \xi$, and maximize the objective $\mu_\xi - \mu_{-\xi}$.

5   Conclusions

We presented an algorithm for checking consistency of and doing inference with uncertainty models based on fully general finitely generated sets of desirable gambles, with gambles specified as linear combinations of propositional sentences. It is designed to avoid a sure exponential computational complexity. As far as we know, it is the first such algorithm presented in the literature.

We have made a preliminary implementation and verified that the algorithm works. Ongoing work consists in improving the implementation and setting up numerical experiments to test the practical efficiency relative to the standard CONEstrip algorithm—for the general problem—and other algorithms—for special cases. Further research directions will be determined by their results.

Acknowledgments. The author wishes to thank Fabio Cozman for useful discussion of ideas at the basis of this research, Marjan van den Akker for her enlightening explanation of linear programming row generation techniques, and the reviewers for their many useful comments.

References

1. Baioletti, M., Capotorti, A., Tulipani, S., Vantaggi, B.: Simplification rules for the coherent probability assessment problem. Annals of Mathematics and Artificial Intelligence 35(1-4), 11–28 (2002)
2. Bansal, N., Raman, V.: Upper bounds for MaxSat: Further improved. In: Aggarwal, A.K., Pandu Rangan, C. (eds.) ISAAC 1999. LNCS, vol. 1741, pp. 247–258. Springer, Heidelberg (1999)
3. Biazzo, V., Gilio, A., Lukasiewicz, T., Sanfilippo, G.: Probabilistic logic under coherence: complexity and algorithms. Annals of Mathematics and Artificial Intelligence 45(1-2), 35–81 (2005)
4. Biere, A., Heule, M., van Maaren, H., Walsh, T. (eds.): Handbook of Satisfiability. Frontiers in Artificial Intelligence and Applications, vol. 185. IOS Press (2009)
5. Cozman, F.G., di Ianni, L.F.: Probabilistic satisfiability and coherence checking through integer programming. In: van der Gaag, L.C. (ed.) ECSQARU 2013. LNCS (LNAI), vol. 7958, pp. 145–156. Springer, Heidelberg (2013)
6. Georgakopoulos, G., Kavvadias, D., Papadimitriou, C.H.: Probabilistic satisfiability. Journal of Complexity 4, 1–11 (1988)
7. Gomes, C.P., Kautz, H., Sabharwal, A., Selman, B.: Satisfiability solvers. In: van Harmelen, F., Lifschitz, V., Porter, B. (eds.) Handbook of Knowledge Representation, ch. 2, pp. 89–134. Elsevier (2008)
8. Hansen, P., Jaumard, B., de Aragão, M.P., Chauny, F., Perron, S.: Probabilistic satisfiability with imprecise probabilities. International Journal of Approximate Reasoning 24(2-3), 171–189 (2000)
9. Manquinho, V., Marques-Silva, J., Planes, J.: Algorithms for weighted boolean optimization. In: Kullmann, O. (ed.) SAT 2009. LNCS, vol. 5584, pp. 495–508. Springer, Heidelberg (2009)
10. Prestwich, S.: CNF encodings. In: Biere, A., Heule, M., van Maaren, H., Walsh, T. (eds.) Handbook of Satisfiability. Frontiers in Artificial Intelligence and Applications, vol. 185, ch. 2, pp. 75–98. IOS Press (2009)
11. Pretolani, D.: Probability logic and optimization SAT: The PSAT and CPA models. Annals of Mathematics and Artificial Intelligence 43(1-4), 211–221 (2005)
12. Quaeghebeur, E.: The CONEstrip algorithm. In: Kruse, R., Berthold, M.R., Moewes, C., Gil, M.A., Grzegorzewski, P., Hryniewicz, O. (eds.) Synergies of Soft Computing and Statistics. AISC, vol. 190, pp. 45–54. Springer, Heidelberg (2013), http://hdl.handle.net/1854/LU-3007274
13. Quaeghebeur, E.: Desirability. In: Coolen, F.P.A., Augustin, T., de Cooman, G., Troffaes, M.C.M. (eds.) Introduction to Imprecise Probabilities. Wiley (2014)
14. Quaeghebeur, E., de Cooman, G., Hermans, F.: Accept & reject statement-based uncertainty models (submitted), http://arxiv.org/abs/1208.4462
15. Walley, P.: Statistical reasoning with imprecise probabilities. Monographs on Statistics and Applied Probability, vol. 42. Chapman & Hall (1991)
16. Walley, P.: Towards a unified theory of imprecise probability. International Journal of Approximate Reasoning 24(2-3), 125–148 (2000)
17. Walley, P., Pelessoni, R., Vicig, P.: Direct algorithms for checking consistency and making inferences from conditional probability assessments. Journal of Statistical Planning and Inference 126(1), 119–151 (2004)

Multinomial Logistic Regression on Markov Chains for Crop Rotation Modelling

Lewis Paton[1], Matthias C.M. Troffaes[1], Nigel Boatman[2],
Mohamud Hussein[2], and Andy Hart[2]

[1] Durham University
[2] The Food and Environmental Research Agency, UK

Abstract. Often, in dynamical systems such as farmer's crop choices, the dynamics are driven by external non-stationary factors, such as rainfall, temperature and agricultural input and output prices. Such dynamics can be modelled by a non-stationary Markov chain, where the transition probabilities are multinomial logistic functions of such external factors. We extend previous work to investigate the problem of estimating the parameters of the multinomial logistic model from data. We use conjugate analysis with a fairly broad class of priors, to accommodate scarcity of data and lack of strong prior expert opinion. We discuss the computation of bounds for the posterior transition probabilities. We use the model to analyse some scenarios for future crop growth.

Keywords: multinomial logistic regression, Markov chain, robust Bayesian, conjugate, maximum likelihood, crop.

1 Introduction

Imagine we wish to model crop distributions, for instance to predict the likely effect of a change of an agricultural policy on a farmer's crop choice. Assume a finite number of crop types can be planted in a particular parcel of land at a particular time. Arable farmers generally grow crops in rotation in order to prevent build-up of pests and diseases, and they aim to maximise yields and profit margins over the period of the rotation. The most advantageous crops to include vary with soil type and climate conditions. The rotation is generally driven by the length of the period required between successive plantings of the most valuable crop that can be grown, in order to allow pests and diseases to decline to non-damaging or readily controllable levels. Rotating crops also spreads risk in the face of weather variability and annual fluctuations in commodity prices.

Future crop rotations may be influenced by climate change, either by affecting the biology of the crops themselves or the pests and diseases that attack them. If changes in climate conditions favour one crop over another, this may lead to changes in profitability of different crops, and hence the balance of the rotation. Such changes on a large scale may have implications for food security as well as landscape and environmental impacts. It is therefore of interest to examine how crop rotations may change in response to future changes in climate, the

A. Laurent et al. (Eds.): IPMU 2014, Part III, CCIS 444, pp. 476–485, 2014.

profitability of crops, or other relevant factors. Although we may have a reasonably sized database, some crop types are quite rare, and so for those there may be very few observations. Moreover, prior expert information can be difficult to obtain. How can we make reasonable inferences about future crop distributions?

In this paper, we investigate the impact of rainfall and crop profit margin on a farmer's decision when faced with a choice of wheat, barley, or anything else. We have already discussed a simpler version of this problem [14]. In this paper we generalise our method for making reliable inferences from any number of categorical observations that are modelled using a Markov chain, where the transition probabilities are influenced by any number of regressors, when both data and expert knowledge are limited. We perform a robust Bayesian analysis using a set of prior distributions that are conjugate to the multinomial logit likelihood. We discuss the computation of bounds for the posterior transition probabilities when we use sets of distributions.

Standard logistic regression uses a generalized linear model [11] with the inverse of the logistic function as link function. Multinomial logit models extend the standard logistic regression model when we need more than two categorical outcomes. The model we consider is the baseline category logit model [1].

When we have limited data, we may want to include expert knowledge into our analysis to improve the quality of our inferences. Bayesian analysis is a useful way to include expert knowledge, where a prior distribution represents the expert's initial uncertainty about the parameters of interest.

However, there are times when specifying a prior distribution is difficult [5,2], which can be problematic, particularly when data is limited. In such cases, inferences may be quite sensitive to the prior. Therefore, robust Bayesian analysis uses a set of prior distributions to more accurately represent prior uncertainty. This results in a set of posterior distributions which is then used for inference, resulting in bounds on probabilities and expectations. There is a large body of work on probability bounding [5,9,2,15]. We follow a similar approach to the imprecise Dirichlet model [16] to make robust inferences from categorical observations, using a near-vacuous set of prior distributions.

Performing a robust Bayesian analysis can often be computationally intensive. We discuss an efficient but approximate approach to estimating the posterior values of the parameters in this paper, using maximum a posteriori (MAP) estimation.

We use conjugate priors in our analysis. Diaconis and Ylvisaker [8] introduced the theory of conjugate priors for exponential families. Chen and Ibrahim [6] used this work to propose a conjugate prior for generalized linear models, including logistic regression. Our approach is inspired by Chen and Ibrahim.

The novel contributions of this paper are:

1. We extend our model to cope with any (finite) number of categories.
2. We also extend the model to allow any (finite) number of regressors.
3. We perform some dynamic simulations of various deterministic scenarios.

This paper is structured as follows. Section 2 introduces the model. Section 3 describes the conjugate distributions, and discusses the parameters of the model. Section 4 explains the posterior inference and explores computation when using sets of distributions. In section 5 we perform some calculations with the model. Section 6 concludes the paper, and details areas of future research.

2   Multinomial Logit Model

We model crop rotations on a particular field as a non-stationary Markov chain, with J states, corresponding to J crop choices. Denote time by k. The crop grown at time k is denoted Y_k. This choice is influenced by regressors $X_{k0}, X_{k1}, \ldots, X_{kM}$, and previous crop choice Y_{k-1}. As usual in a regression analysis, we set $X_{k0} = 1$. Thus, we have the vector of regressors $X_k = (1, X_{k1}, \ldots, X_{kM})$. For example, we will investigate the impact of rainfall prior to sowing (X_{k1}), and the difference in profit margin between wheat and barley prior to sowing (X_{k2}), on a choice between wheat, barley, and any other crops (so $J = 3$). Repeated or heavy rainfall not only delays or prevents sowing at the desirable time period but can also cause soil erosion [4]. The differences between profit levels of wheat and barley would underpin a farmer's economic decision-making. We denote the transition probabilities by:

$$\pi_{ij}(x) := P(Y_{k+1} = j | Y_k = i, X_k = x) \tag{1}$$

The vector x contains the values of the regressors, i represents the farmer's previous crop choice, and j the farmer's current crop choice. We assume a multinomial logistic regression model for $\pi_{ij}(x)$, with $J^2(M+1)$ model parameters β_{ijm}, where $i \in \{1, \ldots, J\}$, $j \in \{1, \ldots, J\}$, $m \in \{0, \ldots, M\}$:

$$\pi_{ij}(x) := \frac{\exp(\beta_{ij}x)}{\sum_{h=1}^{J} \exp(\beta_{ih}x)} \tag{2}$$

where we use the notation $\beta_{ij}x = \sum_{m=0}^{M} \beta_{ijm}x_m$. Without loss of generality we can set $\beta_{iJm} = 0$. We call this the *baseline-category* logit model [1].

3   Parameter Estimation

3.1   Data

We now wish to estimate the parameters of the model, given some data. We have $n_i(x)$ observations where the previous crop was i, and the regressors were x. Obviously $n_i(x)$ will be zero at all but a finite number of $x \in \mathcal{X}$, where $\mathcal{X} := \{1\} \times \mathbb{R}^M$. Of these $n_i(x)$ observations, the crop choice was j in $k_{ij}(x)$ cases. Obviously, $n_i(x) = \sum_{j=1}^{J} k_{ij}(x)$ for each i. Table 1 shows an extract from the data set.

3.2   Likelihood

The likelihood of our model is:

$$p(k|x, n, \beta) = \prod_{i=1}^{J} \prod_{x \in \mathcal{X}} \binom{n_i(x)}{k_{i1}(x), \ldots, k_{iJ}(x)} \prod_{j=1}^{J} \pi_{ij}(x)^{k_{ij}(x)} \tag{3}$$

Table 1. Crop rotation data

previous crop i	rain x_1	profit margin difference x_2	current crop total $n_i(x)$	current wheat $k_{i1}(x)$	current barley $k_{i2}(x)$	current others $k_{i3}(x)$
1	76	93	1	1	0	0
2	15	156	1	1	0	0
1	19	115	1	0	1	0
3	6	129	1	0	1	0
⋮	⋮	⋮	⋮	⋮	⋮	⋮

For conjugate analysis, we rewrite this directly as a function of the parameters:

$$\propto \prod_{i=1}^{J} \exp\left(\sum_{x \in \mathcal{X}} \left[\left(\sum_{j=1}^{J} k_{ij}(x)\beta_{ij}x \right) - n_i(x) \ln \sum_{j=1}^{J} e^{\beta_{ij}x} \right] \right) \quad (4)$$

3.3 Conjugate Prior and Posterior

Following [3, p. 266, Proposition 5.4], we can now simply define a conjugate prior distribution for our model parameters:

$$f_0(\beta|s_0, t_0) \propto \prod_{i=1}^{J} \exp\left(\sum_{x \in \mathcal{X}} s_{0i}(x) \left[\sum_{j=1}^{J} t_{0ij}(x)(\beta_{ij}x) - \ln \sum_{j=1}^{J} e^{\beta_{ij}x} \right] \right) \quad (5)$$

where s_{0i} and t_{0ij} are non-negative functions such that $s_{0i}(x) = t_{0ij}(x) = 0$ for all but a finite number of $x \in \mathcal{X}$, with $0 \le t_{0ij}(x) \le 1$ and $\sum_{j=1}^{J} t_{0ij}(x) = 1$ on those points x where $s_{0i}(x) > 0$.

We can update our prior distribution to the posterior distribution in the usual way:

$$f(\beta|k, n, s_0, t_0) = f_0(\beta|s_n, t_n) \quad (6)$$

where

$$s_{ni}(x) := s_{0i}(x) + n_i(x) \qquad t_{nij}(x) := \frac{s_{0i}(x)t_{0ij}(x) + k_{ij}(x)}{s_{0i}(x) + n_i(x)} \quad (7)$$

Note that the subscript n here simply denotes the fact we are considering posterior hyperparameters, as opposed to the prior hyperparameters which are denoted by subscript 0. This is standard practice in conjugate analysis [3].

4   Inference

4.1   Posterior Transition Probability

To study future crop rotation scenarios, we are mostly interested in the posterior transition probability:

$$\hat{\pi}_{ij}(x) := P(Y_{k+1} = j | Y_k = i, X_k = x, k, n, s_0, t_0) \tag{8}$$

$$= \int_{\mathbb{R}^{J^2 \times (M+1)}} \pi_{ij}(x) f(\beta | k, n, s_0, t_0) \, d\beta \tag{9}$$

$$= \int_{\mathbb{R}^{J \times (M+1)}} \pi_{ij}(x) f(\beta_i | k_i, n_i, s_{0i}, t_{0i}) \, d\beta_i \tag{10}$$

where it is worth recalling that $\pi_{ij}(x)$ is a non-linear function of our model parameters β. We are interested in evaluating Eq. (10). The challenge is the evaluation of the integral. We have a number of options.

We can evaluate the integral numerically, for example using an MCMC method. However, as eventually we want to use sets of distributions, this may not necessarily be the most sensible route to take. We have performed an initial analysis using an adaptive Metropolis-Hastings algorithm, but it was very slow, even for small J and M. Of course, there are advanced MCMC methods which may speed things up, but that goes beyond the scope of this paper. A more detailed analysis, including full MCMC simulation, is planned.

Therefore, we may prefer to rely on faster approximations of the integral. We could approximate the prior by a multivariate normal distribution. Chen and Ibrahim [6] showed that for large sample sizes, in the specific case where $J = 2$, this approximation yields good results. Whilst the mean is easily approximated by the mode, the covariance structure is less obvious (see [6, Theorem 2.3]).

A more crude but very fast approximation is to use the MAP estimate for β, and to assume that all our probability mass is at this estimate. We obtain an estimate β^* of β, and then our estimate of $\hat{\pi}_{ij}$ is simply:

$$\hat{\pi}_{ij}(x) := \frac{\exp(\beta_{ij}^* x)}{\sum_{h=1}^{J} \exp(\beta_{ih}^* x)} \tag{11}$$

This approximation is obviously horribly crude, but we note that it corresponds to the maximum likelihood estimate, where the data has been augmented with pseudo counts. Hence, it reflects current practice quite well. In the rest of this paper we will use the MAP approach.

4.2   Sets of Prior Distributions

We now want to propose sets of prior distributions, in a similar vein to Walley's imprecise Dirichlet model [16]. For now, we study the inferences resulting from a fixed prior function for $s_{0i}(x)$, namely:

$$s_{0i}(x) := \begin{cases} s & \text{if } x \in \mathfrak{X}, \\ 0 & \text{otherwise,} \end{cases} \tag{12}$$

for some finite set $\mathfrak{X} \subset \mathcal{X}$, and a vacuous set $\mathfrak{T}$ of prior functions for t_0. $\mathfrak{X}$ is the set of regressor values where we specify prior beliefs. As in the imprecise Dirichlet model [16, Section. 2.5], smaller values of s typically produce tighter posterior predictive bounds.

Perhaps this deserves some explanation. For conjugate analysis with a likelihood from a full rank exponential families, such as multinomial sampling, the predictive expectation is a convex combination of the natural statistic and its corresponding hyperparameter [8, eq. (2.10)], where s controls the weight on the prior. In such cases, smaller values of s always produce tighter posterior predictive bounds provided that the prior predictive is vacuous. However, our likelihood comes from an exponential family that does not have full rank. Consequently, the predictive MAP estimate depends on the hyperparameters t_0 and s in non-linear and possibly even non-monotone ways. As such, situations where a smaller s value produces wider posterior predictive bounds cannot be excluded. Nevertheless, we have never observed such behaviour in any of our numerical results. An intuitive explanation for this is that s still relates to the prior variance, and hence, still weighs the influence of the prior on the posterior predictive.

4.3 Posterior Transition Probability Bounds

If we can find a MAP estimate for all $t_0 \in \mathfrak{T}$, we obtain a set B^* of solutions β^*, one for each $t_0 \in \mathfrak{T}$. Each member of B^* corresponds to an estimate of the posterior transition probability, as in Eq. (11). Therefore;

$$\underline{\hat{\pi}}_{ij}(x) \approx \inf_{\beta^* \in B^*} \frac{\exp(\beta^*_{ij} x)}{\sum_{h=1}^{J} \exp(\beta^*_{ih} x)} \qquad \overline{\hat{\pi}}_{ij}(x) \approx \sup_{\beta^* \in B^*} \frac{\exp(\beta^*_{ij} x)}{\sum_{h=1}^{J} \exp(\beta^*_{ih} x)} \qquad (13)$$

are the desired lower and upper posterior approximations of the transition probability.

4.4 Choice of $\mathfrak{T}$

The choice of $\mathfrak{T}$ affects the posterior inferences. We follow a similar approach to that of the imprecise Dirichlet model [16]. We are dealing with situations with very little prior expert information, and so we identify a reasonably vacuous set of prior distributions. One such way of achieving this is by choosing:

$$\mathfrak{T}_v := \{t_0 : t_{0ij}(x) = 0 \text{ when } x \notin \mathfrak{X},$$

$$0 < t_{0ij}(x) < 1, \sum_{j=1}^{J} t_{0ij}(x) = 1 \text{ when } x \in \mathfrak{X}\} \qquad (14)$$

for each i. However, for high dimensional problems, this is somewhat computationally involved. We will restrict ourselves to the extreme points of $\mathfrak{T}_v$, namely:

$$\mathfrak{T}'_v := \{t_0 : t_{0ij}(x) = 0 \text{ when } x \notin \mathfrak{X},$$

$$t_{0ij}(x) \in \{0, 1\}, \sum_{j=1}^{J} t_{0ij}(x) = 1 \text{ when } x \in \mathfrak{X}\} \qquad (15)$$

for each i. It is easy to see that even the size of this set can become large very quickly for higher dimensional problems, making the speed of computation even more important. Of course, the inferences we make can be made more precise by choosing a less general set. However, we want to build a model which can still make reasonable inferences when faced with a severe lack of prior information. Hence, from here on in, we proceed with using $\mathfrak{T}'_v$.

5   Example

We now use our model to explore the relationship between crop choice, rainfall, and profit margin difference for a region in England. Our dataset consists of 1000 observations of crop rotations [13], rainfall [10], and the crop profit margin difference between wheat and barley [12].

Firstly, we will use the model to estimate $\hat{\underline{\pi}}_{ij}(x)$ and $\hat{\overline{\pi}}_{ij}(x)$ as in Eq. (13). We use $\mathfrak{T}'_v$, and set $s = 2$. Figures 1 and 2 show the results. The black box represents $\hat{\underline{\pi}}_{ij}(x)$, while the grey box shows $\hat{\overline{\pi}}_{ij}(x)$ for $i = 1, 2$ and $j = 1, 2, 3$. Note, that we have omitted the graph for $i = 3$ due to space requirements. We show results for various levels of rainfall and profit margin difference. For example $x = (1, 10, 70)$ shows the transition probabilities for low rainfall and low profit margin difference, while $x = (1, 10, 140)$ shows transition probabilities for low rainfall and high profit margin difference, and so on.

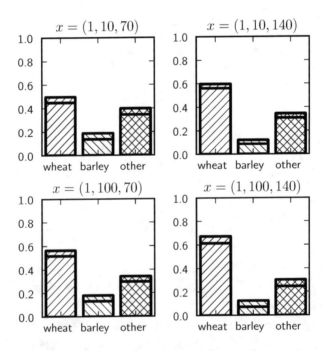

Fig. 1. Transition probabilities when previous crop grown is wheat ($i = 1$)

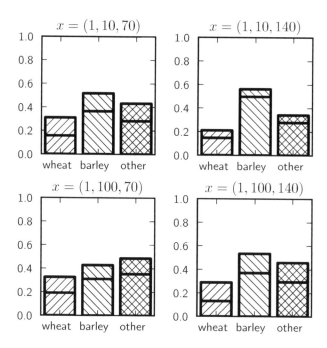

Fig. 2. Transition probabilities when previous crop grown is barley ($i = 2$)

The obvious result to note is that, generally, the transition probability is highest for planting the same crop twice in a row. Changes between transition probabilities at different levels of rainfall and profit margin difference are not as obvious, but they do exist. For example, when the previous crop is wheat, the probability of the next crop being wheat increases with profit margin difference, which is a logical result. We also see that when the previous crop is wheat, the probability of the next crop being wheat increases with rainfall too. Strangely, when the previous crop grown is barley, the model suggests we are more likely to grow barley again even as the profit margin difference increases. This is counter-intuitive, but it perhaps suggests that there are other factors at play other than rainfall and profit, which future work will investigate.

We now consider future crop distributions. For this we use the imprecise Markov chain. Calculations use the methodology for imprecise Markov chains developed in [7]. Our initial distribution is calculated empirically from the data. It is 32% wheat, 13% barley, and 55% others. We also require future scenarios for rainfall, and the profit margin difference between wheat and barley. For ease, we assume the scenarios are constant in time. We consider two scenarios—the rainfall and profit margin difference are low, and the rainfall and profit margin difference are high. Figure 3 shows the results. We see that when the rainfall and profit margin difference remain low, the probability of wheat will decrease. The opposite is true when the rainfall and profit margin difference remain high. The probability of growing barley remains reasonably level. As expected, the probability bounds widen as time increases. Of course, these results

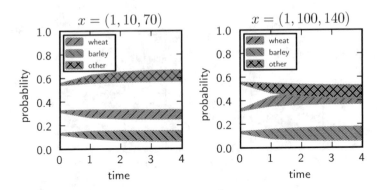

Fig. 3. Future crop distributions under different constant scenarios

aren't particularly realistic, as rainfall and crop profit margins are not stationary in time. Future analysis is planned.

6   Conclusion

We extended previous work to be able to model multiple crop choices and multiple regressors. We proposed a model for multinomial imprecise logistic regression, using sets of conjugate prior distributions, to get bounds on the posterior transition probabilities of growing wheat, barley or anything else, as functions of rainfall and the difference in profit between wheat and barley.

We care about robustness because, for certain rare crop types, very little data is available. By using sets of prior distributions, our approach allows us to make robust inferences even from near-vacuous prior information.

For computational reasons, we use MAP estimation to approximate the actual posterior expectation. Other options, such as MCMC, are computationally far more complex.

The application of this work to crop rotation modelling is still at an early stage. Farmers are faced with more than just three crop choices, and there are other variables that effect their choice. Furthermore, the future is not a fixed constant in time. Indeed, land use in the United Kingdom is strongly influenced by the EU's Common Agricultural Policy (CAP), which affects the relative profitability of different alternative land uses through subsidies. The model can be used to evaluate the impact of recent CAP reforms, such as the decoupling of direct payments from crop production, with farmers no longer required to produce commodities to be entitled to support, but to keep land in good environmental condition. In particular, as the decoupling has aligned EU prices for cereals with world prices, the model can be used to study the effect changes in world prices has on arable rotations. Providing that suitable data is available, the model can also be used to predict the impact of any future CAP reforms, or to analyse how the frequency of crops following wheat varies under different circumstances – wheat being the most valuable crop grown over much of the UK arable area – a question particularly interesting to land use analysts. Future work will investigate issues such as these.

Acknowledgements. The first and second authors are supported by FERA and EPSRC.

References

1. Agresti, A.: Categorical data analysis, 2nd edn. John Wiley and Sons (2002)
2. Berger, J.O.: The robust Bayesian viewpoint. In: Kadane, J.B. (ed.) Robustness of Bayesian Analyses, pp. 63–144. Elsevier Science, Amsterdam (1984)
3. Bernado, J.M., Smith, A.F.M.: Bayesian theory. John Wiley and Sons (1994)
4. Boardman, J., Evans, R., Favis-Mortlock, D.T., Harris, T.M.: Climate change and soil erosion on agricultural land in England and Wales. Land Degradation and Development 2(2), 95–106 (1990)
5. Boole, G.: An investigation of the laws of thought on which are founded the mathematical theories of logic and probabilities. Walton and Maberly, London (1854)
6. Chen, M.-H., Ibrahim, J.G.: Conjugate priors for generalized linear models. Statistica Sinica 13, 461–476 (2003)
7. de Cooman, G., Hermans, F., Quaeghebeur, E.: Imprecise Markov chains and their limit behavior. Probability in the Engineering and Informational Sciences 23(4), 597–635 (2009)
8. Diaconas, P., Ylvisaker, D.: Conjugate priors for exponential families. The Annals of Statistics 7(2), 269–281 (1979)
9. Keynes, J.M.: A treatise on probability. Macmillan, London (1921)
10. Data collected by the Met Office, http://www.metoffice.gov.uk/climate/uk/stationdata/ (accessed: February 11, 2013)
11. Nelder, J.A., Wedderburn, R.W.M.: Generalized linear models. Journal of the Royal Statistical Society. Series A 135(3), 370–384 (1972)
12. Nix, J.: Farm management pocketbook. Agro Business Consultants Ltd. (1993-2004)
13. Data collected by the Rural Payments Agency under the integrated administration and control system for the administration of subsidies under the common agricultural policy
14. Troffaes, M.C.M., Paton, L.: Logistic regression on Markov chains for crop rotation modelling. In: Cozman, F., Denœux, T., Destercke, S., Seidenfeld, T. (eds.) ISIPTA 2013: Proceedings of the Eighth International Symposium on Imprecise Probability: Theories and Applications (Compiègne, France), pp. 329–336. SIPTA (July 2013)
15. Walley, P.: Statistical reasoning with imprecise probabilities. Chapman and Hall, London (1991)
16. Walley, P.: Inferences from multinomial data: Learning about a bag of marbles. Journal of the Royal Statistical Society, Series B 58(1), 3–34 (1996)

Approximate Inference in Directed Evidential Networks with Conditional Belief Functions Using the Monte Carlo Algorithm

Wafa Laâmari, Narjes Ben Hariz, and Boutheina Ben Yaghlane

LARODEC Laboratory, Institut Supérieur de Gestion de Tunis, Tunisia

Abstract. Evidential networks are frameworks of interest commonly used to represent uncertainty and to reason within the belief function formalism. Despite their success in handling different uncertain situations, the exponential computational complexity which occurs when carrying out the exact inference in these networks makes the use of such models in complex problems difficult. Therefore, with real applications reaching the size of several tens or hundreds of variables, it becomes important to address the serious problem of the feasibility of the exact evidential inference. This paper investigates the issue of applying an approximate algorithm to the belief function propagation in evidential networks.

1 Introduction

Bayesian networks [11,13] are powerful tools for representing uncertainty using probability. These probabilistic graphical models are especially effective for describing complex systems when a complete knowledge of their states is available. If not, the use of alternative to probabilistic networks like possibilistic [4], credal [6] and evidential networks [2,22] is more appropriate. Evidential graphical models have gained popularity in the last decade as a tool for modeling uncertain human reasoning in a large variety of applications including reliability analysis [23] and threat assessment [3], and have provided a suitable framework for handling imprecise probabilities [18]. One of the most commonly used models in the evidential setting is the Evidential Network with Conditional belief functions (ENC) [22]. The propagation algorithm in this network is restricted to graphs with only binary relations among variables. In order to deal with n-ary relations between variables, a generalization of ENC, called Directed EVidential Network with conditional belief functions (DEVN) was formalized in [2].

Intuitively, a DEVN is represented by a directed acyclic graph (DAG) G which consists of a set N= $\{N_1, \ldots, N_n\}$ of nodes and a set E= $\{E_1, \ldots, E_m\}$ of edges. Instead of probabilities, uncertainty is expressed in a DEVN by defining a conditional belief function Bel_{N_i} for each node N_i in the context of its parents. The major challenge in this model has been to limit the computational complexity of the following belief chain rule which computes the global belief function $Bel_{N_1, \ldots, N_n}$ by combining all the conditional belief functions Bel_{N_i} in this network using repeated application of Dempster's rule of combination[1]:

---

[1] Dempster's rule of combination is traditionally interpreted as an operator for fusing belief functions (see Section 2).

A. Laurent et al. (Eds.): IPMU 2014, Part III, CCIS 444, pp. 486–497, 2014.

$$\text{Bel}_{N_1,\ldots,N_n} = \text{Bel}_{N_1} \otimes \ldots \otimes \text{Bel}_{N_n} \qquad (1)$$

where symbol $\otimes$ denotes the operator of Dempster's rule of combination (DRC).

One main and significant reason for the popularity of the DEVNs is their ability of modeling cases of partial and complete ignorance.

Despite this advantageous feature, the computational complexity of the exact inference in these models makes sometimes the reasoning process intractable. In fact, to avoid computing explicitly the global belief function using equation (1), the exact methods proposed for inference in these networks are based on the local computation technique [2]. Under these inference methods, DRC is the pivot mechanism for the propagation of belief functions.

In [12], Orponen has proved that the exact computation of a belief function Bel which is the combination of d given belief functions $\text{Bel}^1,\ldots,\text{Bel}^d$ using DRC is a **#P-complete** problem. Since the number of variables in the evidential networks modeling real applications is usually large, and the size of variables' domains grows exponentially with their number, the reasoning process tends to be infeasible. This is due to the exponential explosion problem of the combination operation which is in the core of this process.

Many works have gotten around the computational problem of DRC by considering approximate techniques. Barnett has shown that for a very special case, it is possible to compute the combined belief Bel in linear time [1]. In [7], Barnett's approach has been extended to allow dealing with another case: the hierarchical case. However, Shafer and Logan have shown in [15] that this case can be tackled more efficiently and a generalization of their algorithm was proposed for belief function propagation in Markov trees [17] having a small product space associated with the largest clique, a condition that cannot be always satisfied. In [21], an alternative approach using the Monte Carlo algorithm was developed for an approximate calculation of belief functions combination. The computational complexity of this calculation is linear. Motivated by this result, we present in this paper an algorithm for belief function propagation in DEVNs based on the Monte Carlo method for calculating combined belief functions. To the best of our knowledge, this is the first algorithm dealing with inference in these evidential networks by approximating combination of belief functions to improve the complexity of the propagation algorithm.

The paper is organized as follows: Section 2 reviews the main concepts of the evidence theory and the DEVN. Section 3 presents the Monte Carlo algorithm. The exact inference algorithm in DEVN is presented in section 4 while our proposed approximate algorithm for belief function propagation in these networks is developed in section 5. Experimental tests are presented in section 6.

2   Evidence Theory and Directed Evidential Networks with Conditional Belief Functions: Background

The evidence theory [14,19] is a general framework for reasoning under uncertainty based on the modeling of evidence.

Definition 1. Let $N= \{N_1, \ldots, N_n\}$ be a set of random discrete variables. Each variable N_i in N assumes its values over a finite space Θ_{N_i}, named the *frame of discernment*.

Definition 2. All possible subsets S of Θ_{N_i} are elements of its *power set* which is denoted by $2^{\Theta_{N_i}}$ and formally defined as $2^{\Theta_{N_i}} = \{S : S \subseteq \Theta_{N_i}\}$.

Definition 3. A mass function or a basic belief assignment (bba) m_{N_i} defined over Θ_{N_i} is a mapping from the power set $2^{\Theta_{N_i}}$ onto the interval [0,1] such as:

$$\sum_{S \subseteq \Theta_{N_i}} m_{N_i}(S) = 1 \tag{2}$$

where $m_{N_i}(S)$ is the mass or the degree of belief to support a subset S of Θ_{N_i}.

A subset S of Θ_{N_i} for which the mass $m_{N_i}(S) > 0$ is called a focal element.

Definition 4. A bba can be equivalently represented by a non additive measure: a belief function. Given a bba m_{N_i} associated with a variable N_i, the *belief* in a subset S of Θ_{N_i}, denoted by $\text{Bel}_{N_i}(S)$, represents the total belief committed to S without being also committed to $\bar{S}$. $\text{Bel}_{N_i}(S)$ is obtained by summing the masses of all the subsets Q of S. It is defined as follows:

$$\text{Bel}_{N_i}(S) = \sum_{Q \subseteq S, Q \neq \emptyset} m_{N_i}(Q) \tag{3}$$

Definition 5. Let X and Y be two disjoint subsets of N. Their frames, denoted by Θ_X and Θ_Y, are the Cartesian product of the frames of the variables they include, respectively. Let m_X be a mass function defined over Θ_X. The *vacuous extension* of m_X to $\Theta_X \times \Theta_Y$[2] produces a new mass function m_{XY} defined as follows:

$$m_{X \uparrow XY}(S') = \begin{cases} m_X(S) & \text{if } S' = S \times \Theta_Y, S \subseteq \Theta_X \\ 0 & \text{otherwise} \end{cases} \tag{4}$$

Definition 6. Suppose m_X and m_Y are two mass functions defined on the spaces Θ_X and Θ_Y, respectively. The *combination* of m_X and m_Y into a single mass function m_{XY}, can be done, $\forall S \subseteq \Theta_X \times \Theta_Y$, using the following DRC:

$$m_{XY}(S) = (m_X \otimes m_Y)(S) = \frac{\displaystyle\sum_{S_1 \cap S_2 = S} m_{X \uparrow XY}(S_1) \times m_{Y \uparrow XY}(S_2)}{1 - \displaystyle\sum_{S_1 \cap S_2 = \emptyset} m_{X \uparrow XY}(S_1) \times m_{Y \uparrow XY}(S_2)} \tag{5}$$

where both $m_{X \uparrow XY}$ and $m_{Y \uparrow XY}$ are computed using the equation (4).

Definition 7. Directed evidential networks with conditional belief functions (denoted as DEVNs) combine the evidence theory and graphs for representing uncertain knowledge. A DEVN defined over the set of variables N consists of

---

[2] $\times$ denotes the Cartesian product.

two representational components: (i) a DAG G whose nodes are in one-to-one correspondence with random variables in N and whose edges represent dependency relationships between these variables, and (ii) a set of local parameters expressing the uncertain information in G. A parameter must be specified:

1. for each root node N_i in G. This parameter is an a priori mass function m_{N_i} defined over Θ_{N_i} using the equation (2).
2. for each parent node N_j of each child node N_i in G. This parameter is a conditional mass function $m_{N_i}[N_j]$ defined over N_i conditionally to its parent N_j.

3   The Monte Carlo Algorithm to Approximate the DRC

The Monte Carlo algorithm for approximating the results of the DRC was developed by Wilson in [21]. This algorithm is based on the hint representation, which is used for modeling the uncertainty in the theory of hints [9].

3.1   Theory of Hints

The theory of hints [9] is a mathematical interpretation of the evidence theory based on probability theory. In this theory, uncertain information about a particular variable A is represented by a hint H_A over its frame of discernment.

Let $\Theta_A = \{\theta_A^1, \ldots, \theta_A^z\}$ be the frame of discernment of the variable A. The hint H_A is defined to be a quadruple $(\Omega_A, P_A, \Gamma_A, \Theta_A)$. $\Omega_A = \{\omega_1, \ldots, \omega_x\}$ is a finite set of different possible interpretations, such as for each focal element $S \subseteq \Theta_A$, there is an interpretation $\omega_u \in \Omega_A$. In other words, if $\omega_u \in \Omega_A$ is a correct interpretation, then the correct value taken by A belongs to a subset $\Gamma_A(\omega_u) \subseteq \Theta_A$, where Γ_A is a compatibility function which assigns to each possible interpretation in Ω_A the corresponding focal element. P_A is a probability distribution which assigns a probability $p_A(\omega_u)$ to each possible interpretation $\omega_u \in \Omega_A$. The probability assigned by P_A to each interpretation is in fact the mass of the corresponding focal element.

To better understand the representation of the hint, let us consider a variable A taking three values in $\Theta_A = \{\theta_A^1, \theta_A^2, \theta_A^3\}$.

Its mass function is defined as follows: $m_A(\{\theta_A^1\})=0.2$; $m_A(\{\theta_A^1, \theta_A^2\})=0.1$; $m_A(\{\theta_A^1, \theta_A^2, \theta_A^3\})=0.3$ and $m_A(\{\theta_A^2, \theta_A^3\})=0.4$.

The corresponding hint H_A is expressed by the quadruple $(\Omega_A, P_A, \Gamma_A, \Theta_A)$ where $\Omega_A = \{\omega_1, \omega_2, \omega_3, \omega_4\}$ contains exactly four interpretations corresponding to the four focal elements, the compatibility function Γ_A associates to each interpretation $\omega_u \in \Omega_A$ the corresponding focal element as follows: $\Gamma_A(\omega_1) = \{\theta_A^1\}$; $\Gamma_A(\omega_2) = \{\theta_A^1, \theta_A^2\}$; $\Gamma_A(\omega_3) = \{\theta_A^1, \theta_A^2, \theta_A^3\}$ and $\Gamma_A(\omega_4) = \{\theta_A^2, \theta_A^3\}$, and the probability function P_A associates to each interpretation $\omega_u \in \Omega_A$ the probability $p_A(\omega_u)$ representing the mass of the corresponding focal element. Formally, $p_A(\omega_u) = m_A(\Gamma_A(\omega_u))$. Thus, the probabilities are as follows: $p_A(\omega_1) = m_A(\Gamma_A(\omega_1)) = 0.2$; $p_A(\omega_2) = 0.1$; $p_A(\omega_3) = 0.3$ and $p_A(\omega_4) = 0.4$.

The evidence relative to a given proposition $S \in 2^{\Theta_A}$ is defined in the theory of hints by:

$$m_A(S) = \sum_{\omega_u \in \Omega_A : \Gamma_A(\omega_u) = S} p_A(\omega_u) \tag{6}$$

The corresponding belief function is then defined as follows:

$$\text{Bel}_A(S) = \sum_{\omega_u \in \Omega_A : \Gamma_A(\omega_u) \subseteq S} p_A(\omega_u) \tag{7}$$

For combining d hints $H_A^1, \ldots, H_A^d$ defined over the same frame of discernment Θ_A, where each hint H_A^k ($k=1, \ldots, d$) is expressed by $(\Omega_A^k, P_A^k, \Gamma_A^k, \Theta_A)$, the DRC can be used to obtain the resulting hint H_A as follows:

$$H_A = H_A^1 \otimes \ldots \otimes H_A^d \tag{8}$$

H_A is defined by the quadruple $(\Omega_A, P_A, \Gamma_A, \Theta_A)$, where:
$\Omega_A = \Omega_A^1 \times \ldots \times \Omega_A^d$, Γ_A and P_A assign to each interpretation $\omega = (\omega^1, \ldots, \omega^d)$ of Ω_A the corresponding focal element and probability, respectively, by the following equations:

$$\Gamma_A(\omega) = \bigcap_{k=1}^{d} \Gamma_A^k(\omega^k) \tag{9} \qquad\qquad p_A(\omega) = p'(\omega)/p'(\Omega_A) \tag{10}$$

where $\omega^k \in \Omega_A^k$ and $p'(\omega) = \prod_{k=1}^{d} p_A^k(\omega^k)$.

Let X and Y be two disjoint subsets of the set of variables N, having the frames Θ_X and Θ_Y, respectively. A hint $H_X = (\Omega_X, P_X, \Gamma_X, \Theta_X)$ defined on Θ_X can be extended to $\Theta_{XY} = \Theta_X \times \Theta_Y$ by simply replacing $\Gamma_X(\omega)$, where $\omega \in \Omega_X$, with $\Gamma_{X \uparrow XY}(\omega) = \Gamma_X(\omega) \times \Theta_Y$. The resulting hint $H_{XY} = (\Omega_X, P_X, \Gamma_{X \uparrow XY}, \Theta_{XY})$ is called the *vacuous extension* of H_X from Θ_X to Θ_{XY}.

3.2   The Monte Carlo Algorithm Principle

Let $\text{Bel}_A^1, \ldots, \text{Bel}_A^d$ be d belief functions defined on the frame Θ_A, and let Bel_A be their combination using the Monte Carlo algorithm (MCA).

Using the hint representation (see Section 3.1), each belief function Bel_A^k ($k=1, \ldots, d$) is represented by its corresponding hint H_A^k ($k=1, \ldots, d$) which is defined by the quadruple $(\Omega_A^k, P_A^k, \Gamma_A^k, \Theta_A)$.

Let $H_A = (\Omega_A, P_A, \Gamma_A, \Theta_A)$ denote the hint representation of Bel_A, where $\Omega_A = \Omega_A^1 \times \ldots \times \Omega_A^d$. For $\omega \in \Omega_A$, where $\omega = (\omega^1, \ldots, \omega^d)$ and $\omega^k \in \Omega_A^k$, Γ_A and P_A are defined using equations (9) and (10), respectively.

For a subset $S \subseteq \Theta_A$, $\text{Bel}_A(S)$ is defined in the theory of hints as follows:

$$\text{Bel}_A(S) = p_A(\Gamma_A(\omega) \subseteq S | \Gamma_A(\omega) \neq \emptyset) \tag{11}$$

To compute $\text{Bel}_A(S)$ for a subset $S \subseteq \Theta_A$, the MCA simulates the last equation by repeating a large number of trials T, where for each trial this algorithm:

(1) picks randomly an element ω^k from each set of interpretations $\Omega_{\rm A}^k$, where $k=1,\ldots,d$, to get an element $\omega = (\omega^1,\ldots,\omega^d) \in \Omega_{\rm A}$.
(2) repeats (1) if $\Gamma_{\rm A}(\omega) = \emptyset$, till $\Gamma_{\rm A}(\omega)$ has been a non-empty set.
(3) checks the trial's success: if $\Gamma_{\rm A}(\omega) \subseteq S$, then the trial succeeds and the number of trials that succeed TS[3] is incremented by 1, otherwise it fails.

$\mathrm{Bel}_{\rm A}(S)$ is then estimated by the average value of TS over the T trials. In fact, Wilson has proved that the proportion of trials that succeed converges to $\mathrm{Bel}_{\rm A}(S)$ [21].

Although the use of the MCA for combining beliefs has been suggested in [11], no work has exploited it in evidential networks for the belief function propagation. In the sequel, we use (MC) to denote the combination using the MCA.

4   Exact Inference in Directed Evidential Networks with Conditional Belief Functions

The key problem in DEVNs is to perform the evidential inference, which intuitively means computing the marginal belief $\mathrm{Bel}(N_i)$ of a particular node N_i. Let $N=\{N_1,\ldots,N_n\}$ be a set of variables in the graphical representation G of a DEVN taking their values over the frames of discernment $\Theta_{N_1},\ldots,\Theta_{N_n}$. Reasoning by computing directly the global belief function over the Cartesian product $\Theta_{N_1} \times \ldots \times \Theta_{N_n}$ using equation (1) and then marginalizing it to the frame Θ_{N_i} of the variable N_i is impractical for the computational problems of DRC presented previously. Algorithms for inference in DEVNs were developed in [2] based on the local computation technique to solve these problems. These methods for propagation of beliefs in DEVNs perform inference by using a computational structure namely the modified binary join tree (MBJT) [2] which is an adaptation of the binary join tree (BJT) proposed by Shenoy in [16].

The main idea is to delete all loops from G gathering some variables in a same node. The resulting MBJT is a tree G'=(N',E') where each node in N' is a non-empty subset of N, the set of nodes in the graphical representation G. This tree satisfies the Markovian property according to which, if a variable belongs to two different nodes in N', then it belongs to all the nodes in the path between them. Two kinds of nodes in N' can be defined as:

(i) nodes or clusters formed by a subset S of nodes in G (i.e $S \subseteq N$). These nodes are called joint nodes, and uncertainty in them is given in terms of joint belief functions defined over the frame of discernment Θ_S of the variables in S.
(ii) nodes formed by a subset S of nodes in G, such as S contains a particular child node N_i and its parents in G. These nodes are called conditional nodes since there is a conditional dependency relation between variables in the subset S. Uncertainty is expressed in these nodes in terms of conditional belief functions over the frame of discernment of the child node.

---

[3] The number of trials that succeed TS=0 in the beginning.

In this paper, only the essential elements about a formal treatment on building MBJTs are provided, and the reader is referred to [2] for more details.

For the evidential inference in a DEVN G, a message-passing algorithm between neighboring nodes is carried out in the corresponding MBJT G' and computations are performed locally. This message-passing algorithm has three phases: an initialization phase (IP), a transferring-message phase (TMP) during which an inward and an outward phases are performed in the tree G', and a computation-marginals phase (CMP) during which the algorithm calculates the marginal belief distribution for every desired variable in the graph. The message-passing algorithm in MBJT is more detailed in [2].

5   Approximate Inference in DEVNs by the MCA

Exact reasoning in DEVNs involves the application of the DRC to perform the combination operation which is the pivot mechanism for belief propagation. It is noteworthy to mention that computations using this rule are performed locally only in the joint nodes of a MBJT over the frame of discernment of the variables forming them [2]. We must also mention that conditional nodes showing conditional dependency relationships among the variables forming them do not perform any computation during the TMP and they are not concerned with the application of the DRC. The complexity of the exact inference in a DEVN is exponential in the maximum joint node size of its corresponding MBJT [10].

The inference algorithm we propose is based on the use of the MCA to approximate the combined belief functions of joint nodes. This proposal is motivated by the good accuracy and computational complexity this algorithm has [21]. In general, the amount of computation the MCA requires, increases linearly with the number of belief functions (NBF) being combined and also with the size of the frame of discernment (SF), while the amount of computation the DRC requires increases exponentially with NBF and SF.

The main idea of our approximate algorithm is to reduce the amount of computation resulting from the use of the DRC during the message propagation process in the DEVNs.

The phases that take most of the time in the exact message propagation algorithm are TMP and CMP since only both of them require the application of the DRC. That is why, these two phases are the focus of our approximation.

The approximate inference algorithm works by first transforming the DAG G defining the graphical component of the DEVN into a MBJT G', using the different construction steps of the standard MBJT generation algorithm [2], then by message passing up and down the tree G'. The principle of propagation of messages is similar to that used in the exact inference method. The difference is that the large amount of calculation resulting from the use of the DRC is moved outside the new propagation algorithm which switches from the DRC to the MCA whenever it is necessary to perform the combination operation.

The basic MCA, presented in Section 3.2, combines belief functions defined on the same frame of discernment. So, to allow the application of this algorithm for the belief propagation in DEVNs, we must extend it by allowing the belief functions being combined to be defined on different frames. To do that, if $\text{Bel}^1, \ldots, \text{Bel}^d$, denoting the belief functions to be combined, are defined on the frames $\Theta^1, \ldots, \Theta^d$, respectively, then to (MC)-combine them using the MCA, we must:

(1) represent each belief Bel^k $(k=1, \ldots, d)$ by its hint $H^k = (\Omega^k, P^k, \Gamma^k, \Theta^k)$.

(2) vacuously extend each hint representation H^k defined over Θ^k $(k=1, \ldots, d)$ to the joint frame $\Theta = \Theta^1 \times \ldots \times \Theta^d$ (The vacuous extension of a hint was presented in Section 3.1).

(3) Let Bel denote the combination of $\text{Bel}^1, \ldots, \text{Bel}^d$ and let $H=(\Omega, P, \Gamma, \Theta)$ be its hint representation, where $\Omega = \Omega^1 \times \ldots \times \Omega^d$, and P and Γ are computed using equations (10) and (9), respectively. For each subset $S \subseteq \Theta$, compute $\text{Bel}(S)$ by simulating the equation $\text{Bel}(S) = p(\Gamma(\omega) \subseteq S | \Gamma(\omega) \neq \emptyset)$, where $\omega \in \Omega$, using the three steps of the MCA presented in Section 3.2.

Our approximate message passing algorithm, based on the MCA, has the same skeleton as the exact one. So, in order to distinguish these two algorithms, we call ours MCMP which means Monte Carlo Message Passing algorithm and we refer to the exact one as EMP which means Exact Message Passing algorithm.

The MCMP is divided into 3 phases:

(1) an **initialization phase (IP)** during which the algorithm associates each conditional belief distribution in G with the corresponding conditional node in G' and each a priori one with the corresponding joint node (same as in EMP).

(2) a **transferring-message phase (TMP)** during which the algorithm:

    (a) picks a node in G' and designates it as a root node R (same as in EMP)

    (b) applies an inward phase by collecting messages from the leaves towards R. When a node N'_i in G' receives messages from all its inward neighbors, it updates its marginal belief by (MC)-combining its initial belief with all the received belief functions (i.e. the received messages), then it sends a message to its only neighbor towards the root. This phase ends when R collects from all its neighboring nodes the inward messages, and updates its marginal by (MC)-combining them with its initial belief function.

    (c) applies an outward phase by distributing messages away from R towards each leaf node in G'. Each node waits for the outward message from its neighbor, and upon receiving it, it updates its marginal by (MC)-combining its own belief distribution with all the received messages, and sends a message to each node from which it received a message in step (b).

(3) a **computation-marginals phase (CMP)** during which the approximate algorithm calculates the marginal distribution for every desired variable in the graph by (MC)-combining its own belief function with all the inward messages and also the outward message received from all its neighbors during steps (b) and (c) of the transferring-message phase.

6   Experiments

In this section, we aim to test experimentally the performance of the MCMP and to compare it with the EMP efficiency. To perform the tests, we have used a benchmark made of two networks modeling two classical systems in the reliability field, namely the 2-out-of-3 system and the bridge system [20], with 4 and 12 variables, respectively. The first model is singly-connected while the second one is multiply-connected. Adding to these two networks, we have also used other DEVNs having from 30 to 50 nodes. These networks are defined via random topologies which may either be singly-connected or multiply-connected. The number of states for each variable in the two models representing the two systems is set to 2. For the other random networks, the number of states is randomly picked from the range [4,10]. The maximum number of links per node is fixed to 4.

All the models are quantified by belief functions which are also randomly generated using the algorithm proposed in [5] for random generation of mass functions. For the MCMP, we have made tests on each network using different values for the number of trials T.

Table 1. Experimental results

Networks	NT	MCMP\ EMP	T	CPUT	Distance
The	20		100	19 secs	0.023
2-out-of-3	20	MCMP	1000	30 secs	0.0032
System	20		10000	2 mins	$0.89 * 10e - 003$
Network	-	EMP	-	18 secs	-
The	20		100	28 secs	0.166
Bridge	20	MCMP	1000	45 secs	0.045
System	20		10000	4 mins	0.009
Network	-	EMP	-	28 secs	-
30-50	10		100	20 mins	0.047
Random	10	MCMP	1000	53 mins	0.0012
Singly-Conn	10		10000	2 hours	0.0004
Networks	-	EMP	-	3 hours	-
30-50	10		100	54 mins	0.112
Random	10	MCMP	1000	1 hour	0.0058
Multiply-Conn	10		10000	2 hours	0.0001
Networks	-	EMP	-	4 hours	-

Before commenting on the results in Table 1, note that for each network, the MCMP is tested several times for each fixed number of trials T. The second column shows the number of tests (NT) for each network. The fourth column shows the number of trials (T), and the fifth one shows the average central processing unit time (CPUT).

Based on the results presented in the table, we notice that for the two networks modeling the 2-out-of-3 and the bridge systems, the MCMP is not outperforming the EMP in terms of running time. We believe that this is due to the fact that DRC, on which the EMP is based, is still feasible and provides quick answers to simple problems where the corresponding models consist of small number of variables with relatively small number of states. This is, also, due to the relatively high number of trials (T≥100) which prevents rapid response of the MCA comparing it to the DRC response to such simple problems.

The accuracy of the MCMP was first measured by computing, for each network in the presented benchmark, the dissimilarity between the exact distribution and the approximate one in each node using Jousselme's distance [8], which is one of the most appropriate measures used for calculating the dissimilarity between two mass functions $m^1_{N_i}$ and $m^2_{N_i}$. This distance is defined as:

$$d(m^1_{N_i}, m^2_{N_i}) = \sqrt{\frac{1}{2}(m^1_{N_i} - m^2_{N_i})D(m^1_{N_i}, m^2_{N_i})} \tag{12}$$

where D is the Jaccard index defined by:

$$D(A, B) = \begin{cases} 0 & \text{if } A = B = \emptyset \\ \frac{|A \cap B|}{|A \cup B|} & \forall A, B \in 2^{\Theta_{N_i}} \end{cases} \tag{13}$$

The dissimilarity measure quantifies how much the two distributions are different. The smaller is the dissimilarity, the more accurate is the MCMP. The last column of Table 1 reports these results, i.e. the upper dissimilarity in each network. The results denote a good accuracy. Even more impressive are the results of the MCMP with T=10000.

In most cases, the lower dissimilarity is very close to 0, and in some other cases, it is null which means that approximation is exact for these cases. The experimental results on the whole benchmark show that the dissimilarity is reduced when increasing the number of trials.

Experiments show speedup and good accuracy of the MCMP on the random networks with relatively large number of variables and frames. The results shown in Table 1, for the randomly generated networks, denote that the MCMP takes less time than the EMP even with T= 10000. We believe the main reason for this is the fact that, with large networks, the DRC becomes inefficient in terms of running times. Indeed, the MCMP produces also good accuracy on the whole generated networks. For instance, for the randomly generated singly-connected networks, the upper dissimilarity=0.047 with T=100.

The accuracy of the MCMP was also evaluated by calculating the confidence intervals for the approximate mass functions. We have especially focused in Table 2 on the exact and the estimated beliefs of three randomly chosen nodes of the bridge system: N_6, N_{10} and N_{12}. The experimentation depicted in the second row of the table was conducted on the node N_{10}. The second, the third and the fourth columns report the masses of the exact distribution, the rest of columns present the confidence intervals of the masses of the approximate distribution.

Table 2. Confidence intervals

Nodes	EMP			MCMP		
N_6	0.4685	0.4486	0.0705	$[0.4817, 0.4947]$	$[0.4373, 0.452]$	$[0.0672, 0.0748]$
N_{10}	0.7271	0	0.281	$[0.728, 0.7482]$	$[0, 0]$	$[0.2545, 0.2727]$
N_{12}	0.7084	0.2429	0.0457	$[0.7085, 0.7182]$	$[0.24, 0.2434]$	$[0.044, 0.048]$

The results show clearly that the estimated values computing by the MCMP are closed to the exact ones.

There are cases where the EMP blocks, while the MCMP provides answers. The corresponding experimentation showed that when the frame of discernment of the combined beliefs contains 9 values, then combining just two generated beliefs using the EMP becomes impossible. We increased the number of values to 15, and the MCMP was only limited by the running-time but did not block.

Unfortunately, we identified some cases where the MCMP is not practical and we observed that when the conflict between the randomly generated beliefs is very high, the MCMP becomes unusable.

The preliminary promising results regarding the accuracy and the MCMP tendency to be speedy relatively with the size of the network and the frames, invite practical applications of this algorithm for large DEVNs.

7   Conclusion

We have presented in this paper a new algorithm based on the Monte Carlo approach for approximate evidential networks propagation. The experiments show that the algorithm can deal efficiently with large networks having large number of variables and large frames and provides good accuracy results. As a future work, we intend to test the algorithm on larger networks, and to deal with the conflict management when combining beliefs by using the Markov Chain Monte Carlo algorithm in the DEVNs.

References

1. Barnett, J.A.: Computational Methods for a Mathematical Theory of Evidence. In: IJCAI, pp. 868–875 (1981)
2. Ben Yaghlane, B., Mellouli, K.: Inference in Directed Evidential Networks Based on the Transferable Belief Model. IJAR 48(2), 399–418 (2008)
3. Benavoli, A., Ristic, B., Farina, A., Oxenham, M., Chisci, L.: Threat Assessment Using Evidential Networks. Information Fusion, 1–8 (2007)
4. Borgelt, C., Gebhardt, J., Kruse, R.: Possibilistic graphical models. In: International School for the Synthesis of Expert Knowledge, ISSEK (1998)
5. Burger, T., Destercke, D.: How to randomly generate mass functions. Int. J. of Uncertainty, Fuzziness and Knowledge-Based Systems 21(5), 645–673 (2013)

6. Cozman, F.G.: Credal networks. Artif. Intell. 120(2), 199–233 (2000)
7. Gordon, G., Shortliffe, E.H.: A Method for Managing Evidential Reasoning in a Hierarchical Hypothesis Space. Artif. Intell. 26(3), 323–357 (1985)
8. Jousselme, A.L., Grenier, D., Bossé, E.: A new distance between two bodies of evidence. Information Fusion 2, 91–101 (2001)
9. Kohlas, P., Monney, A.: A Mathematical Theory of Hints. An Approach to the Dempster Shafer Theory of Evidence. Lecture Notes in Economics and Mathematical Systems. Springer (1995)
10. Laâmari, W., Ben Yaghlane, B., Simon, C.: Comparing Evidential Graphical Models for Imprecise Reliability. In: Deshpande, A., Hunter, A. (eds.) SUM 2010. LNCS, vol. 6379, pp. 191–204. Springer, Heidelberg (2010)
11. Murphy, K.: Probabilistic Graphical Models. Michael Jordan (2002)
12. Orponen, P.: Dempster's rule is #P-complete. Artif. Intell. 44, 245–253 (1990)
13. Pearl, J.: Probabilistic Reasoning in Intelligent Systems: Networks of Plausible Inference. Morgan Kaufmann (1988)
14. Shafer, G.: A Mathematical Theory of Evidence. Princeton University Press, Princeton (1976)
15. Shafer, G., Logan, R.: Implementing Dempster's Rule for Hierarchical Evidence. Artifi. Intell. 33, 271–298 (1978)
16. Shenoy, P.P.: Binary Join Trees for Computing Marginals in the Shenoy-Shafer Architecture. Int. J. Approx. Reasoning 17(2-3), 239–263 (1997)
17. Shafer, G., Shenoy, P.P., Mellouli, K.: Propagating Belief Functions in Qualitative Markov Trees. I. J. Approx. Reasoning 1, 349–400 (1987)
18. Simon, C., Weber, P.: Evidential Networks for Reliability Analysis and Performance Evaluation of Systems With Imprecise Knowledge. IEEE Transactions on Reliability 58(1), 69–87 (2009)
19. Smets, P., Kennes, R.: The transferable belief model. Artif. Intell. 66, 191–234 (1994)
20. Villemeur, A.: Reliability, availability, maintainability and safety assessment: methods and techniques. John Wiley and Sons Inc. (1992)
21. Wilson, N.: Justification, Computational Efficiency and Generalisation of the Dempster-Shafer Theory, Research Report 15, Oxford Polytechnic, Artif. Intell. (1989)
22. Xu, H., Smets, P.: Evidential Reasoning with Conditional Belief Functions. In: Heckerman, D., Poole, D., Lopez De Mantaras, R. (eds.) UAI 1994, pp. 598–606. Morgan Kaufmann, San Mateo (1994)
23. Yang, J., Huang, H.Z., Liu, Y.: Evidential networks for fault tree analysis with imprecise knowledge. Int. J. of Turbo & Jet Engines, 111–122 (2012)

A Note on Learning Dependence under Severe Uncertainty

Matthias C.M. Troffaes[1], Frank P.A. Coolen[1], and Sébastien Destercke[2]

[1] Durham University, UK
[2] Université de Technologie de Compiègne, France

Abstract. We propose two models, one continuous and one categorical, to learn about dependence between two random variables, given only limited joint observations, but assuming that the marginals are precisely known. The continuous model focuses on the Gaussian case, while the categorical model is generic. We illustrate the resulting statistical inferences on a simple example concerning the body mass index. Both methods can be extended easily to three or more random variables.

Keywords: bivariate data, categorical data, copula, Gaussian copula, robust Bayesian, imprecise probability.

1 Introduction

Sklar's theorem [10] states that any multivariate distribution of k variables can be expressed through a density on $[0, 1]^k$ with uniform marginals—this density is called a copula—and the marginal distributions of each of the variables. For this reason, copulas [7] have become an indispensible tool to model and learn statistical dependence in multivariate models: they allow of estimation of the dependence structure, separately from the marginal structure.

Estimating dependence requires joint observations, which in many cases are only available in small amounts, while substantial amounts of marginal data may be available. For example, when studying the reliability of a system, it is common to have good information about the reliability of each system component, yet to have only little information about joint failures [11]. Imprecise probabilities provide one possible theoretical basis for dealing with small sample sizes, by representing knowledge as a *set* of distributions [3,13,1], rather than a single distribution necessarily based on somewhat arbitrary assumptions [6].

Copulas and imprecise probabilities have been studied in the literature by various researchers. The Fréchet–Hoeffding copula bounds, which represent completely unknown dependence, are used for instance in probabilistic arithmetic [16] and p-boxes [4,12]. One theoretical difficulty is that there is no straightforward imprecise equivalent of Sklar's theorem, say, expressing any set of joint distributions as a sets of copulas along with a set of marginal distributions [9]: it appears that, when working with sets of distributions, separating the dependence structure from the marginal structure is a lot more difficult in general.

A. Laurent et al. (Eds.): IPMU 2014, Part III, CCIS 444, pp. 498–507, 2014.

In this paper, we propose and investigate a few statistical models for robust dependence learning from limited data. We state an imprecise version of Sklar's theorem when marginal distributions are fully known, separating precise marginals from the imprecise dependence structure. We propose a range of parametric models for the bivariate categorical case. Finally, we demonstrate our findings on a toy example: estimating the body mass index from height and mass data.

Section 2 explores a continuous model, focusing on the multivariate normal model, while section 3 provides a first exploration of a generic categorical model.

2   Robust Bayesian Correlation Learning for Bivariate Normal Sampling

We start with revising a simple and well-studied model: sampling from the bivariate normal distribution. We will derive some new results that are relevant to dependence learning. Our analysis starts from Quaeghebeur and De Cooman's robust Bayesian framework for sampling from the exponential family [8].

2.1   Inference with Known Mean and Unknown Covariance Matrix

Let $Z_i := (Z_{i1}, \ldots, Z_{ik})$ be a multivariate normally distributed random variable with known mean—which we can assume to be zero without loss of generality through translation of the data—but unknown covariance matrix $\Sigma \in \mathbb{R}^{k \times k}$. A particular realisation of Z_i is denoted by a lower case letter $z_i := (z_{i1}, \ldots, z_{ik}) \in \mathbb{R}^k$. The likelihood of an i.i.d. sample $z_1, \ldots, z_n$ is

$$f(z_1, \ldots, z_n \mid \Sigma) \propto |\Sigma|^{-n/2} \prod_{i=1}^{n} \exp\left(-\frac{1}{2} z_i^T \Sigma^{-1} z_i\right) \tag{1}$$

$$= |\Sigma|^{-n/2} \exp\left[-\frac{1}{2} \sum_{i=1}^{n} \operatorname{tr}\left(z_i z_i^T \Sigma^{-1}\right)\right], \tag{2}$$

where the data $z_i \in \mathbb{R}^k$ are organised as row vectors, so $z_i z_i^T$ is the matrix containing $z_{i\ell} z_{i\ell'}$ in row ℓ and column ℓ'.

A family of conjugate priors for this density is the family of inverse Wishart distributions with hyperparameters $\nu_0 > 0$ and $\Psi_0 \in \mathbb{R}^{k \times k}$ positive definite [2]:

$$f(\Sigma \mid \nu_0, \Psi_0) \propto |\Sigma|^{-\frac{\nu_0 + k + 1}{2}} \exp\left[-\frac{1}{2} \operatorname{tr}\left(\Psi_0 \Sigma^{-1}\right)\right]. \tag{3}$$

The posterior distribution is obtained by updating the hyperparameters through

$$\nu_n = \nu_0 + n, \qquad\qquad \Psi_n = \Psi_0 + \sum_{i=1}^{n} z_i z_i^T. \tag{4}$$

The prior expected covariance matrix is given by

$$\mathrm{E}\left(\Sigma \mid \nu_0, \Psi_0\right) = \frac{\Psi_0}{\nu_0 - k - 1} =: \Sigma_0 \tag{5}$$

and therefore, through conjugacy, the posterior expected covariance matrix is

$$\mathrm{E}\left(\Sigma \mid z_1, \ldots, z_n, \nu_0, \Psi_0\right) = \mathrm{E}\left(\Sigma \mid \nu_n, \Psi_n\right) = \frac{\Psi_n}{\nu_n - k - 1} \tag{6}$$

$$= \frac{\Psi_0 + \sum_{i=1}^n z_i z_i^T}{\nu_0 + n - k - 1} \tag{7}$$

$$= \frac{(\nu_0 - k - 1)\Sigma_0 + \sum_{i=1}^n z_i z_i^T}{\nu_0 + n - k - 1} =: \Sigma_n. \tag{8}$$

For robust Bayesian analysis aiming to learn about the dependence between two random variables, we now need to identify a reasonable set of prior distributions, or, in our conjugate setting, a reasonable set of hyperparameters ν_0 and Ψ_0.

The formula for the posterior expected covariance matrix shows that ν_0 determines our learning speed, that is, how many observations n we need before starting to move towards our data. So, ν_0 is similar to the s value in the imprecise Dirichlet model [14]. Here too, we will simply assume ν_0 to be fixed to whatever value is judged to lead to a reasonable learning speed. For fixed ν_0, any particular choice of Ψ_0 corresponds to a prior covariance matrix Σ_0.

Let us now study the bivariate case ($k = 2$) in more detail. We will write X_i for Z_{i1} and Y_i for Z_{i2}. We would choose

$$\Psi_0 = \nu_0' \begin{bmatrix} \sigma_X^2 & \rho_0 \sigma_X \sigma_Y \\ \rho_0 \sigma_X \sigma_Y & \sigma_Y^2 \end{bmatrix} \tag{9}$$

where $\nu_0' = \nu_0 - k - 1 = \nu_0 - 3$, if we had prior standard deviations $\sigma_X > 0$ and $\sigma_Y > 0$ for the two components as well as the prior correlation coefficient $\rho_0 \in [-1, 1]$. For this paper focusing on dependence, we are mostly interested in cases where the marginals are well known, i.e. well known prior σ_X and σ_Y, but unknown prior correlation ρ_0. We will therefore study the set of priors with all parameters fixed, except for ρ_0, which we assume to be vacuous a priori. Without loss of generality, by rescaling, we can assume that $\sigma_X = \sigma_Y = 1$, leaving us with just two hyperparameters: $\nu_0 > 0$ and $\rho_0 \in [-1, 1]$.

The posterior covariance matrix becomes

$$\Sigma_n = \frac{1}{\nu_0' + n} \begin{bmatrix} \nu_0' + \sum_{i=1}^n x_i^2 & \nu_0' \rho_0 + \sum_{i=1}^n x_i y_i \\ \nu_0' \rho_0 + \sum_{i=1}^n x_i y_i & \nu_0' + \sum_{i=1}^n y_i^2 \end{bmatrix}. \tag{10}$$

Provided that the sample variance is approximately equal to the prior variance, i.e.

$$\sum_{i=1}^n x_i^2 \approx n\sigma_X^2 = n, \qquad \sum_{i=1}^n y_i^2 \approx n\sigma_Y^2 = n, \tag{11}$$

our expression for Σ_n becomes

$$\begin{bmatrix} 1 & \rho_n \\ \rho_n & 1 \end{bmatrix},$$ (12)

where

$$\rho_n = \frac{\nu_0' \rho_0 + \sum_{i=1}^n x_i y_i}{\nu_0' + n}.$$ (13)

Equation (12) is the covariance matrix of a bivariate normal with unit marginal variances and correlation coefficient ρ_n. For vacuous prior correlation, $\rho_0 \in [-1, 1]$, we thus get the following posterior bounds on the correlation:

$$\underline{\rho}_n = \frac{-\nu_0' + \sum_{i=1}^n x_i y_i}{\nu_0' + n}, \qquad \overline{\rho}_n = \frac{\nu_0' + \sum_{i=1}^n x_i y_i}{\nu_0' + n},$$ (14)

provided that our observations, and our prior, have unit variance and zero mean (which we can achieve by linear transformation without loss of generality).

This analysis generalises easily to cases with more than two variables, $k > 2$—we leave this to the reader. Essentially, we simply have to deal with more correlation parameters.

2.2   Application to the Body Mass Index Example

We now illustrate our model on the evaluation of the body mass index $R = X/Y^2$, where X is a person's weight in kilograms and Y is his or her height in meters. The body mass index is commonly used to detect under- and over-weight. We aim (i) to assess the dependence between X and Y in a particular population, and (ii) to extract a robust inference about R in this population.

We consider 30 paired observations of heights and weights of girls aged 11 [5, p. 75]. The weight X has sample mean $\bar{x} = 36.2$, sample standard deviation $s_X = 7.7$, with no strong evidence against normality (p-value[1] 0.017). The height Y has sample mean $\bar{y} = 1.448$, sample standard deviation $s_Y = 0.077$, with no evidence against normality whatsoever (p-value 0.711). We will assume that X and Y have known means, equal to $\bar{x}$ and $\bar{y}$. We also assume that, a priori, $\sigma_X = 7.7$ and $\sigma_Y = 0.077$ in eq. (9), but we are vacuous about the prior correlation. For reference, it may be useful to note that the sample correlation between X and Y is in fact 0.742. For the sake of the example, we assume that the sample is drawn from a bivariate normal distribution, although there is reasonably strong evidence against joint normality (p-value 0.00388).

Figure 1 shows the bounds on the correlation of the posterior covariance matrix in eq. (10) with $\nu_0' = 2$ and $\rho_0 \in [-1, 1]$. The two values converge steadily with a final interval $[\underline{\rho}_{30}, \overline{\rho}_{30}] = [0.630, 0.759]$. The expectation of R is bounded by $\underline{E}(R) = 17.10$ and $\overline{E}(R) = 17.16$. Similarly, we may wonder about the probability of R to be in a "healthy" range, which is about $A = [14, 19.5]$ for girls aged 11. We obtain bounds $\underline{P}(R \in A) = 0.66$ and $\overline{P}(R \in A) = 0.71$. Note that bounds were obtained by straightforward numerical optimisation.

---

[1] Throughout, we test for normality using the Shapiro-Wilk test.

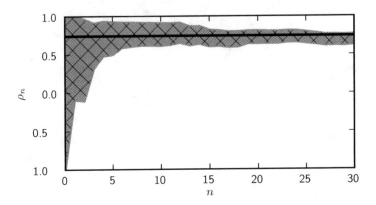

Fig. 1. Lower and upper correlation estimates $\underline{\rho}_n$ and $\overline{\rho}_n$ as a function of the sample size n, with $\nu_0' = 2$. The solid horizontal line denotes the sample correlation for $n = 30$.

3   Robust Bayesian Dependence Learning for Bivariate Categorical Data

3.1   The Model

Consider a bivariate categorical random quantity $Z := (X, Y)$ with X taking values in a finite set $\mathcal{X} = \{1, \dots, m_X\}$, and Y taking values in a finite set $\mathcal{Y} = \{1, \dots, m_Y\}$. The parameters θ_x and ϕ_y determine the marginal distributions:

$$p(x \mid \theta) = \theta_x, \qquad\qquad p(y \mid \phi) = \phi_y. \qquad (15)$$

We assume that $m_X \geq 2$, $m_Y \geq 2$, $\theta_x > 0$ and $\phi_y > 0$.

We are interested in learning the dependence structure of X and Y. One very general way to express the full joint distribution of (X, Y) is by introducing parameters w_{xy} such that

$$p(x, y \mid \theta, \phi, w) = w_{xy}\theta_x\phi_y, \qquad (16)$$

subject to the constraints

$$\sum_{x \in \mathcal{X}} \sum_{y \in \mathcal{Y}} w_{xy}\theta_x\phi_y = 1, \qquad (17)$$

$$\sum_{x \in \mathcal{X}} w_{xy}\theta_x = 1, \qquad (18)$$

$$\sum_{y \in \mathcal{Y}} w_{xy}\phi_y = 1. \qquad (19)$$

Equations (18) and (19) simply follow from

$$\sum_{x \in \mathcal{X}} w_{xy} \theta_x \phi_y = \sum_{x \in \mathcal{X}} p(x, y \mid \theta, \phi, w) = p(y \mid \theta, \phi, w) = \phi_y, \qquad (20)$$

$$\sum_{y \in \mathcal{Y}} w_{xy} \theta_x \phi_y = \sum_{y \in \mathcal{Y}} p(x, y \mid \theta, \phi, w) = p(x \mid \theta, \phi, w) = \theta_x, \qquad (21)$$

respectively. This model specification is over-parametrized, but it allows us to model the marginal distributions and the dependence structure separately, where the matrix w plays precisely a similar role as a copula in general bivariate models for (usually) continuous random quantities. However, a key difference and major difficulty with the above model is that the constraints on w depend on θ and ϕ: the separation is thus not as complete as with copulas, where the dependence structure is parametrised independently of the marginals. For this reason, it seems most natural to consider a two-stage situation where we first learn about the marginal parameters θ and ϕ, followed by learning about the dependence structure w *conditional on what we learnt about θ and ϕ.*

3.2   Inference for Known Marginals

For this reason, as a stepping stone towards general inference about θ, ϕ, and w, here we consider a scenario where the marginal distributions are already fully known, and we only aim at inference about w. While this may appear somewhat restrictive, and perhaps even artificial, there are practical scenarios where one has very substantial information about the probability distributions for the random quantities X and Y separately, but relatively little information about their joint distribution.

There are $(m_X - 1)(m_Y - 1)$ degrees of freedom for the components of w. In case data is limited, to enable sufficiently useful inference, it seems natural to assume a reduced-dimensional parametric form for w, which may naturally correspond to an (assumed) ordering of the categories, as we will illustrate in section 3.3. Let n_{xy} denote the number of observations of $(X, Y) = (x, y)$, with total number of observations $n = \sum_{x \in \mathcal{X}} \sum_{y \in \mathcal{Y}} n_{xy}$ and row and column totals denoted by $n_{x*} = \sum_{y \in \mathcal{Y}} n_{xy}$ and $n_{*y} = \sum_{x \in \mathcal{X}} n_{xy}$, respectively. So, there are n_{x*} observations of $X = x$ and n_{*y} observations of $Y = y$.

Without further restrictions on w, it seems tempting to fit the model to match the non-parametric maximum likelihood estimate

$$\hat{p}(x, y \mid w) = \frac{n_{xy}}{n} \qquad (22)$$

by setting

$$\hat{w}_{xy} = \frac{n_{xy}}{n \theta_x \phi_y}. \qquad (23)$$

A problem is that this estimate will usually violate eqs. (18) and (19). For instance,

$$\sum_{x \in \mathcal{X}} \hat{w}_{xy} \theta_x = \sum_{x \in \mathcal{X}} \frac{n_{xy}}{n \theta_x \phi_y} \theta_x = \frac{n_{*y}}{n \phi_y} \neq 1 \qquad (24)$$

as soon as $\frac{n_{*y}}{n} \neq \phi_y$. A proper maximum likelihood estimate would maximize the likelihood subject to all constraints embodied by eqs. (17) to (19). Solving this optimisation problem poses an interesting challenge.

Bayesian inference for w will face a similar challenge: w lives in a convex subspace of $\mathbb{R}^{m_X \times m_Y}$ determined by eqs. (17) to (19). Application of Bayes's theorem requires numerical integration over this space. Nevertheless, the basic principles behind Bayesian inference for w are simple, and sensitivity analysis is similar to the imprecise Dirichlet model [14]. Specific dimension-reduced models, where we have a much better handle on the parameter space, will be illustrated in more detail in section 3.3.

The likelihood is given by

$$\prod_{x \in \mathcal{X}} \prod_{y \in \mathcal{Y}} (w_{xy} \theta_x \phi_y)^{n_{xy}}, \tag{25}$$

so as conjugate prior we can choose

$$f(w \mid \alpha_0) \propto g(w) \prod_{x \in \mathcal{X}} \prod_{y \in \mathcal{Y}} (w_{xy} \theta_x \phi_y)^{\alpha_{0xy}}, \tag{26}$$

where $\alpha_{0xy} > 0$ and g is some arbitrary non-negative function (as long as the right hand side integrates to a finite value). With $\nu_0 := \sum_{xy} \alpha_{0xy}$, this prior distribution can be interpreted as reflecting prior information equivalent to ν_0 observations of which α_{0xy} were $(X, Y) = (x, y)$. The corresponding posterior distribution is clearly $f(w \mid \alpha_n)$ with $\alpha_{nxy} = \alpha_{0xy} + n_{xy}$.

Sensitivity analysis on this model could then follow an approach similar to Walley's imprecise Dirichlet model [14], by taking the set of all prior distributions for a fixed value of ν_0. In case of an informative set of prior distributions, one may also allow the value of ν_0 to vary within a set to allow prior-data conflict modelling [15].

As already mentioned, the remaining key difficulty is to integrate the conjugate density over the parameter space. For this reason, in the next section, we consider a reduced model.

3.3   Reduced Model

As a first and basic example of a reduced parametric form, consider the case $\mathcal{X} = \mathcal{Y} = \{1, 2, 3\}$ with known $\theta_x = \phi_y = 1/3$ for all x and $y \in \{1, 2, 3\}$. If the categories are pairwise ordered in some natural manner, then it might be quite reasonable to specify

$$w = \begin{bmatrix} 1 + 2\alpha & 1 - \alpha & 1 - \alpha \\ 1 - \alpha & 1 + 2\alpha & 1 - \alpha \\ 1 - \alpha & 1 - \alpha & 1 + 2\alpha \end{bmatrix} \tag{27}$$

with $\alpha \in [0, 1]$. It is easily verified that this model satisfies eqs. (17) to (19): the full matrix sums to 9, and each of the rows and colums sum to 3.

Note that there is no logical requirement to avoid $\alpha \in [-1/2, 0)$, we just use this small example as an illustration. In this model, $\alpha = 0$ corresponds to full independence between X and Y, whereas $\alpha = 1$ corresponds to perfect correlation $X = Y$. Therefore, this corresponds to a scenario where we may suspect positive correlation between X and Y, but we are unsure about the strength of correlation. Note that the actual model reduction is achieved by additionally assuming that $X = Y = x$ has the same probability for all $x \in \{1, 2, 3\}$, and similar for $X = x \cap Y = y$ for all $x \neq y$.

With these assumptions, statistical inference is concerned with learning about the parameter $\alpha \in [0, 1]$. The likelihood function is

$$(1 + 2\alpha)^t (1 - \alpha)^{n-t} \tag{28}$$

with $t = n_{11} + n_{22} + n_{33}$ and $n = \sum_{xy} n_{xy}$ as before. The maximum likelihood estimate is

$$\hat{\alpha} = \begin{cases} \frac{3t-n}{2n} & \text{if } 3t \geq n, \\ 0 & \text{otherwise.} \end{cases} \tag{29}$$

For a Bayesian approach to inference for this model, we can define a conjugate prior

$$f(\alpha \mid \nu_0, \tau_0) \propto (1 + 2\alpha)^{\tau_0} (1 - \alpha)^{\nu_0 - \tau_0} \tag{30}$$

with $\tau_0 \in [0, \nu_0]$, with the possible interpretation that it reflects prior information which is equivalent to ν_0 observations of which τ_0 have $X = Y$.

The posterior distribution is simply $f(\alpha \mid \nu_0 + n, \tau_0 + t)$. Sensitivity analysis is again straightforward by taking the set of prior distributions for $\tau_0 \in [0, \nu_0]$ and a fixed ν_0. For instance, we would get the following robust estimate for the posterior mode of α:

$$\hat{\alpha}_n = \left[\frac{3t - \nu_0 - n}{2(\nu_0 + n)}, \frac{3t + 2\nu_0 - n}{2(\nu_0 + n)} \right] \tag{31}$$

when $3t \geq \nu_0 + n$, with similar formulas when $3t < \nu_0 + n$ (truncating negative values to zero).

3.4   Application to the Body Mass Index Example

To apply the categorical model to our data, we must first discretize them, with the ordering of the categories following the ordering of natural numbers. To obtain three categories with uniform marginals, we simply discretized the 99% prediction intervals of each Gaussian marginals of section 2.2, obtaining $\mathcal{X} = \{[17, 32], [32, 39], [39, 56]\}$ and $\mathcal{Y} = \{[1.24, 1.41], [1.41, 1.47], [1.47, 1.64]\}$.

Figure 2 shows the bounds on the posterior mode $\hat{\alpha}_n$ in eq. (31), with $\nu_0 = 2$. The results are similar to those obtained in section 2.2, showing that even this very simple discretized model can capture the correlation between X and Y, with the bounds on $\hat{\alpha}_{30}$ being $[0.56, 0.66]$. From these values and the bounds of the categories, we can easily obtain bounds on the expectation of R: $\underline{E}(R) = 12.9$

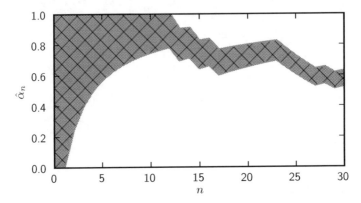

Fig. 2. Bounds on $\hat{\alpha}_n$ as a function of the sample size n, with $\nu_0 = 2$

and $\overline{E}(R) = 22.4$, which gives a much wider interval than in section 2.2. This is due to the very rough discretization: a finer discretization would likely provide a narrower interval. The lower and upper probabilities of $A = [14, 19.5]$ are this time $\underline{P}(R \in A) = 0$ and $\overline{P}(R \in A) = 0.96$, which are almost vacuous and again show that to have meaningful inferences, a finer discretization is needed.

4   Conclusion

In this paper, we have introduced two preliminary models—a continuous one and a discrete one—to model dependence when joint data is limited, but assuming that the marginals are precisely known. The continuous model focused on the very special multivariate normal case. However, already in our simple example, we have seen that joint normality is rarely satisfied in practice. A major challenge is to provide methods for dependence modelling, that are both flexible and computationally tractable, whilst still producing useful inferences.

Even though the models and example studied are very preliminary, we feel that extensions of the discrete model could provide more flexibility, whilst still being easy to learn and to compute with. We see it as a promising path to learn dependency structures with imprecise probabilistic models. In particular, it can be seen as a way to approximate a continuous model, as we did in the example. In the future we plan to work on such extensions and on the identification of parametric matrices of weights more flexible than the reduced one presented here.

Finally, an obvious extension to the present work would be to relax the assumption that marginals are precisely identified, and to work with sets of marginals instead. However, this raises challenging theoretical issues, as defining a well-founded extension or equivalent formulation of Sklar's theorem for imprecise models is far from trivial.

References

1. Augustin, T., Coolen, F.P.A., de Cooman, G., Troffaes, M.C.M.: Introduction to imprecise probabilities. Edited book (submitted to publisher)
2. Bernado, J.M., Smith, A.F.M.: Bayesian Theory. John Wiley and Sons (1994)
3. Boole, G.: An investigation of the laws of thought on which are founded the mathematical theories of logic and probabilities. Walton and Maberly, London (1854)
4. Ferson, S., Kreinovich, V., Ginzburg, L., Myers, D.S., Sentz, K.: Constructing probability boxes and Dempster-Shafer structures. Technical Report SAND2002-4015, Sandia National Laboratories (January 2003)
5. Hand, D.J., Daly, F., McConway, K., Lunn, D., Ostrowski, E.: A handbook of small data sets. CRC Press (1993)
6. Kass, R.E., Wasserman, L.: The selection of prior distributions by formal rules. Journal of the American Statistical Association 91(435), 1343–1370 (1996)
7. Nelsen, R.B.: An introduction to copulas. Springer (1999)
8. Quaeghebeur, E., de Cooman, G.: Imprecise probability models for inference in exponential families. In: Cozman, F.G., Nau, R., Seidenfeld, T. (eds.) ISIPTA 2005: Proceedings of the Fourth International Symposium on Imprecise Probabilities and Their Applications, Pittsburgh, USA, pp. 287–296 (July 2005)
9. Pelessoni, R., Vicig, P., Montes, I., Miranda, E.: Imprecise copulas and bivariate stochastic orders. In: De Baets, B., Fodor, J., Montes, S. (eds.) Proceedings of Eurofuse 2013 Workshop, pp. 217–225 (2013)
10. Sklar, A.: Fonctions de répartition à n dimensions et leurs marges. Publ. Inst. Statist. Univ. Paris 8, 229–231 (1959)
11. Troffaes, M.C.M., Blake, S.: A robust data driven approach to quantifying common-cause failure in power networks. In: Cozman, F., Denœux, T., Destercke, S., Seidenfeld, T. (eds.) ISIPTA 2013: Proceedings of the Eighth International Symposium on Imprecise Probability: Theories and Applications, Compiègne, France, pp. 311–317. SIPTA (July 2013)
12. Troffaes, M.C.M., Destercke, S.: Probability boxes on totally preordered spaces for multivariate modelling. International Journal of Approximate Reasoning 52(6), 767–791 (2011)
13. Walley, P.: Statistical Reasoning with Imprecise Probabilities. Chapman and Hall, London (1991)
14. Walley, P.: Inferences from multinomial data: Learning about a bag of marbles. Journal of the Royal Statistical Society, Series B 58(1), 3–34 (1996)
15. Walter, G., Augustin, T.: Imprecision and prior-data conflict in generalized Bayesian inference. Journal of Statistical Theory and Practice 3, 255–271 (2009)
16. Williamson, R.C., Downs, T.: Probabilistic arithmetic I: Numerical methods for calculating convolutions and dependency bounds. International Journal of Approximate Reasoning 4, 89–158 (1990)

Brain Computer Interface by Use of Single Trial EEG on Recalling of Several Images

Takahiro Yamanoi[1], Hisashi Toyoshima[2], Mika Otsuki[3],
Shin-ichi Ohnishi[1], Toshimasa Yamazaki[4], and Michio Sugeno[5]

[1] Faculty of Engineering, Hokkai-Gakuen University,
Nishi 11, Minami 26, Chuo-ku, Sapporo, 064-0926, Japan
yamanoi@lst.hokkai-s-u.ac.jp, 6512102y@hgu.jp,
onishi@eli.hokkai-s-u.ac.jp
[2] Japan Technical Software, Nisi 3, Kita 21, Kita-ku, Sapporo, 001-0021, Japan
toyoshima@jtsnet.co.jp
[3] Faculty of Health Sciences, Hokkaido University,
Nishi 5, Kita 12, Kita-ku, Sapporo, 060-0812, Japan
lasteroideb612@pop.med.hokudai.ac.jp
[4] Faculty of Computer Science and Systems Engineering,
Kyushu Institute of Technology 680-4 Kawazu, Iizuka, Fukuoka, 820-8502, Japan
t-ymzk@bio.kyutech.ac.jp
[5] European Center for Soft Computing, Edificio Cientifico Technologico
c/o Gonzalo Guterrez Quiros, s/n 33600 Mieres-Asutrias, Spain
michio.sugeno@softcomputing.es

Abstract. In order to develop a brain computer interface (BCI), the present authors investigated the brain activity during recognizing or recalling some images of line drawings. The middle frontal robe is known to be related to the function of central executive system on working memory. It is supposed to be so called headquarters of the higher order function of the human brain.Taking into account these facts, the authors recorded Electroencephalogram (EEG) from subjects looking and recalling four types of line drawings of body part, tetra pod, and further ten types of tetra pods, home appliances and fruits that were presented on a CRT. They investigated a single trial EEGs of the subjects precisely after the latency at 400ms, and determined effective sampling latencies for the discriminant analysis to some types of images. They sampled EEG data at latencies from 400ms to 900ms at 25ms intervals by the four channels such as Fp2, F4, C4 and F8. Data were resampled -1 ms and -2 ms backward. Results of the discriminant analysis with the jack knife (cross validation) method for four type objective variates, the discriminant rates for two subjects were more than 95 %, and for ten objective variates were almost 80%.

Keywords: image recognition, single trial Electroencephalogram, canonical discriminant analysis, brain computer interface.

1   Introduction

According to researches on the human brain, the primer process of visual stimulus is done on V1 in the occipital robe. In the early stage of it, a stimulus from the right visual

A. Laurent et al. (Eds.): IPMU 2014, Part III, CCIS 444, pp. 508–517, 2014.
© Springer International Publishing Switzerland 2014

field is processed on the left hemisphere and a stimulus from the left visual field is processed on the right hemisphere. Then the process goes to the parietal associative area [1].

Higher order process of the brain thereafter has its laterality. For instance, 99% of right-handed people and 70% of left-handed people have their language area on the left hemisphere as the Wernicke's area and the Broca's area [2, 3]. Besides these areas, language is also processed on the angular gyrus (AnG), the fusiform gyrus (FuG), the inferior frontal gyrus (IFG) and the prefrontal area (PFA) [4].

By use of the equivalent current dipole localization (ECDL) method [5] applied to the event related potentials (ERPs), summed and averaged EEGs, some of the present authors have investigated that at first equivalent current dipole (ECD) was localized to the right middle temporal gyrus with arrow symbols, and then they were estimated in areas related to the working memory for spatial perception, such as the right inferior or the right middle frontal gyrus. Further, as with kanji characters, ECD was localized to the prefrontal area and the precentral gyrus [6-9], [11].

However, in the case of the mental translation, activities were observed on areas around same latencies regardless to the Kanji or the arrow. After on the right frontal lobe, which is so called the working memory, ECDs were localized to the de Broca's area which is said to be the language area for speech. Like in our preceding researches, it was found that peak latencies were almost the same but polarities of potentials were reversed (Fig. 1) on the frontal lobe in the higher order process [10].

The middle frontal robe is known to be related to the function of central executive system on working memory from the research on the focus and by fMRI. Functions of the central executive system are to select information from the outer world, to hold memory temporally, to order functions following it, to evaluate these orders and to decide and order for erasing information stored temporally. It is supposed to be so called headquarters of the higher order function of the brain.

Some of the present authors thought that this reversion of EEG potential could play a switch to control a robot. Appling these facts to the brain computer interface (BCI), the authors compared each channel of EEGs and its latency. They found that the channel No.4 (F4), No.6 (C4) and No.12 (F8) according to the International 10-20 system were effective to discriminate the four types of EEGs in mental translation. Each discrimination ratio was more than 80% [10].

Those data to discriminate were off lined and fixed, once it was tried the jack knife statistical method, discriminant ratio fell down to around 50%. Hence, the present paper improved the precedent method by adding an EEG channel No.2 (Fp$_2$), and a number of data were tripled as resampling -1ms and -2ms backward from the precedent data and reassembled [12]. After the results of the discriminant analysis with the jack knife (cross validation) method, the mean of discriminant ratio was 98.40%.

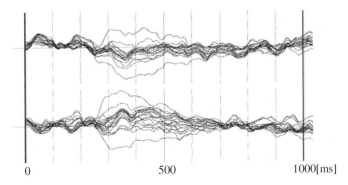

Fig. 1. Comparison between ERPs for rightward (above) and for leftward (below)

2 EEGs Measurements on Recognition and Recalling

Subjects are two university students, that were 22-year-olds, had normal visual acuity, and their dominant hands are the right ones. The subjects put on an electrode cap with 19 active electrodes and watched a 21-inch CRT 30cm in front of them. Each stimulus was displayed on the CRT.

Stimuli to be presented had been stored on the disk of a PC as a file and they were presented in random order. Their heads were fixed on a chin rest on the table. Positions of electrodes on the cap were according to the international 10-20 system and other two electrodes were fixed on the upper and lower eyelids for eye blink monitoring. Impedances were adjusted to less than 10k . Reference electrodes were put on both earlobes and the ground electrode was on the base of the nose.

EEGs were recorded on the multi-purpose portable bio-amplifier recording device (Polymate, TEAC) by means of the electrodes; the frequency band was between 1.0 Hz and 2000 Hz. Output was transmitted to a recording PC. Analog outputs were sampled at a rate of 1 kHz and stored on a hard disk in a PC.

In the experiment, subjects were presented four types and ten types of line drawings of body part, tetrapod, home appliance, that were presented on a CRT. In the first masking period, during 3000ms no stimulus was presented except a gazing point. In the second period (recognizing period), stimulus was presented in the center of CRT during 2000ms, and it was followed by a masking period of 3000ms: the third period. Then in the fourth period during 2000ms (recalling period), visual stimulus was hidden and a subject read the name of stimulus silently. Each stimulus was presented at random, and measurement was repeated thirty times for each stimulus, so the total was 120 times. In these cycles, we measured EEGs during the second and the fourth period during 2000ms (Fig. 2).

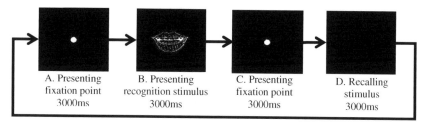

| A. Presenting
fixation point
3000ms | B. Presenting
recognition stimulus
3000ms | C. Presenting
fixation point
3000ms | D. Recalling
stimulus
3000ms |

Repeated A, B, C and D

Fig. 2. Schema of the recognition and recalling image experiment

3 Single trial EEGs Discrimination by Use of Canonical Discriminant Analysis

3.1 Data Sampling from EEGs for Canonical Discriminant Analysis

By use of single trial EEGs data (Fig. 1), that were measured in the experiment with directional symbols, some of the present authors had attempted the canonical discriminant analysis; one of the methods of the multivariate analysis. From the result of our preceding research [12], the pathway goes to the right frontal area at the latency after 400ms. So we sampled EEGs from latency of 400ms to 900ms at 25ms intervals, from 399ms to 899ms at 25ms intervals and from 398ms to 898ms at 25ms intervals.

Electrodes that lie near to the right frontal area are Fp2 (No.2), F4 (No.4), C4 (No.6) and F8 (No.12) (Fig. 4) according to the International 10-20 system, so we chose these four channels among 19 channels. Although the EEGs are time series data, we regarded them as vectors in an 84, i. e. 21 by 4, dimensional space. So the total sample data were 360.

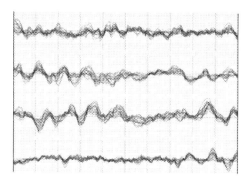

Fig. 3. Single trial EEGs in recalling period for image of body part (mouth, finger, ear and foot from the upper)

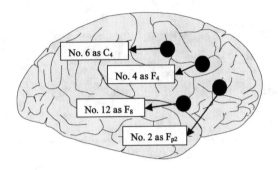

Fig. 4. Position of selected electrodes on right side lateral view

For the use of real time application, it is natural to use a small number of EEGs channels and/or sampling data. Some of the authors have investigated to minimize a number of EEGs channels and a number of sampling data [10]. They investigated the minimal sampling number to obtain complete discriminant ratio (100%) for the same subjects by three channels. However, the sampling interval was 50 ms between the latency 400ms and 900ms. The above analyses were done by use of the statistical software package JUSE-Stat Works/v4.0 MA (Japanese Union of Scientists and Engineers). These results showed a possibility of control in four types of order by use of EEGs. We must note that the discriminant analyses have to be done one by one for each single trial data. So the discriminant coefficients should be determined for each single data for BCI. To improve a single trial discriminant ratio, we adopted the jackknife (cross validation) method.

3.2   Canonical Discriminant Analysis by Learning with Sampling Data

In order to apply the results to BCI, discriminant coefficients should be fixed by some learning process. We grouped each thirty single trial EEGs data into four types, i. e. 120 trials, to play as learning data (Fig. 5).

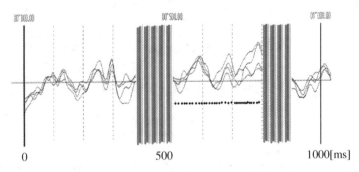

Fig. 5. Selected channels of EEGs and their sampling points: bold lines denote sampling points

3.3   Results of Canonical Discrimination

We gathered each single trial EEGs data to play as learning data. For four type of mental translation (four objective variates), the number of experiments was thirty. These data were resampled three times, in three types of sample timing. Sampling data 1 are taken from latency of 400 ms to 900 ms at 25 ms interval (21 sampling points), sampling data 2 are taken from latency of 399 ms to 899 ms at 25 ms interval and sampling data 3 are taken from latency of 398 ms to 898 ms at 25 ms interval. Each data has one criterion variable i. e. a type of image, and 84 explanatory variates. Because explanatory variates consist of four channels by 21 sampling data, the learning data are 360 with 84 variates. And each criterion variable has four type index, e. g. mouth, finger, ear and foot. We had tried so called the jackknife statistics (cross validation), we took one sample to discriminate, and we used the other samples left as learning data, and the method was repeated.

The subjects were two undergraduate students; however, two samples in recognition period of EEGs were taken twice in two days, so the total number of experiments was six. We denote each experimental data according to the subjects as HF1, HF2, YN1, YN2, HF3, and YN3. We tried to discriminate the four types by 360 samples using the canonical discriminant analysis. As a result, the mean of discriminant ratio was 98.40% (Table 1, 2, 3, 4, 5 and 6). These results are acceptable for an application of BCI.

Table 1. Result of discrimination: recognition of body part (HF1)

Obs./Pred.	Mouth	Finger	Ear	Foot	Total
Mouth	59	0	0	1	60
Finger	0	60	0	0	60
Ear	1	0	57	2	60
Foot	0	0	2	58	60
Total	60	60	59	61	240

Discrimination rate: 97.5%

Table 2. Result of discrimination: recognition of body part (HF2)

Obs./Pred.	Mouth	Finger	Ear	Foot	Total
Mouth	59	0	0	1	60
Finger	0	60	0	0	60
Ear	0	0	60	0	60
Foot	0	1	0	59	60
Total	59	61	60	60	240

Discrimination rate: 99.1%

Table 3. Result of discrimination: recognition of body part (YN1)

Obs./Pred.	Mouth	Finger	Ear	Foot	Total
Mouth	60	0	0	0	60
Finger	0	60	0	0	60
Ear	0	0	60	0	60
Foot	0	0	0	60	60
Total	60	60	60	60	240

Discrimination rate: 100.0%

Table 4. Result of discrimination: recognition of body part (YN2)

Obs./Pred.	Mouth	Finger	Ear	Foot	Total
Mouth	59	0	0	1	60
Finger	2	58	0	0	60
Ear	0	0	60	0	60
Foot	0	0	0	60	60
Total	61	58	60	61	240

Discrimination rate: 98.8%

Table 5. Result of discrimination: recognition of tetra pod (HF)

Obs./Pred.	Dog	Giraffe	Bear	Lion	Total
Dog	59	0	0	1	60
Giraffe	0	56	1	3	60
Bear	0	0	60	0	60
Lion	0	3	0	57	60
Total	59	59	61	61	240

Discrimination rate: 96.7%

Table 6. Result of discrimination: recognition of tetrapod (YN)

Obs./Pred.	Dog	Giraffe	Bear	Lion	Total
Dog	59	0	1	0	60
Giraffe	1	59	0	0	60
Bear	0	0	60	0	60
Lion	0	1	1	58	60
Total	60	60	62	58	240

Discrimination rate: 98.3%

Further the present authors tried to discriminate ten stimuli, those were as tetra pods, home appliances and fruits. The stimuli were also drawn with lines as before. The subjects were two undergraduate students. We denote each experimental data according to the subjects as YS1, YS2 and YN. Two samples in recognition period of

EEGs were taken twice in two days for the subject YS. Tetra pods were Dog (Do), Cow (Co), Horse (Ho), Giraffe (Gi), Bear (Be), Rhino (Rh), Deer (De), Sheep (Sh), Lion (Li), and Camel (Ca). Home appliances were Iron (Ir), Toaster (To), Hair Dryer (Dr), Sewing Machine (Se), Rice Cooker (Ri), Fun, Washing Machine (Wa), Vacuum Cleaner (Va), Electronic Range (Ra) and Refrigerator (Fr). Fruits were Strawberry (Sb), Persimmon (Pe), Cherry (Ch), Water Melon (Wm), Pineapple (Pa), Banana (Ba), Grapefruit (Gf), Melon (Mel), Peach (Pe) and Apple (Ap).   As a result, discriminant ratios were almost 80% (Table 7, 8 and 9). These results are also acceptable for an application of BCI.

Table 7. Result of discrimination: recalling of tetra pod (YS1)

Obs./Pred.	Do	Co	Ho	Gi	Be	Rh	De	Sh	Li	Ca	Total
Do	17	2	1	0	0	1	3	0	0	0	24
Co	0	19	3	1	0	0	1	0	0	0	24
Ho	0	1	20	2	1	0	0	0	0	0	24
Gi	0	1	0	20	1	0	0	0	1	1	24
Be	0	0	0	0	19	0	1	0	4	0	24
Rh	0	0	0	0	1	20	1	1	1	0	24
De	0	1	2	0	3	1	17	0	0	0	24
Sh	0	1	2	0	0	1	2	18	0	0	24
Li	0	2	1	1	0	0	1	0	18	1	24
Ca	0	1	3	0	1	0	1	0	1	17	24
Total	17	28	32	24	26	23	27	19	25	19	240

Discrimination rate: 77.1%

Table 8. Result of discrimination: recalling of home appliance (YS2)

Obs./Pred.	Ir	To	Dr	Se	Ri	Fun	Wa	Va	Ra	Fr	Total
Ir	19	0	2	1	0	0	2	0	0	0	24
To	0	19	1	0	2	0	2	0	0	0	24
Dr	0	0	18	0	3	0	2	0	1	0	24
Se	0	0	1	21	1	0	1	0	0	0	24
Ri	0	0	1	0	18	0	5	0	0	0	24
Fun	0	0	1	1	1	19	1	0	1	0	24
Wa	0	1	0	0	2	1	20	0	0	0	24
Va	0	0	3	0	1	0	1	19	0	0	24
Ra	0	0	1	0	1	1	1	0	20	0	24
Fr	0	0	0	0	2	1	1	0	0	20	24
Total	19	20	28	23	31	21	36	19	22	20	240

Discrimination rate: 80.4%

Table 9. Result of discrimination: recalling of fruits (YN)

Obs./Pred.	Sb	Pe	Ch	Wm	Pa	Ba	Gf	Mel	Pe	Ap	Total
Sb	18	1	1	0	1	1	1	0	0	1	24
Pe	0	18	0	0	2	0	2	0	0	2	24
Ch	1	0	20	0	0	0	0	0	0	3	24
Wm	1	2	0	16	1	1	1	0	1	1	24
Pa	5	0	0	0	17	0	0	0	0	2	24
Ba	0	0	0	0	2	18	3	0	0	1	24
Gf	2	0	0	1	1	0	19	0	1	0	24
Mel	1	1	0	0	0	0	2	19	0	1	24
Pe	1	1	1	0	0	0	1	0	20	0	24
Ap	1	0	0	0	1	0	4	0	0	18	24
Total	30	23	22	17	25	20	33	19	22	29	240

Discrimination rate: 77.1%

Fruits were Strawberry (Sb), Persimmon (Pe), Cherry (Ch), Water Melon (Wm), Pineapple (Pa), Banana (Ba), Grapefruit (Gf), Melon (Mel), Peach (Pe) and Apple (Ap).

4   Concluding Remarks

In this study, the authors investigated a single trial EEGs of the subject precisely after the latency at 400 ms, and determined effective sampling latencies for the canonical discriminant analysis to some four types of image. We sampled EEG data at latency around from 400 ms to 900 ms in three types of timing at 25ms intervals by the four channels Fp_2, F_4, C_4 and F_8. And data were resampled -1 ms and -2 ms backward. From results of the discriminant analysis with jack knife method for four type objective variates, the mean discriminant ratio for two subjects was 96.8%. On recalling four types of images, one could control four type instructions for a robot or a wheel chair i. e. forward, stop, turn clockwise and turn counterclockwise. Furthermore, we tried to discriminate ten types single trial EEGs, the mean discriminant ratio for two subjects was 78.2%. In practical applications to the brain computer interface, it is fairly good that the mean discriminant ratio becomes around 80%. By these ten types instruction, one could control a robot in more complicated movements.

Acknowledgements. This research has been partially supported by the grant from the ministry of education, sports, science and technology to the national project in the High-tech Research Center of Hokkai-Gakuen University ended in March 2013.

References

1. McCarthy, R.A., Warrington, E.K.: Cognitive neuropsychology: a clinical introduction. Academic Press, San Diego (1990)
2. Geschwind, N., Galaburda, A.M.: Cerebral Lateralization, The Genetical Theory of Natural Selection. Clarendon Press, Oxford (1987)
3. Parmer, K., Hansen, P.C., Kringelbach, M.L., Holliday, I., Barnes, G., Hillebrand, A., Singh, K.H., Cornelissen, P.L.: Visual word recognition: the first half second. NeuroImage 22(4), 1819–1825 (2004)
4. Yamanoi, T., Yamazaki, T., Vercher, J.L., Sanchez, E., Sugeno, M.: Dominance of recognition of words presented on right or left eye - Comparison of Kanji and Hiragana. In: Modern Information Processing, From Theory to Applications, pp. 407–416. Elsevier Science B.V., Oxford (2006)
5. Yamazaki, T., Kamijo, K., Kiyuna, T., Takaki, Y., Kuroiwa, Y., Ochi, A., Otsubo, H.: PC-based multiple equivalent current dipole source localization system and its applications. Res. Adv. in Biomedical Eng. 2, 97–109 (2001)
6. Yamanoi, T., Toyoshima, H., Ohnishi, S., Yamazaki, T.: Localization of brain activity tovisual stimuli of linear movement of a circleby equivalent current dipole analysis (in Japanese). In: Proceeding of the 19th Symposium on Biological and Physical Engineering, pp. 271–272 (2004)
7. Yamanoi, T., Toyoshima, H., Ohnishi, S., Sugeno, M., Sanchez, E.: Localization of the Brain Activity During Stereovision by Use of Dipole Source Localization Method. In: The Forth International Symposium on Computational Intelligence and Industrial Application, pp. 108–112 (2010)
8. Hayashi, I., Toyoshima, H., Yamanoi, T.: A Measure of Localization of Brain Activity for the Motion Aperture Problem Using Electroencephalogram. In: Developing and Applying Biologically-Inspired Vision System: Interdisciplinary Concept, ch. 9, pp. 208–223 (2012)
9. Yamanoi, T., Tanaka, Y., Otsuki, M., Ohnishi, S., Yamazaki, T., Sugeno, M.: Spatiotemporal Human Brain Activities on Recalling Names of Bady Parts. Journal of Advanced Computational Intelligence and Intelligent Informatics 17(3) (2013)
10. Yamanoi, T., Toyoshima, H., Yamazaki, T., Ohnishi, S., Sugeno, M., Sanchez, E.: Micro Robot Control by Use of Electroencephalograms from Right Frontal Area. Journal of Advanced Computational Intelligence and Intelligent Informatics 13(2), 68–75 (2009)
11. Toyoshima, H., Yamanoi, T., Yamazaki, T., Ohnishi, S.: Spatiotemporal Brain Activity During Hiragana Word Recognition Task. Journal of Advanced Computational Intelligence and Intelligent Informatics 15(3), 357–361 (2011)
12. Yamanoi, T., Toyoshima, H., Yamazaki, T., Ohnishi, S., Sugeno, M., Sanchez, E.: Brain Computer Interface by use Electroencephalograms from Right Frontal Area. In: The 6th International Conference on Soft Computing and Intelligent Systems, and The 13th International Symposium on Advanced Intelligent Systems, pp. 1150–1153 (2012)

Model Reference Gain Scheduling Control of a PEM Fuel Cell Using Takagi-Sugeno Modelling

Damiano Rotondo, Vicenç Puig, and Fatiha Nejjari

Advanced Control Systems Group (SAC), Universitat Politècnica de Catalunya (UPC), TR11, Rambla de Sant Nebridi, 10, 08222 Terrassa, Spain

Abstract. In this paper, a solution for the oxygen stoichiometry control problem for Proton Exchange Membrane (PEM) fuel cells is presented. The solution relies on the use of a reference model, where the resulting nonlinear error model is brought to a Takagi-Sugeno (TS) form using the nonlinear sector approach. The TS model is suitable for designing a controller using Linear Matrix Inequalities (LMI)-based techniques, such that the resulting closed-loop error system is stable with poles placed in some desired region of the complex plane. Simulation results are used to show the effectiveness of the proposed approach. In particular, the PEM fuel cell can reach asymptotically the oxygen stoichiometry set-point despite all the considered stack current changes.

Keywords: Takagi-Sugeno model, Reference model based control, Gain-scheduling, PEM Fuel Cell, LMIs.

1 Introduction

Proton Exchange Membrane (PEM, also known as Polymer Electrolyte Membrane) fuel cells are one of the most promising technologies to be used, in a near future, as power supply sources in many portable applications. A good performance of these devices is closely related to the kind of control that is used, so a study of different control alternatives is justified [1]. A fuel cell integrates many components into a power system, which supplies electricity to an electric load or to the grid. Several devices, such as DC/DC or DC/AC converters, batteries or ultracapacitors, are included in the system and, in case the fuel cell is not fed directly with hydrogen, a reformer must also be used. Therefore, there are many control loops schemes depending on the devices that must be controlled. The lower control level takes care of the main control loops inside the fuel cell, which are basically fuel/air feeding, humidity, pressure and temperature. The upper control level is in charge of the whole system, integrating the electrical conditioning, storage and reformer (if necessary). Many control strategies have been proposed in literature, ranging from feedforward control [1], LQR [2] or Model Predictive Control [3].

Recently, the complex and nonlinear dynamics of the power generation systems based on fuel cell technology, described in detail in [4], led to the use of linear models that include parameters varying with the operating point (known as LPV models) not only for advanced control techniques [5] but also for model-based fault diagnosis algorithms [6]. The use of Takagi-Sugeno (TS) models [7] is an alternative to the LPV models, as proposed in [8]. This paper will follow this last approach.

In this paper, a solution for the oxygen stoichiometry control problem for PEM fuel cells using TS models is presented. The solution relies on the use of a reference model,

A. Laurent et al. (Eds.): IPMU 2014, Part III, CCIS 444, pp. 518–527, 2014.

where the resulting nonlinear error model is brought to a TS form using the nonlinear sector approach [9]. The TS model is suitable for designing a controller using Linear Matrix Inequalities (LMI)-based techniques, such that the resulting closed-loop error system is stable with poles placed in some desired region of the complex plane [10]. Simulation results are used to show the effectiveness of the proposed approach. In particular, the PEM fuel cell can reach asymptotically the oxygen stoichiometry set-point despite all the considered stack current changes.

The structure of the paper is the following: Section 2 shows how, starting from the nonlinear model of a PEM fuel cell, and using a reference model, a model of the error dynamics suitable for TS modelling can be derived. Section 3 presents the methodology to design a TS controller based on the TS model of the error. Section 4 illustrates the performance of the proposed TS control strategy in simulation. Finally, Section 5 provides the main conclusions and future work.

2 Model Reference Control of the PEM Fuel Cell System

2.1 PEM Fuel Cell Description

A fuel cell is an electrochemical energy converter that converts the chemical energy of fuel into electrical current. It has an electrolyte, a negative electrode and a positive electrode, and it generates direct electrical current through an electrochemical reaction. Typical reactants for fuel cells are hydrogen as fuel and oxygen as oxidant that, once the reaction takes place, produce water and waste heat.

The basic physical structure of a fuel cell consists of an electrolyte layer in contact with a porous anode and cathode electrode plates. There are different kinds of electrolyte layers. Here a PEM (Polymer Electrolyte Membrane or Proton Exchange Membrane) fuel cell is used. The PEM has a special property: it conducts protons but is impermeable to gas (the electrons are blocked through the membrane). Auxiliary devices are required to ensure the proper operation of the fuel cell stack.

2.2 PEM Fuel Cell System Model

The model used in this work has been presented in [4]. The model is widely accepted in the control community as a good representation of the behaviour of a Fuel Cell Stack (FCS) system.

Air Compressor. The air compressor is decomposed into two main parts. One part concerns the electric motor, whereas the other part concerns the compressor box. The compressor motor is modelled using a direct current electric motor model. A compressor flow map is used to determine the air flow rate W_{cp}, supplied by the compressor. The model of the air compressor is given by:

$$\dot{\omega}_{cp} = \frac{\eta_{cp}}{J_{cp}} \frac{k_t}{R_{cm}} (v_{cm} - k_v \omega_{cp}) - \frac{C_p T_{atm}}{J_{cp} \omega_{cp} \eta_{cp}} \left[\left(\frac{p_{sm}}{p_{atm}} \right)^{\frac{\gamma-1}{\gamma}} - 1 \right] W_{cp} \qquad (1)$$

where v_{cm} is the motor supply voltage (V).

Supply Manifold. Manifolds are modelled as a lumped volume in pipes or connections between different devices. The following differential equation is used to model the supply manifold pressure behaviour:

$$\dot{p}_{sm} = \frac{\gamma R_a}{V_{sm}} \left\{ W_{cp} \left[T_{atm} + \frac{T_{atm}}{\eta_{cp}} \left[\left(\frac{p_{sm}}{p_{atm}} \right)^{\frac{\gamma-1}{\gamma}} - 1 \right] \right] - k_{sm.out} \left(p_{sm} - \frac{m_{O_2.ca} R_{O_2} T_{st}}{V_{ca}} \right) T_{sm} \right\} \quad (2)$$

Return Manifold. An equation similar to the one introduced for the supply manifold is used to describe the return manifold behaviour:

$$\dot{p}_{rm} = \frac{R_a T_{rm}}{V_{rm}} \left[k_{ca,out} \left(\frac{m_{O_2.ca} R_{O_2} T_{st}}{V_{ca}} - p_{rm} \right) - k_{rm,out} \left(p_{rm} - p_{atm} \right) \right] \quad (3)$$

Anode Flow Dynamics. The hydrogen supplied to the anode is regulated by a proportional controller. The controller takes the differential pressure between anode and cathode to compute the regulated hydrogen flow:

$$\dot{m}_{H_2,an} = K_1 \left(K_2 p_{sm} - \frac{m_{H_2,an} R_{H_2} T_{st}}{V_{an}} \right) - M_{H_2} \frac{n I_{st}}{2F} \quad (4)$$

Cathode Flow Dynamics. The cathode flow dynamics is described by the following differential equation:

$$\dot{m}_{O_2,ca} = k_{sm,out} p_{sm} - \frac{m_{O_2,ca} R_{O_2} T_{st}}{V_{ca}} (k_{sm,out} + k_{ca,out}) + k_{ca,out} p_{rm} - M_{O_2} \frac{n I_{st}}{4F} \quad (5)$$

Oxygen Stoichiometry. The efficiency optimization of the current system can be achieved by regulating the oxygen mass inflow toward the stack cathode [11]. If an adequate oxidant flow is ensured through the stack, the load demand is satisfied with minimum fuel consumption. In addition, oxygen starvation and irreversible damage are averted. To accomplish such an oxidant flow is equivalent to maintaining at a suitable value the oxygen stoichiometry, defined as:

$$\lambda_{O_2} = \frac{k_{sm,out} \left(p_{sm} - \frac{m_{O_2.ca} R_{O_2} T_{st}}{V_{ca}} \right)}{M_{O_2} \frac{n I_{st}}{4F}} \quad (6)$$

The overall model has five state variables (the compressor speed ω_{cp} (rad/s), the pressure in the supply manifold p_{sm} (Pa), the pressure in the return manifold p_{rm} (Pa), the mass of hydrogen in the anode $m_{H_2,an}$ (kg) and the mass of oxygen in the cathode $m_{O_2,ca}$ (kg)) and three inputs, two of which can be used as control variables (the compressor mass flow W_{cp} (kg/s) and the return manifold outlet orifice constant $k_{rm,out}$ (ms)) while the other (the current in the stack I_{st} (A)) can be considered as a disturbance input that can be included in the reference model in order to generate an appropriate feedforward action and make the feedback loop insensitive to its variations. The values used in this work have been taken from [12], and are listed in Table 1.

2.3   Reference Model

Let us define the following reference model:

$$p_{sm}^{ref} = \frac{\gamma R_a}{V_{sm}} \left\{ W_{cp}^{ref} \left[T_{atm} + \frac{T_{atm}}{\eta_{cp}} \left[\left(\frac{p_{sm}}{p_{atm}} \right)^{\frac{\gamma-1}{\gamma}} - 1 \right] \right] - k_{sm.out} \left(p_{sm}^{ref} - \frac{m_{O_2.ca}^{ref} R_{O_2} T_{st}}{V_{ca}} \right) T_{sm} \right\} \tag{7}$$

$$p_{rm}^{ref} = \frac{R_a T_{rm}}{V_{rm}} \left[k_{ca.out} \left(\frac{m_{O_2.ca}^{ref} R_{O_2} T_{st}}{V_{ca}} - p_{rm}^{ref} \right) - k_{rm.out}^{ref} \left(p_{rm} - p_{atm} \right) \right] \tag{8}$$

$$\dot{m}_{O_2.ca}^{ref} = k_{sm.out} p_{sm}^{ref} - \frac{m_{O_2.ca}^{ref} R_{O_2} T_{st}}{V_{ca}} \left(k_{sm.out} + k_{ca.out} \right) + k_{ca.out} p_{rm}^{ref} - M_{O_2} \frac{n I_{st}}{4F} \tag{9}$$

The reference model provides the state trajectory to be tracked by the real PEM fuel cell, starting from the reference inputs W_{cp}^{ref} and $k_{rm.out}^{ref}$. The values of the reference inputs to be fed to the reference model (feedforward actions) are obtained from steady-state considerations about the fuel cell system, so as to keep the supply manifold pressure and the oxygen stoichiometry at some desired values p_{sm}^{∞} and $\lambda_{O_2}^{ref}$.

2.4   Error Model

By subtracting the reference model equations (7)-(9) and the corresponding system equations (2), (3), (5), and by defining the tracking errors $e_1 \triangleq p_{sm}^{ref} - p_{sm}$, $e_2 \triangleq p_{rm}^{ref} - p_{rm}$, $e_3 \triangleq m_{O_2.ca}^{ref} - m_{O_2.ca}$, and the new inputs $u_1 \triangleq W_{cp}^{ref} - W_{cp}$, $u_2 \triangleq k_{rm.out}^{ref} - k_{rm.out}$, the error model for the PEM Fuel Cell can be brought to the following representation:

$$\dot{e}_1 = -\frac{\gamma R_a}{V_{sm}} k_{sm.out} T_{sm} \left(e_1 - \frac{R_{O_2} T_{st}}{V_{ca}} e_3 \right) + b_{11} \left(p_{sm} \right) u_1 \tag{10}$$

$$\dot{e}_2 = -\frac{R_a T_{rm} k_{ca.out}}{V_{rm}} \left(e_2 - \frac{R_{O_2} T_{st}}{V_{ca}} e_3 \right) + b_{22} \left(p_{rm} \right) u_2 \tag{11}$$

$$\dot{e}_3 = k_{sm.out} e_1 + k_{ca.out} e_2 - \left(k_{sm.out} + k_{ca.out} \right) \frac{R_{O_2} T_{st}}{V_{ca}} e_3 \tag{12}$$

with:

$$b_{11} \left(p_{sm} \right) = \frac{\gamma R_a}{V_{sm}} \left[T_{atm} + \frac{T_{atm}}{\eta_{cp}} \left[\left(\frac{p_{sm}}{p_{atm}} \right)^{\frac{\gamma-1}{\gamma}} - 1 \right] \right] \tag{13}$$

$$b_{22} \left(p_{rm} \right) = -\frac{R_a T_{rm}}{V_{rm}} \left(p_{rm} - p_{atm} \right) \tag{14}$$

Table 1. List of parameters and values

Variable	Description	Value and Unit
η_{cp}	Compressor efficiency	0.8
γ	Specific heat capacity of gas	1.4
R_a	Air gas constant	$286.9 J/(kgK)$
R_{O_2}	Oxygen gas constant	$259.8 J/(kgK)$
R_{H_2}	Hydrogen gas constant	$4124.3 J/(kgK)$
V_{sm}	Supply manifold volume	$0.02 m^3$
V_{ca}	Cathode volume	$0.01 m^3$
V_{rm}	Return manifold volume	$0.005 m^3$
V_{an}	Anode volume	$0.005 m^3$
T_{atm}	Air temperature	$298.15 K$
T_{st}	Temperature in the stack	$350 K$
T_{sm}	Supply manifold temperature	$300 K$
T_{rm}	Return manifold temperature	$300 K$
p_{atm}	Air pressure	$101325 Pa$
$k_{sm.out}$	Supply manifold outlet flow constant	$0.3629 \cdot 10^{-5} kg/sPa$
$k_{ca.out}$	Cathode outlet flow constant	$0.2177 \cdot 10^{-5} kg/sPa$
M_{H_2}	Hydrogen molar mass	$2.016 \cdot 10^{-3} kg/mol$
M_{O_2}	Oxygen molar mass	$32 \cdot 10^{-3} kg/mol$
n	Number of cells in the fuel cell stack	381
F	Faraday constant	$96485 C/mol$
K_1	Proportional gain	2.1
K_2	Nominal pressure drop coefficient	0.94
J_{cp}	Combined inertia of motor and compressor	$5 \cdot 10^{-5} kgm^2$
k_t	Torque constant	$0.0153 Nm/A$
R_{cm}	Resistance	0.82Ω
k_v	Motor constant	$0.0153 V s/rad$
C_p	Specific heat capacity of air	$1004 J/kgK$

3   Controller Design Scheme

From the previous section, the PEM Fuel Cell error model can be expressed in a TS form, as follows:

$$
\begin{aligned}
&IF\ \vartheta_1(k)\ is\ M_{i1}\ AND\ \vartheta_2(k)\ is\ M_{i2} \\
&THEN\ \begin{cases} e_i(k+1) = Ae(k) + B_i u(k) \\ y_i(k) = Ce(k) \end{cases} \quad i = 1,\dots,N
\end{aligned} \tag{15}
$$

where $e \in \mathbb{R}^{n_e}$ is the error vector, $e_i \in \mathbb{R}^{n_e}$ is the error update due to the i-th rule of the fuzzy model, $u \in \mathbb{R}^{n_u}$ is the input vector, and $\vartheta_1(k)$, $\vartheta_2(k)$ are premise variables (in this paper, $\vartheta_1(k) = b_{11}(p_{sm}(k))$ and $\vartheta_2(k) = b_{22}(p_{rm}(k))$).

The entire fuzzy model of the error system is obtained by fuzzy blending of the consequent submodels. For a given pair of vectors $e(k)$ and $u(k)$, the final output of the fuzzy system is inferred as a weighted sum of the contributing submodels:

$$e(k+1) = \frac{\sum\limits_{i=1}^{N} w_i(\vartheta(k))[Ae(k) + B_iu(k)]}{\sum\limits_{i=1}^{N} w_i(\vartheta(k))} = \sum_{i=1}^{N} h_i(\vartheta(k))[Ae(k) + B_iu(k)] \quad (16)$$

$$y(k) = Ce(k) \quad (17)$$

where $w_i(\vartheta(k))$ and $h_i(\vartheta(k))$ are defined as follows:

$$w_i(\vartheta(k)) = M_{i1}(\vartheta_1(k))M_{i2}(\vartheta_2(k)) \qquad h_i(\vartheta(k)) = \frac{w_i(\vartheta(k))}{\sum_{i=1}^{N} w_i(\vartheta(k))} \quad (18)$$

where $M_{i1}(\vartheta_1(k))$ and $M_{i2}(\vartheta_2(k))$ are the grades of membership of $\vartheta_1(k)$ and $\vartheta_2(k)$ in M_{i1} and M_{i2}, respectively, and $h_i(\vartheta(k))$ is such that:

$$\sum_{i=1}^{N} h_i(\vartheta(k)) = 1 \quad h_i(\vartheta(k)) \geq 0 \quad i = 1,\dots,N \quad (19)$$

The error submodels in (15) are controlled through TS error-feedback control rules:

$$\begin{array}{l} IF\ \vartheta_1(k)\ is\ M_{i1}\ AND\ \vartheta_2(k)\ is\ M_{i2} \\ THEN\ u_i(k) = K_ie(k) \quad i = 1,\dots,N \end{array} \quad (20)$$

such that the overall controller output is inferred as the weighted mean:

$$u(k) = \sum_{i=1}^{N} h_i(\vartheta(k))K_ie(k) \quad (21)$$

Since the vector of premise variables $\vartheta(k)$ is a function of the state variables p_{sm} and p_{rm}, (21) represents a nonlinear gain-scheduled control law. The goal of the controller design is to find the matrices K_i such that the resulting closed-loop error system is stable with the poles of each subsystem in some desired region of the complex plane.

In this paper, both stability and pole clustering are analyzed within the quadratic Lyapunov framework, where a single quadratic Lyapunov function is used to assure the desired specifications. Despite the introduction of conservativeness with respect to other existing approaches, where the Lyapunov function is allowed to be parameter-varying, the quadratic approach has undeniable advantages in terms of computational complexity.

In particular, the TS error system (16), with the error-feedback control law (21), is quadratically stable if and only if there exist $X_s = X_s^T > 0$ and matrices K_i such that [13]:

$$\begin{pmatrix} -X_s & (A + B_jK_i)X_s \\ X_s(A + B_jK_i)^T & -X_s \end{pmatrix} < 0 \quad i,j = 1,\dots,N \quad (22)$$

On the other hand, pole clustering is based on the results obtained by [14], where subsets $\mathscr{D}$ of the complex plane, referred to as *LMI regions*, are defined as:

$$\mathscr{D} = \{z \in \mathbb{C} : f_{\mathscr{D}}(z) < 0\} \quad (23)$$

where $f_{\mathscr{D}}$ is the *characteristic function*, defined as:

$$f_{\mathscr{D}}(z) = \alpha + z\beta + \bar{z}\beta^T = [\alpha_{kl} + \beta_{kl}z + \beta_{lk}\bar{z}]_{k,l \in [1,m]} \quad (24)$$

where $\alpha = \alpha^T \in \mathbb{R}^{m \times m}$ and $\beta \in \mathbb{R}^{m \times m}$. Hence, the TS error system (16), with the error-feedback control law (21), has its poles in $\mathcal{D}$ if there exist $X_{\mathcal{G}} = X_{\mathcal{G}}^T > 0$ and matrices K_i such that:

$$\left[\alpha_{kl} X_{\mathcal{G}} + \beta_{kl} \left(A + B_j K_i\right) X_{\mathcal{G}} + \beta_{lk} X_{\mathcal{G}} \left(A + B_j K_i\right)^T \right]_{k,l \in [1,m]} < 0 \qquad i,j = 1,\dots,N \quad (25)$$

Two issues arising when using (22) and (25) for design:

- conditions (22) and (25) are Bilinear Matrix Inequalities (BMIs), since the products of the variables K_i by the matrices X_s and $X_{\mathcal{G}}$ appear. In order to reduce the BMIs to Linear Matrix Inequalities (LMIs), a common Lyapunov matrix $X_s = X_{\mathcal{G}} = X$ is chosen, and the change of variables $\Gamma_i \triangleq K_i X$ is introduced;
- in TS systems where only the input matrix B changes according to the considered subsystem, the solution provided by (22) and (25) exhibits too conservatism, due to the fact that a given K_i has to guarantee stability/pole clustering for all the possible B_j. In order to reduce such conservatism, a gridding approach is considered for obtaining the TS model, and the design conditions are written at the grid points only. Even though the stability and the pole clustering specification are theoretically guaranteed only at the grid points, from a practical point of view such specifications should be guaranteed by choosing a grid of points dense enough.

Hence, the conditions to be used for finding the gains K_i are the following:

$$\begin{pmatrix} -X & AX + B_i \Gamma_i \\ XA^T + \Gamma_i^T B_i^T & -X \end{pmatrix} < 0 \qquad i = 1,\dots,N \qquad (26)$$

$$\left[\alpha_{kl} X + \beta_{kl} \left(AX + B_i \Gamma_i\right) + \beta_{lk} \left(XA^T + \Gamma_i^T B_i^T\right) \right]_{k,l \in [1,m]} < 0 \qquad i = 1,\dots,N \qquad (27)$$

These LMIs can be solved efficiently using available software, e.g. the YALMIP toolbox [15] with SeDuMi solver [16].

4   Simulation Results

The TS control design technique described in Section 3 has been applied to the error model of the PEM Fuel Cell presented in Section 2, where the state matrix is given as follows:

$$A = \begin{pmatrix} -21.8644 & 0 & 1.9881 \cdot 10^8 \\ 0 & -37.4749 & 3.4076 \cdot 10^8 \\ 3.6290 \cdot 10^{-6} & 2.1770 \cdot 10^{-6} & -52.7940 \end{pmatrix}$$

By considering that the supply and the return manifold pressures p_{rm} and p_{sm} can take values in given intervals:

$$p_{rm} \in \left[1.3 \cdot 10^5, 13 \cdot 10^5\right] \qquad p_{sm} \in \left[1.3 \cdot 10^5, 13 \cdot 10^5\right]$$

it is obtained that the elements of the input matrix vary in the indicated ranges:

$$B = \begin{pmatrix} b_{11}(p_{sm}) & 0 \\ 0 & b_{22}(p_{rm}) \\ 0 & 0 \end{pmatrix} \in \begin{pmatrix} \left[1.3 \cdot 10^5, 13 \cdot 10^5\right] & 0 \\ 0 & \left[1.3 \cdot 10^5, 13 \cdot 10^5\right] \\ 0 & 0 \end{pmatrix}$$

The nonlinear sector approach has been applied to obtain a TS model by dividing these intervals into 30 points each. This leads to a grid of 900 pairs (b_{11}, b_{22}), i.e., submodels.

Using this model, a TS controller with the structure (20) has been designed using (26) to assure stability and using (27) to achieve pole clustering in a circle of radius 0.4 and center $(0.599, 0)$. This controller needs an observer to estimate the error between the reference and the real states of the PEM fuel cell. Even though the observer is implemented as a TS observer (see [10] for more details about TS observers) due to the variability of the values $b_{11}(p_{sm})$ and $b_{22}(p_{rm})$, the design of the observer can be performed using an LTI exact pole placement technique, since both the A and the C matrices of the TS PEM Fuel Cell error model are constant. In particular, in this work the error observer eigenvalues have been put in $\{0.3, 0.25, 0.2\}$.

The results shown in this paper refer to a simulation which lasts $300\,s$, where abrupt change in the stack current $I_{st}(t)$ and the desired oxygen excess ratio $\lambda_{O2}^{ref}(t)$ were introduced. The PEM fuel cell initial states have been chosen as follows:

$$\begin{bmatrix} p_{sm}(0) \\ p_{rm}(0) \\ m_{H2,an}(0) \\ m_{O2,ca}(0) \end{bmatrix} = \begin{bmatrix} 1.6 \cdot 10^5\,Pa \\ 1.6 \cdot 10^5\,Pa \\ 5 \cdot 10^{-4}\,kg \\ 0.01\,kg \end{bmatrix}$$

A Gaussian noise with zero mean and standard deviation equal to 5 ‰ of the measurement has been considered for both the available sensors (state variables p_{sm} and p_{rm}). Fig. 1 shows the evolution of the stack current I_{st} during the simulation and the tracking of the desired oxygen excess ratio. It can be seen that the reference is correctly followed independently of the values taken by the stack current. This is done by changing the compressor mass flow W_{cp} and the return manifold outlet constant $k_{rm,out}$, taking into account both the feedforward and the feedback control law, as shown in Fig. 2. Finally, the values taken by the state variables are shown in Fig. 3.

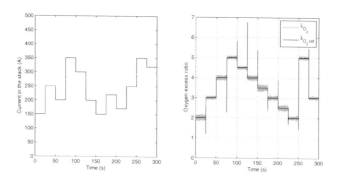

Fig. 1. Current in the stack and oxygen excess ratio

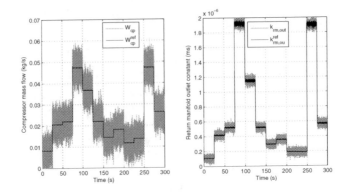

Fig. 2. Input variables

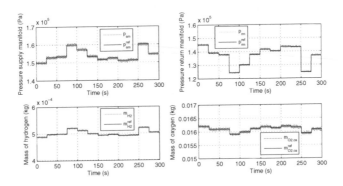

Fig. 3. State variables

5   Conclusions

In this paper, the problem of controlling the oxygen stoichiometry for a PEM fuel cell
has been solved. The proposed solution relies on the use of a reference model that
describes the desired behaviour. The resulting nonlinear error model is brought to a
TS form that is used for designing a TS controller using LMI-based techniques. The
results obtained in simulation environment have demonstrated the effectiveness of the
proposed technique. As future work, the proposed TS approach will be compared with
the LPV approach in order to see the advantages and disavantages of each one of these
two techniques.

Acknowledgement. This work has been funded by the Spanish MINECO through
the project CYCYT SHERECS (ref. DPI2011-26243), by the European Commission
through contract i-Sense (ref. FP7-ICT-2009-6-270428) and by AGAUR through the
contract FI-DGR 2013 (ref. 2013FIB00218).

References

1. Pukrushpan, J.T., Stefanopoulou, A.G., Peng, H.: Control of Fuel Cell Power Systems: Principles, Modeling, Analysis and Feedback Design. Series Advances in Industrial Control. Springer (2004)
2. Niknezhadi, A., Allué, M., Kunusch, C., Ocampo-Martínez, C.: Design and implementation of LQR/LQG strategies for oxygen stoichiometry control in PEM fuel cells based systems. Journal of Power Sources 196(9), 4277–4282 (2011)
3. Ziogou, C., Papadopoulou, S., Georgiadis, M.C., Voutetakis, S.: On-line nonlinear model predictive control of a PEM fuel cell system. Journal of Process Control 23(4), 483–492 (2013)
4. Pukrushpan, J.T., Peng, H., Stefanopoulou, A.G.: Control-oriented modeling and analysis for automotive fuel cell systems. ASME Journal of Dynamic Systems, Measurement and Control 126(1), 14–25 (2004)
5. Bianchi, F.D., Kunusch, C., Ocampo-Martínez, C., Sánchez-Peña, R.S.: A gain-scheduled LPV control for oxygen stoichiometry regulation in PEM fuel cell systems. IEEE Transactions on Control Systems Techonology (in preview, 2014)
6. de Lira, S., Puig, V., Quevedo, J., Husar, A.: LPV observer design for PEM fuel cell system: Application to fault detection. Journal of Power Sources 196(9), 4298–4305 (2011)
7. Takagi, T., Sugeno, M.: Fuzzy Identification of Systems and Its Applications to Modeling and Control. IEEE Transactions on Systems, Man, and Cybernetics SMC-15(1), 116–132 (1985)
8. Olteanu, S., Aitouche, A., Oueidat, M., Jouni, A.: PEM fuel cell modeling and simulation via the Takagi-Sugeno Fuzzy model. In: International Conference on Renewable Energies for Developing Countries (REDEC), pp. 1–7 (2012)
9. Tanaka, K., Wang, H.O.: Fuzzy control systems design and analysis: A linear matrix inequality approach. John Wiley and Sons, Inc. (2001)
10. Korba, P., Babuška, R., Verbruggen, H., Frank, P.: Fuzzy gain scheduling: controller and observer design based on Lyapunov method and convex optimization. IEEE Transactions on Fuzzy Systems 11(3), 285–298 (2003)
11. Kunusch, C., Puleston, P., Mayosky, M.: Sliding-Mode control of PEM fuel cells. Springer, London (2012)
12. Aitouche, A., Yang, Q., Ould Bouamama, B.: Fault detection and isolation of PEM fuel cell system based on nonlinear analytical redundancy. The European Physical Journal Applied Physics 54(2) (2011)
13. Wang, H., Tanaka, K., Griffin, M.: An approach to fuzzy control of nonlinear systems: stability and design issues. IEEE Transactions on Fuzzy Systems 4(1), 14–23 (1996)
14. Chilali, M., Gahinet, P.: H_∞ Design with Pole Placement Constraints: An LMI Approach. IEEE Transactions on Automatic Control 41(3), 358–367 (1996)
15. Löfberg, J.: YALMIP: A toolbox for modeling and optimization in MATLAB. In: Proceedings of the CACSD Conference (2004)
16. Sturm, J.F.: Using SeDuMi 1.02, a MATLAB toolbox for optimization over symmetric cones. Optimization Methods and Software 11(12), 625–653 (1999)

Non-quadratic Stabilization for T-S Systems with Nonlinear Consequent Parts

Hoda Moodi[1], Jimmy Lauber[2], Thierry Marie Guerra[2],
and Mohamad Farrokhi[1]

[1] Department of Electrical Engineering, Iran University of Science and Technology,
Narmak, 16846-13114, Tehran, Iran
{moodi,farrokhi}@iust.ac.ir
[2] LAMIH UMR CNRS 8201, University of Valenciennes and Hainaut-Cambrsis,
Le Mont Houy, 59313, Valenciennes, France
{lauber,guerra}@univ-valenciennes.fr

Abstract. This paper deals with nonlinear systems, which are modeled by T-S fuzzy model containing nonlinear functions in the consequent part of the fuzzy IF-THEN rules. This will allow modeling a wider class of systems with smaller modeling errors. The consequent part of each rule is assumed to contain a linear part plus a sector-bounded nonlinear term. The proposed controller guarantees exponential convergence of states by utilizing a new non-quadratic Lyapunov function for Lyapunov stability analysis and Linear Matrix Inequality (LMI) formulation. Moreover, new relaxation methods are introduced for further reduction of conservativeness and maximizing the region of attractions. Numerical examples illustrate effectiveness of the proposed method.

Keywords: Fuzzy Sugeno, Nonlinear Subsystem, Non-Quadratic Lyaponuv.

1 Introduction

Takagi-Sugeno (T-S) fuzzy model is a well-known tool for nonlinear system modeling with increasing interest in recent years. As T-S model is a universal approximator, it can model any smooth nonlinear system with any degree of accuracy. Furthermore, the local linear subsystems of this model allow one to use powerful linear systems tools, such as Linear Matrix Inequalities (LMIs), to analyze and synthesize T-S fuzzy systems.

As complexity of the system increases, the number of rules in the fuzzy model and hence, the number and dimensions of LMIs (used for the stability analysis) increase and become harder to solve. Many works in literature are devoted to decrease the conservativeness of these LMIs in order to apply them to a wider class of systems [1–3].

One different possible solution is to use nonlinear local subsystems for the T-S model. A priori it seems that this method increases the complexity of the fuzzy model, whereas it decreases the number of rules and at the same time increases

A. Laurent et al. (Eds.): IPMU 2014, Part III, CCIS 444, pp. 528–538, 2014.
© Springer International Publishing Switzerland 2014

the model accuracy. The key idea of using nonlinear terms in the subsystems is to use some kind of nonlinearity, which is less complicated than the nonlinearities of the main system. One main class of T-S system with nonlinear consequent is the polynomial fuzzy systems, which has attracted the attention of researchers in recent years [4–6]. Another possible form of nonlinear consequent part is a linear subsystem plus a nonlinear term, which is used by Dong *et al.* [7] and [8] and Dong [9], where they have employed sector-bounded functions in the subsystems.

In this paper, a similar form of the Dong's model is employed. In other words, every subsystem in the Sugeno model contains a linear term plus a nonlinear term in the consequent part of the fuzzy IF-THEN rules. However, to avoid conservativeness introduced by the sector conditions, a novel non-quadratic Lyapunov function is introduced.

The reminder of the paper is organized as follows. In Section 2, nonlinear Sugeno model is described. In Section 3, existing methods on stabilization of nonlinear Sugeno system as well as new proposed methods are introduced. Simulation results are given in Section 4. Section 5 concludes the paper.

2 Problem Statement

Consider a class of nonlinear systems described by

$$\dot{x}(t) = f_a\big(x(t)\big) + f_b\big(x(t)\big)\varphi\big(x(t)\big) + g\big(x(t)\big)u(t)$$
$$y(t) = f_{ya}\big(x(t)\big) + f_{yb}\big(x(t)\big)\varphi\big(x(t)\big) \tag{1}$$

where $x(t) \in \mathbb{R}^{n_x}$ is the state, $u(t) \in \mathbb{R}^{n_u}$ is the control input, $y(t) \in \mathbb{R}^{n_y}$ is the measurable output, $f_n\big(x(t)\big): n \in [a, b, ya, yb]$ and $g\big(x(t)\big) \in \mathbb{R}^{(n_x \times n_u)}$ are nonlinear functions and $\varphi\big(x(t)\big) \in \mathbb{R}^{(n_x \times n_\varphi)}$ is a vector of sector-bounded nonlinear functions satisfying

$$\varphi_i\big(x(t)\big) \in co\{E_{Li}x(t), E_{Ui}x(t)\}, 1 \le i \le s \tag{2}$$

or without loss of generality

$$\varphi_i\big(x(t)\big) \in co\{0, E_i x(t)\}, 1 \le i \le s \tag{3}$$

which results $\varphi_i\big(x(t)\big)\Big(E_i x(t) - \varphi_i\big(x(t)\big)\Big) \ge 0$. Considering $\Gamma > 0$, it immediately results that

$$\varphi^T\big(x(t)\big)\Gamma^{-1}Ex(t) - \varphi^T\big(x(t)\big)\Gamma^{-1}\varphi\big(x(t)\big) \ge 0. \tag{4}$$

where Γ is diagonal.

2.1   Nonlinear Sugeno Model

The system (1) can be represented by a T-S fuzzy system with local nonlinear models as follows:

> Plant Rule i:
> IF $z_1(t)$ is $M_{i1}(z)$, ..., and $z_p(t)$ is $M_{ip}(z)$ THEN:
> $$\dot{x}(t) = A_i x(t) + G_{xi}\varphi(x(t)) + B_i u(t) + D_{1i}\nu(t)$$
> $$y(t) = C_i x(t) + G_{yi}\varphi(x(t)) + D_{2i}\nu(t)$$

(5)

where $A_i \in \mathbb{R}^{(n_x \times n_x)}$, $B_i \in \mathbb{R}^{(n_x \times n_u)}$, $C_i \in \mathbb{R}^{(n_y \times n_x)}$, $G_{xi} \in \mathbb{R}^{(n_x \times n_\varphi)}$, $G_{yi} \in \mathbb{R}^{(n_y \times n_\varphi)}$, $D_{1i} \in \mathbb{R}^{(n_x \times n_\nu)}$ and $D_{2i} \in \mathbb{R}^{(n_y \times n_\nu)}$ $(i = 1, \ldots, r)$ are constant matrices, in which r is the number of rules, n_x is the number of states, n_u is the number of inputs, n_y is the number of outputs, n_φ is the number of nonlinear functions in $\varphi(x)$ vector and n_ν is the dimension of ν. Moreover, $z_1(t), \ldots, z_p(t)$ are the premise variables, M_{ij} denote the fuzzy sets and $\nu(t)$ is a band-limited white noise. In this case, the whole fuzzy system can be represented as

$$\dot{x}(t) = \sum_{i=1}^{r} w_i(z)\big[A_i x(t) + G_{xi}\varphi(x(t)) + B_i u(t) + D_{1i}\nu(t)\big]$$

$$y(t) = \sum_{i=1}^{r} w_i(z)\big[C_i(x(t)) + G_{yi}\varphi(x(t)) + D_{2i}\nu(t)\big]$$

(6)

where

$$w_i(z) = \frac{h_i(z)}{\sum_{k=1}^{r} h_k(z)}, \quad h_i(z) = \Pi_{j=1}^{p}\mu_{ij}(z).$$

(7)

and $\mu_{ij}(z)$ is the grade of membership of z_j in M_{ij}.

3   Controller Design

The control scheme used in this paper is as follows:

> Controller Rule i:
> IF $z_1(t)$ is $M_{i1}(z)$, ..., and $z_p(t)$ is $M_{ip}(z)$ THEN:
> $$u(t) = K_{ai}x(t) + K_{bi}\varphi(x(t)).$$

(8)

From (5) and (8), the closed-loop fuzzy system can be obtained as follows:

$$\dot{x}(t) = (A_i + B_i K_{ai})x(t) + (G_{xi} + B_i K_{bi})\varphi(x(t)) + D_{1i}\nu(t)$$

$$y(t) = C_i x(t) + G_{yi}\varphi(x(t)) + D_{2i}\nu(t)$$

(9)

In this section, the conditions for asymptotic convergence of the states of the system (5) by the controller (8) will be given. The following lemmas are used in this paper.

Lemma 1. *[10] If the following conditions hold:*

$$\Xi_{ii} < 0, \ \ 1 < i < r$$

$$\frac{1}{r-1}\Xi_{ii} + \frac{1}{2}(\Xi_{ij} + \Xi_{ji}) < 0, \ \ 1 < i \neq j < r \tag{10}$$

then, the following inequality holds:

$$\sum_{i=1}^{r}\sum_{j=1}^{r}\alpha_i\alpha_j\Xi_{ij} < 0 \tag{11}$$

where $0 \leq \alpha_i \leq 1$ and $\sum_{i=1}^{r}\alpha_i = 1$.

Lemma 2. *Matrix inversion lemma [11]: For matrices of the correct size, the following property holds:*

$$(A + BCD)^{-1} = A^{-1} - A^{-1}B(C^{-1} + DA^{-1}B)^{-1}DA^{-1} \tag{12}$$

In the following theorem, sufficient conditions for stability of the system dynamic (9) will be given.

Theorem 1. *[8] If there exist matrices $P = P^T > 0$, X_{ai} and X_{bi}, $1 \leq i \leq r$, and diagonal matrix $\Gamma > 0$ such that (10) is satisfied with Ξ_{ij} defined as*

$$\Xi_{ij} = \begin{bmatrix} He(A_iP + B_iX_{aj}) & * \\ \Gamma G_{xi}^T + X_{bj}^T B_i^T + EP & -2\Gamma \end{bmatrix} \tag{13}$$

then, fuzzy system (5) is asymptotically stable using controller (8) with

$$K_{ai} = X_{ai}P^{-1} \ \ K_{bi} = X_{bi}\Gamma^{-1} \ \ 1 \leq i \leq r. \tag{14}$$

*In (13), $He(A)$ denotes the Hermit of matrix A and * indicates the symmetric term.*

Corollary 1. *If there exist matrices $P = P^T > 0$, X_{ai} and X_{bi}, $1 \leq i \leq r$, and diagonal matrix $\Gamma > 0$ such that (10) is satisfied with Ξ_{ij} defined as*

$$\Xi_{ij} = \begin{bmatrix} He(A_iP + B_iX_{aj}) + \beta P & * \\ \Gamma G_{xi}^T + X_{bj}^T B_i^T + EP & -2\Gamma \end{bmatrix} \tag{15}$$

then, fuzzy system (5) is asymptotically stable via controller (8) with decay rate β. I.e., if $V(x(t))$ is a Lyapunov function for system (9) then $\dot{V}(x(t)) < -\beta V(x(t)))$

Proof. Based on the Theorem 1, the proof is easily obtained.

3.1 Non-quadratic Stabilization

Based on Theorem 1, the nonlinear term is treated as a disturbance that must be compensated. This may yield a more conservative solution to the problem. The following theorem reduces this conservativeness.

Theorem 2. *If there exist matrices $P = P^T > 0$, X_{ai} and X_{bi}, $1 \leq i \leq r$, and diagonal matrix $\Gamma > 0$ such that (10) holds with Ξ_{ij} defined as below*

$$\Xi_{ij} = \begin{bmatrix} He(A_i P + B_i X_{aj}) & * \\ \Xi_{ij}^{21} & \Xi_{ij}^{22} \end{bmatrix} \tag{16}$$

where

$$\begin{aligned} \Xi_{ij}^{21} &= \Gamma G_i^T + X_{bj}^T B_i^T + EP + \alpha E(A_i P + B_i X_{aj}) \\ \Xi_{ij}^{22} &= He(-\Gamma + \alpha E(G_i \Gamma + B_i X_{bj})) \end{aligned} \tag{17}$$

then, fuzzy system (5) is asymptotically stable via controller (8) with K_{ai} and K_{bi} defined in (14), if the Jacobian matrix of the vector $\varphi(x(t))$ is symmetric.

Proof. Let define the following non-quadratic Lyapunov function:

$$V(x, \varphi(x)) := x^T(t) P^{-1} x(t) + 2\alpha \int_0^x \varphi^T(y) \Gamma^{-1} E \, dy. \tag{18}$$

Then, if the Jacobian matrix of the vector $\varphi(x(t))$ is symmetric, the integral in (18) is path independent and hence, the time derivative of (18) becomes

$$\begin{aligned} \dot{V}(x, \varphi(x)) &= 2\dot{x}^T(t) P^{-1} x(t) + 2\alpha \varphi^T(x(t)) \Gamma^{-1} E \dot{x}(t) \\ &= \begin{bmatrix} x(t) \\ \varphi(x(t)) \end{bmatrix}^T \begin{bmatrix} He(P^{-1} A_i + P^{-1} B_i K_{ai}) & * \\ \tilde{\Xi}_{ij}^{21} & \tilde{\Xi}_{ij}^{22} \end{bmatrix} \begin{bmatrix} x(t) \\ \varphi(x(t)) \end{bmatrix} \end{aligned} \tag{19}$$

where

$$\begin{aligned} \tilde{\Xi}_{ij}^{21} &= (G_{xi} + B_i K_{bj})^T P^{-1} + \alpha \Gamma^{-1} E(A_i + B_i K_{aj}) \\ \tilde{\Xi}_{ij}^{22} &= \alpha(\Gamma^{-1} E(G_i + B_i K_{bj}) + *). \end{aligned} \tag{20}$$

Substituting (14) into (16) and pre- and post-multiplying it by $\text{diag}[P^{-1}, \Gamma^{-1}]$ and its transpose, and based on (19) and (4) it follows that $\dot{V}(x, \varphi(x)) < 0$, which completes the proof.

Remark 1. To show that the selected $V(x, \varphi(x))$ is a Lyapunov function, it is necessary to investigate its positiveness. It is obvious that $V(x, \varphi(x))$ for $x(t) = 0$ is equal to zero. Based on (4), it yields

$$\varphi^T(x(t)) \Gamma^{-1} E x(t) > 0 \tag{21}$$

and as a result

$$\int_0^x \varphi^T(y) \Gamma^{-1} E \, dy > 0. \tag{22}$$

Therefore, if $\alpha \geq 0$, then $V(x, \varphi(x)) > 0$. For $\alpha < 0$ a bound on the value of this integral should be found. To this aim, note that if $\varphi(x(t))$ is bounded to the sector $[0\ Ex(t)]$, then its integral from 0 to $x(t)$ is less than the area of this sector. In other words

$$\int_0^x \varphi^T(y)\Gamma^{-1}E\,dy < \frac{1}{2}x^T(t)E^T\Gamma^{-1}Ex(t) \tag{23}$$

Therefore, to guarantee the positiveness of the Lyapunov function, the inequlity $P^{-1} + \alpha E^T\Gamma^{-1}E > 0$ must be satisfied. This can be converted into an LMI as

$$\begin{bmatrix} P & * \\ EP & -\frac{1}{\alpha}\Gamma \end{bmatrix} > 0. \tag{24}$$

Remark 2. If by changing Ξ_{ij} in (17) as

$$\begin{aligned} \Xi_{ij}^{22} &= He(\alpha E(G_i\Gamma + B_iX_{bj})) \\ \Xi_{ij}^{21} &= \Gamma G_i^T + X_{bj}^T B_i^T + \alpha E(A_iP + B_iX_{aj}) \end{aligned} \tag{25}$$

$\Xi_{ij} < 0$ still holds, then the condition of sector boundedness of $\varphi(x(t))$ is not necessary for making $\dot{V}(x, \varphi(x)) < 0$. However, to guarantee positiveness of the Lyapunov function, a milder condition on $\varphi(x(t))$ must be satisfied. That is, $\varphi^T(x(t))Ex(t) > 0$.

Corollary 2. *If there exist matrices $P = P^T > 0$, X_{ai} and X_{bi}, $1 \leq i \leq r$, and diagonal matrix $\Gamma > 0$ such that (10) holds with Ξ_{ij} defined as below*

$$\Xi_{ij} = \begin{bmatrix} He(A_iP + B_iX_{aj}) + \beta P & * & * \\ \beta EP & -\frac{\beta}{\alpha}\Gamma & 0 \\ \Xi_{ij}^{21} & 0 & \Xi_{ij}^{22} \end{bmatrix} \tag{26}$$

where Ξ_{ij}^{21} and Ξ_{ij}^{22} are defined in (17), then, fuzzy system (5) is asymptotically stable via controller (8) with decay rate β.

Proof. Based on (23) and the proof of Theorem 2, proof can be easily obtained.

3.2   Maximum Region of Attraction

One of the property of the Lyapunov's method is that it can be used to find the region of attraction of states or an estimate of it. This region is usually estimated by the largest ball $V(x(t)) = c$ contained in the domain of the system definition, where $\dot{V}(x(t)) < 0$ [12]. The invariant ellipsoid $\varepsilon = \{x \in \mathbb{R}^n,\ x^TP^{-1}x \leq 1\}$ is contained in polytope $\mathbb{P} = \{x \in \mathbb{R}^n,\ a_k^Tx \leq 1,\ k = 1, \ldots, q\}$ if the following set of linear inequalities are satisfied:

$$a_k^T P^{-1} a_k < 1, \quad k = 1, \ldots, q. \tag{27}$$

Applying the Schur complement, (27) can be rewritten as

$$\begin{bmatrix} 1 & a_k^T \\ a_k & P \end{bmatrix} > 0. \tag{28}$$

On the other hand, The largest invariant ellipsoid ε in $\mathbb{P}$ can be found by the following optimization problem [13]:

$$\max \lambda \tag{29}$$
$$\text{subject to: } P > \lambda I$$

where $\lambda > 0$ and I is an identity matrix. Since Lyapunov function (18) is not an ellipsoid, it is rather difficult to find its maximum value by a convex optimization problem. Instead, two surrounding ellipsoids can be determined. To this aim, the following lemma is stated.

Lemma 3. *Lyapunov function (18) with $\alpha > 0$ is bounded as*

$$x^T P^{-1} x < V(x, \varphi(x)) < x^T P^{-1} x + \alpha x E^T \Gamma^{-1} E x \tag{30}$$

Proof. based on (22) and (23), the proof is straightforward. Next, the following theorem can be stated.

Theorem 3. *In Theorem 2, the maximum region of attraction of states can be found by the following optimization problem:*

$$\max \lambda$$
$$\text{subject to: } \begin{bmatrix} P - \lambda I & PE^T \\ EP & (1/\alpha)\Gamma + EPE^T \end{bmatrix} > 0 \tag{31}$$
$$and \quad \begin{bmatrix} 1 & a_k^T \\ a_k & P \end{bmatrix} > 0$$

Proof. Based on (3), the Lyapunov function is always outside the ellipsoid $\varepsilon_1 = \{x \in \mathbb{R}^n, \ x^T(P^{-1} + \alpha E^T \Gamma^{-1} E)x \leq 1\}$ and inside the ellipsoid $\varepsilon_2 = \{x \in \mathbb{R}^n, \ x^T P^{-1} x \leq 1\}$. In order to guarantee that the region $V(x(t)) < 1$ is inside the polytope $\mathbb{P}$, it suffices to check whether ε_2 satisfies this condition, which is equal to $a_k^T P^{-1} a_k < 1, k = 1, \ldots, q$. And in order to maximize the region of attraction, it is enough to maximize ε_1, which can be stated as the following optimization problem:

$$\max \lambda \tag{32}$$
$$\text{subject to: } (P^{-1} + \alpha E^T \Gamma^{-1} E)^{-1} > \lambda I.$$

Based on the matrix inversion lemma, the condition in (32) can be written as

$$P + PE^T(-\frac{1}{\alpha}\Gamma - EPE^T)^{-1}EP - \lambda I > 0 \tag{33}$$

which is equal to

$$\begin{bmatrix} P - \lambda I & PE^T \\ EP & (\frac{1}{\alpha})\Gamma + EPE^T \end{bmatrix} > 0. \tag{34}$$

4   Simulations

In this section, simulations are given to illustrate the effectiveness of the new non-quadratic Lyapunov function. It is shown that the proposed method can design fuzzy controllers with less conservatism than the conventional methods with a larger region of attraction. All LMIs are solved using YALMIP [14] as a parser and SeDuMi as a solver.

Example 1. Consider the following system (Example 1 in [15] with minor modifications):

$$\dot{x}(t) = \begin{bmatrix} a + bx_2^2 & -1 \\ 2 & c + dx_2^2 \end{bmatrix} x(t) + \begin{bmatrix} 1 \\ -x_2^2 \end{bmatrix} u(t) + \begin{bmatrix} b & 0 \\ 1 & d \end{bmatrix} \begin{bmatrix} x_1^3 \\ \sin(x_2) \end{bmatrix} \tag{35}$$

with $a = -0.2363$, $b = 0.0985$, $c = -0.7097$ and $d = 0.3427$. Suppose $x_1 \in [-1 \ 1]$ and $x_2 \in [-2 \ 2]$. By selecting x_2^2 as the premise variable and $\varphi(x(t)) = \left[x_1^3 \ \sin(x_2)\right]^T$, this system can be modeled by sector nonlinearity approach as follows:

$$A_1 = \begin{pmatrix} a & -1 \\ 2 & c \end{pmatrix}, A_2 = \begin{pmatrix} a + 4b & -1 \\ 2 & c + 4d \end{pmatrix} B_1 = \begin{pmatrix} 1 \\ -4 \end{pmatrix},$$

$$B_2 = \begin{pmatrix} 1 \\ 0 \end{pmatrix} G_{x1} = G_{x2} = \begin{pmatrix} b & 0 \\ 1 & d \end{pmatrix}, E = \begin{pmatrix} 1 & 0 \\ 0 & 1 \end{pmatrix}. \tag{36}$$

Part1: *Comparing Nonlinear T-S vs. Linear T-S*

To show the effectiveness of modeling a system using nonlinear Sugeno model, this system is modeled with a traditional Sugeno model with linear local subsystems as follows:

$$A_1 = \begin{pmatrix} a & -1 \\ 1 & c \end{pmatrix}, A_2 = \begin{pmatrix} a + 4b & -1 \\ 1 & c + 4d \end{pmatrix} A_3 = \begin{pmatrix} a + b & -1 \\ 1 & c \end{pmatrix},$$

$$A_4 = \begin{pmatrix} a + 5b & -1 \\ 1 & c + 4d \end{pmatrix} A_5 = \begin{pmatrix} a & -1 \\ 3 & c \end{pmatrix}, A_6 = \begin{pmatrix} a + 4b & -1 \\ 3 & c + 4d \end{pmatrix}$$

$$A_7 = \begin{pmatrix} a + b & -1 \\ 3 & c \end{pmatrix}, A_8 = \begin{pmatrix} a + 5b & -1 \\ 3 & c + 4d \end{pmatrix} \tag{37}$$

$$B_1 = \begin{pmatrix} 1 \\ -4 \end{pmatrix}, B_2 = \begin{pmatrix} 1 \\ 0 \end{pmatrix}, B_3 = \begin{pmatrix} 1 \\ -4 \end{pmatrix}, B_4 = \begin{pmatrix} 1 \\ 0 \end{pmatrix}$$

$$B_5 = \begin{pmatrix} 1 \\ -4 \end{pmatrix}, B_6 = \begin{pmatrix} 1 \\ 0 \end{pmatrix}, B_7 = \begin{pmatrix} 1 \\ -4 \end{pmatrix}, B_8 = \begin{pmatrix} 1 \\ 0 \end{pmatrix}.$$

Note that in this case, the system model and hence the controller, have eight rules. Moreover, note that x_1 is one of the premise variables now. Figure 1 compares the feasibility area of these models by changing parameters b and d. The

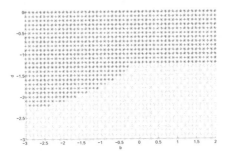

Fig. 1. Feasible area of Example 1 for parameters b and d: linear T-S [16] (Star) and non-linear T-S (Theorem 1) (Circle)

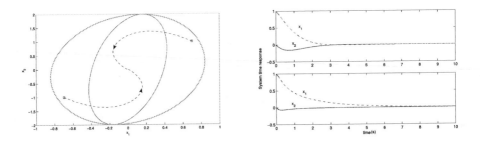

Fig. 2. Left: Stability area for Theorem 1 (blue line) and Theorem 2 (green line). Right: System Time response for Theorem 2 (Top) and Theorem 1 (Bottom).

simulation results show that the nonlinear Sugeno model can control a larger class of systems.

Part2: *Comparing Non-Quadratic Lyapunov Function vs. Quadratic one in Nonlinear T-S*
The convergence region is compared in Figure 2 for Theorem 2 ($\alpha = 0.5$) and Theorem 1. It should be mentioned that conditions of Theorem 3 have also been applied in this case. The dash line shows the state trajectories for some initial conditions for Theorem 3. Figure 2 also shows system time responses for Theorems 1 and 3. The decay rates are shown in Figure 3, which shows improvement of the decay rate by changing from quadratic (Corollary 1) to non-quadratic (Corollary 2) Lyapunov function. In addition to these benefits, the non-quadratic Lyapunov function increases the feasibility area. As an example, for $b = 0.1$ and $d = 1$, no result can be obtained for Theorem 1 while Theorem 2 provides a feasible solution with $\alpha = -1$. The Lyapunov function is also shown in Figure 3 to show its positivity while α is negative.

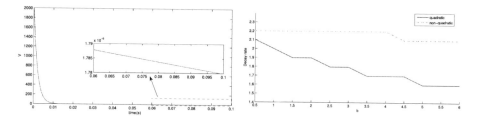

Fig. 3. Left: Non-quadratic Lyapunov function for Example 1 ($\alpha = -1$). Right: Decay rate (β) based on Corollary 1 (solid line) and Corollary 2 (dotted line).

5   Conclusion

In this paper, T-S model with nonlinear consequent part was considered to reduce the number of rules in a fuzzy system. In this model, each sub-system is supposed to be linear plus a sector-bounded nonlinearity. This may recall of a traditional linear T-S plus disturbance but a major difference is that the controller for these systems also includes the nonlinear term. This resulted in introducing a non-quadratic Lyapunov function considering the nonlinear term, which can improve the decay-rate and the Lyapunov level.

Acknwoledgement. Thanks our support CISIT (Intern. Campus on Safety and Intermodality in Transportation), Regional Delegation for Research and Technology, Ministry of Higher Education and Research, Region Nord Pas de Calais and CNRS.

References

1. Guerra, T.M., Bernal, M., Guelton, K., Labiod, S.: Non-quadratic local stabilization for continuous-time Takagi-Sugeno models. Fuzzy Sets and Systems 201, 40–54 (2012)
2. Abdelmalek, I., Golea, N., Hadjili, M.L.: A new fuzzy Lyapunov approach to non-quadratic stablization of Takagi-Sugeno fuzzy models. International Journal of Applied Mathematics and Computer Science 17(1), 39–51 (2007)
3. Bernal, M., Husek, P.: Non-quadratic performance design for Takagi-Sugeno fuzzy systems. International Journal of Applied Mathematics and Computer Science 15(3), 383–391 (2005)
4. Tanaka, K., Yoshida, H., Ohtake, H., Wang, H.O.: A sum-of-squares approach to modeling and control of nonlinear dynamical systems with polynomial fuzzy systems. IEEE Transaction on Fuzzy Systems 17(4), 911–922 (2009a)
5. Sala, A.: On the conservativeness of fuzzy and fuzzy-polynomial control of nonlinear systems. Annual Reviews in Control 33(1), 48–58 (2009)
6. Bernal, M., Sala, A., Jaadari, A., Guerra, T.M.: Stability analysis of polynomial fuzzy models via polynomial fuzzy lyapunov functions. Fuzzy Sets and Systems 185(1), 5–14 (2011)

7. Dong, J., Wang, Y., Yang, G.H.: Control synthesis of continuous-time t-s fuzzy systems with local nonlinear models. IEEE Transaction on System, Man and Cybernatics, B. Cybern. 39(5), 1245–1258 (2009)

8. Dong, J., Wang, Y., Yang, G.H.: Output feedback fuzzy controller design with local nonlinear feedback laws for discrete-time nonlinear systems. IEEE Transaction on System, Man and Cybernatics, B. Cybern. 40(6), 1447–1459 (2010)

9. Dong, J.: H_∞ and mixed H_2/H_∞ control of discrete-time T-S fuzzy systems with local nonlinear models. Fuzzy Sets and Systems 164(1), 1–24 (2011)

10. Tuan, H.D., Apkarian, P., Narikiyo, T., Yamamoto, Y.: Parameterized linear matrix inequality techniques in fuzzy control system design. IEEE Transactions on Fuzzy Systems 9(2), 324–332 (2001)

11. Henderson, H.V., Searle, S.R.: On deriving the inverse of a sum of matrices. SIAM Review 23(1), 53–60 (1981)

12. Khalil, H.: Nonlinear Systems. Prentice Hall (2002)

13. Boyd, S., Ghaoui, L.E., Feron, E., Balakrishnan, V.: Linear matrix inequalities in system and control theory. SIAM Studies in Applied Mathematics, Philadelphia (1994)

14. Löfberg, J.: Yalmip: A toolbox for modeling and optimization in MATLAB. In: IEEE International Symposium on Computer Aided Control Systems Design, Taipei, Taiwan (2004)

15. Bernal, M., Guerra, T.M.: Strategies to exploit non-quadratic local stability analysis. International Journal of Fuzzy Systems 14(3), 372–379 (2012)

16. Wang, H.O., Tanaka, K., Griffin, M.: An approach to fuzzy control of nonlinear systems: Stability and design issues. IEEE Trans. on Fuzzy Systems 4(1), 14–23 (1996)

Model Following Control of a Unicycle Mobile Robot via Dynamic Feedback Linearization Based on Piecewise Bilinear Models

Tadanari Taniguchi[1], Luka Eciolaza[2], and Michio Sugeno[2]

[1] IT Education Center, Tokai University,
Hiratsuka, Kanagawa 2591292, Japan
taniguchi@tokai-u.jp
[2] European Centre for Soft Computing,
33600 Mieres, Asturias, Spain
luka.eciolaza@softcomputing.es, michio.sugeno@gmail.com

Abstract. We propose a dynamic feedback linearization of a unicycle as a non-holonomic system with a piecewise bilinear (PB) model. Input-output (I/O) dynamic feedback linearization is applied to stabilize PB control system. We propose a method for nonlinear model following controller to the unicycle robot. Although the controller is simpler than the conventional I/O feedback linearization controller, the tracking performance based on PB model is the same as the conventional one. Examples are shown to confirm the feasibility of our proposals by computer simulations.

Keywords: Piecewise bilinear model, dynamic feedback linearization, non-holonomic system, model following control.

1 Introduction

This paper deals with the model following control of a unicycle robot using dynamic feedback linearization based on piecewise bilinear (PB) models. Wheeled mobile robots are completely controllable. However they cannot be stabilized to a desired position using time invariant continuous feedback control [1]. The wheeled mobile robot control systems have a non-holonomic constraint. Non-holonomic systems are much more difficult to control than holonomic ones. Many methods have been studied for the tracking control of unicycle robots. The backstepping control methods are proposed in (e.g. [2], [3]). The sliding mode control methods are proposed in (e.g., [4], [5]), and also the dynamic feedback linearization methods are in (e.g., [6], [7]). For non-holonomic robots, it is never possible to achieve exact linearization via static state feedback [8]. It is shown that the dynamic feedback linearization is an efficient design tool to solve the trajectory tracking and the setpoint regulation problem in [6], [7].

In this paper, we consider PB model as a piecewise approximation model of the unicycle robot dynamics. The model is built on hyper cubes partitioned in

A. Laurent et al. (Eds.): IPMU 2014, Part III, CCIS 444, pp. 539–548, 2014.

state space and is found to be bilinear (bi-affine) [9], so the model has simple nonlinearity. The model has the following features: 1) The PB model is derived from fuzzy if-then rules with singleton consequents. 2) It has a general approximation capability for nonlinear systems. 3) It is a piecewise nonlinear model and second simplest after the piecewise linear (PL) model. 4) It is continuous and fully parametric. The stabilizing conditions are represented by bilinear matrix inequalities (BMIs) [10], therefore, the design of the stabilizing controller needs long time computing. To overcome these difficulties, we have derived stabilizing conditions [11], [12], [13] based on feedback linearization, where [11] and [13] apply input-output linearization and [12] applies full-state linearization.

We propose a dynamic feedback linearization for PB control system and the model following control [14] for a unicycle robot system. The control system has the following features: 1) Only partial knowledge of vertices in piecewise regions is necessary, not overall knowledge of an objective plant. 2) These control systems are applicable to a wider class of nonlinear systems than conventional I/O linearization. 3) Although the controller is simpler than the conventional I/O feedback linearization controller, the tracking performance based on PB model is the same as the conventional one.

This paper is organized as follows. Section 2 introduces the canonical form of PB models. Section 3 presents PB modeling of the unicycle mobile robot. Section 4 proposes a model following controller design using dynamic feedback linearization based on PB model. Section 5 shows examples demonstrating the feasibility of the proposed methods. Section 6 summarizes conclusions.

2   Canonical Forms of Piecewise Bilinear Models

2.1   Open-Loop Systems

In this section, we introduce PB models suggested in [9]. We deal with the two-dimensional case without loss of generality. Define vector $d(\sigma, \tau)$ and rectangle $R_{\sigma\tau}$ in two-dimensional space as $d(\sigma, \tau) \equiv (d_1(\sigma), d_2(\tau))^T$,

$$R_{\sigma\tau} \equiv [d_1(\sigma), d_1(\sigma+1)] \times [d_2(\tau), d_2(\tau+1)]. \tag{1}$$

σ and τ are integers: $-\infty < \sigma, \tau < \infty$ where $d_1(\sigma) < d_1(\sigma+1), d_2(\tau) < d_2(\tau+1)$ and $d(0, 0) \equiv (d_1(0), d_2(0))^T$. Superscript T denotes a *transpose* operation.

For $x \in R_{\sigma\tau}$, the PB system is expressed as

$$\begin{cases} \dot{x} = \sum_{i=\sigma}^{\sigma+1} \sum_{j=\tau}^{\tau+1} \omega_1^i(x_1)\omega_2^j(x_2)f_o(i, j), \\ x = \sum_{i=\sigma}^{\sigma+1} \sum_{j=\tau}^{\tau+1} \omega_1^i(x_1)\omega_2^j(x_2)d(i, j), \end{cases} \tag{2}$$

where $f_o(i, j)$ is the vertex of nonlinear system $\dot{x} = f_o(x)$,

$$
\begin{cases}
\omega_1^\sigma(x_1) = (d_1(\sigma + 1) - x_1)/(d_1(\sigma + 1) - d_1(\sigma)), \\
\omega_1^{\sigma+1}(x_1) = (x_1 - d_1(\sigma))/(d_1(\sigma + 1) - d_1(\sigma)), \\
\omega_2^\tau(x_2) = (d_2(\tau + 1) - x_2)/(d_2(\tau + 1) - d_2(\tau)), \\
\omega_2^{\tau+1}(x_2) = (x_2 - d_2(\tau))/(d_2(\tau + 1) - d_2(\tau)),
\end{cases}
\tag{3}
$$

and $\omega_1^i(x_1), \omega_2^j(x_2) \in [0, 1]$. In the above, we assume $f(0,0) = 0$ and $d(0,0) = 0$ to guarantee $\dot{x} = 0$ for $x = 0$.

A key point in the system is that state variable x is also expressed by a convex combination of $d(i, j)$ for $\omega_1^i(x_1)$ and $\omega_2^j(x_2)$, just as in the case of $\dot{x}$. As seen in equation (3), x is located inside $R_{\sigma\tau}$ which is a rectangle: a hypercube in general. That is, the expression of x is polytopic with four vertices $d(i, j)$. The model of $\dot{x} = f(x)$ is built on a rectangle including x in state space, it is also polytopic with four vertices $f(i, j)$. We call this form of the canonical model (2) parametric expression.

Representing $\dot{x}$ with x in Eqs. (2) and (3), we obtain the state space expression of the model found to be bilinear (biaffine) [9], so the derived PB model has simple nonlinearity. In PL approximation, a PL model is built on simplexes partitioned in state space, triangles in the two-dimensional case. Note that any three points in three-dimensional space are spanned with an affine plane: $y = a + bx_1 + cx_2$. A PL model is continuous. It is, however, difficult to handle simplexes in the rectangular coordinate system.

2.2 Closed-Loop Systems

We consider a two-dimensional nonlinear control system.

$$
\begin{cases}
\dot{x} = f_o(x) + g_o(x)u(x), \\
y = h_o(x).
\end{cases}
\tag{4}
$$

The PB model (5) is constructed from a nonlinear system (4).

$$
\begin{cases}
\dot{x} = f(x) + g(x)u(x), \\
y = h(x),
\end{cases}
\tag{5}
$$

where

$$
\begin{cases}
f(x) = \sum_{i=\sigma}^{\sigma+1} \sum_{j=\tau}^{\tau+1} \omega_1^i(x_1)\omega_2^j(x_2)f_o(i, j), \quad g(x) = \sum_{i=\sigma}^{\sigma+1} \sum_{j=\tau}^{\tau+1} \omega_1^i(x_1)\omega_2^j(x_2)g_o(i, j), \\
h(x) = \sum_{i=\sigma}^{\sigma+1} \sum_{j=\tau}^{\tau+1} \omega_1^i(x_1)\omega_2^j(x_2)h_o(i, j), \quad x = \sum_{i=\sigma}^{\sigma+1} \sum_{j=\tau}^{\tau+1} \omega_1^i(x_1)\omega_2^j(x_2)d(i, j),
\end{cases}
\tag{6}
$$

and $f_o(i, j)$, $g_o(i, j)$, $h_o(i, j)$ and $d(i, j)$ are vertices of the nonlinear system (4). The modeling procedure in region $R_{\sigma\tau}$ is as follows:

1. Assign vertices $d(i,j)$ for $x_1 = d_1(\sigma)$, $d_1(\sigma + 1)$, $x_2 = d_2(\tau)$, $d_2(\tau + 1)$ of state vector x, then partition state space into piecewise regions (see Fig. 1).
2. Compute vertices $f_o(i,j)$, $g_o(i,j)$ and $h_o(i,j)$ in equation (6) by substituting values of $x_1 = d_1(\sigma)$, $d_1(\sigma + 1)$ and $x_2 = d_2(\tau)$, $d_2(\tau + 1)$ into original nonlinear functions $f_o(x)$, $g_o(x)$ and $h_o(x)$ in the system (4). Fig. 1 shows the expression of $f(x)$ and $x \in R_{\sigma\tau}$.

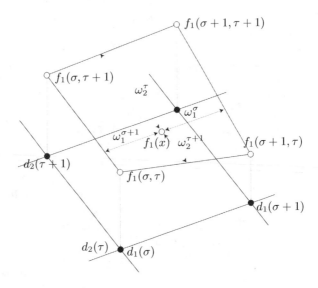

Fig. 1. Piecewise region ($f_1(x) = \sum_{i=\sigma}^{\sigma+1} \sum_{j=\tau}^{\tau+1} \omega_1^i \omega_2^j f_1(i,j)$, $x \in R_{\sigma\tau}$)

The overall PB model is obtained automatically when all vertices are assigned. Note that $f(x)$, $g(x)$ and $h(x)$ in the PB model coincide with those in the original system at vertices of all regions.

3   Dynamic Feedback Linearization of Unicycle Mobile Robot [7]

We consider a unicycle mobile robot model.

$$\begin{pmatrix} \dot{x} \\ \dot{y} \\ \dot{\theta} \end{pmatrix} = \begin{pmatrix} \cos\theta \\ \sin\theta \\ 0 \end{pmatrix} v + \begin{pmatrix} 0 \\ 0 \\ 1 \end{pmatrix} \omega, \tag{7}$$

where x and y are the position coordinates of the center of the wheel axis, θ is the angle between the center line of the vehicle and the x axis. The control inputs are the forward velocity v and the angular velocity ω. In this case, we consider $\eta = (x,y)^T$ as the output, the time derivative of η is calculated as

$$\dot{\eta} = \begin{pmatrix} \dot{x}_1 \\ \dot{x}_2 \end{pmatrix} = \begin{pmatrix} \cos x_3 & 0 \\ \sin x_3 & 0 \end{pmatrix} \begin{pmatrix} v \\ \omega \end{pmatrix},$$

where $x_1 = x$, $x_2 = y$, $x_3 = \theta$. The linearized system of (7) at any points $(x_1, x_2, x_3) = (x, y, \theta)$ is clearly not controllable and the only v affects $\dot{\eta}$. To proceed, we need to add an integrator of the forward velocity input v. The new input a is the acceleration of the unicycle. The time derivative of $\dot{\eta}$ is obtained as

$$\ddot{\eta} = a \begin{pmatrix} \cos x_3 \\ \sin x_3 \end{pmatrix} + v\omega \begin{pmatrix} -\sin x_3 \\ \cos x_3 \end{pmatrix} = \begin{pmatrix} \cos x_3 & -v \sin x_3 \\ \sin x_3 & v \cos x_3 \end{pmatrix} \begin{pmatrix} a \\ \omega \end{pmatrix}$$

and the matrix multiplying the modified input (a, ω) is nonsingular if $v \neq 0$. Since the modified input is obtained as (a, ω), the integrator with respect to the input v is added to the original input (v, ω). Finally, the stabilizing controller of the unicycle system (7) is presented as a dynamic feedback controller.

$$\begin{cases} \dot{v} = u_1 \cos x_3 + u_2 \sin x_3, \\ \omega = (-u_1 \sin x_3 + u_2 \cos x_3)/v, \end{cases} \tag{8}$$

where (u_1, u_2) is the linearized controller of the unicycle system (7).

4 PB Modeling and Model Following Controller Design of the Unicycle Model

4.1 PB Model of the Unicycle Model

We construct PB model of the unicycle system. The state space of θ in the unicycle model (7) is divided by the 9 vertices $x_3 \in \{-2\pi, -7\pi/4, \ldots, 2\pi\}$. The PB model is constructed as

$$\begin{pmatrix} \dot{x}_1 \\ \dot{x}_2 \\ \dot{x}_3 \end{pmatrix} = \begin{pmatrix} f_1 \\ f_2 \\ 0 \end{pmatrix} v + \begin{pmatrix} 0 \\ 0 \\ 1 \end{pmatrix} \omega, \tag{9}$$

$$f_1 = \sum_{i=\sigma}^{\sigma+1} w^i(x_3) f_1(d_3(i)), \quad f_2 = \sum_{i=\sigma}^{\sigma+1} w^i(x_3) f_2(d_3(i))$$

$$w^\sigma(x_3) = \frac{d_3(\sigma+1) - x_3}{d_3(\sigma+1) - d_3(\sigma)}, \quad w^{\sigma+1}(x_3) = \frac{x_3 - d_3(\sigma)}{d_3(\sigma+1) - d_3(\sigma)},$$

where $x_3 \in [d_3(\sigma), d_3(\sigma+1)]$, $w^i(x_3) \geq 0, i = \sigma, \sigma+1$ and $w^\sigma(x_3) + w^{\sigma+1}(x_3) = 1$. Table 1 shows a part of the PB model of the unicycle system. We can construct PB model with respect to x_3 and v. The PB model structure is independent of the vertex position v since v is the linear term. This paper constructs the PB model with respect to the nonlinear term of x_3.

Table 1. PB model of the unicycle system

	$d_3(1) =$ -2π	$d_3(2) =$ $-7\pi/4$	$d_3(3) =$ $-3\pi/2$	$d_3(4) =$ $-5\pi/4$	$d_3(5) =$ $-\pi$	$\cdots$ $\cdots$	$d_3(8) =$ $7\pi/4$	$d_3(9) =$ 2π
$f_1(d_3(i))$	1	0.707	0	-0.707	1	$\cdots$	0.707	1
$f_2(d_3(i))$	0	0.707	1	0.707	0	$\cdots$	-0.707	0

4.2   Model Following Controller Design Using Dynamic Feedback Linearization Based on PB Model

We define the output as $\eta = (x_1, x_2)^T$ in the same manner as the previous section, the time derivative of η is calculated as

$$\dot{\eta} = \begin{pmatrix} \dot{x}_1 \\ \dot{x}_2 \end{pmatrix} = \sum_{i=\sigma}^{\sigma+1} w^i(x_3) \begin{pmatrix} f_1(d_3(i)) & 0 \\ f_2(d_3(i)) & 0 \end{pmatrix} \begin{pmatrix} v \\ \omega \end{pmatrix} = v \sum_{i=\sigma}^{\sigma+1} w^i(x_3) \begin{pmatrix} f_1(d_3(i)) \\ f_2(d_3(i)) \end{pmatrix}$$

The time derivative of η doesn't contain the control input ω. An integrator on the forward velocity input v is considered as $\dot{v} = a$. We continue to calculate the time derivative of $\dot{\eta}$ then we get

$$\ddot{\eta} = a \sum_{i=\sigma}^{\sigma+1} w^i(x_3) \begin{pmatrix} f_1(d_3(i)) \\ f_2(d_3(i)) \end{pmatrix} + v\dot{x}_3 \frac{1}{\Delta d_3} \begin{pmatrix} f_1(d_3(\sigma+1)) - f_1(d_3(\sigma)) \\ f_2(d_3(\sigma+1)) - f_2(d_3(\sigma)) \end{pmatrix} = G \begin{pmatrix} a \\ \omega \end{pmatrix},$$

where $\omega = \dot{x}_3$, $\Delta d_3 = d_3(\sigma+1) - d_3(\sigma)$,

$$G = \begin{pmatrix} G_{11} & G_{12} \\ G_{21} & G_{22} \end{pmatrix} = \begin{pmatrix} \sum_{i=\sigma}^{\sigma+1} w^i(x_3) f_1(d_3(i)) & \dfrac{v(f_1(d_3(\sigma+1)) - f_1(d_3(\sigma)))}{\Delta d_3} \\ \sum_{i=\sigma}^{\sigma+1} w^i(x_3) f_2(d_3(i)) & \dfrac{v(f_2(d_3(\sigma+1)) - f_2(d_3(\sigma)))}{\Delta d_3} \end{pmatrix}.$$

If $v \neq 0$, we can derive the input (a, ω): $(a, \ \omega)^T = G^{-1}(u_1, \ u_2)^T$ so as to obtain $\ddot{\eta} = (u_1, \ u_2)^T$. Note that $v = 0$ means that the unicycle doesn't move.

The I/O linearized system can be formulated as

$$\begin{cases} \dot{z} = Az + Bu, \\ y = Cz, \end{cases} \tag{10}$$

where $z = (x_1, \ x_2, \ \dot{x}_1, \ \dot{x}_2) \in \Re^4$,

$$A = \begin{pmatrix} 0 & 0 & 1 & 0 \\ 0 & 0 & 0 & 1 \\ 0 & 0 & 0 & 0 \\ 0 & 0 & 0 & 0 \end{pmatrix}, \quad B = \begin{pmatrix} 0 & 0 \\ 0 & 0 \\ 1 & 0 \\ 0 & 1 \end{pmatrix}, C = \begin{pmatrix} 1 & 0 \\ 0 & 1 \\ 0 & 0 \\ 0 & 0 \end{pmatrix}^T.$$

In this case, the state space of the unicycle system is divided into 17 vertices. Therefore the system has 16 local PB models. Note that all the linearized systems of these PB models are the same as the linear system (10).

In the same manner of (8), the dynamic feedback linearizing controller of the PB system (9) is obtained

$$
\begin{cases}
\dot{v} = \dfrac{1}{G_{11}G_{22} - G_{12}G_{21}} \left(u_1 G_{22} - u_2 G_{12} \right), \\[4mm]
\omega = \dfrac{1}{G_{11}G_{22} - G_{12}G_{21}} \left(-u_1 G_{21} + u_2 G_{11} \right),
\end{cases}
\tag{11}
$$

$$
u = \left(u_1, u_2 \right)^T = -Fz.
\tag{12}
$$

The stabilizing linear controller $u = -Fz$ of the linearized system (10) can be obtained so that the transfer function $C(sI - A)^{-1}B$ is Hurwitz.

Note that the dynamic controller (11) based on PB model is simpler than the conventional one (8). Since the nonlinear terms of controller (11) are not the original nonlinear terms (e.g., $\sin x_3$, $\cos x_3$) but the piecewise approximation models.

4.3   Model Following Control for PB System

We apply a model reference control [14] to the unicycle model (7). Consider the following reference signal model

$$
\begin{cases}
\dot{x}_r = f_r \\
\eta_r = h_r
\end{cases}
$$

The controller is designed to make the error signal $e = (e_1, e_2)^T = \eta - \eta_r \to 0$ as $t \to \infty$. The time derivative of e is obtained as

$$
\dot{e} = \dot{\eta} - \dot{\eta}_r = v \sum_{i=\sigma}^{\sigma+1} w^i(x_3) \begin{pmatrix} f_1(d_3(i)) \\ f_2(d_3(i)) \end{pmatrix} - \begin{pmatrix} h_{r1} \\ h_{r2} \end{pmatrix}.
$$

Furthermore the time derivative of $\dot{e}$ is calculated as $\ddot{e} = \ddot{\eta} - \ddot{\eta}_r = -H + G(a, \, \omega)^T$, where $H = (H_1, \, H_2)^T = (\dot{h}_{r_1}, \, \dot{h}_{r_2})^T$, The model following controller

$$
\begin{cases}
\dot{v} = \dfrac{1}{G_{11}G_{22} - G_{12}G_{21}} \left((u_1 + H_1)G_{22} - (u_2 + H_2)G_{12} \right), \\[4mm]
\omega = \dfrac{1}{G_{11}G_{22} - G_{12}G_{21}} \left(-(u_1 + H_1)G_{21} + (u_2 + H_2)G_{11} \right)
\end{cases}
\tag{13}
$$

yields the linear closed loop system $\ddot{e} = u = (u_1, u_2)^T$. The linearized system and controller $u = -Fz$ are obtained in the same manners as (10) and (12). The coordinate transformation vector is $z = (e_1, \, e_2, \, \dot{e}_1 - h_{r_1}, \, \dot{e}_2 - h_{r_2})^T$. Note that the dynamic controller (13) based on PB model is simpler than the conventional one on the same reason of Section 4.2.

5   Simulation Results

We consider two nonlinear systems as the nonlinear reference models. Although the controller is simpler than the conventional I/O feedback linearization controller, the tracking performance based on PB model is the same as the conventional one. In addition, the controller is capable to use a nonlinear system with chaotic behavior as the reference model.

5.1   Eight-Shaped Reference Trajectory

Consider an eight-shaped reference trajectory as the reference model.

$$\begin{pmatrix} x_{r_1} \\ x_{r_2} \end{pmatrix} = \begin{pmatrix} \sin \frac{t}{10} \\ \sin \frac{t}{20} \end{pmatrix} \tag{14}$$

The feedback gain is calculated as

$$F = \begin{pmatrix} 1 & 0 & 1.7321 & 0 \\ 0 & 1 & 0 & 1.7321 \end{pmatrix},$$

which stabilizes the linearized system (10). To substitute the feedback gain for (13), we get the model following controller. The initial positions are set at $(x, y) = (-1, 0)$ and $(x_r, y_r) = (0, 0)$. Fig. 2 shows the simulation result. The solid line is the reference signal and the dotted line is the signal of PB model.

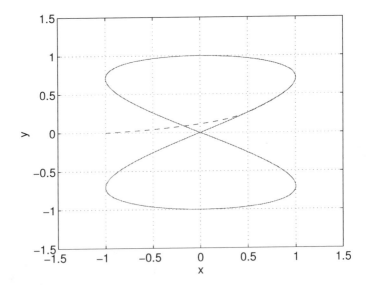

Fig. 2. Eight-shaped reference trajectory

5.2   Van der Pol Oscillator

Consider the following nonlinear system as a reference model.

$$\begin{pmatrix} \dot{x}_{r_1} \\ \dot{x}_{r_2} \end{pmatrix} = \begin{pmatrix} x_{r_2} \\ \alpha(1 - x_{r_1}^2)x_{r_2} - x_{r_1} \end{pmatrix} \tag{15}$$

where $\alpha = 2$ is the constant. The feedback gain F is the same as the previous example. We select the initial positions $(x, y) = (1, 1)$ and $(x_r, y_r) = (1, 1)$. Fig. 3 shows the trajectories of this simulation result. The solid line is the reference signal and the dotted line is the signal of PB model.

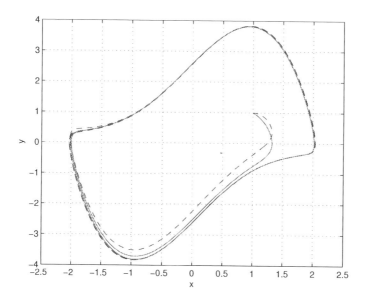

Fig. 3. Van del pol reference trajectory

6   Conclusions

We have proposed a dynamic feedback linearization of a unicycle as a non-holonomic system with a PB model. The approximated model is fully parametric. I/O dynamic feedback linearization is applied to stabilize PB control system. PB modeling with feedback linearization is a very powerful tool for analyzing and synthesizing nonlinear control systems. We have proposed a method for nonlinear model following controller to the unicycle robot. Although the controller is simpler than the conventional I/O feedback linearization controller, the tracking performance based on PB model is the same as the conventional one. Examples have been shown to confirm the feasibility of our proposals by computer simulations.

Acknowledgments. This project was supported by a URP grant from the Ford Motor Company, which the authors thankfully acknowledge. The authors also wish to thank Dr. Dimitar Filev and Dr. Yan Wang of Ford for their invaluable comments and discussion. This work was supported by Grant-in-Aid for Scientific Research (C:26330285) of Japan Society for the Promotion of Science.

References

1. d'Andréa-Novel, B., Bastin, G., Campion, G.: Modeling and control of non holonomic wheeled mobile robots. In: The 1991 IEEE International Conference on Robotics and Automation, pp. 1130–1135 (1991)
2. Fierro, R., Lewis, F.L.: Control of a nonholonomic mobile robot: backstepping kinematics into dynamics. In: The 34th Conference on Decision and Control, pp. 3805–3810 (1995)
3. Lee, T.C., Song, K.T., Lee, C.H., Teng, C.C.: Tracking control of unicycle-modeled mobile robots using a saturation feedback controller. IEEE Transactions on Control Systems Technology 9(2), 305–318 (2001)
4. Guldner, J., Utkin, V.I.: Stabilization of non-holonomic mobile robots using lyapunov functions for navigation and sliding mode control. In: The 33rd Conference on Decision and Control, pp. 2967–2972 (1994)
5. Yang, J., Kim, J.: Sliding mode control for trajectory tracking of nonholonomic wheeled mobile robots. IEEE Transactions on Robotics and Automation, 578–587 (1999)
6. d'Andréa-Novel, B., Bastin, G., Campion, G.: Dynamic feedback linearization of nonholonomic wheeled mobile robot. In: The 1992 IEEE International Conference on Robotics and Automation, pp. 2527–2531 (1992)
7. Oriolo, G., Luca, A.D., Vendittelli, M.: WMR control via dynamic feedback linearization: Design, implementation, and experimental validation. IEEE Transaction on Control System Technology 10(6), 835–852 (2002)
8. Luca, A.D., Oriolo, G., Vendittelli, M.: Stabilization of the unicycle via dynamic feedback linearization. In: The 6th IFAC Symposium on Robot Control (2000)
9. Sugeno, M.: On stability of fuzzy systems expressed by fuzzy rules with singleton consequents. IEEE Trans. Fuzzy Syst. 7(2), 201–224 (1999)
10. Goh, K.C., Safonov, M.G., Papavassilopoulos, G.P.: A global optimization approach for the BMI problem. In: Proc. the 33rd IEEE CDC, pp. 2009–2014 (1994)
11. Taniguchi, T., Sugeno, M.: Piecewise bilinear system control based on full-state feedback linearization. In: SCIS & ISIS 2010, pp. 1591–1596 (2010)
12. Taniguchi, T., Sugeno, M.: Stabilization of nonlinear systems with piecewise bilinear models derived from fuzzy if-then rules with singletons. In: FUZZ-IEEE 2010, pp. 2926–2931 (2010)
13. Taniguchi, T., Sugeno, M.: Design of LUT-controllers for nonlinear systems with PB models based on I/O linearization. In: FUZZ-IEEE 2012, pp. 997–1022 (2012)
14. Taniguchi, T., Eciolaza, L., Sugeno, M.: Look-Up-Table controller design for nonlinear servo systems with piecewise bilinear models. In: FUZZ-IEEE 2013 (2013)

An Anti-windup Scheme for PB Based FEL

Luka Eciolaza[1], Tadanari Taniguchi[2], and Michio Sugeno[1]

[1] European Centre for Soft Computing, Mieres, Asturias, Spain
luka.eciolaza@softcomputing.es, michio.sugeno@gmail.com
[2] Tokai University, Hiratsuka, Kanagawa, 2591292, Japan
taniguchi@tokai-u.jp

Abstract. Anti-windup methods deal with saturation problems in systems with known actuator limits. Part of the controller design is devoted to constraint handling in order to avoid pernicious behaviour of the systems. Feedback error learning is an on-line learning strategy of inverse dynamics. It sequentially acquires an inverse model of a plant through feedback control actions. In previous works we proposed an approach to implement the FEL control scheme through Piecewise Bilinear models. In this paper, a PB based FEL implementation scheme is proposed, where saturations provoked by the input constraints are taken into account for the inverse model learning algorithm.

1 Introduction

Actuators which deliver the control signal in physical applications are in the majority of cases subject to limits in their magnitude or rate. Apart from restricting the achievable performance, if these limits are not treated carefully, considering them appropriatedly in the controller design, they could result in a pernicious behaviour of the system.

The study of anti-windup probably began in industry where practitioners noticed performance degradation in systems where saturation occurred [1]. The term *windup* refers to the saturation in systems with integral controllers. During saturation, the integrator's value builds up of charge and the subsequent dissipation of this charge causes long settling times and excessive overshoot, degrading substantially system's performance. Modifications to the controller which avoided this charge build-up were often termed as *anti-windup*.

There are two main approaches to avoid saturation problems in systems which are known to have actuator limits [1]. The *one-step* approach, where a single controller attempts to ensure that all nominal performace specifications are met and also handles the saturation constraits imposed by the actuators. This approach has often been criticized due to its conservatism.

On the other hand, the *anti-windup* approach separates the controller design in two steps where one part is devoted to achieving nominal performance and the other part is devoted to constraint handling. A controller which does not explicitly take into account the saturation constraints is first designed, using standard design tools. Then, the *anti-windup* compensator is designed to handle the saturation constraints.

In the conventional form of anti-windup, no restriction is placed upon the nominal controller design and, assuming no saturation is encountered, this controller alone dictates the behaviour of the linear closed loop. It is only when saturation is encountered

A. Laurent et al. (Eds.): IPMU 2014, Part III, CCIS 444, pp. 549–558, 2014.

that the anti-windup compensator becomes active and acts to modify the closed-loops behaviour such that it is more resilient to saturation.

In this paper, we want to analyze the influence of input constraints for the implementation of Feedback Error Learning (FEL) scheme. We will propose a new inverse model learning approach where the saturation will be taken into account.

This article follows on with our previous work [2] where we proposed the approach to implement FEL based on Piecewise Bilinear (PB) models.

FEL scheme, proposed by [3], was originally inspired on biological control systems which usually are systems with time delays and strong nonlinearities. Fast and coordinated limb movements cannot be executed solely under feedback control, since biological feedback loops are slow and have small gains. Thus, the brain uses inverse dynamical models in order to calculate necessary feed-forward commands from desired trajectory information, see [4]. FEL and the inverse model identification scheme represent an important role for the quick and smooth motions of the limbs in human motor control. The FEL has the feature that the learning and control can be performed simultaneously.

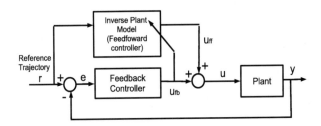

Fig. 1. Feedback error learning architecture

A typical schematic of the FEL control system is shown in Fig. 1. It consists of the combination of a feedback controller that ensures the stability of the system and an adaptive feedforward controller that improves control performance. An inverse model as a feedforward controller is learned by FEL so as to minimize the square error between u_{ff}, the output of a feedforward controller and u_0, the ideal feedforward control based on an inverse model. Using a conventional steepest descent method for the minimization, the parameters of a feedforward controller can be sequentially updated in proportion to $(u_{ff} - u_0)$. However since the ideal u_0 is unknown, this error signal is approximated with u_{fb} called feedback error in FEL. As it is shown in Fig. 1, u (the control input to a plant) is equal to $u_{ff} + u_{fb}$, and $u_{ff} - u_0$ is replaced with $u_{ff} - u = u_{fb}$. After learning is complete, i.e., $y(t) = r(t)$, then u_{fb} tends to zero and feedback control is partially replaced by feedforward control. Thus, $u = u_{ff}$ the feedforward controller should be serving as the inverse of the original plant.

In [2] we concluded the PB model represents the most convenient model to implement FEL with regard to nonlinear modeling, control objective and on-line learning capability. Previously the FEL scheme was normally implemente via neural networks [5], [6], [7], however they are not convenient for control purpose. On the other hand,

FEL was analyzed by control system community, [8], [9], [10], however these analysis were only made for linear systems.

The PB model, proposed in [11], is very effective for modeling and control of non-linear systems. It is a fully parametric model to represent Linear/Nonlinear systems. PB model has a big number of parameters to tune, but [12] showed that the global tuning of PB models can be made in a very simple and efficient way. PB models are very convenient for control purpose as they are interpretable (for instance as LUT), easy to compute and simple to handle.

The rest of the paper is organized as follows. Sections 2 introduces how we can implement FEL using PB models. Section 3 presents the approacht proposed in this paper to implement FEL with Anti-Windup. Section 4 presents some simulations and section 6 concludes the paper with some discussion and final remarks of the results.

2   PB Model-Based FEL

As is explained in the previous section, Fig. 1 illustrates the feedback error learning architecture. It consists of an objective plant to be controlled with a feedback controller and, in addition, a feedforward controller to be learned by FEL. The objective of control is to minimize the error e between the reference signal r and the plant output y.

In the FEL approach, the feedback controller action is converted into motor command error and used to learn the feedforward controller. By FEL, eventually the feedback control is replaced by feedforward control. FEL includes the features of simultaneous learning and control, making it an adaptive controller.

2.1   PB Models

The PB model is a fully parametric model to represent Linear/Nonlinear systems. It is designed to be easily applicable for control purpose. In the model, bilinear functions are used to regionally approximate any given function. The obtained model is built on piecewise rectangular regions, and each region is defined by four vertices partitioning the state space. A bilinear function is a nonlinear function of the form $y = a + bx_1 + cx_2 + dx_1x_2$, where any four points in the three dimensional space are spanned with a bi-affine plane.

If a general case of an affine two-dimensional nonlinear control system is considered,

$$\begin{cases} \dot{x}_1 = & f_1(x_1, x_2) \\ \dot{x}_2 = f_2(x_1, x_2) + g(x_1, x_2) \cdot r \\ y = & h(x_1, x_2) \end{cases} \tag{1}$$

where r is the input (control, reference or both). For the PB representation of a state-space equation, a coordinate vector $d(\sigma, \tau)$ of the state space and a rectangle $\Re_{ij}$ must be defined as,

$$d(i, j) \equiv (d_1(i), d_2(j))^T \tag{2}$$

$$R_{ij} \equiv [d_1(i), d_1(i+1)] \times [d_2(j), d_2(j+1)] \tag{3}$$

where $i \in (1, \dots, n_1)$ and $j \in (1, \dots, n_2)$ are integers, and $d_1(i) < d_1(i + 1)$, $d_2(j) < d_2(j + 1)$. The PB models are formed by matrices of size $(n_1 \times n_2)$, where n_1 and n_2 represent the number of partitions of dimension x_1 and x_2 respectively. Each value in the matrix is referred to as a vertex in the PB model. The operational region of the system is divided into $(n_1 - 1 \times n_2 - 1)$ piecewise regions that are analyzed independently.

The PB model was originally derived from a set of fuzzy if-then rules with singleton consequents [11] such that

$$\text{if } x \text{ is } W^{\sigma\tau}, \text{ then } \dot{x} \text{ is } f(\sigma, \tau) \tag{4}$$

which in a two-dimensional case, $x \in \Re^2$ is a state vector, $W^{\sigma\tau} = (w_1^\sigma(x_1), w_2^\tau(x_2))^T$ is a membership function vector, $f(\sigma, \tau) = (f_1(\sigma, \tau), f_2(\sigma, \tau))^T \in \Re$ is a singleton consequent vector, and $\sigma, \tau \in Z$ are integers ($1 \le \sigma \le n_1, 1 \le \tau \le n_2$) defined by,

$$\sigma(x_1) = d_1(\max(i)) \text{ where } d_1(i) \le x_1, \tag{5}$$

$$\tau(x_2) = d_2(\max(j)) \text{ where } d_2(j) \le x_2. \tag{6}$$

The superscript T denotes transpose operation.

For $x \in \Re_{\sigma\tau}$, the PB model that approximates $f_1(x_1, x_2)$ in (1) is expressed as,

$$f_1(x_1, x_2) = \sum_{i=\sigma}^{\sigma+1} \sum_{j=\tau}^{\tau+1} w_1^i(x_1) w_2^j(x_2) f_1(i, j), \tag{7}$$

where

$$\begin{cases} w_1^\sigma(x_1) & = 1 - \alpha, \\ w_1^{\sigma+1}(x_1) & = \alpha, \\ w_2^\tau(x_2) & = 1 - \beta, \\ w_2^{\tau+1}(x_2) & = \beta, \end{cases} \tag{8}$$

and

$$\alpha = \frac{x_1 - d_1(\sigma)}{d_1(\sigma + 1) - d_1(\sigma)} \tag{9}$$

$$\beta = \frac{x_2 - d_2(\tau)}{d_2(\tau + 1) - d_2(\tau)} \tag{10}$$

in which case $w_1^i, w_2^j \in [0, 1]$. PB models representing $f_2(x_1, x_2)$, $g(x_1, x_2)$ and $h(x_1, x_2)$ in (1) have the same form. In every region of the PB models, i.e.: $f_1(x_1, x_2)$, the values are computed through bilinear interpolation of the corresponding four vertexes.

2.2   FEL: On-Line Sequential Learning Algorithm

Let us express a plant model for the sake of simplicity as $y = p(u)$ where y is output and u control input. We also denote its inverse in a similar manner as $u = p^{-1}(y)$, assuming that a plant is invertible. We note that an inverse model used in FEL is usually written as $u_{ff} = p^{-1}(r)$. We consider a feedforward controller expressed as $u_{ff} = p^{-1}(r, \dot{r})$, where r is a desired output y_d and $\dot{r}$ is $\dot{y}_d$.

In what follows, we deal with a feedforward controller with two inputs r, $\dot{r}$ and single output u_{ff}. We will initially assume that an objective plant is unknown and its feedback controller is given. In a realistic nonlinear system control scenario, both plant identification and controller design could also be performed through PB models.

Let u_0 be an ideal feedforward control based on a pseudo-inverse model. We design a feedforward controller so that $u_{ff} = u_0$. That is, we learn u_{ff}, i.e., identify the feedforward controller parameters, to minimize the performance index

$$I = \frac{(u_{ff} - u_0)^2}{2} \tag{11}$$

where the PB representation of the feedforward controller is,

$$u_{ff} = p^{-1}(r, \dot{r}) = \sum_{i=\sigma}^{\sigma+1} \sum_{j=\tau}^{\tau+1} w_1^i(r) w_2^j(\dot{r}) V(i, j) \tag{12}$$

I can be sequentially minimized using the derivative of (11)

$$\frac{\partial I}{\partial V} = \frac{\partial u_{ff}}{\partial V}(u_{ff} - u_0). \tag{13}$$

However, the error $(u_{ff} - u_0)$ is not available since u_0 is unknown. Therefore Kawato [3] suggested to use u_{fb} for $(u_{ff} - u_0)$ since $u = u_{fb} + u_{ff}$. This is why u_{fb} is called a feedback error playing a role as the error $(u_{ff} - u_0)$. FEL is a learning scheme based on a signal u_{fb}, a feedback error signal.

Then we have

$$\frac{\partial I}{\partial V} = \frac{\partial u_{ff}}{\partial V} u_{fb} = \frac{\partial p^{-1}(r, \dot{r})}{\partial V} u_{fb} \tag{14}$$

The sequential learning of each vertex of a region is made using the following algorithm:

$$V_{new(i,j)} = V_{old(i,j)} - \delta \frac{\partial u_{ff}}{\partial V(i, j)} u_{fb} \tag{15}$$

where δ is an adjustable parameter as a learning rate. This is the conventional steepest descent algorithm to minimize a performance index. If learning is successfully completed, i.e., $V_{new} = V_{old}$, then u_{fb} must become zero, and only u_{ff} works.

In the case of a two dimensional PB model, if we develop (12), with (8) (9) and (10), we have:

$$u_{ff} = p^{-1}(r, \dot{r}) = (1 - \alpha)(1 - \beta)V_{(\sigma,\tau)} + (1 - \alpha)\beta V_{(\sigma,\tau+1)}$$
$$+ \alpha(1 - \beta)V_{(\sigma+1,\tau)} + \alpha\beta V_{(\sigma+1,\tau+1)}. \tag{16}$$

$(V_{(\sigma,\tau)}, V_{(\sigma,\tau+1)}, V_{(\sigma+1,\tau)}, V_{(\sigma+1,\tau+1)})$ refer to the values of 4 vertexes of a region and as the function is linear, the calculation of partial derivatives (∇p^{-1}) is straightforward.

3 Anti-windup

Anti-windup techniques are used to deal with stability and performance degradation problems for systems with saturated inputs. There have bee several studies around this phenomenon as explained in [1], [13]. Anti-windup techniques have evolved from many diverse sources. Modern techniques provide LMI conditions for the synthesis of anti-windup compensators with which guarantees on stability can be made.

In this paper we focus on more conventional anti-windup methods. We want to show that PB model based FEL implementation can be made even for systems with actuator constraints. The saturations provoked by these constraints are taken into account for the inverse model learning algorithm.

Windup was initially observed in PID controllers designed for SISO control systems with a saturated actuator. If the error is positive for a substantial time, the control signal u_s gets saturated at the high limit u_{max}.

$$u_s = S(u) = \begin{cases} u_{min}, & if u < u_{min} \\ u, & if u_{min} \le u \le u_{max} \\ u_{max}, & if u > u_{max} \end{cases} \qquad (17)$$

The integrator continues accumulation the error while the error remains positive and actuator saturated. The control signal remains saturated at this point, due to the large integral value, until the error becomes negative and remains negative for a sufficiently long time to make the integral value small again. This effect causes large overshoots and sometimes instability in the output.

To avoid windup, the actuators output u_s is measured and en extra feedback path is provided in the controller. The difference between the controller output u and the acuator output u_s is used as error signal to correct the windup effect $e_{aw} = u_s - u$.

In the *conventional anti-windup* scheme, shown in Fig. 2, the compensation is provided by feeding the difference $u_s - u$ through a high gain X to the controller input e. When the actuator saturates, the feedback signal $u_s - u$ attemps to drive the difference to zero. This prevents the integrator from winding up.

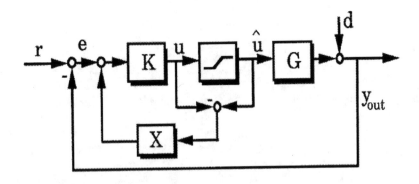

Fig. 2. Conventional Anti-windup

4   FEL with Anti-windup for Actuator Constraints

Based on the conventional anti-windup approach, we have implemented the FEL control scheme for a plant with input constraints as shown in Fig. 3.

From Fig. 3, we have $u_s = S(u)$ and $u = u_{fb} + u_{ff}$, where S is a saturator to model input constraints, u an input to the saturator, u_s its output, u_{fb} feedback control, and u_{ff} feedforward control.

Suppose that u_s is an ideal feedforward control which should be realized by the inverse model. Then an error signal to learn the inverse model can be chosen as,

$$u_e = u_s - u_{ff}, \tag{18}$$

where u_{ff} is a current feedforward control based on an inverse model under learning.

Since $u = u_{fb} + u_{ff}$, we obtain that

$$u_e = u_s - u_{ff} \Leftrightarrow u_e = u_s - (u - u_{fb}) \Leftrightarrow u_e = u_{fb} + u_s - u. \tag{19}$$

The signal (u_{fb}) used for learning the inverse model in (15) will be modified to

$$V_{new(i,j)} = V_{old(i,j)} - \delta \frac{\partial u_{ff}}{\partial V(i,j)} (u_{fb} + u_s - u) \tag{20}$$

When actuator is saturated the inverse model output (u_{ff}) is limited to the saturation limits (u_{min}, u_{max}).

We note that if there is no input contraint, then we have, since $u_s = u$,

$$u_e = u_{fb} \tag{21}$$

which is the conventional error signal used in FEL. And so, when controller is performing within actuator saturation limits, the FEL will work as explained in seciton 2.

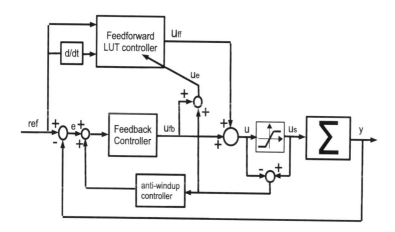

Fig. 3. Implemented FEL with anti-windup scheme

5   Simulations

For the simulation of the FEL with Anti-Windup scheme proposed in previous section and shown in Fig. 3, we have used a linear second order plant controlled by a PID.

The plant P is described by the state-space matrices (A_p, B_p, C_p, D_p)

$$A_p = \begin{bmatrix} 0 & 1 \\ -10 & -10 \end{bmatrix}, B_p = \begin{bmatrix} 0 \\ 10 \end{bmatrix}, C_p = \begin{bmatrix} 1 & 0 \end{bmatrix}, D_p = \begin{bmatrix} 0 \end{bmatrix}. \tag{22}$$

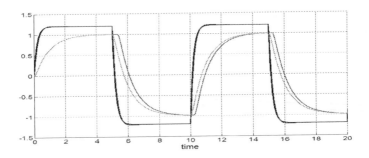

Fig. 4. Reference in black. Blue: Plant value without limitations and a PID controller. Red: Plant Saturated without Anti-Windup action. Green: Plant Saturated with Anti-Windup action.

The PID controler implemented to control the plant has the following constants, $K_P = 20$, $K_I = 30$ $K_D = 2.5$. K_I is very big in order to make sure that windup effect occurs when actuator constraints are used.

Fig. 4 shows the performance of the plant with only the feedback loop controlled by the PID. The reference signal, in black, is superposed by the plant output with no constraints, in blue. In this case tracking is perfect. After that an acutator contraint is set to 1.2. For this case, the red line, shows the performance of the plant without the Anti-Windup scheme. In this case the plant can not meet the reference signal, and the windup effect can be observed clearly each time reference signal has a sudden change. Finally the green line shows the performance of the plant with the mentioned actuator constraints and Anti-Windup scheme implemented. In this case the performance clearly improves with respect to the red line and the windup effect is avoided. The Anti-Windup gain used is $K_{AW} = 5$.

In Fig. 5, the performance of the plant implementing the FEL scheme is shown without actuator constraints. A sinusoidal reference signal with variable amplitude is used. It can be observed that u_{fb} action tends to zero while the inverse model of the plant is learned. In this case, the learning rate has been set to $\delta = 0.16 \times sr$, where sr refers to the sample rate.

Finally, in Fig. 6, the performance of the plant controlled under the FEL scheme with actuator constraints is shown. In this case we consider the inverse model has been already learned and the main control action u is defined by u_{ff} while u_{fb} is close to

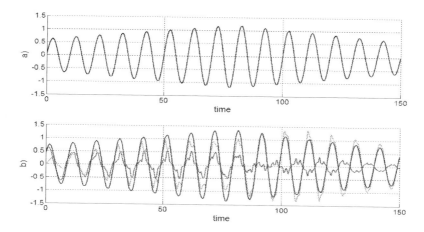

Fig. 5. FEL scheme with no saturation. a) refernce signal, in black, and the plant output, in blue, are superposed. b) Feedback action of controller in red, the feedforward in green, and the total $u = u_{fb} + u_{ff}$ in blue.

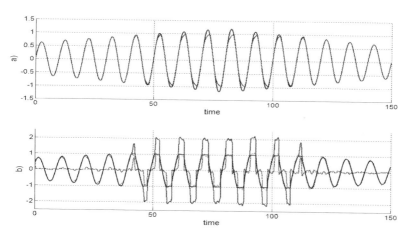

Fig. 6. FEL with actuator constraints and Anti-Windup action, once after the inverse model learning is done. a) refernce signal, in black, and the plant output, in blue. b) Feedback action of controller in red, the feedforward in green, and the total $u = u_{fb} + u_{ff}$ in blue.

zero. Using the same, sinusoidal reference signal with variable amplitude, it can be observed that at some point the actuator gets saturated and the plant can not meet the refernce input. At this point, u_{fb} gets to saturation limits and the learning of the inverse model is limited, preventing u_{ff} from windup effect. However, when the reference signal gets back to achievable values, the plant can track the reference almost perfectly and main control action u is once again defined by u_{ff} while u_{fb} is close to zero.

6   Conclusions

In this paper, a PB model based FEL implementation scheme is proposed, where saturations provoked by the actuator constraints are taken into account for the inverse model learning algorithm. The windup effect when learning the inverse model is avoided using the difference between the controller output u and the constrained actuator output u_s in order to limit the inverse model learning.

Acknowledgment. The authors acknowledge Ford Motor Company for their financial support this work under the URP program.

References

1. Tarbouriech, S., Turner, M.: Anti-windup design: an overview of some recent advances and open problems. IET Control Theory & Applications 3(1), 1–19 (2009)
2. Eciolaza, L., Taniguchi, T., Sugeno, M., Filev, D., Wang, Y.: Piecewise bilinear models for feedback error learning: On-line feedforward controller design. In: IEEE International Conference on Fuzzy Systems (FuzzIEEE). IEEE (2013)
3. Kawato, M., Furukawa, K., Suzuki, R.: A hierarchical neural-network model for control and learning of voluntary movement. Biological Cybernetics 57(3), 169–185 (1987)
4. Kawato, M.: Internal models for motor control and trajectory planning. Current Opinion in Neurobiology 9(6), 718–727 (1999)
5. Jung, S., Kim, S.: Control experiment of a wheel-driven mobile inverted pendulum using neural network. IEEE Transactions on Control Systems Technology 16(2), 297–303 (2008)
6. Ruan, X., Chen, J.: On-line nnac for a balancing two-wheeled robot using feedback-error-learning on the neurophysiological mechanism. Journal of Computers 6(3), 489–496 (2011)
7. Nakamura, Y., Morimoto, K., Wakui, S.: Experimental validation of control for a positioning stage by feedback error learning. In: 2011 International Conference on Advanced Mechatronic Systems (ICAMechS), pp. 11–16. IEEE (2011)
8. Miyamura, A., Kimura, H.: Stability of feedback error learning scheme. Systems & Control Letters 45(4), 303–316 (2002)
9. Miyamura Ideta, A.: Stability of feedback error learning method with time delay. Neurocomputing 69(13), 1645–1654 (2006)
10. AlAli, B., Hirata, K., Sugimoto, K.: Generalization of feedback error learning (fel) to strictly proper mimo systems. In: American Control Conference, 6 p. IEEE (2006)
11. Sugeno, M.: On stability of fuzzy systems expressed by fuzzy rules with singleton consequents. IEEE Transactions on Fuzzy Systems 7(2), 201–224 (1999)
12. Eciolaza, L., Sugeno, M.: On-line design of lut controllers based on desired closed loop plant: Vertex placement principle. In: IEEE International Conference on Fuzzy Systems (FuzzIEEE), pp. 1–8. IEEE (2012)
13. Kothare, M.V., Campo, P.J., Morari, M., Nett, C.N.: A unified framework for the study of anti-windup designs. Automatica 30(12), 1869–1883 (1994)

Some Aggregation Operators
of General Misinformation

Doretta Vivona and Maria Divari

Facoltá di Ingegneria
"Sapienza" - Università di Roma
Roma, Italy
vivona@uniroma1.it, maria.divari@alice.it

Abstract. On setting of general information (i.e.without probability) we define, by axiomatic way, general misinformation of an event. We give a class of measures of misinformation, solving a sistem of functional equations, given by the properties of the misinformation.

Then, we propose some aggregation operators of these general misinformation.

Keywords: Information, functional equation, aggegation operator.

1 Introduction

On setting of the axiomatic theory of general information (i.e.without probability) [4, 5, 6, 7, 8, 9], we define a measure of misinformation which we link to every information of a crisp event.

For example, misinformation could be given by a bad reliability of the source or by a poor ability to take into a good testing the received information.

This misinformation is called by us *general*, because it is defined without using probability measure.

The paper develops in the following way: in Sect.2 we recall some preliminaires, in Sect.3 we present the axiomatic definition of misinformation without probability and we propose a class of its measures.

In Sect.4 we propose some aggregation operators for the introduced measures. Sect.5 is devoted to the conclusion.

2 Preliminaires

Let X be an abstract space and $\mathcal{A}$ a $\sigma-$algebra of all subsets of X, such that $(X, \mathcal{A})$ is measurable.

We recall that measure $J(\cdot)$ of general information (i.e. without probability) is a mapping $J(\cdot) : \mathcal{A} \to [0, +\infty]$ such that $\forall\, A, A' \in \mathcal{A}$:

(i) $A' \supset A \Rightarrow J(A') \leq J(A)$,

(ii) $J(\emptyset) = +\infty$,

(iii) $J(X) = 0$.

A. Laurent et al. (Eds.): IPMU 2014, Part III, CCIS 444, pp. 559–564, 2014.

3   General Misinformation

Now, we present the following

Definition 1 *Fixed any measure of general information J, measure of general misinformation M_J is a mapping*

$$M_J(\cdot) : \mathcal{A} \to [0, +\infty]$$

such that $\forall\ A, A' \in \mathcal{A}$:
   (j) $A' \supset A \Rightarrow M_J(A') \le M_J(A)$,
   (jj) $M_J(\emptyset) = +\infty$,
   (jjj) $M_J(X) = 0$.

We suppose that misinformation of an event $A \in \mathcal{A}$ depends on measure of general information $J(A)$ and on measure of reliability of the source of information $\rho(A)$ and our capacity of take into a good testing the received information $\tau(A)$. For the previuos functions ρ and τ, we propose that

$$\rho : \mathcal{A} \to [0, +\infty]   ,$$

such that:
   (a) $A' \supset A \Rightarrow \rho(A') \le \rho(A)$, as $J(A') \le J(A)$,
   (b) $\rho(\emptyset) = +\infty$,
   (c) $\rho(X) = 0$,

   and

$$\tau : \mathcal{A} \to [0, +\infty]   ,$$

such that
   (a') $A' \supset A \Rightarrow \tau(A') \le \tau(A)$, as $J(A') \le J(A)$,
   (b') $\tau(\emptyset) = +\infty$,
   (c') $\tau(X) = 0$.

Now, we are expressing $M_J(A)$ as a function Φ of $J(A)$, $\rho(A)$, $\tau(A)$, in the following way:

$$M_J(A) = \Phi\left(J(A),\ \rho(A),\ \tau(A) \right) \tag{1}$$

where $\Phi : H \to [0, +\infty]$ with
   $H = \left\{ (x, y, z) : x = J(A), y = \rho(A), z = \tau(A),\ x, y, z \in [0, +\infty] \right\}$.
   Putting $J(A) = x, J(A) = x', \rho(A) = y, \rho(A') = y', \tau(A) = z, \tau(A') = z'$, the (1) becomes

$$M_J(A) = \Phi\left(x, y, z \right). \tag{2}$$

The properties of ρ and τ above and (2) take to the following:

$$\begin{cases} (I) \ \Phi(x, y, z) \leq \Phi(x', y', z') \\ \qquad\qquad \forall \ x \leq x', \ y \leq y', z \leq z', \\ \\ (II) \ \Phi(+\infty, +\infty, +\infty) = +\infty, \\ \\ (III) \ \Phi(0, 0, 0) = 0. \end{cases}$$

As regards a class of functions $\Phi(x, y, z)$ which satisfies the previous conditions, we can give the following:

Theorem 1. *If $h : [0, +\infty] \to [0, +\infty]$ is an continuous strictly increasing function with $h(0) = 0$ and $h(+\infty) = +\infty$, the function*

$$\Phi\left(x, y, z\right) = h^{-1}\left(h(x) + h(y) + h(z)\right) \qquad (3)$$

verifies the conditions $(I), (II), (III)$.

Proof. The (I) is valid as h is monotone, the (II) and (III) are satisified by the values of h.

4 Some Aggregation Operators of General Misinformation

Many authors have studied aggregation operators, for example, [3, 10, 2]. In [12, 13] we have applied the aggregation operators to general conditional information. We shall use again the same procedure for the characterization of some forms of aggregation operators of general misinformation.

Let $\mathcal{M}$ be the family of the misinformation $M_J(\cdot)$. The aggregation operator $L : \mathcal{M} \to [0, K], \ 0 \leq K \leq [0, +\infty]$ of $n \in [0, +\infty)$ misinformation $M_J(A_1), ..., M_J(A_i), ..., M_J(A_n)$, with $A_i \in \mathcal{A}, i = 1, ..., n$, has the following properies as in [12, 13]:

(I) *idempotence* : $M_J(A_i) = \lambda, \forall i = 1, ..., n \Longrightarrow$

$$L\underbrace{\left(\lambda, ..., \lambda\right)}_{n \ times} = \lambda;$$

(II) *monotonicity* : $M_J(A_1) \leq M_J(A_1') \Longrightarrow$
$$L\left(M_J(A_1), ..., M_J(A_i), ..., M_J(A_n)\right) \leq$$
$$L\left(M_J(A_1'), ..., M_J(A_i), ..., M_J(A_n)\right),$$
$A_1', A_i \in \mathcal{A}, i = 1, ..., n;$

(III) *continuity from below* : $M_J(A_{1,m}) \nearrow M_J(A_1) \Longrightarrow$

$$L\left(M_J(A_{1,m}), ..., M_J(A_i), ..., M_J(A_n)\right) \quad \nearrow \quad L\left(M_J(A_1), ..., M_J(A_i), ...,\right.$$
$$\left. M_J(A_n)\right) A_1', A_i \in \mathcal{A}, i = 1, ..., n.$$

Putting $M_J(A_i) = x_i, i = 1, ...n$, $M_J(A_1') = x_1'$, $M_J(A_{1,m}) = x_{1,m}$, with $x_i, i = 1, ...n$, x_1', $x_{1,m} \in [0,1]$, we obtain the following system of functional equations:

$$\begin{cases} (I') \ L\underbrace{\left(\lambda, ..., \lambda\right)}_{n\ times} = \lambda, \\[2em] (II') \ x_1 \le x_1' \Longrightarrow L\left(x_1, ..., x_n\right) \le L\left(x_1', ...x_n\right), \\[2em] (III') \ x_{1,m} \nearrow x_1 \Longrightarrow \\ \quad L\left(x_{1,m}, ..., x_n\right) \nearrow L\left(x_1, ...x_n\right). \end{cases}$$

For the solution of the system $[(I') - (III')]$, we propose the following:

Theorem 2. *Two natural solutions of the system $[(I') - (III')]$ are*

$$L(x_1, ..., x_n) = \bigwedge_{i=1}^{n} x_i$$

and

$$L(x_1, ..., x_n) = \bigvee_{i=1}^{n} x_i.$$

Proof. The proof is immediate.

Theorem 3. *A class of solution of the system $[(I') - (III')]$ is*

$$L(x_1, ..., x_n) = h^{-1}\left(\frac{h(x_1) + ... + h(x_n)}{n}\right),$$

where $h : [0,1] \to [0, K](0 \le K \le +\infty)$ is a continuous, strictly increasing function with $h(0) = 0$ and $h(1) = K$.

Proof. The proof is immediate.

5   Remark

1) If the function h is linear, then the aggregation operator L is the aritmetica mean.

2) All these results can be extended to fuzzy sets [11, 14].

6   Conclusion

First, we have given the definition of measure of general misinformation $M_J(A)$ of a set A, second we have proposed it as a function of the general information $J(A)$, reliability of the source $\rho(A)$, and our capacity $\tau(A)$ of taking into a good testing. We have obtained the following form:

$$M_J(A) = h^{-1}\left(h(J(A)) + h(\rho(A)) + h(\tau(A))\right) \ ,$$

where $h : [0,1] \to [0,K]$ is any continuous strictly increasing function with $h(0) = 0$ and $h(1) = K$,

Finally, we have proposed some classes of aggregation operators of misinformation solving a suitable system of functional equations:

$$1) \ L\left(M_J(A_1), ..., M_J(A_n)\right) = \bigwedge_{i=1}^{n} M_J(A_i) \ ,$$

$$2) \ L\left(M_J(A_1), ..., M_J(A_n)\right) = \bigvee_{i=1}^{n} M_J(A_i) \ ,$$

$$3) \ L\left(M_J(A_1), ..., M_J(A_n)\right) = h^{-1}\left(\frac{\Sigma_{i=1}^{n} h(M_(A_i))}{n}\right),$$

where $h : [0,1] \to [0,K](0 \leq K \leq +\infty)$ is a continuous, strictly increasing function with $h(0) = 0$ and $h(1) = K$.

References

[1]   Aczel, J.: Lectures on functional equations and their applications. Academic Press, New York (1966)

[2]   Benvenuti, P., Vivona, D., Divari, M.: Aggregation operators and associated fuzzy mesures. Int. Journ. of Uncertainty, Fuzziness and Knownledge-Based Systems 2, 197–204 (2001)

[3]   Calvo, T., Mayor, G., Mesiar, R.: Aggregation operators. In: New Trend and Applications. Physisc-Verlag, Heidelberg (2002)

[4]   Forte, B.: Measures of information: the general axiomatic theory, R.A.I.R.O., pp. 63–90 (1969)

[5]   Kampé De Fériet, J.: Mesures de l'information par un ensemble d'observateures. C.R. Acad. Sc. Paris 269 A, 1081–1086 (1969)

[6]   Kampé De Fériet, J.: Mesures de l'information fornie par un evénement. Coll. Int. Centre Nat. de La Rech. Scien. 186, 191–221 (1970)

[7]   Kampé De Fériet, J., Benvenuti, P.: Forme générale de l'operation de composition continue d'une information. C.R. Acad. Sc. Paris 269, 529–534 (1969)

[8]   Kampé De Fériet, J., Benvenuti, P.: Sur une classe d'informations. C.R. Acad. Sc. Paris 269A, 97–101 (1969)

[9]   Kampé De Fériet, J., Forte, B.: Information et Probabilité. C.R. Acad. Sc. Paris 265, 110–114, 142–146, 350–353 (1967)

[10] Klement, E.P.: Binary aggregation operators which are bounded by the minimum (a mathematical overview). In: Proc. AGOP 2005, pp. 13–16 (2005)

[11] Klir, G.J., Folger, T.A.: Fuzzy sets, Uncertainty, and Information. Prentice-Hall International editions (1988)

[12] Vivona, D., Divari, M.: Aggregation operators for conditional information without probability. In: Proc. IPMU 2008, pp. 258–260 (2008)

[13] Vivona, D., Divari, M.: Aggregation operators for conditional crispness. In: Proc. IFSA-EUSFLAT 2009, pp. 88–91 (2009)

[14] Zadeh, L.A.: Fuzzy sets. Inf. and Control 8, 338–353 (1965)

Author Index